Optical Fiber
Telecommunications
VIA

Components and Subsystems

Dedication

To the memory of Dr. Tingye Li
(July 7, 1931–December 27, 2012)

A pioneer, luminary, friend, mentor, and champion of our field.
We will miss him dearly.

From the optical communications community

题词

光通信领域的同仁们：
请向我们的先驱、引路人、朋友、导师、倡导者和勇士，

厉鼎毅博士
(1931年7月7日－2012年12月27日)

致以深切的怀念！

Optical Fiber Telecommunications VIA

Components and Subsystems

Sixth Edition

Ivan P. Kaminow

Tingye Li

Alan E. Willner

AMSTERDAM • BOSTON • HEIDELBERG • LONDON
NEW YORK • OXFORD • PARIS • SAN DIEGO
SAN FRANCISCO • SINGAPORE • SYDNEY • TOKYO

Academic Press is an Imprint of Elsevier

Academic Press is an imprint of Elsevier
The Boulevard, Langford Lane, Kidlington, Oxford OX5 1GB, UK
225 Wyman Street, Waltham, MA 02451, USA

Sixth edition 2013

Library of Congress Cataloging-in-Publication Data
A catalog record for this book is availabe from the Library of Congress

British Library Cataloguing in Publication Data
A catalogue record for this book is available from the British Library

ISBN: 978-0-12-396958-3

For information on all Academic Press publications
visit our web site at books.elsevier.com

13 14 15 16 17   10 9 8 7 6 5 4 3 2 1

Contents

CHAPTER 9 Multi-Core Optical Fibers 321

Tetsuya Hayashi

CHAPTER 10 Plastic Optical Fibers and Gb/s Data Links 353

Yasuhiro Koike and Roberto Gaudino

CHAPTER 11 Integrated and Hybrid Photonics for High-Performance Interconnects 377

Nikos Bamiedakis, Kevin A. Williams, Richard V. Penty, and Ian H. White

CHAPTER 16 All-Optical Regeneration of Phase
Encoded Signals... 589

*Joseph Kakande, Radan Slavík, Francesca Parmigiani,
Periklis Petropoulos, and David Richardson*

PHASE SENSITIVE OPTICAL REGENERATION 589

Preface—Overview of OFT VI A & B

Optical Fiber Telecommunications VI (*OFT VI*) is the sixth installment of the *OFT* series. Now 34 years old, the series is a compilation by the research and development community of progress in the field of optical fiber communications. Each edition reflects the current state of the art at the time. As editors, we started with a clean slate and selected chapters and authors to elucidate topics that have evolved since *OFT V* or that have now emerged as promising areas of research and development.

SIX EDITIONS

Installments of the series have been published roughly every 5–8 years and chronicle the natural evolution of the field:

- In the late 1970s, the original *OFT* (Chenoweth and Miller, 1979) was concerned with enabling a simple optical link, in which reliable fibers, connectors, lasers, and detectors played the major roles.
- In the late 1980s, *OFT II* (Miller and Kaminow, 1988) was published after the first field trials and deployments of simple optical links. By this time, the advantages of multi-user optical networking had captured the imagination of the community and were highlighted in the book.
- *OFT III* (Kaminow and Koch, 1997) explored the explosion in transmission capacity in the early-to-mid-1990s, made possible by the erbium-doped fiber amplifier (EDFA), wavelength-division-multiplexing (WDM), and dispersion management.
- By 2002, *OFT IV* (Kaminow and Li, 2002) dealt with extending the distance and capacity envelope of transmission systems. Subtitle nonlinear and dispersive effects, requiring mitigation or compensation in the optical and electrical domains, were explored.
- *OFT V* (Kaminow, Li, and Willner, 2008) moved the series into the realm of network management and services, as well as employing optical communications for ever-shorter distances. Using the high-bandwidth capacity in a cost-effective manner for customer applications started to take center stage.
- The present edition, *OFT VI* (Kaminow, Li, and Willner, 2013), continues the trend of photonic integrated circuits, higher-capacity transmission systems, and flexible network architectures. Topics that have gained much interest in increasing performance include coherent technologies, higher-order modulation formats, and space-division-multiplexing. In addition, many of the topics from earlier volumes are brought up to date and new areas of research which show promise of impact are featured.

Although each edition has added new topics, it is also true that new challenges emerge as they relate to older topics. Typically, certain devices may have adequately solved transmission problems for the systems of that era. However, as systems become more complex, critical device technologies that might have been considered a "solved problem" would now have new requirements placed upon them and need a fresh technical treatment. For this reason, each edition has grown in sheer size, i.e. adding the new and, if necessary, re-examining the old.

An example of this circular feedback mechanism relates to the fiber itself. At first, systems simply required low-loss fiber. However, long-distance transmission enabled by EDFAs drove research on low-dispersion fiber. Further, advances in WDM and the problems of nonlinear effects necessitated development of non-zero-dispersion fiber. Cost and performance considerations today drive research in plastic fibers, highly bendable fibers, few-mode fibers, and multicore fibers. We believe that these cycles will continue.

Perspective of the past 5 years

OFT V was published in 2008. At that point, our field was still emerging from the unprecedented upheaval circa 2000, at which time worldwide telecom traffic ceased being dominated by the slow-growing voice traffic and was overtaken by the rapidly growing Internet traffic. The *irrational* investment exuberance and subsequent depression-like period of oversupply (i.e. the "bubble-and-bust") wreaked havoc on our field. We are happy to say that, by nearly all accounts, the field continues to gain strength again and appears to have entered a stage of *rational* growth. Demand for bandwidth continues to grow at a very healthy rate. Capacity needs are real, and are expected to continue in the future.

We note that optical fiber communications is firmly entrenched as part of the global information infrastructure. For example: (i) there would be no Internet as we know it if not for optics, (ii) modern data centers may have as many as 1,000,000 lasers to help interconnect boards and machines, and (iii) Smartphones would not be so smart without the optical fiber backbone.

A remaining question is how deeply will optical fiber penetrate and complement other forms of communications, e.g. wireless, access and on-premises networks, Interconnections, satellites, etc. The odds are that, indeed, optics will continue to play a significant role in assisting all types of future communications. This is in stark contrast to the voice-based future seen by *OFT*, published in 1979, which occurred before the first commercial intercontinental or transatlantic cable systems deployed in the 1980s. We now have Tbit/s systems for metro and long-haul networks. It is interesting and exciting to contemplate what topics, concerns, and innovations might be contained in the next edition of the series, *OFT VII*.

In this edition, *OFT VI*, we have tried to capture the rich and varied technical advances that have occurred in our field. Innovations continue to abound! We hope our readers learn and enjoy from all the chapters.

We wish sincerely to thank Tim Pitts, Charlie Kent, Susan Li, Jason Mitchell of Elsevier and Hao Huang of USC for their gracious and invaluable support throughout the publishing process. We are also deeply grateful to all the authors for their laudable efforts in submitting their scholarly works of distinction. Finally, we wish to thank the many people whose insightful suggestions were of great assistance.

Below are brief highlights of the different chapters in the two volumes.

OFT VI Volume A: Components and Subsystems
1A. Advances in Fiber Distributed-Feedback Lasers
Michalis N. Zervas

This chapter covers advances in fiber distributed-feedback (DFB) lasers and their potential use in modern coherent optical telecommunication systems. In particular, it describes novel DFB cavity designs and configurations and considers their impact on the laser performance. Special emphasis is given to the fiber parameters that define the power scalability and stability, the polarization performance, as well as the linewidth and phase-noise characteristics. The wavelength coverage and tunability mechanisms are also discussed. The chapter finally reviews the use of fiber DFB lasers in non-telecom applications, such as advanced optical fiber sensors, and concludes with an outlook of the fiber laser technologies and their future prospects.

2A. Semiconductor Photonic Integrated Circuit Transmitters and Receivers
Radhakrishnan Nagarajan, Christopher Doerr, and Fred Kish

This chapter covers the field of semiconductor photonic integrated circuits (PIC) used in access, metro, long-haul, and undersea telecommunication networks. Although there are many variants to implementing optical integration, the focus is on monolithic integration where multiple semiconductor devices, up to many hundreds in some cases, are integrated onto the same substrate. Monolithic integration poses the greatest technical challenge and the biggest opportunity for bandwidth and size scaling. The PICs discussed here are based on the two most popular semiconductor material systems: Groups III–V indium phosphide-based devices and Group IV silicon-based devices. The chapter also covers the historical evolution of the technology from the decades-old original proposal to the current-day Tbit/s class, coherent PICs.

3A. Advances in Photodetectors and Optical Receivers
Andreas Beling and Joe C. Campbell

This chapter reviews the significant advances in photodetectors that have occurred since *Optical Fiber Telecommunications V*. The quests for higher-speed p-i-n detectors and lower-noise avalanche photodiodes (APDs) with high gain-bandwidth product remain.

To a great extent, high-speed structures have coalesced to evanescently coupled wave-guide devices; bandwidths exceeding 140 GHz have been reported. A primary APD breakthrough has been the development of Ge on Si separate-absorption-and-multiplication devices that achieve long-wavelength response with the low-noise behavior of Si. For III–V compound APDs, ultra-low noise has been achieved by strategic use of complex multilayer multiplication regions that provide a more deterministic impact ionization. However, much of the excitement and innovation have focused on photodiodes that can be incorporated into InP-based integrated circuits and photodetectors for Si photonics.

4A. Fundamentals of Photonic Crystals for Telecom Applications— Photonic Crystal Lasers

Susumu Noda

Photonic crystals, in which the refractive index changes periodically, provide an exciting tool for the manipulation of photons and have made substantial progresses in recent years. This chapter first introduces research activities that are geared toward realizing the ultimate nanolaser using the photonic bandgap effect. Important aspects of this effort are in the achievement of spontaneous emission suppression and strong optical confinement using a photonic nanocavity. During the process of implementation of this goal, interesting phenomena, which can be classified as Quantum Anti-Zeno effect, have been observed. The rest of the chapter focuses on the current state of research in the field of broad-area coherent photonic crystal lasers using the band-edge effect, which occupies a position opposite to that of nanolasers discussed above. The main characteristics of these lasers will be discussed, including their high-power operation, the generation of tailored beam patterns, the surface-emitting laser operation in the blue-violet region, and even the beam-steering functionality.

5A. High-Speed Polymer Optical Modulators

Raluca Dinu, Eric Miller, Guomin Yu, Baoquan Chen, Annabelle Scarpaci, Hui Chen, and Corey Pilgrim

Recent advances in thin-film-polymer-on-silicon (TFPS) technology have provided the foundation to support commercial devices manufactured at production levels. A fundamental understanding of the material systems and fabrication techniques has been demonstrated, and will provide a stable platform for future developments to support next-generation applications. The chapter focuses on high-speed polymer-based optical modulators and on the molecular engineering of chromophores. The design of electron donor, bridge, electron acceptor, and isolating groups are discussed. Finally, the current commercial technologies are presented.

6A. Nanophotonics for Low-Power Switches

Lars Thylen, Petter Holmström, Lech Wosinski, Bozena Jaskorzynska, Makoto Naruse, Tadashi Kawazoe, Motoichi Ohtsu, Min Yan, Marco Fiorentino, and Urban Westergren

Switches and modulators are key devices in ubiquitous applications of photonics: telecom, measurement equipment, sensor, and the emerging field of optical interconnects in high-performance computing systems. The latter could accomplish a breakthrough in offering a mass market for these switches. This chapter deals with photonic switches and the quest for the partly interlinked properties of low-power dissipation in operation and nanostructured photonics. It first summarizes some of the most important existing and emerging materials for nanophotonics low-power switches, and describes their physical mechanisms, operation mode, and characteristics. The chapter then focuses on basic operation and power dissipation issues of electronically controlled switches, which in many important cases by using a simple model are operated by charging and discharging capacitors and thus changing absorption and/or refraction properties of the medium between the capacitor plates. All optical switches are also discussed and some present devices are presented.

7A. Fibers for Short-Distance Applications

John Abbott, Scott Bickham, Paulo Dainese, and Ming-Jun Li

This chapter first reviews the current use of multimode fibers (MMF) with short-wavelength VCSELs for short-distance applications. Standards are in place for 100 Gbit/s applications based on 10 Gbit/s optics and are being developed for ~25 Gbit/s optics. Then it briefly introduces the theory of light propagation in multimode fibers. The actual performance of an MMF link (the bit error rate and inter-symbol interference) depends on both the fiber and the laser. Effective model bandwidth, which includes both fiber and laser effects, will be discussed, and the method of characterizing fiber with the differential-mode-delay measurement and the laser with the encircled flux measurement will be summarized as well. Bend-insensitive multimode fiber is then presented, explaining how the new fiber achieves high bandwidth with low bend loss. New fibers for short-distance consumer applications and home networking are discussed. Finally, fibers designed for high-performance computing are reviewed, including multicore fibers for optical interconnects.

8A. Few-Mode Fiber Technology for Spatial Multiplexing

David W. Peckham, Yi Sun, Alan McCurdy, and Robert Lingle Jr.

This chapter gives an overview of design and optimization of few-mode optical fibers (FMF) for space-division-multiplexed transmission. The design criteria are outlined, along with performance limitations of the traditional step-profile and graded-index profiles. The trade-offs between number of usable optical modes (related to total channel capacity), differential group delay, differential mode attenuation, mode coupling, and the impact on multiple-input and multiple-output (MIMO) receiver complexity are outlined. Improved fiber designs are analyzed which maximize channel capacity with foreseeable next-generation receiver technology. FMF measurement technology is overviewed.

9A. Multi-Core Optical Fibers
Tetsuya Hayashi

Spatial division multiplexing attracts lots of attention for tackling the "capacity crunch," which is anticipated as a problem in the near future, and therefore various types of optical fibers and multiplexing methods have been intensively researched in recent years. This chapter introduces the multi-core fibers for spatial division multiplexed transmission. It describes various characteristics specific to the multi-core fibers, which have been elucidated theoretically and experimentally in recent years. Though there are many important factors, many pages are devoted especially to the description of inter-core crosstalk, which is crucial when signals are transmitted over each core independently. The chapter also describes other characteristics related to the improvement of core density.

10A. Plastic Optical Fibers and Gb/s Data Links
Yasuhiro Koike and Roberto Gaudino

As high-speed data processing and communication systems are required, plastic optical fibers (POFs) become promising candidates for optical interconnects as well as optical networking in local area networks. This chapter presents an overview of the evolution of POF, reviewing the technical achievements of both fiber design and system architectures that today allow using POF for Gb/s data links. In particular, the chapter presents the different POF materials such as polymethyl methacrylate (PMMA), perfluorinated polymers, types such as step-index POF and graded-index POF, as well as the POF production process, describing the resulting optical characteristics in terms of attenuation, dispersion, and bandwidth. The main applications of POF in industrial automation, home networking, and local area networks are also discussed.

11A. Integrated and Hybrid Photonics for High-Performance Interconnects
Nikos Bamiedakis, Kevin A. Williams, Richard V. Penty, and Ian H. White

Optical interconnection technologies are increasingly deployed in high-performance electronic systems to address challenges in connectivity, size, bandwidth, latency, and cost. Projected performance requirements lead to formidable cost and energy efficiency challenges. Hybrid and integrated photonic technologies are currently being developed to reduce assembly complexity and to reduce the number of individually packaged parts. This chapter provides an overview of the important challenges that photonics currently face, identifies the various optical technologies that are being considered for use at the different interconnection levels, and presents examples of demonstrated state-of-the-art optical interconnection systems. Finally, the prospects and potential of these technologies in the near future are discussed.

12A. CMOS Photonics for High-Performance Interconnects
Jason Orcutt, Rajeev Ram, and Vladimir Stojanović

For many applications, multicore chips are primarily constrained by the latency, bandwidth, and capacity of the external memory system. One of the most significant challenges is how to effectively connect on-chip processors to off-chip memories. This chapter introduces optical interconnects as a possible solution to the emerging performance wall in high-density supercomputer applications, arising from limited bandwidth and density of on-chip interconnects and chip-to-chip (processor-to-memory) electrical interfaces. The chapter focuses on the translation of system- and link-level performance metrics to photonic component requirements. The topics to be developed include network topology, photonic link components, circuit and system design for photonic links.

13A. Hybrid Silicon Lasers
Brian R. Koch, Sudharsanan Srinivasan, and John E. Bowers

The term "hybrid silicon laser" refers to a laser that has a silicon waveguide and a III–V material that is in close optical contact. In this structure, the optical confinement can be easily transferred from one material to the other and intermediate modes exist for which the light is contained in both materials simultaneously. In hybrid silicon lasers, the optical gain is provided by the electrically pumped III–V material and the optical cavity is ultimately formed by the silicon waveguide. This type of laser can be heterogeneously integrated with silicon components that have superior performance compared to III–V components. These lasers can be fabricated in high volumes as components of complex photonic integrated circuits, largely with CMOS-compatible processes. These traits are expected to allow for highly complex, non-traditional photonic integrated circuits with very high yields and relatively low manufacturing costs. This chapter discusses the theory of hybrid silicon lasers, wafer-bonding techniques, examples of experimental results, examples of system demonstrations based on hybrid silicon lasers, and prospects for future devices.

14A. VCSEL-Based Data Links
Julie Sheridan Eng and Chris Kocot

Vertical cavity surface emitting laser (VCSEL)-based data links are attractive due to their low-power dissipation and low-cost manufacturability. This chapter reviews the foundations for this technology, as well as the device and module design challenges of extending the data rate beyond the current level. The chapter begins with a review of data communications from the business perspective, and continues with a brief discussion of the current and future standards. This is followed by a survey of recent advances in VCSELs, including data links operating at 28 Gbit/s. Recent efforts on ultra-fast data links are reviewed and the advantages of the different approaches are discussed. The chapter also examines key design aspects of optical transceiver modules and focuses on novel applications in high-performance computing using both multi-mode and single-mode fiber optics. The importance of the device/component-level and system-level modeling is highlighted, and some modeling examples are shown with comparison to measured data. The chapter concludes with a comparison of the VCSEL-based data links with other competing technologies, including silicon photonics and short-cavity edge-emitting lasers.

15A. Implementation Aspects of Coherent Transmit and Receive Functions in Application-Specific Integrated Circuits
Andreas Leven and Laurent Schmalen

One of the most challenging components of an optical coherent communication system is the integrated circuits (ICs) that process the received signals or condition the transmit signals. This chapter discusses implementation aspects of these ICs and their main building blocks, as data converters, baseband signal processing, forward error correction, and interfacing. This chapter also highlights selected implementation details for some baseband signal processing blocks of a coherent receiver. The latest generation of coherent ICs also supports advanced forward error correction techniques based on soft decisions. The circuits for encoding and decoding low-density parity-check (LDPC) codes are introduced and evaluation of different forward error correction schemes based on a set of recorded measurement data is presented in this chapter.

16A. All-Optical Regeneration of Phase-Encoded Signals
Joseph Kakande, Radan Slavík, Francesca Parmigiani, Periklis Petropoulos, and David Richardson

This chapter reviews the general principles and approaches used to regenerate phase-encoded signals of differing levels of coding complexity. It first reviews different approaches and nonlinear processes that may be used to perform the regeneration of phase-encoded signals. The primary focus is on parametric effects, which as explained previously can operate directly on the optical phase. The chapter then proceeds to review progress on regenerating the simplest of phase modulation formats, namely DPSK/BPSK- and for which the greatest progress has been made to date. In the following, the progress in regenerating more complex modulation format signals—in particular (D)QPSK and other M-PSK signals—is discussed. The chapter also reviews the choice of nonlinear components available to construct phase regenerators. Finally, it reviews the prospects for regenerating even more complex signals including QAM and mixed phase-amplitude coding variants.

17A. Ultra-High-Speed Optical Time Division Multiplexing

Leif Katsuo Oxenløwe, Anders Clausen, Michael Galili, Hans Christian Hansen Mulvad, Hua Ji, Hao Hu, and Evarist Palushani

The attraction of optical time division multiplexing (OTDM) technology is the promise of achieving higher bit rates per channel than electronics could provide, thus alleviating the so-called electronic speed bottleneck. In this chapter, the state-of-the-art OTDM systems are presented, with a focus on experimental demonstrations. This chapter especially highlights demonstrations at 640–1280 Gbaud per polarization based on a variety of materials and functionalities. Many essential network functionalities are available today using a plethora of available materials, so now it is time to look at new network scenarios that take advantage of the serial nature of the data, e.g. try to come up with practical schemes for ultra-high bit rate optical data packets in supercomputers or within data centers.

18A. Technology and Applications of Liquid Crystal on Silicon (LCoS) in Telecommunications

Stephen Frisken, Ian Clarke, and Simon Poole

Liquid crystal is now the dominant technology for flat-screen displays and has been used in telecom systems since the late 1990s. More recently, the adoption of liquid crystals in Wavelength Selective Switches—with the control of light on a pixel-by-pixel basis—has been enabled by developments in Liquid Crystal on Silicon (LCoS) backplane technologies derived from projection displays. This chapter presents the principles of operation of liquid crystals, focusing in particular on how they operate within an LCoS chip. It then explains in detail the design and operation of an LCoS-based wavelength selective switch (WSS), with particular emphasis on the key optical parameters that determine performance in an optical communications network. In the final section, the chapter briefly describes the broad scope of new opportunities that arise from the intrinsic performance and flexibility of LCoS as a switching medium.

OFT VI Volume B: Systems and Networks

1B. Fiber Nonlinearity and Capacity: Single-Mode and Multimode Fibers

René-Jean Essiambre, Robert W. Tkach, and Roland Ryf

This chapter presents the trends in optical network traffic and commercial system capacity, discusses fundamentals of nonlinear capacity of single-mode fibers, and indicates that improvements in the properties of single-mode fibers only moderately increase the nonlinear fiber capacity. This leads to the conclusion that fiber capacity can be most effectively grown by increasing the number of spatial modes. This chapter also discusses nonlinear propagation in multimode fiber, a complex field still largely unexplored. It gives a basic nonlinear propagation equation derived from the Maxwell equation, along with simplified propagation equations in the weak- and strong-coupling approximations, referred to as generalized Manakov equations. Finally, the chapter presents experimental observations of two inter-modal nonlinear effects, inter-modal cross-phase modulation, and inter-modal four-wave mixing, over a few-km-long few-mode fiber. Important differences between intra-modal and inter-modal nonlinear effects are also discussed.

2B. Commercial 100-Gbit/s Coherent Transmission Systems
Tiejun J. Xia and Glenn A. Wellbrock

This chapter provides a global network service provider's view on technology development and product commercialization of 100-Gbit/s for optical transport networks. Optical channel capacity has been growing over the past four decades to address traffic demand growth and will continue this trend for the foreseeable future to meet ever-increasing bandwidth requirements. In this chapter, optical channels are reclassified into three basic design types. Commercial 100-Gbit/s channel development experienced all three types of channel designs before eventually settling on the single-carrier polarization-multiplexed quadrature-phase-shift keying (PM-QPSK) format using coherent detection, which appears to be the optimal design in the industry. A series of 100-Gbit/s channel related field trials was performed in service providers' networks to validate the technical merits and business advantages of this new capacity standard before its deployment. Introduction of the 100-Gbit/s channel brings new opportunities to boost fiber capacity, accommodates increases in client interface speed rates, lowers transmission latency, simplifies network management, and speeds up the realization of next-generation optical add/drop functions.

3B. Advances in Tb/s Superchannels
S. Chandrasekhar and Xiang Liu

Optical superchannel transmission, which refers to the use of several optical carriers combined to create a channel of desired capacity, has recently attracted much research and development in an effort to increase the capacity and cost-effectiveness of wavelength-division multiplexing (WDM) systems. Using superchannels avoids the electronic bottleneck via optical parallelism and provides high per-channel data rates and better spectral utilization, especially in transparent mesh optical networks. This chapter reviews recent advances in the generation, detection, and transmission of optical superchannels with channel data rates on the order of Tbit/s. Multiplexing schemes such as optical orthogonal-frequency-division-multiplexing (O-OFDM) and Nyquist-WDM are described, in conjunction with modulation schemes such as OFDM and Nyquist-filtered single-carrier modulation. Superchannel transmission performance is discussed. Finally, networking implications brought by the use of superchannels, such as flexible-grid WDM, are also discussed.

4B. Optical Satellite Communications
Hamid Hemmati and David Caplan

Current satellite-based communication systems are increasingly capacity-limited. Based on radio frequency or microwave technologies, current state-of-the-art satellite communications (Satcom) are often constrained by hardware and spectrum allocation limitations. Such limitations are expected to worsen due to the use of more sophisticated data-intensive sensors in future interplanetary, deep-space, and manned missions, an increased demand for information, and the demand for a bigger return on space-exploration investment. This chapter presents the recent advances in optical satellite communications technologies. Lasercom link budgets, the first step in designing a lasercom system, are discussed. The chapter then reviews the major challenges facing laser beam propagation through the atmosphere, including atmospheric attenuation, scattering, radiance, and turbulence. It also discusses mitigation approaches. The rest of the chapter focuses on optical transceiver technologies for satellite communications systems. Finally, space and ground terminals in optical satellite communications are discussed.

5B. Digital Signal Processing (DSP) and its Application in Optical Communication Systems
Polina Bayvel, Carsten Behrens, and David S. Millar

The key questions in current optical communications research are how to maximize both capacity and transmission distance in future optical transmission networks by using spectrally efficient modulation formats with coherent detection and how digital signal processing can aid in this quest. There is a clear trade-off between spectral efficiency and transmission distance, since the more spectrally efficient modulation formats are more susceptible to optical fiber nonlinearities. This chapter illustrates the application of nonlinear back-propagation to mitigate both linear and nonlinear transmission impairments in a range of modulation formats at varying symbol rates, wavelength spacing, and signal bandwidth. The basics of coherent receiver structure and digital signal processing (DSP) algorithms for chromatic dispersion compensation, equalization, and phase recovery of different modulation formats employing amplitude, phase, and polarization are reviewed and the effectiveness of the nonlinearity compensating DSP based on digital back-propagation is explored. This chapter includes a comprehensive literature review of the key experimental demonstrations of nonlinearity compensating DSP.

6B. Advanced Coding for Optical Communications
Ivan B. Djordjevic

This chapter represents an overview of advanced coding techniques for optical communication. Topics include the following: codes on graphs, coded modulation, rate-adaptive coded modulation, and turbo equalization. The main objectives of this chapter are as follows: (i) to describe different classes of codes on graphs of interest for optical communications, (ii) to describe how to combine multilevel modulation and channel coding, (iii) to describe how to perform equalization and soft-decoding jointly, and (iv) to demonstrate efficiency of joint demodulation, decoding, and equalization in dealing with various channel impairments simultaneously. The chapter describes both binary and nonbinary LDPC codes, their design, and decoding algorithms. A field-programmable gate array (FPGA) implementation of decoders for binary LDPC codes is discussed. In addition, this chapter demonstrates that an LDPC-coded turbo equalizer is an excellent candidate to simultaneously mitigate chromatic dispersion, polarization mode dispersion, fiber nonlinearities, and I/Q-imbalance. In the end, the information capacity study of optical channels with memory is provided for completeness of presentation.

7B. Extremely Higher-Order Modulation Formats
Masataka Nakazawa, Toshihiko Hirooka, Masato Yoshida, and Keisuke Kasai

This chapter reviews recent progress on coherent quadrature amplitude modulation (QAM) and orthogonal frequency-division multiplexing (OFDM) transmission with higher-order multiplicity, which is aiming at ultra-high spectral efficiency approaching the Shannon limit. Key technologies are the coherent detection with a frequency-stabilized fiber laser and an optical PLL circuit. Single-carrier 1024 QAM and 256 QAM-OFDM transmissions are successfully achieved, demonstrating a spectral efficiency exceeding 10 bit/s/Hz. Such an ultra-high spectrally efficient transmission system would also play a very important role in increasing the total capacity of WDM systems and improving the tolerance to chromatic dispersion and polarization mode dispersion as well as in reducing power consumption. The chapter also describes a novel high-speed, spectrally efficient transmission scheme that combines the OTDM and QAM techniques, in which a pulsed local oscillator (LO) signal obtained with an optical phase-lock loop (OPLL) enables precise demultiplexing and demodulation simultaneously. An optimum OTDM and QAM combination would provide the possibility for realizing long-haul Tbit/s/channel transmission with a simple configuration, large flexibility, and low-power consumption.

8B. Multicarrier Optical Transmission
Xi Chen, Abdullah Al Amin, An Li, and William Shieh

This chapter is an overview of multicarrier transmission and its application to optical communication. Starting with an introduction to historical perspectives in the development of optical multicarrier technologies, the chapter presents different variants of optical multicarrier transmission, including electronic and optical fast Fourier transform (FFT)-based realizations. In the next section, several problems of fiber nonlinearity in optical multicarrier transmission systems are highlighted and an analysis of fiber capacity under nonlinear impairments is presented. The applications of multicarrier techniques to long-haul systems, access networks, and free-space optical communication systems are also discussed. Finally, this chapter summarizes several possible directions for research into the implementation of multicarrier technologies in optical transmission.

9B. Optical OFDM and Nyquist Multiplexing
Juerg Leuthold and Wolfgang Freude

New pulse shaping techniques allow for optical multiplexing with the highest spectral efficiencies. This chapter introduces the general theory of orthogonal pulse shaping followed by a discussion that places more emphasis on the orthogonal frequency-division multiplexing (OFDM) and Nyquist frequency-division multiplexing schemes. Subsequently, the chapter shows that the rectangular-shaped pulses used for OFDM can mathematically be treated by the Fourier transform. This leads to the theory of the time-discrete Fourier transform (DFT) and to a discussion of practical implementations of the DFT and its inverse in the optical domain. The chapter concludes with exemplary implementations of OFDM transceivers that either rely on direct pulse shaping or use the DFT approaches.

10B. Spatial Multiplexing Using Multiple-Input Multiple-Output Signal Processing
Peter J. Winzer, Roland Ryf, and Sebastian Randel

In order to further scale network capacities and to avoid a looming "capacity crunch," *space* has been identified as the only known physical dimension yet unexploited for optical modulation and multiplexing. Space-division multiplexing (SDM) may use uncoupled or coupled cores of multi-core fiber, or individual modes of multimode waveguides. If crosstalk rises to levels where it cannot be treated as a transmission impairment any more, multiple-input multiple-output (MIMO) digital signal processing (DSP) techniques have to be used to manage crosstalk in highly integrated SDM systems. This chapter reviews the fundamentals and practical experimental aspects of MIMO-SDM. First, it discusses the importance of selectively addressing all modes of a coupled-mode SDM channel at transmitter and receiver in order to achieve reliable capacity gains. It shows that reasonable levels of mode-dependent loss (MDL) are acceptable without much loss of channel capacity. The chapter then introduces MIMO-DSP techniques as an extension of familiar algorithms used in polarization-division multiplexed (PDM) digital coherent receivers and discusses their functionality and scalability. Finally, the design of mode multiplexers that allows for the mapping of the individual transmission signals onto an orthogonal basis of waveguide mode is reviewed and its performance in experimental demonstrations is discussed.

11B. Mode Coupling and its Impact on Spatially Multiplexed Systems

Keang-Po Ho and Joseph M. Kahn

Mode coupling is the key to overcoming challenges in mode-division multiplexed transmission systems in multimode fiber. This chapter provides an in-depth description of mode coupling, including its physical origins, its effect on modal dispersion (MD) and mode-dependent loss (MDL) or gain, and the resulting impact on system performance and implementation complexity. Strong mode coupling reduces the group delay spread from MD, minimizing the complexity of digital signal processing used for compensating MD and separating multiplexed signals. Likewise, strong mode coupling reduces the variations of MDL that arise from transmission fibers and inline optical amplifiers, thus maximizing average channel capacity. When combined with MD, strong mode coupling creates frequency diversity, which reduces the probability of outage caused by MDL and enables outage capacity to approach average capacity. The statistics of strongly coupled MD and MDL depend only on the number of modes and the variances of MD or MDL, and they can be derived from the eigenvalue distributions of certain random matrices.

12B. Multimode Communications Using Orbital Angular Momentum

Jian Wang, Miles J. Padgett, Siddharth Ramachandran, Martin P.J. Lavery, Hao Huang, Yang Yue, Yan Yan, Nenad Bozinovic, Steven E. Golowich, and Alan E. Willner

Laser beams with a helical phase front, such as Laguerre-Gaussian beams, carry orbital angular momentum (OAM). Based on the fact that different OAM beams can be inherently orthogonal with each other, OAM multiplexing was introduced to provide an additional degree of freedom in optical communications, and further increase the capacity and spectral efficiency in combination with advanced multilevel modulation formats and conventional multiplexing technologies. This chapter provides a comprehensive review of multimode communications using OAM technologies. The fundamentals of OAM are introduced first, followed by the techniques for OAM generation, multiplexing/demultiplexing, and detection. The chapter then presents recent research into free-space communication links and fiber-based transmission links using OAM multiplexing with optical signal processing using OAM (data exchange, add/drop, multicasting, monitoring, and compensation). Future challenges for OAM communications are then discussed.

13B. Transmission Systems Using Multicore Fibers

Yoshinari Awaji, Kunimasa Saitoh, and Shoichiro Matsuo

As the simplest form of space-division multiplexing (SDM), multi-core fiber (MCF) transmission technologies have been widely studied. Many types of MCFs exist, but the most common is "Uncoupled MCF" in which each individual core is assumed to be an independent optical path. The key issue in these systems is how to suppress the inter-core crosstalk and the coupling/decoupling mechanism. Currently, many MCF varieties, coupling methods, splicing techniques, and transmission schemes have been proposed and demonstrated, and despite the fact that many of the component technologies are still in the development stage, MCF systems already present the capability for huge transmission capacities. In this chapter, these component technologies and the early experimental trials of MCF transmission are reviewed. First, an overview of medium- to long-haul MCF transmission and theories is provided. Second, coupling technologies between MCF-SMF and MCF-MCF are reviewed. Finally, several experimental demonstrations, including transmission exceeding 100 Tbit/s and over 1000 km, are described.

14B. Elastic Optical Networking
Ori Gerstel and Masahiko Jinno

Service provider (SP) networks are undergoing major changes. These changes imply that the optical layer will have to be low-cost, flexible, and reconfigurable. To properly address this challenge, flexible and adaptive networks equipped with flexible transceivers and network elements that can adapt to the actual traffic demands are needed. The combination of adaptive transceivers, a flexible grid, and intelligent client nodes enables a new "elastic" networking paradigm, allowing SPs to address the increasing needs of the network without frequently overhauling it. This chapter starts by looking at the challenges faced by the optical layer in the future. These challenges are fueled by the insatiable appetite for more bandwidth, coupled with a reduced ability to forecast and plan for such growth. Different enabling technologies, including flexible spectrum reconfigurable optical add/drop multiplexers (ROADM), bit rate variable transceivers, and the extended role of network control systems are reviewed. The concept of elastic optical network (EON) is envisioned and the benefits are highlighted by further comparing the EON to a fixed WDM system.

15B. ROADM-Node Architectures for Reconfigurable Photonic Networks
Sheryl L. Woodward, Mark D. Feuer, and Paparao Palacharla

The deployment of reconfigurable optical add/drop multiplexers (ROADMs) is gradually transforming a transport layer made of point-to-point optical links into a highly interconnected, reconfigurable photonic mesh. To date, the widespread use of ROADMs has been driven by the cost savings and operational simplicity they provide to quasi-static networks (i.e. networks in which new connections are frequently set up but rarely taken down). However, new applications exploiting the ROADMs' ability to dynamically reconfigure a photonic mesh network are now being investigated. This chapter reviews the attributes and limitations of today's ROADMs and other node hardware. It also surveys proposals for future improvements, including colorless, non-directional, and contentionless add/drop ports. The application of reconfigurable networks is also discussed with emphasis on the backbone network of a major communications service provider. Finally, the chapter assesses which of these new developments is most likely to bring added value in the short and long future.

16B. Convergence of IP and Optical Networking
Kristin Rauschenbach and Cesar Santivanez

Rapidly increasing network demand based on unpredictable services has driven research into methods to provide intelligent provisioning, efficient restoration and recovery from failures, and effective management schemes that reduce the amount of "hands-on" activity to plan and run the network. Integrating the service-oriented IP layer together with the efficient transport capabilities of the optical layer is a cornerstone of this research. Converged IP-optical networks are being demonstrated in large multi-carrier and multi-vendor venues. Research is continuing on making this convergence more efficient, flexible, and scalable. This chapter reviews the current key technologies that contribute to the convergence of IP and optical networks, and describes control and management plane technologies, techniques, and standards in some detail. Current research challenges and future research directions are also discussed.

17B. Energy-Efficient Telecommunications
Daniel C. Kilper and Rodney S. Tucker

For many years, advances in telecommunications have been driven by the need for increased capacity and reduced cost. Recently, however, concerns about the rising energy use of telecommunications networks have brought the issue of energy efficiency into the mix for both equipment vendors and network operators. This chapter provides an overview of energy consumption in telecommunications networks. This chapter identifies the key contributors to energy consumption and the trends in the growth of energy consumption. The chapter also compares the performance of state-of-the-art equipment with theoretical lower bounds on energy consumption and points to opportunities for improving the energy efficiency of core metro and access networks. The potential of significantly improving energy efficiency in telecommunications is envisioned.

18B. Advancements in Metro Regional and Core Transport Network Architectures for the Next-Generation Internet
Loukas Paraschis

The expanding role of Internet-based service delivery, and its underlying infrastructure of internetworked data centers, is motivating an evolution to an IP next-generation network architecture with a flatter hierarchy of more densely interconnecting networks. This next-generation Internet is required to cost-effectively scale to Zettabytes of bandwidth with improved operational efficiency, in an environment of increasing traffic variability, dynamism, forecast unpredictability, and uncertainty of future traffic types. This chapter explores the implications of this change in the metro regional and core transport network architectures, and the important advancements in optical, routing, and traffic engineering technologies that are enabling this evolution. The chapter accounts particularly for the increasingly important role of optical transport, and photonics technology innovations.

19B. Novel Architectures for Streaming/Routing in Optical Networks
Vincent W.S. Chan

Present-day networks are being challenged by dramatic increases in the data rate demands of emerging applications. New network architectures for streaming/routing large "elephant" transactions will be needed to reduce costs and improve power efficiency. This chapter examines a number of possible optical network transport mechanisms, including optical packet switching, burst switching, and flow switching and describes the necessary physical layer, routing, and transport layer architectures for these transport mechanisms. Performance comparisons are made based on capacity utilization, scalability, costs, and power consumption.

20B. Recent Advances in High-Frequency (>10 GHz) Microwave Photonic Links
Charles H.Cox, III and Edward I.Ackerman

The transmission of multi-band radio signals through optical fibers has attracted great attention recently due to its potential for cellular backhaul networks, mobile cloud computing, and wireless local area networks. As wireless services and technologies evolve into multi-gigabit radio access networks, their speed is increased, but the wireless coverage of a single access point is inevitably and dramatically reduced. As a result, the importance of >10 GHz radio-over-fiber techniques has been emphasized for the capability of expanding wireless coverage feasibility, and in the meantime reducing system complexity and operation expenditure, especially in the high-speed millimeter-wave regime. This chapter introduces the radio-over-fiber technique and its challenge to handle optical millimeter-wave generation, transmission, and converged multi-band systems. By exploring real-world system implementation and characterization, the unique features and versatile applications of radio-over-fiber technologies are investigated and reviewed to reach next-generation converged optical and wireless access networks.

21B. Advances in 1-100 GHz Microwave Photonics: All-Band Optical Wireless Access Networks Using Radio Over Fiber Technologies
Gee-Kung Chang, Yu-Ting Hsueh, and Shu-Hao Fan

With the growing bandwidth demand for the last mile and last meter in the access network, radio-over-fiber (RoF) technology at millimeter-wave (mmW) band has been viewed as one of the most promising solutions to providing ubiquitous multi-gigabit wireless services with simplified and cost-effective base stations (BSs) and low-loss, bandwidth-abundant fiber optic networks. This chapter first outlines the general methods and types of optical mmW generation, and summarizes their advantages and disadvantages. Owing to ultra-wide bandwidth and protocol transparent characteristics, a RoF system can be utilized to simultaneously deliver wired and multi-band wireless services for both fixed and mobile users. In the rest of this chapter, several multi-band 60-GHz RoF systems are reviewed, including mmW with baseband, microwave, mmW with commercial wireless services in low RF regions, and 60-GHz sub-bands.

22B. PONs: State of the Art and Standardized
Frank Effenberger

This chapter aims to describe the current state of the passive optical network (PON) technology, including both state-of-the-art systems that are currently under research in the laboratory and "standardized" systems that have been or soon will be described as an industry norm. A short introduction to the PON topic is given, to set the scene and provide the basic motivation for why PON is so important to fiber access. Then, each of the major technologies is reviewed, including time division multiplexing, video overlay, wavelength-division multiplexing, frequency-division multiplexing, and hybrid multiplexing. The focus of each review is at a system level to present a wide view of the whole range, and comparisons are made to different technologies.

23B. Wavelength-Division-Multiplexed Passive Optical Networks (WDM PONs)
Y.C. Chung and Y. Takushima

Wavelength-division multiplexed passive optical network (WDM PON) has long been considered as an ultimate solution for a future optical access network capable of providing practically unlimited bandwidth to each subscriber. On the other hand, it is still considered to be too expensive for mass deployment. To solve this problem and to meet the ever-increasing demand for bandwidth, there have been numerous efforts to improve the competitiveness of WDM PON. This chapter reviews the current status and future direction of these WDM PON technologies. It first reviews various colorless light sources, which are critical for the cost-effective implementation of the optical network units (ONUs), and several representative network architectures proposed for WDM PONs. The chapter then reviews the recent research activities for the realization of high-speed (>10 Gb/s) and long-reach WDM PONs. Various fault-monitoring and protection techniques are also reviewed, as they may be increasingly important in future high-capacity WDM PONs.

24B. FTTX Worldwide Deployment
Vincent O'Byrne, Chang Hee Lee, Yoon Kim, and Zisen Zhao

Since the early 2000s, Fiber-to-the-X, where X refers to different meanings for to different operators, has taken off around the world and is seen as the main method to meet the continued growth in the broadband needs of residential and business customers. This chapter covers two types of architectures, including the shared network among many users and the point-to-point network, and the standing of the various technologies for access space. The status of FTTX and some of the issues that operators are facing around the world are discussed. The chapter then reviews technologies that have been deployed to date and the new technologies that are under consideration to meet their customers' residential and business needs in the future.

25B. Modern Undersea Transmission Technology
Jin-xing Cai, Katya Golovchenko, and Georg Mohs

Much progress has been made over the last few years in undersea optical fiber telecommunication systems. Most importantly, coherent receivers have become practical, enabling polarization multiplexing and higher-order modulation formats with increased spectral efficiency. This chapter provides an overview of the progress in undersea transmission technology. After a brief general introduction to undersea systems and their unique challenges and design constraints, the principles of coherent transmission technologies are outlined. These include polarization multiplexing, linear equalizers, and multiple bits per symbol. The chapter then describes the use of strong optical filtering to help to improve spectral efficiency, and it reviews the techniques to mitigate the effects of inter-symbol interference. Higher-order modulation formats that can further increase spectral efficiency by increasing the number of bits per symbol are then introduced. The implications of the receiver sensitivity degradation and the mitigation techniques are discussed.

Ivan P. Kaminow
(Bell Labs, retired)
University of California, Berkeley, CA, USA

Tingye Li
(Bell Labs and AT&T Labs, deceased)
Boulder, CO, USA

Alan E. Willner
University of Southern California,
Los Angeles, CA, USA

Advances in Fiber Distributed-Feedback Lasers

1

Michalis N. Zervas

Optoelectronics Research Centre, University of Southampton,
Southampton SO17 1BJ, UK

1.1 INTRODUCTION

The recent re-introduction of optical coherent techniques has revolutionized modern telecom communications and has resulted in the most spectrally efficient optical system demonstrations to date [1]. In addition to record spectral efficiencies, coherent communication systems offer other advantages in terms of system flexibility, reduced signal-to-noise ratio requirements, increased resilience against fiber chromatic dispersion and in many aspects simplify the system design. However, coherent detection systems, such as, for example, coherent-optical orthogonal-frequency-division-multiplexed (CO-OFDM) systems, and advanced modulation formats, such as quadrature amplitude modulation (QAM), impose much more stringent requirements on source linewidth, stability, and phase/frequency noise characteristics. The development of inexpensive laser sources, with high coherence and stability, is expected to replace expensive external-cavity semiconductor lasers and allow their use as local oscillators similar to the way local oscillators are used in today's radio communication systems. The performance of CO-OFDM is in general 3 dB better than that of incoherent OFDM since no optical power is allocated for the carrier. However, CO-OFDM requires a laser at the receiver to generate the carrier locally. It is, therefore, more sensitive to phase noise and the main challenge of CO-OFDM is that the phase noise of the local oscillator must be compensated for. It has been shown that the influence of phase noise can be reduced by using lasers with narrow linewidth [2,3].

So far, practical commercial telecommunication systems are almost entirely based on semiconductor-based lasers as their optical sources, because of their compact size, high speed and reliability, and direct electronic control. Over the past decade, continuous-wave and pulsed fiber lasers, covering a quite large spectral window, have experienced a remarkable progress and have penetrated successfully quite diverse industrial sectors. Also, a number of different fiber lasers, such as ultrashort, ultrafast, or frequency stabilized ring lasers, operating in the telecom window have already been used extensively in a number of coherent system demonstrators [4]. In addition, fiber distributed-feedback (DFB) laser technology has matured considerably

Optical Fiber Telecommunications VIA. http://dx.doi.org/10.1016/B978-0-12-396958-3.00001-9

1

and various sectors, such as the optical sensor sector, have benefited enormously from their unique characteristics, such as their robustness, electromagnetic-radiation immunity, power scalability, narrow linewidth, and exceptional phase-noise performance. A review of the high-performance fiber DFB lasers and their potential use in future high spectral efficiency telecom systems, in particular, is timely. This chapter covers advances in fiber DFB lasers and their potential use in modern telecom systems. It considers different designs and configurations and their impact on the laser performance. Special emphasis is given to power scalability and stability, linewidth, and phase/frequency noise characteristics. A brief comparison with other technologies in the telecom and non-telecom applications space is also given. The chapter concludes with a summary and an outlook of the fiber laser technologies and the future prospects.

1.2 FIBER DFB LASERS

1.2.1 Introduction

DFB lasers were first proposed and demonstrated by Kogelnik and Shank [5,6]. In DFB laser cavities; feedback is provided by distributed Bragg structures, instead of classic discrete mirrors, resulting in compact and frequency-stable devices. The first demonstration involved a UV-exposed periodic structure on a gelatin film soaked in a dye (rhodamine 6G). This was followed by demonstrations of optically and electrically pumped DFB lasers with semiconductor (GaAs) active materials and ion-milled corrugated waveguides [7,8]. The first demonstrations involved waveguide corrugations; corresponding to uniform periodic effective-index variations (see Figure 1.1a). The mode spectrum of such periodic structures exhibits a gap around the Bragg wavelength, with equal threshold modes λ_1 and λ_2 occurring symmetrically on either side of the gap. This threshold degeneracy results in frequency instabilities and compromises the use of such devices in telecom and other high-performance applications. To overcome this problem, Haus and Shank [9] suggested the introduction of an antisymmetric variation of the periodic index modulation. This can be practically achieved by the insertion of a quarter-wavelength (QW) section ($\lambda_0/4n_{eff} = \Lambda/2$)—or equivalently π-phase-shift—between two equal uniform DFB structures (see Figure 1.1b). This "defect" allows the existence of a mode within the stopband of the uniform sections, with the field decaying exponentially from the center into the gratings on either side. Compared to uniform DFB structures, this mode has lower threshold and higher stability when used in a laser cavity. This quarter-wavelength phase slip in the middle of the RI modulation profile essentially turns the uniform DFB structure into an effective Fabry-Perot cavity of length L_{eff}. Due to symmetry, the symmetric QW-shifted DFB is bidirectional with almost equal powers emitted from both laser ends. Unidirectionality can be achieved by moving the QW shift closer to one end [10], as shown schematically in Figure 1.1c. In this case, the length of the effective FP cavity remains largely unchanged and the overall

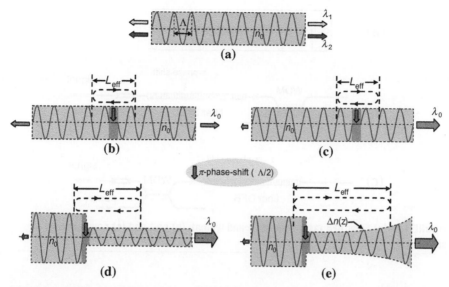

FIGURE 1.1 Symmetric and asymmetric DFB laser refractive index profiles: (a) uniform variation, (b) symmetric π-phase-shifted, (c) asymmetric π-phase-shifted, (d) step-apodized, asymmetric π-phase-shifted, and (e) optimized-apodization asymmetric π-phase-shifted.

power extraction efficiency remains almost constant. However, the forward/backward power ratio improves considerably.

1.2.2 Fiber π-phase-shifted DFB lasers

The basic DFB fiber laser configuration is shown schematically in Figure 1.2a. It consists of a length of doped fiber within which a phase-shifted Bragg grating is inscribed. Each end of the doped fiber is spliced to a passive fiber and the DFB laser is pumped through a WDM coupler in a co-directional (Figure 1.2b) or counter-directional (Figure 1.2c) pumping scheme.

Early fiber DFB lasers focused on the important $1.5\,\mu m$ telecom window. The first fiber DFB laser demonstration used a single Bragg grating at $1.5\,\mu m$ written directly into a 2 cm-long Er^{3+}/Yb^{3+}-doped fiber. Robust single-frequency operation was achieved by locally heating the center of the grating to create the necessary phase shift [11]. Robust single-mode operation was achieved with an Er^{3+}-doped DFB fiber laser based on a 36 mm-long Bragg grating with a permanently UV-induced $\pi/2$ phase-shift, yielding an output power of 5.4 mW and a linewidth of 15 kHz [12]. A number of different DFB lasers followed using either Er^{3+} or Er^{3+}/Yb^{3+}-codoped fibers [13,14]. In order to avoid clustering, Er^{3+}-doped fibers are usually of low concentration and result in low efficiencies (a few %) and relatively low output powers

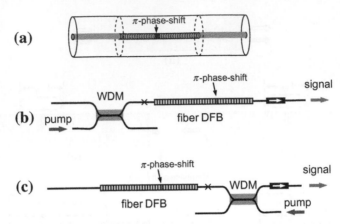

FIGURE 1.2 (a) Fiber DFB laser schematic and typical, (b) co-, and (c) counter-pumping schemes.

(a few mW). Yb^{3+} codoping results in much more efficient pump absorption and gives increased efficiencies (tens of %) and larger output powers (tens of mW).

1.2.2.1 Grating writing techniques

Grating writing techniques are essential in developing robust, high-performance fiber DFB lasers. Their modal and polarization performance depends critically on the writing beam characteristics. Fiber DFB laser fabrication is based primarily on well-established ultraviolet (UV) light fiber-Bragg grating writing techniques [15]. They use predominantly Argon-ion 244 nm lasers, because of their superior stability and beam quality, to either directly scan phase masks—with all the apodization and phase characteristics designed in—or multiple "exposure-and-shift" techniques, which provide more flexibility and versatility [16]. The aforementioned techniques require specially prepared and photo-sensitized fiber and somehow limit the range of fibers that can be ultimately used. Lately other grating inscribing techniques have been developed which rely on femtosecond pulsed lasers in the near infrared and do not require special photosensitive fibers [17], opening up almost endless possibilities in terms of candidate fiber host materials and geometries.

1.2.2.2 Novel fiber DFB cavity designs

Despite the performance improvements achieved by the introduction of the center or off-center π-phase-shift, uniform DFB lasers are still sub-optimum, in that the gain medium is not utilized effectively throughout its length and the power extraction efficiency as well as the side-mode suppression are reduced. The directionality and power extraction efficiency can be improved further by using novel step-apodized, π-phase-shifted refractive-index profiles [18] or optimized π-phase-shifted structures with ultimate efficiency [19], where the extraction efficiency is maximized over the entire laser length.

1.2.2.2.1 Step-apodized, asymmetric π-phase-shifted designs

Propagation in an active fiber with periodically modified core refractive index $n(z) = n_0 + \Delta n(z) \cos[(2\pi/\Lambda)z + \phi]$, with period Λ and refractive index change profile $\Delta n(z)$ (see Figure 1.1), is described by the standard coupled-mode equations for the two counter-propagating field amplitudes $R_+(z)$ and $R_-(z)$ [19], namely:

$$\frac{dR_+(z)}{dz} = +\alpha(z)R_+(z) + \kappa(z)R_-(z)\exp(+i\Phi z), \qquad (1.1)$$

$$\frac{dR_-(z)}{dz} = -\alpha(z)R_-(z) + \kappa(z)R_+(z)\exp(-i\Phi z). \qquad (1.2)$$

In the equations above, $\alpha(z)$ is the gain distribution, $\kappa(z) \approx (\pi/\lambda)\Delta n(z)$ is the coupling constant, $\Phi = 2(\beta - \pi/\Lambda)$, β is the propagation constant, and λ is the free-space wavelength. In a uniform grating, when $\alpha \ll \kappa$ (as is the case in practical fiber DFBs) the effective penetration depth at the Bragg wavelength ($\lambda_B = 2n_0\Lambda$) is approximated by $d \approx \tanh(\kappa L)/(2\kappa)$. By changing the grating strengths κ_1 and κ_2 and lengths L_1 and L_2 of the uniform gratings on either side of the phase shift, as shown schematically in Figure 1.1d, the effective cavity length d_T, defined as the sum of effective penetration depths ($d_T = d_1 + d_2$), is increased and, in addition to uni-directionality, the asymmetric DFB laser shows enhanced efficiency. Figure 1.3a shows that in comparison with the uniform asymmetric phase-shift design, when the phase shift is moved toward the input end, the step-apodized effective cavity length increases by $\sim$23% and the output power increases by $\sim$33%. The proportionally larger increase in output power is due to the fact that in addition to the effective cavity length increase, the step-apodized designs result in better overall saturation and larger extraction efficiency.

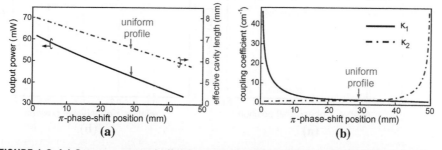

FIGURE 1.3 (a) Output power and effective cavity lengths and (b) required grating strengths as a function of π-phase-shift position (the arrows indicate the parameters of the optimum uniform design).

The active fiber length is kept constant in all cases. Figure 1.3b plots the required variation of the grating coupling constants showing that the maximum improvements require extremely strong gratings on the input side. Such strengths are not possible with common grating writing techniques and achievement of the maximum efficiencies might require use of butt-coupled mirrors or spliced external passive gratings [20].

Figure 1.4a and b shows the threshold gain for the fundamental mode and gain margins with respect to the second-order side modes, respectively, for step-apodized fiber DFBs, as a function of π-phase-shift position. They also include the corresponding parameters for a uniform DFB without phase shift and with optimized π-phase-shift. All lasers have the same length and optimum coupling constants. It is shown that compared to uniform structures with asymmetrically placed π-phase-shift, the step-apodized designs show lower threshold for the fundamental lasing mode and substantially increased gain margins. The uniform DFB structure with no phase shift shows the lowest threshold, but as already mentioned it is bidirectional and dual-wavelength and, therefore, of limited practical use. The gain margins of the step-apodized design are larger than the other two configurations.

Figure 1.5a and b compares the internal gain and signal power distribution for the conventional uniform asymmetric and step-apodized π-phase-shifted designs. The corresponding grating strengths are shown in Figure 1.3b. It is evident that the step-apodized design shows a larger saturated area and it is expected to have higher efficiency. Figure 1.5c, on the other hand, shows the corresponding pump power distribution and the positions of effective reflections and delineates the areas of the resulting effective cavities. It is realized that in the case of the step-apodized design not only the effective cavity is longer (7.4 mm as opposed to 6.6 mm in the case of uniform profile), but there is also more pump power delivered at the effective cavity area. Finally, due to the higher saturation more pump power is absorbed. The combination of all these effects results in the aforementioned efficiency increase.

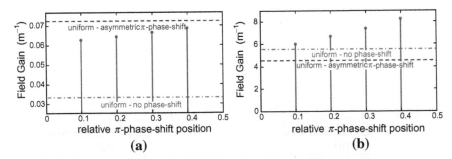

FIGURE 1.4 (a) Fundamental mode threshold gain and (b) threshold gain margin as a function of π-phase-shift position. Uniform DFB without phase shift and with optimized π-phase-shift are also shown for comparison.

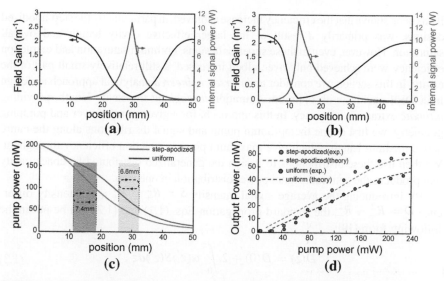

FIGURE 1.5 Internal gain and signal power distribution for (a) conventional uniform asymmetric and (b) step-apodized π-phase-shifted designs, (c) corresponding pump power distribution and the positions of effective reflections and effective cavities, and (d) experimental results (from [18]).

We applied experimentally the step-apodized design approach in an Er^{3+}/Yb^{3+}-codoped fiber and compared it to a conventional uniform asymmetric design. The alumino-phospho-silicate fiber core was surrounded by a photosensitive germano-silicate ring doped with boron, matching the refractive index of the silica cladding. The core radius of the fiber was 2.3 μm and the cut-off wavelength was 1150 nm, giving NA = 0.19. The refractive index grating was written in the photosensitive ring by exposing the fiber to ultraviolet interference pattern produced by a phase mask. The optimum coupling coefficient and phase-shift position for maximum unidirectional output of the uniform profile were experimentally found to be 1.53 cm^{-1} and 29 mm, respectively, for a 50 mm-long device. These experimental results are in very good agreement with the simulation predictions. Using these values for a 50 mm-long step-apodized DFB profile, the optimum phase-shift position was calculated to be ~12.5 mm. Figure 1.5d compares the experimental and theoretical output power variation with input pump power for two fiber DFB lasers. It is shown that the step-apodized design results in ~33% efficiency increase, in close agreement with the theory.

1.2.2.3 Inverse-engineered designs with ultimate efficiency

So far, the optimum values of phase shift and grating coupling constant are found through a *parametric optimization* process of an a priori defined cavity by varying the parameters over predetermined ranges, either by simulation or by experimentation.

It has been shown that the efficiency increase achieved, in particular by the step-apodized designs, was primarily a result of the larger effective cavity length and overall saturation. However, even in these superior designs, optimum saturation and extraction efficiency were achieved only over, albeit increased, a still relatively small part of the cavity. In this section, we consider an *inverse-engineered* analytical approach to design fiber DFB cavities with optimum saturation *throughout* the cavity length, providing *ultimate* extraction efficiency. In this approach, for a given pump power and pumping geometry, we first define the optimum pump and signal distributions along the entire active medium length, which give maximum power extraction efficiency throughout. We then use these distributions to design the generalized distributed-feedback cavity that produces the optimum signal power distribution in one step.

By introducing the average signal intensity $S = R_+^2 + R_-^2$ and intensity difference $D = R_+^2 - R_-^2$, the standard propagation Eqs. (1.1) and (1.2) can be put in the following form [19]:

$$D(z) = D(0) + 2 \int_0^z \alpha(z')S(z')dz', \tag{1.3}$$

$$\kappa(z) = \frac{1}{\sqrt{S(z)^2 - D(z)^2}} \left(\frac{1}{2}\frac{dS(z)}{dz} + D(z)\alpha(z) \right) \left\{ \cos\left[\left(2\beta - \frac{2\pi}{\Lambda(z)} \right) \right] \right\}^{-1}. \tag{1.4}$$

Equations (1.3) and (1.4) constitute the basis of the new design method. Given $S(z)$, $\alpha(z)$, and $\Lambda(z)$, we can use Eq. (1.3) to calculate $D(z)$ and then the required (z) distribution can be calculated through Eq. (1.4). In the case of a real and physically realizable active medium, the local gain distribution is a function of the pump and total signal intensity, as defined by the appropriate rate equations. Therefore, for a given pump power distribution, the choice of $S(z)$ implicitly defines $\alpha(z)$. The $D(z)$ variation is fully determined by the laser boundary conditions and defines the outputs at both laser ends as well as the device length.

In order to design a laser cavity that provides the maximum possible efficiency, we should first define the optimum $S(z)$. At any point in an active medium the energy transfer from pump to signal gives the local conversion efficiency, which when maximized results in a laser with ultimate efficiency. The net signal generated per unit length $\Delta S = 2\alpha(z)S(z)$ and the absorbed pump per unit length $\Delta P = 2\alpha_p(z)P(z)$ where α_p is the absorption coefficient for pump field and P is the pump power. The local pump-to-signal conversion efficiency $\eta(S, P) = \Delta S(S, P)/\Delta P(S, P)$ is a function of signal S and pump intensity P and it is determined by the rate equations and various loss mechanisms of the active medium. This implies that for a given pump intensity P, there exists certain signal intensity S_{opt} such that the conversion efficiency η is maximum. Figure 1.6 shows (a) the S_{opt} and the corresponding optimum gain α and (b) the resultant generated signal (ΔS) and absorbed pump power (ΔP) for an Er^{3+}/Yb^{3+}-codoped fiber for pump powers up to 200 mW at 976 nm and signal wavelength at 1550 nm. Figure 1.6c shows the maximum expected efficiency (left) and output power (right) for the ultimate efficiency design. The curves in Figure 1.6 can be considered as the master curves, which for a given active fiber enable the design of maximum efficiency DFB lasers. Firstly, the input pump power and pumping scheme are defined, then the spatial distributions of pump power $P(z)$, signal power $S_{opt}(z)$, and corresponding gain $\alpha(z)$ are calculated using the results of Figure 1.6 in an iterative way. Assuming a co-pumped configuration, $P(0)$ is launched from the left-hand side and the majority of the output signal power is expected to exit from the right-hand side. At $z=0$ we find $S_{opt}(0)$ and calculate $\Delta P(0)$. The pump power exiting the infinitesimal segment at δz is then $P(\delta z)=P(0)-\Delta P(0)\delta z$. This is used to calculate $S_{opt}(\delta z)$, $\alpha(\delta z)$, and $\Delta P(\delta z)$, and proceed to $z=2\delta z$. Repeating these steps until the pump power reduces to zero gives the entire spatial distributions $S_{opt}(z)$ and $\alpha(z)$ (see Figure 1.6a and b). $D(z)$ can then be calculated using Eq. (1.3).

The boundary conditions at $z=0$ require a sharp transition between the $S_{opt}(0)$ and the small unwanted output power. This results in an extremely large delta-like function κ at $z=0$, which is beyond any grating writing technology capability. Therefore, we should define a smooth, physically realizable transition between a relatively small $D(0)$ and S_{opt}. $S(z)$ in such transition regions can be described by a $\cosh(mz)$ function (shown by dashed line in Figure 1.7b), which can be shown to correspond to an almost constant (z) (see Figure 1.7c). We then calculate $D(z)$ and find the position $z=L$ where the second boundary condition for $D(L)$ is met, giving the total device length. In the

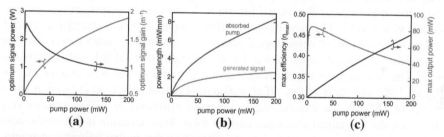

FIGURE 1.6 (a) Optimum signal power, (b) power extraction, and (c) maximum efficiency.

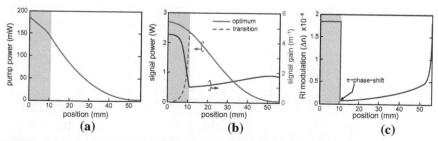

FIGURE 1.7 (a) Pump power, (b) signal power and local field gain, and (c) refractive index modulation as a function of position for a DFB laser with maximum efficiency.

co-directional pumping scheme $S_{opt}(L)$ is equal to the output power at $z=L$. Finally, substituting $S(z)$, $\alpha(z)$, $D(z)$ in Eq. (1.1) we can calculate the required apodization profile $\kappa(z)$, as shown in Figure 1.7c.

The grating apodization profile will depend on the pumping direction also as this will change the balance between optimum pump and signal distributions. Figures 1.8a and b show the refractive-index modulation profile for counter- and bi-directional pumping, respectively, for the same total pump power of 185 mW at 977 nm. It is shown that the pump direction affects the required optimum design, with the counter- and bi-directional designs requiring longer active fiber lengths and the bi-directional pumping scheme requiring three π-phase-shifts. Figure 1.8c compares the signal output power as a function of pump power for the three inverse-engineered DFB laser designs and contrasts them to the uniform, asymmetric π-phase-shifted configuration of the same active fiber length. It is shown that using the new designs, co-directional pumping results in a slightly smaller efficiency because it requires the largest unoptimized transition region. Bi-directional pumping, on the other hand, results in the best efficiency because the available pump power is split equally between the two ends. Smaller local pump intensities result in higher local efficiencies, with a positive net effect on the overall efficiency. All output power curves are shown to be very linear up to the design pump power of ∼185 mW after which they start to saturate.

Figure 1.9a compares the experimental and simulation results for the 45 mm-long new design with a co-pumped 50 mm-long standard optimized design. Under the same operating conditions, a dramatic 57% increase in output power and efficiency is achieved. The new design is 10% shorter, shows a reduced threshold, and provides 63 mW of output signal power with 230 mW of pump power. This we believe is the highest output power for the shortest fiber DFB length reported to date. The standard optimized design uses the pump power less efficiently (see Figure 1.9b) and starts rolling off at ∼30 mW of output power. At the design pump power of 185 mW, the experimentally observed efficiency of 33% compares very well with the theoretical efficiency limit of 39%. This deviation is due to the non-optimized transition region at the input, imposed by the grating writing constraints.

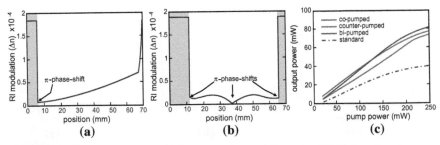

FIGURE 1.8 Inverse-engineered designs with (a) counter pumping, (b) bi-directional pumping, and (c) output power comparison.

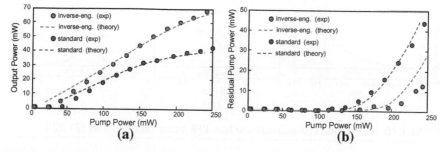

FIGURE 1.9 Comparison of (a) output powers and (b) residual pump powers from standard and new design DFB lasers (adapted from [19]).

It is also observed that the input/output linearity deteriorates and residual pump increases above the 185 mW design pump level. The new design operated in a single longitudinal mode and showed ~25 dB output power directionality at all the pump power up to 250 mW.

1.2.3 Optical performance of fiber DFB lasers

The effective integration of laser sources in high-performance systems/subsystems depends on a number of characteristics, such as polarization and wavelength stability as well as phase/frequency noise and/or relative intensity noise (RIN). These performance characteristics can be controlled accurately by proper fiber design.

1.2.3.1 Output power/polarization

The control of output signal polarization is extremely important since advanced lasers are invariably interfaced with a number of polarization-sensitive devices, such as modulators, isolators, and filters. The output polarization of fiber DFB lasers can be influenced by a number of fiber and cavity parameters, such as fiber birefringence and twist, polarization-dependent grating strength and/or phase shift as well as grating chirp [21,22]. The polarization mode competition and overall performance are influenced by the magnitude of the resulting polarization hole-burning and the local as well as global spatial hole-burning effects.

Figure 1.10a and b shows the normalized grating strength requirements as a function of polarization-dependent grating strength difference and phase-shift difference, respectively, in order to achieve single polarization operation. In this case, $\Delta \kappa = \kappa_x - \kappa_y$ is the grating strength difference along the x and y polarizations, $\Delta \phi = \Delta \phi_y - \pi$ is the phase error in the y-polarization alone, while always $\Delta \phi_x = \pi$. It is shown that in the absence of sufficient grating strength difference and/or phase-shift difference, the fiber DFB lasers operate in two polarizations. It is also shown that the stronger the fiber grating, the larger the required $\Delta \kappa$ and $\Delta \phi$ in order to achieve single polarization operation. The behavior in the presence of fiber twist is

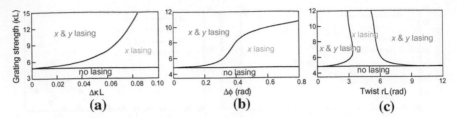

FIGURE 1.10 Polarization characteristics of fiber DFB lasers (adapted from [21,22]).

more complicated. Figure 1.10c shows a "window" of single polarization operation for moderate fiber twist, with dual polarization operation on either side. The width of the single polarization window depends on the grating strength.

Single polarization operation has been achieved by a number of techniques, such as twisting ordinary low-bi fibers [23], controlled UV exposure resulting in polarization-dependent phase shift and/or grating strength [24–27], or controlled-polarization back-reflection injection locking [28]. In most applications, single polarization DFB lasers are required to be subsequently spliced to hi-bi passive fibers in order to maintain polarization. Birefringent axes in hi-bi fibers, on the other hand, are known to rotate along the fiber length, with different rotation rates. It has been shown that single-mode, single-polarization fiber DFB lasers written directly in active hi-bi fibers exhibit elliptical polarization output. This is primarily due to inherent internal, frozen-in birefringence-axes rotation in hi-bi fibers. In this case, application of external birefringence-axes twist can be used to manipulate and control accurately the output state of polarization.

Figure 1.11a shows the schematic of a hi-bi fiber DFB laser, spliced to a matching passive hi-bi fiber. The passive fiber was angle-cleaved to minimize back-reflections. A quarter-wave plate and a high extinction ratio polarizer were used at the output in order to measure the polarization extinction ratio of the collimated laser output. The

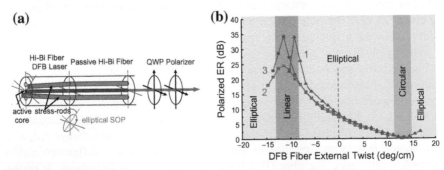

FIGURE 1.11 (a) Twisted fiber DFB laser characterization set-up, (b) PER as a function of DFB fiber twist [29].

polarization extinction ratio is defined as the max/min throughput power ratio (in dB), corresponding to two orthogonal polarizer orientations.

The active hi-bi fiber under test was measured to have an average internal axis twist of about +6°/cm. Figure 1.11b shows the output polarization extinction ratio (PER) as a function of applied DFB fiber external twist. Without any external twist, the PER was ~10 dB, which corresponds to an elliptical SOP. For twist rates of about −10 to −12°/cm the output SOP became linear, with max PER of ~35 dB. It is worth noting that the external twist required to achieve maximum PER is about two times the internal one. This is probably required in order to counteract the effect of the extra twist-induced stresses in addition to the initial geometrical twist. For opposite external twist rates of about +12°/cm the PER was ~0 dB, which corresponds to circular output SOP. For larger external twists, the PER is shown to increase again. This demonstrates an effective way of controlling and aligning the fiber DFB laser output SOP. It also shows that twisted hi-bi fiber DFB lasers can be potentially used as optical twist sensors.

1.2.3.2 Wavelength coverage

An advantage of fiber lasers in general is their ability to emit over broad bandwidths in different spectral regions by simply using different active dopants and slightly varying glass host compositions.

Using advanced grating writing techniques, the wavelength of operation in fiber DFB lasers is set by the Bragg grating period to <50 pm accuracy. Figure 1.12 shows the outputs of different π-phase-shifted fiber DFB lasers covering the entire C band. The fiber core is Er^{3+}/Yb^{3+}-codoped and the grating was written in a photosensitive boron codoped germano-silicate ring. The typical output power was 10 dBm and the wavelength stability was better than 0.14 nm over 70 °C. The optical signal-to-noise ratio, with 0.1 nm resolution, was better than 70 dB. L-band coverage has also been demonstrated using very strong, asymmetric π-phase-shifted gratings in high gain Er^{3+}-doped fibers [30,31]. Single-frequency, single-polarization outputs with power

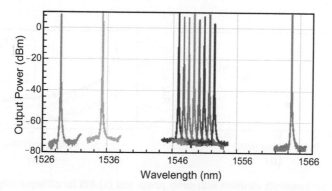

FIGURE 1.12 Er^{3+}/Yb^{3+} fiber DFB laser wavelength coverage in the C-band.

in excess of 80 mW were demonstrated with 980 nm pumping. These devices showed a rather large threshold of 16–20 mW.

In addition to more traditional Er^{3+} or Er^{3+}/Yb^{3+} fiber DFB lasers operating at telecom related wavelengths, fiber DFB lasers have also been demonstrated in Yb^{3+}-doped fibers operating at 970 nm [32] and in the 1.0 µm window [33], as well as thulium-doped fibers emitting in the 1735 nm [34] to 1943 nm [35] wavelength range. Fiber DFB lasers using Brillouin [36] and Raman amplification as the gain mechanism have also been demonstrated [37–39], opening up the possibilities for single-frequency sources in extended wavelength ranges not covered by active dopants. In addition to core pumping, cladding-pumped fiber DFB lasers have also been demonstrated as a means of scaling up the output power to 160 mW and simplifying the pumping configuration [40].

Typically Er^{3+} or Er^{3+}/Yb^{3+} are pumped at the 976–980 nm Er^{3+} and Yb^{3+} pump absorption peaks to keep the total length to a minimum. Pumping at the 520 nm [14] and 1480 nm Er^{3+} absorption bands has also been demonstrated. More complex intracavity pumping schemes have also been used for more efficient use of the pump power [41].

1.2.3.3 Linewidth and RIN performance

In addition to single stable longitudinal mode, showing no mode hopping, and single polarization output fiber DFB lasers show extremely narrow optical linewidth. Typical variation of the measured linewidth with increasing pump power is shown in Figure 1.13a for a 5 cm-long Er^{3+}/Yb^{3+} uniform, asymmetric fiber DFB laser pumped with an FBG-stabilized pump diode at 976.8 nm. The optical linewidth was measured using a delayed self-heterodyned interferometer with 5 km optical path imbalance.

At low pump powers, a linewidth of ~16 kHz is measured, but instead of the expected $1/P_{out}$ dependence, the linewidth increases with pump power. It has been shown that two noise sources are mainly responsible for the observed behavior, namely fundamental thermal noise at low pump power levels and temperature fluctuations induced by pump intensity noise at higher powers [42]. Potential techniques

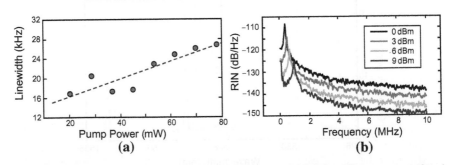

FIGURE 1.13 (a) Linewidth variation with pump power and (b) RIN for different output signal powers.

to overcome these limitations are the usage of a low noise pump or materials with a temperature-insensitive refractive index, e.g. by using specially tailored phosphate glasses. This has the added advantage of reducing both the pump fluctuation-induced linewidth broadening and the fundamental thermal noise. Varying the length of the grating, while keeping the grating strength constant, also modifies the fundamental thermal noise. Similarly, the fiber core radius can be enlarged to increase the mode volume and thereby decrease the thermal effects. Another approach to reduce the linewidth deterioration by pump noise is to use the DFB laser at low power in a master oscillator power amplifier (MOPA) to boost the power.

The relative intensity noise (RIN) measurement is plotted in Figure 1.13b for different signal output powers. It is shown that the RIN reduces as the output power increases and approaches −150 dB/Hz beyond 10 MHz. This is accompanied by a very low peak RIN of −125 dB/Hz at the relaxation frequency of 1 MHz. This is achieved without the added complication of feedback noise-reduction circuits.

Using a low concentration Er^{3+}-doped fiber, an asymmetric DFB laser has been demonstrated with sub-kHz (~250 Hz) linewidth, without applying any external stabilization technique. The DFB laser length was 17 mm and the phase shift was introduced by UV post-processing. Due to short length and low Er^{3+} concentration, however, the laser showed relatively low optical efficiency [43].

1.2.3.4 Phase/frequency noise performance

As already mentioned, optical phase is used lately to encode data in advanced optical communications systems, resulting in record spectral efficiencies [44]. The phase characteristics of optical sources are also shown to play an important role in a number of other advanced applications, such as high-performance interferometric optical sensors, ultrasensitive interferometer-based phase measurements in optical component response, and ultrahigh resolution optical spectroscopy. In all the aforementioned applications the absolute frequency stability, spectral linewidth, and spectrally/temporally dependent phase/frequency noise characteristics of the lasers used are critical to understanding/optimizing the system performance. Theoretical studies have predicted that the phase noise and linewidth of the laser source can affect significantly the transmission penalties in coherent-optical systems [45,46].

Figure 1.14a shows the frequency noise spectrum of a free-running fiber DFB laser compared to other laser sources [47,48]. The theoretical thermal noise floor of a 10 mm-long fiber cavity is shown as a dashed line. Measured noise levels for other commonly used laser sources, such as a Nd:YAG ring laser, a Hitachi HLP 1400 semiconductor laser, and a commercial semiconductor DFB laser with 1.6 MHz linewidth, are included for comparison. The estimated frequency noise level of a bulk glass Er^{3+}/Yb^{3+} laser is indicated with a diamond. Figure 1.14b compares the frequency noise spectrum of commercially available fiber DFB (Basik) with the ones of a fiber-based Fabry-Perot laser (ROCK), a hybrid integrated-optics/semiconductor laser (RIO), and a virtual-ring waveguide laser (Orbit) [49]. It is shown that the fiber DFB and the virtual-ring waveguide lasers show better frequency noise performance, with the fiber laser having the edge especially at low frequencies.

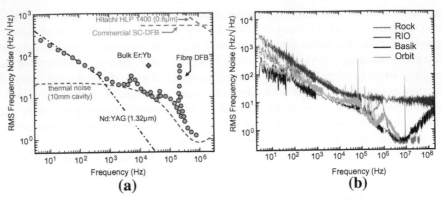

FIGURE 1.14 RMS frequency noise measurements (a) of a free-running fiber DFB laser [48] and (b) comparison of commercially available lasers [49].

A fundamental source of laser frequency noise is the thermodynamically defined thermal energy within the fiber. For a fiber DFB laser with a 1 cm-long effective cavity length, the expected thermal noise floor decreases from about $25\,dBHz/\sqrt{Hz}$ at $1\,kHz$ to about $-5\,dBHz/\sqrt{Hz}$ at $1\,MHz$. Below $50\,kHz$ the noise increase above the thermal noise floor is believed to be caused by acoustical vibrations. Above $1\,kHz$ and away from the relaxation oscillation frequency (f_{ro}), the measured frequency noise floor varies from 0 to $6\,dB$ above the expected thermal noise floor. The origin of excess frequency noise above $f = 50\,kHz$ is not understood in detail. The use of longer cavities is proven to lower the frequency noise contribution from fundamental thermal fluctuations, and therefore the previously mentioned step-apodized and inverse-engineered designs, characterized by substantially longer effective cavities, are expected to show lower frequency noise. For frequencies below $1\,kHz$ the rms frequency noise follows a $1/f$ dependence, the origin of which has not been fully identified. Around f_{ro} the optical frequency noise and RIN are found to be strongly interrelated and associated by the linewidth enhancement factor [47].

1.2.3.5 Tunability

Tunable lasers are of paramount importance in modern telecom systems as they can be used to populate different channel slots and reduce laser inventory. In addition, they can also provide a route to reconfigurable WDM systems. The response times are in the $0.1\,ms–1\,s$ range, depending on the actuator and tuning mechanism utilized. The lower limit is imposed by the acoustic velocity of strain waves in the fibers. Fiber DFB lasers can be tuned effectively by heating, stretching, and/or compressing uniformly the grating cavity. Continuous tuning over $27\,nm$ has been demonstrated in Er^{3+}/Yb^{3+} all-fiber, DFB lasers using a simple tuning mechanism for axial extension and compression. The demonstrated devices operated with powers up to $10\,dBm$ and remain operating in single mode over the full tuning range [50].

1.2.3.6 Power scaling—master-oscillator power amplifiers (MOPAs)

Optimized step-apodized and inverse-engineered designs provide maximum efficiency and output powers in the 10s of mW range. Higher powers are potentially needed if the output of a single DFB laser is to be split in multiple channels. Also, output powers should be scaled up to higher levels, in order to make fiber DFB lasers suitable for medical and other industrial applications. A single-frequency narrow-linewidth laser at 1552 nm was demonstrated with an output power of 83 W using a MOPA configuration. A fiber DFB seed laser with a linewidth of 13 kHz was polarization scrambled, in order to avoid the onset of SBS, and boosted through a chain of fiber amplifiers [51]. Using a MOPA configuration, the output of an Yb^{3+}-doped all-fiber DFB laser was boosted to 400 mW [52]. The DFB fiber laser was pumped by a 976 nm amplified spontaneous emission (ASE) source based on an Yb^{3+}-doped jacketed-air-clad fiber pumped by a 915 nm multimode laser diode source. The total output power response was approximately linear and the overall performance was limited by the available pump power. The spectral characteristics and signal-to-noise ratio remained similar to the master-oscillator DFB laser over the entire output power range.

1.2.4 Multi-wavelength fiber DFB lasers and fiber DFB laser arrays

In addition to single-wavelength outputs, fiber DFB lasers can be designed to provide multi-wavelength outputs. Moire-type fiber DFB lasers operating simultaneously on two-wavelength channels around 1.55 μm at room temperature have been realized. The demonstrated lasers have channel separations of 25, 50, and 100 GHz with identical output powers of 1 mW [53]. Multi-wavelength fiber lasers were also obtained by writing two superimposed chirped fiber 8 cm-long Bragg gratings in a photosensitive Er^{3+}/Yb^{3+}-codoped optical fiber. The chirped gratings create a distributed Fabry-Pérot structure in which the resonating fields of the different laser lines are spatially separated, thus, reducing gain cross-talk. Multi-wavelength lasers emitting simultaneously over 8 and 16 lines spaced by 50 GHz have been demonstrated [54].

Despite their simplicity, in multi-wavelength DFB laser sources all channels are intertwined and co-exist inside the fiber cavity. Although this might be beneficial for sensor applications, possible use in telecom systems requires separation and individual access of all the different channels. To this end, fiber DFB laser arrays with pump redundancy have been demonstrated [55]. Figure 1.15a shows a schematic of the 16-channel transmitter module. This module consists of eight pump diodes, a pump redundancy unit, and sixteen 5 cm-long asymmetric fiber DFB lasers with isolators on the output end. The pump redundancy module splits the powers from the pumps equally between the fiber DFB lasers, such that each laser still receives the pump power corresponding to that provided by one pump diode. The particular module is made up of twenty 2×2 3 dB couplers at 980 nm and splits equally the eight inputs to 16 outputs. The total insertion loss for each pump input channel in this 16-channel module is ~1.5 dB.

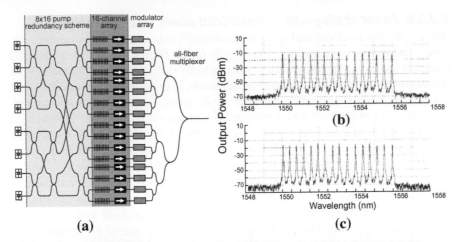

FIGURE 1.15 (a) Schematic of 16-channel fiber DFB laser array with pump redundancy. 50 GHz WDM transmitter module output with (b) all pumps, (c) four pumps only [55].

Sixteen fiber DFB lasers separated in frequency by 50 GHz were fabricated, operating in a single polarization mode with a purity of 40 dB and with a single-sided output power ratio of 50:1. The slope efficiency of the lasers was ~25% and using the pump redundancy scheme with 50 mW power from each pump diode resulted in output powers of about −10 dBm ± 0.25 dBm for all channels. The outputs of the 16 lasers were combined in an all-fiber multiplexer consisting of fifteen 1550 nm 3 dB splitters with a total insertion loss of ~10 dB. Noise measurements on the individual channels showed RIN < −160 dB/Hz for frequencies above 10 MHz and RIN < −165 dB/Hz for frequencies larger than 30 MHz, indicating performance well suited for high-speed communication systems.

DFB fiber laser arrays with one shared pump have also been used as cost-effective and wavelength selectable optical sources. A large number of wavelengths can be selected via optical space switches, and this system is a good candidate for use as a wavelength selectable, backup transmitter for wavelength division multiplexed (WDM) systems [56].

1.2.5 Optical transmission system experiments

Fiber DFB lasers have been used successfully in a number of optical transmission system experiments. A DFB fiber laser operating at 1607 nm having 75.4 dB OSNR and RIN below −150 dB/Hz was used in a 10 Gbit/s transmission experiment over 72 km of standard single mode fiber with no observable difference when compared to an external-cavity semiconductor source [30].

WDM transmission and dispersion compensation at 40 Gbit/s over 200 km standard fiber has been demonstrated on a 100 GHz grid using four high-power single-polarization single-sided output fiber DFB laser-based transmitters and a single four-channel WDM chirped fiber-Bragg grating dispersion compensator [57]. In-line fiber DFB lasers have also been used as orthogonally polarized pump sources to provide phase conjugation, by fiber four-wave mixing, and spectral inversion for efficient mid-span dispersion compensation in optical links. This technique provides polarization-independent operation in a simple all-fiber configuration without the need for externally injected pumps [58].

Finally, fiber DFB lasers can be used in microwave photonics applications. Dual polarization mode fiber DFB lasers fabricated in an elliptical-core Er-doped fiber have been used for efficient 40 GHz optical millimeter wave generation with 3 dB beat linewidth of 900 Hz [59]. Such millimeter-wave sources can be advantageous for applications involving phased-array antennas or fiber-fed radio systems.

1.2.6 Fiber DFB laser in non-telecom applications

Due to the unique combination of output power and polarization stability, narrow linewidth and extremely low intensity and phase/frequency noise, fiber DFB lasers have already found widespread usage outside the telecom field. It should be first emphasized that the only laser in the National Ignition Facility (NIF), the largest laser system in the world, is a fiber DBF laser [60]. The 1.8 MJ energy, 500 MW peak power at the output of the NIF laser system starts from a single-fiber DFB oscillator, before it splits in 192 chains of multiple solid-state high-power amplifiers. The Yb-doped oscillator produces $\sim$20 mW of output power at 1053.01 ± 0.01 nm in a single longitudinal mode when pumped with 130 mW from a 980 nm laser diode.

Fiber DFB lasers have been used in a number of sensor applications such as high-performance seeds for sensor arrays [61] enabling remote measurements from a large-scale interferometric optical sensor system, using a 500 km optical transmission link between interrogator and sensor array. A phase noise floor of -80 dB re 1 rad/$\sqrt{\text{Hz}}$ peak was achieved (equivalent to 1 mPa/$\sqrt{\text{Hz}}$).

DFB fiber lasers have also been used as the sensing elements [62]. Radio-frequency (RF) beat frequencies between two longitudinal modes and two polarization modes of a birefringent dual-longitudinal-mode moire distributed-feedback fiber laser have been employed to measure strain and temperature simultaneously. Operating entirely in the RF domain, this approach potentially allows one to employ low-cost and precise RF measuring techniques. The achieved sensor accuracy was $\pm 15\,\mu\varepsilon$ and $\pm 0.2\,°\text{C}$ [63]. They have also been used as sensors to characterize temperature distribution along other DFB lasers [64].

1.3 SUMMARY AND CONCLUDING REMARKS—OUTLOOK

This chapter has reviewed the latest developments in high-performance fiber lasers, and their potential use in modern high spectral efficiency coherent optical communication systems. We have presented results on asymmetric, step-apodized, and inverse-engineered fiber DFB laser designs with high ultimate efficiency. We have also discussed the output power scalability, wavelength coverage, tunability, polarization properties, output power stability and linewidth, as well as frequency noise characteristics of these lasers.

It has been shown that modern optical communication systems using nonbinary modulation formats and coherent detection provide maximum spectral efficiency and improve tolerances to other transmission impairments [65]. In has been predicted that compared to other coherent schemes, 16-QAM modulation with coherent detection results in 4 bits/symbol efficiency with minimum possible optical signal-to-noise-ratio (OSNR) requirements. However, in coherent optical systems the increased sensitivity to phase noise and frequency stability sets extremely stringent requirements on laser oscillator linewidth. It has also been shown that other effects, like dispersion-enhanced phase noise, induce severe degradations to advanced optical systems performance, such as CO-OFDM [47]. In this case, to reduce the OSNR requirements for transmission of 56 Gbaud DP-QPSK (224 Gb/s) over 3200 km requires lasers with linewidth of a few 10–100 kHz. Undoubtedly, transmission over longer distances and higher repetition rates will further increase the requirements on laser linewidth and phase noise. We have shown that fiber DFB lasers can exhibit record sub-kHz linewidth without the need of external controls and offer themselves as cheap and versatile replacements of currently used expensive and complicated external-cavity semiconductor counterparts [66]. We have also shown frequency noise comparison showing superior fiber DFB laser performance over other fiber and semiconductor lasers, especially in the low frequency regime. This feature is proving to be particularly advantageous in other photonic applications, such as interferometric sensor arrays.

Although fiber lasers show a number of advantages over other technologies in terms of linewidth, phase noise, power scalability, stability, and simplicity more research is required to address the remaining issues affecting the ultimate linewidth limits and phase noise performance. Work is required to further understand the impact of host material and cavity design, output power level, and pump characteristics on the aforementioned fundamental limits. Also more effort is needed to study the parameters that affect the single polarization performance and polarization control. Finally, further work should be carried out on alternative gain mechanisms such as Raman and Brillouin DFBs to increase the wavelength coverage.

References

[1] M. Nakazawa, T. Hirooka, M. Yoshida, K. Kasai, Ultrafast coherent optical transmission, J. Sel. Top. Quant. Electron. 18 (2012) 363.

[2] S.L. Jensen, I. Morita, H. Tanaka, 10 Gb/s OFDM with conventional DFB lasers, ECOC, Paper Tu.2.5.2 2007.

[3] W-R. Peng, Analysis of laser phase noise effect in direct-detection optical OFDM transmission, J. Lightwave Technol. 28 (2010) 2526.

[4] M. Nakazawa, Recent progress on ultrafast/ultrashort/frequency-stabilized erbium-doped fiber lasers and their applications, Front. Optoelectron. China 3 (2010) 38.

[5] H. Kogelnik, C.V. Shank, Stimulated emission in a periodic structure, Appl. Phys. Lett. 18 (1971) 152.

[6] H. Kogelnik, C.V. Shank, Coupled-wave theory of distributed feedback lasers, J. Appl. Phys. 43 (1972) 2327–2335.

[7] M. Nakamura, A. Yariv, H.W. Yen, S. Somekh, H.L. Garvin, Optically pumped GaAs surface laser with corrugation feedback, Appl. Phys. 22 (1973) 515.

[8] D.R. Scifres, R.D. Burnham, W. Streifer, Distributed feedback single heterojunction GaAs diode laser, Appl. Phys. Lett. 25 (1974) 203.

[9] H.A. Haus, C.V. Shank, Antisymmetric taper of distributed feedback lasers, J. Quant. Electron. 12 (1976) 532.

[10] K. Utaka, S. Akiba, K. Sakai, Y. Matsushima, $\lambda/4$-shifted InGaAsP/InP DFB lasers, J. Quant. Electron. 22 (1986) 1042.

[11] J.T. Kringlebotn, J.L. Archambault, L. Reekie, D.N. Payne, "Er^{3+}-Yb^{3+}-codoped fiber distributed-feedback laser, Opt. Lett. 19 (24) (1994) 2101.

[12] M. Sejka, P. Varming, J. Hubner, M. Kristensen, Distributed-feedback Er^{3+}-doped fiber laser, Electron. Lett. 31 (1995) 1445.

[13] W.H. Loh, R.I. Laming, 1.55 μm phase-shifted distributed feedback fiber laser, Electron. Lett. 31 (1995) 1440.

[14] W.H. Loh, S.D. Butterworth, W.A. Clarkson, Efficient distributed feedback erbium-doped germanosilicate fiber laser pumped in 520 nm band, Electron. Lett. 32 (1996) 2088.

[15] G. Meltz, W.W. Morey, W.H. Glenn, Formation of Bragg gratings in optical fibers by a transverse holographic method, Opt. Lett. 14 (1989) 823.

[16] M.J. Cole, W.H. Loh, R.I. Laming, M.N. Zervas, S. Barcelos, Moving fibre/phase mask-scanning beam technique for enhanced flexibility in producing fibre gratings with a uniform phase mask, Electron. Lett. 31 (1995) 1488.

[17] S.J. Mihailov, C.W. Smelser, D. Grobnic, R.B. Walker, L. Ping Lu, D. Huimin, J. Unruh, Bragg gratings written in all-SiO_2 and Ge-doped core fibers with 800-nm femtosecond radiation and a phase mask, J. Lightwave Technol. 22 (2004) 94.

[18] K. Yelen, L.M.B. Hickey, M.N. Zervas, A new design approach for fiber DFB lasers with improved efficiency, J. Quant. Electron. 40 (2004) 711.

[19] K. Yelen, M.N. Zervas, L.M.B. Hickey, Fiber DFB lasers with ultimate efficiency, J. Lightwave Technol. 23 (2005) 32.

[20] W.H. Loh, B.N. Samson, L. Dong, G.J. Cowle, K. Hsu, High performance single frequency fiber grating-based erbium:ytterbium-codoped fiber laser, J. Lightwave Technol. 16 (1998) 114.

[21] E. Ronnekleiv, M.N. Zervas, J.T. Kringlebotn, Modeling of polarization mode competition in fiber DFB lasers, J. Quant. Electron. 34 (1999) 1559.

[22] E. Ronnekleiv, M.N. Zervas, J.T. Kringlebotn, Corrections to modeling of polarization mode competition in fiber DFB lasers, J. Quant. Electron. 35 (1999) 1097.

[23] Z.E. Harutjunian, W.H. Loh, R.I. Laming, D.N. Payne, Single polarization twisted DFB laser, Electron. Lett. 32 (1996) 346.

[24] L.B. Fu, M. Ibsen, P.W. Turner, D.J. Richardson, D.N. Payne, Keyed axis single-polarization all-fiber DFB laser, Electron. Lett. 38 (2002) 1537.

[25] H. Storøy, B. Sahlgren, R. Stubbe, Single polarization fiber DFB laser, Electron. Lett. 33 (1997) 56.

[26] J.I. Philipsen, M.O. Berendt, P. Varming, V.C. Lauridsen, J.H. Povlsen, J. Hübner, M. Kristensen, B. Pálsdóttir, Polarization control of DFB fiber laser using UV-induced birefringent phase-shift, Electron. Lett. 34 (1998) 678–679.

[27] M. Ibsen, E. Rønnekleiv, G.J. Cowle, M.O. Berendt, O. Hadeler, M.N. Zervas, R.I. Laming, Robust high-power (>20 mW) all-fiber DFB lasers with unidirectional and truly single polarization outputs, in: Proceedings of the Conference on Lasers and Electro-Optics (CLEO) 1999, Baltimore, MD, 1999, CWE4, p. 245.

[28] S. Yamashita, G.J. Cowle, Single-polarization operation of fiber distributed feedback lasers by injection locking, J. Lightwave Technol. 17 (1999) 509–513.

[29] M.N. Zervas, R. Wilmshurst, L.M.B. Walker, Twisted Hi-Bi fiber DFB lasers with controllable output polarization, in: Proceedings of the Conference on Lasers and Electro-Optics (CLEO), Paper CF3N.5, 2012.

[30] H.N. Poulsen, P. Varming, A. Buxens, A.T. Clausen, I. Munoz, P. Jeppesen, C.V. Poulsen, J.E. Pedersen, L. Eskildsen, 1607 nm DFB fibre laser for optical communications in the L-band, in: European Conference on Optical Communication, ECOC '99, Nice, France, September 1999, Paper MoB2.1.

[31] P. Varming, V.C. Lauridsen, J.H. Povlsen, J.B. Jensen, M. Kristensen, B. Palsdottir, Design and fabrication of Bragg grating based DFB fiber lasers operating above 1610 nm, in: Optical Fiber Communication Conference, OSA Technical Digest Series (Optical Society of America, 2000), Paper ThA6.

[32] L.B. Fu, M. Ibsen, D.J. Richardson, D.N. Payne, 977-nm all-fiber DFB laser, Photon. Technol. Lett. 16 (2004) 2442.

[33] A. Asseh, H. Storøy, J.T. Kringlebotn, W. Margulis, B. Sahlgren, S. Sandgren, R. Stubbe, G. Edwall, 10 cm Yb^{3+} DFB fibre laser with permanent phase shifted grating, Electron. Lett. 31 (1995) 969.

[34] S. Agger, J. Hedegaard Povlsen, P. Varming, Single-frequency thulium-doped distributed-feedback fiber laser, Opt. Lett. 29 (2004) 1503.

[35] Z. Zhang, D.Y. Shen, A.J. Boyland, J.K. Sahu, W.A. Clarkson, M. Ibsen, High-power Tm-doped fiber distributed-feedback laser at 1943 nm, Opt. Lett. 33 (2008) 2059.

[36] K.S. Abedin, P.S. Westbrook, J.W. Nicholson, J. Porque, T. Kremp, X. Liu, Distributed feedback fiber laser employing Brillouin gain, in: Proceedings of the Conference on Lasers and Electro-Optics (CLEO), Paper CF3N, 2012.

[37] V.E. Perlin, H.G. Winful, Distributed feedback fiber Raman laser, J. Quantum Electron. 37 (2001) 38.

[38] Y. Hu, N.G.R. Broderick, Improved design of a DFB Raman fibre laser, Opt. Commun. 282 (2009) 3356.

[39] P.S. Westbrook, K.S. Abedin, J.W. Nicholson, T. Kremp, J. Porque, Raman fiber distributed feedback lasers, Opt. Lett. 36 (2011) 2895.

[40] A. Schülzgen, L. Li, D. Nguyen, Ch. Spiegelberg, R. Matei Rogojan, A. Laronche, J. Albert, N. Peyghambarian, Distributed feedback fiber laser pumped by multimode pump diodes, Opt. Lett. 33 (2008) 614.

[41] W.H. Loh, B.N. Samson, Z.E. Harutjunian, R.I. Laming, Intracavity pumping for increased output power from a distributed feedback erbium fiber laser, Electron. Lett. 32 (1996) 1204.

[42] P. Horak, N.Y. Voo, M. Ibsen, W.H. Loh, Pump-noise-induced linewidth contributions in dfb fiber lasers, Photon. Technol. Lett. 18 (2006) 998.

[43] A.C.L. Wong, W.H. Chung, H.Y. Tam, C. Lu, Ultra-short distributed feedback fiber laser with sub-kilohertz linewidth for sensing applications, Laser Phys. 21 (2011) 163.

[44] A.H. Gnauck, P.J. Winzer, Optical phase-shift-keyed transmission, J. Lightwave Technol. 23 (2005) 115.

[45] W. Shieh, K.-P. Ho, Equalization-enhanced phase noise for coherent-detection systems using electronic digital signal processing, Opt. Express 16 (2008) 15718.

[46] Q. Zhuge, M. Morsy-Osman, D.V. Plant, Analysis of dispersion-enhanced phase noise in CO-OFDM systems with RF-pilot phase compensation, Opt. Express 19 (2011) 24030.

[47] E. Ronnekleiv, Frequency and intensity noise of single frequency fiber Bragg grating lasers, Opt. Fiber Technol. 7 (2001) 206.

[48] E. Rønnekleiv, S.W. Løvseth, J.T. Kringlebotn, Er-doped fiber distributed feedback lasers: properties, applications and design considerations, Proc. SPIE 4943 (2003) 69.

[49] R. Slavik, Y. Liao, E. Austin, P. Petropoulos, D.J. Richardson, Full characterization and comparison of phase properties of narrow linewidth lasers operating in the C-band, in: 21st International Conference on Optical Fiber Sensors, Proc. SPIE, vol. 7753, 2011, p. 775338.

[50] M. Ibsen, S.Y. Set, G.S. Goh, K. Kikuchi, Broad-band continuously tunable all-fiber DFB lasers, Photon. Technol. Lett. 14 (2002) 21.

[51] C. Alegria, Y. Jeong, C. Codemard, J.K. Sahu, J.A. Alvarez-Chavez, L. Fu, M. Ibsen, J. Nilsson, 83-W single-frequency narrow-linewidth MOPA using large-core erbium-ytterbium co-doped fiber, Photon. Technol. Lett. 16 (2004) 1825.

[52] C.A. Codemard, L.M.B. Hickey, K. Yelen, D.B.S. Soh, R. Wixey, M. Coker, M.N. Zervas, J. Nilsson, 400 mW, 1060 nm ytterbium doped fiber DFB laser, Proc. SPIE, vol. 5335, 2004, p. 56.

[53] M. Ibsen, E. Ronnekleiv, G.J. Cowle, M.N. Zervas, R.I. Laming, Multiple wavelength all-fibre DFB lasers, Electron. Lett. 36 (2000) 143.

[54] R. Slavik, I. Castonguay, S. LaRochelle, S. Doucet, Short multiwavelength fiber laser made of a large-band distributed Fabry-Pérot structure, Photon. Technol. Lett. 16 (2004) 1017.

[55] M. Ibsen, S. Alam, M.N. Zervas, A.B. Grudinin, D.N. Payne, 8- and 16-Channel all-fiber DFB laser WDM 8- and 16-channel all-fiber DFB laser WDM, Photon. Technol. Lett. 11 (1999) 1114.

[56] X. Zheng, P.J.S. Pedersen, P. Varming, A. Buxens, Y. Qian, P. Jeppesen, Cost-effective wavelength selectable light source using DFB fibre laser array, Electron. Lett. 36 (2000) 620.

[57] M. Ibsen, A. Fu, H. Geiger, R.I. Laming, Fibre DFB lasers in a 4×10 Gbit/s WDM link with a single Sinc-sampled fibre grating dispersion compensator, in: 24th European Conference on Optical Communication, vol. 3, 20–24 September, 1998, pp. 107–111.

[58] S. Yamashita, S.Y. Set, R.I. Laming, Polarization independent, all-fiber phase conjugation incorporating inline fiber DFB lasers, Photon. Technol. Lett. 10 (1998) 1407.

[59] W.H. Loh, J.P. de Sandro, G.J. Cowle, B.N. Samson, A.D. Ellis, 40 GHz optical-millimeter wave generation with a dual polarization distributed feedback fiber laser, Electron. Lett. 33 (1997) 594.

[60] D.F. Browning, G.V. Erbert, Distributed feedback fiber laser: the heart of the national ignition facility, Report UCRL-ID-155446, 2003.

[61] E. Austin, P. Nash, Q. Zhang, S. Alam, M.N. Zervas, R. Slavik, P. Petropoulos, D.J. Richardson, 500 km remote interrogation of optical sensor arrays, Proc. SPIE 7753 (2011) 77532M.

[62] G.A. Cranch, G.M.H. Flockhart, C.K. Kirkendall, Distributed feedback fiber laser strain sensors, IEEE Sens. J. 8 (2008) 1161.

[63] O. Hadeler, M. Ibsen, M.N. Zervas, Distributed-feedback fiber laser sensor for simultaneous strain and temperature measurements operating in the radio-frequency domain, Appl. Opt. 40 (2001) 3169.

[64] O. Hadeler, M.N. Zervas, Application of a DFB fibre laser temperature sensor for characterizing pump induced temperature distributions along another DFB fibre laser, in: SPIE Proceedings of the International Conference on Optical Fiber Sensors, 4185, 2000, p. 142.

[65] J.M. Kahn, K.P. Ho, Spectral efficiency limits and modulation/detection techniques for DWDM systems, J. Sel. Top. Quant. Electron. 10 (2004) 259.

[66] E. Ip, J.M. Kahn, D. Anthon, J. Hutchins, Linewidth measurements of MEMS-based tunable lasers for phase-locking applications, Photon. Technol. Lett. 17 (2005) 2029.

Semiconductor Photonic Integrated Circuit Transmitters and Receivers

2

Radhakrishnan Nagarajan[a], Christopher Doerr[b], and Fred Kish[a]

[a]*Infinera Corporation, 140 Caspian Court, Sunnyvale, CA 94089, USA,*
[b]*Acacia Communications, 1715 Route 35N, Suite 207, Middletown, NJ 07748, USA*

2.1 INTRODUCTION

In early fiber-optic systems, the cost of the transmitters and receivers, i.e. transceivers, was only a small part of the total cost. Today, with wavelength-division multiplexing (WDM) and advanced modulation formats, transmission network cost for long-haul and metropolitan distances is dominated by the transceivers. This is because there are many transceiver pairs per link, sometimes 80 or more, and transceivers are becoming more and more complicated. For example, a typical 100-Gb/s coherent communication link used in today's state-of-the-art long-haul networks is shown in Figure 2.1. The transmitter contains 23 optical components with 29 inter-component connections, and the receiver contains 13 optical components with 13 inter-component connections. Next generation transceivers continue to compound this problem: a 1-Tb/s transceiver comprises multiple 100s of components and interconnections. The rising costs for these complex transceivers are crippling network operators as they are expected to constantly provide more bandwidth at roughly the same connection price. Carriers must constantly reduce their price per bit per second while installing more and more expensive equipment.

A solution to this dilemma is optical integration. Optical integration has many levels, from putting multiple discrete optics on the same line card, to putting multiple optical elements on the same chip. In this chapter, we focus on the latter, because this is where the greatest opportunity for simultaneous bandwidth scaling and cost reduction exists. Such integration is referred to as monolithic integration. The multiple optical elements are fabricated at once on the same chip with many chips per wafer. This greatly reduces cost and makes transceivers scalable, because the elements are automatically optically aligned to each other and can be tested all at once rather than one at a time. In fact, some optical integration schemes can be "self testing" in that the testing sources and receivers can be integrated on the chip. Reliability increases dramatically because there are fewer mechanically separate piece parts. Without optical integration, scaling a line card to 1 Tb/s and beyond may not even be feasible commercially.

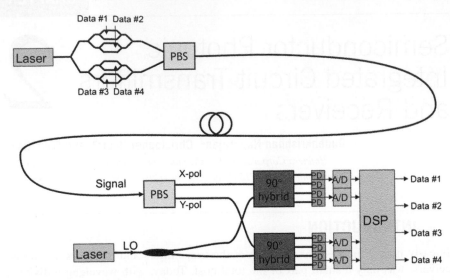

FIGURE 2.1 An optical coherent communication link.

Furthermore, monolithic optical integration greatly reduces footprint. Specifically, integration can reduce footprint on line cards by combining packages and it can also reduce faceplate footprint by having smaller transceivers, by having fibers route directly to optical-to-electronic elements, and/or by employing a more compact connection density.

Integrated optical components are categorized into "active" and "passive." There is a large range of what is considered to be an active optical component. At one end of the range are devices with only optical gain and at the other are devices with any dynamic or electronic behavior. Here we use the term "active" for any optical component that involves a dynamic interaction between light and matter. There are many types of optical active components including amplifying, electrooptic, thermooptic, acousto-optic, mechanical, piezoelectric, and liquid crystal.

We use the following terminology: planar lightwave circuit (PLC) for optical integrated devices that are fully passive (except thermooptic), and photonic integrated circuit (PIC) for all other optical integrated devices. Because this chapter focuses on transceivers, only PICs are discussed here.

PICs are often compared to electronic integrated circuits (EICs). Trends based on generic EICs have been predicted for PICs, such as acquiring billions of elements. However, it is better to compare PICs to analog EICs, as we do in Section 2.5. Unlike analog EICs, digital EICs are formed by the application of a repeated unit cell with varying interconnections to implement a circuit functionality. Analog EIC performance is often linked closely to custom fabrication processes which maximize performance whereas digital EICs tend to more universally utilize a standard fabrication process (foundry process) wherein the value of the integration scale outweighs compromises made in the performance of the integrated devices. PICs, which tend to have high diversity in architectures and designs and require high performance, are much more comparable to analog EICs.

It is generally assumed that PICs perform significantly worse than their discrete-optic counterparts. The argument is often made that because each discrete part can be optimized independently, a discrete optics solution can achieve superior performance to an integrated part in which tradeoffs must be made because all components are made at once. However, this analogy is based on a digital EIC comparison and is faulty. The performance of PICs can be nominally equivalent to discretes as demonstrated in the commercial system performance of PIC-based transport systems as described in Sections 2.3.1.2 and 2.4.3.1. Furthermore, at some point optimization and alignment of discrete optics becomes too difficult, especially when skews between devices must be carefully managed, and then the raw performance of a PIC can potentially exceed that of discretes. Thus for something highly complex as a coherent transceivers, a PIC may outperform its discrete optic counterpart for some of the critical performance parameter.

In this chapter we cover PICs used for metro, long-haul, and undersea telecommunication systems with some coverage of access and short-reach applications. The two most popular semiconductor PIC material systems for these applications are Group III–V materials which comprise mostly indium phosphide (InP) along with gallium (Ga), arsenic (As) and aluminum (Al); and Group IV materials which comprise mostly silicon (Si) along with germanium (Ge), oxygen (O), and nitrogen (N).

2.2 TECHNOLOGY

2.2.1 Group III–V PICs

For metro, long-haul, ultra long-haul, and submarine optical communications transmission applications, III–V InP-based photonic integrated circuits have achieved the highest level of integration and have been the only PICs to enjoy commercial success to date. The success has been predicated on the fact that InP-based III–V materials (including InGaAsP and InAlGaAs alloys) are direct bandgap semiconductor materials capable of being grown lattice-matched on high-quality substrates, and possess a tunable bandgap and heterostructures that enable a full-spectrum of active and passive devices to be realized over the low-loss spectrum of optical fiber.

Typical waveguides made in InP and its related materials are shown in Figure 2.2. A rich diversity of structures in the low-loss wavelength regime for fiber-optic transmission ($\sim$1.3–1.6 μm) is feasible due to the ability to form lattice-matched ternary and quaternary compounds of InGaAsP as well as substitute Al for P to form InAlGaAs. Waveguide types range from buried channels to deep ridges. By changing the bandgap of the materials via composition one can make efficient optical amplifiers, modulators, and photodetectors all in the same platform.

The successful development (over the last >40 years) and commercialization (over the last >20 years) of PICs has been the result of numerous advances in materials,

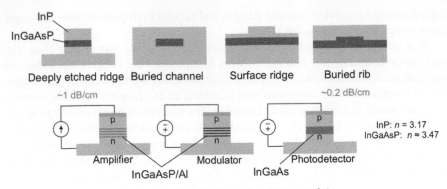

InP
InGaAsP

Deeply etched ridge   Buried channel   Surface ridge   Buried rib

~1 dB/cm   ~0.2 dB/cm

Amplifier   Modulator   Photodetector

InGaAsP/Al   InGaAs

InP: $n = 3.17$
InGaAsP: $n \approx 3.47$

FIGURE 2.2 The Group III–V InP family of integrated optical materials.

processing, and device capability. Fundamental advancements that have enabled PIC technology include the following: the realization of the semiconductor laser [1–4], the development of: heterojunction lasers and cw laser operation [5,6], the realization of quantum-well lasers [7], the development of distributed feedback (DFB) lasers [8,9], the development of pseudomorphic materials (including strained quantum wells) [10–12], and the development of long-wavelength InP-based semiconductor lasers in the low-loss spectrum of the optical fiber [13]. Furthermore, many key advances in technology have enabled the commercialization of PICs, including the following: the availability of high-quality, low-defect density 50–100 mm diameter InP substrates, the development of metalorganic chemical vapor deposition (MOCVD) as a viable means for the growth of high-precision lasers and optoelectronics devices [14–16] in multi-wafer reactors, the development of precision dry-etch technologies for low-loss waveguides and highly reliable devices, and fine line lithography. The culmination of these and other advances over more than 50 years has resulted in a capability for InP PICs that can now integrate all required functions for the most sophisticated transmitters and receivers (on a scale of hundreds of elements per chip). As a result, InP-based PICs are now widely utilized for transmitter and receiver solutions in the optical communications industry as they have brought significant benefits to both component and network scaling.

Unlike electrical ICs, each different component in a photonic IC generally requires a different semiconductor layer stack (with varying bandgap, doping, and/or heterostructures) to enable high-performance operation. A number of technologies have been developed to realize these structures in InP-based PICs. The predominantly employed methods consist of one or more combinations of: butt-joint regrowth [17,18], selective-area growth [19–22], quantum well disordering/layer intermixing [23–26], and etch-back of multiple vertical device layers [27–31]. After realization of the multi-device/bandgap regions within a wafer, standard wafer fabrication techniques similar to those utilized in heterostructure bipolar transistor (HBT) integrated circuits [32] are utilized to: define active and passive waveguides, form inter-device and inter-channel electrical isolation, establish DC and RF contacts/bondpads, and passivate the devices and circuits within a chip.

Figure 2.3 graphically shows the progression of non-coherent on-off keying (OOK) PIC complexity. The details of the devices presented in Figure 2.1 are covered in Refs. [6,33–49] (the references are next to the devices in Figure 2.3). For the first decade or so after the demonstration of the CW laser in the GaAs system, InP lasers started to mature. In the mid-1980s there was active work in the area of opto-electronic integrated circuits (OEICs) where the integration of electronic devices such as heterojunction bipolar transistors (HBTs) and field effect transistors (FETs) with laser diodes and photodetectors was pursued. In the late 1980s, three-section tunable distributed Bragg reflector (DBR) lasers were introduced. This was also when electro-absorption modulators (EAM) integrated with distributed feedback (DFB) lasers were demonstrated. The trend continued with more complicated widely tunable laser sources which were also integrated with an EAM or a semiconductor optical amplifier (SOA). The next step was the demonstration of the arrayed waveguide grating (AWG) or PHASAR (phased array) router integrated with photodetectors for multi-channel receivers or with gain regions and EAM for multi-frequency lasers and multi-channel modulated sources. At this stage the most sophisticated laboratory devices still integrated less than 20 functions and the most sophisticated commercially deployed devices integrated at most four functions.

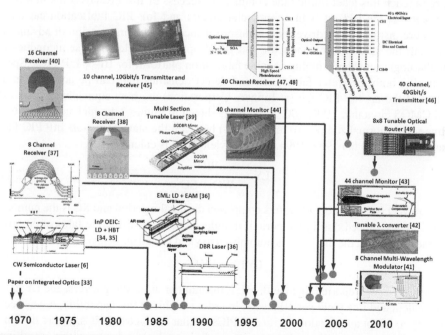

FIGURE 2.3 Historical trend and timeline for monolithic, photonic integration on InP for OOK (non-coherent) transmitters and receivers (Refs. are next to the description of the devices in the figure). We have not included vertical cavity InP devices. The trend shows an exponential growth in PIC complexity in recent years.

Steps towards a significantly larger integrated chip were reported in 2003. ThreeFive Photonics reported a 40-channel WDM monitor chip (a receiver), integrating 9 AWGs with 40 detectors. MetroPhotonics reported a 44-channel power monitor based on an echelle grating demultiplexer. The commercial development of both chips was subsequently discontinued. The first successful long-term deployment of a commercial large-scale photonic integrated chip was made in 2004 when Infinera Corporation introduced a 10-channel transmitter, with each channel operating at 10 Gb/s. This device with an integration of more than 50 individual functions was the first large-scale PIC device deployed in the field to carry live network traffic. This was quickly followed in 2006 by a demonstration of 40-channel monolithic InP transmitter, each channel operating at 40 Gb/s (for an aggregate data rate of 1.6 Tb/s) integrating more than 240 functions, and a complementary 40-channel receiver PIC. As a further step in complexity, the 40-channel receiver PIC also had an integrated, polarization independent, multi-channel SOA at the input. These levels of complexity are still the benchmark for monolithic integration for InP PICs based on OOK. In 2009, the University of California, Santa Barbara (UCSB), reported an 8×8 monolithic tunable router with a component count of about 200.

A fundamental premise to the success of the IC is that the economic value derived from the integrated component must outweigh the cost of the integration itself. Failure to reach this tenet limited the commercial success of InP PICs over the first few decades as the lack of both commercial drivers and InP PIC fabrication capability were insufficient to drive their widespread success. The progression of advances in both the silicon semiconductor and III–V optical and electrical semiconductor technologies enabled a highly capable fabrication platform for III–V InP PICs. Specifically, both the wafer fabrication and defect densities that can be achieved in a state-of-the-art InP PIC fab (at Infinera Corporation) are now equivalent to that of Si CMOS circa mid-1990s [50–52]. This is shown in Figure 2.4 which compares the line yield (normalized per 10 mask layers) between the state-of-the-art InP PIC fab and a Si CMOS fab 1989–1994. The line yield (LY) is calculated as

$$LY = \frac{WO}{WO + SC},\tag{2.1}$$

where WO is the number of wafers completed during the period and SC is the number of wafers scrapped during the period. In order to compare disparate fabrication processes and flows, the line yield is normalized to yield per 10 mask levels by the relation

$$LY10 = LY^{\frac{10}{ML}},\tag{2.2}$$

where LY is the reported line yield, ML is the number of masking layers, and LY10 is the calculated line yield per 10 layers. In order to achieve this capability, it was necessary to adopt a methodology for InP PICs similar to electronic ICs, where designers are given a fixed (limited) tool set (e.g. design rules) to design within, resulting in a manufacturable (cost-effective) device. As a result, both highly capable fabrication and device performance (as to be shown in Sections 2.3.1 and 2.4.2) have been simultaneously achieved.

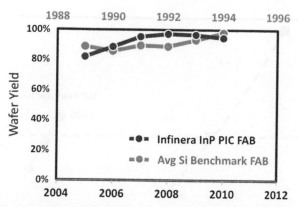

FIGURE 2.4 Wafer fabrication yield (normalized to 10 mask levels) for a state-of-the-art InP PIC fab (Infinera Corporation) and the average Si CMOS fab (1989–1994) [50].

Since their inception [53,54], the scaling of electronic integrated circuits has progressed at an exponential rate as predicted by Moore [55]. This scaling is in stark contrast to photonic integrated circuits, wherein the pace of integration has been dwarfed by that of electronic ICs. This dramatic difference is mainly due to the inability for the value created by a PIC to outweigh the cost of integrating optical elements monolithically onto a chip. One of the main costs of integration is directly related to the ability to integrate an economically significant number of functions onto the chip at a high yield. This yield relies upon fabrication processes that possess sufficiently low killer defect densities that enable integration scaling. As a result of the aforementioned advances, state-of-the-art PIC fabrication technology now enables sufficiently low killer defect densities to economically integrate devices of significant scale. Figure 2.5 shows killer random defect densities versus time for Si CMOS integrated circuits [51,52] in comparison with those achieved in the state-of-the-art InP PIC fabrication fab at Infinera Corporation. The killer defect densities for the InP PICs are obtained via yield modeling theory wherein the rate at which yield decreases with increasing die size is used to determine both systematic and random yield contributions, and allows the killer random defect density to be quantified [51,52,56,57]. The defect density was obtained using the Murphy yield model for random defects which assumes a symmetrical triangular distribution for the defect density probability

$$Y_r = \left(\frac{1 - e^{-AD_0}}{AD_0} \right)^2,$$ (2.3)

where Y_r is the yield due to random defects, A is the die size area, and D_0 is the average defect density per unit area. Yield calculations and fits to the data were performed on average production DC test data for Infinera Corporation's first generation 100-Gb/s InP transmitter PICs (solid circles) as well as on early data from pilot production for our third generation 500-Gb/s transmitter PICs (open circle) and 500-Gb/s receiver PICs

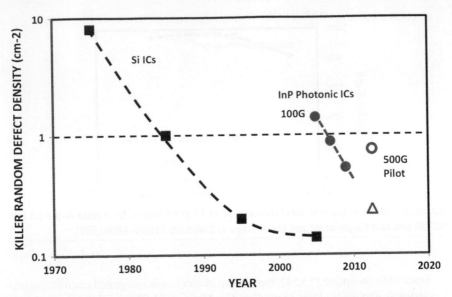

FIGURE 2.5 Trends for killer random defect reduction versus time for Si ICs and InP large-scale transmitter PICs for both 100-Gb/s transmitter products (solid circles) and 500-Gb/s transmitter (open circle) and receiver (open triangle) products. The 100-Gb/s data points represent average production performance whereas the 500-Gb/s data points are based on sampling of better performing wafers during early production.

(open triangle) as shown in Figure 2.5. Larger PIC chip areas were generated by analyzing chip multiples as an effectively larger chip [58]. The methodology utilized herein uses an overall yield and total chip area for determination of the killer random defect density and assumes an effective average defect density arising from different failure modes (which each have their own defect density) for a given chip architecture. This effective average defect density is appropriate to measure the rate of defect reduction (learning) as well as project how a given architecture will scale (e.g. scaling the number of channels, assuming a constant active/passive device area ratio on the PIC).

As a result of the deployment of a wide breadth of semiconductor IC and III–V optoelectronic and electronic fabrication advances, as well as continued learning, rigorous manufacturing controls, and process improvements, a state-of-the-art InP fabrication capability at Infinera Corporation has been developed with a killer random defect density of 0.5 cm^{-2} for 100-Gb/s PICs as shown in Figure 2.5. These defect density numbers compare well with those in the Si industry circa 1988 [51,52], and have a rate of decrease (defect reduction learning) similar to what the Si industry achieved in 1975–1995. In addition, data from a sampling of better performing wafers for initial production of 500-Gb/s PICs is indicative of defect densities $< 1 \text{ cm}^{-2}$. We would expect with continued learning and process improvement for defect reduction, similar defect reduction versus time should be obtainable on average 500-Gb/s PIC wafers.

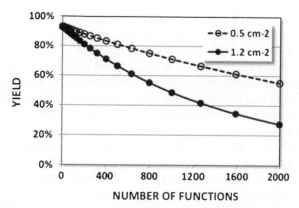

FIGURE 2.6 Yield vs. number of functions for different defect densities for large-scale InP PICs. This data assumes chip areas and area density that corresponds to a 500-Gb/s transmitter PIC.

A key difference between EIC's and PIC's driving the data in Figure 2.5 is that the random killer defect size is very different for Si ICs than InP PICs. The killer random defect size in InP PICs is typically on the order of ~0.5 μm (corresponding to the wavelength of light in the semiconductor materials), compared to that of Si ICs where the killer defect size is on the order of the linewidth of the current technology node (presently ~32 nm, going to 22 nm). Despite these differences, the implications for the scaling of ICs to higher integration levels are the same. Figure 2.6 shows the functional yield for various defect densities obtained with the Murphy model (Eq. (2.3)) for the chip areas and an areal density that correspond to that of a 500-Gb/s transmitter PIC. At a conservative defect density <1.2 cm^{-2}, it is possible to integrate ~450 functions with ~70% functional yield. Furthermore, expected defect reduction over time (to <0.5 cm^{-2}) should enable continued scaling of devices to incorporate >1000 functions with >70% yield and hence achieve even higher bandwidth per chip.

The field of InP-based optoelectronic devices has matured from a field of discrete devices to one that is capable of realizing large-scale integrated devices that are economically viable. These developments coupled with the superior device performance capability for many functions currently make InP an excellent platform for many PIC applications. As shown in Figures 2.5 and 2.6, the platform is capable of high wafer fabrication yields and to scaling to over a thousand functions per chip with reasonable functional yields.

2.2.2 Group IV PICs

The primary advantages of the Si material system are the abundance of Si, the high mechanical strength of Si, and Si's high-quality oxide SiO$_2$. Si makes up 27% of the mass of the Earth's crust. Its companions O and N are also plentiful. Ge makes up only 1.4 ppm of the Earth's crust, but fortunately the amount of Ge used in Si PICs is usually very small. Si's high mechanical strength allows for large wafers, the industry currently focusing on 300-mm-diameter wafers for EICs. PICs tend to use previous-generation EIC technology, 200-mm-diameter wafers. Si has a high quality oxide, SiO_2, that is facile to make. When Si is put in an oxygen-rich atmosphere at a high temperature the Si at the surface slowly turns into a high quality, dense oxide. Oxidized Si has extremely low optical loss and is an excellent electrical insulator. For instance, a very small core waveguide of Si_3N_4 embedded in SiO_2 on a Si wafer, in which most of the light resides in the oxide cladding, achieved a waveguide loss of 0.003 dB/cm [59]. The oxide can be molecularly bonded to another wafer also with an oxide layer with high yield. This allows for the creation of waveguides without requiring epitaxy, ultimately increasing the yield. The thickness of a 200-mm Si wafer is typically 725 μm.

The oxide furthermore allows Si optical waveguides to have high index contrast both vertically and horizontally, whereas Group III–V waveguides are generally limited to horizontal high index contrast. High vertical index contrast in InP waveguides has been achieved by using bonded [60] or suspended structures [61], but these are in early research stages. High index contrast in both dimensions allows for very small bend radii (<5 μm) and very strong gratings (making grating couplers possible).

A non-exhaustive display of the Group IV family of materials is shown in Figure 2.7. It invariably involves a Si substrate. On that Si substrate there can be SiO_2 (silicon dioxide, often referred to as just oxide), more Si, Si_3N_4 (silicon nitride), SiON (silicon oxynitride), Ge, and various metals such as Al, Cu, and W. For the Si, there can be crystalline Si (c-Si), amorphous Si (a-Si), and polycrystalline Si (p-Si). The distinction between a-Si and p-Si is blurry, p-Si having larger single-crystal domains. a-Si can be converted to p-Si by annealing. The refractive indices of Ge, a-Si, p-Si, c-Si, Si_3N_4, and SiO_2 at 1550-nm wavelength are approximately 4.2, 3.8, 3.5, 3.48, 2.0, and 1.45, respectively.

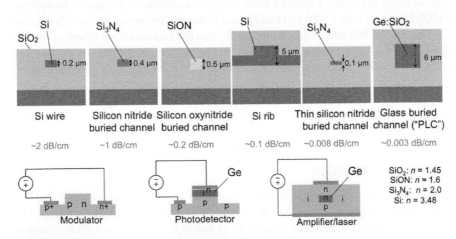

FIGURE 2.7 The Group IV family of integrated optic materials.

The most common starting platform for silicon photonics is a silicon-on-insulator (SOI) wafer. An SOI wafer consists of a Si substrate, a layer of oxide on top of that [called a buried oxide layer (BOX)] and a thin layer of c-Si on top of that. An SOI wafer is fabricated by taking two Si wafers, oxidizing their surfaces to a certain depth and then molecularly bonding the two wafers. The BOX thickness is typically $2\,\mu$m for an optics SOI wafer and typically $0.05\,\mu$m for a high-performance microprocessor SOI wafer. The c-Si thickness is typically $0.22\,\mu$m. SOI allows the waveguide core to be completely surrounded by oxide, providing a very high index contrast waveguide in both dimensions.

From right to left in Figure 2.7 the effective lateral waveguide index contrast increases but the waveguide propagation loss also increases. The waveguide on the far left is a Si wire waveguide, providing very high index contrast, and the waveguide on the far right is a typical PLC Ge-doped silica waveguide with very low index contrast. Typical losses in Si wire waveguides $0.5\,\mu$m wide are 2 dB/cm, whereas losses of 0.003 dB/cm have been achieved in PLC waveguides [62]. However, a Si wire waveguide can be bent with a radius as small as $3\,\mu$m, whereas a PLC waveguide requires a radius larger than 2 mm.

Some useful intermediate materials are silicon oxynitride (SiON) [63] and silicon nitride Si_3N_4 when stoichiometric. The intermediate refractive index results in optical filters with lower crosstalk, and its reduced thermooptic coefficient results in less temperature dependence. The thermooptic coefficients of Si and Si_3N_4 are $\sim$0.1 and 0.02 nm/°C, respectively. The presence of H in Si_3N_4 causes high absorption around 1520 nm, though, and must be eliminated by high temperature annealing or other techniques.

Si is a very weak electrooptic material. Fortunately, the index contrast of Si waveguides is high enough that the waveguide cross section can be made small enough to allow one to move carriers in and out of the light beam passing through the Si in a short enough time to reach 40-Gb/s modulation [64]. Some typical modulators in Si are shown in Figure 2.8. Figure 2.8a shows a carrier injection modulator. While this effect is quite strong ($V_\pi L$ of 2.5 V-mm is possible), it is also slow (typically below 5 GHz). Speeds up to 10 Gb/s have been achieved when embedded in a photonic crystal, but require electronic pre-emphasis [65]. Figure 2.8b shows a carrier depletion modulator [65]. The effect is much weaker ($V_\pi L$ of 25 V-mm is typical) but speeds in excess of 40 GHz are possible. 50-Gb/s modulation has been reported [66]. Figure 2.8c shows a metal-oxide-semiconductor (MOS) modulator [67,68]. Actually, it is a semiconductor-oxide-semiconductor modulator but the electronics industry has

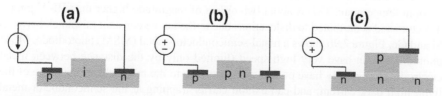

FIGURE 2.8 Various Si modulators: (a) carrier injection, (b) carrier depletion, and (c) MOS.

stayed with the term MOS for such structures. The thin layer of oxide allows a larger carrier density change. The effect is both strong ($V_\pi L$ of 2.5 V-mm is possible) and fast, but the insertion loss and capacitance are high.

The real and imaginary parts of the refractive index change of Si with carrier density at $1.55\,\mu m$ wavelength are given by

$$\Delta n_r = -8.8 \times 10^{-22} N_e - 8.5 \times 10^{-18} N_h^{0.8}, \tag{2.4}$$

$$\Delta n_i = 2.5 \times 10^{-26} N_e^{1.2} + 2.2 \times 10^{-24} N_h^{1.08}, \tag{2.5}$$

where N_e and N_h are the electron and hole densities, respectively, per cubic cm. These equations are rough curve fits to the data presented in [64]. The hole density has a stronger effect on the real part of the index change than the electron density yet the absorption is approximately the same for hole and electron concentrations. For this reason, Si modulators are typically designed to have a larger optical mode overlap with the p-doped region.

One can instead use a non-group IV material deposited on the Si to make a modulator. Sullivan et al. [69] report a modulator in which an electrooptic polymer has been deposited in a slot in a Si waveguide, and Liu et al. [70] report a one-atom-layer-thick film of graphene on a Si waveguide. The graphene modulator works on the principle that the graphene changes from an insulator to a semimetal by putting a voltage on it. The polymer modulator requires high-cost polymers and may have reliability concerns, and the graphene modulator may have a low saturation power limit.

Si does not absorb enough light at 1550 nm to be used as a photodetector. However, when the optical power is high enough (typically greater than 10 mW in a Si wire waveguide), two-photon absorption occurs. The generated carriers can be collected, operating as a nonlinear photodetector [71]; however, this is mainly suitable for an optical monitor. The nonlinearity is deleterious to receive the signal for most communication applications.

Instead, one can use Ge as a photodetector. Crystalline Ge is 4% lattice-mismatched to Si. Despite this mismatch, it can be grown on Si using a buffer layer of SiGe with a sufficiently low defect density to result in high-responsivity operation. Ge photodetectors work fine in the C-band, but unless the Ge is heavily strained, the Ge absorption decreases significantly at wavelengths greater than ~1580 nm. Some typical Ge photodetectors in Si are shown in Figure 2.9. Figure 2.9a shows a Ge p-i-n photodiode. Such photodetectors can achieve both high responsively and high speed: 0.89 A/W and 40 Gb/s were achieved in [72]. Typical dark currents are 100 nA at −1 V at room temperature. This is about two orders of magnitude higher than III–V photodetectors and is attributed to dislocations due to the lattice constant mismatch between Si and Ge. Figure 2.9b shows a metal-semiconductor-metal (MSM) photodiode. MSM photodiodes can have very high speed (limited only by the distance between metal electrodes) but usually have poor responsivity due to the metal absorbing some of the light, high dark current, and can exhibit carrier trapping at the semiconductor-metal interface, giving rise to long tails in the impulse response. A performance of 0.07 A/W

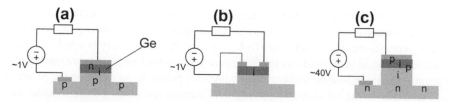

FIGURE 2.9 Various Si/Ge photodetectors: (a) p-i-n, (b) MSM, and (c) APD.

and 40 Gb/s were achieved, as reported in [73]. Figure 2.9c shows an avalanche photodiode (APD). The avalanche region is in Si, which has a lower avalanche noise than in III–V materials. This holds promise for a Si–Ge APD that has higher sensitivity than an InGaAsP APD [74].

Recently there has been publication of a Ge optical amplifier, which was used to make an electrically pumped Ge laser [75], as shown in the lower right portion of Figure 2.7. Ge has an indirect band gap at the Γ point, so it ordinarily provides very inefficient optical emission and gain. This is because there is another valley in the conduction band with a lower energy, and most of the electrons pile up there. To make a transition to the valence band the electrons need a momentum change, which is an unlikely transition. To solve this problem, the Ge is very heavily n-doped with phosphorus. This results in band filling of the L-valley and making it more likely for electrons to remain in the higher-energy Γ-valley and thus have a direct transition to the valence band. However, the doping requirement is so high it must be incorporated in the Ge during the crystal growth. Unlike a photodiode, a Ge amplifier/laser cannot tolerate significant dislocations and so the Ge must be grown in a narrow trench allowing it to grow as a perfect crystal with reduced stress.

There is current interest in Ge-only photonics for sensors [76]. Ge is transparent above ∼1600 nm wavelength. While useful for sensors, these wavelengths are currently too lossy in optical fibers and so will not be covered in this chapter.

2.2.3 Hybrid integration of Groups III–V and IV

One would ideally like to combine the best of both Groups III–V and IV into one platform. Because Group IV has the less expensive substrate and more mature fabrication tools and larger wafer diameters, the focus has been on putting Group III–V semiconductors onto Group IV substrates. Epitaxy of III–V on IVs has had very limited success so far in optics; however, in electronics significant progress is being made. The main difficulty is that the lattice mismatch between InP and Si is very large. In electronics, one can first grow or mechanically bond (wafer bond) Ge on the Si (this is called Ge-on-insulator, GOI) and then grow GaAs, which is closely lattice matched to Ge, on the Ge, provided one takes care to manage the anti-phase domains. In optics, one generally needs InP, not GaAs, and Ge is absorbing in the telecommunications window. There has been some progress in growing InP on off-axis-polished Si substrates, but it is difficult to make this electrically pumped and efficient.

Most of the progress for hybrid integration has been made in mechanical attachment of already-grown III–V layers to Si. The coupling can be evanescent, adiabatic, grating-coupler-based, or butt. The III–V chips can be molecularly bonded, glued, or soldered to the Si or an oxide on the Si. Figure 2.10 shows four of the most successful options. The molecular-bonding approach involves polishing an oxide on the Si wafer and an oxide on the InP chip to high degree of flatness and placing them together and annealing, causing the two oxide layers to form together as a single oxide. One cannot generally bond an entire InP wafer to an entire Si wafer at once, because currently the largest available InP wafer diameter is 150 mm, and thus small pieces must be bonded across the Si wafer. After bonding, the InP chips are thinned and then processed along with the Si wafer. The coupling between the InP and Si waveguides can be either evanescent or adiabatic. In evanescent coupling [77], some of the light remains in the Si waveguide, whereas in adiabatic coupling the light is completely transferred to the InP waveguide. The III–V waveguides can be lasers, optical amplifiers, modulators, or photodetectors. Using benzocyclobutene (BCB), a spin-on liquid that cures into a hard plastic, InP ring lasers have been coupled to Si waveguides [78]. In Ref. [79] a laser assembly is soldered to a Si chip and is optically coupled via a lens, 45° mirror, and a grating coupler, as in Figure 2.10c. As in Figure 2.10d an InP gain chip is butt-coupled to a Si PIC containing ring resonators to create a tunable laser [80].

2.2.4 Comparison of PIC technologies

There is no simple comparison of Group III–V and Group IV materials for optical devices because of the large variety of optical components in each platform and also because the platforms are still rapidly evolving. Nevertheless, with today's technology, there are some relatively definite statements that can be made.

Group III–V devices provide superior amplifying and electrooptic components, mainly because of the direct bandgap of III–V materials and because of the availability of heterogeneous, lattice-matched materials, such as InGaAsP and AlGaInAs. These are significant advantages, and to date have resulted in commercial PICs being primarily implemented in III–V materials (and exclusively implemented in III–V materials for long-haul applications). Furthermore, as described in Section 2.2.1,

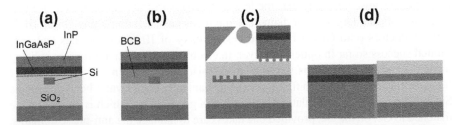

FIGURE 2.10 Some hybrid integration schemes: (a) evanescently coupled using molecular bonding, (b) evanescently coupled using benzocyclobutene (BCB) adhesive, (c) grating-coupler coupled, and (d) butt-coupled.

III–V PIC fabrication capability has improved drastically over the last several decades and is now capable of integrating hundreds if not a thousand devices onto a single chip with high yield.

However, Si photonics has been an area of active development over the last decade as Si has more optical foundries available and offers potentially higher integration densities as a result of the more mature Si manufacturing capabilities. As a result, Si PICs are emerging in a number of applications as described in Sections 2.3.2 and 2.4.2, especially in short and potentially intermediate reach applications. Also Si is a more easily obtained, stronger material with a more robust oxide.

Hybrid integration of Groups III–V and IV materials promises the best of both worlds, but significant issues remain such as yield/cost and thermal management. Table 2.1 shows a high-level comparison between device characteristics for Group III–V and Group IV materials.

Table 2.1 Comparison of Groups III–V and IV PICs.

Characteristic	Group III–V PICs	Group IV PICS
Combine transistors and optical functions monolithically	Excellent for low-integration density	Excellent for high-integration density although expensive for small node size
Integrate different passive optical functions	Very good compact elements	Extremely compact elements. Can integrate with PLC materials
Typical commercial wafer diameter	50–100 mm	200–300 mm
Laser/optical amplifier performance	Excellent	Not viable at present time
Modulator performance	Excellent	$\sim$10 $\times$ higher $V_\pi L$ for low-loss modulation
Photodetector performance	Excellent 1200–1620 nm	Good 1000–1580 nm. High dark current
Consolidation of electrical connections	Very good due to integrated active sources	Very good/excellent due to integrated electronics
Consolidation of optical connections, including fiber coupling	Excellent	Very good. Needs integrated gain elements
Testing consolidation	Excellent due to integration of active sources	No integrated source but wafer-level optic access via grating couplers
Size/footprint	Excellent for PICs with active sources. Very good for passives	Good but lack of active sources. Excellent due to high index contrast for passives
Reliability improvement vs. discretes	Excellent	Excellent for passives. Emerging for photodetectors and modulators
Power consumption vs. discretes	Excellent due to low modulator voltage and integration of gain elements	Very good due to efficient thermooptics

2.3 DEVICES BASED ON ON-OFF KEYING (OOK)

2.3.1 Group III–V PICs for OOK transmission

2.3.1.1 Group III–V single-channel PICs for OOK transmission

The first transmitter PIC was the electroabsorption-modulated laser (EML) developed in 1987 [81]. This PIC integrated two devices: a distributed feedback (DFB) laser and an electroabsorption modulator onto a single monolithic chip as shown in Figure 2.11. A key design criterion for this device that drives the required integration technologies is detuning the active region of the electroabsorption modulator to a more transparent region (bandgap) than that of the lasing wavelength. Typically, the bandgap of this region must be detuned 20–150 nm from the lasing wavelength [19,82] in order to provide a good trade-off between modulator extinction ratio, loss, and drive voltage. As a result, EMLs have conventionally utilized multiple regrowths [17,18] or selective-area growth [23–26] to integrate the devices monolithically. Photonic ICs based on externally modulated lasers utilizing Mach-Zehnder modulators have also been developed [39,83]. However, the EML was also the first widely deployed commercial PIC as a result of its better performance and cost structure at the time of introduction. The need and success of this device was driven by the demand for high bandwidth (>2.5 Gb/s) and high performance (overcoming transmission impairments for intermediate and long-haul applications) while simultaneously providing a better cost than comparable discrete components. This device was commercially deployed at 2.5 Gb/s in 1996 and 10 Gb/s in 1998 [84].

The continued development of PIC design, integration, and fabrication capabilities led to the next major advancement in PIC transmitter devices: the widely tunable laser with a tuning capability over the full C- or L-Band. The development of this device was originally driven in the late 1990s by the vision to enable dynamic WDM optical networks wherein the tunable laser would enable wavelength switching and routing in conjunction with reconfigurable optical add/drop multiplexers

First Transmitter PIC

Electroabsorption-Modulated Laser

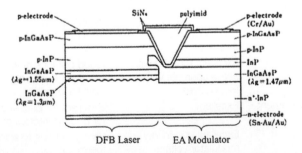

FIGURE 2.11 Schematic diagram of the first transmitter PIC: the electroabsorption-modulated laser [81].

(ROADMs). However, the development of these network architectures was deferred due to network overcapacity in the early 2000s. Consequently, the main application of these devices to date has been in inventory reduction and sparing of lasers and optical transmission system line cards by virtue of replacing fixed frequency devices. Many different widely tunable laser technologies have been developed over the last 30 years (see [85] for a comprehensive review). However, two main PIC technologies have met with commercial success. The first is the sampled-grating distributed Bragg reflector (DBR) laser which employs four distinct sections to tune the gain, phase, and front and back sampled gratings that are slightly detuned from each other [86]. A sampled grating is a conventional grating with grating elements removed periodically which leads to reflection spectra with periodic maxima in a limited wavelength region. These devices operate by having the two widely-spaced and independently current-tuned sampled grating reflection combs at each end of the cavity select a single mode within the gain bandwidth of the laser (due to a small difference in periodicity between the reflection spectra). This is commonly referred to as the Vernier effect. The phase tuning section is utilized to tune the mode to the resonance condition selected by the overlap of the sampled gratings. The differing comb pitches of the front and back sampled gratings lead to large changes in lasing wavelength with tuning current relative to standard DBR lasers. Thus, this device architecture results in a wide-tunability (full C- or L-Band) with an excellent side mode suppression ratio. An analogous device called a DS-DBR laser employs similar physics [87].

In order to improve output power, further integration of a SOA has been employed in these devices [88,89]. Commercial devices incorporating this structure are capable of >13 dBm output power [90,91]. Furthermore, a fully integrated transmitter has been developed by the additional integration of a modulator. First generation versions of this device utilized an EAM [92,93] and were first deployed in live networks in 2003. However, due to the limited operating wavelength bandwidth of the EAM, second generation devices have integrated Mach-Zehnder modulators to achieve an integrated transmitter PIC capable of tuning over the entire C-Band at 10 Gb/s [85,94]. A schematic and photograph of this device are shown in Figure 2.12.

Despite their wide tunability and excellent mode suppression ratio, SG-DBR lasers suffer from the need to employ complex control schemes to assure wavelength stability and mode-hop free tuning [85]. Consequently, an alternate tunable laser based on PICs has been developed. This device employs a 12-channel multiwavelength distributed feedback (DFB) array. The wavelength is tuned by selecting the appropriate wavelength DFB (coarse grid) and then thermally tuning the device to the desired wavelength of interest. The light from the appropriate DFB is collected and coupled to an optical fiber via a MEMS (micro-electrical mechanical systems) tilt mirror [95]. The MEMS mirror eliminates the $1/N$ combiner loss (and associated need for an integrated amplifier) of previous tunable laser research demonstrations that incorporated a selectable DFB array with a WDM combiner and SOA [96]. Like the SG-DBR lasers, these devices have met with significant commercial success and are capable of delivering >13 dBm output power in product form [97].

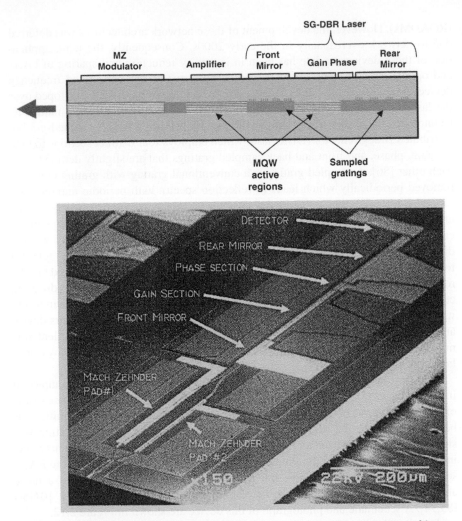

FIGURE 2.12 Schematic diagram (a) and photograph (b) of a SG-DBR tunable laser with an integrated Mach-Zehnder modulator [98].

In addition to intermediate and long-haul applications, single-channel PICs have also been developed for network access applications. In order to minimize cost and complexity, a single epi-step process has been developed that facilitates integration of lasers and passive designs [28–31,94]. Figure 2.13 shows an example of a optical network unit (ONU) transceiver PIC that incorporates a 1310-nm DFB laser operating for upstream communications, a 1310-nm/1490-nm wavelength splitter, and a 1490-nm waveguide detector (with an optional SOA) [94]. These devices are currently being sampled for commercial applications by OneChip Photonics.

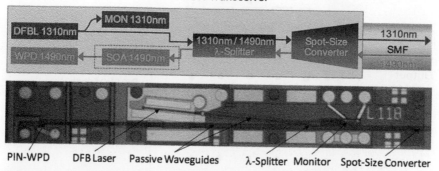

FIGURE 2.13 Schematic diagram (top) and chip photograph (bottom) of an InP diplexer for FTTH applications. The device incorporates a 1310 nm upstream DFB laser and a downstream 1490 nm waveguide photodetector integrated onto a single chip [94].

Their commercial success will be largely dictated by the ability of this platform to be significantly lower cost than the discrete devices it seeks to replace, as the access network is very cost sensitive.

2.3.1.2 Group III–V multichannel PICs for OOK transmission

The need to continue scale bandwidth in the network has led to a multi-decade demand to similarly increase the bandwidth per transmitter/receiver. The scaling of the optical components provides simultaneous benefits of component and system cost, size, power, and reliability. From the beginning of the first optical network until 2004, this scaling was achieved by scaling the bandwidth of a single transmitter/receiver optical channel. However, in the late 1990s bandwidth scaling beyond 10 Gb/s stalled as a result of increased fiber transmission impairments. In 2004, this bottleneck was broken with the introduction of the first commercial large-scale DWDM (dense-wavelength division multiplexed) 100-Gb/s PIC transmitter and receiver chips and modules which were enabled by simultaneous serial (within a channel) and parallel (multiple channels) integration [45].

The architecture for the multi-channel transmitter PIC is shown in Figure 2.14. The active train of each monolithically integrated channel includes an individually thermally tunable distributed feedback (DFB) laser, a high-speed (>10 Gb/s) electroabsorption modulator (EAM), an optical power monitor (OPM) photodiode at the back of the laser, and per-channel variable optical attenuators (VOAs). Second generation devices add a per-channel semiconductor optical amplifier (SOA) to provide high output power [99,100] and enable ultra-long haul transmission and tunability [100,101]. In this case, the VOAs are modified to also be optical power monitor photodiodes to enable the per-channel control of output power. Each individual channel is multiplexed into a single output channel via a monolithically integrated arrayed

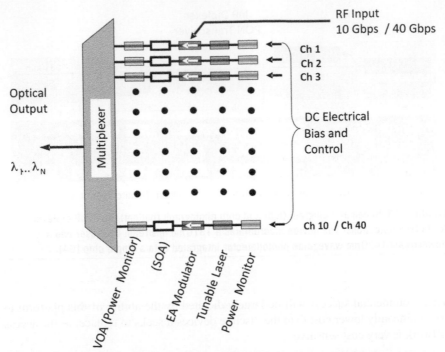

FIGURE 2.14 Schematic of the architecture of multi-channel OOK DWDM transmitter PICs. Commercial devices integrate 10 channels operating at 10 Gb/s to realize a 100-Gb/s monolithic transmitter PIC [45,99–101]. Research devices have integrated 40 channels operating at 40 Gb/s to realize a PIC capable of 1.6 Tb/s OOK transmission [46].

waveguide grating (AWG) which is terminated in a spot size converter for optimized fiber coupling in the package. First generation 100-Gb/s commercial devices consisted of 10 channels operating at 10 Gb/s per channel and integrated 51 functions per chip [45]. Second generation 100-Gb/s commercial devices added semiconductor optical amplifiers, resulting in 61 functions per chip [99–101]. Research demonstrations of this architecture have integrated 40 channels operating up to 40 Gb/s per channel (1.6 Tb/s per PIC) and integrated >240 functions per chip [102].

The multi-channel OOK receiver PIC is complementarily architected as shown in Figure 2.15 [45]. It consists of a single optical input channel/spot-size converter routed through a waveguide to a polarization-independent demultiplexer. Ten high-speed (>10 Gb/s) PIN photodiodes (PDs) are monolithically integrated into the waveguides at the output of each AWG port. Optionally, a polarization independent (PI) SOA is integrated at the input after the spot-size converter [103]. Commercial devices operating at 100 Gb/s consist of 10 channels operating at 10 Gb/s per channel [45]. Research demonstrations of this architecture have integrated 40 channels operating at 12.5 Gb/s per channel (500 Gb/s per PIC) [47].

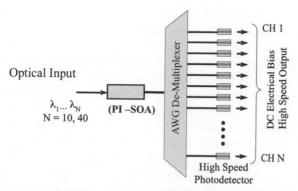

FIGURE 2.15 Schematic of the architecture of multi-channel OOK DWDM receiver PICs. Commercial devices integrate 10 channels operating at 10 Gb/s to realize a 100-Gb/s monolithic receiver PIC [45]. Research devices have integrated 40 channels [103] as well as an optional polarization-independent SOA [104].

Commercial versions of the multi-channel DWDM OOK transmitter and receiver PICs operating at 100 Gb/s have been deployed commercially since 2004 [45]. The 100-Gb/s commercial PICs are packaged in state-of-the-art commercial optical packages. A photograph of a module with the lid removed for a second generation 100-Gb/s transmitter (with SOAs) [99,100] is shown in Figure 2.16. The package is hermetic and contains over 150 electrical leads with over 600 electrical interconnects within the package. A single-mode lensed fiber is coupled to the output waveguide of the transmitter PIC. A 10-channel monolithic SiGe modulator-driver ASIC array is electrically connected in a hybrid fashion to the PIC modulators. A thermoelectric cooler is utilized to maintain the requisite temperature control of the PIC for DWDM operation. The commercial 100-Gb/s receiver PIC module packaging is similar to that of the transmitter, and includes a lensed fiber coupling to the input of the receiver

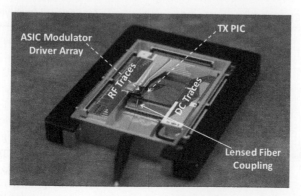

FIGURE 2.16 Photograph of a second generation 100-Gb/s PIC transmitter module with the hermetic lid removed [100].

de multiplexer and 10-channel monolithic SiGe transimpedance amplifier (TIA) array electrically connected in a hybrid fashion to the PDs on the PIC (not shown).

The normalized, fiber-coupled output power spectrum of a 40-channel transmitter PIC based on OOK is shown in Figure 2.17 (left). This device employs per-channel SOAs to level the output powers from the individual channels [99–101,103]. Each channel shows an average spacing of 50 GHz between the channels, thus spanning ~8 nm for 40 channels (a quarter of the traditional C-Band). Note that for commercial devices, the AWG filter functions are aligned on a 200 GHz ITU wavelength grid. In a deployed network a grid spacing of 50 GHz or 25 GHz is achieved by multi-stage interleaving PIC-based linecards [100]. The transmission spectrum for a 40-channel receiver PIC based on OOK is shown in Figure 2.17 (right). This device employs a single PI-SOA with channels spaced at 50 GHz (spanning 8 nm) and achieves a power flatness across all 40 channels better than ±0.75 dB. Note that for commercial devices, the AWG filter functions are aligned on a 200-GHz ITU wavelength grid. In a deployed network a grid spacing of 50 GHz or 25 GHz is achieved by multi-stage (de)interleaving PIC-based linecards [100]. The 3 dB optical bandwidth for all 40 channels ranges between 36 GHz and 40 GHz with an adjacent channel cross-talk better than 20 dB. The spectral characteristics and performance of these multi-channel PICs are suitable for deployment in long-haul and ultra long-haul systems [45,99,100].

The inclusion of the SOA on the 2nd generation 100-Gb/s transmitter PIC has enabled the devices to be tunable, reducing sparing costs for customers [101]. The spectra remain unaffected over the tuning range of 150 GHz. The transmitter PIC per-channel output power is maintained across the tuning range by compensating for reduced output power and channel skew with the per-channel SOA. Tunable (second generation) receiver PICs utilize the same architecture as first generation devices; however, they have sufficient margin to enable their performance also to be tunable

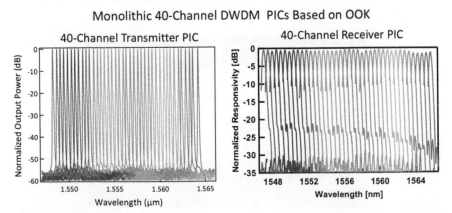

FIGURE 2.17 Spectra for 40-channel DWDM transmitter (left) and receiver (right) PICs based on OOK.

over 150 GHz [101]. In order to maintain the requisite over life margins for transmitter and receiver PICs, the tunability performance is limited to 75 GHz in commercial systems, enabling a tunable PIC-module to reduce sparing costs by 4×. This sparing cost reduction is achieved by virtue of the system implementation which utilizes multi-stage interleaving of PIC wavelengths (spaced 200 GHz on-chip) to achieve a 25 GHz-spaced system [100] enabling a capacity of 1.6 Tb/s in the C-Band.

Data is encoded for optical transmission on each channel of the multi-channel transmitter PIC via electro-absorption modulators (EAMs). The high-speed transmission performance for 40-channel transmitter PICs is shown in Figure 2.18 for devices operating at 12.5 Gb/s (500 Gb/s per PIC) at left and 40 Gb/s (1.6 Tb/s per PIC) at right. The data lines for EAM operation at 12.5 Gb/s (left) are driven using a broadband amplifier and exhibit very uniform performance (left) across all 40 channels and are comparable to the commercial 10 channel 100 Gb/s PIC at similar data rates [45]. The data for transmitter PICs utilizing EAMs designed for 40 Gb/s operation is shown in Figure 2.18 (right) [46]. Bandwidth limitations in the measurement instruments result in some filtering and transmitter eye closure. The eye diagrams are very uniform over all 40 channels demonstrating a robust design and a high degree of control. The extinction ratio for all channels operating at 40 Gb/s is in the range of 6 dB–8 dB. Despite the high level of integration, the transmission performance of the commercial large-scale PICs is nominally equivalent to that of discrete devices (at the same data rates). Second generation transmitter PICs using 10 Gb/s OOK per channel have been deployed in submarine links spanning 6200 km without regeneration [105].

An essential requirement for the commercialization of transmitter and receiver PICs for telecommunications applications is long life (>20 years). Photonic integration offers significant potential improvements in reliability over discrete devices in its ability to eliminate many complex opto-mechanical packages which typically limit the reliability performance of these optical transmission devices. First and second generation commercial 100 Gb/s large-scale transmitter and receiver PICs and modules based on OOK have undergone extensive reliability testing and successfully completed Telcordia GR-468 qualification. These transmitter and receiver PICs result in over a 20–30× reduction in the number of fiber-couplings compared

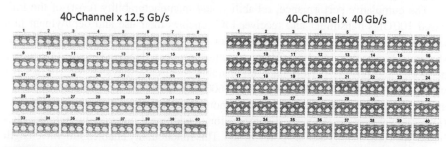

FIGURE 2.18 Eye diagrams for 40-channel DWDM transmitter PICs based on OOK. Data at 12.5 Gb/s per channel (500 Gb/s per PIC) is shown at left and 40 Gb/s (1.6 Tb/s per PIC) is shown at right.

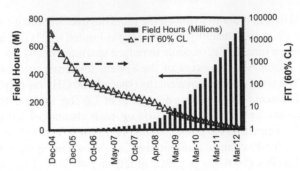

FIGURE 2.19 Reliability data for field-deployed 100 Gb/s first generation large-scale PIC module pairs (transmitter and receiver). No failures are observed after >727 h corresponding to a 1.3 PIC-pair FIT rate.

to a conventional system comprised of discrete optical components. Accordingly, the drastic reduction in fiber connections results in substantial improvement in the reliability of the overall system. Furthermore, these PICs are designed with sufficient controls and performance margins to maintain reliability performance for over 20 years of life. The cumulative effect of the benefits of photonic integration for these devices is shown in the reliability results shown in Figure 2.19. This data is based on field operating hours reported periodically after the first product shipment for 1st generation 100 Gb/s PIC-pairs (pairs of a transmitter and receiver). The vertical bars represent the integrated PIC-pair field operating hours up to the reported date. The dashed curve represents the PIC-pair failure-in-time (FIT) rate assuming a random failure probability model and calculated at each report date using a 60% confidence level. 1 FIT = 1 failure in 10^9 h of operation. As of May 2012, there have been over 727 Mh of PIC-pair field operation with no observed PIC-caused field failures. Accordingly, the failure rate is limited by field hours and has continuously dropped to the last reported level of 1.3 FIT. This performance is comparable to the best reported submarine pump laser performance [106,107], yet for a pair of PICs integrating 62 optical functions.

The cumulative performance, reliability, and manufacturability (cost) of the InP-based 100 Gb/s transmitter and receiver PICs enables their use and deployment in a commercial optical telecommunications transport system by Infinera Corporation as shown in Figure 2.20 [45,100]. These systems have been deployed since 2004 in live networks and possess performance comparable to discrete devices for 10 Gb/s per wavelength systems: 100 Gb/s per FRU, >6000 km reach (25 GHz channel spacing), and 1.6 Tb/s C-Band fiber capacity. Furthermore, the advantages of photonic integration (lower power, size, cost), enable the integration of additional functionality into the PIC-based line card: namely a cross-point switch. This functionality enables sub-wavelength bandwidth management within the system (at ODU1, 2.5 Gb/s granularity) in addition to the traditional transport functions. Thus, a variety of client interfaces of any protocol or speed (up 100 Gb/s) may be virtualized onto the transmission bandwidth of the system.

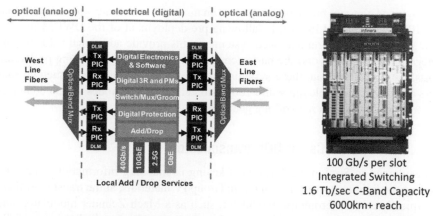

FIGURE 2.20 Schematic diagram (left) and picture (right) of a PIC-based "digital" ROADM node with integrated bandwidth management.

This functionality is enabling relative to traditional architectures based on discrete optics that map a fixed service to a wavelength. The term digital ROADM is utilized for these PIC-based optical-electrical-optical (OEO) systems as the bandwidth is managed digitally by electronics integrated into the PIC-based line cards and system. This architecture allows 3R regeneration, switching/muxing/grooming, digital protection, and add/drop all in the same node, all performed digitally. Accordingly, these features allow the digital ROADM to support colorless, directionless, and contentionless operation without wavelength blocking. In addition to the enhanced functionality and capability, this architecture also enables an improved bandwidth efficiency of 20–40% [108] compared to conventional systems (which use a transponder operating at the per wavelength bit rate). Hence, significant capital savings is incurred from not only the reduced cost of the systems, but also their increased network bandwidth efficiency.

The PIC-based OOK transport systems also exhibits lower power dissipation and reduced operating costs compared to equivalent systems based on discrete optics. The 100 Gb/s PIC-based systems can facilitate >35% reduction in network power consumption [109] as a result of lower PIC component power consumption as well as a reduced volume of total equipment required to implement a network (arising from the aforementioned network efficiency improvements). The reduced PIC power consumption results from a number of different contributions in the integrated PIC-module. For example, on-chip couplings between functional elements are much less lossy than the typical fiber coupling loss in discrete optical components, resulting in reduced power consumption. Furthermore, the 100 Gb/s PIC-based systems also enable a significant reduction in operating expenses. A case study for the US nation-wide deployment of a DWDM system for a major carrier showed a 25–200% saving in operating expenses compared to conventional systems for a PIC-based system [110]. The key factors contributing to this saving were: reduced fiber terminations/couplings, adding capacity in 100 Gb/s increments (thus requiring fewer line cards to deploy), and constant bit-error-rate testing on non-service bearing wavelengths from the 100 Gb/s deployed line cards (allowing pre-deployment of the DWDM line for new services). The aforementioned system benefits provide additional value for the integration of PICs beyond the basic economic advantage of eliminating packaging costs. As a result, to date these systems have met with significant commercial success and have been deployed with >100 customers on >9500 nodes in >55 countries accounting for >$2B in networking gear revenue.

2.3.2 Group IV PICs for OOK transmission

As explained in Section 2.3.1.1, on-off keying is encoding information simply by turning the light on or off for each bit. In Group IV materials, at the transmitter there is typically an interferometric modulator, such as a Mach-Zehnder modulator, and at the receiver there is typically a Ge photodetector. All of today's commercially deployed Si-photonic transmitters use only on-off keying.

2.3.2.1 Group IV single-channel PICs for OOK transmission

There are currently few commercial Si photonics products designed for single channel OOK applications. This is because there is no practical monolithically integratable laser on the Si platform yet and single-channel OOK transceivers are simple and hence do not require much integration. The few devices that are commercially sold include fast variable optical attenuators to place an identifying tone on a signal pioneered by Kotura [111], which use carrier injection to create loss by free-carrier absorption, and 10-Gb/s Mach-Zehnder OOK modulators, which use MOS modulators pioneered by Lightwire [68] and Intel [67] to have a low driving voltage, at the expense of a high capacitance per unit length.

2.3.2.2 Group IV multi-channel PICs for OOK transmission

Wavelength-division multiplexed (WDM) devices require optical filtering and are best off using a mixture of SiO_2-family materials and Si-family materials. That is because the compound materials tend to have a lower refractive index and lower index change with temperature. One example is the 8-channel polarization-independent coarse wavelength-division multiplexed (CWDM) receiver on a silicon substrate shown in Figure 2.21 [112]. The input spot-size converter, AWG, output waveguides, and photodetectors comprise SiO_2, Si_3N_4, Si, and Ge waveguides, respectively. Si_3N_4 has an index change with temperature about five times lower than that of Si, making the AWG passbands move about five times less in wavelength for than Si for a given temperature change, making it well suited for non-temperature-controlled operation. The total insertion loss from fiber to photodetector, including the photodetector responsivity, is less than 5.6 dB for channels 1–6.

The spot-size converter design used in Figure 2.21 is shown in detail in Figure 2.22 [113]. It uses an inverse taper (i.e. a waveguide that narrows down to widen the mode instead of widening to widen the mode) inside of a suspended glass beam formed by etching the glass around it and the silicon underneath it and backfilling with a fluid

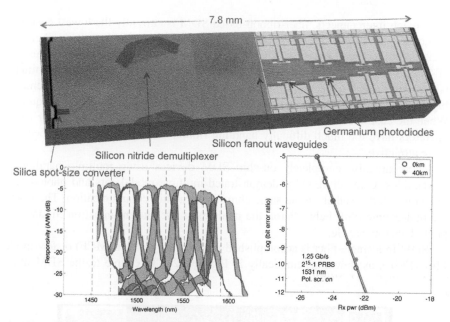

FIGURE 2.21 Chip photograph of 8-channel CWDM receiver, measured spectral response over all input polarizations, and measured bit-error rate vs. received optical power.

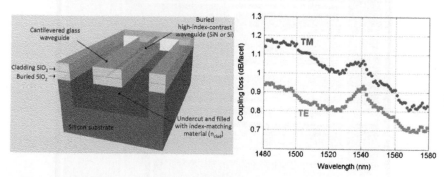

FIGURE 2.22 Si$_3$N$_4$ cantilever spot-size converter.

with a refractive index slightly lower than that of the glass. It can achieve less than 1.2 and 2.0 dB loss for both polarizations coupling to a standard single-mode fiber from Si$_3$N$_4$ and Si wire waveguides, respectively.

2.3.2.3 Space-division multiplexed devices

Space-division multiplexing (SDM) can serve two main purposes: to increase the transmission capacity over a single fiber and/or increase the optical connection density. The most common form of SDM is to use multiple fibers in a fiber ribbon. For short reach interconnects, fiber-ribbon-based SDM is currently the lowest cost way to increase the bit rate. Figure 2.23 shows an 8-fiber Si transceiver (four fibers for each direction) from Luxtera with 10 Gb/s OOK on each fiber for a total transmission of 40 Gb/s [114]. Only one III–V laser, which is power split to the four transmitters, is required in each transceiver. this laser is hybridly integrated using the scheme of Figure 2.10c. If WDM was used instead of SDM, then four lasers would be required. When the cost of short-reach transceivers falls below that of the link cost, or when operators have too much difficulty handling ribbon fiber, then WDM will become more attractive.

There are also completely on-chip SDM devices using multiple parallel Si waveguides. Urino et al. [115] demonstrated multiple modulators and photodetectors connected together on the same chip for intra-chip communication. The lasers are butt-coupled InP Fabry-Perot, the modulators are current injection Si, and the photodetectors are Ge.

SDM in a single fiber is accomplished with multicore fiber (MCF) or few-mode fiber (FMF), as shown schematically in Figure 2.24. To couple to the SDM fiber,

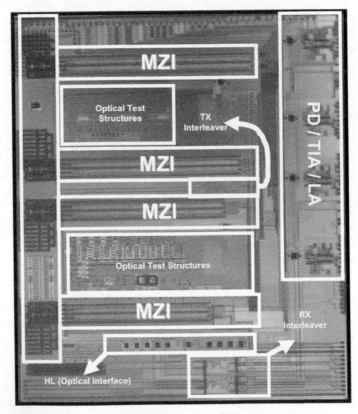

FIGURE 2.23 4 × 10 Gb/s SDM transceiver in Si.

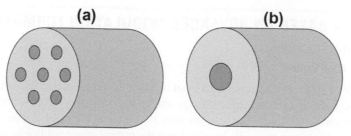

FIGURE 2.24 SDM fibers: (a) multicore fiber (MCF) and (b) few-mode fiber (FMF).

one traditionally uses free-space micro optics. However, using integrated optics has the promise of lower cost and large-volume manufacturability. However, it is very difficult to couple SDM fiber to the facet of a PIC efficiently because of the 2-D arrangement of the modes that can cover thousands of μm^2. A more practical way to couple a single-fiber SDM fiber to a PIC is to couple to the top surface of the PIC using grating couplers.

Figure 2.25 shows a Si polarization-independent, 7-core, 2-wavelength CWDM receiver [116]. 1-D grating couplers are arranged in a triangular lattice, matching the MCF core distribution. The MCF is tilted at the appropriate angle such that transverse-electric (TE) polarized light couples to the left and transverse magnetic (TM) light propagates to the right, forming a polarization-diversity receiver. There are also length-imbalanced Mach-Zehnder interferometers serving as coarse wavelength-division multiplexed (CWDM) demultiplexers.

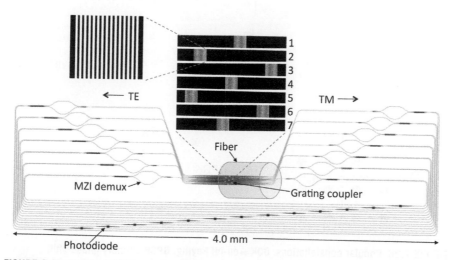

FIGURE 2.25 7-Core CWDM receiver PIC.

2.4 PICs BASED ON ADVANCED MODULATION FORMATS

2.4.1 Introduction

2.4.1.1 Overview

In the face of continued increase in channel bandwidth and reduction in channel spacing, OOK has begun to asymptote in its ability to scale fiber capacity because of its low spectral efficiency. One can improve the amount of bit/s/Hz transmitted by using multiple magnitude levels, by also using the optical phase, and/or by using both polarizations [117]. The location of the data points in the space of the optical signal is called a constellation. This space can have three dimensions (in-phase, quadrature, and polarization), but usually only two dimensions are shown. Some of the most popular constellations are shown in Figure 2.26. From top to bottom the number of bits per symbol is increased. The transitions between the symbols are not shown, for clarity.

The fundamental building block of a transmitter for advanced modulation formats is a Mach-Zehnder modulator (MZM). A single stage MZM, depending on how it is biased, may be used to implement the OOK (on-off keying, where the amplitude of the optical carrier is turned on and off) or the BPSK (binary phase shift keying, where phase of the optical carrier is modulated between 0 and π radians) modulation formats. As shown in Figure 2.27, if two such single stage MZ phase modulators are combined in quadrature (with a relative $\pi/2$ phase shift), you obtain a two-stage

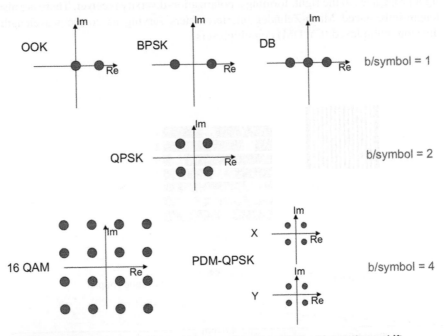

FIGURE 2.26 Popular constellations. OOK = on-off keying, BPSK = binary phase-shift keying, DB = duobinary, QPSK = quadrature phase-shift keying, QAM = quadrature amplitude modulation, and PDM = polarization-division multiplexed.

Coherent Transmitter Architecture
PM-QPSK

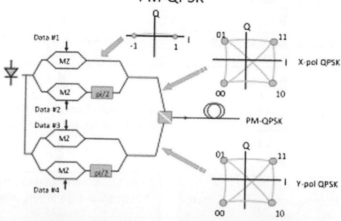

FIGURE 2.27 The I/Q MZM for the generation of polarization multiplexed QPSK signal. Two BPSK signals are combined in quadrature to generate the QPSK signal. Output of a single laser is split to generate the two polarization components. One of the outputs is rotated in orientation and then combined with the other output to generate the polarization multiplexed signal.

MZM (two single stage MZMs nested within a larger MZM) capable of generating four complex states. In principle, all higher order amplitude-phase modulated signals (QAM, quadrature amplitude modulation) may be decomposed into their real and imaginary (I, in-phase and Q, quadrature) components. Hence the two stage, I/Q MZM is the basic building block for implementing QPSK or QAM. In the case of a generalized QAM signal, the I and Q components would be linear multi-level signals unlike the QPSK format where the I and Q components are binary.

A straightforward way to double the capacity of the optical carrier is to carry modulated data in both of the fiber's polarization states. This is called polarization multiplexing (PM). At the receiver in order to demodulate a polarization-multiplexed signal, the polarization state of the light must be accurately tracked or equivalently a polarization diverse receiver must be implemented. This is true even if only one polarization state of the carrier is modulated with data. So, in principle, there is little additional complexity in the receiver to recover polarization multiplexed signals. At the transmitter, the output from a single laser is split into two, and these are then independently modulated, one is polarization rotated, and then they are polarization combined, as shown in Figure 2.27.

Direct detection cannot detect base-band advanced modulation formats. Both the phase and polarization need to be detected. Interference with a reference is required to detect phase, the reference being either another time portion of the signal itself or another signal. Differential QPSK (DQPSK) uses differential coding at the transmitter or the receiver. In other words, the information is now in the phase difference between adjacent symbols rather than the absolute phase of each symbol. DQPSK can be demodulated using a delay line interferometer. In this DQPSK format, the data is encoded as a phase change relative to the phase state of the previous bit.

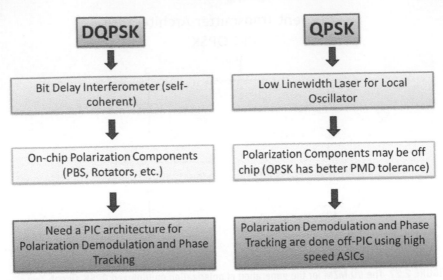

FIGURE 2.28 Differences in the implementation of DQPSK and QPSK receivers. One of the major differences is that the DQPSK format does not require an absolute phase reference at the receiver. In DQSK format after the delay line interferometer, the phase information is discarded. In the QPSK format the phase information is preserved through the detection process and subsequently used for further signal processing.

The major differences between the QPSK and DQPSK formats in terms of the receiver implementation are shown in Figure 2.28.

Figure 2.29a shows a differential receiver for QPSK, in which one symbol is interfered with the previous symbol in a 90° hybrid. (b) shows a coherent receiver, in which the signal is interfered with a cw signal. The advantage of (a) is that it does not require a LO, high-speed analog-to-digital converter, or digital signal processing. The advantage of (b) is that it has ~2.5-dB better sensitivity for QPSK, it can work with any modulation format, and the DSP can compensate many impairments. Both (a) and (b) assume a single-polarization signal. (a) can be used with a randomly varying input polarization, but (b) works only when the signal is co-polarized with the LO. Later we show configurations that can handle both polarizations simultaneously for both differential and coherent.

It is common to use differential encoding in a coherent receiver. If one does not use either differential detection or encoding, then one must know the absolute phase of the received QPSK signal. If the phase were to suddenly jump by more than 90°, called a cycle slip, then the absolute phase would no longer be accurately known, and all the received data after that would be in error. Differential encoding, like differential detection, measures the difference in phase between successive symbols, except that differential detection takes the difference before the symbol decision is made, and differential encoding takes the difference after the symbol decision is made. Differential encoding results in a doubling of the BER, which is about a 1-dB penalty at 10^{-3} BER.

In addition, polarization-multiplexed systems, typically used to double the spectral efficiency, require real-time polarization tracking and demodulation. In QPSK or other

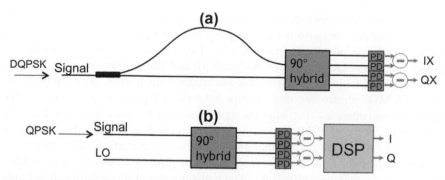

FIGURE 2.29 Single-polarization: (a) differential and (b) coherent detection.

coherent systems using a local oscillator this can be done using digital signal processing algorithms [118,119]. Unlike QPSK, the DQPSK modulation format is less tolerant to polarization mode dispersion so for a PIC implementation, the polarization components need to be integrated on the chip. In the early demonstrations of non-LO based PM DQPSK systems, the incoming signal was polarization demultiplexed manually using a polarization controller followed by a polarization beam splitter (PBS) [120,121]. Recently, methods for automatic polarization tracking using external active optics have been reported [122,123]. In a modified version of the PM DQPSK format reported recently, the signal is time/polarization multiplexed, with half-symbol time interleaving, at the transmitter, and then detected using decision circuitry operating at twice the symbol rate without the need for explicit optical polarization demultiplexing [124]. However, the time interleaving results in reduced spectral efficiency.

Coherent communication, which involves interfering the signal with a local oscillator (LO) laser, provides a way to measure the entire complex optical field. Optical coherent systems were heavily researched in the 1980s. The main reason was not because of coherent's ability to detect the phase of light but instead because coherent provides effective optical gain, through the beating of the signal with the LO in the photodetectors. Once the Er-doped fiber amplifier (EDFA) was invented, this advantage of coherent detection mostly evaporated and work on coherent decreased significantly. It picked up again at the turn of the century, when higher bit rates and spectral efficiency put advanced modulation formats in great demand.

Coherent detection can be divided into three categories based on the frequency difference between the signal and LO, as shown in Figure 2.30. Original coherent

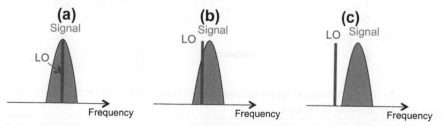

FIGURE 2.30 Coherent detection types: (a) homodyne, (b) intradyne, and (c) heterodyne.

systems were either homodyne or heterodyne. However, homodyne detection is extremely difficult because it requires frequency and phase locking the LO to a signal which has no carrier (most advanced modulation formats do not have a carrier left in the spectrum after modulation), and heterodyne detection requires extremely high-frequency photodiodes and receiver electronics. With the improvement of CMOS electronics, one can now have enough processing power to do the frequency and phase locking, after the signal has been received, by post-processing in real time. For a 100-Gb/s signal, this processing is done at about 1 Tb/s, assuming 5 bits of resolution per symbol and two times oversampling. This allows for intradyne detection, which is the most common type of coherent detection used in today's telecommunication links. The real-time signal processing can not only perform frequency and phase locking, but it can also perform polarization demultiplexing, chromatic dispersion compensation, polarization-mode dispersion compensation, many transmitter and receiver imperfections compensation, and other impairments. Research is ongoing to even compensate nonlinear fiber transmission impairments.

There are varying degrees of coherent receiver sophistication. Figure 2.31 shows the evolution of coherent receivers. Early coherent receivers detected only a single quadrature and were mainly used for heterodyne detection. Today's coherent receivers detect both polarizations and quadratures.

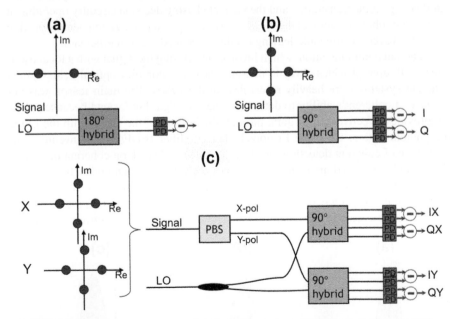

FIGURE 2.31 Coherent receiver types: (a) single quadrature, (b) dual quadrature, and (c) dual polarization, dual quadrature.

The hybrids in coherent receivers consist of optical couplers. Some examples of hybrids are shown in Figure 2.32. A 180° hybrid is simply a 2 × 2, 3-dB coupler. A 90° hybrid is a 2 × 4 coupler. Not all 2 × 4 couplers are 90° hybrids, though.

A more detailed coherent receiver is shown in Figure 2.33. A LO (A) is mixed with the incoming signal to establish the phase reference for demodulation. A balanced photodetector pair is used for signal detection. In this case, the real and imaginary parts of the incoming complex signal are recovered. The relevant field product terms are shown in Figure 2.33. The 90° hybrid is used to mix the LO with the I and Q components of the signal. The transfer matrix for the 90° hybrid is shown in Eq. (2.6). The math is then done for the top pairs of PDs. Expanding out the terms one obtains the equations for the desired coherent component and the undesired common mode noise. The degree to which the PDs are balanced in their responsivities determines the extent to which the common mode noise is rejected. With perfect balanced detection, the squared signal and LO terms cancel out. In practice, there is always some amount of asymmetry between the PDs. The degree to which these are balanced and the noise is rejected is called the common mode rejection ratio

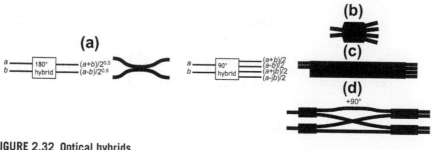

FIGURE 2.32 Optical hybrids.

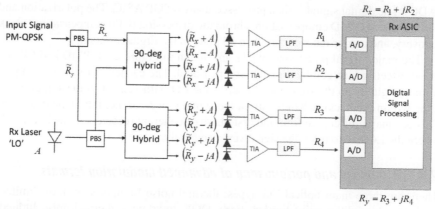

FIGURE 2.33 Dual-polarization, dual-quadrature coherent receiver.

(CMRR). This is expressed as the ratio of the difference and the sum of the respective PD responsivities [125]. The math can then be repeated for the remaining three terms of the complex polarization multiplexed signal.

$$
\begin{bmatrix} \text{OUT1} \\ \text{OUT2} \end{bmatrix} = \begin{bmatrix} 1 & j \\ j & 1 \end{bmatrix} \begin{bmatrix} \tilde{R}_x \\ A \end{bmatrix} = \begin{bmatrix} \tilde{R}_x + jA \\ j\tilde{R}_x + A \end{bmatrix}, \tag{2.6}
$$

$$
\begin{bmatrix} \text{OUT3} \\ \text{OUT4} \end{bmatrix} = \begin{bmatrix} 1 & j \\ j & 1 \end{bmatrix} \begin{bmatrix} \tilde{R}_x \\ jA \end{bmatrix} = \begin{bmatrix} \tilde{R}_x - A \\ j\tilde{R}_x + jA \end{bmatrix}, \tag{2.7}
$$

$$
\left| \tilde{R}_x + A \right|^2 = \left| \tilde{R}_x \right|^2 + A^2 + 2\Re[\tilde{R}_x], \tag{2.8}
$$

$$
\tilde{R}_x = R_1 + jR_2, \tag{2.9}
$$

$$
\text{balanced detection} = \rho_1 \left| \tilde{R}_x + A \right|^2 - \rho_2 \left| \tilde{R}_x - A \right|^2, \tag{2.10}
$$

$$
= 2(\rho_1 + \rho_2)R_1 + (\rho_1 - \rho_2)\left(\left| \tilde{R}_x \right|^2 + A^2 \right), \tag{2.11}
$$

$$
= 4\rho R_1 \text{ for } \rho_1 = \rho_2 = \rho, \tag{2.12}
$$

$$
\frac{\text{undesired signal}}{\text{desired signal}} = \left(\frac{\rho_1 - \rho_2}{\rho_1 + \rho_2} \right) \left(\frac{\left| \tilde{R}_x \right|^2 + A^2}{2R_1} \right), \tag{2.13}
$$

$$
\text{CMRR} = \left| \frac{\rho_1 - \rho_2}{\rho_1 + \rho_2} \right|, \tag{2.14}
$$

where the responsivities of photodiodes 1 and 2 are ρ_1 and ρ_2, respectively. The detected signal is then sampled and digitized using an analog-to-digital converter (ADC). The digital signal is then processed using a DSP ASIC. The polarization and phase tracking, PMD compensation, chromatic dispersion (CD) compensation, clock recovery, and carrier recovery are then all done in the digital domain.

The major signal processing blocks are shown in Figure 2.34. PMD and CD are linear effects which can be digitally compensated. The carrier recovery is also performed digitally so the frequency offset of the LO with respect to the signal needs to be within the tracking bandwidth of the signal processor, i.e. intradyne detection. This is a major difference from the first phase of coherent receivers in the late 1980s where the LO had to track the signal very accurately.

2.4.1.2 Devices and performance of advanced modulation formats

There are three main optical link types: thermal-noise limited, shot-noise limited, and optical-amplifier-noise limited. Early OOK links were thermal-noise limited. The thermal noise comes from the receiver electronics and determines the sensitivity

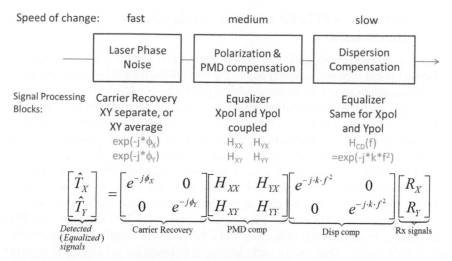

FIGURE 2.34 Major signal processing blocks in the coherent receiver. The order of operations may vary from implementation to implementation.

limit when the received optical power is very low. In early coherent systems (in the 1980s) the gain of the LO resulted in a near shot-noise limited system. With the advent of the EDFA, many OOK systems, especially most long-reach systems, became optical-amplifier-noise limited, often called optical signal-to-noise ratio (OSNR) limited. Still today, though, most short and intermediate OOK links are not optically amplified (to save cost) and thus are thermal-noise limited.

Coherent systems typically have a high optical power due to high LO power, and thus are almost always either shot-noise limited or optical-amplifier-noise limited. Since coherent is usually used in optically amplified systems, we will focus on the optical-amplifier-noise-limited case here.

The theoretical relationship between bit-error rate (BER) and OSNR for PDM-QPSK, assuming Gaussian noise, is

$$ \text{OSNR} = [\text{erfc}^{-1}(2\text{BER})]^2 \frac{R}{25 \text{ GHz}}, \tag{2.15} $$

where R is the bit rate (not symbol rate). This assumes that the OSNR (in linear units) is measured using a 0.1-nm (12.5-GHz) resolution bandwidth. If differential encoding is used, which it usually is, then the required BER must be divided by 2, because in differential encoding for QPSK with appropriate Gray coding, every time there is an error, two bit errors are generated. For PDM-16-QAM, the relationship is

$$ \text{OSNR} = 5 \left[\text{erfc}^{-1} \left(\frac{8}{3}\text{BER} \right) \right]^2 \frac{R}{50 \text{ GHz}}. \tag{2.16} $$

Again, R is the bit rate. At a BER of 10^{-3}, 16-QAM has a $\sim$4-dB penalty compared to QPSK.

2.4.2 Group III–V PICs for advanced modulation format transmission

There was a brief period in the mid to late 1980s when there was a lot of interest in coherent optical communication [126]. In the mid-2000s, the field of coherent optical communication was revived by the availability of high-speed Si ASICs and advanced digital signal processing algorithms which eliminated the requirement for ultra-stable optical sources and analog phase/frequency/polarization tracking of the optical signal at the receiver. This is reflected in the coherent PIC development timeline of Figure 2.35. Devices reported in Refs. [127–137] are shown in Figure 2.35.

Early coherent receiver PICs were all single channel. They were designed for the binary phase shift keying (BPSK) modulation format. BPSK is similar to QPSK except that there is no data in the quadrature component of the signal. A simple, single-stage MZM may be used to generate the BPSK signal. A BPSK signal has a lower spectral efficiency, but a better noise margin for longer transmission distances. There were early attempts to integrate a LO on the receiver PIC as well.

A multi-channel PIC with *I/Q* MZM integrated with an optical source was reported in 2008. There have been a number of variants on the DQPSK and QPSK (with external LO) receiver PICs reported since then. The DQPSK PICs also have

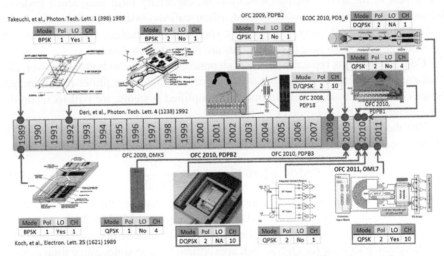

FIGURE 2.35 Brief timeline for the development of coherent PICs. There is a gap between the early 1990s, when EDFAs were first introduced, and late 2000s when coherent communication saw practical deployment. Key: Mode = BPSK, QPSK; Pol = number of polarizations detected; LO = whether a LO was integrated into the PIC; CH = number of channels integrated onto a PIC. Most of these are receivers, with exception of the entry for 2008 when a 10 channel transmitter PIC was reported which included an *I/Q* modulator integrated with an optical source, for each channel, on the same substrate.

the polarization components integrated onto the same substrate. The first multichannel, dual polarization, QPSK PIC with an integrated LO per wavelength was reported in 2011.

2.4.2.1 III–V single-channel PICs for advanced modulation format transmission

Because of their complexity and multiple electrical inputs/outputs, advanced modulation format transmitters and receivers are ideal candidates for optical integration. III–V materials are especially suitable for high-baud-rate transmitters and receivers because of their efficient and high-speed modulators and receivers.

NTT demonstrated a III–V traveling-wave InP *I-Q* modulator, which achieved 80-Gb/s QPSK operation [138]. More recently, Cogo demonstrated another InP *I-Q* modulator using a capacitively loaded traveling-wave structure [139]. As an example of a higher-order modulation format III–V device, a research-stage demonstration of a III–V, single-channel, advanced modulation format transmitter, an InP 16-QAM modulator, is shown in Figure 2.36 [140]. It used four electroabsorption modulators (EAMs), instead of the conventional nested Mach-Zehnder modulators, to generate the constellation. Each EAM was driven with an independent, but synchronized, binary data stream. The advantage is a very compact size, because of the strength of the quantum-confined Stark effect for electroabsorption in InGaAsP quantum wells. The disadvantage is high insertion loss and a residual carrier due to a residual optical DC component. Unlike traditional LiNbO$_3$ modulators, III–V modulators do not have a drifting bias voltage, simplifying their operation.

A research-stage 86-Gb/s InP single-polarization differential receiver for QPSK is shown in Figure 2.37 [140]. It consists of a Mach-Zehnder delay interferometer (MZDI; the longer arm configured in a loop) connected to a 2×4 star coupler acting as 90° hybrid connected to InGaAs photodetectors. It uses deep ridge waveguides. It successfully received an 86-Gb/s single-polarization DQPSK signal. Low

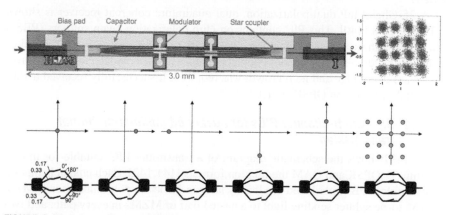

FIGURE 2.36 InP 16-QAM modulator.

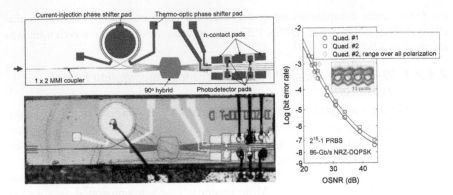

FIGURE 2.37 InP DQPSK receiver.

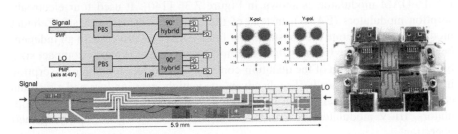

FIGURE 2.38 InP coherent receiver.

polarization dependence in the MZDI was achieved by using a polarization dependent phase shifter on the MZDI in concert with a polarization-independent phase shifter.

A 25-Gbaud InP dual-polarization, dual quadrature coherent receiver is shown in Figure 2.38 [140]. This PIC contains on-chip polarization beam splitters (PBSs) that separate the signal and LO into two polarizations. These proceed to 90° hybrids, which in this case are 2 × 4 MMI couplers, and InGaAs photodetectors. The waveguides are deep ridge and no regrowth was required. It successfully received a 100-Gb/s dual-polarization QPSK signal.

2.4.2.2 III–V multi-channel PICs for advanced modulation format transmission

Figure 2.39 shows the schematic diagram of a transmitter PIC suitable for implementing (D)QPSK and QAM modulation formats [141,142]. Each channel consists of a tunable DFB (distributed feedback) laser, a backside power monitor (PM), a TE/TM-to-be splitter sending light to a nested pair of MZMs in every path, and two AWGs separately combine the TE/TM-to-be channels into two output waveguides

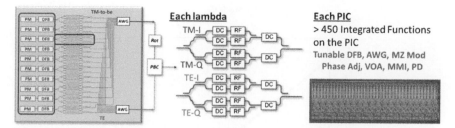

FIGURE 2.39 10-Wavelength, polarization-multiplexed QPSK transmitter PIC.

all integrated monolithically on a single chip. The polarization rotator and polarization beam combiner (PBC) are implemented off-chip via discrete optical components. Figure 2.39 also shows a photograph of the active block which consists of all the elements except the spectral multiplexers. The density of the elements is significantly higher than those of the previous generation PICs used for on-off keying modulation format. With the higher density, the PIC size scaled only by a factor of ~2, while the number of functions scaled by more than a factor of nine. The active block indicates the complexity of the device layout and optical and electrical routing. The active block contains 10 tunable DFBs (bottom), 20 nested modulators comprising 40 total MZMs (top) as well as all of the sense and control elements required for the PIC.

The light-current (L-I) curves for the DFBs on the transmitter PICs are shown in Figure 2.40 (left). The curves were taken using the optical power monitors (PMs) at the rears of the DFB lasers. The threshold current for all channels is <20 mA. Both the thresholds and slope efficiencies are very uniform across the array. The inset shows the DFB spectrum for all 10 channels that exhibit a side mode suppression ratio of >48 dB. There are no deleterious reflections from the downstream integrated optical circuitry that impact the laser performance (the MZMs are held in their transmission state to maximize any reflected light back into the laser during the linewidth measurement).

Low phase noise lasers are a critical requirement for coherent transmitter and receivers. The DFB phase noise spectra were measured using a delay-line interferometer (DLI) with a 10-GHz free spectral range (FSR) as a frequency discriminator. The lasers were biased under nominal operating conditions, and the center frequency was stabilized at the quadrature frequency of the transmission window of the DLI, converting the signal frequency fluctuations into intensity fluctuations at the output. The power spectrum of the output signal from the DLI was measured on an electrical spectrum analyzer over 100 kHz to 10 GHz. A typical measured spectra for 10 lasers in a PIC is shown in Figure 2.40 (right). The notch at 10 GHz is due to the FSR of the DLI. Above the ~1–10 MHz region, the spectrum is almost white and is used to determine the spectral linewidth of the laser source. The laser linewidth is ~300 kHz (π power spectral density between 100 MHz and 5 GHz [143]).

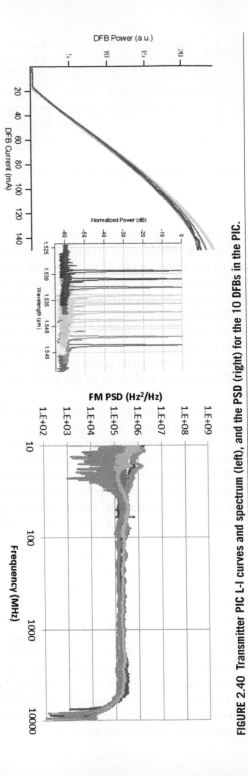

FIGURE 2.40 Transmitter PIC L-I curves and spectrum (left), and the PSD (right) for the 10 DFBs in the PIC.

Optical power splitters and combiners, connected in various ways, are common in this PIC layout. These are typically built either as directional couplers [144] or multi-mode interference (MMI) [145] couplers. Directional couplers require very good fabrication control of the width of the gap between the coupler waveguides. On the other hand, for acceptable performance, MMI couplers require very good control of the dimensions of the multi-mode section. An MMI-based architecture is most often chosen for PICs.

The DC performance of the sub-MZMs is characterized by scanning the RF electrode biases along the RF path and measuring the power transfer function. Excellent uniformity is demonstrated in Figure 2.41 where we show the superimposed power transfer functions of all 40 sub-MZMs on one PIC. A V_π of 2.6 V and DC extinction ratios derived from Figure 2.41 in excess of 32 dB across all 40 sub-MZMs are achieved with a median ER > 38 dB. These high extinction ratios imply a power imbalance between MZ arms of <0.5 dB which has negligible impact on system performance. Overall, the results indicate uniform and high extinction ratio performance for integrated InP modulators that compare favorably with conventional discrete modulators.

Figure 2.42 shows the architecture of the PM-QPSK receiver [146]. The H and V polarization components are split off-chip. As in the PM-DQPSK receiver PIC (details to be discussed later), the polarization components may also be integrated on the PIC. The coherent PIC receives fiber coupled signals with the polarization axes of the fibers oriented such that both H and V inputs are launched in the TE orientation on-chip. There is an input TE polarizer on both paths to strip any remnants of the TM component. Each signal arm also has a VOA to power balance the input signals. The V and H signal components are then wavelength demultiplexed using a single AWG.

Each demultiplexed output is then mixed with its own local oscillator (LO) using a 90° optical hybrid. These optical hybrids may be constructed out of a series of 2×2 couplers or made of a single 2×4 coupler [147]. LOs are tunable DFB lasers integrated on the same InP substrate, and tuned to the incoming signal frequencies.

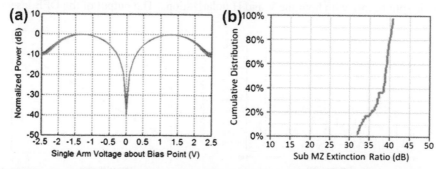

FIGURE 2.41 (left) 40 MZM DC power transfer functions vs. the push-pull voltage on one arm about the central bias point and (right) cumulative distribution plot of all 40 sub-MZM extinction ratios, each exceeding 30 dB on the Tx PIC.

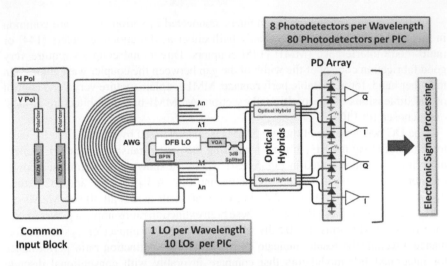

FIGURE 2.42 Multichannel coherent receiver PIC architecture.

A single LO (output controlled by a VOA) is split to mix with both the V and H polarizations. The outputs of the optical hybrid are then fed to high-speed, balanced photodetector (HSPD) pairs. When packaged, the HSPD outputs are wire bonded to a high-speed TIA (transimpedance amplifier) array off-PIC. As shown in Figure 2.42, the linear TIA outputs are then sampled using high-speed ADCs and are processed in real time using a DSP ASIC.

Figure 2.43 shows the normalized input responsivity curve for the PIC, which essentially maps out the spectral response of the demultiplexer at the input. The peak frequency plot in Figure 2.43 shows that the channel separation is 25 GHz. The 10-wavelength block occupies only a 250-GHz bandwidth [148].

The LOs are temperature tunable DFB lasers integrated on the same InP substrate, and are tuned to the incoming signal frequencies. Within a channel, a single LO is split to mix with both the V and H polarizations. The output of the DFB lasers

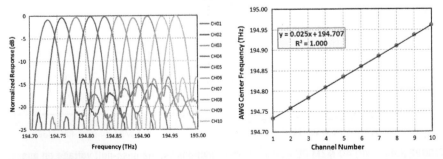

FIGURE 2.43 (left) Normalized receiver spectral response, and (right) peak frequency as a function of channel number.

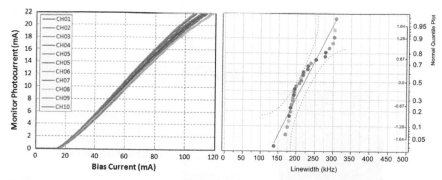

FIGURE 2.44 (left) L-I curves of the LO DFB lasers, and (right) local oscillator linewidth distribution.

(LO), are measured on-PIC using the integrated back PIN (BPIN) PD. Figure 2.44 shows that each DFB is capable of greater than 20 mW output power.

The linewidth for the receiver PIC is measured on test cells, on the PIC wafer, where we can route part of the DFB output to the facet. Figure 2.44 shows the distribution of the LO linewidth for a sample of DFB lasers over multiple wafers. Like the transmitter, the linewidth is inferred from noise power spectral density (PSD) measurements. The median linewidth of the integrated DFB laser is 200 kHz which is more than narrow enough for the coherent detection of QPSK signals at 14 Gbaud.

Commercial systems require the signal processing associated with coherent detection to be done in real time. Currently, single-channel ASICs integrated with FEC are available for data rates up to 100 Gbit/s [149]. For the PIC-based system implementation, we have a real time DSP and FEC ASICs that, together with the 10 wavelength transmitter and receiver PICs, have been used to implement a 570 Gbit/s (500 Gbit/s data throughput plus the FEC and other overheads) which is the heart of an Infinera Corporation commercial superchannel transport system available for deployment today [100,142]. The resultant system is capable of long-haul reach consistent with discrete components. A photograph of the system is shown in Figure 2.45 (right). The system exhibits a total fiber capacity of 8 Tb/s in the C-Band, and 5 Tb/s per chassis of DWDM transport capacity and non-blocking OTN switching. The feasibility of implementing 5 Tb/s of non-blocking switching and WDM transmission in a single bay is made feasible by the reduced power and size of the 500-Gb/s PICs. The system exhibits similar types of benefits afforded by the earlier generation OOK PIC-based system described in Section 2.3.1.2 in terms of increased network bandwidth efficiency, reduced power consumption, reduced capital and operating costs. However, these advantages are even more prevalent in this system as mismatch between the transmission capacity per line card and the required client services is significantly higher for high-capacity, high-spectral efficiency coherent-based systems.

The commercial 500 Gb/s (14.3 Gbaud symbol rate) PICs have been optimized for higher data rate operation up to 1 Tbit/s (28 Gbaud symbol rate). The MZM lengths

Commercial Transport System Based on 500 Gb/s Coherent PIC Superchannels

1732km Error Free Transmission

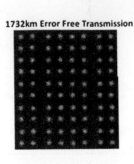

500 Gb/s Superchannel
500 Gb/s per slot
5 Tb/s per Chassis
Converged DWDM / OTN

FIGURE 2.45 Dual polarization, 14.25 Gbaud constellation diagrams for error-free 500 Gb/s superchannel transmission of a field trial of an Infinera Corporation commercial grade system using 500 Gb/s PICs over 1732 km (left). Photograph of a fully populated chassis for a commercial system long-haul transport system (right) based in 500-Gb/s coherent PM-QPSK PICs that is capable of: 500 Gb/s superchannels, 5 Tb/s per chassis capacity, and integrated a converged DWDM transport and OTN switch with 5 Tb/s of non-blocking capacity.

were made shorter and the capacitance was reduced for high baud rate operation. The high-speed PD has excess bandwidth that enables a higher-speed operation [142]. The receiver PIC was packaged as previously described and the transmitter PIC was on a high-speed probe station. Unlike the commercial product, although the data was driven via sampled in real time, the signal processing for data recovery was done offline.

In the experimental setup, pseudo-random bit sequences of length $(2^{15} - 1)$ are used to modulate the transmitter whose output is a polarization multiplexed QPSK signal with a baud rate of 28 Gbaud. At the receiver, the incoming signal is scrambled. A polarization beam splitter is used to split the polarization-scrambled incoming signal to the TE/TM inputs of the integrated receiver. Each on-PIC LO is tuned to its respective transmitter wavelength, to obtain intradyne reception. Using a real-time oscilloscope, the four RF outputs from each wavelength were sampled at 50 GSa/s. The channels were measured one at a time (thus not impacted by adjacent channel crosstalk here), and the acquired data was processed offline to obtain the constellations. Figure 2.46 shows I/Q, TE/TM, QPSK constellation diagrams for all 10 channels under test. The signal was not noise loaded. The BER (bit error rate) of the received signal was within the FEC (forward error correction) limit up to 28 Gbaud. With all 10 channels operating at 28 Gbaud in the PM QPSK mode, the PIC pair is capable of a 1.12 Tbit/s superchannel.

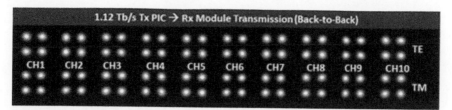

FIGURE 2.46 28-Gbaud, PM-QPSK constellation diagrams for 10-channel Tx-to-Rx PIC transmission for an aggregate data rate of 1.12 Tbit/s.

Figure 2.47 shows a four-wavelength coherent receiver in InP using a different scheme than described above. The wavelength separation, polarization splitting, and 90° hybrids are all accomplished by an interleave-chirped AWG [150]. An interleave-chirped AWG is similar to a conventional AWG demultiplexer except that the grating arms are slightly adjusted in path length with a repeating pattern from arm to arm. In this case, every 2nd grating arm has its path length shortened by $\lambda/8$ and every 4th arm has its path length increased by $3\lambda/8$, where λ is the wavelength in the waveguide. The advantages of this design are that it is robust to fabrication changes and has no waveguide crossings. The disadvantage is that the AWG requires four times as many grating arms as a conventional AWG with the same demultiplexing characteristics.

Figure 2.48 shows the architecture of a DQPSK receiver PIC with electronic polarization and phase tracking [151]. At the PIC input is the polarization processing block that is common to all 10 wavelengths. The input signal is first split into its TE and TM components using a polarization beam splitter (PBS). The TM output

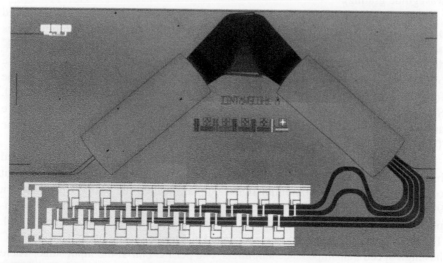

FIGURE 2.47 Four-channel coherent receiver in InP.

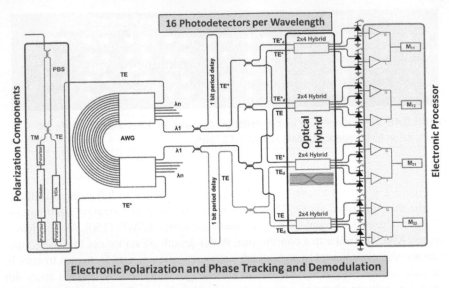

FIGURE 2.48 DQPSK receiver PIC architecture.

of the PBS then passes through a polarization rotator that converts the signal to the TE polarization (labeled TE*). The adjacent arm with the original TE component has a variable optical attenuator (VOA) to compensate for the insertion loss of the rotator, and power balance the two outputs of the polarization processing block. The TE and TE* signals are then wavelength demultiplexed using a single AWG (array waveguide grating).

The demultiplexed outputs of the AWG are then fed to the network composed of 1-bit delay interferometers, and 90 optical hybrids. The circuit combination is repeated for each demultiplexed wavelength channel. In a conventional DQPSK decoder the TE and TM components would be separately processed by mixing the original signal with its delayed component (TEd). In this architecture we create the following four combinations, TE + TEd, TE* + TEd, TE + TE*d, and TE* + TE*d. Since they are orthogonal, combinations of TE and TM signals would not produce any signal output. The polarization rotator at the input which converts the TM to the TE makes this architecture possible.

The four outputs of the optical hybrid are terminated in two pairs of balanced high-speed photodetectors (PD). The two inputs to a balanced PD pair are 180° out of phase, and create a differential signal output. Further, the two PD pairs have a 90° phase offset between them, and this phase offset is used to separate the I (in-phase) and Q (quadrature) components of the quadrature phase modulated signal. The PD outputs are then fed to a co-packaged electronic processor ASIC. This ASIC has high-speed transimpedance amplifiers, one per PD pair at the input stage. There are 16 PDs per channel (wavelength) for a total of 160 PDs on the PIC. It is almost impossible to build such a receiver reliably out of discrete components, and it is in realizing architectures like these that monolithic photonic integration triumphs.

This section describes the mathematical foundation for the PM-DQPSK demodulation, in particular the signal processing which enables polarization tracking,

$$
\begin{aligned}
D_{\text{out}} &= PD_1 - PD_2 + jPD_3 - jPD_4 \\
&= |E_{k-1} + E_k|^2 - |E_{k-1} - E_k|^2 + j|E_{k-1} + jE_k|^2 - j|E_{k-1} - jE_k|^2 \\
&= (|E_{k-1}|^2 + 2\Re[E_{k-1}^* E_k] + |E_k|^2) - (|E_{k-1}|^2 - 2\Re[E_{k-1}^* E_k] + |E_k|^2) + \\
&\quad\ j(|E_{k-1}|^2 + 2\Im[E_{k-1}^* E_k] + |E_k|^2) - (|E_{k-1}|^2 - 2\Im[E_{k-1}^* E_k] + |E_k|^2) \\
&= 4E_{k-1}^* E_k.
\end{aligned}
\tag{2.17}
$$

The purpose of the two balanced receivers, mixing the bit-delayed and current optical signals, is to extract electrically the difference in phase between two sequential bits. Equation (2.17) demonstrates how the electrical outputs can represent the complex phase between the two signals. Practically, the two electrical signals carry the in-phase and quadrature portions of the phase on two distinct wires. It is convenient to represent the two signals mathematically as one complex value.

For phase modulated formats, the electrical field at the transmitter is $E_k = e^{j\alpha_k}$, representing the electric field in the optical domain. In order to extract the phase change between two sequential bits in pol-muxed DQPSK, the circuit can be represented mathematically using Eqs. (2.18) and (2.19). In this case, the two orthogonal polarization states generated at the transmitter are shown as having arrived at the receiver without any rotations. The matrix represents an ideal polarization splitter, which yields two (complex) electrical outputs representing the phase change for horizontal and vertical polarizations

$$
\begin{bmatrix} e^{-i\alpha_{k+1}} & e^{-i\beta_{k+1}} \end{bmatrix}
\begin{bmatrix} 1 & 0 \\ 0 & 0 \end{bmatrix}
\begin{bmatrix} e^{i\alpha_k} \\ e^{i\beta_k} \end{bmatrix},
\tag{2.18}
$$

$$
\begin{bmatrix} e^{-i\alpha_{k+1}} & e^{-i\beta_{k+1}} \end{bmatrix}
\begin{bmatrix} 0 & 0 \\ 0 & 1 \end{bmatrix}
\begin{bmatrix} e^{i\alpha_k} \\ e^{i\beta_k} \end{bmatrix}.
\tag{2.19}
$$

The polarization rotation through a lossless fiber can be represented by the Jones matrix shown in Eq. (2.20)

$$
R = \begin{bmatrix} e^{i\phi}\cos\theta & -e^{i\psi}\sin\theta \\ e^{i\psi}\sin\theta & -e^{i\phi}\cos\theta \end{bmatrix}.
\tag{2.20}
$$

For light transmitted through an arbitrary polarization rotation, the corresponding signals seen electrically can be described using Eqs. (2.21) and (2.22)

$$
\begin{bmatrix} e^{-i\alpha_{k+1}} & e^{-i\beta_{k+1}} \end{bmatrix} R^{-1}
\begin{bmatrix} 1 & 0 \\ 0 & 0 \end{bmatrix} R
\begin{bmatrix} e^{i\alpha_k} \\ e^{i\beta_k} \end{bmatrix},
\tag{2.21}
$$

$$
\begin{bmatrix} e^{-i\alpha_{k+1}} & e^{-i\beta_{k+1}} \end{bmatrix} R^{-1} \begin{bmatrix} 0 & 0 \\ 0 & 1 \end{bmatrix} R \begin{bmatrix} e^{i\alpha_k} \\ e^{i\beta_k} \end{bmatrix}. \tag{2.22}
$$

Expanding the $R^{-1} \begin{bmatrix} 1 & 0 \\ 0 & 0 \end{bmatrix} R$ matrix by itself yields Eqs. (2.23) and (2.26). A conventional technique for receiving pol-mux DQPSK signals would be to place an optical polarization tracker before the optical receiver. However, a real-time polarization tracker is difficult to realize in optics. If the TE and TM signals are also mixed together (using a 90° polarization rotation on one of the polarization states), additional information is obtained about the incoming data stream. While additional photodiodes are required, this enables polarization tracking of the data sequence through signal processing. For simplicity, the Jones matrix angles are simplified with the notation $c = \cos(2\theta)$ and $s = \sin(2\theta)$,

$$
R^{-1} \begin{bmatrix} 1 & 0 \\ 0 & 0 \end{bmatrix} R = \frac{1}{2} \begin{bmatrix} 1+c & -se^{-i\phi-i\psi} \\ -se^{i\phi+i\psi} & 1+c \end{bmatrix}, \tag{2.23}
$$

$$
R^{-1} \begin{bmatrix} 0 & 1 \\ 0 & 0 \end{bmatrix} R = \frac{1}{2} \begin{bmatrix} se^{-i\phi+i\psi} & (1+c)e^{-2i\phi} \\ -(1-c)e^{2i\psi} & -se^{-i\phi+i\psi} \end{bmatrix}, \tag{2.24}
$$

$$
R^{-1} \begin{bmatrix} 0 & 0 \\ 1 & 0 \end{bmatrix} R = \frac{1}{2} \begin{bmatrix} se^{i\phi-i\psi} & -(1-c)e^{-2i\psi} \\ (1+c)e^{2i\phi} & -se^{i\phi-i\psi} \end{bmatrix}, \tag{2.25}
$$

$$
R^{-1} \begin{bmatrix} 0 & 0 \\ 0 & 1 \end{bmatrix} R = \frac{1}{2} \begin{bmatrix} 1+c & se^{-i\phi-i\psi} \\ se^{i\phi+i\psi} & 1-c \end{bmatrix}. \tag{2.26}
$$

In order to demonstrate that the data is sufficient for demodulating the incoming streams electrically without regard to the polarization state, the 16 values shown in Eqs. (2.23–2.26) are rearranged into a 4×4 matrix M

$$
M = \frac{1}{2} \begin{bmatrix} 1+c & -se^{-i\phi-i\psi} & -se^{i\phi+i\psi} & 1-c \\ se^{-i\phi+i\psi} & (1+c)e^{-2i\phi} & -(1-c)e^{2i\psi} & -se^{-i\phi+i\psi} \\ se^{i\phi-i\psi} & -(1-c)e^{-2i\psi} & (1+c)e^{2i\phi} & -se^{i\phi-i\psi} \\ 1-c & se^{-i\phi-i\psi} & -se^{i\phi+i\psi} & 1+c \end{bmatrix}. \tag{2.27}
$$

M is nonsingular, which is required for polarization tracking, and its inverse is

$$
M^{-1} = \frac{1}{2} \begin{bmatrix} 1+c & se^{i\phi-i\psi} & se^{-i\phi+i\psi} & 1-c \\ -se^{i\phi+i\psi} & (1+c)e^{2i\phi} & -(1-c)e^{2i\psi} & se^{i\phi+i\psi} \\ -se^{-i\phi-i\psi} & -(1-c)e^{-2i\psi} & (1+c)e^{-2i\phi} & se^{-i\phi-i\psi} \\ 1-c & -se^{i\phi-i\psi} & -se^{-i\phi+i\psi} & 1+c \end{bmatrix}. \tag{2.28}
$$

M^{-1} represents the coefficients which weight the outputs from the eight differential TIA outputs to generate the eight mixtures of Eqs. (2.23–2.26). However, only the mixing of the bits within the same polarization is of importance. This is represented by the first and last row of M^{-1}, rewritten as S in Eq. (2.29):

$$S = \frac{1}{2} \begin{bmatrix} 1+c & se^{i\phi-i\psi} & se^{-i\phi+i\psi} & 1-c \\ 1-c & -se^{i\phi-i\psi} & -se^{-i\phi+i\psi} & 1+c \end{bmatrix}. \tag{2.29}$$

In order to realize this receiver, a multiple-input, multiple-output signal processing structure is needed which can make linear combinations of the input signals to produce a single output signal.

Figure 2.49 shows a schematic representation of this approach for four possible states of polarization at the input. For the case where the transmit polarization aligns with the polarization splitter at the receiver, polarization demultiplexing appears reasonably conventional, with the splitter separating the two polarizations and a bit-delay interferometer performing the DQPSK demultiplexing. In Figure 2.49a, the highlighted waveguides show one of the four data streams' paths. The demultiplexed signal comes from a single balanced photodiode pair, as shown by the arrow. In Figure 2.49b, a 90° rotation occurs and the same data stream ends up on a different set of photodiodes. In Figure 2.49c, the polarization is circularly polarized such that the signal must be reconstructed from four independent photodiodes, as shown by the arrows. For a linear polarization at 45° from the transmit (Figure 2.49d), a similar combination restores the original data stream; only the contribution from the polarization mixed components needs to change.

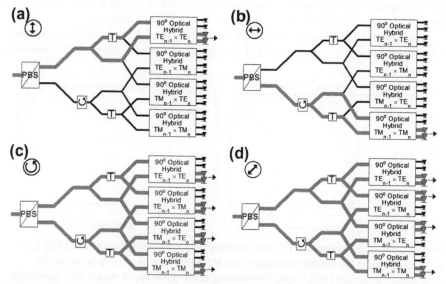

FIGURE 2.49 Signal paths required to reconstruct four example states of polarization.

The signal processing can be performed in either the analog or digital domain. Moreover, adaptation of the receiver can be achieved using conventional least mean square (LMS) adaptation. By employing a signal-processing-based adaptation, rapid polarization transients can be tracked and optical response time does not limit tracking capability [152].

A benefit of the LMS adaptation is that crosstalk occurring between the outputs is always minimized. Variation in the response of photodiode pairs will be tracked by the gain cells without any intervention. DQPSK has excellent phase noise tolerance, such as to cross-phase modulation often present in WDM systems. To first order, sensitivity to chromatic and polarization dispersion depends on the baud rate. Chromatic dispersion needs to be compensated to within a window that is similar to other 10-Gbaud modulation formats, such as single-polarization DQPSK. Simulation and experiment have not shown any particular sensitivity that the tracking introduces to chromatic dispersion tolerance.

Figure 2.50 shows the polarization extinction ratio (PER) performance of the PBS at the input of the PIC. The PBS is based on an asymmetric Mach-Zehnder interferometer (MZI) structure [153]. PBS devices based on directional coupler geometries have also been successfully demonstrated [154]. In the MMI-based PBS structure, the TM mode index is preferentially changed in one of the arms of the MZI. When we induce a π-phase change preferentially for the TM polarization, we get the response shown in Figure 2.50. The PER is measured between the TM cross and TE cross or between TM bar and TE bar states and is better than 20 dB.

The wavelength response of the PBS is very flat over a large wavelength range. The FSR (free spectral range) of 16 nm in the standalone MZI test structure, shown in Figure 2.50, is due to the deliberate design modification on a test chip that was used to characterize the PER. In the integrated PBS device, the path lengths of the MZI arms are matched for performance over a wide wavelength range.

The TE and TM outputs of the PBS have orthogonal electric field orientations. The rotator is used to convert the TM output (which has its electric field primarily

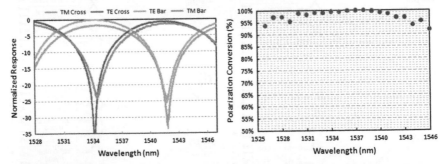

FIGURE 2.50 Performance of a PBS test structure over wavelength (upper). The PBS FSR of 16 nm is due to the deliberate design of the MZI test structure used to measure the polarization extinction ratio. (right) Performance of the polarization rotator with wavelength (lower).

oriented perpendicular to the plane of the substrate) to the same orientation as the TE output (which has its electric field primarily oriented parallel to the plane of the substrate). Even symmetric optical waveguides have some birefringence in general, i.e. they have different propagation constants for the TE and TM modes. Beat length refers to the propagation distance in the waveguide over which a phase difference accumulates between the TE and TM modes. Cross-sectional asymmetries or compositional variations may be used to increase the birefringence in waveguides, and shorten the beat length.

Polarization rotators are commonly made of asymmetric waveguides, e.g. waveguides with one sloped sidewall and one vertical sidewall [155–157]. The waveguide design is such that the eigenmodes of the structure are oriented at an angle of 45° to the TE/TM eigenmodes of the input and output waveguides with vertical sidewalls. After propagating a beat length inside the rotator, the electric field of the input optical mode is effectively rotated to its orthogonal orientation. A rotator built on this principle is a periodic structure, and has to be terminated at its beat length. Otherwise the input optical field will continue to evolve inside the rotator. Rotators may also be built out of waveguides with trenches [158,159] or tight optical bends [160,161] to provide the required asymmetry.

We used an asymmetric waveguide design similar to [157] for the rotator. Figure 2.50 (right) shows performance of the polarization rotator. In the wavelength range shown the TM/TE conversion is well over 90%, typically in excess of 20 dB extinction of the unwanted polarization. The insertion loss is less than 0.5 dB.

A TE polarizer is a device with high insertion loss for the TM polarization, and operates with minimal insertion loss for the TE polarization. The TM polarizer does the opposite. There is a TM polarizer at the TM output of the PBS just before the rotator. Although the PBS has a PER in excess of 25 dB, the TM polarizer further cleans up the signal path by stripping away any residual TE signal. There is a TE polarizer after the rotator in the TM path and after the VOA in the TE path. This ensures that any residual TM signal (capable of causing coherent crosstalk) has been completely eliminated from the signal paths. The polarizers are metal clad waveguides that preferentially affect the TE or TM signal states.

Figure 2.51 shows the performances of the TE and TM polarizers as a function of wavelength. They both have PER in excess of 25 dB over a wide wavelength range.

Figure 2.52 shows the 40, 11.4-Gb/s eye diagrams of the demodulated DQPSK signal. The BER performance of all the channels is well below the FEC correctable limit. Most of the channels were error-free for the duration of the test. Thus, the package is capable of a total data rate of 456 Gb/s. Visually, some of the eye diagrams may look better than others. This is mostly due to variations in loss between the individual paths. The module was tested with the TEC nonoperational, i.e. without strict temperature control. This is possible because the resulting phase variation in the 1-bit delay was automatically tracked by the electronics. To date, this PIC architecture has not been commercialized as it has less ability to correct for polarization mode dispersion (PMD) and chromatic dispersion (CD) than a coherent receiver. However, the analog MIMO signal processing enables significantly lower power dissipation,

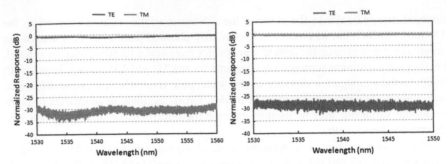

FIGURE 2.51 TE polarizer and (right) TM polarizer performance over wavelength. They both show a better than 25 dB extinction ratio.

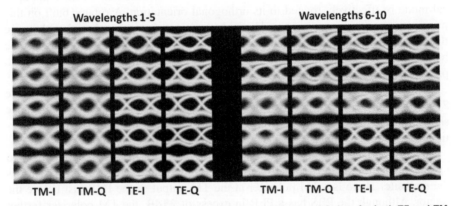

FIGURE 2.52 40, 11.4 Gbit/s eye diagrams comprising of the I and Q data for both TE and TM polarizations from a packaged receiver.

making this technology potentially very interesting for intermediate reach applications, which have less PMD and CD.

2.4.3 Group IV PICs for advanced modulation format transmission

2.4.3.1 Group IV single-channel PICs for advanced modulation format transmission

Previously we discussed two ways to receive a single-polarization phase-shift keyed signal: with differential or coherent detection as shown in Figure 2.29. To receive a dual-polarization phase-shift keyed signal, one must somehow unscramble the two polarization signals after transmission through the fiber, which contains randomly changing birefringence. For differential detection, one can use an optical polarization tracker, and for coherent one can use a dual-polarization coherent receiver and a DSP, as shown in Figure 2.53. Yet another method is to use differential detection with a DSP [162] as just discussed in Section 2.4.2.

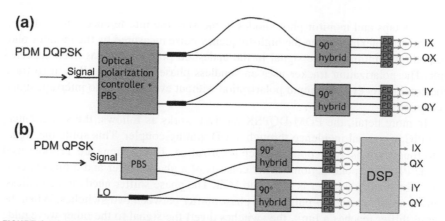

FIGURE 2.53 Two methods of receiving a PDM-QPSK signal. (a) Differential detection with an optical polarization tracker and (b) coherent detection with a digital polarization tracker.

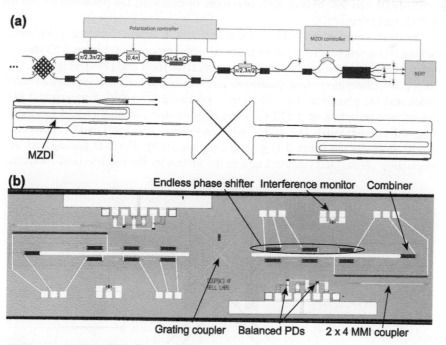

FIGURE 2.54 PDM-DQPSK receiver in silicon.

Figure 2.54 shows a demonstration of a Si PIC with two integrated polarization trackers and two DQPSK receivers [163]. Having two independent polarization trackers allows this PIC to handle the case of PDL in the transmission. The PIC contains taps and monitor photodiodes to measure the interference after polarization demultiplexing. When the high-frequency noise measured by the monitor photodiode is minimized, the signal in the undesired polarization has been canceled out. The polarization tracker uses an "endless phase shifter" to allow it to track the continually evolving input polarization without ever needing to interrupt signal reception.

In more detail, the PDM-DQPSK receiver works as follows: the signal enters the PIC at normal incidence through a 2-D grating coupler. This splits the signal into four portions, two copies of each polarization. The portions are TE-polarized in the PIC. The polarization tracker consists of a phase shifter and a tunable coupler. The tunable coupler is a tunable MZI. The phase shifter needs to be endless. This is accomplished by putting the phase shifter between two switches. When the phase shifter reaches a limit, the switches divert the signal to the other waveguide temporarily while the phase shifter resets. The switching is done when the two waveguides have the same phase. After the polarization controller there is a tap with a Ge photodiode for feedback control of the polarization tracker. Finally there is the MZDI and 90° hybrid with two pairs of balanced Ge photodiodes for the DQPSK reception.

The most common method to track polarization is to use electronic polarization tracking, via a coherent receiver with digital signal processing. Figure 2.55 shows a dual-polarization, dual-quadrature coherent receiver front-end [164]. (Figure 2.38 is also a dual-polarization, dual-quadrature receiver, but in InP.) It comprises Si waveguides and Ge photodetectors. The entire packaged assembly demonstrated high performance reception of a 112-Gb/s polarization-division multiplexed quadrature phase-shift keyed signal. The device works as follows: the signal and LO ports enter the PIC from the top using 2-D grating couplers [165]. The 2-D grating acts as a polarization splitter and rotator. Light in the fiber with the electric field oscillating

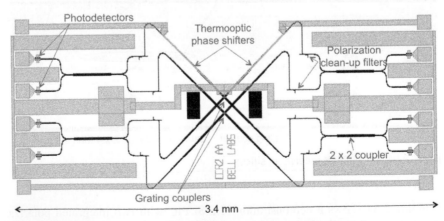

FIGURE 2.55 Layout of Si dual-polarization, dual-quadrature coherent receiver.

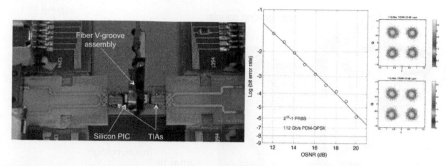

FIGURE 2.56 Packaged Si dual-polarization, dual-quadrature coherent receiver.

along one axis of the grating couples into the waveguides normal to that axis as TE light, and light in the fiber with the electric field oscillating along the orthogonal axis couples to the other set of waveguides, as TE light. Thus all the light in the PIC is TE polarized, making the PIC significantly simpler to design. Like the PDM-DQPSK receiver, the fiber is oriented perfectly normal to the PIC surface, and thus the light splits equally in both directions for a given polarization. This gives high coupling efficiency, a broader bandwidth, and allows the grating coupler to also act as a 3-dB coupler. The four portions from the signal interfere with the four portions from the LO in four 2×2 couplers and proceed to four Ge photodiode pairs. There are thermooptic phase shifters that set the relative phase between the two interferences for each polarization to 90°. These four Ge photodiode pairs are wire bonded to four transimpedance amplifiers (TIAs) and form the coherent receiver. Figure 2.56 shows the silicon coherent receiver PIC packaged with TIAs and experimental results with 112-Gb/s PDM-QPSK.

2.4.4 Space-division multiplexing PICs

We discussed a Si PIC which can receive OOK signals from an MCF in Section 2.3.2.3. Here we show a Si PIC that can spatially multiplex/demultiplex signals from an FMF.

An FMF with a ring-shaped core can be designed to be multimode azimuthally but single-mode radially. Such a fiber is well suited for coupling to a PIC, because it allows all the modes to be accessed via a planar geometry without requiring waveguide crossings. The calculated fiber modes of a 10-mode ring-core fiber are shown in Figure 2.58. Such a coupler is shown in Figure 2.57 [166]. It consists of a circular grating coupler [167] connected with equal-length waveguides to a star coupler. Each port of the star coupler couples to a different angular momentum mode at the circular grating coupler, e.g. the center star coupler port couples to the grating coupler mode in which all the waveguides are in phase around, giving an angular momentum of zero.

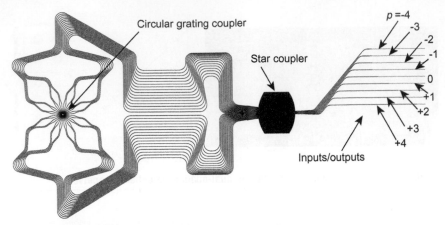

FIGURE 2.57 Multimode ring-core fiber coupler silicon PIC.

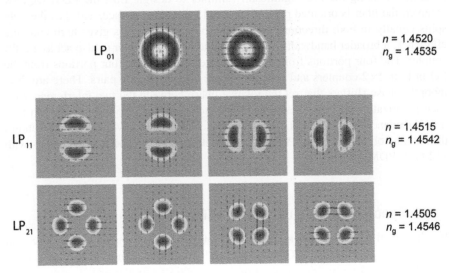

FIGURE 2.58 Calculated modes of a 5-mode (10 modes if include polarization) ring-core fiber.

The adjacent star coupler port excites a mode with a 2π linear phase change around the grating coupler, resulting in an angular momentum mode of ± 1. The angular momentum modes are linear combinations of fiber modes, as shown in Figure 2.59. This particular device works for only one polarization, due to the strong polarization dependence of the grating coupler.

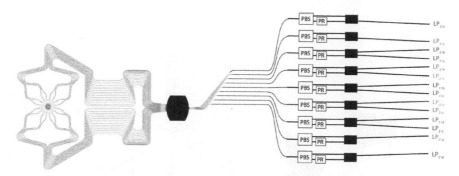

FIGURE 2.59 Physical configuration mapping of angular momentum modes to ring-core fiber modes.

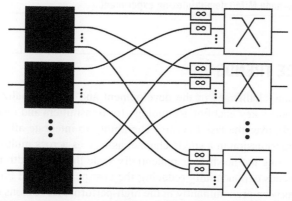

FIGURE 2.60 Configuration for optical MIMO demultiplexing. The solid rectangles are power splitters, the rectangles containing infinity signs are infinite phase shifters, and the switches are adjustable power combiners.

If there is coupling between the modes in the SDM fiber, then when the light is received, each mode will contain a linear combination of all the launched signals. Because the signals will remain essentially orthogonal, then the original signals can be recovered by adding together a linear combination of the fields in each mode. This addition is typically done in a DSP [168]. However, such a DSP may consume very high power. One can instead perform the addition optically [169], saving significant power consumption, provided that the modal dispersion is low. An architecture for optical MIMO demultiplexing is shown in Figure 2.60 emanating from an FMF, and an experimental demonstration that demultiplexes one mode from six modes is shown in Figure 2.61 [170].

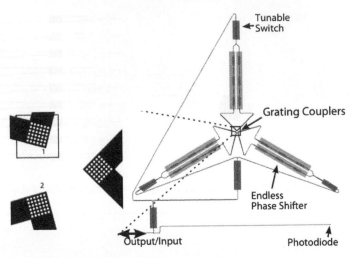

FIGURE 2.61 Six-mode MIMO demultiplexing experiment.

2.5 FUTURE TRENDS

Despite the many challenges in the development and commercialization of PICs, PICs have become a key enabling technology for transmitters and receivers in commercial networks over the last 15 years. The ability to integrate all of the requisite functions for transmission in a single channel as well as across multiple channels is essential to meeting the needs to exponentially scale the bandwidth in the network while simultaneous exponentially reducing the cost per deployed bit. The application of PICs has been predominantly in the high-performance regions of the network (intermediate through ultra-long-reach) as these portions have the highest optical components costs and leverage the enabling features of integration at the component, system, and network levels. However, as the costs of integration have decreased, PICs are starting to make inroads into shorter-reach applications. Accordingly, decreased cost as well as reduced power consumption will likely drive adoption over time in these applications.

Figure 2.62 shows the scaling of the data capacity per chip for commercial transmitter devices employed in long-haul telecommunications networks. Over time, the data capacity per transmitter chip has doubled every ~2.2 years (dashed line, fit to data). The first PICs deployed in optical fiber networks began with electroabsorption modulated lasers in 1996 [84]. This ultimately enabled the realization of devices with a capacity up to 10 Gb/s per chip. The next significant progression occurred with the development of large-scale PICs where over 50 functions were integrated onto a monolithic chip to realize a 100-Gb/s transmitter chip (10 channels × 10 Gb/s) based on on-off keying (OOK) [45]. This advance represented an order of magnitude increase in data capacity per chip

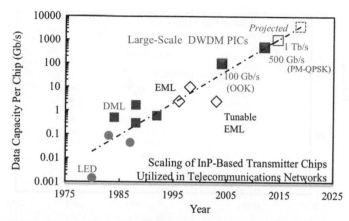

FIGURE 2.62 Scaling of the data capacity/chip for commercial InP-based transmitter chips utilized in optical telecommunications networks. The data capacity per chip has doubled an average of every 2.2 years to date and continuing at this rate with the introduction of commercial 500 Gb/s devices in 2012 based on coherent PM-QPSK PIC technology.

and integration level compared to existing commercial devices. Next-generation large-scale PIC technologies based on coherent modulation (PM-QPSK) are being deployed in 2012 with capacities of 500 Gb/s [100] and integrate over 450 functions onto a single chip. The continued scaling of data capacity per chip as shown in Figure 2.62 will be driven by bandwidth demands in the network and the need to continue to reduce the system cost per Gb/s while simultaneously scaling the total fiber capacity. Assuming the scaling trends continue to hold, PICs capable of >1 Tb/s are expected to be commercialized and deployed in the next ~3 years. In order to achieve this scaling, PICs are likely to utilize a combination of higher order modulation formats (e.g. 16-QAM), higher baud rate, and more transmission channels.

As bandwidth demands in the network continue to grow, PICs will likely become more prevalent in the intermediate and short-reach portions of the network (metro, access, premise). This will be enabled by higher bandwidth demands and reduced PIC costs that are obtained by leveraging the previous generations of PICs in the case of InP-based devices as well as potentially the emergence of more applications for Si-based photonics. The key to the success of the latter will be the simultaneous cost and performance competitiveness with both discrete and existing InP-based PIC solutions.

The ability to continue to scale the bandwidth of PICs for transmission applications is predicated on the ability to integrate a higher number and diversity of functions. Figure 2.63 shows the scaling of the functions per chip of InP-based transmitter PICs utilized in commercial long-haul telecommunications networks. For this figure, we define a function as any of the following: light emission, light detection, laser frequency tuning, optical amplitude amplification, optical amplitude

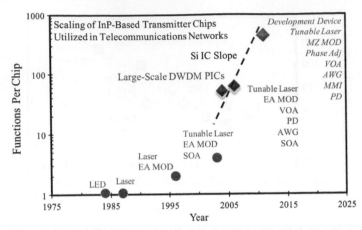

FIGURE 2.63 Scaling of the number of functions per chip for InP-based transmitter PICs utilized in commercial telecommunications networks.

attenuation, optical phase tuning, optical modulation, optical multiplexing, and optical demultiplexing. The first generation 100 Gb/s transmitter PICs introduced in 2004 resulted in an inflection point in scaling of functions per chip. The 500 Gb/s transmitter PIC integrated over 450 functions into a single chip with a resulting new scaling rate of $\sim 100\times$ functions per decade. This rate is approximately $10\times$ slower than the rate of scaling of transistors per electronic Si IC [171], but is consistent with the rate of scaling of the functions per Si electronic IC [172] as shown in the dashed line in Figure 2.63 (the rates cited for Si ICs are circa 1960s, around the invention of the IC). We believe this is a more representative metric for scaling as many of the elements integrated on a PIC have a higher level of functionality that typically require circuit implementations (e.g. multiple transistors and other components) to realize their electronic equivalent (e.g. multiplexing/demultiplexing of multiple data streams, frequency tuning, etc.). Furthermore, the cascaded components in each channel of today's telecommunications PICs require implementations that are more circuit-like than the transistor-like logic implementations seen in digital electrical ICs. The long-haul, high-performance applications (as shown in Figure 2.63) are likely to be the drivers of the highest integration densities. PICs with reduced element counts are likely to become more prevalent in the short and intermediate reach portions of the network, especially as bandwidth continues to grow. The adoption of higher-order modulation formats will drive the integration of more functions per channel, and hence will be another driver for integration density as well.

Two of the key limitations of the development of PICs are the fundamentally larger size required for photonic devices as well as the larger set of diverse of building blocks. The latter places requirements and limits on how the integrated

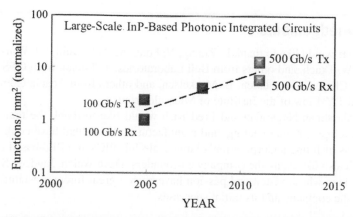

FIGURE 2.64 Scaling of the density of normalized functions (per mm²) for large-scale PICs.

components may be usefully interconnected. Consequently, a key differentiation between photonic ICs and electronic ICs is the density of integration. Figure 2.64 shows the functions per mm^2 (normalized) for various generations of large-scale PICs. A key to enabling the economics of next-generation devices has been the ramp of this density. Despite these advances, the integration density is approximately five orders of magnitude lower than electronic ICs which have circuit function densities of $10^6/mm^2$ for SRAM circuits [173]. Although it is unlikely that photonic ICs will approach the integration density of electronic ICs, future generations of PICs will likely utilize multi-level interconnects (electrical and optical) to continue to enable the progression of this scaling. Furthermore, the development and integration of additional functions and optoelectronic device structures such as the transistor-laser [174,175] reconfigurable filters [176], multi-function modulators [177], and multi-function splitters/routers [177] may provide important paths to future PIC scaling.

Over the last 15 years, PICs have moved from a research interest to important commercial devices that are at the heart of enabling the exponential scaling of the network bandwidth while simultaneously exponentially reducing cost per bit. Photonic ICs are poised to continue to enable this scaling over many generations of devices, and across the entire breadth of the network (access, premise, metro, long-haul, ultra-long-haul, and submarine), and hence, will be a growing and essential part of the optical communications infrastructure. These trends are enabled by the fact that PICs benefit from the same semiconductor manufacturing infrastructure and learning curves as electronic ICs. Future generations of PICs are likely to expand for both InP-based devices as well as Si-based devices. Over time, we would expect to see virtually all discrete optical functions integrated onto PICs. This should enable new applications for these devices as well as expanding their impact in the communications infrastructure.

Acknowledgements

Chris Doerr would like to thank L. Zhang, N. Fontaine, L.L. Buhl, P.J. Winzer, Y.-K. Chen, R.W. Tkach, and others from Bell Laboratories, T. Taunay from OFS Laboratories, L. Chen, D. Vermeulen, B. Mikkelsen, and others from Acacia Communications, and T.Y. Liow of the Institute of Microelectronics.

Radhakrishnan Nagarajan and Fred Kish would like to thank the many people on the development, engineering, and manufacturing teams that have contributed to the success of Infinera Corporation's large-scale InP PICs and PIC-based products. We would also like thank the company co-founders: Dave Welch, Jagdeep Singh and Drew Perkins whose vision and passion have been a great force in establishing the basis for the company and its initial successes.

References

[1] R.N. Hall, G.E. Fenner, J.D. Kingsley, T.J. Soltys, R.O. Carlson, Coherent light emission from GaAs junctions, Phys. Rev. Lett. 9 (1962) 366–368.

[2] N. Holonyak, Jr S.F. Bevacqua, Coherent (visible) light emission from P_x junctions, Appl. Phys. Lett. 1 (1962) 82–83.

[3] M.I. Nathan, W.P. Dumke, G. Burns, F.H. Dill, G. Lasher, Stimulated emission of radiation from GaAs p-n junctions, Phys. Rev. Lett. 1 (1962) 62–64.

[4] T.M. Quist, Semiconductor laser of GaAs, Appl. Phys. Lett. 1 (1962) 91–92.

[5] Z.I. Alfërov, Injection lasers based on heterojunctions in the AlAs-GaAs system with low threshold at room temperature, Fiz: Tekh Polupr. 3 (1969) 1328–1332.

[6] I. Hayashi, M.B. Panish, P.W. Foy, S. Sumski, Junction lasers which operate continuously at room temperature, Appl.Phys. Lett. 17 (1970) 109–111.

[7] E. Rezek, N. Holonyak, B.A. Vojak, G. Stillman, J. Rossi, D. Keune, J. Fairing, LPE $In_{1-x}Ga_xP_{1-z}As_z$ ($x \sim 0.12$, $z \sim 0.26$) DH laser with multiple thin-layer (500 A) active region, Appl. Phys. Lett. 31 (1977) 288–290.

[8] D. Scifres, R. Burnham, W. Streifer, Distributed feedback single heterojunction GaAs diode laser, Appl. Phys. Lett. 25 (1974) 203–206.

[9] K. Utaka, Room-temperature CW operation of distributed-feedback buried-heterostructure InGaAsP/InP lasers emitting at $1.57\,\mu m$, Electron. Lett. 17 (1981) 961–963.

[10] A.R. Adams, Band-structure engineering for low-threshold high-efficiency semiconductor lasers, Electron. Lett. 22 (1986) 249–250.

[11] E. Yablonovitch, E. Kane, Reduction of lasing threshold current density by the lowering of valence band effective mass, J. Lightwave Technol. 4 (1986) 504–506.

[12] W.D. Laidig, P.J. Caldwell, Y.F. Lin, C. Peng, Strained-layer quantum-well injection laser, Appl. Phys. Lett. 44 (1984) 653–655.

[13] J.J. Hsieh, J. Rossi, J. Donnelly, Room-temperature CW operation of GaInAsP/InP double-heterostructure diode lasers emitting at $1.1\,\mu m$, Appl. Phys. Lett. 28 (1976) 709.

[14] H.M. Manasevit, Single crystal GaAs on insulating substrates, Appl. Phys. Lett. 12 (1968) 156–159.

[15] H.M. Manaevit, W.I. Simpson, The use of metalorganics in the preparation of semiconductor materials II: IIG VI compounds, J. Electrochem. Soc. 118 (1971) 644–647.

[16] R.D. Dupuis, P.D. Dapkus, Room temperature operation of $Ga_{1-x}Al_xAs$ GaAs double-heterostructure lasers grown by metalorganic chemical vapor deposition, Appl. Phys. Lett. 31 (1977) 466–468.

[17] Y. Kawamura, K. Wakita, Y. Yoshikuni, Y. Itaya, H. Asahi, Monolithic integration of a DFB laser and an MQW modulator in the 1.5 μm wavelength range, J. Quant. Electron 23 (1987) 915–918.

[18] J. Binsma, P. Thijs, T. VanDongen, E. Jansen, A. Staring, G. Van-DenHoven, L. Tiemeijer, Characterization of butt-joint InGaAsP waveguides and their application to 1310 nm DBR-Type MQW gain-clamped semiconductor optical amplifiers, IEICE Trans. Electron. E80-C (1997) 675–681.

[19] M. Aoki, H. Sano, M. Suzuki, M. Takahashi, K. Uomi, A. Takai, Novel structure MQW electroabsorption modulator/DFB-laser integrated device fabricated by selective area MOCVD growth, Electron. Lett. 27 (1991) 2138–2140.

[20] C.H. Joyner, S. Chandrasekhar, J.W. Sulhoff, A.G. Dentai, Extremely large band gap shifts for MQW structures by selective epitaxy on SiO_2 masked substrates, IEEE Photon. Technol. Lett. 4 (1992) 1006–1009.

[21] Y.D. Galeuchet, P. Roentgen, Selective area MOVPE of GaInAs/InP heterostructures on masked and nonplanar (1 0 0) and (1 1 1) substrates, J. Cryst. Growth 107 (1991) 147–150.

[22] M. Aoki, M. Suzuki, H. Sano, T. Kawano, T. Ido, T. Taniwatari, K. Uomi, A. Takai, InGaAs/InGaAsP MQW electroabsorption modulator integrated with a DFB laser fabricated by bandgap energy control selective area MOCVD, IEEE J. Quant. Electron 27 (1993) 2281–2295.

[23] W.D. Laidig, N. Holonyak, Jr, M.D. Camras, K. Hess, J.J. Coleman, P.D. Dapkus, J. Bardeen, Disorder of an AlAs-GaAs superlattice by impurity diffusion, Appl. Phys. Lett. 38 (1981) 776–778.

[24] D. Deppe, N. Holonyak Jr, Atom diffusion and impurity-induced layer disordering in quantum well III–V semiconductor heterostructures, J. Appl. Phys. 64 (1988) R93–R113.

[25] S. Charbonneau, E.S. Koteles, P.J. Poole, J.J. He, G.C. Aers, J. Haysom, M. Buchanan, Y. Feng, A. Delage, F. Yang, M. Davies, R.D. Goldberg, P.G. Piva, I.V. Mitchell, Photonic integrated circuits fabricated using ion implantation, IEEE J. Sel. Top. Quant. Electron. 4 (1998) 772–793.

[26] E. Skogen, J. Barton, S. DenBaars, L. Coldren, A quantum-well intermixing process for wavelength-agile photonic integrated circuits, IEEE J. Sel. Top. Quant. Electron. 8 (2002) 863–869.

[27] M.L. Masanovic, V. Lal, J.A. Summers, J.S. Barton, E.J. Skogen, L.G. Rau, L.A. Coldren, D.J. Blumenthal, Widely-tunable monolithically-integrated all-optical wavelength converters in InP, J. Lightwave Technol. 23 (2005) 1350–1363.

[28] Y. Suematsu, M. Yamada, K. Hayashi, Integrated twin-guide AlGaAs laser with multiheterostructure, IEEE J. Quant. Electron. QE-11 (1975) 457–460.

[29] R.J. Deri, O. Wada, Impedance matching for enhanced waveguide/photodetector integration, Appl. Phys. Lett. 55 (1989) 2712–2714.

[30] P.V. Studenkov, M.R. Gokhale, S.R. Forrest, Efficient coupling in integrated twin-waveguide lasers using waveguide tapers, IEEE Photon. Technol. Lett. 11 (1999) 1096–1098.

[31] S. Kuntze, V. Tolstikhin, F. Wu, Y. Logvin, C. Watson, K. Pimenof, R. Moore, A. Moore, J. Wang, T. Oogarah, Transmitter and receiver solutions for regrowth-free multi-guide vertical integration in InP, in: Integrated Photonics Research (IPR), 2010.

[32] M. Sokolich, M.Y. Chen, R.D. Rajavel, D.H. Chow, Y. Royter, S. Thomas, III C.H. Fields, B. Shi, S.S. Bui, J.C. Li, D.A. Hitko, K.R. Elliott, InP HBT integrated circuit technology with selectively implanted subcollector and regrown device layers, IEEE Solid-State Circuits 39 (2004) 1615–1621.

[33] S. Miller, Integrated optics: an introduction, Bell Syst. Technol. J. 48 (1969) 2059–2069.

[34] J. Shibata, I. Nakao, Y. Sasai, S. Kimura, N. Hase, H. Serizawa, Monolithic integration of an InGaAsP/InP laser diode with heterojunction bipolar transistors, Appl. Phys. Lett. 45 (1984) 191–193.

[35] T.S.O. Wada, T. Nakagami, Recent progress in optoelectronic integrated circuits (OEICs), IEEE J. Quant. Electron. QE-22 (1986) 805–823.

[36] T.L. Koch, U. Koren, Semiconductor photonic integrated circuits, IEEE J. Quant. Electron. QE-27 (1991) 641–653.

[37] M. Zirngibl, C.H. Joyner, L. Stulz, WDM receiver by monolithic integration of an optical preamplifier, waveguide grating router and photodiode array, Electron. Lett. 31 (1995) 581–582.

[38] C.A.M. Steenbergen, C. Van Dam, A. Looijen, C.G.P. Herben, M. De Kok, M.K. Smit, J.W. Pedersen, I. Moerman, R.G.F. Baets, B. Verbeek, Compact low loss 8 × 10 GHz polarisation independent WDM receiver, in: European Conference on Optical, Communication, 1996, pp. 129–132.

[39] L.A. Coldren, Monolithic tunable diode lasers, IEEE J. Sel. Top. Quant. Electron. 6 (2000) 988–999.

[40] Y. Yoshikuni, Semiconductor arrayed waveguide gratings for photonic integrated devices, J. Sel. Top. Quant. Electron. 8 (2002) 1102–1114.

[41] Y. Suzaki, K. Asaka, Y. Kawaguchi, S. Oku, Y. Noguchi, S. Kondo, R. Iga, H. Okamoto, Multi-channel modulation in a DWDM monolithic photonic integrated circuit, in: International Conference on Indium Phosphide and Related Materials, 2002, pp. 681–683.

[42] M.L. Masanovic, V.L. Lal, J.S. Barton, E.J. Skogen, L.A. Coldren, D.J. Blumenthal, Monolithically integrated Mach-Zehnder interferometer wavelength converter and widely tunable laser in InP, IEEE Photon. Technol. Lett. 15 (2003) 1117–1119.

[43] V.I. Tolstikhin A. Densmore, Y. Logvin, K. Pimenov, F. Wu, S. Laframboise, 44-channel optical power monitor based on an echelle grating demultiplexer and a waveguide photodetector array monolithically integrated on an InP substrate, in: OFC, vol. 1, 2003, pp. 2–4.

[44] X. Leijtens, M. Smit, Developments in arrayed wavguide grating devices for photonic integrated circuits, in: Proceedings of the SPIE 7218, Integrated Optics: Devices, Materials, and Technologies XIII, 72180C February 09, 2009. http://dx.doi.org/10.1117/12.810217.

[45] R. Nagarajan, C.H. Joyner, R.P. Schneider, J.S. Bostak, T. Butrie, A.G. Dentai, V.G. Dominic, P.W. Evans, M. Kato, M. Kauffman, D.J.H. Lambert, S.K. Mathis, A. Mathur, R.H. Miles, M.L. Mitchell, M.J. Missey, S. Murthy, A.C. Nilsson, F.H. Peters, S.C. Pennypacker, J.L. Pleumeekers, R.A. Salvatore, R.K. Schlenker, R.B. Taylor, H.-S. Tsai, M.F.V. Leeuwen, J. Webjorn, M. Ziari, D. Perkins, J. Singh, S.G. Grubb, M.S. Reffle, D.G. Mehuys, F.A. Kish, D.F. Welch, Large-scale photonic integrated circuits, IEEE J. Sel. Top. Quant. Electron. 11 (2005) 50–65.

[46] M. Kato, J. Pleumeekers, P. Evans, D. Lambert, A. Chen, V. Dominic, A. Mathur, P. Chavarkar, M. Missey, A. Dentai, S. Hurtt, J. Back, R. Muthiah, S. Murthy, R. Salvatore, S. Grubb, C. Joyner, J. Rossi, R. Schneider, M. Ziari, F. Kish, D.F. Welch, Single-chip 40-channel InP transmitter photonic integrated circuit capable of aggregate data rate of 1.6 Tbit/s, Electron. Lett. 42 (2006) 771–773.

[47] M. Kato, R. Nagarajan, J. Pleumeekers, P. Evans, A. Chen, A. Mathur, A. Dentai, S. Hurtt, D. Lambert, P. Chavarkar, M. Missey, J. Back, R. Muthiah, S. Murthy, R. Salvatore, C. Joyner, J. Rossi, R. Schneider, M. Ziari, F. Kish, D. Welch, 40-channel transmitter and receiver photonic integrated circuits operating at per channel data rate 12.5 Gbit/s, Electron. Lett. 43 (2007) 468–469.

[48] R. Nagarajan, M. Kato, S. Hurtt, A. Dentai, J. Pleumeekers, P. Evans, M. Missey, R. Muthiah, A. Chen, D. Lambert, P. Chavarkar, A. Mathur, J. Bäck, S. Murthy, R. Salvatore, C. Joyner, J. Rossi, R. Schneider, M. Ziari, F. Kish, D. Welch, Monolithic 10 and 40 channel InP receiver photonic integrated circuits with on-chip amplification, in: OFC PDP32, 2007, pp. 40–42.

[49] S. Nicholes, M. Mašanović, B. Jevremović, E. Lively, L. Coldren, D. Blumenthal, The world's first InP 8 × 8 monolithic tunable optical router (MOTOR) operating at 40 Gbps line rate per port, in: OFC, Paper PDPB1, 2009.

[50] R.C. Leachman, D.A. Hodges, Benchmarking semiconductor manufacturing, IEEE Trans. Semicond. Manuf. 9 (1996) 158–169.

[51] Integrated Circuit Engineering Corporation, "Yield and yield management," Cost Effective IC Manufacturing, Integrated Circuit Engineering Corporation, Scottsdale, AZ, 1997 (Chapter 3).

[52] <http://www.icknowledge.com/trends/defects.pdf>.

[53] J.S. Kilby, Minaturized electronic circuits, 1964.

[54] R.N. Noyce, Semiconductor Device-And-Lead Structure, U.S. Patent No. 2,981, 877, 1961.

[55] G.E. Moore, Cramming more components onto integrated circuits, Electronics 38 (1965) 114–117.

[56] R. Ross, N. Atchinson, Yield analysis, TI Technical J. (1998) 58–103.

[57] C.H. Stapper, LSI yield modeling and process monitoring, IBM J. Res. Develop. 44 (2000) 58–103.

[58] C.H. Stapper, F.M. Armstrong, K. Saji, Integrated circuit yield statistics, Proc. IEEE 71 (1983) 453–470.

[59] J.F. Bauters, M.J.R. Heck, D. John, M.-C. Tien, A. Leinse, R.G. Heideman, D. Blumenthal, J.E. Bowers, Ultra-low loss silica-based waveguides with millimeter bend radius, in: 36th European Conference and Exhibition on Optical Communication, 2010, pp. 1–3.

[60] F. Bordas, G. Roelkens, R. Zhang, Compact passive devices in InP membrane on silicon, in: ECOC, vol. 2, 2009, pp. 3–4.

[61] L. Chen, L. Zhang, C.R. Doerr, N. Dupuis, N.G. Weimann, R.F. Kopf, Efficient membrane grating couplers on InP, IEEE Photon. Technol. Lett. 22 (2010) 890–892.

[62] T. Kominato, Y. Hida, M. Itoh, H. Takahashi, S. Sohma, Extremely low-loss (0.3 dB/m) and long silica-based waveguides with large width and clothoid curve connection, in: ECOC, 2004, pp. 5–6.

[63] B. Little, A VLSI photonics platform, in: Fiber Communications Conference 2003, OFC, vol. 2, 2003, pp. 444–445.

[64] R. Soref, B. Bennett, Electrooptical effects in silicon, IEEE J. Quant. Electron. 23 (1987) 123–129.

[65] (a) H.C. Nguyen, Y. Sakai, M. Shinkawa, N. Ishikura, T. Baba, "10 Gb/s operation of photonic crystal silicon optical modulators," Opt. Express 19 (14) (2011) 13000. (b) A. Liu, L. Liao, D. Rubin, H. Nguyen, High-speed optical modulation based on carrier depletion in a silicon waveguide, Opt. Express 15 (2007) 660–668.

[66] D.J. Thomson, F.Y. Gardes, J.-M. Fedeli, S. Zlatanovic, Y. Hu, B. Ping, P. Kuo, E. Myslivets, N. Alic, S. Radic, G.Z. Mashanovich, G.T. Reed, 50-Gb/s silicon optical modulator, IEEE Photon. Technol. Lett. 24 (2012) 234–236.

[67] A. Liu, R. Jones, L. Liao, D. Samara-Rubio, D. Rubin, A high-speed silicon optical modulator based on a metal G oxide G semiconductor capacitor, Nature 427 (2004) 615–618.

[68] R. Montgomery, M. Ghiron, P. Gothoskar, High-Speed Silicon-Based Electro-Optic Modulator, US Patent, 2005.

[69] P. Sullivan, J.-M. Fedeli, M. Fournier, L. Dalton, Demonstration of a low V sub/sub L modulator with GHz bandwidth based on electro-optic polymer-clad silicon slot waveguides, Optics 18 (2010) 15618–15623.

[70] M. Liu, X. Yin, E. Ulin-Avila, B. Geng, T. Zentgraf, L. Ju, F. Wang, X. Zhang, A graphene-based broadband optical modulator, Nature 474 (2011) 64–67.

[71] I.-W. Hsieh, H. Rong, Two-photon-absorption-based optical power monitor in silicon rib waveguides, in: Group IV Photonics (GFP), 2010, pp. 326–328.

[72] D. Ahn, C.-Y. Hong, J. Liu, W. Giziewicz, M. Beals, L.C. Kimerling, J. Michel, J. Chen, F.X. Kärtner, High performance, waveguide integrated {Ge} photodetectors, Opt. Express 15 (2007) 3916–3921.

[73] S. Assefa, F. Xia, S. Bedell, Y. Zhang, CMOS-integrated 40 GHz germanium waveguide photodetector for on-chip optical interconnects, in: Optical Fiber Communication Conference, 2009, pp. 10–12.

[74] Y. Kang, H. Liu, M. Morse, M. Paniccia, Monolithic germanium/silicon avalanche photodiodes with 340 GHz gain G bandwidth product, Nat. Photon. 3 (2008) 59–63.

[75] J. Michel, R.E. Camacho-Aguilera, Y. Cai, N. Patel, J.T. Bessette, B.R. Dutt, L.C. Kimerling, An electrically pumped Ge-on-Si laser, in: OFC, 2012, pp. 5–7.

[76] R. Soref, Mid-infrared photonics in silicon and germanium, Nature 4 (2010) 495–497.

[77] A. Fang, H. Park, O. Cohen, R. Jones, M. Paniccia, Electrically pumped hybrid AlGaInAs-silicon evanescent laser, Opt. Express 14 (2006) 9203–9210.

[78] G. Roelkens, D. Van Thourhout, R. Baets, R. Nötzel, M. Smit, Laser emission and photodetection in an InP/InGaAsP layer integrated on and coupled to a silicon-on-insulator waveguide circuit, Opt. Express 14 (2006) 8154–8159.

[79] C. Gunn, CMOS photonics for high-speed interconnects, Micro IEEE (2006) 58–66.

[80] N. Fujioka, T. Chu, M. Ishizaka, Compact and low power consumption hybrid integrated wavelength tunable laser module using silicon waveguide resonators, J. Lightwave Technol. 28 (2010) 3115–3120.

[81] Y.I.Y. Kawamura, K. Wakita, Y. Yoshikuni, H. Asahi, Unknown, J. Quant. Electron QE-23 (1987) 915–918.

[82] W. Kobayashi, M. Arai, T. Yamanaka, N. Fujiwara, T. Fujisawa, T. Tadokoro, K. Tsuzuki, Y. Kondo, F. Kano, Design and fabrication of 10-/40-Gb/s uncooled electroabsorption modulator integrated DFB laser with butt-joint structure, J. Lightwave Technol. 28 (2010) 164–170.

[83] J.E. Zucker, K.L. Jones, M.A. Newkirk, R.P. Gnall, B.I. Miller, M.G. Young, U. Koren, C.A. Burrus, B. Tell, Quantum well interferometric modulator monolithically

integrated with a 1.55 μm tunable distributed bragg reflector laser, Electron. Lett. 28 (1992) 1888–1889.

[84] R.C. Alferness, H. Kogelnik, T.H. Wood, The evolution of optical systems: optics everywhere, Bell Labs. Technol. J. 5 (2000) 188–202.

[85] L.A. Coldren, G.A. Fish, Y. Akulova, J.S. Barton, L. Johansson, C.W. Coldren, Tunable semiconductor lasers: a tutorial, J. Lightwave Technol. 22 (2004) 193–202.

[86] V. Jayaraman, Z.-M. Chuang, L. Coldren, Theory, design, and performance of extended tuning range semiconductor lasers with sampled gratings, IEEE J. Quant. Electron. 29 (1993) 1824–1834.

[87] D. Robbins, D.C.J. Reid, A.J. Ward, N.D. Whitbread, P.J. Williams, G. Busico, A.C. Carter, A.K. Wood, N. Carr, J.C. Asplin, M.Q. Kearley, W.J. Hunt, A novel broadband DBR laser for DWDM networks with simplified quasi-digital wavelength selection, in: Proceedings of Optical Fiber Communication OFC Conference, 2002, pp. 541–543.

[88] B. Mason, J. Barton, G.A. Fish, S.P. DenBaars, L.A. Coldren, Design of sampled grating DBR lasers with integrated semiconductor optical amplifiers, IEEE Photon. Technol. Lett. 12 (2000).

[89] A.J. Ward, D.J. Robbins, G. Busico, E. Barton, L. Ponnampalam, J.P. Duck, N.D. Whitbread, P.J. Williams, D.C.J. Reid, A.C. Carter, M.J. Wale, Widely tunable DS-DBR laser with monolithically integrated SOA: design and performance, IEEE J. Quant. Electron. 11 (2005) 149–156.

[90] JDSU, CW-TOSA Integrable Tunable Laser Assemblies 5205-T/5206-T ITLA Data Sheet 5205T–5206TITLA.DS.CMS.AE, 2011.

[91] Oclaro, LambdaFLEXG iTLA TL5000DCJ Integrable Tunable Laser Assembly Data Sheet D00255-PB, 2011.

[92] S.B. Mason, G.A. Fish, S.P. DenBaars, L.A. Coldren, Widely tunable sampled grating DBR laser with integrated electroabsorption modulator, IEEE Photon. Technol. Lett. 11 (1999) 638–640.

[93] Y.A. Akulova, G.A. Fish, P.-C. Koh, C.L. Schow, P. Kozodoy, A.P. Dahl, S. Nakagawa, M.C. Larson, M.P. Mack, T.A. Strand, C.W. Coldren, E. Hegblom, S.K. Penniman, T. Wipiejewski, L.A. Coldren, Widely tunable electroabsorption-modulated sampled-grating DBR laser transmitter, IEEE J. Sel. Top. Quant. Electron. 8 (2002) 1349–1356.

[94] V. Tolstikhin, Regrowth-free multi-guide vertical integration in InP for optical communications, in: IPRM, 2011.

[95] B. Pezeshki, E. Vail, J. Kubicky, G. Yoffe, S. Zou, J. Heanue, P. Epp, S. Rishton, D. Ton, B. Faraji, M. Emanuel, X. Hong, M. Sherback, V. Agrawal, C. Chipman, T. Razazan, 20-mW widely tunable laser module using DFB array and MEMS selection, IEEE Photon. Technol. Lett. 14 (2002) 1457–1459.

[96] N. Natakeyama, K. Naniwae, K. Kudo, N. Suzuki, S. Sudo, S. Ae, Y. Muroya, K. Yashiki, S. Satoh, T. Morimoto, K. Mori, T. Sasaki, Wavelength-selectable microarray light sources for S-, C-, and L-band WDM systems, IEEE Photon. Technol. Lett. 15 (2003) 903–905.

[97] Neophotonics, <http://www.neophotonics.com/_upload/pro/2011-10-09/NLW%20 ITLA%20R1.pdf>.

[98] J.S. Barton, E.J. Skogen, M.L. Masanovic, S.P. DenBaars, L. Coldren, Widely-tunable high-speed transmitters using integrated SGDBRs and Mach-Zehnder modulators, IEEE J. Sel. Top. Quant. Electron. 9 (2003) 1113–1117.

[99] S. Murthy, Large-scale photonic integrated circuit transmitters with monolithically integrated semiconductor optical amplifiers, in: OFC, 2008, p. Paper OTuN1.

[100] F. Kish, D. Welch, R. Nagarajan, J. Pleumeekers, V. Lal, M. Ziari, A. Nilsson, M. Kato, S. Murthy, P. Evans, S. Corzine, M. Mitchell, P. Samra, M. Missey, S. DeMars, R. Schneider, M. Reffle, T. Butrie, J. Rahn, M. Van Leeuwen, J. Stewart, D. Lambert, R. Muthiah, H. Tsai, J. Bostak, A. Dentai, K. Wu, H. Sun, D. Pavinski, J. Zhang, J. Tang, J. McNicol, M. Kuntz, V. Dominic, B. Taylor, R. Salvatore, M. Fisher, A. Spannagel, E. Strzelecka, P. Studenkov, M. Raburn, W. Williams, D. Christini, K. Thomson, S. Agashe, R. Malendevich, G. Goldfarb, S. Melle, C. Joyner, M. Kaufman, S. Grubb, Current status of large-scale InP photonic integrated circuits, IEEE J. Sel. Top. Quant. Electron. 17 (2011) 1470–1489.

[101] M. Missey, M. Kato, S. Murthy, V. Lal, J. Zhang, B. Taylor, M. Ziari, J. Stewart, A. Mathur, P. Evans, J. Pleumeekers, R. Muthiah, V. Dominic, M. Fisher, A. Nilson, S. Agashe, A. Chen, R. Salvatore, P. Liu, J. Bäeck, C. Joyner, J. Rossi, R. Schneider, M. Reffle, F. Kish, D.F. Welch, Tunable 100 Gb/s photonic integrated circuit transmitter and receiver, Technology (2008) 9–10.

[102] R. Nagarajan, Single-chip 40-channel InP transmitter photonic integrated circuit capable of aggregate data rate of 1.6 Tbit/s, Electron. Lett. 42 (2006) 771–773.

[103] R. Nagarajan, M. Kato, J. Pleumeekers, P. Evans, S. Corzine, S. Hurtt, A. Dentai, S. Murthy, M. Missey, R. Muthiah, R. Salvatore, C. Joyner, R. Schneider, Jr M. Ziari, F. Kish, D. Welch, InP photonic integrated circuits, J. Sel. Top. Quant. Electron. 16 (2010) 1113–1125.

[104] M. Kato, 40-channel transmitter and receiver photonic integrated circuits operating at per channel data rate 12.5 Gbit/s, Electron. Lett. 43 (2007).

[105] S. Grubb, High capacity upgrade of existing submarine wet plant using photonic integrated circuit based systems (2010).

[106] H.-U. Pfeiffer, S. Arlt, M. Jacob, C.S. Harder, I.D. Jung, F. Wilson, T. Oldroyd, Reliability of 980 nm pump lasers for submarine, long-haul terrestrial, and low cost metro applications, in: OFC, 2002, pp. 483–484.

[107] JDSU, <http://www.jdsu.com/News-and-Events/news-releases/Pages/jdsu-introduces-most-powerful-pump-laser-for-undersea-optical-networks.aspx>.

[108] S. Melle, V. Vusirikala, Network planning and architecture analysis of wavelength blocking in optical and digital ROADM networks, in: OFC, 2007, p. NTuC.

[109] Infinera Corporation internal network modeling data.

[110] S. Melle, Building agile optical networks, in: OFC, 2008, p. NME.

[111] D. Zheng, B. Smith, Improved efficiency Si-photonic attenuator, Opt. Express 16 (2008) 16754–16765.

[112] C.R. Doerr, L. Chen, L.L. Buhl, Y.-K. Chen, Eight-Channel SiON$_4$ {/Si/Ge} CWDM Receiver, IEEE Photon. Technol. Lett. 23 (2011) 1201–1203.

[113] L. Chen, C.R. Doerr, Y.-K. Chen, T.-Y. Liow, Low-loss and broadband cantilever low-loss and broadband cantilever couplers between standard cleaved fibers and high-index-contrast Si$_3$N$_4$ or Si, IEEE Photon. Technol. Lett. 22 (2010) 1744–1746.

[114] A. Narasimha, B. Analui, Y. Liang, S. Member, T.J. Sleboda, S. Abdalla, E. Balmater, S. Gloeckner, D. Guckenberger, M. Harrison, R.G.M.P. Koumans, D. Kucharski, A. Mekis, S. Mirsaidi, D. Song, T. Pinguet, A fully integrated 4 × 10-Gb/s DWDM optoelectronic transceiver implemented in a standard 1.3 μm CMOS SOI technology, IEEE J. Solid-State Circuits 42 (2007) 2736–2744.

[115] Y. Urino, T. Shimizu, M. Okano, N. Hatori, M. Ishizaka, T. Yamamoto, T. Baba, T. Akagawa, S. Akiyama, D. Okamoto, M. Miura, M. Noguchi, J. Fujikata, D. Shimura, H. Okayama, T. Tsuchizawa, T. Watanabe, K. Yamada, S. Itabashi, E. Saito, T. Nakamura, Y. Arakawa, First demonstration of high density optical interconnects integrated with lasers, optical modulators, and photodetectors on single silicon substrate, Opt. Express 19 (2011) B159–65.

[116] C.R. Doerr, T.F. Taunay, Silicon photonics core-, wavelength-, and polarization-diversity receiver, IEEE Photon. Technol. Lett. 23 (2011) 597–599.

[117] P.J. Winzer, Advanced optical modulation formats, in: Proc. IEEE, 94, 2006.

[118] P.J. Winzer, Modulation and multiplexing in optical communication systems, IEEE LEOS Newsletter (2009).

[119] P.J. Winzer, R.-J. Essiambre, Advanced optical modulation formats, in: Optical Fiber Telecommunications V, vol. B, 2008, pp. 23–94.

[120] D. Van Den Borne, S.L. Jansen, E. Gottwald, P.M. Krummrich, G.D. Khoe, H. De Waardt, 1.6-b/s/Hz spectrally efficient 40 85.6-Gb/s transmission over 1700 km of SSMF using POLMUX-RZ-DQPSK, in: Optical Fiber Communication Conference and the National Fiber Optic Engineers Conference, vol. 25, 2006, pp. 1–3.

[121] A.H. Gnauck, G. Charlet, P. Tran, P.J. Winzer, C.R. Doerr, J.C. Centanni, E.C. Burrows, T. Kawanishi, T. Sakamoto, K. Higuma, 25. 6-Tb/s WDM transmission of polarization-multiplexed RZ-DQPSK signals, J. Lightwave Technol. 26 (1) (2008) 79–84.

[122] M. Yagi, S. Satomi, S. Ryu, Field trial of 160-Gbit/s, polarization-division multiplexed RZ-DQPSK transmission system using automatic polarization control, in: OFC, 2008, p. OTuT7.

[123] B. Koch, R. Noé, V. Mirvoda, D. Sandel, V. Filsinger, K. Puntsri, 40-krad/s polarization tracking in 200-Gb/s PDM-RZ-DQPSK transmission over 430 km, IEEE Photon. Technol. Lett. 22 (2010) 613–615.

[124] S. Chandrasekhar, X. Liu, A. Konczykowska, F. Jorge, J. Dupuy, J. Godin, Direct detection of 107-Gb/s polarization-multiplexed RZ-DQPSK without optical polarization demultiplexing, IEEE Photon. Technol. Lett 20 (2008) 1878–1880.

[125] G. Abbas, V. Chan, Y. Ting, A dual detector optical heterodyne receiver for local oscillator noise suppression, J. Lightwave Technol. 3 (1985) 1110.

[126] T. Okoshi, Recent advances in coherent optical fiber communication systems, J. Lightwave Technol. 5 (1987) 44–52.

[127] H. Takeuchi, K. Kasaya, Y. Kondo, H. Yasaka, K. Oe, Y. Imamura, Monolithic integrated coherent receiver on InP substrate, IEEE Photon. Technol. Lett. 1 (1989) 398.

[128] T. Koch, U. Koren, R. Gnall, F. Choa, F. Hernandez-Gil, C. Burrus, M. Young, M. Oron, B. Miller, GaInAs/GaInAsP multiple-quantum-well integrated heterodyne receiver, Electron. Lett. 25 (1989) 1621.

[129] R.J. Deri, E. Pennings, A. Scherer, A. Gozdz, C. Caneau, N. Andreadakis, V. Shah, L. Curtis, R. Hawkins, J. Soole, J. Song, Ultracompact monolithic integration of balanced, polarization diversity photodetectors for coherent lightwave receivers, IEEE Photon. Technol. Lett. 4 (1992) 1238.

[130] S. Corzine, P. Evans, M. Kato, G. He, M. Fisher, M. Raburn, A. Dentai, I. Lyubomirsky, A. Nilsson, J. Rahn, R. Nagarajan, C. Tsai, J. Stewart, D. Christini, M. Missey, V. Lal, H. Dinh, A. Chen, J. Thomson, W. Williams, P. Chavarkar, S. Nguyen, D. Lambert, S. Agashe, J. Rossi, P. Liu, J. Webjorn, T. Butrie, M. Reffle, R. Schneider, M. Ziari, C. Joyner, S. Grubb, F. Kish, D. Welch, 10-channel × 40Gb/s per channel DQPSK

monolithically integrated InP-based transmitter PIC, in: OFC/NFOEC, 2008, p. PDP18.

[131] H. Bach, A. Matiss, C. Leonhardt, R. Kunkel, D. Schmidt, M. Schell, A. Umbach, Monolithic 90° hybrid with balanced PIN photodiodes for 100 Gbit/s PM-QPSK receiver applications, in: OFC, 2009, p. OMK5.

[132] C. Doerr, P. Winzer, S. Chandrasekhar, M. Rasras, M. Earnshaw, J. Weiner, D. Gill, Y.K. Chen, Monolithic silicon coherent receiver, in: OFC/NFOEC, 2009, p. PDPB2.

[133] C. Doerr, L. Zhang, P. Winzer, Monolithic InP multi-wavelength coherent receiver, in: OFC, 2010, p. PDPB1.

[134] R. Nagarajan, M. Kato, J. Pleumeekers, D. Lambert, V. Lal, A. Dentai, M. Kuntz, J. Rahn, H. Tsai, R. Malendevich, G. Goldfarb, J. Tang, J. Zhang, T. Butrie, M. Raburn, B. Little, A. Nilsson, M. Reffle, F. Kish, D. Welch, 10 channel, 45.6 Gb/s per channel, polarization multiplexed DQPSK InP receiver photonic integrated circuit, in: OFC/NFOEC, 2010, p. PDPB2.

[135] A. Matiss, R. Ludwig, J.-K. Fischer, L. Molle, C. Schubert, C. Leonhardt, H.-G. Bach, R. Kunkel, A. Umbach, Novel integrated coherent receiver module for 100G serial transmission, in: OFC, 2010, p. PDPB3.

[136] C.R. Doerr, L. Chen, Monolithic PDM-DQPSK receiver in silicon, ECOC (2010) 1–3.

[137] R. Nagarajan, D. Lambert, M. Kato, V. Lal, G. Goldfarb, J. Rahn, M. Kuntz, J. Pleumeekers, A. Dentai, H. Tsai, R. Malendevich, M. Missey, K. Wu, H. Sun, J. McNicol, J. Tang, J. Zhang, T. Butrie, A. Nilsson, M. Reffle, F. Kish, D. Welch, 10 Channel, 100 Gbit/s per channel, dual polarization, coherent QPSK, monolithic InP receiver photonic integrated circuit, in: Optical Fiber Communication Conference, 2011, p. OML7.

[138] N. Kikuchi, H. Sanjoh, 80-Gbit/s InP DQPSK modulator with an npin structure, in: 33rd ECOC 2007, vol. 2, 2007, pp. 8–9.

[139] K. Prosyk, T. Brast, M. Gruner, M. Hamacher, D. Hoffmann, R. Millett, K.-O. Velthaus, Tunable InP-based Optical IQ Modulator for 160Gb/s–OSA Technical Digest (CD), in: 37th European Conference and Exposition on Optical Communications, Optical Society of America, 2011, p. Th.13.A.5.

[140] (a) C.R. Doerr, P.J. Winzer, L. Zhang, L.L. Buhl, N.J. Sauer, Monolithic InP 16-QAM Modulator (2008); 20–22.(b) C. R. Doerr, L. Zhang, L. L. Buhl, J. H. Sinsky, A. H. Gnauck, P. J. Winzer, A. L. Adamiecki, N. J. Sauer, "High-speed InP DQPSK receiver," in: PDP23, OFC 2008; (c) C.R. Doerr, L. Zhang, P.J. Winzer, N. Weimann, V. Houtsma, T.-C. Hu, N.J. Sauer, L.L. Buhl, D.T. Neilson, S. Chandrasekhar, Y.K. Chen, "Monolithic InP Dual-Polarization and Dual-Quadrature Coherent Receiver," IEEE Photon. Technol. Lett. 23 (11) (2011) 694–696.

[141] P. Evans, 1.12 Tb/s superchannel coherent PM-QPSK InP transmitter photonic integrated circuit (PIC), Opt. Express 19 (2011) B154–B158.

[142] R. Nagarajan, D. Lambert, M. Kato, V. Lal, G. Goldfarb, J. Rahn, J. McNicol, K.-T. Wu, M. Kuntz, J. Pleumeekers, A. Dentai, H.-S. Tsai, R. Malendevich, M. Missey, J. Tang, J. Zhang, O. Khayam, H.S.T. Butrie, A. Nilsson, V. Dangui, M. Mitchell, M. Reff, F. Kish, D. Welch, Five-channel, 114 Gbit/s per channel, dual carrier, dual polarisation, coherent QPSK, monolithic InP receiver photonic integrated circuit, Electron. Lett. 47 (2011) 555–556.

[143] K. Kikuchi, Effect of 1/f-type FM noise on semiconductor-laser linewidth residual in high-power limit, IEEE J. Quant. Electron. 25 (1989) 684.

[144] R. Alferness, Guided-wave devices for optical communication, IEEE J. Quant. Electron. 17 (1981) 946–959.

[145] L.B. Soldano, E.C.M. Pennings, Optical multi-mode interference devices based on self-imaging: principles and applications, J. Lightwave Technol. 13 (1995) 615–627.

[146] R. Nagarajan, 10 channel, 100Gbit/s per channel, dual polarization, coherent QPSK, monolithic InP receiver photonic integrated circuit, in: OFC, 2011, p. OML7.

[147] R. Kunkel, No Title, in: IPRM, 2009, p. TuB2.2.

[148] M. Kato, R. Malendevich, D. Lambert, M. Kuntz, A. Damle, V. Lal, A. Dentai, O. Khayam, R. Nagarajan, J. Tang, J. Zhang, H.-S. Tsai, T. Butrie, M. Missey, J. Rahn, D. Krause, J. McNicol, K.-T. Wu, H. Sun, M. Reffle, F. Kish, D. Welch, 10 channel, 28 Gbaud PM-QPSK, monolithic InP Terabit Superchannel receiver PIC, in: IEEE Photonic Society 24th Annual Meeting vol. 3, 2011, pp. 340–341.

[149] NTT, 2012. <http://www.ntt-electronics.com/en/news/2012_02_29.html>.

[150] C.R. Doerr, L. Zhang, P.J. Winzer, Monolithic InP multiwavelength coherent receiver using a chirped arrayed waveguide grating, J. Lightwave Technol. 29 (2011) 536–541.

[151] R. Nagarajan, J. Rahn, M. Kato, J. Pleumeekers, D. Lambert, V. Lal, H. Tsai, A. Nilsson, A. Dentai, M. Kuntz, R. Malendevich, J. Tang, J. Zhang, T. Butrie, M. Raburn, B. Little, W. Chen, G. Goldfarb, V. Dominic, B. Taylor, M. Reffle, F. Kish, D. Welch, 10 Channel, 45.6 Gb/s per channel, polarization-multiplexed DQPSK, InP receiver photonic integrated Circuit, J. Lightwave Technol. 29 (2011) 386–395.

[152] J. Rahn, G. Goldfarb, H.-S. Tsai, W. Chen, S. Chu, B. Little, J. Hryniewicz, F. Johnson, W. Chen, T. Butrie, J. Zhang, M. Ziari, J. Tang, A. Nilsson, S. Grubb, I. Lyubomirsky, J. Stewart, R. Nagarajan, F. Kish, D. Welch, Low-power, polarization tracked 45.6 GB/s per wavelength PM-DQPSK receiver in a 10-channel integrated module, in: OFC, 2010, p. OThE2.

[153] E.M.L. Soldano, A. de Vreede, M. Smit, B. Verbeek, F. Groen, Mach-Zehnder interferometer polarization splitter in InGaAsP/InP, IEEE Photon. Technol. Lett. 6 (1994) 402–405.

[154] L.M. Augustin, J.J.G.M.V.D. Tol, R. Hanfoug, W.J.M.D. Laat, M.J.E.V.D. Moosdijk, P.W.L.V. Dijk, Y.-S. Oei, M.K. Smit, A single etch-step fabrication-tolerant polarization splitter (2007)

[155] J. van der Tol, F. Hakimzadeh, J. Pedersen, D. Li, H. van Brug, A new short and low-loss passive polarization converter on InP, IEEE Photon. Technol. Lett 7 (1995) 32–34.

[156] H. El-Refaei, D. Yevick, T. Jones, Slanted-rib waveguide InGaAsP-InP polarization converters, J. Lightwave Technol. 22 (2004) 1352–1357.

[157] L. Augustin, J. van der Tol, E. Geluk, M. Smit, Short polarization converter optimized for active-passive integration in InGaAsP-InP, IEEE Photon. Technol. Lett. 19 (2007) 1673–1675.

[158] M. Kotlyar, L. Bolla, M. Midrio, L. O'Faolain, T. Krauss, Compact polarization converter in InP-based material, Opt. Express 13 (2005) 5040–5045.

[159] S.-H. Kim, R. Takei, Y. Shoji, T. Mizumoto, Single-trench waveguide TE-TM mode converter, Opt. Express 17 (2009) 11267–11273.

[160] C. van Dam, L. Spiekman, F. van Ham, F. Groen, J. van der Tol, I. Moerman, W. Pascher, M. Hamacher, H. Heidrich, C. Weinert, M. Smit, Novel compact InP-based polarization converters using ultra short bends, in: Conf. Int. Photonics Res., 1996, p. IWC5.

[161] S. Obayya, A. Rahman, K. Grattan, H. El-Mikati, Improved design of a polarization converter based on semiconductor optical waveguide bends, Appl. Opt. 40 (2001) 5395–5401.

[162] J. Rahn, G. Goldfarb, H.-S. Tsai, W. Chen, S. Chu, B. Little, J. Hryniewicz, F. Johnson, W. Chen, T. Butrie, J. Zhang, M. Ziari, J. Tang, A. Nilsson, I. Lyubomirsky, J. Stewart,

R. Nagarajan, F. Kish, D.F. Welch, Low-power polarization tracked 45.6 GB/s per wavelength PM-DQPSK receiver in a 10-channel integrated module in: Optical Fiber Communication Conference, 2010, pp. 10–12.

[163] C.R. Doerr, N.K. Fontaine, L.L. Buhl, PDM DQPSK silicon receiver with integrated monitor and minimum number of controls, IEEE Photon. Technol. Lett. 24 (8) (2012) 697–699.

[164] C.R. Doerr and Others, Packaged Monolithic Silicon 112-{Gb}/s Coherent Receiver, IEEE Photon. Technol. Lett. 23 (2011) 762–764.

[165] D. Taillaert, P. Borel, L. Frandsen, R. De La Rue, R. Baets, A compact two-dimensional grating coupler used as a polarization splitter, IEEE Photon. Technol. Lett. 15 (2003) 1249–1251.

[166] C.R. Doerr, N. Fontaine, M. Hirano, T. Sasaki, L. Buhl, P. Winzer, Silicon photonic integrated circuit for coupling to a ring-core multimode fiber for space-division multiplexing, in: ECOC, 2011, p. Th.13.A.3.

[167] C.R. Doerr, L.L. Buhl, Circular grating coupler for creating focused azimuthally and radially polarized beams, Opt. Lett. 36 (2011) 1209–1211.

[168] R. Ryf et al., Space-division multiplexing over 10 km of three-mode fiber using coherent 6 × 6 MIMO processing, in: Optical Fiber Communication Conference and Exposition OFC/NFOEC and the National Fiber Optic Engineers Conference, CA, Los Angeles, March 6–10 2011, pp. 1–3.

[169] C.R. Doerr, Proposed architecture for MIMO optical demultiplexing using photonic integration, IEEE Photon. Technol. Lett. 23 (2011) 1573–1575.

[170] N. Fontaine, C. Doerr, M. Mestre, Space-division multiplexing and all-optical MIMO demultiplexing using a photonic integrated circuit, Opt. Fiber (2012) 2–4.

[171] G.E. Moore, No exponential is forever: but G forever G can be delayed! in: IEEE Intl Solid State Circuits Conf., 2003, pp. 20–23.

[172] J.A. Cunningham, Using the learning curve as a management tool, IEEE Spectrum (1980) 45–48.

[173] S. Natarajan, A 32 nm logic technology featuring 2nd-generation high-k + metal-gate transistors, enhanced channel strain and $0.171\,\mu m^2$ SRAM cell size in a 291 Mb array, in: Electron Devices Meeting, 2008, pp. 1–3.

[174] G. Walter, N. Holonyak, Jr M. Feng, R. Chan, Laser operation of a heterojunction bipolar light-emitting transistor, Appl. Phys. Lett. 85 (2004) 4769–4770.

[175] M. Feng, H.W. Then, N. Holonyak, Jr G. Walter, A. James, Resonance-free frequency response of a semiconductor laser, Appl. Phys. Lett. 95 (2009) 3–5.

[176] Submitted for publication: L.A. Coldren, S.C. Nicholes, L.A. Johansson, S. Ristic, R.S. Guzzon, E.J. Norberg, U. Krishnamachari, G High Performance InP-based Photonic ICsGa Tutorial, J. Lightwave Technol.

[177] C.R. Doerr, Monolithic InP photonic integrated circuits for transmitting or receiving information with augmented fidelity or spectral efficiency, in: High Spectral Density Optical Communication Technologies, Springer-Verlag, 2010 (Chapter 4).

Advances in Photodetectors and Optical Receivers

Andreas Beling and Joe C. Campbell

University of Virginia, Department of Electrical and Computer Engineering,
Charlottesville, VA 22904, USA

3.1 INTRODUCTION

The landscape for advances in photodetectors and optical receivers has experienced significant change since Optical Fiber Telecommunications V. The quests for higher speed p-i-n detectors and lower noise avalanche photodiodes (APDs) with high gain-bandwidth product remain. To a great extent, high-speed structures have coalesced to evanescently-coupled waveguide devices; bandwidths exceeding 140 GHz have been reported. A primary APD breakthrough has been the development of Ge on Si separate-absorption-and-multiplication devices that achieve long-wavelength response with the low-noise behavior of Si. For III–V compound APDs ultra-low noise has been achieved by strategic use of complex multilayer multiplication regions that provide more deterministic impact ionization. However, much of the excitement and innovation have focused on photodiodes that can be incorporated into InP-based integrated circuits and photodetectors for Si photonics. Optical receivers have followed the wave to higher and higher bit rates; 112 Gb/s has been achieved. A number of receiver architectures, particularly those for coherent detection, now incorporate balanced detectors, which are being pushed to higher bandwidths. Finally, there have been concerted efforts to develop detectors for integrated receivers for advanced modulation formats such as DPSK, DQPSK, and DP-QPSK.

We begin with a very brief review of photodetector fundamentals. The most basic function of a photodetector is to convert light to current; this is evaluated in terms of the responsivity expressed in A/W or the external quantum efficiency, the percentage of free carriers collected relative to the number of incident photons. In an ideal photodetector, each incident photon would result in the charge of one electron flowing in the external circuit. In practice, there are several physical effects, such as incomplete absorption, recombination, reflection from the semiconductor surface, coupling losses, and contact shadowing, which reduce the responsivity. Once the photons have entered the semiconductor, the photodetection process consists of (1) absorption and (2) collection of the photogenerated electron-hole pairs.

Structurally, photodetectors can be classified as normal-incidence or side-illuminated. Figure 3.1a shows a common structure for normal-incidence

Optical Fiber Telecommunications VIA. http://dx.doi.org/10.1016/B978-0-12-396958-3.00003-2

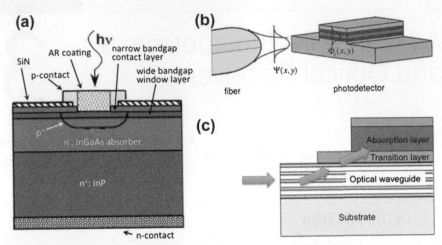

FIGURE 3.1 Illustrations of (a) normal-incidence, (b) butt-coupled waveguide, and (c) evanescently-coupled waveguide photodiodes.

telecommunications p-i-n photodiodes. Light is absorbed in a fully depleted $In_{0.53}Ga_{0.47}As$ layer. In the rest of this chapter $In_{0.53}Ga_{0.47}As$ will be referred to as InGaAs unless a different composition is explicitly stated. These photodiodes usually also include a wide-bandgap (e.g. InGaAsP or InP) window layer to reduce surface recombination and an InGaAs contact layer that is removed in the input area. For normal-incidence, carriers flow parallel to the incident light flux, which gives rise to a trade-off between the responsivity and bandwidth. High responsivity requires a relatively thick absorption layer. The absorption coefficient, α, in InGaAs is $\sim 10^4 \, cm^{-1}$, which means that approximately $2 \, \mu m$ is needed to absorb 88% of the light. However, the thicker the absorption region, the longer it takes to collect the carriers. This is the origin of the transit-time limit to the bandwidth. For waveguide photodiodes, on the other hand, light is coupled in perpendicular to carrier transport and the absorption layers are much thinner than for normal-incidence devices. Even though the confinement factor, i.e. the spatial overlap of the optical field and the absorbing layer, in the waveguide may be much less than unity, producing an effective absorption coefficient less than that of the bulk semiconductor, the interaction length is typically a few tens of microns and thus the internal quantum efficiency can easily approach 100%. Since the photogenerated carriers transit perpendicular to the thin absorption/depletion region, high bandwidths can be achieved. The most common implementations for side-illumination utilize input by butt coupling (Figure 3.1b) or an evanescently-coupled waveguide (Figure 3.1c). Normal-incidence has the advantages of high input coupling efficiency, uniform current flow, and relaxed packaging constraints. The waveguide structure, as noted, decouples the responsivity from the transit-time-limited bandwidth and it is compatible with photonic integrated circuits. For normal-incidence photodiodes the responsivity, R, at wavelength, λ, can be written as

$$R = \frac{I_{\text{ph}}}{P_{\text{opt}}} = \frac{\eta_{\text{ext}}\lambda\ [\mu\text{m}]}{1.24} \text{ and } \eta_{\text{ext}} = (1 - R_{\text{surface}})(1 - e^{-\text{ad}}), \quad (3.1)$$

where I_{ph} is the photocurrent, P_{opt} is the incident optical power, η_{ext} is the external quantum efficiency, R_{surface} is the power reflection coefficient, α is the absorption coefficient, and d is the thickness of the absorbing region. The external quantum efficiency of waveguide photodiodes is multiplied by the confinement factor, $\Gamma(x)$, which may vary along the length of the detector.

For most photodetectors the bandwidth is determined by the RC time constant and the electron and hole transit times. Typically, the junction resistance in parallel with the diode capacitance, C_{pd}, is large and can be neglected, which leaves the diode series resistance, R_s, and the load resistance, R_l. The RC-limited 3 dB bandwidth is given by

$$f_{\text{RC}} = \frac{1}{2\pi\ R_{\text{eff}}C_{\text{pd}}} \quad (3.2)$$

with $R_{\text{eff}} = R_s + R_l$. If an impedance matching resistor, usually $50\,\Omega$, is employed, $R_{\text{eff}} = R_s + R_l \cdot 50\,\Omega/(R_l + 50\,\Omega)$. The time constants required for carrier transport through the active region prior to collection by the contacts vary with the device structure and the electric field profile. For the case that the active region is fully depleted and assuming the following: (1) the electric field is sufficient for carriers to achieve saturation velocity, (2) equal electron and hole velocities, $\bar{v}$, and (3) uniform photogeneration in the absorber with thickness d, the transit time-limited bandwidth can be approximated as [1]:

$$f_t = \frac{1}{2\pi\tau_t} \approx \frac{3.5\bar{v}}{2\pi d}. \quad (3.3)$$

An important exception is the uni-traveling-carrier (UTC) photodiode [2,3] that has been widely used for applications that require high power-bandwidth product [4,5]. The UTC structure utilizes an undepleted p-layer to absorb light and inject electrons into a wide-bandgap non-absorbing drift region. Having only electrons in the depletion region greatly suppresses the space-charge effect, which is the primary source of compression at high photocurrent levels [6–9] in conventional p-i-n photodiodes [8]. In addition, the higher velocity of electrons relative to that of holes results in higher bandwidth compared to conventional p-i-n photodiodes. The transit time for electrons has two components corresponding to the undepleted absorber and the drift region with widths W_A and W_C, respectively. The total transit time is [10]

$$\tau_t = \frac{W_C}{3.5\bar{v}} + \left[\frac{W_A^2}{3D_e} + \frac{W_A}{v_{\text{th}}}\right], \quad (3.4)$$

where D_e is the electron diffusion coefficient and v_{th} is the electron thermal velocity. $W_A^2/3D_e$ is the diffusion transit time and W_A/v_{th} is the correction factor associated with the finite thermal velocity.

The 3 dB bandwidth is given by the expression:

$$f_{3\,\text{dB}} \approx \sqrt{\frac{1}{\frac{1}{f_{\text{RC}}^2} + \frac{1}{f_t^2}}}. \tag{3.5}$$

The product of Eq. (3.5) and the efficiency in Eq. (3.1) yields the bandwidth-efficiency product. By proper scaling of the detector area, it is possible to achieve transit-time-limited response.

The p-i-n photodiodes discussed above typically consist of a depleted absorbing layer, the "i" region, sandwiched between heavily doped p^+ and n^+ quasi-neutral layers. They operate without gain. The other type of photodetector that has been widely used for telecommunications is the avalanche photodiode (APD). Gain is achieved by impact ionization in a high electric field region in the APD. If the noise associated with impact ionization is less than the noise of the following circuitry, the internal gain of the APD can provide higher sensitivity in optical receivers than p-i-n photodiodes [11–14]. Since the noise of amplifiers increases with bit rate, the sensitivity margin provided by APDs relative to p-i-ns also increases with bit rate up to the gain-bandwidth limit of the APD. Consequently, a major thrust in APD research has been to increase the bandwidth and the gain-bandwidth product. According to the local-field model that has been widely used to design and characterize APDs [15–17], both the noise- and the gain-bandwidth product are determined by the electron, α, and hole, β, ionization coefficients of the semiconductor in the multiplication region, or more specifically, their ratio, $k = \beta/\alpha$. The noise power spectral density, ϕ, for mean gain, $\langle M \rangle$, and mean photocurrent, $\langle I_{\text{ph}} \rangle$, are given by the expression $\phi = 2q\langle I_{\text{ph}} \rangle \langle M \rangle 2F(M)$. $F(M)$ is the excess noise factor, which arises from the stochastic nature of impact ionization. Under the conditions of uniform electric fields and single carrier injection, the excess noise factor is:

$$F(M) = \langle M^2 \rangle / \langle M \rangle^2 = k\langle M \rangle + (1 - k)(2 - 1/\langle M \rangle). \tag{3.6}$$

Equation (3.6) has been derived under the condition that the ionization coefficients are in local equilibrium with the electric field, hence, the designation "local-field" model. This model assumes that the ionization coefficients at a specific position are determined solely by the electric field at that position irrespective of the carrier energy. It is clear from Eq. (3.6) that lower noise is achieved when $k \ll 1$. The gain-bandwidth product results from the time required for the avalanche process to build up or decay: the higher the gain, the higher the associated time constant and, thus, the lower the bandwidth. Emmons [18] has shown that for $M_o > \alpha\beta$, where M_o is the dc gain, the frequency response is characterized by a constant gain bandwidth-product that increases as k decreases. In recent years it has been shown that low excess noise and high gain-bandwidth product can also be achieved by submicron scaling of the high-field multiplication region [19–39] or by incorporating new materials and impact ionization engineering (I^2E) with appropriately designed heterostructures [40–47]. These concepts will be discussed in more detail in Section 3.5.

3.2 HIGH-SPEED WAVEGUIDE PHOTODIODES

Carrier transit times and the resistance-capacitance (RC) time constant determine the bandwidth of conventional, lumped element photodiodes. For these photodiodes it is imperative to reduce the depletion width and device area to minimize transit times and capacitance while simultaneously maintaining low series resistance. In surface-normal photodiodes this approach has yielded devices with bandwidths greater than 100 GHz [48], however, at the expense of low responsivity. To overcome the bandwidth-efficiency trade-off, waveguide photodiodes (WGPDs) have been successfully employed. The benefit of this type of photodiode is that high quantum efficiency and short carrier transit times can be achieved simultaneously [49]. Furthermore, by adopting a traveling-wave structure the requirement to compromise between the RC- and the transit-time-limited bandwidth can effectively be evaded [50–52].

3.2.1 Side-illuminated and evanescently-coupled waveguide photodiodes

To overcome the bandwidth-efficiency trade-off waveguide photodiodes have been developed for p-i-n [53,54], metal-semiconductor-metal (MSM) [55], uni-traveling-carrier (UTC) [56,57], and avalanche photodiodes [58,59]. As illustrated in Figure 3.1, waveguide photodiodes can be classified as side-illuminated or evanescently-coupled WGPDs. In the side-illuminated WGPD the light is butt-coupled from the fiber into the absorber. A prominent example for this type of WGPD is the mushroom-mesa photodiode, which achieved a record bandwidth-efficiency product of 55 GHz demonstrated by Kato et al. [60]. The disadvantages of side-illumination are the limited high-power capability due to non-uniform carrier distribution along the optical path and the low tolerance to lateral and vertical displacement of the input signal. The latter implies the use of additional optics or a tapered fiber in order to efficiently illuminate the small active region. Recently, coupling tolerances at 1 dB extra loss of $\pm 1.3\,\mu$m and $\pm 0.85\,\mu$m in the horizontal and vertical directions, respectively, were reported for a 5 μm-wide tapered waveguide photodiode [61]. This photodiode had an absorption layer of only 100 nm thickness embedded within thicker depleted transparent layers to reduce capacitance and balance electron and hole drift times. The reported bandwidth was 42 GHz with responsivity of 1.08 A/W.

In contrast to side-illuminated devices the evanescently-coupled WGPD consists of a photodiode located on top of a passive waveguide. The light couples evanescently from the input waveguide to the photodiode mesa, which ensures a more uniform absorption profile along the absorber length and leads to an improved high-power capability [62]. This WGPD is well suited for monolithic integration with planar lightwave circuits in photonic integrated circuits and, independent of the active device, a mode field transformer (taper) can be integrated in order to improve the fiber-chip coupling efficiency. This enables the use of a cleaved fiber instead of a tapered/lensed fiber, which simplifies the fiber-chip coupling process and also provides large alignment tolerances. Using a cleaved fiber 1-dB alignment tolerances of $\pm 2.5\,\mu m$ and $\pm 3.5\,\mu m$ in the vertical and horizontal directions, respectively, were demonstrated for a WGPD with vertically tapered mode field transformer [53]. Using a similar input taper a highly efficient p-i-n photodetector suitable for detection up to data rates of 160 Gbit/s was reported in Refs. [63,64]. The photodetector chip comprised an evanescently-coupled InGaAs/InP photodiode with an active area of $100\,\mu m^2$, a vertically tapered mode field transformer, a biasing network, and a $50\,\Omega$ load resistor. An optimized impedance of the electrical output line of the detector enabled a cut-off frequency of 100 GHz. More recently, a fully packaged photodiode of this type capable of providing a flexible dc offset voltage at its RF output achieved a bandwidth of 90 GHz. The on-chip bias network of this photodetector was modified to supply a voltage to the postamplifier or demux integrated circuit without the need of an external bias tee [65]. The responsivity of the photodetector module was 0.53 A/W with a polarization-dependent loss (PDL) of only 0.1 dB.

The reduction of absorber thickness and photodiode length enabled higher bandwidth of similar p-i-n photodiodes in Refs. [66,67]. By downscaling the photodiode length to $7\,\mu m$ the reduced capacitance of these small-area photodiodes resulted in transit-time-limited bandwidths at $25\,\Omega$ effective load of 100 GHz, 120 GHz, and 145 GHz for 430 nm-, 350 nm-, and 200 nm-thick absorbers, respectively (Figure 3.2). Owing to an optimized evanescent coupling scheme the fiber-coupled responsivities were 0.51 A/W, 0.48 A/W, and 0.35 A/W, respectively. To mitigate the decrease in responsivity, which generally follows reduction in absorber length, the n-contact layer was extended by a well-defined length toward the single-mode input waveguide. As this portion of the waveguide becomes multimode, mode-beating effects can be exploited in order to facilitate an efficient coupling from the single-mode input waveguide into the absorber and maximize the confinement factor Γ_{xy} [68,69]. Simulations of the optimized structure showed that the confinement factor was particularly

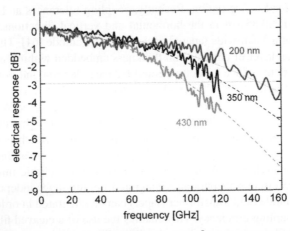

FIGURE 3.2 Measured frequency responses for $5 \times 7\,\mu m^2$ p-i-n photodiodes with 430 nm, 350 nm, and 200 nm thick absorbers.

enhanced in the first few microns of the absorber [70]. Since in multimode WGPDs the confinement factor Γ_{xy} is generally a function of z, the propagation direction, the quantum efficiency can be estimated using the relation:

$$\eta_{eWGPD} = \eta_c \left[1 - \exp \left(-\alpha \int_0^{l_{abs}} \Gamma_{xy}(z)dz \right) \right], \qquad (3.7)$$

where η_c is the fiber-chip input coupling efficiency, α is the absorption coefficient, and l_{abs} is the photodiode length. It should be noted that designing a WGPD with a non-constant confinement factor can be beneficial for high-power applications [71]. To achieve a more uniform absorption profile and current density, the confinement factor should increase toward the end of the WGPD.

By using a short multimode input waveguide it has been shown that high fiber-coupled efficiencies can be achieved without the need of a mode field transformer [72]. The photodiode in [73] utilized a planar diluted waveguide and two optical matching layers designed to provide a gradual increase of the optical refractive index from the diluted waveguide to the absorbing layer, which resulted in a significant enhancement in the quantum efficiency. For an optimal input waveguide length the responsivity was 1.07 A/W with a PDL of less than 0.5 dB when using a lensed fiber. The reported bandwidth was 48 GHz. Recently, dry-etching of the waveguide input facet has been demonstrated to precisely control the length of the planar multimode waveguide. In addition, etching a lensed facet improved horizontal fiber-chip 1-dB alignment tolerances to >20 µm [74]. The reported lensed facet waveguide UTC photodiodes achieved 0.55 A/W and a bandwidth of >50 GHz [75].

Lateral tapering of diluted input waveguides was demonstrated in Ref. [76] using an asymmetric twin-waveguide technology [77]. In this photodiode the incident light is collected by a single-mode diluted waveguide and transferred via a taper to a thinner coupling waveguide from where it couples evanescently into the absorber. Using a lensed fiber for input coupling a responsivity of ∼1 A/W and a bandwidth 42 GHz were demonstrated. Using a similar waveguide taper Rouvalis et al. reported UTC waveguide photodiodes with a bandwidth of 110 GHz and 0.32 A/W [78]. The photodiode-active area and absorber thickness were $4 \times 15 \, \mu m^2$ and 70 nm, respectively. For sub-THz applications similar photodiodes with traveling-wave electrodes and integrated resonant antennas achieved a maximum extracted power of 150 µW at 460 GHz.

3.2.2 Distributed and traveling-wave photodetectors

The traveling-wave photodetector (TWPD) is a distributed structure in that optical and photogenerated electrical signals co-propagate along the device length. The electrical contacts are designed as a transmission line with its characteristic impedance matched to that of the external load. The bandwidth of TWPDs is determined by the carrier transit times, microwave losses, and velocity mismatch. Velocity mismatch results from the difference between the optical and electrical signal velocities as they propagate along the structure [51]. The fact that both, the characteristic impedance and the electrical signal velocity, depend on the capacitance of the photodiode makes a velocity match difficult. Thus, it has been suggested that in practical devices a residual velocity mismatch is inevitable, which can lead to bandwidth degradations in very long devices [81,79]. TWPDs were first demonstrated for GaAs-based p-i-n photodiodes that operated at short wavelengths [80–82]. In [81], a 7 μm-long WGPD achieved a bandwidth of 172 GHz and a record-high bandwidth-efficiency product of 76 GHz. More recently, the traveling-wave principle was also applied to evanescently-coupled InGaAs/InGaAsP/InP partially depleted absorber photodiodes [83]. Bandwidths up to 110 GHz with output power levels of +4.5 dBm at 110 GHz were reported. Owing to a double-stage mode field transformer, a responsivity of 0.45 A/W was achieved.

In a second type of TWPD, referred to as periodic TWPD, discrete photodiodes located on top of an optical waveguide are combined in a traveling-wave structure [84,85]. The variation of the spacing between the photodiodes provides an additional degree of freedom in the device design and thus, an impedance and velocity match can be achieved simultaneously. Assuming that the electrical signal wavelength is on the order of the device length, the periodic TWPD can be treated as a capacitively loaded, synthetic transmission line [86]. This approach is valid up to the Bragg frequency [87]. To eliminate the reflections of the backward propagating electrical signal, periodic TWPDs with a termination resistor at the input of the transmission line have been demonstrated. Using three $2 \times 5\,\mu m^2$ InAlGaAs/InGaAs UTC photodiodes and an integrated termination resistor, Hirota et al. reported a periodic TWPD with 115 GHz bandwidth and 0.15 A/W responsivity [86]. In [88], Murthy et al. reported a multisection coplanar strip transmission line to avoid the internal resistor and its associated 6 dB RF power loss. In their device multisection coplanar strips with step-reduced impedances connected three discrete photodiodes. The design of the transmission line discontinuities was optimized to cancel the backward propagating electrical waves. The measured bandwidth, linear dc photocurrent, and responsivity were 38 GHz, 12 mA, and 0.24 A/W, respectively. The lengths of the photodiodes (8, 10, and 20 μm) were designed to achieve uniform photocurrent distribution.

An efficient way to achieve uniform photocurrent distribution and thus higher saturation photocurrents is to divide the optical signal with a power splitter to several discrete photodiodes that are connected by an output transmission line [89]. In [90], the monolithic integration of a multimode interference (MMI) power splitter and a TWPD with four 80 μm-long p-i-n photodiode was demonstrated. Linear dc photocurrent of 52 mA and unsaturated RF power of 9 dBm at 10 GHz signal frequency were achieved. A high-speed TWPD comprising a mode field converter, a 1×4 MMI power splitter, and four evanescently coupled $4 \times 7\,\mu m^2$ p-i-n photodiodes was reported in [67]. In order to eliminate electrical reflections at the input of the transmission line, a 50 Ω matching resistor was integrated (Figure 3.3). Impedance and velocity match were verified experimentally to be >80%. Figure 3.4 compares the measured frequency responses from a chip-based and a fully packaged TWPD (inset) indicating bandwidths of 80 GHz and 40 GHz, respectively [91]. Despite some additional losses in the packaging, a TWPD module with a responsivity of 0.24 A/W and 0.2 dB PDL was successfully operated in 80 Gbit/s back-to-back experiments [91].

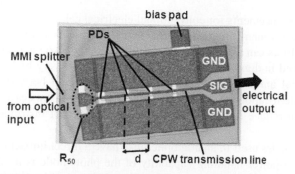

FIGURE 3.3 TWPD chip with termination resistor R_{50} and $d=90\,\mu m$. The input signal is fed from the left via the mode field converter and rib waveguide (not shown) into the MMI splitter. The RF output pads are in ground-signal-ground configuration.

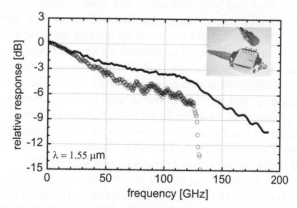

FIGURE 3.4 Frequency response of TWPD chip (line) and module (open circles). The response of the TWPD module (inset) is limited by the 1 mm coaxial connector around 125 GHz [91].

3.3 HIGH-POWER PHOTODIODES

While digital links constitute the majority of fiber optic links for optical transmission, there are a growing number of applications for analog optic links such as cable TV (CATV), local oscillator (LO) distribution for radio telescopes, beamforming networks for phased array antennas, and antenna remoting for military radar. The Atacama Large Millimeter/Submillimeter Array (ALMA) in Chile, one of the largest radio telescopes in the world, uses analog optic links to distribute the photonic LO over 100 GHz [92]. Analog links are also being deployed in microwave transmission equipment at base stations of cellular phone systems [93]. Analog links can be viewed as replacements for conventional electrical cables or waveguides, which are impractical for many applications due to their high loss and limited bandwidth. Analog optic links can also overcome the limitations of A/D and D/A conversion that are required in digital transmission and can achieve very high speed. Signals with substantial amount of RF power can be transmitted between a center station and remote locations (e.g. antenna feeds) over long distances via analog optic links. This can reduce the complexity and maintenance of instruments at remote locations.

The photodiodes used in analog optic links have a direct impact on the link performance. High-power handling capability of the photodiode is needed to achieve high link gain. High-speed operation of the link is impossible without photodiodes that have correspondingly large bandwidth. It is also important for the photodiodes to have high linearity in order to minimize signal distortion and maintain large spurious-free dynamic range (SFDR).

The two primary limitations on the power handling capability of photodiodes are the space-charge effect and heat dissipation. The space-charge effect has its origin in the spatial distribution of the photogenerated carriers as they transit the depletion layer [6–9]. The electric field generated by the free carriers, referred to as the space-charge field, opposes the field established by ionized dopants and the applied bias voltage. As a result, the total electric field can collapse in some portion of the depletion region at high current densities. Once this happens, the carrier transit time increases significantly and RF power output suffers from compression/saturation as illustrated in Figure 3.5. Also significant at high current levels, the voltage drop across the load resistor reduces the effective bias voltage, which also pushes the photodiode toward saturation. In the following "saturation current" will denote the photocurrent at which the RF output power is compressed by 1 dB.

Another primary limitation on the RF output power is thermally induced catastrophic failure. Watt-level DC power is dissipated in an active area less than $10^{-4}\,cm^2$ for a high-power photodiode. Depending on factors such as heat conductance of the semiconductor layers, the photodiode geometry, and the heat sink design, the junction temperature can exceed 500 K. The need for better thermal management is made more evident by the fact that higher bias voltage is often used to improve photodiode saturation and linearity but this results in more joule heating.

The photodiode linearity measures the extent that the electrical output of the photodiode follows its optical input in a linear fashion. Third-order intermodulation distortion is of particular interest because the associated frequencies can be close to the fundamental modulation frequencies. The third-order intercept point (IP3), which is illustrated in Figure 3.5, is a standard figure of merit third-order intermodulation. Virtually all physical mechanisms are nonlinear to various degrees. Nonlinear optical transmission, carrier generation, and transport can all contribute to photodiode nonlinearities. The dominant nonlinear mechanisms are usually different for photodiodes that operate at different power levels, frequency ranges, and bias voltages. Trade-offs have often to be made between conflicting requirements for high linearity and other metrics in photodiode design.

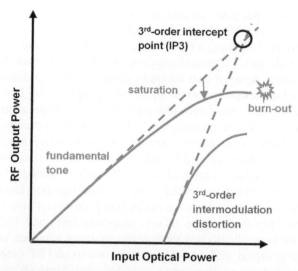

FIGURE 3.5 Output RF power versus input optical power. This graph illustrates saturation, thermal failure, and the impact of intermodulation distortion.

3.3.1 Normal-incidence uni-traveling-carrier photodiodes

In the frequency domain the space-charge effect is observed as a decrease in the bandwidth with increasing photocurrent [96–98]. The first photodiode structure to demonstrate significantly higher saturation current than conventional p-i-n photodiodes was the uni-traveling-carrier (UTC) structure [3,10,94,95]. The absorbing region of the UTC-photodiode is a narrow-bandgap (typically InGaAs) p-type layer adjacent to an undoped (or a lightly n-type doped) wide-bandgap (typically InP) depleted drift layer. Since the absorbing layer is undepleted, the photogenerated excess hole density decays within the dielectric relaxation time. The electrons, on the other hand, move by both drift and diffusive forces through the InGaAs absorber and are injected into the InP collector. Hence, the photocurrent is purely electron transport. This provides two distinct advantages compared to the p-i-n structure, in which both electrons and holes are generated in and drift through the depletion layer. The first is higher speed; frequency response up to 457 GHz has been demonstrated using a traveling-wave configuration [79]. While both carriers contribute to the photocurrent in a p-i-n, since the hole velocity is much lower than that of electrons, it is the holes that limit the bandwidth and exacerbate the space-charge effect [96].

High-power operation of UTC photodiodes has been achieved by modifying the doping profile in the InP drift layer. Using a charge-compensated UTC (CC-UTC) structure with 0.45 A/W responsivity, Li et al. [97] reported 25 GHz bandwidth and large-signal 1-dB compression current greater than 90 mA; the RF output power at 20 GHz was 20 dBm. A smaller ∼100 μm² photodiode exhibited a bandwidth of 50 GHz and large-signal 1-dB compression current greater than 50 mA.

The maximum RF output power at 40 GHz was 17 dBm. Chtioui et al. have achieved high responsivity (0.83 A/W), 24 GHz bandwidth, 80 mA saturation current, and IP3 of 30 dBm at 10 GHz [98] using a UTC with a non-uniform doping in the collector.

In normal-incidence UTC photodiodes, since the photogenerated electrons transit to the wide-bandgap depleted drift region by diffusion in the p-doped absorber, in order to achieve high-speed operation, the absorbing region must be relatively thin, which results in an efficiency-bandwidth trade-off. Back illumination is frequently employed to increase the responsivity; reflection from the top contact effectively increases the thickness of the absorber. Another approach is to replace the wide bandgap drift layer with a thin depleted InGaAs absorber, which also results in higher responsivity. This structure is referred to as a Partially Depleted Absorber (PDA) photodiode. The homojunction aspect facilitates carrier transport since there are no band discontinuities at the p-i or i-n interfaces to impede carrier mobility. PDA photodiodes having 1 μm-thick p-type InGaAs absorbing layer and 700-nm depleted InGaAs collector layer illuminated through the substrate achieved responsivity of 1.0 A/W. The large-signal modulation saturation currents and RF output powers for 28 μm-, 34 μm-, and 40 μm-diameter photodiodes were 70 mA, 80 mA, and 90 mA and 17 dBm, 18.5 dBm, and 19.5 dBm, respectively. Williams et al. [99] reported 1-dB small-signal compression currents for a 34 μm-diameter PDA photodiode of 700 mA, 620 mA, and 260 mA at 300 MHz, 1 GHz, and 6 GHz, respectively. A disadvantage of the PDA structure is that the presence of both electrons and holes in a portion of the drift region can degrade the saturation characteristics. In addition, the poor thermal conductivity of the thicker InGaAs layer also contributes to thermal failure.

Jun et al. [100] merged the UTC and PDA structures by inserting an undoped InGaAs layer between the InP drift layer and the p: InGaAs absorption region of a UTC. This Modified UTC (MUTC) has achieved higher responsivity-saturation current performance than either the UTC or the PDA. At 5 GHz an MUTC photodiode exhibited saturation current of 260 mA and RF power of 29 dBm; the bandwidth was 7 GHz [101]. Wang et al. [102] have reported a 20 μm-diameter charge-compensated (CC) MUTC photodiode with bandwidth, saturation current, and RF output power of 30 GHz, 70 mA, and 18.6 dBm, respectively. At a bias of 6 V a 28 μm-diameter MUTC exhibited 23-GHz bandwidth, 90-mA saturation current, and 83 mW RF output power. The responsivity at $\lambda = 1550$ nm was 0.75 A/W. Chioui et al. have carried out a direct comparison of UTC and MUTC photodiodes [103]. Both device structures had 1.5 μm-thick absorption regions followed by a 0.5 μm-thick InP collector layer. Devices from both structures exhibited 0.92 A/W responsivity at 1.55 μm and saturation current >100 mA at 10 GHz. Owing in part to a partially depleted absorber the MUTC photodiode demonstrated higher bandwidth (>20 GHz at high current), while the UTC photodiode achieved higher saturation current. The lower saturation current of the MUTC is most likely due to field collapse in the depleted absorber.

For the MUTC photodiodes at very high optical power levels, the field collapses in the depleted portion of the InGaAs absorber. It has been shown that this can be mitigated with an optimized charge layer between the depleted InGaAs layer and the charge-compensated InP layer in CC-MUTC photodiodes [104,105].

(a)

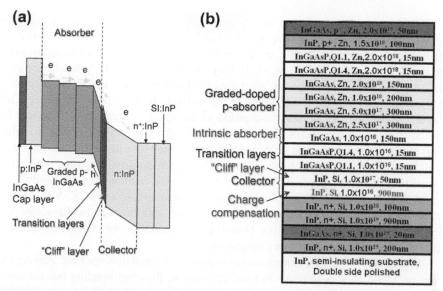

(b)

	InGaAs, p+ , Zn, 2.0x10¹⁷, 50nm	
	InP, p+ , Zn, 1.5x10¹⁸, 100nm	
	InGaAsP,Q1.1, Zn,2.0x10¹⁸, 15nm	
	InGaAsP,Q1.4, Zn,2.0x10¹⁸, 15nm	
Graded-doped p-absorber	InGaAs, Zn, 2.0x10¹⁸, 150nm	
	InGaAs, Zn, 1.0x10¹⁸, 200nm	
	InGaAs, Zn, 5.0x10¹⁷, 300nm	
	InGaAs, Zn, 2.5x10¹⁷, 300nm	
Intrinsic absorber	InGaAs, 1.0x10¹⁶, 150nm	
Transition layers	InGaAsP,Q1.4, 1.0x10¹⁶, 15nm	
"Cliff" layer	InGaAsP,Q1.1, 1.0x10¹⁶, 15nm	
Collector	InP, Si, 1.0x10¹⁷, 50nm	
Charge compensation	InP, Si, 1.0x10¹⁶, 900nm	
	InP, n+, Si, 1.0x10¹⁸, 100nm	
	InP, n+, Si, 1.0x10¹⁹, 900nm	
	InGaAs, n+, Si, 1.0x10¹⁷, 20nm	
	InP, n+, Si, 1.0x10¹⁹, 200nm	

InP, semi-insulating substrate, Double side polished

FIGURE 3.6 (a) Band structure and (b) layer structure of a CC-MUTC with cliff layer.

Figure 3.6 shows (a) the band structure and (b) a schematic cross-section of the layer structure of a CC-MUTC with a cliff layer. The step-graded *p*-type doping in the InGaAs absorber provides a quasi-electric field to aid electron transport [110]. The role of the cliff layer is to increase the electric field in the depleted InGaAs layer with a concomitant decrease in the InP drift layer. The simulated electric field profiles in Figure 3.7a assume bias voltage of 5V and 235mW optical power. Without the cliff layer the field collapses in the InGaAs but remains high with the cliff layer.

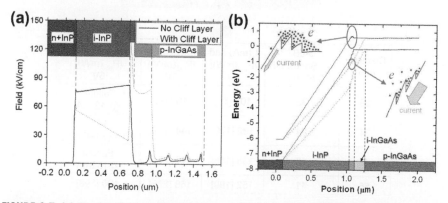

FIGURE 3.7 (a) Electric field profile and (b) band structure of CC-MUTC photodiodes with and without cliff layer. The bias voltage is 5V and the optical power level is 235mW.

This effect can also be seen in the energy diagram (Figure 3.7b). The energy band of the device without a cliff layer has flattened out. This has two deleterious effects, reduced electron velocity and, more significant, many of the electrons are trapped at the heterojunction interfaces where they recombine before they can be injected into the InP collector. However, after the cliff layer is added, electron transport is enhanced due to higher electric field in the intrinsic InGaAs layer. While the cliff layer renders the MUTC less susceptible to the space-charge effect, it should be noted that this is accompanied by a decrease in the electric field in the depleted InP layer. When the electric field in this region is very low, the frequency response will be impaired. Simulations show that performance is optimized when the charge density in the cliff layer is $\sim 5 \times 10^9 \, cm^{-2}$. Figure 3.8 summarizes the bandwidth, saturation current, and RF output power for various diameters of a CC-MUTC with cliff layer. The 34 μm-diameter device at 6V bias exhibits bandwidth-saturation current product of 3456 mA GHz. Note that in some cases the devices experience thermal failure prior to reaching saturation.

Higher RF output power necessitates improved thermal management. Among the various techniques to improve thermal dissipation, flip-chip bonding has achieved the best results. Kuo et al. reported a cascaded-two-diode photodetector flip-chip bonded onto AlN with an output power of 63 mW at 95 GHz [106]. Itakura et al. have demonstrated a maximum output power of 790 mW at 5 GHz using a flip-chip bonded 4-diode array with a monolithically integrated Wilkinson power combiner circuit on AlN [107]. Li et al. have flip-chip bonded the MUTC in Figure 3.6 to AlN [108]. Figure 3.9 shows the RF output power versus bandwidth. The inset is a schematic cross-section of the flip-chip bond configuration. A 40 μm diameter

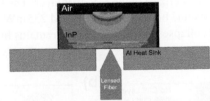

Diameter (μm)	3-dB Bandwidth (GHz)	I_s(mA) and [P_{out}(mW)] 4V	I_s(mA) and [P_{out}(mW)] 5V	I_s(mA) and [P_{out}(mW)] 6V
28	32	103 [83]	110 [126]	Failure @112 mA
34	24	132 [126]	144 [187]	145 [204@ 140 mA]
40	18	135 [155]	146 [234]	Failure @152mA
56	11	152 [190]	168 [275]	190 [398]

FIGURE 3.8 Illumination configuration and summary of bandwith, saturation current, and RF output power for various diameters of the CC-MUTC in Figure 3.6.

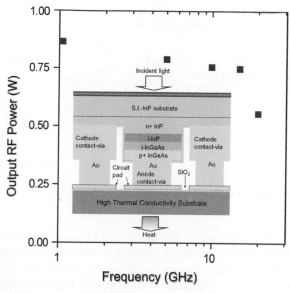

FIGURE 3.9 RF output power versus frequency for the CC-MUTC in Figure 3.6 flip-chip bonded to AIN [108].

photodiode operated at 11 V bias achieved 180 mA saturation current without saturation. The 3-dB bandwidth was >15 GHz when the photocurrent was greater than 40 mA. The bandwidth increased with increasing photocurrent, which can be attributed to carrier acceleration by the enhanced self-induced field in the gradated p-type absorber [100]. At 15 GHz the RF power from a single photodiode was 0.75 W. The maximum power dissipated by the photodiode is ∼1.5 W, which is 70% higher than the back-illuminated configuration in Figure 3.8.

The frequency response of UTC photodiodes can be further enhanced by velocity overshoot of electrons [109,110] in the depletion layer. It is well known that the velocity of electrons reaches a maximum >2.5×10^7 cm/s for electric field strengths in the range 10–15 kV/cm and decreases to <1×10^7 cm/s at high fields. It follows that in order to achieve the highest operating bandwidths, it is advantageous to bias a UTC photodiode in the velocity overshoot regime. However, at low fields the space-charge effect is most severe. Shimizu et al. successfully utilized velocity overshoot to achieve 76 mA average current and 159 GHz bandwidth (25 Ω load resistor) with a 20-μm^2 UTC photodiode [110]. In order to accommodate higher bias voltages and further suppress high current field screening, Shi et al. have developed the near-ballistic UTC (NBUTC) [111]; this photodiode utilizes a p^+ delta-doped layer in the collector of a conventional UTC to achieve electron transport near the maximum overshoot velocity under high bias voltage. The band diagram of the NBUTC is shown in Figure 3.10. Optimizing the charge in the p^+ delta-doped layer enables electrons to sustain high velocity throughout most of the collector layer.

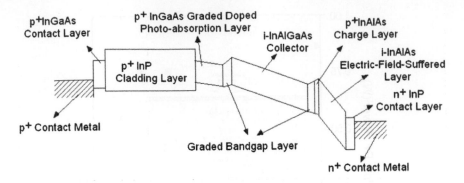

FIGURE 3.10 Band diagram of near-ballistic UTC photodiode [113].

An advantage of this approach is that the reduced transit time permits the use of thicker depleted collector layers, which results in lower capacitance and, thus, relaxes the scaling required to achieve a given bandwidth. In essence the NBUTC structure yields higher bandwidth per unit area or for a specific bandwidth higher saturation current. Wu and Shi reported a back-illuminated $64\,\mu m^2$ NBUTC having 24.6 mA photocurrent and 120 GHz bandwidth with a 25 Ω load. Even higher saturation current was obtained by flip-chip bonding the normal-incidence NBUTC photodiodes to AlN [112]. For a $144\,\mu m^2$ photodiode at 110 GHz the saturation current was 37 mA, which yields a saturation current-bandwidth product of 4070 GHz mA; the responsivity was 0.15 A/W at 1550 nm. Smaller area NBUTC photodiodes have achieved bandwidths as high as 250 GHz (50 Ω load) with 17 mA saturation current [113]. In order to obtain such high bandwidth it was necessary to reduce the thicknesses of the InGaAs absorption and InP collector regions to 150 nm and 120 nm, respectively, which reduced the responsivity to 0.08 A/W.

3.3.2 High-power WG photodiodes

The re-emergence of coherent receivers coupled with the upsurge of photonic integrated circuits has created a need for waveguide photodiodes that can operate at high current with high linearity at GHz frequencies. Also, as illustrated in the previous section, even with structures that utilize graded doping in the absorber and velocity overshoot in the collector, very high speed is accomplished at the price of reduced responsivity. Also, for high-power normal-incidence photodiodes, it is critical to achieve spatially uniform illumination, which is usually accomplished by using a graded-index lens [114] or by increasing the distance from the input fiber to the photodiode. However, this also reduces responsivity. The waveguide structure, on the other hand, decouples transit-time-limited bandwidth from responsivity, which provides higher bandwidth-responsivity product. In addition, as discussed above, waveguide photodiodes are more compatible with monolithic integration and can be designed for traveling-wave operation. However, it should be noted that waveguide photodiodes have lower alignment tolerances than normal-incidence structures although this has been alleviated to some extent by using evanescent coupling techniques [74,115] (Figure 3.1c). Thermal management such as flip-chip bonding is more difficult to implement with waveguide devices.

A primary determining factor for the power handling capability of waveguide photodiodes is spatial uniformity of absorption. A high degree of optical confinement in the absorbing layer minimizes the diode length and, thus, the active area, which is beneficial for achieving a low RC time constant. However, this also results in a highly non-uniform exponential absorption profile and, consequently, poor saturation characteristics. Klamkin et al. [116] addressed this issue by fabricating UTC waveguide photodiodes with reduced optical confinement; the saturation current was 80.5 mA at a frequency of 1 GHz. An alternative approach to a more uniform absorption profile is evanescent coupling. Using evanescent coupling from a dilute multimode waveguide, Demiguel et al. [73] reported a responsivity of 1.02 A/W, a bandwidth of 48 GHz, and a saturation current of 11 mA at 40 GHz. Another method to obtain uniform generation of photocarriers is a distributed-absorption waveguide [117]. One method to accomplish this is to incorporate a lateral-taper mode size transformer to vary the optical mode dimensions and the confinement along the length of the device as illustrated in Figure 3.11a. At the device input the mode diameter is large, which facilitates low input fiber coupling loss, while also limiting absorption, which diminishes front-end saturation. The simulated absorption profiles with and without variable absorption are compared in Figure 3.11b. It is clear that tapering the photodiode spreads the absorption more uniformly along the length. Klamkin et al. [71] have combined a UTC photodiode with a variable confinement optical waveguide structure that employs slab coupled waveguide technology [118]. Two structures were reported. A photodiode designed for high-speed and moderate saturation current demonstrated bandwidth of 12.6 GHz, saturation current of 34 mA at 10 GHz, and responsivity of 0.7 A/W. A lower bandwidth design exhibited saturation current >100 mA at 1 GHz and responsivity of 0.8 A/W. For very high-speed performance an

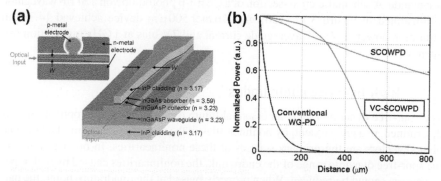

FIGURE 3.11 (a) UTC waveguide photodiode with variable coupling (b) simulation of absorption profiles of a conventional waveguide photodiode and one with variable coupling.

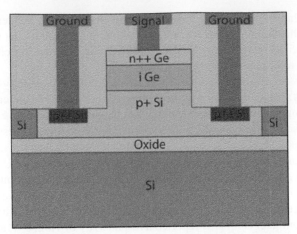

FIGURE 3.12 Schematic cross-section of Ge photodiode integrated with Si rib waveguide [120].

evanescently-coupled NBUTC photodiode has been developed [112]. The responsivity was 1.14 A/W and at 40 GHz the saturation current was 30 mA with 12 dBm RF output power.

As with normal-incidence photodiodes, thermal issues can significantly limit the saturation current and RF output power. In fact, thermal degradation is typically more acute for waveguide photodiodes. For InP-based photodiodes, an InGaAsP layer is usually located below the InGaAs absorber to enhance the coupling efficiency in addition to reducing carrier trapping at the absorber-collector interface. The thermal conductivity of these layers is very low (~0.05 W/cm K [119]). From the perspective of more efficient heat dissipation, using a Si substrate is very attractive because the thermal conductivity of Si is more than twice that of InP [127]. In order to achieve photoresponse at telecomm wavelengths, a Ge photodiode can be integrated with a Si waveguide. A schematic cross-section of a Ge n-i-p photodiode on a Si rib waveguide is illustrated in Figure 3.12 [120]. A 7.4 μm × 500 μm device achieved 14.2 dBm saturation power at 60 mA average dc current and 7 V bias at 1 GHz modulation frequency [120].

3.3.3 High-linearity photodiodes

The spur-free dynamic range, an essential figure of merit for analog optical links, is determined, to a great extent, by nonlinearities in the system. Optical modulators and photodetectors are the primary sources of these nonlinearities. In order to improve the spur-free dynamic range of the entire link, the nonlinearities caused by both types of devices should be reduced. When properly biased at the quadrature point, the harmonics and intermodulation products produced by the modulators can be effectively minimized, though not completely eliminated [121]. However, unlike modulators, there is not any explicit expression for the transfer function of photodiodes; their deviation from a linear response is a complicated function of reverse bias, photocurrent, and temperature [122]. Frequently, the linearity of the photodetectors is the limiting factor for the spur-free dynamic range in high-performance analog optical links, especially when the optical power is high [123]. When considering nonlinearities in photodiodes, third-order intermodulation distortions (IMD3) are particularly significant, since their frequencies can be close to the fundamental modulation frequencies. The key figure of merit to characterize IMD3 is the third-order output intercept point (OIP3); it is defined as the extrapolated intercept point of the power of fundamental frequency and IMD3, assuming that the fundamental power has a perfect slope of 1 and the power of IMD3 has a perfect slope of 3, as shown in Figure 3.2. OIP3 can be calculated from the measured fundamental power and the power of IMD3 using the expression [8]:

$$OIP3 = P_f + \frac{1}{2}(P_f - P_{IMD3}), \tag{3.8}$$

where P_f is the power of the fundamental frequency and P_{IMD3} is the power of the IMD3. OIP3 is typically measured with a three-tone measurement setup and procedure similar to those described in Ref. [124].

At low frequencies, photodiodes with very good linearity have been demonstrated. A dual depletion region (DDR) photodiode has been reported to have a third-order harmonic output intercept point (HOIP3) of 54 dBm at 829 MHz [115] and a partially depleted absorber (PDA) photodiode has achieved an HOIP3 of 51 dBm at 3 GHz [125], which is equivalent to OIP3 values of 49.2 dBm and 46.2 dBm, respectively, as estimated from the cubic dependence of the third-order nonlinearity terms [129]. For waveguide UTC photodiode 40.9 dBm OIP3 at 1 GHz has been reported by Klamkin et al. [116]. However, the OIP3 of previously reported photodiodes exhibited significant roll-off with frequency and thus high OIP3 has been difficult to obtain at high frequencies [126]. OIP3 values of 35 dBm and 36 dBm at 20 GHz were reported for a uni-traveling-carrier (UTC) photodiode [127] and a charge-compensated modified uni-traveling-carrier (CC-MUTC) photodiode [128], respectively.

Various measurement-based models have been developed to explain photodiode nonlinearities [124,129–132]. Using bias modulation the voltage-dependent responsivity has been identified as the primary factor that determines nonlinearities at low frequencies (<3 GHz). As the frequency increases, the nonlinear capacitance becomes significant and dominates [136]. The origin of the voltage-dependent responsivity appears to be the Franz-Keldysh effect and, if any of the absorber is depleted, impact ionization [133].

In order to reduce the voltage dependence of the capacitance in MUTC photodiodes and increase OIP3 at high frequencies, Pan et al. [131] have incorporated a highly doped p-type absorber (referred to as HD-MUTC). At optimized photocurrent and bias conditions, the OIP3 of the HD-MUTC is 55 dBm at low frequencies (<1 GHz) and remains as high as 47.5 dBm at 20 GHz. Figure 3.13a compares the frequency response of a conventional MUTC with the HD-MUTC. At low frequency

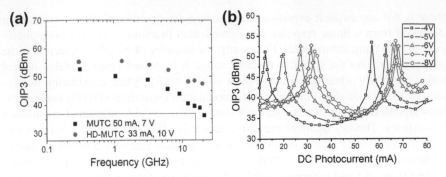

FIGURE 3.13 (a) Measured OIP3 of MUTC (■) and HD-MUTC (●) photodiodes. (b) OIP3 of
HD-MUTC versus photocurrent for various bias voltages, measurement frequency = 300 MHz,
and 40 μm diode diameter [131].

the difference is small but at 20 GHz the HD-MUTC provides 10 dB higher OIP3.
As illustrated in Figure 3.13b the OIP3 of the HD-MTC photodiodes exhibits strong
photocurrent dependence at low frequencies. The peaks shift to higher photocurrent
values for higher voltage and lower temperature. Fu et al. have developed a model
that explains the variation of OIP3 with photocurrent, bias voltage, and temperature [134]. The primary physical effects are impact ionization and the Franz-Keldysh
effect in the InP collector and the depleted InGaAs absorber. Also, the space-charge
effect is important in the depleted portion of the absorber. The DC photocurrent
through the space-charge effect and the DC bias both affect the device temperature

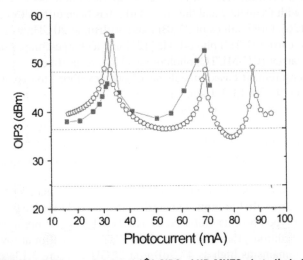

FIGURE 3.14 Measured (■) and simulated (○) OIP3 of HD-MUTC photodiode [134].

and the electric field in the depleted InGaAs absorber. The AC photocurrent and voltage swing across the photodiode alter the electric field profile. The temperature and the internal electric field strongly influence impact ionization and the Franz-Keldysh effect, which, in turn, give rise to nonlinear responsivity. Changes in the responsivity feed back to the photocurrent, electric field, and AC voltage and photocurrent. For MUTC photodiodes it was found that the primary contribution to the voltage dependence of the responsivity is the Franz-Keldysh effect in the depleted InGaAs absorber with a minor contribution from impact ionization in that layer. Figure 3.14 shows the measured and simulated OIP3 at 3 GHz for an HD MUTC photodiode at 9 V bias.

3.3.4 **High-power balanced detectors**

Balanced receivers, which consist of anti-parallel photodiodes, are widely used in coherent receivers owing to their capacity to suppress laser relative intensity noise (RIN) and the amplified spontaneous emission noise (ASE) from erbium-doped fiber amplifiers. This enables analog links to achieve shot noise-limited performance at high optical powers, which significantly improves the link gain, spurious-free dynamic range, and noise figure. To realize these advantages, it is important to have balanced photodetectors with high saturation photocurrents. Since optimal performance is achieved when both receiver channels are perfectly matched electrically and optically, it is beneficial to monolithically integrate the receiver components, particularly the two photodiodes.

   Islam et al. fabricated balanced receivers using velocity-matched distributed photodetectors (VMDP) [79]. This type of balanced receiver with metal-semiconductor-metal VMDP detectors had bandwidth of 13.8 GHz and 26 mA average current under small-signal modulation of 1.55-μm light [135]. A similar structure comprised of p-i-n VMDPs achieved 31 mA per channel at 8 GHz with 43 dB common-mode rejection [136]. Using balanced UTC photodiodes with a tunable multi-mode interference coupler Klamkin et al. obtained >40 mA saturation current per diode at 1 GHz [137]. The OIP3 was estimated to be 49 dBm at 20 mA photocurrent and −8 V bias. The common-mode rejection was 20 dB at 1 GHz and increased to 40 dB at higher frequencies. Using 34 μm-diameter MUTC photodiodes with a cliff layer [110], Li et al. [138] reported a balanced receiver with 0.82 A/W responsivity, 11 GHz bandwidth, saturation current of 136 mA per diode, and 35 dB common-mode rejection up to 15 GHz. Very high output power of 1 W at 2 GHz has been attained with a monolithic balanced pair of UTC photodiodes with on-chip capacitors; the output of the balanced receiver was 6 dB higher power than discrete diodes with the same junction area. The responsivity was 0.65 A/W, the OIP3 was 48 dBm for a single diode of the pair, and the common-mode rejection was >40 dB [139].

3.3.5 **Photodetector arrays**

Often very high RF output power is linked to lower bandwidth. To operate at high speed the photodiode has to have low capacitance and short carrier transit times.

This usually leads to a small active area in conjunction with thin drift layer thickness. However, for a given optical input power, the reduction in photodiode area results in higher photocurrent densities and earlier onset of saturation. This can be mitigated, to an extent, by increasing the bias voltage but the concomitant increase in power dissipation is accompanied by a thermal penalty. One approach to circumventing the trade-off between high speed and large saturation current is to distribute the optical signal to several photodiodes and combine their photocurrents by means of a transmission line. In this configuration the optical signal is split by a power divider and fed into several discrete photodiodes, which are connected by an output transmission line [89]. The photodiode arrays can be combined as lumped elements or embedded in a transmission line to form a traveling-wave detector. Both normal-incidence and waveguide arrays have been reported.

An early approach to combining the output of a linear array of photodiodes in phase utilized the VMDP configuration [144]. However, the initial serially fed structures achieved limited saturation current because most of the input power was absorbed in the first diode in the array, which became the point of thermal runaway. This was addressed by using lateral injection into each element using a 1×4 MMI coupler [90]. The combined responsivity was 0.2 A/W. The saturation current was 52 mA and the bandwidth was 9 GHz. Beling et al. [140] have also used lateral input except with normal-incidence photodiodes. This traveling-wave detector consisted of two monolithically integrated back-illuminated MUTC photodiodes with an integrated termination resistor to reduce microwave reflections at the input of the coplanar waveguide transmission line. Light was input through a two-channel lensed fiber. The advantages of normal-incidence include negligible polarization dependence, high saturation current, and tolerant input coupling. At 17 GHz saturation photocurrent at 1-dB RF power compression was 114 mA and the maximum output power delivered to a 50 Ω load was 13 dBm. Compared to a single 40 μm-diameter photodiode this represents an improvement of 60 mA and 5 dB, respectively. Kuo et al. have used a two-channel fiber input with a pair of NBUTC connected in series [112]. This configuration achieved maximum current and RF output power of 75 mA and 18 dBm (50 Ω load), respectively, at 95 GHz, which yields a high current-bandwidth product of 7500 mA-GHz.

The outputs of photodiode arrays can also be efficiently collected using microwave power combiners. Itakura et al. have incorporated a four-element PDA photodiode array into a hybrid integrated module with a beam splitter that divides the signal from a single-mode fiber into four beams with identical powers and a two-stage Wilkinson RF power combiner [141]. At 5 GHz the maximum RF output power was 1.0 W when the total optical input power was 1.1 W [107]. Ideally, the output of the array should be four times that of a single photodiode. In this work, the relative increase was 2.6, which was attributed to phase shifts in the RF combiner and variations in the temperatures of the photodiodes. An array also yields 10 log 10(N) dB improvement in OPI3, where N is the number of diodes in the array. OIP3 for the four-element array was 32.5 dBm. Fu et al. have monolithically integrated four HD-MUTC photodiodes with a Wilkinson power combiner. Optical input to the

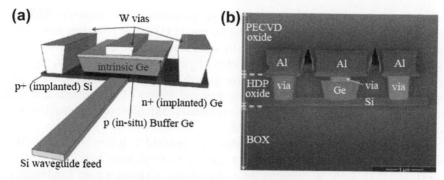

FIGURE 3.15 (a) Schematic of Ge waveguide n-i-p PD. (b) SEM cross-section of selective area epitaxially grown Ge photodiode with electrical contacts [167].

back-illuminated photodiodes was done with an array of four lensed fibers. This receiver circuit achieved 400 mA operating current and 20.5 dBm RF output power at 15 GHz. The OIP3 of the discrete photodiode was 40 dBm while the OIP3 of the photodiode array was ~46 dBm at 15 GHz.

3.4 LONG-WAVELENGTH PHOTODIODES ON SILICON

Over the last two decades InP with its lattice-matched compounds has proved itself to be the most appropriate material system for the fabrication of high-performance photodiodes in the C and L bands. Owing to its good material quality, high absorption efficiency, and high carrier drift velocities, the InGaAs/InP photodiode has become a standard solution for today's high-speed applications.

However, concomitant with the rapid advances in Si photonics there has been considerable research toward the integration of high-speed photodiodes on silicon-on-insulator (SOI)/Si substrates. Although Si is transparent at wavelengths >1.1 μm numerous approaches have been reported to enable efficient light detection at 1.55 μm wavelength, including ion-implanted all-silicon [142], InGaAs/GaAs growth on Si [143,144], polycrystalline Ge films [145,146], Si-Ge hetero-epitaxy [147–149], and III–V on Si bonding [150].

3.4.1 High-speed Ge photodiodes

Germanium is a viable absorber material for high-performance photodiodes as it provides sufficient absorption up to ~1550 nm wavelengths. Depending on growth conditions and strain, absorption coefficients between $1000 \, cm^{-1}$ and $4000 \, cm^{-1}$ at 1.55 μm wavelength have been reported [151,152]. Ge can be grown on Si; however, the main difficulty is the ~4% lattice mismatch between Ge and Si which can lead to high defect densities in the Ge [159]. These defects may cause high dark current, which would compromise the sensitivity of the photodiode. The amount of dark current that can be tolerated depends on the bit rate [153] and application; however, it has been suggested that <1 μA is typically sufficient for high-speed receivers [154]. Growth techniques including the deposition of graded SiGe buffer layers, high/low temperature growth [148], area-selective growth [155,156], and cyclic annealing have been shown to effectively improve material quality [157]. Several low dark current Ge on Si photodiodes have been presented [158]; however, the problem with high dark current persists to some extent.

Recently, surface-normal photodiodes that reached a bandwidth of 36 GHz, responsivity of 0.47 A/W, and low dark current <100 nA have been demonstrated [158]. In [159], a 10 μm-diameter n-i-p photodiode achieved a high bandwidth of 49 GHz. This structure was grown by a two-step technique using MBE. First, a silicon buffer followed by a very thin Ge virtual substrate was grown at low temperature for lattice mismatch accommodation. Then, a highly boron-doped Ge layer followed by a sharp doping transition to the 330 nm intrinsic Ge layer was grown at 300 °C. The n-contact layers were formed by a 200 nm Ge layer and a thin Si cap layer, both highly Sb-doped. The bandwidth and responsivity were 49 GHz and 39 GHz and 0.05 A/W and 0.04 A/W at 2 V and zero-bias, respectively. Since the absorption coefficient in Ge is ~50% lower at 1.55 μm compared to InGaAs, surface-normal Ge photodiodes typically exhibit lower bandwidth-efficiency product when compared to their InGaAs/InP counterparts.

Waveguide photodiodes can achieve higher bandwidth-efficiency products and to date several high-performance Ge waveguide photodiodes have been demonstrated [160,161]. By their nature they have become important devices in SOI/Si photonic integrated circuits. Recently, research has focused on CMOS-compatible processing techniques for Ge waveguide photodiodes [154,162]. The goal is to enable large-scale photonic-electronic integration using the available CMOS infrastructure. For CMOS integration one critical issue is the thermal budget of the Ge epitaxy and postgrowth annealing [163]. On one hand, it is useful to anneal the Ge at very high temperatures (>900 °C) to minimize the density of dislocations. On the other hand, this anneal can degrade the CMOS-device performance, if the Ge photodiodes are integrated after the formation of the CMOS electronics. Another requirement arises from the fact that both detector and the electronic circuit should operate on a single power supply <1.5 V [164]. Ideally, zero-bias operation is desired, not only to minimize dark current but also to reduce power consumption.

Utilizing a lateral seeded crystallization method a CMOS-compatible evanescently-coupled Ge MSM waveguide photodetector was reported in [165]. For lateral seeded crystallization a thin SiON layer was deposited on top of the Si waveguide and a small seed window was etched down to Si layer. After growth of a thin SiGe buffer and a 150 nm-thick Ge layer, the Ge waveguide was patterned and then melted during rapid thermal annealing (RTA). The Si touching the Ge at the window served as a seed for lateral crystallization growth of Ge. The crystallization front starting from the seed window propagates along the waveguide leaving a high-quality single-crystalline SiGe strip with less than 10% total concentration of Si. Outside of the crystallization window, the SiON blocks intermixing of Ge with Si and provides electrical isolation between the Ge and Si. The MSM detector was formed by adding 150 nm-wide interdigitated metal fingers on top of the Ge waveguide with separation distance of 300 nm; this creates a series of MSM diodes along the length of the Ge waveguide. Owing to its intrinsic small capacitance and small area ($0.7 \times 20\,\mu m^2$), the reported photodetector reached a 3 dB bandwidth of 40 GHz [166]. Internal responsivity at 1.3 μm and dark current, both measured at 1 V, were 0.42 A/W and 90 μA, respectively.

An ultra-compact n-i-p Ge waveguide photodiode with an active area of only $1.3 \times 4\,\mu m^2$ was recently demonstrated using selective growth of Ge in an oxide window on top of a Si pedestal [167]. As dislocations that form are much closer to the window edge they can terminate on this surface, which can effectively reduce dislocation density and dark current. The selective Ge epitaxy consisted of a low temperature buffer layer grown at 400 °C and in situ doped with boron followed by growth at 600 °C to fill and overgrow the oxide trench. The device was then planarized by chemical-mechanical polishing to a final Ge thickness of 0.6 μm. Phosphorous was implanted to form the *n*-type contact layer followed by the deposition of a capping oxide. Implants were activated by rapid thermal annealing at 630 °C for 30 s. Prior to growth the Si pedestal was implanted with boron and activated to form the *n*-contact layer. Figure 3.15 shows (a) the device schematic and (b) a scanning electron micrograph cross-section of a completed photodiode. Due to the photodiode's low intrinsic capacitance ($\sim$1 fF) high bandwidths of 37 GHz and 45 GHz at zero-bias and 0.8 V, respectively, were measured. The dark current and responsivity were 3 nA at 1 V and 0.6 A/W at 1.55 μm, respectively.

Compared to evanescent coupling butt coupling can lead to higher efficiencies in short photodiodes. In this configuration the absorber is butt-coupled to the input waveguide, making the detector an extended part of the waveguide and thus providing a good overlap between the guided modes in the SOI waveguide and the Ge layer. Butt-coupled photodiodes typically require recess etching into the waveguide layer followed by selective growth of the absorber material. Using this approach Vivien et al. presented a high responsivity of 1 A/W at 1.55 μm wavelength for a 15 μm-long device [168]. At 4 V bias voltage they measured dark current and bandwidth of 4 μA and 42 GHz, respectively.

Feng et al. demonstrated a lateral Ge photodetector butt-coupled with a large cross-section SOI waveguide [169]. The tapered SOI waveguide had a cross-section of 3 μm $\times$ 3 μm at its input and fiber-chip coupling loss of <1.2 dB when measured with a lensed fiber. To reduce the carrier transit times, a horizontal p-i-n configuration with an intrinsic Ge width of 650 nm was designed. The demonstrated photodetector had an active area of $0.8 \times 10\,\mu m^2$ and achieved a bandwidth of 32 GHz, responsivity of 0.8 A/W at 1.55 μm wavelength, and dark current of 1.3 μA at 1 V.

Recently, a butt-coupled lateral p-i-n photodiode with >110 GHz bandwidth was reported [170]. A $10 \times 10\,\mu m^2$ silicon recess was etched at the end of the 500 nm-wide Si waveguide and Ge was selectively grown by RP-CVD. Boron and phosphorus were implanted to form a horizontal p-i-n junction with a nominal intrinsic Ge width of 500 nm. The responsivity was 0.8 A/W at 1.55 μm and open eye diagrams at 40 Gb/s were obtained under zero-bias operation.

3.4.2 Heterogeneously integrated III–V photodiodes on Si

Wafer bonding of III–V material on Si has been shown to be a viable technique to integrate dissimilar materials without compromising their properties [150,171–175]. While this approach is particularly interesting for silicon transmitters [176], it may also lead to low dark current photodiodes with high efficiencies beyond 1.55 μm. Furthermore, heterogeneous integration has the potential to fully exploit bandgap engineering available in III–V materials to design more complex detector heterostructures.

Park et al. reported an evanescently-coupled waveguide photodetector utilizing AlGaInAs quantum wells bonded to a SOI waveguide [177]. The III–V epitaxial structure was grown on an InP substrate with the absorbing region consisting of both compressively and tensile strained quantum wells and a total thickness of the undoped quantum well region of ~0.15 μm. This III–V structure was then transferred to the patterned silicon wafer through low temperature oxygen plasma-assisted wafer bonding with 300 °C annealing temperature under vacuum [178]. After removal of the InP substrate, 12 μm-wide mesas were formed by dry-etching the p-type layers followed by wet-etching of the quantum well layers to the n-type layers. Metal contacts were then deposited onto the exposed n-type InP layer and on the center of the mesas of the absorber region. After proton implantation on the two sides of the p-type mesa, probe pads were deposited on SiNx (Figure 3.16). The photodetector had fiber-coupled responsivity of 0.31 A/W and 0.23 A/W at 1.55 μm and 1.65 μm wavelengths, respectively, an internal quantum efficiency of 90% at 1.55 μm, and a dark current <100 nA at 2 V. Using the same epitaxial layer structure similar photodiodes have been integrated with hybrid optical amplifiers [179]. The pre-amplified receiver exhibited responsivity of 5.7 A/W and receiver sensitivity of −17.5 dBm at 2.5 Gb/s.

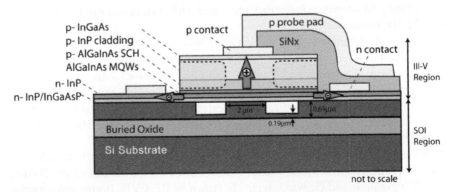

FIGURE 3.16 Device cross-section of a hybrid silicon evanescent photodetector [175].

In these devices the bandwidth was primarily limited by the quantum well valence band offset, which causes hole trapping. An improved bandwidth of 16 GHz has been reported for an InGaAs/InP p-i-n photodiode [172,180]. The $12\,\mu m \times 120\,\mu m$ photodiode with $0.5\,\mu m$ intrinsic layer thickness was part of a triplexer with different epitaxial layers being bonded on a single Si chip. Recently, InP-based modified uni-traveling-carrier photodiodes (MUTC photodiodes) heterogeneously integrated onto SOI/Si waveguides were fabricated and characterized [181]. The MUTC photodiode was previously demonstrated to achieve high saturation current and high linearity [131], and a schematic of the layer stack is shown in Figure 3.17. A $700\text{-}\mu m^2$ area waveguide photodiode had a dark current of 100 nA at 3 V, 0.85 A/W internal responsivity, and bandwidth of 15 GHz. Saturation currents reached 24 mA and 42 mA for single photodiodes and 2-element photodiode arrays, respectively [182].

Using small InP-based membrane p-i-n photodetectors on SOI a bandwidth of 33 GHz was demonstrated by Binetti et al. [183]. In this approach the photodiode structure consisted of an InP membrane input waveguide to couple the light out of a Si photonic wire waveguide into the p-i-n junction, similar to a vertical directional coupler [184]. III–V on Si integration was achieved by direct molecular bonding of InP dies using a 300 nm-thick SiO_2 interface layer. The photodiode with 700 nm absorber thickness and $50\,\mu m^2$ mesa footprint had a junction capacitance below 10 fF and

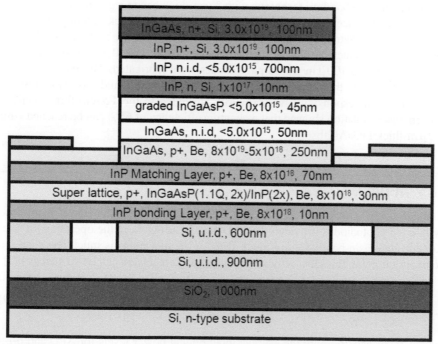

FIGURE 3.17 Layer stack of MUTC PD on SOI. Doping concentrations in cm^{-3} [182].

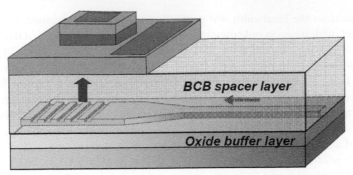

FIGURE 3.18 Vertical coupling scheme for III–V photodetector bonded to SOI waveguide [186].

achieved dark current below 2 nA at 4 V. Measurements with a 20%-efficiency grating coupler [185] for fiber input coupling indicated an internal responsivity of 0.45 A/W.

In [186], Roelkens et al. proposed a grating coupler to diffract light from an SOI waveguide into a photodetector. A schematic cross-section is shown in Figure 3.18. The 10 μm-long grating with a grating period of 610 nm and a duty cycle of 50% was designed to diffract the light from the waveguide toward the InP/InGaAsP photodiode. The InP die of 1 cm² was bonded onto the SOI waveguide structure using an adhesive bonding technique with 3 μm-thick benzocyclobutene (BCB) as a bonding layer. Since this vertical coupling scheme allows thick bonding layers, the requirements on flatness, roughness, and cleanliness of the two surfaces that have to be mated were somewhat relaxed. After curing the BCB at 250 °C, the InP substrate was removed and 100 μm² mesa photodiodes were fabricated. The measured dark current at 1 V and the responsivity at 1.55 μm were 0.3 nA and 0.02 A/W, respectively. The low responsivity was primarily attributed to the 120 nm-thin InGaAsP absorber as simulation showed that a quantum efficiency of 65% can be reached with a 2 μm-thick InGaAs absorber.

3.5 APDs

Optical receivers that utilize APDs can operate with lower input power than those with p-i-n photodiodes provided the APD gain is sufficient at the operating bit rate, i.e. adequate gain-bandwidth product. If the p-i-n shot noise is much less than the total circuit noise, the sensitivity advantage of APD receivers can be written as [12]

$$\frac{\overline{P}_{APD}}{\overline{P}_{PIN}} = \frac{1}{M}\left[1 + \frac{qI_1QM^2F(M)B}{\langle i^2\rangle_c^{1/2}}\right], \quad (3.9)$$

where $\overline{P}_{APD}$ and $\overline{P}_{PIN}$ are the average input optical power levels for APD and p-i-n receivers, respectively, that are required for a given bit error rate, B. The Personick integral, I_1, is a normalized noise-bandwidth integral [202], Q is the signal-to-noise ratio ($Q = 6$ for bit error rate $= 10^{-9}$), and $\langle i^2 \rangle_c^{1/2}$ is the total circuit noise current. If the APD noise is less than the circuit noise, an APD receiver requires $\sim 1/M$ less input optical power than the same receiver with a p-i-n. Typically, up to 10 Gb/s, APD receivers achieve ~ 10 dB higher sensitivity [14]. In essentially all cases the circuit noise is much greater than the shot noise of a p-i-n photodiode. As the bias voltage on an APD is increased to provide higher gain, the noise that arises from impact ionization increases but there is minimal increase in the overall receiver noise until the APD noise is comparable to the circuit noise. At higher gain values, the receiver sensitivity is APD-noise limited.

3.5.1 SACM APDs

First generation optical fiber communication systems operated in the wavelength range 800–900 nm. The optical receivers utilized Si p-i-n photodiodes or Si APDs [187]. In subsequent generations the operating wavelengths migrated to 1300 nm and 1550 nm, which resulted in the development of "long-wavelength" photodetectors. The most straightforward approach would have been to develop $In_{0.53}Ga_{0.47}As$ homojunction APDs [188,189] but this was prevented by excessive dark current caused by tunneling at the electric fields required for impact ionization [204]. To circumvent this limitation, separate absorption and multiplication (SAM) APD structures similar to Si reach-through APDs [190,191] were investigated [192]. In these APDs the p-n junction and thus the high-field multiplication region is located in a wide-bandgap semiconductor such as InP where tunneling is insignificant and absorption occurs in an adjacent InGaAs layer. By properly controlling the charge density in the multiplication layer, it is possible to maintain a high enough electric field to achieve good avalanche gain while keeping the field low enough to minimize tunneling and impact ionization in the InGaAs absorber. However, the frequency response of the InP/InGaAs SAM APDs, as originally implemented, was very poor owing to accumulation of photogenerated holes at the absorption/multiplication heterojunction interface [193]. Several methods to eliminate the slow release of trapped holes were reported; however, the approach that has been most widely adopted utilizes a transition region consisting of one or more intermediate-bandgap $In_xGa_{1-x}As_{1-y}P_y$ layers [209,194,195]. A second modification to the original SAM APD structure has been the inclusion of a high-low doping profile in the multiplication region [196–198]. In this structure the wide-bandgap multiplication region consists of a lightly doped (usually unintentionally doped) layer where the field is high and an adjacent, doped charge layer or field control region. This type of APD, which is frequently referred to as the SACM structure with the "C" representing the charge layer, decouples the thickness of the multiplication region from the charge density constraint in the SAM APD. At present, most commercial SACM APDs are planar, as opposed to mesa structures, with lateral guard rings to suppress edge breakdown. Figure 3.19 shows a schematic cross-section of an InP/InGaAsP/InGaAs SACM APD with a double-diffused floating guard ring [199]. The adjacent graph shows the electric field profiles

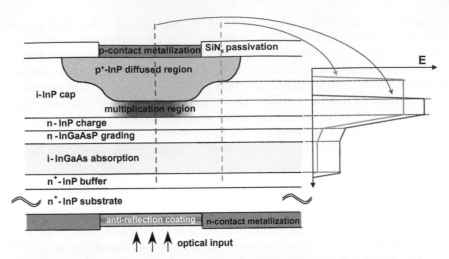

FIGURE 3.19 Schematic cross-section of InP/InGaAsP/InGaAs SACM APD with double-diffused floating guard ring configuration. The adjacent graph shows electric field profiles normal to the surface through the active and the guard regions [199].

normal to the surface through the active and the guard ring regions and illustrates how the charge layer is used to tailor the relative fields in the multiplication and absorption layers.

3.5.2 Low-noise APDs

There are three documented methods to achieve low excess noise in an avalanche photodiode. The best-known and most straightforward approach is to select a material that has k, the ratio of ionization coefficients, $\ll 1$. Recently, HgCdTe [200–202] and InAs [203,204] APDs have demonstrated excess-noise-free operation. However, these materials have bandgaps that correspond to mid-wavelength infrared and require cryogenic (77 K) cooling. The low-noise characteristics of Si are well documented [205–208]; however, the bandgap of Si renders it transparent at telecommunications wavelengths. One approach to utilizing the excellent multiplication characteristics of Si for telecommunications is to combine narrow bandgap absorber materials such as InGaAs or Ge with a Si multiplication region. Si/InGaAs APDs have been fabricated by wafer bonding [209–211]. These APDs have achieved low dark current (4×10^{-5} A/cm^2 @ $M = 50$) [210], excess noise levels comparable to those of Si homojunction devices ($k \sim 0.02$) [209,40], and bandwidths up to 4.8 GHz [40]. However, these APDs have not displaced InP/InGaAs SACM APDs owing to materials issues related to the bonded interface between InGaAs and Si. Recently, more promising results have been reported for Ge on Si SACM APDs [212]. Epitaxial layers of Ge were grown on (100) Si substrates by chemical vapor deposition [42]. A schematic cross-section of the device structure is shown in Figure 3.20. The wafer

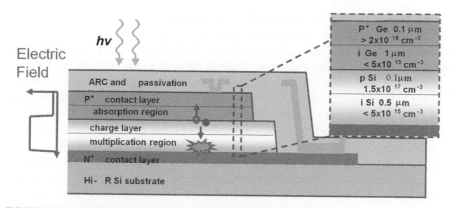

FIGURE 3.20 Schematic cross-section of a Ge/Si SACM APD [212].

has a $1\,\mu$m-thick unintentionally doped Ge absorption layer and a $0.5\,\mu$m-thick Si multiplication layer. The multiplication layer is also unintentionally doped to achieve a relatively constant high electric field. The charge layer ($0.1\,\mu$m) is p-doped with a concentration of $1.52 \times 10^{-17}\,cm^{-3}$. Figure 3.20 also shows the electric field profile. The current-voltage characteristics are typical of SACM APDs with a clear punch-through voltage, corresponding to the voltage at which the depletion region penetrates into the germanium of approximately -22V and breakdown at -25V. The responsivity was 5.88 A/W at 1310 nm. The breakdown voltage thermal coefficient, defined as $\delta = (\Delta V_{bd}/V_{bd})/\Delta T$, was 0.05%/°C over a temperature range from 200 K to 380 K. This value is approximately one third to one half of that obtained from a typical InGaAs/InP APD with a similar multiplication layer thickness [213,214]. Of particular note, these APDs achieved a gain-bandwidth product of 340 GHz, which is two to three times higher than InP/InGaAs APDs. The effective noise factor k was only 0.09. Optical receivers built with these APDs demonstrated a sensitivity of -28 dBm at 10 Gbps and a bit error rate (BER) of 10^{-12} (-28.5 dBm at BER of 10^{-11}).

It has been well established with device simulations [19–26] and experimental work [19–39] that reducing the thickness of the multiplication region can significantly reduce the excess noise of an APD even as the ratio of the ionization coefficients approaches unity. This apparent contradiction with early APD theories [15–18] is due to the inherent non-local nature of impact ionization. Unless they have been "pre-heated" in a high-field region characterized by high threshold energy, carriers injected into the high-field region do not have sufficient energy to initiate impact ionization and require a certain distance to attain the requisite energy [215]. This also applies to carriers immediately after ionization because their final states are typically near the band edge. The distance in which essentially no impact ionization occurs is frequently referred to as the "dead space." If the multiplication region is thick, the dead space can be neglected and the local-field model provides an accurate description of APD characteristics. However, for thin multiplication layers the nonlocal nature of impact ionization has a profound impact. Further, materials with lower bulk excess noise characteristics benefit the most from the dead-space effect. For example, the bulk value of k for InP is ~0.45. However, for a multiplication thickness of 0.25–0.3 μm the excess noise is reduced to the equivalent of $k=0.3$ [27,29]. The bulk value for $In_{0.52}Al_{0.48}As$ (simplified to InAlAs in the following) is ~0.3 and for multiplication layer of 0.2 μm the excess noise is characterized by $k=0.18$ [227,216]. Note that this does not mean that the actual ratio of the ionization coefficients has been reduced, only that the excess noise follows a curve that would correspond to a certain k value plot using the local-field model. In fact, Ong et al. have demonstrated that the ratio of hole to electron ionization events increases toward unity as the multiplication region decreases [217]. We quote an equivalent k factor solely for reference because the k value has become a widely used figure of merit for excess noise. While decreasing the thickness of the multiplication layer is a viable approach to reducing the excess noise of an APD, band-to-band tunneling imposes a lower limit on the thickness. For InAlAs the minimum thickness appears to be in the range from 100 to 150 nm.

Another approach to achieving low noise utilizes new materials and impact ionization engineering (I^2E) with appropriately designed heterostructures [45,40–42,218,44,46,47] that can be employed to achieve even lower noise. The structures that have achieved the lowest excess noise, to date, utilize multiplication regions in which electrons are injected from a wide-bandgap semiconductor into adjacent low-bandgap material. InGaAlAs/InP implementations that operate at the telecommunications wavelengths have been reported. Duan et al. reported an InAlAs/InGaAlAs/InGaAs I^2E SACM APD that exhibited excess noise characterized by k factor of 0.12 and gain-bandwidth product of 160 GHz at 1550 nm. Figure 3.21a and b shows the multiplication region layer structure and band structure of an optimized I^2E APD that achieved even lower noise. Between the InAlGaAs/InAlAs I^2E multiplication region and the InGaAs absorber are InAlGaAs transition regions to prevent carrier trapping at the heterojunction interfaces. Figure 3.22a shows the dark current and photocurrent for a 50 μm-diameter device. The dark current is relatively high near breakdown, which is probably indicative of the onset of tunneling; however, it is still acceptable for high-bit-rate applications owing to the high-frequency noise of transimpedance amplifiers. The measured (■) and Monte Carlo simulated (□) excess noise is plotted in Figure 3.22b. There is good agreement at low gain. At higher gain impact ionization of holes becomes significant. At low gain, the excess noise is below the $k=0$ line. That does not mean that k is negative, which is physically unreasonable, but it does reflect the fact that the local-field model, which is used to generate the constant k curves, is inappropriate for this type of APD.

Impact ionization engineering (I^2E) has been shown to be a viable approach to reduce excess noise in III–V compound APDs. An enhancement of this approach is to cascade multiple I^2E multiplication cells all operated at relatively low gain [219–221]. Proper design of the connection between two adjacent multiplication cells can significantly enhance impact ionization of the carrier type with higher ionization rate and suppress impact ionization of the carrier with lower ionization rate.

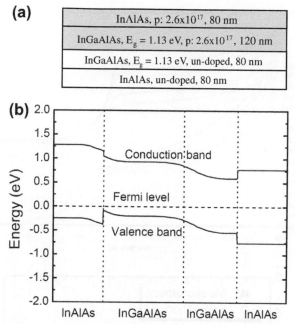

(a)

InAlAs, p: 2.6x10^{17}, 80 nm
InGaAlAs, E$_g$ = 1.13 eV, p: 2.6x10^{17}, 120 nm
InGaAlAs, E$_g$ = 1.13 eV, un-doped, 80 nm
InAlAs, un-doped, 80 nm

FIGURE 3.21 (a) Layer structure and (b) band structure of InAlAs/InGaAlAs/InGaAs I^{2}E SACM APD.

A.S. Huntington et al. have reported a five-stage I^{2}E InAlAs/InGaAs tandem APD that achieved $k < 0.1$ for gain, M, <20 [222]. Clark et al. have successfully demonstrated a three-stage tandem I^{2}E InAlAs/InGaAs APD [223]. Both structures have been designed to enhance electron ionization while suppressing that of holes. The structure of each I^{2}E multiplication cell in Ref. [223] contains a 100 nm InAlAs hole relaxation layer, a 100 nm p-type InAlAs layer in which the electric field increases, a 200 nm InAlAs "acceleration" layer, a 10 nm InAlAs layer with decreasing electric field, and a 100 nm InAlGaAs (band-gap = 0.92 eV) impact ionization layer. The band structure is illustrated in Figure 3.23. The excess noise versus gain for the three-cell tandem APD is plotted in Figure 3.24; APDs with InP and thin InAlAs multiplication regions are included for comparison. Very low noise is achieved for $M < 10$. At higher gain the noise asymptotes toward the $k = 0.1$ curve. This is due to transport across the valance band discontinuity (labeled "barrier for hole" in Figure 3.23b). At low electric field, the barrier effectively blocks hole transport owing to the relatively low energy of holes. With increasing field, more holes can surmount the barrier and initiate impact ionization events. This is illustrated by the spatial distribution of ionization events plotted in Figure 3.25 [224]. As gain increases from 10 (Figure 3.25a) to 100 (Figure 3.25b), there is a significant increase in hole-initiated impact ionization as a result of the fact that an increasing number of holes can cross the barrier and impact ionization. This will cause an increase in the excess noise. To suppress

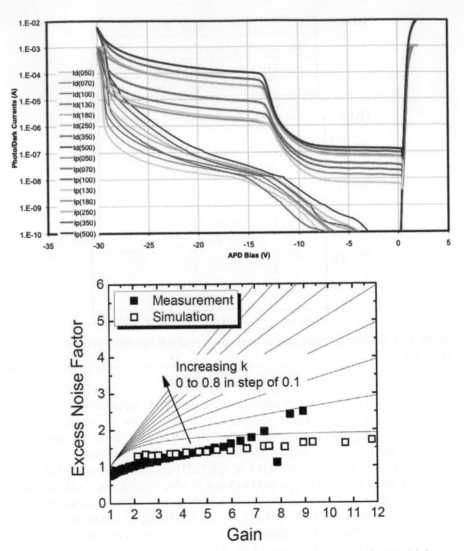

FIGURE 3.22 (a) Dark current and photocurrent and (b) measured (■) and simulated (□) excess noise of InAlAs/InGaAlAs/InGaAs I²E SACM APD.

hole-initiated ionization, Sun et al. [224] have used Monte Carlo simulations to show that increasing the thickness of the relaxation layer in the multiplication cells [225] has the benefit of allowing the holes to cool before they enter the InAlGaAs multiplication layer. For the excess noise simulation in Figure 3.26, the thickness was increased from 100 nm (■) to 500 nm (●). The k factor for the thicker hole relaxation layer is less than 0.05 for gain values up to ~100.

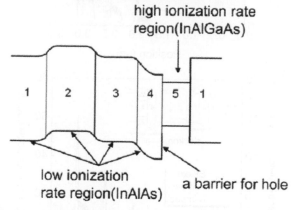

InAlAs	uid	100 nm	1
InAlAs	p	100 nm	2
InAlAs	uid	200 nm	3
InAlAs	n	10 nm	4
InAlGaAs	uid	100 nm	5

high ionization rate region(InAlGaAs)

1 2 3 4 5 1

low ionization rate region(InAlAs)

a barrier for hole

FIGURE 3.23 Single-cell layer structure and energy band diagram of I^{2}E InAlAs/InGaAs tandem APD [223].

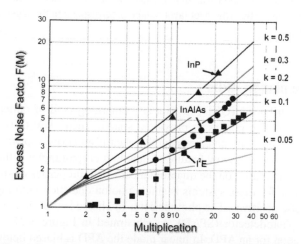

FIGURE 3.24 Measured excess noise versus gain for three-cell I^{2}E InAlAs/InGaAs tandem APD (■); InP (▲) and thin InAlAs (●) APDs are included for Ref. [223].

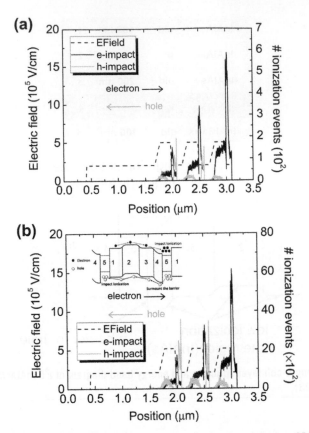

FIGURE 3.25 Simulated spatial distribution of impact ionization of tandem APD at (a) gain = 10 and (b) gain = 100. The inset layer numbers correspond to the layer structure in Figure 3.23 [224].

3.5.3 Single photon APDs

Quantum key distribution (QKD) is a rapidly emerging field of optical fiber communications [226]. The goal is to provide a shared, secret key to two authorized parties who desire to communicate securely, even if an eavesdropper has access to all of the message traffic. A key feature of quantum cryptography is that it provides the ultimate security based on the quantum mechanical properties of single photons. A critical function for these systems is single photon detection with high efficiency and minimal false positives. The photodetector of choice is the single photon counting avalanche photodetector (SPAD). As illustrated in Figure 3.27, there are two modes of operation for an APD. In linear mode the APD is biased below breakdown and the output photocurrent is directly proportional to the input optical signal level. Operation above the breakdown voltage is fundamentally different; the APD acts

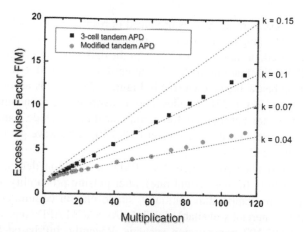

FIGURE 3.26 Simulated excess noise of I^{2}E InAlAs/InGaAs tandem APD with 100 nm (■) and 500 nm (●) relaxation layers [224].

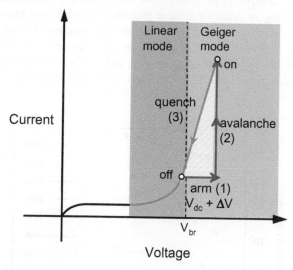

FIGURE 3.27 Current-voltage characteristics of an avalanche photodiode illustrating linear mode and single photon counting mode.

as a trigger element similar to a Geiger-Muller counter of nuclear radiation. Consequently, this mode of operation is frequently referred to as Geiger-mode. For single photon detection the APD is biased above its breakdown voltage with an excess bias, ΔV, to arm it for single photon detection ((1) in Figure 3.27). In this state a single carrier (electron or hole) can trigger a self-sustaining avalanche breakdown ((2) in Figure 3.27), which transitions the APD to its high current "on" state. Only the device series resistance and the excess voltage limit the current. The net result is that a macroscopic current pulse is produced in response to a single carrier in the depletion layer. The single carrier can be photogenerated, by which single photon detection is achieved, or may have its origin in the dark current, giving rise to a dark count. The APD will remain in the "on" state until it is returned to its "off" state by reducing the bias below breakdown ((3) in Figure 3.27). This is accomplished with a quenching circuit; Ref. [227] provides a broad overview of quenching techniques. The methods that are appropriate for communications include gated quenching [228], passive quenching with active reset [229,230], sine wave gating [231–233], and self-differencing [234].

The key performance parameters for SPADs are photon detection efficiency, dark count probability (or dark count rate), after-pulsing probability, jitter, spectral range, and operating temperature. Initially, single photon communication experiments utilized commercially available InP/InGaAs SACM APDs that were designed for OC-48 and OC-192 transmission systems. Recently, InP-based SACM APDs have been designed specifically for single photon counting applications [235–237]. While Si SPADs can achieve low dark counts at room temperature, the higher dark current of III–V compound photodetectors that operate at telecomm wavelengths necessitates cooling in order to achieve acceptable dark count rates. Figure 3.28 shows the dark count rate (DCR) versus the photon detection efficiency (PDE) of a 40 µm-diameter InP/InGaAs SACM SPAD for temperature in the range 180–300 K [238]. The PDE is >35% for all temperatures and reaches 50% at 180 K. However, the DCR increases from mid-10^3 s^{-1} to high 10^6 s^{-1} as the temperature increases

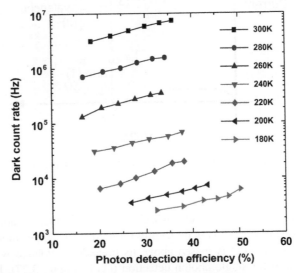

FIGURE 3.28 Dark count rate (DCR) versus photon detection efficiency (PDE) for 40 µm-diameter InP/InGaAs SACM SPAD at 1310 nm wavelength and temperature in the range 180–300 K [238].

from 180 K to 300 K. At this point, the detection efficiencies of telecomm SPADs are significantly higher than those of photomultiplier tubes and the dark counts are acceptable. It would be desirable to have lower dark counts but a more pressing issue is to suppress afterpulsing.

During an avalanche event, as current flows through the high-field multiplication layer of a SPAD, carriers are captured by deep-level traps. It should be noted that trapping is less severe for Si SPADs than those fabricated from III–V compounds. The trapped carriers are subsequently thermally emitted. However, if the SPAD is rearmed by biasing above breakdown before all the carriers have been emitted, escaping carriers can trigger avalanches. This effect, referred to as afterpulsing, is observed as an increase in the dark count rate at higher frequencies. At present, afterpulsing is the primary limitation on transmission rates. It has been shown that higher charge flow during an avalanche event, which is proportional to the gain, results in more afterpulsing [229]. Afterpulsing can be reduced with a delay time between incident photons that exceeds the detrapping times or using quenching techniques that restrict charge flow. Since the detrapping time constants are typically in the range 100 ns to 1 μs, reducing the total charge flow per avalanche event is the only practical approach. Recently, three improved quenching techniques, narrow pulse width gating [239], passive quenching with active reset [229,230], and sinusoidal gating [231–233], have achieved significantly reduced afterpulsing. Figure 3.29 shows that

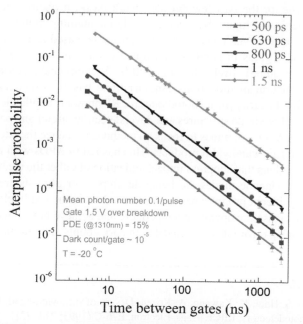

FIGURE 3.29 Afterpulse probability versus the delay time between incident light pulses for 0.5 ns, 0.63 ns, 1.0 ns, and 1.5 ns pulse widths [240].

decreasing the gate width from 1.5 ns to 0.5 ns reduces afterpulsing by a factor of 85 [240]. The challenge for this approach is synchronization of the incident photons with the narrow bias pulse and cancelation of capacitive transients at the leading and trailing edges of the pulse [227]. Sine wave gating is similar to pulsed gating in that the SPAD is periodically biased above breakdown. Its success in achieving high data transmission rate, with low DCR and reduced afterpulsing, is attributable to two factors. First, a high-frequency sinusoidal signal generates short sub-nanosecond gates that are beneficial for reducing total charge flow and afterpulsing. Second, high signal-to-noise detection is facilitated by the fact that the SPAD output consists of two components, a sinusoidal response at the bias frequency, which can be effectively eliminated with one or more narrow band filters, and a short temporal pulse with a broad frequency spectrum that is relatively unperturbed by the filter. While QKD field experiments [241–246] have focused on practical aspects such as long-term stability and immunity from hacking, a primary performance trend has been toward higher transmission rates. Tanaka et al. have reported 1 Mb/s transmission at 1550 nm over 50 km of fiber using an InP/InGaAs SACM SPAD receiver [247]. The SPAD was sine wave gated at 1.244 GHz. The afterpulse probability was 0.61%; the dark count probability was 0.7×10^{-6}, and the photon detection efficiency was 11%.

3.6 CONCLUSION

As has been true for the past few decades, work on p-i-n photodetectors has concentrated on achieving higher bandwidths while maintaining high responsivity and low-dark current. However, recently, there has been increased emphasis in achieving these performance goals in photonic integrated circuits in InP- and Si-based platforms. This has required novel column IV and III–V compound materials efforts and device-level structural modifications. Another trend has been increased interest in high-power, high-linearity p-i-n photodiodes for RF photonic applications, which has resulted in new photodiode structures and novel detection circuit and packaging techniques. Work on APDs continues to focus on methods to reduce the excess noise factor and increase gain-bandwidth products. This has led to APDs with multiple-layer multiplication regions that suppress impact ionization of either the electrons or holes. Single photon avalanche diodes are being developed for ultra-secure communications. While significant progress has been achieved since the last book in this series, challenges to increase the operating temperature and achieve high detection efficiencies while reducing error rates associated with dark counts and afterpulsing remain.

References

[1] K. Kato, S. Hata, K. Kawano, A. Kozen, Design of ultrawide-band, high sensitivity p-i-n photodetectors, IEICE Trans. Electron. E76-C (1993) 214–221.

[2] G.A. Davis, R.E. Weiss, R.A. LaRue, K.J. Williams, R.D. Esman, A 920–1650 nm high current photodetector, IEEE Photon. Technol. Lett. 8 (10) (1996) 1373–1375.

[3] T. Ishibashi, N. Shimizu, S. Kodama, H. Ito, T. Nagatsuma, T. Furuta, Uni-traveling-carrier photodiodes, Tech. Dig. Ultrafast Electronics and Optoelectronics, 1997, pp. 83–87.

[4] H. Ito, T. Nagatsuma, High-speed and high-output-power unitraveling-carrier photodiodes, Proc. SPIE – Int. Soc. Opt. Eng. 5246 (1) (2003) 465–479.

[5] S. Kodama, H. Ito, UTC-PD-based optoelectronic components for high-frequency and high-speed applications, IEICE Trans. Electron. E90-C (2) (2007) 429–435.

[6] M. Dentan, B.D. de Cremoux, Numerical simulation of a p-i-n photodiode under high illumination, J. Lightwave Technol. 8 (8) (1990) 1137–1144.

[7] P.-L. Liu, K.J. Williams, M.Y. Frankel, R.D. Esman, Saturation characteristics of fast photodetectors, IEEE Trans. Microwave Theory Tech. 47 (7) (1999) 1297–1303.

[8] K.J. Williams, R.D. Esman, Design considerations for high current photodetectors, J. Lightwave Technol. 17 (1999) 1443–1454.

[9] K.J. Williams, R.D. Esman, M. Dagenais, Effects of high space-charge fields on the response of microwave photodetectors, IEEE Photon. Technol. Lett. 6 (5) (1994) 639–641.

[10] T. Ishibashi, T. Furtua, H. Fushimi, H. Ito, Photoresponse characteristics of uni-traveling-carrier photodiodes, Proc. SPIE – Int. Soc. Opt. Eng. 4283 (2001) 469–479.

[11] S.D. Personick, Receiver design for digital fiber-optic communication systems, Parts I and II, Bell Syst. Tech. J. 52 (1973) 843–886.

[12] R.G. Smith, S.D. Personick, Receiver design for optical fiber communications systems, in: Semiconductor Devices for Optical Communication, Springer-Verlag, New York, 1980 (Chapter 4)

[13] S.R. Forrest, Sensitivity of avalanche photodetector receivers for high-bit-rate long-wavelength optical communication systems, Semiconductors and Semimetals, Lightwave Communication Technology, vol. 22, Academic Press, Orlando, FL, 1985 (Chapter 4)

[14] B.L. Kasper, J.C. Campbell, Multigigabit-per-second avalanche photodiode lightwave receivers, J. Lightwave Technol. LT-5 (1987) 1351.

[15] R.J. McIntyre, Multiplication noise in uniform avalanche diodes, IEEE Trans. Electron Dev. 13 (1) (1966) 154–158.

[16] R.J. McIntyre, B. John, The distribution of gains in uniformly multiplying avalanche photodiodes: theory, IEEE Trans. Electron Dev. ED-19 (1972) 703–713.

[17] R.J. McIntyre, Factors affecting the ultimate capabilities of high speed avalanche photodiodes and a review of the state-of-the-art, Tech. Dig. International Electron Dev. Mtg., 1973, pp. 213–216.

[18] R.B. Emmons, Avalanche-photodiode frequency response, J. Appl. Phys. 38 (9) (1967) 3705–3714.

[19] M.M. Hayat, B.E.A. Saleh, M.C. Teich, Effect of dead space on gain and noise of double-carrier multiplication avalanche photodiodes, IEEE Trans. Electron Dev. 39 (3) (1992) 546–552.

[20] R.J. McIntyre, A new look at impact ionization – Part 1: A theory of gain, noise, breakdown probability and frequency response, IEEE Trans. Electron Dev. 48 (8) (1999) 1623–1631.

[21] X. Li, X. Zheng, S. Wang, F. Ma, J.C. Campbell, Calculation of gain and noise with dead space for GaAs and $Al_xGa_{1-x}As$ avalanche photodiodes, IEEE Trans. Electron Dev. 49 (2002) 1112–1117.

[22] B. Jacob, P.N. Robson, J.P.R. David, G.J. Rees, Fokker-Planck model for nonlocal impact ionization in semiconductors, J. Appl. Phys. 90 (3) (2001) 1314–1317.

[23] A. Spinelli, A.L. Lacaita, Mean gain of avalanche photodiodes in a dead space model, IEEE Trans. Electron Dev. 43 (1) (1996) 23–30.

[24] G.M. Dunn, G.J. Rees, J.P.R. David, S.A. Plimmer, D.C. Herbert, Monte Carlo simulation of impact ionization and current multiplication in short GaAs p^+in^+ diodes, Semicond. Sci. Technol. 12 (1997) 111–120.

[25] D.S. Ong, K.F. Li, G.J. Rees, G.M. Dunn, J.P.R. David, P.N. Robson, A Monte Carlo investigation of multiplication noise in thin p^+in^+ avalanche photodiodes, IEEE Trans. Electron Dev. 45 (8) (1998) 1804–1810.

[26] S.A. Plimmer, J.P.R. David, D.S. Ong, K.F. Li, A simple model including the effects of dead space, IEEE Trans. Electron Dev. 46 (4) (1999) 769–775.

[27] K.F. Li, S.A. Plimmer, J.P.R. David, R.C. Tozer, G.J. Rees, P.N. Robson, C.C. Button, J.C. Clark, Low avalanche noise characteristics in thin InP p^+-i-n^+ diodes with electron initiated multiplication, IEEE Photon. Technol. Lett. 11 (1999) 364–366.

[28] J.C. Campbell, S. Chandrasekhar, W.T. Tsang, G.J. Qua, B.C. Johnson, Multiplication noise of wide-bandwidth InP/InGaAsP/InGaAs avalanche photodiodes, J. Lightwave Technol. 7 (3) (1989) 473–477.

[29] P. Yuan, C.C. Hansing, K.A. Anselm, C.V. Lenox, H. Nie, A.L. Holmes Jr. B.G. Streetman, J.C. Campbell, Impact ionization characteristics of III–V semiconductors for a wide range of multiplication region thicknesses, IEEE J. Quant. Electron. 36 (2000) 198–204.

[30] M.A. Saleh, M.M. Hayat, P.O. Sotirelis, A.L. Holmes, J.C. Campbell, B. Saleh, M. Teich, Impact-ionization and noise characteristics of thin III–V avalanche photodiodes, IEEE Trans. Electron Dev. 48 (2001) 2722–2731.

[31] K.F. Li, D.S. Ong, J.P.R. David, R.C. Tozer, G.J. Rees, S.A. Plimmer, K.Y. Chang, J.S. Roberts, Avalanche noise characteristics of thin GaAs structures with distributed carrier generation, IEEE Trans. Electron Dev. 47 (5) (2000) 910–914.

[32] K.F. Li, D.S. Ong, J.P.R. David, G.J. Rees, R.C. Tozer, P.N. Robson, R. Grey, Avalanche multiplication noise characteristics in thin GaAs p^+-i-n^+ diodes, IEEE Trans. Electron Dev. 45 (10) (1998) 2102–2107.

[33] C. Hu, K.A. Anselm, B.G. Streetman, J.C. Campbell, Noise characteristics of thin multiplication region GaAs avalanche photodiodes, Appl. Phys. Lett. 69 (24) (1996) 3734–3736.

[34] C.H. Tan, J.C. Clark, J.P.R. David, G.J. Rees, S.A. Plimmer, R.C. Tozer, D.C. Herbert, D.J. Robbins, W.Y. Leong, J. Newey, Avalanche noise measurements in thin Si p^+-i-n^+ diodes, Appl. Phys. Lett. 76 (26) (2000) 3926–3928.

[35] C.H. Tan, J.P.R. David, J. Clark, G.J. Rees, S.A. Plimmer, D.J. Robbins, D.C. Herbert, R.T. Carline, W.Y. Leong, Avalanche multiplication and noise in submicron Si p-i-n diodes, in: Proc. SPIE, Silicon-based Optoelectronics II, vol. 3953, 2000, pp. 95–102.

[36] S.A. Plimmer, J.P.R. David, G.J. Rees, R. Grey, D.C. Herbert, D.R. Wright, A.W. Higgs, Impact ionization in thin $Al_xGa_{1-x}As$ ($x = 015 - 0.30$) p-i-n diodes, J. Appl. Phys. 82 (3) (1997) 1231–1235.

[37] B.K. Ng, J.P.R. David, G.J. Rees, R.C. Tozer, M. Hopkinson, R.J. Riley, Avalanche multiplication and breakdown in $Al_xGa_{1-x}As$ ($x < 0.9$), IEEE Trans. Electron Dev. 49 (12) (2002) 2349–2351.

[38] B.K. Ng, J.P.R. David, R.C. Tozer, M. Hopkinson, G. Hill, G.H. Rees, Excess noise characteristics of $Al_{0.8}Ga_{0.2}As$ avalanche photodiodes, IEEE Trans. Electron Dev. 48 (10) (2001) 2198–2204.

[39] C.H. Tan, J.P.R. David, S.A. Plimmer, G.J. Rees, R.C. Tozer, R. Grey, Low multiplication noise thin $Al_{0.6}Ga_{0.4}As$ avalanche photodiodes, IEEE Trans. Electron Dev. 48 (7) (2001) 1310–1317.

[40] P. Yuan, S. Wang, X. Sun, X.G. Zheng, A.L. Holmes, Jr., J.C. Campbell, Avalanche photodiodes with an impact-ionization-engineered multiplication region, IEEE Photon. Technol. Lett. 12 (2000) 1370–1372.

[41] O.-H. Kwon, M.M. Hayat, S. Wang, J.C. Campbell, A.L. Holmes, Jr B.E.A. Saleh, M.C. Teich, Optimal excess noise reduction in thin heterojunction $Al_{0.6}Ga_{0.4}As$–GaAs avalanche photodiodes, IEEE J. Quant. Electron. 39 (10) (2003) 1287–1296.

[42] C. Groves, C.K. Chia, R.C. Tozer, J.P.R. David, G.J. Rees, Avalanche noise characteristics of single $Al_xGa_{1-x}As(0.3 < x < 0.6)$–GaAs heterojunction APDs, IEEE J. Quant. Electron. 41 (1) (2005) 70–75.

[43] S. Wang, R. Sidhu, X.G. Zheng, X. Li, X. Sun, A.L. Holmes, Jr J.C. Campbell, Studies on electro-polymerizations, Physica C 13 (2001) 1346.

[44] S. Wang, F. Ma, X. Li, R. Sidhu, X.G. Zheng, X. Sun, A.L. Holmes, Jr J.C. Campbell, Ultra-low noise avalanche photodiodes with a "centered-well" multiplication region, IEEE J. Quant. Electron. 39 (2003) 375–378.

[45] M.M. Hayat, O.-H. Kwon, S. Wang, J.C. Campbell, B.E.A. Saleh, M.C. Teich, Boundary effects on multiplication noise in thin heterostructure avalanche photodiodes: theory and experiment, IEEE Trans. Electron Dev. 49 (2002) 2114–2123.

[46] S. Wang, J.B. Hurst, F. Ma, R. Sidhu, X. Sun, X.G. Zheng, A.L. Holmes, Jr J.C. Campbell, A. Huntington, L.A. Coldren, Low-noise impact-ionization-engineered avalanche photodiodes grown on InP substrates, IEEE Photon. Technol. Lett. 14 (2002) 1722–1724.

[47] N. Duan, S. Wang, F. Ma, N. Li, J.C. Campbell, C. Wang, L.A. Coldren, High-speed and low-noise SACM avalanche photodiodes with an impact-ionization engineered multiplication region, IEEE Photon. Technol. Lett. 17 (8) (2005) 1719–1721.

[48] Y.-G. Wey, K. Giboney, J. Bowers, M. Rodwell, P. Silvestre, P. Thiagarajan, G. Robinson, 110-GHz GaInAs/InP double heterostructure p-i-n photodetectors, J. Lightwave Technol. 13 (1995) 1490–1499.

[49] J.E. Bowers, C.A. Burrus, High-speed zero-bias waveguide photodetectors, Electron. Lett. 22 (1986) 905–906.

[50] H.F. Taylor, O. Eknoyan, C.S. Park, K.N. Choi, K. Chang, Traveling wave photodetectors, Proc. SPIE 1217 (1990) 59–63.

[51] K.S. Giboney, M.J.W. Rodwell, J.E. Bowers, Traveling-wave photodetectors, IEEE Photon. Technol. Lett. 4 (1992) 1363–1365.

[52] K.S. Giboney, M.J.W. Rodwell, J.E. Bowers, Traveling-wave photodetector theory, IEEE Trans. Microwave Theory Tech. 45 (1997) 1310–1319.

[53] A. Umbach, D. Trommer, R. Steingrüber, S. Seeger, W. Ebert, G. Unterbörsch, Ultrafast, high-power $1.55 \mu m$ side-illuminated photodetector with integrated spot size converter, in: Tech. Dig. Opt. Fiber Commun. Conf. (OFC'00), Baltimore, MD, 2000, pp. 117–119.

[54] T. Takeuchi, T. Nakata, K. Makita, T. Torikai, High-power and high-efficiency photodiode with an evanescently coupled graded-index waveguide for 40 Gb/s applications, in: Tech. Dig. Opt. Fiber Commun. Conf. (OFC'01), Anaheim, CA, vol. 3, 2001, WQ2-1-3.

[55] E. Dröge, E.H. Böttcher, D. Bimberg, O. Reimann, R. Steingrüber, 70 GHz InGaAs MSM photodetectors for polarization-insensitive operation, Electron. Lett. 34 (1998) 1421–1422.

[56] Y.-S. Wu, P.-H. Chiu, J.-W. Shi, High-speed and high-power performance of a dual-step evanescently-coupled uni-traveling-carrier photodiode at 1.55 μm wavelength, in: Tech. Dig. Opt. Fiber Commun. Conf. (OFC'07), Anaheim, CA, 2007, Paper OThG1.

[57] Y. Muramoto, K. Kato, M. Mitsuhara, O. Nakajima, Y. Matsuoka, N. Shimizu, T. Ishibashi, High output voltage, high speed, high efficiency uni-travelling carrier waveguide photodiode, Electron. Lett. 34 (1998) 122–123.

[58] C. Cohen-Jonathan, L. Giraudet, A. Bonzo, J.P. Praseuth, Waveguide AlInAs/GaAlInAs avalanche photodiode with a gain-bandwidth product over 160 GHz, Electron. Lett. 33 (1997) 1492–1493.

[59] J. Wei, F. Xia, S.R. Forrest, A high-responsivity high-bandwidth asymmetric twin-waveguide coupled InGaAs-InP-InAlAs avalanche photodiode, IEEE Photon. Technol. Lett. 14 (2002) 1590–1592.

[60] K. Kato, A. Kozen, Y. Muramoto, Y. Itaya, T. Nagatsuma, M. Yaita, 110-GHz, 50%-efficiency mushroom-mesa waveguide p-i-n photodiode for a 1.55-μm wavelength, IEEE Photon. Technol. Lett. 6 (1994) 719–721.

[61] J.-W. Park, High-responsivity and high-speed waveguide photodiode with a thin absorption region, IEEE Photon. Technol. Lett. 22 (13) (2010) 975–977.

[62] G. Unterbörsch, D. Trommer, A. Umbach, R. Ludwig, H.-G. Bach, High-power performance of a high-speed photodetector, in: Proc. 24th Europ. Conf. Opt. Commun. (ECOC'98), Madrid, Spain, 1998, pp. 67–68.

[63] H.-G. Bach, A. Beling, G.G. Mekonnen, R. Kunkel, D. Schmidt, W. Ebert, A. Seeger, M. Stollberg, W. Schlaak, InP-based waveguide-integrated photodetector with 100-GHz bandwidth, IEEE J. Sel. Top. Quant. Electron. 10 (2004) 668–672.

[64] A. Beling, PIN photodiode modules for 80 Gbit/s and beyond, in: Tech. Dig. Opt. Fiber Commun. Conf. (OFC'06), Anaheim, CA, 2006, Paper OFI1.

[65] H.-G. Bach, R. Kunkel, G.G. Mekonnen, R. Zhang, A. Sigmund, D. Schmidt, C. Sakkas, D. Pech, C. Schubert, Novel 107 Gb/s bias-feeding photodetector OEIC for efficient low-cost photoreceiver co-packaging, in: 35th European Conference on Optical Communication, 2009 (ECOC '09), 20–24 September 2009.

[66] A. Beling, H.-G. Bach, G.G. Mekonnen, R. Kunkel, D. Schmidt, Miniaturized waveguide-integrated p-i-n photodetector with 120-GHz bandwidth and high responsivity, IEEE Photon. Technol. Lett. 17 (2005) 2152–2154.

[67] A. Beling, H.-G. Bach, G.G. Mekonnen, R. Kunkel, D. Schmidt, High-speed miniaturized photodiode and parallel-fed traveling-wave photodetectors based on InP, IEEE J. Sel. Top. Quant. Electron. 13 (2007) 15–21.

[68] R.J. Hawkins, R.J. Deri, O. Wada, Optical power transfer in vertically integrated impedance-matched waveguide/photodetectors: physics and implications for diode-length reduction, Opt. Lett. 16 (1991) 470–472.

[69] R.J. Deri, W. Döldissen, R.J. Hawkins, R. Bhat, J.B.D. Soole, L.M. Schiavone, M. Seto, N. Andreadakis, Y. Silberberg, M.A. Koza, Efficient vertical coupling of photodiodes to InGaAsP rib waveguides, Appl. Phys. Lett. 58 (1991) 2749–2751.

[70] A. Beling, InP-based 1.55 μm waveguide-integrated photodetectors for high-speed applications, Proc. SPIE 6123 (2006) 156–167.

[71] J. Klamkin, S.M. Madison, D.C. Oakley, A. Napoleone, F. O'Donnell, J. Frederick, M. Sheehan, L.J. Missaggia, J.M. Caissie, J.J. Plant, P.W. Juodawlkis, Uni-traveling-carrier variable confinement waveguide photodiodes, Opt. Express 19 (11) (2011) 10199–10205.

[72] M. Achouche, V. Magnin, J. Harari, D. Carpentier, E. Derouin, C. Jany, F. Blanche, D. Decoster, Design and fabrication of a p-i-n photodiode with high responsivity and large alignment tolerances for 40 Gb/s applications, IEEE Photon. Technol. Lett. 18 (2006) 556–558.

[73] S. Demiguel, N. Li, X. Li, X. Zheng, J. Kim, J.C. Campbell, H. Lu, A. Anselm, Very high-responsivity evanescently coupled photodiodes integrating a short planar multimode waveguide for high-speed applications, IEEE Photon. Technol. Lett. 15 (2003) 1761–1763.

[74] M. Achouche, C. Cuisin, E. Derouin, F. Pommereau, J.Y. Dupuy, F. Blache, P. Berdaguer, M. Riet, H. Gariah, S. Vuye, D. Carpentier, 43 Gb/s balanced photoreceiver using monolithic integrated lensed facet waveguide dual-UTC photodiodes, in: Joint Conference of the 2008 Opto-Electronics and Communications Conference, and the Australian Conference on Optical Fibre Technology (OECC/ACOFT 2008), 7–10 July, 2008, pp. 1–2.

[75] G. Glastre, D. Carpentier, F. Lelarge, B. Rousseau, B. Blache, M. Achouche, High-linearity and high responsivity UTC photodiodes for multi-level formats applications, in: 35th European Conference on Optical Communication (ECOC 2009), 20–24 September, 2009, Paper 9.2.5.

[76] F. Xia, J.K. Thomson, M.R. Gokhale, P.V. Studenkov, J. Wei, W. Lin, S.R. Forrest, An asymmetric twin-waveguide high-bandwidth photodiode using a lateral taper coupler, IEEE Photon. Technol. Lett. 13 (2001) 845–847.

[77] V.M. Menon, F. Xia, S.R. Forrest, Integration using asymmetric twin-waveguide (ATG) technology: Part II – Devices, IEEE J. Sel. Top. Quant. Electron. 11 (11) (2005) 30–42.

[78] E. Rouvalis, C.C. Renaud, D.G. Moodie, M.J. Robertson, A.J. Seeds, Traveling-wave uni-traveling carrier photodiodes for continuous wave THz generation, Opt. Express 18 (11) (2010) 11105–11110.

[79] L.Y. Lin, M.C. Wu, T. Itoh, T.A. Vang, R.R. Muller, D.L. Sivco, A.Y. Cho, High-power high-speed photodetectors—design, analysis, and experimental demonstration, IEEE Trans. Microwave Theory Tech. 45 (1997) 1320–1331.

[80] V.M. Hietala, A. Vawter, T.M. Brennan, B.E. Hammons, Traveling-wave photodetectors for high-power, large-bandwidth applications, IEEE Trans. Microwave Theory Tech. 43 (1995) 2291–2298.

[81] K.S. Giboney, R.L. Nagarajan, T.E. Reynolds, S.T. Allen, R.P. Mirin, M.J.W. Rodwell, J.E. Bowers, Travelling-wave photodetectors with 172-GHz bandwidth and 76-GHz bandwidth-efficiency product, IEEE Photon. Technol. Lett. 7 (1995) 412–414.

[82] Y.-J. Chiu, S.B. Fleischer, J.E. Bowers, High-speed low-temperature-grown GaAs p-i-n traveling-wave photodetector, IEEE Photon. Technol. Lett. 10 (1998) 1012–1014.

[83] A. Stöhr, S. Babiel, P.J. Cannard, B. Charbonnier, F. van Dijk, S. Fedderwitz, D. Moodie, L. Pavlovic, L. Ponnampalam, C.C. Renaud, D. Rogers, V. Rymanov, A. Seeds, A.G. Steffan, A. Umbach, M. Weiss, Millimeter-wave photonic components for broadband wireless systems, IEEE Trans. Microwave Theory Tech. 58 (11) (2010) 3071–3082.

[84] M.S. Islam, S. Murthy, T. Itoh, M.C. Wu, D. Novak, R.B. Waterhouse, D.L. Sivco, A.Y. Cho, Velocity-matched distributed photodetectors with p-i-n photodiodes, IEEE Trans. Microwave Theory Tech. 49 (2001) 1914–1920.

[85] E. Dröge, E.H. Böttcher, St. Kollakowski, A. Strittmatter, D. Bimberg, O. Reimann, R. Steingrüber, 78 GHz distributed MSM photodetector, Electron. Lett. 34 (1998) 2241–2242.

[86] Y. Hirota, T. Ishibashi, H. Ito, 1.55 μm wavelength periodic traveling-wave photodetector fabricated using unitraveling-carrier photodiode structures, J. Lightwave Technol. 19 (2001) 1751–1758.

[87] M.J.W. Rodwell, S.T. Allen, R.Y. Yu, M.G. Case, U. Bhattacharya, M. Reddy, E. Carman, M. Kamegawa, Y. Konishi, J. Pusl, R. Pullela, Active and nonlinear wave propagation devices in ultrafast electronics and optoelectronics, Proc. IEEE 82 (1994) 1037–1059.

[88] S. Murthy, S.-J. Kim, T. Jung, Z.-Z. Wang, W. Hsin, T. Itoh, M.C. Wu, Backward-wave cancellation in distributed traveling-wave photodetectors, J. Lightwave Technol. 21 (2003) 3071–3077.

[89] C.L. Goldsmith, G.A. Magel, R.J. Baca, Principles and performance of traveling-wave photodetector arrays, IEEE Trans. Microwave Theory Tech. 45 (1997) 1342–1350.

[90] S. Murthy, M.C. Wu, D. Sivco, A.Y. Cho, Parallel feed traveling wave distributed pin photodetectors with integrated MMI couplers, Electron. Lett. 38 (2002) 78–79.

[91] A. Beling, J.C. Campbell, H.-G. Bach, G.G. Mekonnen, D. Schmidt, Parallel-fed traveling wave photodetector for >100-GHz applications, J. Lightwave Technol. 26 (2008) 16–20.

[92] W. Shillue, W. Grammer, C. Jacques, R. Brito, J. Meadows, J. Castro, J. Banda, Y. Masui, The ALMA local oscillator system, in: General Assembly and Scientific Symposium, 2011 XXXth URSI, 13–20 August 2011, pp. 1–4.

[93] D. Wake, Trend and prospects for radio over fibre picocells, in: Proc. Int. Top. Meeting Microw. Photon., 2002, pp. 21–24.

[94] H. Ito, H. Fushimi, Y. Muramoto, T. Furuta, T. Ishibashi, High-power photonic microwave generation at K- and Ka-bands using a uni-traveling-carrier photodiode, J. Lightwave Technol. 20 (8) (2002) 1500–1505.

[95] H. Ito, S. Kodama, Y. Muramoto, T. Furuta, T. Nagatsuma, T. Ishibashi, High-speed and high-output InP-InGaAs uni-traveling-carrier photodiodes, IEEE J. Sel. Top. Quant. Electron. 10 (4) (2004) 709–727.

[96] T. Furuta, H. Ito, T. Ishibashi, Photocurrent dynamics of uni-traveling-carrier and conventional pinphotodiodes, Inst. Phys. Conf. Ser. 166 (2000) 419–422.

[97] N. Li, X. Li, S. Demiguel, X. Zheng, J.C. Campbell, D.A. Tulchinsky, K.J. Williams, T.D. Isshiki, G.S. Kinsey, R. Sudharsansan, High-saturation-current charge-compensated InGaAs/InP uni-traveling-carrier photodiode, Photon. Tech. Lett. 16 (3) (2004) 864–866.

[98] M. Chtioui, D. Carpentier, S. Bernard, B. Rousseau, F. Lelarge, F. Pommereau, C. Jany, A. Enard, M. Achouche, Thick absorption layer uni-traveling-carrier photodiodes with high responsivity, high speed and high saturation power, IEEE Photon. Technol. Lett. 21 (7) (2009) 429–431.

[99] K.J. Williams, D.A. Tulchinsky, J.B. Boos, Doewon Park, P.G. Goetz, High-power photodiodes, in: 2006 Digest of the LEOS Summer Topical Meetings, IEEE Cat. No. 06TH8863C, 2006, pp. 50–51.

[100] D.-H. Jun, J.-H. Jang, Ilesanmi Adesida, J.-I. Song, Improved efficiency-bandwidth product of modified uni-traveling carrier photodiode structures using an undoped photo-absorption layer, Jpn. J. Appl. Phys. 45 (4B) (2006) 3475–3478.

[101] S. Itakura, K. Sakai, T. Nagatsuka, T. Akiyama, Y. Hirano, E. Ishimura, M. Nakaji, T. Aoyagi, High-current backside-illuminated InGaAs/InP p-i-n potodiode, in: Microwave Photonics, 2009, MWP '09, International Topical Meeting, October 2009.

[102] Xin Wang, Ning Duan, Hao Chen, J.C. Campbell, InGaAs/InP photodiodes with high responsivity and high saturation power, IEEE Photon. Technol. Lett. 19 (16) (2007) 1272–1274.

[103] M. Chtoiui, F. Lelarge, A. Enard, F. Pommereau, D. Carpentier, A. Marceaux, F. Van Dijk, M. Achouche, High responsivity and high power UTC and MUTC GaInAs-InP photodiodes, IEEE Photon. Technol. Lett. 24 (4) (2012) 318–320.

[104] N. Shimizu, N. Watanabe, T. Furuta, T. Ishibashi, InP-InGaAs uni-traveling-carrier photodiode with improved 3-dB bandwidth of over 150 GHz, IEEE Photon. Technol. Lett. 10 (3) (1998) 412–414.

[105] Naofumi Shimizu, Noriyuki Watanabe, Tomofumi Furuta, Todao Ishibashi, Improved response of uni-traveling-carrier photodiodes by carrier injection, Jpn. J. Appl. Phys. 37 (March 1998) 1424–1426.

[106] F.-M. Kuo, M.-Z. Chou, J.W. Shi, Linear-cascaded near-ballistic unitraveling-carrier photodiodes with an extremely high saturation current-bandwidth product, IEEE J. Lightwave Technol. 29 (4) (2011) 432–438.

[107] Shigetaka Itakura, K. Sakai, T. Nagatsuka, T. Akiyama, Y. Hirano, E. Ishimura, M. Kakaji, T. Aoyagi, High-current backside-illuminated photodiode array module for optical analog links, IEEE J. Lightwave Technol. 28 (6) (2010) 965–971.

[108] Zhi Li, Fu Yang, M. Piels, Huapu Pan, A. Beling, J.E. Bowers, J.C. Campbell, High-power high-linearity flip-chip bonded modified uni-traveling carrier photodiode, Opt. Express 19 (26) (2011) B385–B390.

[109] T. Ishibashi, High Speed Heterostructure Devices, Semiconductors and Semimetals vol. 41, Academic Press, San Diego, 1994 p. 333 (Chapter 5).

[110] Y.-S. Wu, J.-W. Shi, P.-H. Chiu, Analytical modeling of a high-performance near-ballistic uni-traveling-carrier photodiode at a 1.55-μm wavelength, IEEE Photon. Technol. Lett. 18 (8) (2006) 938–940.

[111] J.-W. Shi, Y.-S. Wu, C.-Y. Wu, P.-H. Chiu, C.-C. Hong, High-speed, high-responsivity, and high-power performance of near-ballistic uni-traveling-carrier photodiode at 1.55-μm wavelength, IEEE Photon. Technol. Lett. 17 (9) (2005) 1929–1931.

[112] J.-W. Shi, F.-M. Kuo, C.-J. Wu, C.L. Chang, C.-Y. Liu, C.Y. Chen, J.-I. Chyi, Extremely high saturation current-bandwidth product performance of a near-ballistic uni-traveling-carrier photodiode with a flip-chip bonding structure, IEEE J. Quant. Electron. 40 (1) (2010) 80–86.

[113] J.-W. Shi, F.-M. Kuo, J.C. Bowers, Design and analysis of ultra-high-speed near-ballistic uni-traveling-carrier photodiodes under 50-Ω load for high-power performance, IEEE Photon. Technol. Lett. 24 (7) (2012) 533–535.

[114] A. Joshi, S. Datta, D. Becker, GRIN lens-coupled top-illuminated highly linear InGaAs photodiodes, IEEE Photon. Technol. Lett. 20 (17) (2008) 1500–1502.

[115] L. Giraudet, F. Banfi, S. Demiguel, G. Herve-Gruyer, Optical design of evanescently coupled waveguide-fed photodiodes for ultrawide-band applications, IEEE Photon. Technol. Lett. 11 (1) (1999) 111–113.

[116] J. Klamkin, A. Ramaswamy, N. Nunoya, L.A. Johansson, J.E. Bowers, S.P. DenBaars, L.A. Coldren, Uni-traveling-carrier waveguide photodiodes with >40 dBm OIP3 for up to 80 mA of photocurrent, in: Device Research Conference, 2009.

[117] S. Jasmin, N. Vodjdani, J. Renaud, A. Enard, Diluted- and distributed-absorption microwave waveguide photodiodes for high efficiency and high power, IEEE Trans. Microwave Theory Tech. 45 (8) (1997) 1337–1341.

[118] S.M. Madison, J.J. Plant, D.C. Oakley, A. Napoleone, P.W. Juodawlkis, Slab-coupled optical waveguide photodiode, in: Conference on Lasers and Electro-Optics/Quantum Electronics and Laser Science Conference and Photonic Applications Systems Technologies, OSA Technical Digest (CD), Optical Society of America, 2008, paper CWF4.

[119] S. Adachi, Lattice thermal conductivity of group-IV and III–V semiconductor alloys, J. Appl. Phys. 102 (6) (2007) 063502-1–063502-7.

[120] A. Ramaswamy, M. Piels, N. Nunoya, T. Yin, J.E. Bowers, High power silicon-germanium photodiodes for microwave photonic applications, IEEE Trans. Microwave Theory Tech. 58 (11) (2010) 3336–3342.

[121] Z. Sen Lin, P.M. Lane, J.J. O'Reilly, Assessment of the nonlinearity tolerance of different modulation schemes for millimeter-wave fiber-radio systems using MZ modulators, IEEE Trans. Microwave Theory Tech. 45 (8) (1997) 1403–1409.

[122] K.J. Williams, R.D. Esman, M. Dagenais, Nonlinearities in p-i-n microwave photodetectors, J. Lightwave Technol. 14 (1) (1996) 84–96.

[123] K.J. Williams, L.T. Nichols, R.D. Esman, Photodetector nonlinearity limitations on a high-dynamic range 3 GHz fiber optic link, J. Lightwave Technol. 16 (2) (1998) 192–199.

[124] H. Pan, Z. Li, A. Beling, J.C. Campbell, Characterization of high-linearity modified uni-traveling carrier photodiodes using three-tone and bias modulation techniques, J. Lightwave Technol. 28 (9) (2010) 1316–1322.

[125] K.J. Williams, D.A. Tulchinsky, A. Hastings, High-power and high-linearity photodiodes, in: 21st Annual Meeting of the IEEE Lasers and Electro-Optics Society, LEOS 2008, Newport Beach, California, November 2008, pp. 290–291.

[126] H. Jiang, D.S. Shin, G.L. Li, T.A. Vang, D.C. Scott, P.K.L. Yu, The frequency behavior of the third-order intercept point in a waveguide photodiode, IEEE Photon. Technol. Lett. 12 (5) (2000) 540–542.

[127] M. Chtioui, A. Enard, D. Carpentier, S. Bernard, B. Rousseau, F. Lelarge, F. Pommereau, M. Achouche, High-power high-linearity uni-traveling-carrier photodiodes for analog photonic links, IEEE Photon. Technol. Lett. 20 (3) (2008) 202–204.

[128] Andreas Beling, Huapu Pan, Hao Chen, Joe C. Campbell, Linearity of modified uni-traveling carrier photodiodes, J. Lightwave Technol. 26 (August) (2008) 2373–2378.

[129] M.N. Draa, J. Bloch, W.S. Chang, P.K.L. Yu, D.C. Scott, S.B. Chen, N. Chen, K.J. Williams, Voltage-dependent nonlinearities in uni-traveling carrier directional coupled photodiodes, in: 2010 IEEE Topical Meeting on Microwave Photonics (MWP), 5–9 October 2010, pp. 15–18.

[130] A.S. Hastings, D.A. Tulchinsky, K.J. Williams, Photodetector nonlinearities due to voltage-dependent responsivity, IEEE Photon. Technol. Lett. 21 (21) (2009) 1642–1644.

[131] H. Pan, Z. Li, A. Beling, J.C. Campbell, Measurement and modeling of high-linearity modified uni-traveling carrier photodiode with highly-doped absorber, Opt. Express 17 (2009) 20221–20226.

[132] A. Beling, H. Pan, H. Chen, J.C. Campbell, Measurement and modelling of high-linearity partially depleted absorber photodiode, Electron. Lett. 44 (24) (2008) 1419–1420.

[133] A.S. Hastings, D.A. Tulchinsky, K.J. Williams, H. Pan, A. Beling, J.C. Campbell, Minimizing photodiode nonlinearities by compensating voltage-dependent responsivity effects, J. Lightwave Technol. 28 (22) (2010) 3329–3333.

[134] Yang Fu, Huapu Pan, Zhi Li, Andreas Beling, Joe C. Campbell, Characterization and modeling nonlinear intermodulation distortions in uni-traveling carrier photodiodes, J. Quant. Electron. 47 (10) (2011) 1312–1319.

[135] M.S. Islam, T. Jung, T. Itoh, M.C. Mu, A. Nespola, D.L. Sivco, A.Y. Cho, High power and highly linear monolithically integrated distributed balanced photo detectors, J. Lightwave Technol. 20 (February 2002) 285–295.

[136] M.S. Islam, M.C. Wu, Recent advances and future prospects in high-speed and high-saturation-current photodetectors, Proc. SPIE 5246 (1) (2003) 448–457.

[137] J. Klamkin, L.A. Johansson, A. Ramaswamy, H. Chou, M.N. Sysak, J.W. Raring, N. Parthasrathy, S.P. Denbaars, J.E. Bowers, L.A. Coldren, Monolithically integrated balanced uni-traveling-carrier photodiode with tunable MMI coupler for microwave photonic circuits, in: Conference on Optoelectronic and Microelectronic Materials and Devices (COMMAD), Perth, Australia, December 2006.

[138] Zhi Li, Hao Chen, Huapu Pan, Andreas Beling, J.C. Campbell, High-power integrated balanced photodetector, IEEE Photon. Technol. Lett. 24 (21) (2009) 1858–1860.

[139] V. Houtsma, T. Hu, N.G. Weimann, R. Kopf, A. Tate, J. Frackoviak, R. Reyes, Y.K. Chen, L. Zhang, High-power linear balanced InP photodetectors for coherent analog optical links, in: 2011 IEEE Avionics, Fiber-Optics and Photonics Technology Conference, 2011, pp. 95–96.

[140] Andreas Beling, Hao Chen, Huapu Pan, Joe C. Campbell, High-power monolithically integrated traveling wave photodiode array, IEEE Photon. Technol. Lett. 21 (24) (2009) 1813–1815.

[141] S. Itakura, K. Sakai, T. Nagatsuka, E. Ishimura, M. Nakaji, H. Otsuka, K. Mori, Y. Hirano, High-current backside-illuminated photodiode array module for optical analog links, J. Lightwave Technol. 28 (6) (2010) 965–971.

[142] M.W. Geis, S.J. Spector, M.E. Grein, R.T. Schulein, J.U. Yoon, D.M. Lennon, S. Deneault, F. Gan, F.X. Kaertner, T.M. Lyszczarz, CMOS-compatible all-Si high-speed waveguide photodiodes with high responsivity in near-infrared communication band, IEEE Photon. Technol. Lett. 19 (3) (2007) 152–154.

[143] M. Zirngibl, J.C. Bischoff, M. Ilegems, J.P. Hirtz, B. Bartenian, P. Beaud, W. Hodel, High speed 1.3pm InGaAs/GaAs superlattice on Si photodetector, Electron. Lett. 26 (1990) 1027–1029.

[144] Y. Gao, Z. Zhong, S. Feng, Y. Geng, H. Liang, A.W. Poon, K.M. Lau, High-speed normal-incidence p-i-n InGaAs photodetectors grown on silicon substrates by MOCVD, IEEE Photon. Technol. Lett. 24 (4) (2012) 237–239.

[145] G. Masini, L. Colace, F. Galluzzi, G. Assanto, Advances in the field of poly-Ge on Si near infrared photodetectors, Mater. Sci. Eng. B 69 (2000) 257–260.

[146] L. Colace, G. Masini, A. Altieri, G. Assanto, Waveguide photodetectors for the near-infrared in polycrystalline germanium on silicon, IEEE Photon. Technol. Lett. 18 (9) (2006) 1094–1096.

[147] S. Luryi, A. Kastalsky, J. Bean, New infrared detector on a silicon chip, IEEE Trans. Electron Dev. 31 (1984) 1135–1139.

[148] L. Colace, G. Masini, F. Galluzzi, G. Assanto, G. Capellini, L. Di Gaspare, E. Palange, F. Evangelisti, Metal-semiconductor-metal near-infrared light detector based on epitaxial Ge/Si, Appl. Phys. Lett. 72 (1998) 3175–3178.

[149] J. Osmond, G. Isella, D. Chrastina, R. Kaufmann, M. Acciarri, H.V. Kanel, Ultralow dark current Ge/Si(100) photodiodes with low thermal budget, Appl. Phys. Lett. 94 (2009) 201106:1–201106:3.

[150] D. Liang, G. Roelkens, R. Baets, J.E. Bowers, Hybrid integrated platforms for silicon photonics, Materials 3 (2010) 1782–1802.

[151] S.J. Koester, J.D. Schaub, G. Dehlinger, J.O. Chu, Germanium-on-SOI infrared detectors for integrated photonic applications, IEEE J. Sel. Top. Quant. Electron. 12 (6) (2006) 1489–1502.

[152] J. Liu, D.D. Cannon, K. Wada, Y. Ishikawa, S. Jongthammanurak, D.T. Danielson, J. Michel, L.C. Kimerling, Tensile strained Ge p-i-n photodetectors on Si platform for C and L band telecommunications, Appl. Phys. Lett. 87 (2005) 011110–011112.

[153] T.V. Muoi, Receiver design for high-speed optical-fiber systems, J. Lightwave Technol. 2 (3) (1984) 243–264.

[154] K.-W. Ang, T.-Y. Liow, M.-B. Yu, Q. Fang, J. Song, G.-Q. Lo, D.-L. Kwong, Low thermal budget monolithic integration of evanescent-coupled Ge-on-SOI photodetector on Si CMOS platform, IEEE J. Sel. Top. Quant. Electron. 16 (1) (2010) 106–113.

[155] H.-Y. Yu, J.-H. Park, A.K. Okyay, K.C. Saraswat, Selective-area high-quality germanium growth for monolithic integrated optoelectronics, IEEE Electron Dev. Lett. 33 (4) (2012) 579–581.

[156] H.-C. Luan, D.R. Lim, K.K. Lee, K.M. Chen, J.G. Sandland, K. Wada, L.C. Kimerling, High-quality Ge epilayers on Si with low threading-dislocation densities, Appl. Phys. Lett. 75 (19) (1999) 2909–2911.

[157] Z. Huang, J. Oh, J.C. Campbell, Back-side-illuminated high-speed Ge photodetector fabricated on Si substrate using thin SiGe buffer layers, Appl. Phys. Lett. 85 (2004) 3286–3289.

[158] D. Suh, S. Kim, J. Joo, G. Kim, 36-GHz high-responsivity Ge photodetectors grown by RPCVD, IEEE Photon. Technol. Lett. 21 (10) (2009) 672–674.

[159] S. Klinger, M. Berroth, M. Kaschel, M. Oehme, E. Kasper, Ge-on-Si p-i-n photodiodes with a 3-dB bandwidth of 49 GHz, IEEE Photon. Technol. Lett. 21 (13) (2009) 920–922.

[160] D. Ahn, C.-Y. Hong, J. Liu, W. Giziewicz, M. Beals, L.C. Kimerling, J. Michel, J. Chen, F.X. Kärtner, High performance, waveguide integrated Ge photodetectors, Opt. Express 15 (7) (2007) 3916–3921.

[161] T. Yin, R. Cohen, M.M. Morse, G. Sarid, Y. Chetrit, D. Rubin, M.J. Paniccia, 31 GHz Ge n-i-p waveguide photodetectors on silicon-on-insulator substrate, Opt. Express 15 (21) (2007) 13965–13971.

[162] P. De Dobbelaere, B. Analui, E. Balmater, D. Guckenberger, M. Harrison, R. Koumans, D. Kucharski, Y. Liang, G. Masini, A. Mekis, S. Mirsaidi, A. Narasimha, M. Peterson, T. Pinguet, D. Rines, V. Sadagopan, S. Sahni, T.J. Sleboda, Y. Wang, B. Welch, J. Witzens, J. Yao, S. Abdalla, S. Gloeckner, G. Capellini, Demonstration of first WDM CMOS photonics transceiver with monolithically integrated photo-detectors, in: Proc. 34th Europ. Conf. Opt. Commun. (ECOC'08), Brussels, Belgium, Paper Tu.3.C.1, 2008.

[163] S.J. Koester, C.L. Schow, L. Schares, G. Dehlinger, J.D. Schaub, F.E. Doany, R.A. John, Ge-on-SOI-detector/Si-CMOS-amplifier receivers for high-performance optical-communication applications, J. Lightwave Technol. 25 (1) (2007) 46–57.

[164] G. Masini, S. Sahni, G. Capellini, J. Witzens, C. Gunn, High-speed near infrared optical receivers based on Ge waveguide photodetectors integrated in a CMOS process, Adv. Opt. Technol. (2008) Article ID 196572

[165] S. Assefa, F. Xia, S.W. Bedell, Y. Zhang, T. Topuria, P.M. Rice, Y.A. Vlasov, CMOS-integrated high-speed MSM germanium waveguide photodetector, Opt. Express 18 (5) (2010) 4986–4999.

[166] S. Assefa, F. Xia, S.W. Bedell, Y. Zhang, T. Topuria, P.M. Rice, Y.A. Vlasov, CMOS-Integrated 40 GHz germanium waveguide photodetector for on-chip optical interconnects, in: Tech. Dig. Opt. Fiber Commun. Conf. (OFC'09), Los Angeles, CA, Paper OMR 4, 2009.

[167] C.T. DeRose, D.C. Trotter, W.A. Zortman, A.L. Starbuck, M. Fisher, M.R. Watts, P.S. Davids, Ultra compact 45 GHz CMOS compatible germanium waveguide photodiode with low dark current, Opt. Express 19 (25) (2011) 24897–24904.

[168] L. Vivien, J. Osmond, J.-M. Fédéli, D. Marris-Morini, P. Crozat, J.-F. Damlencourt, E. Cassan, Y. Lecunff, S. Laval, 42 GHz p.i.n germanium photodetector integrated in a silicon-on-insulator waveguide, Opt. Express 17 (8) (2009) 6252–6257.

[169] D. Feng, S. Liao, P. Dong, N.-N. Feng, H. Liang, D. Zheng, C.-C. Kung, J. Fong, R. Shafiiha, J. Cunningham, A.V. Krishnamoorthy, M. Asghari, High-speed Ge photodetector monolithically integrated with large cross-section silicon-on-insulator waveguide, Appl. Phys. Lett. 95 (2009) 261105–261107.

[170] L. Vivien, A. Polzer, D. Marris-Morini, J. Osmond, J.M. Hartmann, P. Crozat, E. Cassan, C. Kopp, H. Zimmermann, J.M. Fédéli, Zero-bias 40 Gbit/s germanium waveguide photodetector on silicon, Opt. Express 20 (2) (2012) 1096–1101.

[171] D. Pasquariello, K. Hjort, Plasma-assisted InP-to-Si low temperature wafer bonding, IEEE J. Sel. Top. Quant. Electron. 8 (1) (2002) 118–131.

[172] H.-H. Chang, Y.-H. Kuo, H.-W. Chen, R. Jones, A. Barkai, M.J. Paniccia, J.E. Bowers, Integrated triplexer on hybrid silicon platform, in: Tech. Dig. Opt. Fiber Commun. Conf. (OFC'10), Los Angeles, CA, Paper OThC4, 2010.

[173] M.J.R. Heck, H.-W. Chen, A.W. Fang, B.R. Koch, D. Liang, H. Park, M.N. Sysak, J.E. Bowers, Hybrid silicon photonics for optical interconnects, IEEE J. Sel. Top. Quant. Electron. 17 (2) (2011) 333–346.

[174] T. Spuesens, F. Mandorlo, P. Rojo-Romeo, P. Régreny, N. Olivier, J.-M. Fédeli, D. Van Thourhout, Compact integration of optical sources and detectors on SOI for optical interconnects fabricated in a 200 mm CMOS pilot line, J. Lightwave Technol. 30 (11) (2012) 1764–1770.

[175] J. Brouckaert, G. Roelkens, D. Van Thourhout, R. Baets, Thin-film III–V photodetectors integrated on silicon-on-insulator photonic ICs, J. Lightwave Technol. 25 (4) (2007) 1053–1060.

[176] H. Park, M.N. Sysak, H.-W. Chen, A.W. Fang, D. Liang, L. Liao, B.R. Koch, J. Bovington, Y. Tang, K. Wong, M. Jacob-Mitos, R. Jones, J.E. Bowers, Device and integration technology for silicon photonic transmitters, IEEE J. Sel. Top. Quant. Electron. 17 (3) (2011) 671–688.

[177] H. Park, A.W. Fang, R. Jones, O. Cohen, O. Raday, M.N. Sysak, M.J. Paniccia, J.E. Bowers, A hybrid AlGaInAs-silicon evanescent waveguide photodetector, Opt. Express 15 (10) (2007) 6044–6052.

[178] A.W. Fang, H. Park, O. Cohen, R. Jones, M.J. Paniccia, J.E. Bowers, Electrically pumped hybrid AlGaInAs-silicon evanescent laser, Opt. Express 14 (2006) 9203–9210.

[179] H. Park, Y.-H. Kuo, A.W. Fang, R. Jones, O. Cohen, M.J. Paniccia, J.E. Bowers, A hybrid AlGaInAs-silicon evanescent preamplifier and photodetector, Opt. Express 15 (21) (2007) 13539–13546.

[180] H.-H. Chang, Y.-H. Kuo, R. Jones, A. Barkai, J.E. Bowers, Integrated hybrid silicon triplexer, Opt. Express 18 (23) (2010) 23891–23899.

[181] A. Beling, Y. Fu, Z. Li, H. Pan, Q. Zhou, A. Cross, M. Piels, J. Peters, J. E. Bowers, J.C. Campbell, Modified uni-traveling carrier photodiodes heterogeneously integrated

on SOI, in: Integrated Photonics Research, Silicon and Nano Photonics (IPR 2012) Topical Meeting, Paper IM2A.2, 2012.

[182] A. Beling, M. Piels, A.S. Cross, Y. Fu, Q. Zhou, J. Peters, J.E. Bowers, J.C. Campbell, High-power InP-based waveguide photodiodes and photodiode arrays heterogeneously integrated on SOI, in: 24th International Conference on Indium Phosphide and Related Materials (IPRM 2012), Santa Barbara, CA, 27–30 August 2012, Postdeadline Paper.

[183] P.R.A. Binetti, X.J.M. Leijtens, T. de Vries, Y.S. Oei, L. Di Cioccio, J.-M. Fedeli, C. Lagahe, J. Van Campenhout, D. Van Thourhout, P.J. van Veldhoven, R. Nötzel, M.K. Smit, InP/InGaAs photodetector on SOI photonic circuitry, IEEE Photon. J. 2 (3) (2010) 299–305.

[184] P.R.A. Binetti, R. Orobtchouk, X.J.M. Leijtens, B. Han, T. de Vries, Y.-S. Oei, L. Di Cioccio, J.-M. Fedeli, C. Lagahe, P.J. van Veldhoven, R. Nötzel, M.K. Smit, InP-based membrane couplers for optical interconnects on Si, IEEE Photon. Technol. Lett. 21 (5) (2009) 337–339.

[185] W. Bogaerts, D. Taillaert, B. Luyssaert, P. Dumon, J. Van Campenhout, P. Bienstman, D. Van Thourhout, R. Baets, V. Wiaux, S. Beckx, Basic structures for photonic integrated circuits in silicon-on-insulator, Opt. Express 12 (8) (2004) 1583–1591.

[186] G. Roelkens, J. Brouckaert, D. Taillaert, P. Dumon, W. Bogaerts, D. Van Thourhout, R. Baets, Integration of InP/InGaAsP photodetectors onto silicon-on-insulator waveguide circuits, Opt. Express 13 (25) (2005) 10102–10108.

[187] H. Melchior, A.R. Hartman, D.P. Schinke, T.E. Seidel, Planar epitaxial silicon avalanche photodiode, Bell Syst. Tech. J. 57 (1978) 1791–1807.

[188] S.R. Forrest, M. DiDomenico, Jr R.G. Smith, H.J. Stocker, Evidence of tunneling in reverse-bias III–V photodetector diodes, Appl. Phys. Lett. 36 (1980) 580–582.

[189] H. Ando, H. Kaaba, M. Ito, T. Kaneda, Tunneling current in InGaAsP and optimum design for InGaAs/InP avalanche photo-diodes, Jpn. J. Appl. Phys. 19 (1980) 1277–1280.

[190] P.P. Webb, R.J. McIntyre, A silicon avalanche photodiode for 1.06 μm radiation, in: Solid State Sensors Symposium, Minneapolis, MN, June 1970.

[191] H.W. Ruegg, An optimized avalanche photodiode, IEEE Trans. Electron Dev. ED-14 (May) (1967) 239–251.

[192] K. Nishida, K. Taguchi, Y. Matsumoto, InGaAsP heterojunction avalanche photodiodes with high avalanche gain, Appl. Phys. Lett. 35 (1979) 251–253.

[193] S.R. Forrest, O.K. Kim, R.G. Smith, Optical response time of In $_{0.53}$Ga$_{0.47}$As avalanche photodiodes, Appl. Phys. Lett. 41 (1982) 95–98.

[194] J.C. Campbell, A.G. Dentai, W.S. Holden, B.L. Kasper, High-performance avalanche photodiode with separate absorption, grading, and multiplication regions, Electron. Lett. 18 (1983) 818–820.

[195] Y. Matsushima, A. Akiba, K. Sakai, K. Kushirn, Y. Node, K. Utaka, High-speed response InGaAs/InP heterostructure avalanche photodiode with InGaAsP buffer layers, Electron. Lett. 18 (1982) 945–946.

[196] F. Capasso, A.Y. Cho, P.W. Foy, Low-dark-current low-voltage 1.3–1.6 μm avalanche photodiode with high-low electric field profile and separate absorption and multiplication regions by molecular beam epitaxy, Electron. Lett. 20 (15) (1984) 635–637.

[197] P. Webb, R. McIntyre, J. Scheibling, M. Holunga, A planar InGaAs APD fabricated using Si implantation and regrowth techniques, in: Tech. Digest of 1990 Opt. Fiber Conf., New Orleans, 1988.

[198] L.E. Tarof, Planar InP-InGaAs avalanche photodetectors with n-multiplication layer exhibiting a very high gain-bandwidth product, IEEE Photon. Technol. Lett. 2 (1990) 643–645.

[199] M.A. Itzler, K.K. Loi, S. McCoy, N. Codd, N. Komaba, Manufacturable planar bulk-InP avalanche photodiodes for 10 Gb/s applications, in: Proc. LEOS'99, San Francisco, CA, November 1999.

[200] J.D. Beck, C.-F Wan, M.A. Kinch, J.E. Robinson, MWIR HgCdTe avalanche photodiodes, Proc. SPIE 4454 (2001) 188–197.

[201] M. Kinch, J. Beck, C. Wan, F. Ma, J. Campbell, HgCdTe electron avalanche photodiodes, J. Electron. Mater. 23 (6) (2004) 630–639.

[202] J. Beck, C. Wan, M. Kinch, J. Robinson, P. Mitra, R. Scritchfield, F. Ma, J. Campbell, The HgCdTe electron avalanche photodiode, J. Electron. Mater. 35 (6) (2006) 1166–1173.

[203] A.R.J. Marshall, C.H. Tan, M.J. Steer, J.P.R. David, Electron dominated impact ionization and avalanche gain characteristics in InAs photodiodes, Appl. Phys. Lett. 93 (11) (2008) 111107.

[204] A.R.J. Marshall, P.J. Ker, A. Krysa, J.P.R. David, C.H. Tan, High speed InAs electron avalanche photodiodes overcome the conventional gain-bandwidth product limit, Opt. Express 19 (23) (2011) 23341–23349.

[205] C.A. Lee, R.A. Logan, R.L. Batdorf, J.J. Kleimack, W. Weigmann, Ionization rates of holes and electrons in silicon, Phys. Rev. 134 (1964) A761–A773.

[206] J. Conradi, The distributions of gains in uniformly multiplying avalanche photodiodes: experimental, IEEE Trans. Electron Dev. ED-19 (6) (1972) 713–718.

[207] W.N. Grant, Electron and hole ionization rates in epitaxial silicon at high electric fields, Solid-State Electron. 16 (1973) 1189–1203.

[208] T. Kaneda, H. Matsumoto, T. Yamaoka, A model for reach-through avalanche photodiodes (RAPD's), J. Appl. Phys. 47 (7) (1976) 3135–3139.

[209] A.R. Hawkins, T.E. Reynolds, D.R. England, D.I. Babic, M.J. Mondry, K. Streubel, J.E. Bowers, Silicon heterointerface photodetector, Appl. Phys. Lett. 70 (1996) 303–305.

[210] Y. Kang, P. Mages, A.R. Clawson, P.K.L. Yu, M. Bitter, Z. Pan, A. Pauchard, S. Hummel, Y.H. Lo, Fused InGaAs-Si avalanche photodiodes with low-noise performances, IEEE Photon. Lett. 14 (2002) 1593–1595.

[211] M. Bitter, Z. Pan, S. Kristjansson, L. Boman, R. Gold, A. Pauchard, InGaAs-on-Si photodetectors for high-sensitivity detection, in: Proc. SPIE Infrared Tech. and Applications XXX, vol. 5406, 2004, pp. 1–12.

[212] Yimin. Kang, Han.-Din. Liu, Mike. Morse, Mario J. Paniccial, Moshe Zadka, Stas Litski, Gadi Sarid, Alexandre Pauchard, Ying-Hao Kuo, Hui-Wen Chen, Wissem Sfar Zaoui, John E. Bowers, Andreas Beling, Dion C. McIntosh, Xiaoguang Zheng, Joe C. Campbell, Monolithic germanium silicon avalanche photodiodes with 340 GHz gain–bandwidth product, Nat. Photon. 3 (2009) 59–63.

[213] C.L.F. Ma, M.J. Dean, L.E. Tarof, J.C.H. Yu, Temperature dependence of breakdown voltages in separate absorption, grading, charge, and multiplication InP/InGaAs avalanche photodiodes, IEEE Trans. Electron Dev. 42 (1995) 810–818.

[214] K.-S. Hyun, C.-Y. Park, Breakdown characteristics in InP/InGaAs avalanche photodiode with p-i-n multiplication layer structure, J. Appl. Phys. 81 (1997) 974–984.

[215] Y. Okuto, C.R. Crowell, Ionization coefficients in semiconductors: a nonlocalized property, Phys. Rev. B 10 (1974) 4284–4296.

[216] W.R. Clark, K. Vaccaro, W.D. Waters, InAlAs-InGaAs based avalanche photodiodes for next generation eye-safe optical receivers, Proc. SPIE 6796 (2008) pp. 6792H 1–15, Photonics North 2007

[217] D.S. Ong, K.F. Li, G.J. Rees, G.M. Dunn, J.P.R. David, P.N. Robson, Monte Carlo estimation of excess noise factor in thin p^+-i-n^+ avalanche photodiodes, in: Proceedings of the IEEE Twenty-Fourth Internat. Symp. Compound Semic., 1998, pp. 631–634.

[218] S. Wang, R. Sidhu, X.G. Zheng, X. Li, X. Sun, A.L. Holmes, Jr J.C. Campbell, Low-noise avalanche photodiodes with graded impact-ionization-engineered multiplication region, IEEE Photon. Technol. Lett. 13 (2001) 1346.

[219] J.P. Gordon, R.E. Nahory, M.A. Pollack, J.M. Worlock, Low-noise multistage avalanche photodetector, IEEE Electron. Lett. 15 (17) (1979) 518–519.

[220] S. Rakshit, N.B. Charkraborti, Multiplication noise in multi-heterostructure avalanche photodiodes, Solid State Electron. 26 (10) (1983) 999–1003.

[221] W. Clark, United States Patent 6,747,296, 2004.

[222] G.M. Williams, M.A. Compton, A.S. Huntington, High-speed photon counting with linear-mode APD receivers, Proc. SPIE 7320 (2009) 732012 (p. 9).

[223] W.R. Clark, K. Vaccaro, W.D. Waters, InAlAs-InGaAs based avalanche photodiodes for next generation eye-safe optical receivers, Proc. SPIE 6796 (2007) 67962H.

[224] W. Sun, X. Zheng, Z. Lu, J.C. Campbell, Monte Carlo simulation of InAlAs/InAlGaAs tandem avalanche photodiodes, J. Quant. Electron. 48 (4) (2012) 528–532.

[225] A. Huntington, M. Compton, S. Coykendall, G. Soli, G.M. Williams, Linear-mode single-photon-sensitive avalanche photodiodes for GHz-rate near-infrared quantum communications, in: Military Communications Conference, 2008, pp. 1–6.

[226] W.P. Risk, D.S. Bethune, Quantum cryptography, Opt. Photon. News 13 (7) (2002) 26–32.

[227] S. Cova, M. Ghioni, A. Lacaita, C. Samori, F. Zappa, Avalanche photodiodes and quenching circuits for single-photon detection, Appl. Opt. 35 (1996) 1956–1976.

[228] D.S. Bethune, W.P. Risk, An autocompensating fiber-optic quantum cryptography system based on polarization splitting of light, IEEE J. Quant. Electron. 36 (3) (2000) 340–347.

[229] Hu Chong, Xiaoguang Zheng, Joe C. Campbell, Bora M. Onat, Xudong Jiang, Mark A. Itzler, Characterization of an InGaAs/InP-based single-photon avalanche diode with gated-passive quenching with active reset circuit, J. Mod. Opt. 58 (2011) 201–209.

[230] Mingguo Liu, Hu Chong, Joe C. Campbell, Zhong Pan, Mark M. Tashima, A novel quenching circuit to reduce afterpulsing of single photon avalanche diodes, IEEE J. Quant. Electron. 44 (5) (2008) 430–434.

[231] N. Namekata et al., 800 MHz single-photon detection at 1550-nm using an InGaAsInP avalanche photodiode operated with a sine wave gating, Opt. Express 14 (2006) 10043–10049.

[232] N. Namekata et al., High-speed single-photon detection using 2-GHz sinusoidally gated InGaAsInP avalanche photodiode, Quant. Commun. Quant. 23 (2010) 34–38.

[233] J. Zhang et al., Practical fast gate rate InGaAs/InP single-photon avalanche photodiodes, Appl. Phys. Lett. 95 (2009) 091103.

[234] Z.L. Yuan et al., Multi-gigahertz operation of photon counting InGaAs avalanche photodiodes, Appl. Phys. Lett. 96 (2010) 071101.

[235] K.K. Forsyth, J.C. Dries, Variations in the photon-counting performance of InGaAs/ InP avalanche photodiodes, in: Proceedings of IEEE LEOS Annual Conference, vol. 2, 2003, p. 777.

[236] K.A. McIntosh, J.P. Donnelly, D.C. Oakley, A. Napoleon, S.D. Calawa, L.J. Mahoney, K.M. Molvar, E.K. Duerr, S.H. Groves, D.C. Shaver, InGaAsP/InP avalanche photodiodes for photon counting at 1.06 μm, Appl. Phys. Lett. 81 (14) (2002) 2505–2507.

[237] Mark Itzler, Rafael Ben-Michael, Chia-Fu Hsu, Krystyna Slomkowsk, Alberto Tosi, Sergio Cova, Franco Zappa, Radu Ispasoiu, in: Single-Photon Workshop (SPW) 2005: Sources, Detectors, Applications and Measurement Methods, Teddington, UK, 24–26 October, 2005.

[238] Chong Hu, Mingguo Liu, J.C. Campbell, Improved passive quenching with active reset circuit, in: Proceedings of the SPIE – The International Society for Optical Engineering, Advanced Photon Counting Techniques, vol. 7320, April 2009.

[239] M.A. Itzler, M. Entwistle, Xudong Jiang, High-rate photon counting with Geiger-mode APDs, in: Proceedings of the IEEE Photonic Society 24th Annual Meeting, 2011, pp. 348–349.

[240] Alessandro Restelli, J. Bienfang, Avalanche discrimination and high-speed counting in periodically gated single-photon avalanche diodes, in: Proceedings of the SPIE – The International Society for Optical Engineering, Advanced Photon Counting Techniques, Baltimore, MD, vol. 8375, April 2012, p. 11.

[241] C. Elliott, A. Colvin, D. Pearson, O. Pikalo, J. Schlafer, H. Yeh, Current status of the DARPA quantum network, Proc. SPIE: Quant. Inf. Comput. III 5815 (February 2005) 138–149.

[242] M. Peev, C. Pacher, R. Alléaume, C. Barreiro, J. Bouda, W. Boxleitner, T. Debuisschert, E. Diamanti, M. Dianati, J.F. Dynes, S. Fasel, S. Fossier, M. Fürst, J.-D. Gautier, O. Gay, N. Gisin, P. Grangier, A. Happe, Y. Hasani, M. Hentschel, H. Hübel, G. Humer, T. Länger, M. Legré, R. Lieger, J. Lodewyck, T. Lorünser, N. Lütkenhaus, A. Marhold, T. Matyus, O. Maurhart, L. Monat, S. Nauerth, J.-B. Page, A. Poppe, E. Querasser, G. Ribordy, S. Robyr, L. Salvail, A.W. Sharpe, A.J. Shields, D. Stucki, M. Suda, C. Tamas, T. Themel, R.T. Thew, Y. Thoma, A. Treiber, P. Trinkler, R. Tualle-Brouri, F. Vannel, N. Walenta, H. Weier, H. Weinfurter, I. Wimberger, Z.L. Yuan, H. Zbinden, A. Zeilinger, The SECOQC quantum key distribution network in Vienna, New J. Phys. 11 (7) (2009) pp. 075001-1–075001-37

[243] Swiss Quantum (Online). <http://www.swissquantum.com/>.

[244] S. Wang, W. Chen, Z.-Q. Yin, Y. Zhang, T. Zhang, H.-W. Li, F.-X. Xu, Z. Zhou, Y. Yang, D.-J. Huang, L.-J. Zhang, F.-Y. Li, D. Liu, Y.-G. Wang, G.-C. Guo, Z.-F. Han, Field test of wavelength-saving quantum key distribution network, Opt. Lett. 35 (14) (2010) 2454–2456.

[245] M. Sasaki, M. Fujiwara, H. Ishizuka, W. Klaus, K. Wakui, M. Takeoka, S. Miki, T. Yamashita, Z. Wang, A. Tanaka, K. Yoshino, Y. Nambu, S. Takahashi, A. Tajima, A. Tomita, T. Domeki, T. Hasegawa, Y. Sakai, H. Kobayashi, T. Asai, K. Shimizu, T. Tokura, T. Tsurumaru, M. Matsui, T. Honjo, K. Tamaki, H. Takesue, Y. Tokura, J.F. Dynes, A.R. Dixon, A.W. Sharpe, Z.L. Yuan, A.J. Shields, S. Uchikoga, M. Legré, S. Robyr, P. Trinkler, L. Monat, J.-B. Page, G. Ribordy, A. Poppe, A. Allacher, O. Maurhart, T. Länger, M. Peev, A. Zeilinger, Field test of quantum key distribution in the Tokyo QKD Network, Opt. Express 19 (11) (2011) 10387–10409.

[246] A. Tanaka, M. Fujiwara, K.-I. Yoshino, S. Takahashi, Y. Nambu, A. Tomita, S. Miki, T. Yamashita, Z. Wang, M. Sasaki, A. Tajima, High-speed quantum key distribution system for 1-Mbps real-time key generation, IEEE J. Quant. Electron. 48 (4) (2012) 542–550.

[247] Y. Nambu, S. Takahashi, K. Yoshino, A. Tanaka, M. Fujiwara, M. Sasaki, A. Tajima, S. Yorozu, A. Tomita, Efficient and low-noise single-photon avalanche photodiode for 1.244-GHz clocked quantum key distribution, Opt. Express 19 (21) (2011) 20531–20541.

Fundamentals of Photonic Crystals for Telecom Applications—Photonic Crystal Lasers

4

Susumu Noda

Department of Electronic Science and Engineering, Kyoto University,
Kyoto 615-8510, Japan

4.1 INTRODUCTION

Photonic crystals are optical nanostructures with refractive indices that vary periodically, and are characterized by a photonic bandgap in terms of the energy of photons. Light with wavelengths that lie in the photonic bandgap cannot exist in the crystal, giving rise to completely different optical phenomena than in free space. These characteristics enable various interesting methods of optical control including materials from which light emission is inhibited [1–5], trapping of photons at a microscopic point defect (known as a photonic nanocavity) [6–9], realization of nanolasers [10–12], and key elements for quantum information processing [13–15]. It is also expected that photonic crystals can be designed to allow the dynamic control of optical nanocavities [16,17] and the control of the dispersion characteristics of waveguides [18,19], which may enable the propagation of light to be slowed down to the limit and ultimately halted. The attractive potentials of photonic crystals also include use of the standing wave state at the band edge, where the group velocity becomes zero, in order to realize broad-area coherent laser operation [20–25]. These characteristics ensure that photonic crystals are currently attracting much attention.

The first half of this chapter introduces research activities that are geared toward realizing the ultimate nanolaser [11] using the photonic bandgap effect. Important aspects of this effort are in the achievement of spontaneous emission suppression and strong optical confinement using a photonic nanocavity. During the process of implementation of this goal, interesting phenomena, which can be classified as Quantum Anti-Zeno effect [26,27], have been observed [14].

The second half of the chapter focuses on the current state of research in the field of broad-area coherent photonic crystal lasers using the band-edge effect [20–25], which occupies a position opposite to that of nanolasers discussed above. The main characteristics of these lasers will be discussed, including their high-power operation, the generation of tailored beam patterns, the surface-emitting laser operation in the blue-violet region, and even the beam-steering functionality.

Optical Fiber Telecommunications VIA. http://dx.doi.org/10.1016/B978-0-12-396958-3.00004-4

4.2 ULTIMATE NANOLASERS

Semiconductor lasers typically suffer from unnecessary spontaneous emission before the laser oscillation starts. This degrades the laser characteristics, for example giving rise to an increased lasing threshold and noise. Therefore, the possibility of realizing a semiconductor laser in which the threshold is ultimately minimized (the so-called thresholdless laser), by eliminating spontaneous emission as far as possible, is attracting attention as one type of the ultimate semiconductor lasers. The development of photonic crystal nanocavities and their integration with quantum dots is accelerating research activities that aim to achieve such ultimate lasers. In the following section we discuss the current status of this area of nanolaser research.

4.2.1 Toward realizing ultimate nanolasers

The spontaneous emission from a semiconductor laser occurs due to interactions between the resonant and leakage modes in the laser cavity with the gain medium. Among these many optical modes, one of the resonant modes undergoes a transition to a laser oscillation, whereas the other modes generate unnecessary spontaneous emission that degrades the laser characteristics. The following three requirements should thus be addressed in order to minimize spontaneous emission and to realize a thresholdless laser:

(i) Elimination, as far as possible, of unnecessary optical modes from the region of the photonic crystal where lasing occurs.

(ii) Utilization of a single-cavity mode with a sufficiently high Q factor and small loss as a laser mode, the modal volume of which should also be as small as possible in order to maximize the interaction with the gain medium.

(iii) Coupling of the gain medium only with the single optical mode described in (ii).

Regarding requirement (i), whereas modes other than the lasing mode eventually generate unnecessary spontaneous emission, progress in photonic crystal research is making it possible to minimize these unnecessary modes. It has been demonstrated both theoretically and experimentally that the two-dimensional photonic crystal slab structure shown in Figure 4.1, which enables pseudo-three-dimensional optical control, eliminates optical modes on the surface [4,5,28], leading to the suppression of 94% of the unnecessary spontaneous emission [29].

Requirement (ii) can be addressed by introducing a point defect, which can be envisaged as a disturbance in the periodic structure, in the middle of the two-dimensional photonic crystal slab; this leads to the formation of a microscopic single photonic mode (photonic nanocavity) with a modal volume V. If the Q factor of this mode is sufficiently large, its associated emission rate can be increased due to the so-called Purcell effect, which enhances the emission rate by a factor of Q/V. This increases the average number of photons in the cavity at the same excitation level, thereby reducing the lasing threshold (on the condition that the average number of photons exceeds 1). The large Q/V also contributes to making the lasing operation

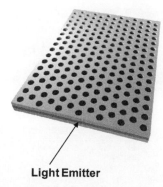

Light Emitter

FIGURE 4.1 Schematic illustration of two-dimensional photonic crystal slab structure.
The two-dimensional photonic bandgap effect occurring in the plane of a slab with a periodic structure inhibits the existence of slab modes that are confined in the slab. A large difference in refractive index at the air/semiconductor interface strongly confines light in the up/down direction, which inhibits most of the slab modes and allows the existence of only a few leakage modes. As a result, more than 94% of the spontaneous emission is suppressed.

appear thresholdless because the emission rate for the nanocavity mode becomes much larger than the emission rate associated with the spontaneous emission that cannot be completely eliminated by photonic crystal design considerations, as described above. Therefore, it is essential to increase the Q factor of the optical nanocavity. Dramatic progress in this area has recently been achieved; a Q factor of 50,000 has been demonstrated by shifting the air holes at the edges of the cavity (see Figure 4.2) [7]. Furthermore, a Q factor in excess of 3 million [30] has recently been attained by introducing a photonic heterostructure [8].

Focusing now on requirement (iii), the most suitable gain medium that can be introduced into an optical nanocavity is the quantum dot [31,32]. Quantum dots enable three-dimensional carrier confinement, avoiding the effects of non-radiative centers that are inevitably introduced in the fabrication of photonic crystals and thereby restricting the coupling of the gain medium to the lasing mode. Furthermore, because quantum dots become transparent above a very weak level of excitation, only limited lowering of the cavity Q factor due to optical absorption by the quantum dots before lasing occurs, which would otherwise inhibit the enhanced emission effect described above. Quantum dots are thus expected to be effective in lowering the threshold of the laser. However, quantum dots with the best characteristics are currently fabricated using a self-formation method, which takes advantage of lattice mismatch introduced during crystal growth [32] and inevitably yields dots with random geometries and exciton resonance wavelengths. A quantum dot is expected to contribute to thresholdless operation only if the exciton resonance wavelength is the same as that of the resonant wavelength of the nanocavity, and when the position is located where the electric field of the nanocavity is strongest. When these

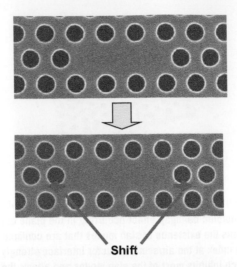

Shift

FIGURE 4.2 Conceptual method for increasing the *Q* factor of a photonic crystal optical nanocavity. A slight shift of the lattice points at the ends of a point defect, which is formed by the removal of three linearly adjacent lattice points in a two-dimensional, circular-hole, triangular-lattice photonic crystal, greatly increases the *Q* factor.

conditions are not met, there is a concern that a quantum dot may make threshold-less operation even more difficult by inhibiting the Purcell effect. Although recent developments in fabrication technologies are beginning to realize systems in which a single quantum dot and photonic nanocavity are aligned under the condition of low quantum dot density [33], the Purcell effect is still inhibited when the wavelengths of the quantum dot and the nanocavity do not match. Precise tuning between the two seems thus essential. However, a phenomenon that is contrary to intuition was recently reported whereby an optical nanocavity mode generates strong emission despite large detuning between the resonant mode of the nanocavity and the exciton resonance wavelength of the quantum dot [12,14]. The explanation of this new emission mechanism is generating debate and will be discussed in the following section [26,27,34–36]. This mechanism significantly relaxes the requirement for precise wavelength tuning and is accelerating progress toward the realization of an ultimate nanolaser [11,12,37].

4.2.2 Quantum anti-Zeno effect in a nanocavity and quantum dot system

Figure 4.3 schematically illustrates the emission mechanism for a system comprised of a photonic nanocavity and quantum dot. In this model, the optical nanocavity and the two-level electron system interact with one another, eventually dissipating into

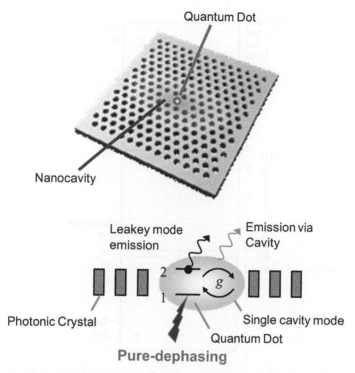

Quantum Dot

Nanocavity

Leakey mode emission

Emission via Cavity

Photonic Crystal

Single cavity mode

Quantum Dot

Pure-dephasing

FIGURE 4.3 Analytical model for integrating a two-dimensional photonic crystal optical nanocavity and a quantum dot.

free space. Two dissipation processes are considered in this model: a process where the two-level electron system directly relaxes, emitting photons to free space, and a process where photons are first emitted to the nanocavity before relaxation to free space. An emission spectrum can be calculated by evaluating the time evolution of the entire system using the master equation, assuming an initial state where only the two-level electron system is excited. We carried out this calculation as a function of the energy difference between the two-level electron system and the nanocavity; the results are presented in Figure 4.4a. An emission peak is observed at the cavity energy when the energies of the electron system and photonic nanocavity match. However, no emission from the optical nanocavity occurs when the energies do not match, in accordance with the intuition expressed in Section 4.2.1. However, because in reality the quantum dots are buried in the semiconductor matrix, a phase disturbance process is caused by factors such as surrounding electron systems and phonons that are not associated with the quantum dots themselves. This process is a pure phase relaxation, and when introduced in the model, it becomes apparent that strong emission can be generated from the nanocavity even when there is detuning in the energies (Figure 4.4b). Here, the pure phase relaxation process refers to a relaxation only in

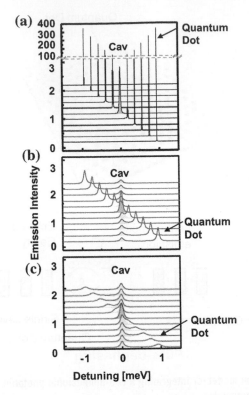

FIGURE 4.4 Emission spectra of integrated two-dimensional photonic crystal optical nanocavity and quantum dot systems. QD indicates emission peaks from the quantum dots, and Cav indicates emission peaks from the nanocavity. Spectra are shown for no pure phase relaxation (a), pure phase relaxation rate = 33 μeV (b), and 94 μeV (c).

the non-diagonal terms of the so-called density matrix, which corresponds to a disturbance process where only the phases of dipoles in the electron system interact with the surrounding environment. When the degree of pure phase relaxation is increased, as shown in Figure 4.4c, emission from the nanocavity becomes extremely strong. This result strongly suggests that pure phase relaxation provides the dominant contribution to emission from the cavity when the cavity and electron system energies do not match. This is an example of a general physical phenomenon known as the anti-Zeno effect, which is opposite to the well-known quantum Zeno effect. In the quantum Zeno effect, observation or decoherence inhibits changes of state, whereas in the anti-Zeno effect, observation or decoherence promotes changes of state. A possible explanation specific to this system is that the energy of the electron system is slightly broadened by the effects of either pure phase relaxation or decoherence; the trailing edge of the electron system energy distribution then coincides with the energy of the cavity mode, enabling emission from cavities whose energies are detuned from those of the electron system. However, the calculation results in Figure 4.4b and c suggest that the broadening of the electron system energy is extremely small, and it does not appear that the trailing edge of the energy distribution overlaps with the energy of the cavity. Emission from the cavity thus appears to be counterintuitive, which led us to investigate further why strong emission is possible for cavities whose energies are far from that of the electron system. It turns out that this phenomenon is specific to a system comprised of an optical nanocavity and quantum dots. This finding is based on the following effects: (a) The photonic bandgap effect significantly inhibits emission relaxation processes in which the electron system couples into leakage modes. (b) The Purcell effect is strongly suppressed due to the detuning effect: The path by which the electron system forces the nanocavity to oscillate in order to relax to free space is strongly suppressed. Thus, even when the energies of the electron system and cavity are detuned, energy is transferred through the trailing edge of the broadening of the cavity energy and then emitted to free space. This represents the major relaxation process and as a result, strong emission is observed at the cavity energy. This also implies that the conditions regarding wavelength detuning are significantly relaxed when quantum dots are used as the gain medium in an optical nanocavity, making the realization of a thresholdless laser more viable. Further discussion of this issue can be found in Ref. [27]. These results provide a direction for future research into coupled quantum dot and nanocavity systems, and are particularly important for the eventual realization of devices such as nanolasers.

4.3 BROAD-AREA COHERENT LASERS

Whereas interactions between photons and electrons in nanocavities were discussed in the previous section, we now focus on the opposite situation: extremely large-area coherent laser oscillation.

It is well known that semiconductor distributed feedback lasers possess a one-dimensional lattice, and that the forward-propagating wave undergoes Bragg reflection due to this grating being diffracted to the opposite direction. The resulting forward- and backward-propagating waves couple with each other to generate a standing wave, forming a cavity. This is equivalent to the fact that in a one-dimensional photonic crystal the cavity loss is smallest at the band edges, which are at both ends of the photonic bandgap, giving rise to a state that causes oscillation. When this idea is extended to photonic crystals with two-dimensional gratings, one can make use of the coupling of optical waves due to Bragg reflection within the two-dimensional plane in order to form a standing wave state that covers the entire surface of the plane [20,21,38]. As a result, it becomes possible to obtain an oscillation mode with an electromagnetic field distribution that is perfectly defined at each grating point in the two-dimensional crystal. The optical output can be diffracted in the direction perpendicular to the plane of the crystal, thus realizing a surface-emitting characteristic. Two-dimensional photonic crystals hence enable the construction of surface-emitting lasers in which not only the longitudinal mode of lasing is defined, but also the beam pattern, usually referred to as the transverse mode. Furthermore, it becomes possible to realize a novel laser that oscillates in a single longitudinal and transverse mode, no matter how large the surface area is, which surpasses a conventional concept in the field of laser research. The rigorous and precise methods to analyze the operation of this laser have been developed recently based on 3D coupled wave theory [39,40].

The first semiconductor laser to be based on this principle was realized in 1999 [19]. Since then, in addition to the demonstration of room-temperature continuous lasing [41], it has been shown that two-dimensional photonic crystals can generate beams with controlled polarization [22] and patterns [23]; for example, a dough-nut-shaped beam can be formed, which is expected to be focusable to sizes smaller than the wavelength [42]. The other notable recent developments using this principle include the realization of a current-injection-type blue-violet region surface-emitting laser [24] and the demonstration of electronically beam steering functionality [25]. We discuss the current state of the art in the following sections.

4.3.1 Broad-area coherent operation

Figure 4.5 shows an example of a laser based on the two-dimensional photonic crystal band-edge effect. This laser consists of two wafers, A and B; wafer A includes an active layer for the injection of electrons and holes, and a photonic crystal as the uppermost layer. The integration of wafers A and B results in the photonic crystal being sandwiched to complete the device. As shown in the insert of Figure 4.5, this photonic crystal has a square-lattice structure and is designed such that the periodicity in the Γ-X direction matches the emission wavelength in the active layer. In this

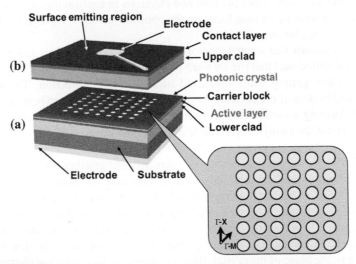

FIGURE 4.5 Schematic picture of a broad-area coherent operation photonic crystal laser based on the band-edge effect.

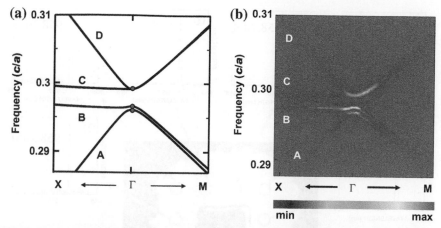

FIGURE 4.6 Band diagram of the photonic crystal laser shown in Figure 4.5. (a) Theoretical photonic band structure. The resonant mode is formed at the band edges, indicated by red dots. (b) Measured band structure. (For interpretation of the references to color in this figure legend, the reader is referred to the web version of this book.)

design, light propagating in a certain Γ-X direction is Bragg diffracted to the opposite (−180°) direction, as well as to the −90° and 90° directions; the four equivalent light waves propagating in the Γ-X direction then couple to form a two-dimensional cavity. More precise explanation on the lasing mechanism can be seen in Refs. [39,40]. Figure 4.6a shows the photonic band structure of this cavity. The lasing mode occurs at the band edges indicated by the red dots at the Γ-points of the four bands, A, B, C, and D. Detailed analysis indicates that the band edge in band A yields the highest Q factor, and that lasing oscillation most readily occurs there [40,43,44].

Figure 4.6b shows the measured band structure of a fabricated device based on the semiconductor GaAs, which is in good agreement with the calculated band structure in Figure 4.6a. Figure 4.7 shows the near-field pattern measured when the device was lasing, as well as lasing spectra at various points. Despite the large lasing area of 150×150 μm, single-wavelength operation was achieved across the device. Comparison between the measured lasing wavelength and the band structure in Figure 4.6b confirmed that the lasing oscillation occurred at the Γ-point of band A as theoretically predicted, demonstrating that this device indeed operates at the band edges of the two-dimensional photonic band structure, and that large-area single-mode operation is possible.

4.3.2 Beam pattern control by designing lattice points

The pattern of the surface-emitted beam from a photonic crystal laser can be determined by the Fourier transformation of its two-dimensional electromagnetic

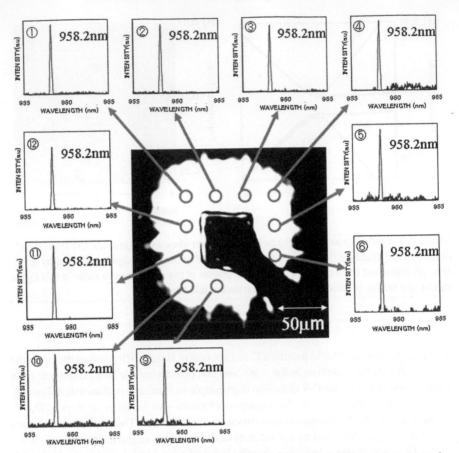

FIGURE 4.7 Near-field pattern of the photonic crystal laser shown in Figure 4.5, measured while the device was lasing. Lasing spectra at various points are shown.

distribution. This implies that the beam pattern can be tailored by varying the electromagnetic distribution in the two-dimensional plane, that is, by changing the coupling state of the light that propagates in various directions in the two-dimensional plane. One effective method of achieving this is to vary the shapes and spacing of the lattice points in the photonic crystal. Figure 4.8a and b shows the electromagnetic field distribution in the unit lattice of a crystal when the holes placed at the lattice points are circles and equilateral triangles, respectively. Changing the shape of the holes from circular to triangular removes the fourfold rotational symmetry in the electromagnetic field distribution; there is no symmetry in the x-direction for triangular holes. Figure 4.8c–g shows the electromagnetic field distributions over the entire crystal in cases where shifts of the lattice points were introduced in order to increase the lattice spacing in either the longitudinal or transverse directions. Figure 4.8c

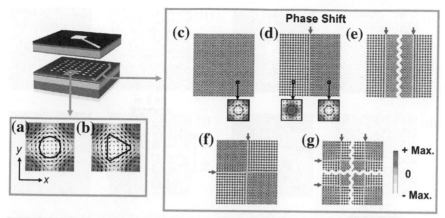

FIGURE 4.8 Electromagnetic field distribution in various photonic crystal lasers. Black arrows represent the electric field, and shading indicates the magnetic field in the direction perpendicular to the paper. Field distributions around the lattice points are shown for (a) circular lattice points and (b) triangular-lattice points. Field distributions over the entire crystal are shown for circular lattice points with (c) no phase shift, (d) one phase shift in the x-direction, (e) two phase shifts in the x-direction, (f) one phase shift in both directions, and (g) two phase shifts in both directions. The phase in each region was shifted by π across the boundary where the phase shift was introduced.

represents the case with no shift, whereas Figure 4.8d–g represents increasing numbers of shifts. It is apparent that shifting the lattice spacing reverses the polarity of the electromagnetic field distribution at the position of the shift. Further increasing the number of shifts repeats the reversal of the electromagnetic field. It is clear that the electromagnetic field distribution in the plane can be controlled in various ways by appropriate design of the photonic crystal.

Based on the above considerations, we fabricated devices with various different photonic crystal structures, as shown in Figure 4.9a–f. All of these devices exhibited lasing oscillation at room temperature with a stable single mode. The right-hand panels of Figure 4.9a–f show the corresponding measured beam patterns. An interesting array of patterns was obtained ranging from a single doughnut shape to twofold doughnut, fourfold doughnut, and regular circular shapes. The beam divergence was extremely narrow, reflecting the fact that these are large-area coherent laser oscillations.

The various beam patterns can be explained as follows. First, the device in Figure 4.9a has regular circular holes, and the corresponding electromagnetic field distribution exhibits well-defined rotational symmetry as shown in Figure 4.8a. When the laser light corresponding to this electromagnetic field distribution is output to free space, the electromagnetic field at the center of the beam cancels out to yield a doughnut-shaped beam. When a shift of the lattice period is introduced, as shown in Figure 4.9b, a polarity (+ and −) reversal of the electromagnetic field

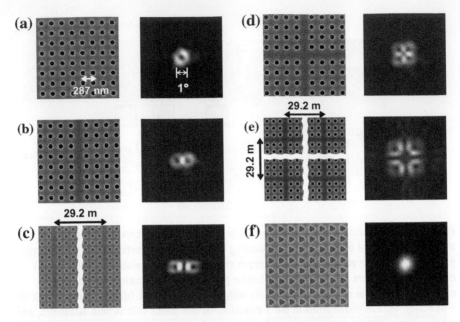

FIGURE 4.9 Electron micrographs (left-hand panels) and obtained beam patterns (right-hand panels) of fabricated photonic crystals used to construct lasers. Photonic crystals with circular lattice points are shown with (a) no shift of the lattice period, (b) one lattice shift in one direction, (c) two lattice shifts in one direction, (d) one lattice shift in both directions, and (e) two lattice shifts in both directions. (f) Triangular-lattice points with no phase shift.

distribution occurs at the location of the shift, as shown in Figure 4.8d. This changes the nature of the interference of the light emitted to free space, yielding two doughnut beams. As the number of lattice shifts is increased, changes in the interference conditions are repeated, yielding a variety of other doughnut-shaped beams as shown in Figure 4.9b–e. In particular, the doughnut beam in Figure 4.9a is expected to possess interesting characteristics such as the ability to manipulate non-transparent substances by acting as a pair of optical tweezers, and focusing characteristics that exceed the wavelength limit.

In contrast, triangular-lattice holes (Figure 4.9f) remove the rotational symmetry of the electromagnetic field distribution, as shown in Figure 4.9b. The cancelation effect at the center of the beam in Figure 4.9a is also lost, yielding a clean circular pattern. In this case, the polarization is also different, being linear. Introducing such a non-symmetrical effect is a key factor to achieve high optical output power by enabling a greater optical extraction efficiency in the perpendicular direction. A device constructed using equilateral triangular air holes exhibited continuous room-temperature optical output in excess of 100 mW [45]. Furthermore, we have also fabricated devices with a higher degree of non-symmetry, and succeeded in achieving a maximum output of more than 35 W in pulsed drive mode [46].

4.3.3 Extension to the blue-violet region

It is anticipated that extension of the lasing wavelength of photonic crystal surface-emitting lasers to the blue-violet region will greatly broaden the number of potential applications. For example, such lasers could act as a light source for later generation high-density optical disks, and as a light source for the observation and manipulation of microscopic objects. It should also be possible to control the polarization of high-power blue-violet surface-emitting lasers operating in a single longitudinal and transverse mode, and even to arrange them in two-dimensional arrays. These blue-violet lasers are expected to be key light sources in a variety of fields including information storage and processing, optical manipulation, and nano-bio applications. (see Figure 4.10)

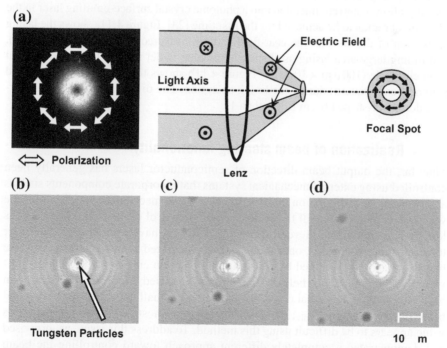

FIGURE 4.10 (a) Example of polarized doughnut beam from a photonic crystal laser, and schematic picture of the focusing method. (b)–(d) Optical trapping of a tungsten particle using a doughnut-shaped focal point. The particle was firmly held at the focal point even when the glass slide was removed.

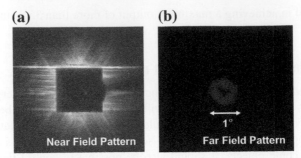

FIGURE 4.11 Near-field pattern (a) and far-field pattern (beam shape) (b) of a blue-violet photonic crystal laser during lasing oscillation.

Novel photonic crystal growth technology, recently developed for GaN and known as air-hole retained over-growth (AROG), which takes advantage of crystal growth characteristics specific to the GaN system, has enabled the fabrication of high-quality GaN/air two-dimensional photonic crystals inside a laser. This has recently allowed current injection into a photonic crystal surface-emitting laser in the blue-violet region to be achieved for the first time [23]. Figure 4.11a shows the lasing oscillation of this device (the near-field pattern measured after lasing had started), confirming large-area lasing oscillation in the blue-violet region. The central region is an electrode (100 μm × 100 μm). Figure 4.11b shows the corresponding far-field pattern (beam pattern) during the lasing operation. The divergence of the characteristic doughnut-shaped beam is less than 1°.

4.3.4 Realization of beam steering functionality

Thus far, the output beam direction of semiconductor lasers has generally been controlled using external mechanical systems that incorporate components such as a galvanometer and polygon mirrors. Because these mechanical systems limit the speed, size, and lifetime of the system, on-chip control of the beam direction has long remained a dream. Several approaches [47–49] have been proposed in order to reach this goal, based on the concept of integrated twin-stripe lasers whose relative phases are changed by adjusting the injection currents to each laser individually; the emitted far-field angle can thus be varied. However, the maximum beam steering angle $\delta\theta$ that could be obtained was small compared to the original beam divergence angle θ_{div}: the ratio $\delta\theta\theta_{div}$ was at most 2–3, and increasing this value appears to be difficult using this method. To address this issue, we proposed and demonstrated a completely different approach toward controlling the beam direction of semiconductor lasers: the use of 2D photonic crystals enables a much larger ratio of $\delta\theta\theta_{div} > 30$. More specifically, we have achieved a maximum beam steering angle of more than ±30° while maintaining a narrow beam divergence angle of ~1°.

In order to achieve the function of beam steering, it is important to be able to arbitrarily shift the emission direction away from the normal direction. It is thus necessary to make the lasing band edge deviate from the Γ-point. More specifically, if a lasing band edge can be formed that deviates from the Γ-point by a wavenumber δk, the emission direction is shifted by an angle of $\delta\theta = \pm\sin^{-1}(\delta k/k_0)$ from the normal direction, where $k_0 = 2\pi\lambda_0$ (λ_0 is the wavelength in vacuum). Moreover, if δk can be tuned arbitrarily, the emission direction can be controlled as desired, which leads to the realization of beam steering functionality. We proposed a laser based on a composite photonic crystal composed of square- and rectangular-lattice structures, as shown in Figure 4.12a. The square-lattice crystal has a lattice constant of a in both the Γ-X_1 and Γ-X_2 directions, while the rectangular-lattice crystal has a lattice constant of a' in the Γ-X_1 direction and a in the Γ-X_2 direction. The composite photonic crystal structure then has two fundamental reciprocal lattice vectors $G_1(|G_1| = 2\pi/a)$ and $G_1'(|G_1'| = 2\pi/a')$ in the Γ-X_1 direction, and a single fundamental reciprocal lattice vector $G_2(|G_2| = 2\pi/a)$ in the Γ-X_2 direction. The reciprocal lattice vectors G_1 and G_1' are expected to modify the diffraction of light inside the crystal and produce a new band edge that can be shifted from the Γ-point. Figure 4.12b shows the calculated photonic band structure for a composite photonic crystal with $a = 297$ nm and $a' = 446$ nm. It is clearly seen that a new band edge is formed (indicated by the red circle)[1] at a point deviating from the Γ-point by a wavenumber of $\delta k = |(G_1 - G_1')/2| = \pi(1/a - 1/a') = 0.167[2\pi/a]$.

Based on the above prediction, we developed a composite photonic crystal consisting of square- and rectangular-lattice structures. The square-lattice constant was basically fixed at a value a but slightly modified (see the details in supplementary

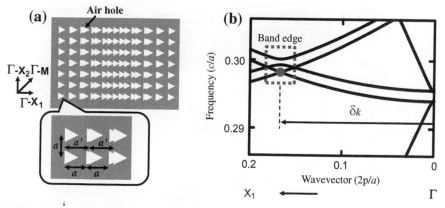

FIGURE 4.12 (a) Composite photonic crystal composed of square- and rectangular-lattice structures, and (b) calculated photonic band structure for a composite photonic crystal with $a=297$ nm and $a'=446$ nm.

---

[1]For interpretation of color in Figure 4.12, the reader is referred to the web version of this book.

information), while the rectangular-lattice constant a' in the Γ-X_1 direction was varied continuously in order to tune the value of $\delta k = |(G_1 - G'_1)/2| = \pi 1/a - 1/a'$. Figure 4.13a shows a scanning microscope image of a portion of the composite photonic crystal with an a lattice constant of 294 nm and an a' lattice constant of approximately 328 nm. The composite photonic crystal was formed on an active layer; further processing gave the device shown schematically in Figure 4.13b. The top p-electrode consisted of multiple elements, each with dimensions $17 \times 50\ \mu m^2$ and separated by intervals of 3 μm, while the bottom n-electrode consisted of a single element. In total, 30–40 p-electrode elements were formed on a single chip in the Γ-X_1 direction. The net length of the device was $\sim$1000 μm in the Γ-X_1 direction and $\sim$300 μm in the Γ-X_2 direction.

The strategy to achieve the beam steering function is twofold. The first step is to excite adjacent pairs of p-electrode elements sequentially. Each pair selects a portion of the device area with a certain δk. Therefore, scanning across all element pairs of

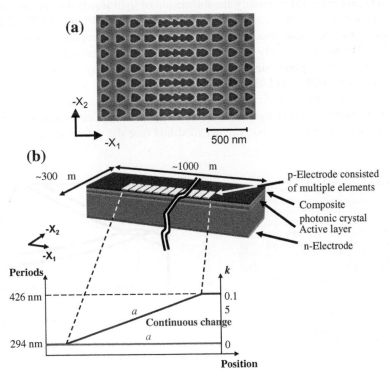

FIGURE 4.13 (a) Scanning microscope image of a portion of the composite photonic crystal with an *a* lattice constant of 294 nm and an *a'* lattice constant of approximately 328 nm, and (b) finished device structure containing the photonic crystal. The inset shows the detailed cross-sectional structure, including the electrodes and both the active and photonic-crystal layers.

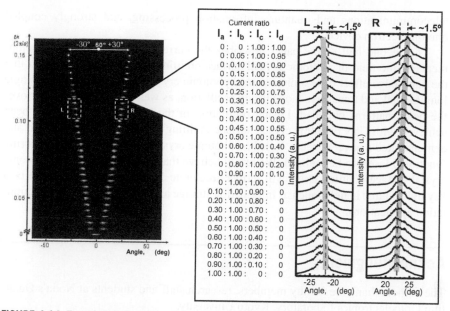

FIGURE 4.14 Experimental results of the beam steering operation.

the device should display discrete (or coarse) changes of the emission angle across the device's entire range. The second step is to enable continuous tuning of δk (and therefore the emission angle) by balancing the current injection levels of multiple adjacent p-electrode elements. This is possible because of the continuous, smooth change of the rectangular-lattice constant a' throughout the entire device. Figure 4.14 shows the experimental results of the beam steering operation. As can be seen in the figure, the coarse operation, namely discrete, $\sim1°$ changes to the emission angle, is clearly seen. As also shown in the inset, the emission angle has been changed continuously and smoothly between the discrete emission angles of $\sim1°$ by exciting two to three adjacent p-electrodes and changing the balance of the current injection levels of individual electrodes. These results clearly indicate beam steering operation has been successfully achieved [25].

4.4 CONCLUSION

The first half of this review shows that the advent of the ultimate nanolaser is coming close to reality due to utilization of the photonic bandgap effect, progress in optical nanocavities, and the integration of nanocavities with quantum dots. It is expected that this type of ultimate nanolaser will finally be realized by focusing research on developing an effective method for current injection [50–53]. Such nanolasers will have a huge impact on developments such as single-photon devices for quantum telecommunications and quantum information processing, and strongly coupled photon-electron devices.

The second half of the paper describes the current status and recent developments in the field of photonic crystal surface-emitting lasers. We have shown that the band-edge effect of two-dimensional photonic crystals enables large-area single longitudinal and transverse mode lasing oscillation, as well as complete control over the beam patterns obtained. Lasing oscillation in the blue-violet region has been achieved with GaN photonic crystal surface-emitting lasers. We have proposed and demonstrated a novel concept, based on photonic crystals, that allows beam steering functional lasers to be realized. We firmly believe that the development of photonic crystal surface-emitting lasers will further accelerate and that they will prove to be a key light source in fields such as information storage and processing, optical manipulation applications.

Acknowledgments

The authors thank the faculty members, research staff and students at Noda's Quantum Optoelectronics Laboratory, Kyoto University.

References

[1] E. Yablonovitch, Phys. Rev. Lett. 58 (1987) 2059.
[2] S. Ogawa, M. Imada, S. Yoshimoto, M. Okano, S. Noda, Science 305 (2004) 227.
[3] P. Lodahl et al., Nature 430 (2004) 654.
[4] M. Fujita, S. Takahashi, Y. Tanaka, T. Asano, S. Noda, Science 308 (2005) 1296.
[5] S. Noda, M. Fujita, T. Asano, Nature Photon. 1 (2007) 449.
[6] S. Noda, M. Imada, A. Chutinan, Nature 289 (2000) 606.
[7] Y. Akahane, T. Asano, B.-S. Song, S. Noda, Nature 425 (2003) 944.
[8] B.-S. Song, S. Noda, T. Asano, Y. Akahane, Nature Mater. 4 (2005) 207.
[9] T. Tanabe, M. Notomi, E. Kuramochi, A. Shinya, H. Taniyama, Nature Photon 1 (2006) 49.
[10] O. Painter et al., Science 284 (1999) 1819.
[11] S. Noda, Science 314 (2006) 206.
[12] S. Strauf et al., Phys. Rev. Lett. 96 (2006) 127404.
[13] T. Yoshie et al., Nature 432 (2004) 200.
[14] K. Hennessy et al., Nature 445 (2007) 896.
[15] D. Englund et al., Nature 450 (2007) 857.
[16] Y. Tanaka et al., Nature Mater. 6 (2007) 862.
[17] Y. Sato et al., Nature Photon. 6 (2012) 56.
[18] M.F. Yanik, S. Fan, Phys. Rev. Lett. 92 (2004) 083901.
[19] T. Baba, Nature Photon. 2 (2008) 465.
[20] M. Imada et al., Appl. Phys. Lett. 75 (1999) 316.
[21] M. Meier, A. Mekis, A. Dodabalapur, A. Timko, R.E. Slusher, J.D. Joannopoulos, O. Nalamasu, Appl. Phys. Lett. 74 (1999) 7.

[22] S. Noda, M. Yokoyama, M. Imada, A. Chutinan, M. Mochizuki, Science 293 (2001) 1123.

[23] E. Miyai et al., Nature 441 (2006) 946.

[24] H. Matsubara et al., Science 319 (2008) 445.

[25] Y. Kurosaka et al., Nature Photon. 4 (2010) 447.

[26] M. Yamaguchi, T. Asano, S. Noda, Opt. Express 16 (2008) 18067.

[27] M. Yamaguchi, T. Asano, S. Noda, Rep. Prog. Phys. 75 (2012) 096401.

[28] D. Englund et al., Phys. Rev. Lett. 95 (2005) 013904.

[29] K. Kounoike et al., Electron. Lett. 41 (2005) 1402.

[30] Y. Taguchi, Y. Takahashi, Y. Sato, T. Asano, S. Noda, Opt. Express 19 (2011) 11916.

[31] Y. Arakawa, H. Sakaki, Appl. Phys. Lett. 40 (1982) 939.

[32] M. Tabuchi, S. Noda, A. Sasaki, in: S. Namba, C. Hamaguchi, T. Ando (Eds.), Science and Technology of Mesoscopic Structures, Springer, Tokyo, 1992, pp. 379.

[33] A. Badolato et al., Science 308 (2005) 1158.

[34] M. Yamaguchi, T. Asano, K. Kojima, S. Noda, Phys. Rev. B 80 (2009) 155326.

[35] M. Yamaguchi, T. Asano, S. Noda, in: The 8th International Conference on Physics of Light-Matter Coupling in Nanostructures, Tokyo, Japan, 7–11 April 2008.

[36] A. Naesby, T. Suhr, P.T. Kristensen, J. Mørk, Phys. Rev. A 78 (2008) 045802.

[37] M. Nomura, N. Kumagai, S. Iwamoto, Y. Ohta, Y. Arakawa, Opt. Express 18 (2009) 15975.

[38] M. Imada, A. Chutinan, S. Noda, M. Mochizuki, Phys. Rev. B 65 (2002) 195306.

[39] Y. Liang, C. Peng, K. Sakai, S. Iwahashi, S. Noda, Phys. Rev. B 84 (2011) 195119.

[40] Y. Liang, C. Peng, K. Sakai, S. Iwahashi, S. Noda, Opt. Express 20 (2012) 15945.

[41] D. Ohnishi, T. Okano, M. Imada, S. Noda, Opt. Express 12 (2004) 1562.

[42] R. Dorn, S. Quabis, G. Leuchs, Phys. Rev. Lett. 91 (2003) 233901.

[43] K. Sakai et al., IEEE J. Sel. Areas Commun. 23 (2005) 1335.

[44] K. Sakai, E. Miyai, S. Noda, IEEE J. Quantum Electron. 46 (2010) 788.

[45] T. Sakaguchi et al., Technical Digest of CLEO/IQEC2009, CTuH1.

[46] T. Sakaguchi, S. Noda, et al. (unpublished).

[47] D.R. Scifres, W. Streifer, R.D. Burnham, Appl. Phys. Lett. 33 (1978) 702.

[48] S. Mukai et al., Opt. Quantum Electron. 17 (1985) 431.

[49] T. Ide et al., Jpn J. Appl. Phys. 38 (1999) 1966.

[50] A. Sugitatsu, S. Noda, in: Int. Symp. Photon. Electromag. Cryst. Struct. (PECS) V, Kyoto International Conference Hall, Kyoto, Japan, March 2004.

[51] H.-G. Park et al., Science 305 (2004) 1444.

[52] B. Ellis et al., Nature Photon. 5 (2011) 297.

[53] S. Matsuo et al., Opt. Express 20 (2012) 3773.

High-Speed Polymer Optical Modulators

5

Raluca Dinu, Eric Miller, Guomin Yu, Baoquan Chen, Annabelle Scarpaci, Hui Chen, and Corey Pilgrim

5.1 INTRODUCTION

5.1.1 The advantages of EO polymer

Optical networks are the telecommunication backbone enabling computers and portable devices to exchange information almost instantly. As ever higher bandwidth is required from data centers and telecom networks, electro-optical (EO) components are challenged to meet the ever increasing performance demands such as higher power efficiency, smaller device size, and lower cost. The vast majority of EO modulators for telecom networks are currently built using lithium niobate technology. Fundamental limitations with lithium niobate devices prevent viable technology roadmaps to support "next generation" devices. EO polymer materials have proven to deliver broadband EO devices to meet the datarates required for many generations to come due to ultra fast response time in range of femto seconds, low drive voltage, and ease of integration and scaleability. These materials provide the foundation for terahertz bandwidth devices with small form factors as compared to lithium niobate, low cost manufacturing, and the ability to support current and future complex modulation demands.

5.1.2 Requirements for commercial applications

EO polymers must meet several requirements for commercial applications. These include a high EO coefficient r_{33} greater than 80 pm/V to support modulators with low $V\pi$, low optical loss of <3 dB/cm at the typical telecommunication wavelength range of 1300–1550 nm, appropriate refractive index to ensure low coupling loss and a high confinement factor, a low dielectric constant to exhibit large bandwidth up to 60 GHz. In addition, a commercially viable EO material must withstand standard Telcordia HTOL (high temperature operation lifetime) testing at 85 °C for 2000 h. This requires stable thermal, photochemical, and mechanical properties. Also, good film forming properties, compatibility with cladding layers, ability to support a wave guide forming processes using photolithography and plasma etch, and a relatively low cost material synthesis are desired to support easy scale up of production and long-term repeatability of the product.

Optical Fiber Telecommunications VIA. http://dx.doi.org/10.1016/B978-0-12-396958-3.00005-6

5.2 MATERIAL DESIGN

EO polymers consist of small molecules with EO properties called chromophores incorporated in a polymer backbone. The primary challenge to design EO polymers is the development of suitable chromophores. They have to present high hyperpolarizability at the molecular level, and good chemical, thermal and photochemical stability to survive harsh processing conditions for modulator fabrication and have a long life-time device. Chromophores are typically composed of three different blocks: a strong electron-donor moiety and a strong electron-acceptor moiety connected by a π-conjugated bridge ensuring efficient electronic polarization between the donor and the acceptor.

Another challenge is to ensure the incorporation of chromophores into a polymer and translating the nonlinear optical (NLO) properties from the molecular level to the macroscopic level. The common method to translate these properties is pol-ing which consists in heating the material at its T_g (glass transition temperature) and to apply an electric field to orient the chromophore and generate a non-centro-symmetric order. To incorporate chromophore into a polymer, the simplest and most common way consists in guest-host system where the chromophores are doped into a host polymer. The limitation of this approach is the possibility of phase separa-tion in films causing scattering or diffraction and high optical loss. Therefore the guest-host system must feature good solubility and good compatibility. Another classical issue is the relaxation of the chromophore after poling. A solution is to work with high T_g polymers such as polycarbonates, polysulfones, and polyquino-lines with highly stable chromophores. Peripheral secondary groups are introduced on the chromophore to improve simultaneously poling efficiency, solubility, com-patibility with the host polymer, and thermal, chemical and photochemical stabil-ity. Thermally crosslinkable guest-host systems, where the guest and the host have functional groups reacting together after orientation of the chromophores lead to EO polymers with high r_{33} value (up to 263 pm/V at 1.3 μm) being stable at 85 °C for 500 h [1]. Devices using guest-host-based EO polymer with sub-1-volt halfwave voltage and bandwidths larger than devices using lithium niobate (110 Vs. 70 GHz) were demonstrated [2]. Also, to the best of our knowledge materials using guest-host system are the only ones currently employed for industrial production of EO modulators.

Other strategies to improve thermal stability of the EO properties have been studied such as side-chain or main-chain crosslinkable polymers [3] where the chromophore is covalently attached to the host polymer and crosslinking occurs after orientation of the chromophores. However, the poling efficiency typically decreases. Multi-chromophore dendrimers showed to be an efficient strategy to improve poling efficiency [4]. Indeed, Dalton's group showed theoretically and experimentally that the globular shape of dendrimer could allow up to a threefold increase of the r_{33} value due to an increase of chromophore number density [5]. To date, this type of material has not yet transferred from academic research to commercial application.

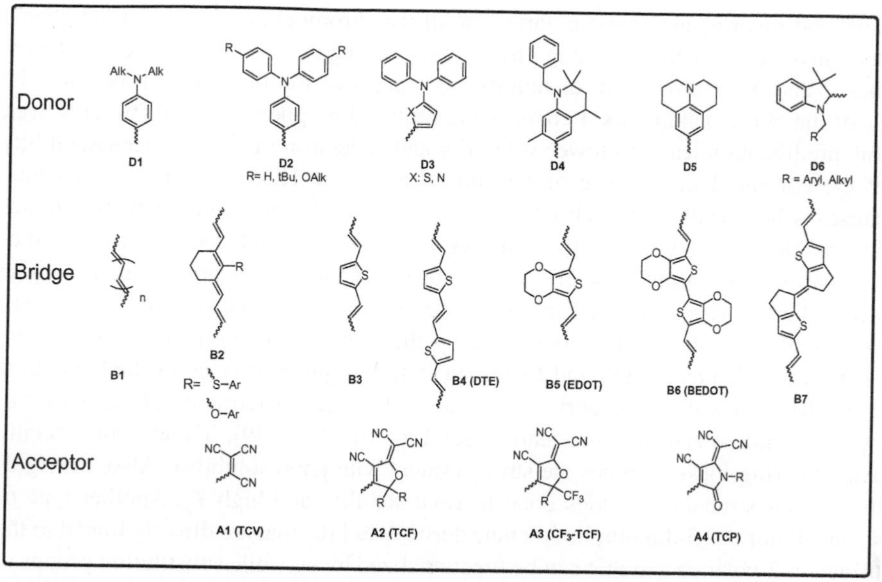

FIGURE 5.1 Structures of efficient electron-donors, bridges, and electron-acceptors.

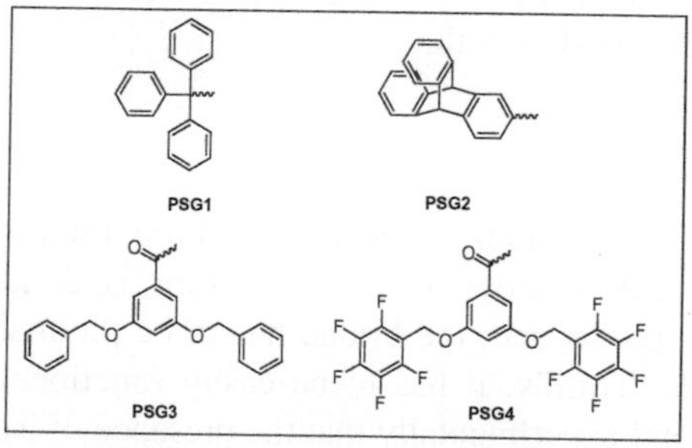

FIGURE 5.2 Structures of peripheral secondary groups (PSGs).

The following paragraphs will be focused on the molecular engineering of chromophores. The design of electron donor, bridge, electron acceptor, and isolating groups will be discussed (Figures 5.1 and 5.2). Finally, the current status of commercial technologies will be presented.

5.2.1 Design and development of EO polymer
5.2.1.1 Electron donor
The most common class of donors is aminophenyl derivatives as shown in Figure 5.1. Their strong donor strength results from the presence of the nitrogen non-bonding lone electron pair destabilizing the HOMO (Highest Occupied Molecular Orbital) energy

level and favoring electronic polarization of the chromophore. Dialkylaminophenyl derivatives present the advantage to combine strong donor character, ease of synthesis, and great possibility of tailoring. To improve the thermal stability and the T_g of the chromophore, alkyl chains were replaced by phenyl groups [6]. However, this modification leads to lower solubility and reduction of the hyperpolarizability of the chromophore because of a stabilization of the HOMO energy level which causes a blue shift of the absorption band of the chromophore. Inductive donor groups such as tertbutyl groups or alkoxy groups were introduced in para- or ortho-position to compensate these effects [7]. The direct functionalization of diaryl amine with a heterocycle connected to the bridge such as thiophene or pyrrole increases the donor strength [8]. This is caused by the reduction of aromatic stabilization. However, their preparation and functionalization requires very controlled synthetic work. Ring-locked aniline derivatives such as 1-R-2,2,4,7-tetramethyl-1,2,3,4-tetra-hydro-quinoline and judoline feature good intermediates [9]. These commercially available ring-locked amines are strong donors with good solubility. Also, the rigidity of their structure provides good thermal stability and high T_g. Another type of strong donor is 3,3-dimethylindolenine derivatives [10] that are directly linked to the bridge in 2-position and that can be functionalized by versatile substitution groups in 1-position in one step. For example, the chromophore CPO-1 reported by Andraud's group using this donor, a ring-locked tetraene bridge, and the TCF acceptor present a high $\mu\beta$ value of $31,000 \times 10^{-48}$ esu at 1.9 μm which is comparable to the CLD chromophore and show good r_{33} value.

5.2.1.2 Bridge

The bridge plays a very important role in chromophore. First, it has to present good planarity to provide ideal electronic polarization from the donor to the acceptor and high hyperpolarizability. Second, the bridge has to be thermally, photochemically, and chemically stable. Thirdly, it has to be easily functionalized. Indeed, it was proven theoretically and experimentally that the presence of isolating groups on the bridge has a greater effect toward reducing chromophore-chromophore interactions than on the donor or on the acceptor moiety.

Polyene bridges such as tetraene can provide CLD-like chromophores with good electronic polarizability and high hyperpolarizability [2], but they are not suitable for commercial applications because of their lack of chemical, thermal and photochemical stability. These limitations can be improved to a certain extent by using a ring-locked tetraene bridge functionalized with a thiol or ether group [11]. A more stable category of bridge includes aromatic groups such as the divinylbenzyl bridge [12]. However, hyperpolarizability is typically reduced because of aromatic stabilization. A good trade-off combining good stability and electronic polarization is the divinylthienyl bridge [13]. The large variety of synthetic routes to prepare thiophene and its chemical stability allow different types of functionalization. The replacement of the thiophene group by the ethylenedioxythiophene (EDOT) unit has the advantage to induce a red-shift of the chromophore absorption band without elongating the bridge.

This shift could be due to intramolecular sulfur-oxygen interactions [14] resulting in rigidification of the structure and a better planarity of the chromophore. The effect of rigidification of thiophene-based bridges using covalent bonding or intramolecular interactions on hyperpolarizability was studied using bridges such as dithienylethylene (DTE), bis-EDOT (BEDOT) [15]. Covalent rigidification is a powerful approach to limit rotational and vibrational disorder and improves thermal stability and NLO properties of chromophores. Rigidification of the bridge using intramolecular interactions, e.g. in the case of BEDOT, provides another efficient way of enhancing the NLO properties of chromophores by improving planarity of the bridge.

5.2.1.3 Electron acceptor

The challenge of acceptor design is to deliver good electron-withdrawing power, which comes from both inductive and resonance effects. Through several decades of effort in the community, heterocyclic structures as shown below stand out the choice of acceptor for NLO chromophores. Especially, 2-dicyanomethylene-3-cyano-4,5,5-trimethyl-2,5-dihydrofuran (TCF) [16] and 3-methyl-4-cyano-5-dicyanomethylene-2-oxo-3-pyrroline (TCP) [17] give very high β chromophores compared to other acceptors like tricyanovinyl (TCV) [18]. 2-Dicyanomethylene-3-cyano-4-methyl-5-phenyl-5-trifluoromethyl-2,5-dihydrofurane (CF3-TCF) [19] is exceptionally strong acceptor because of the powerful inductive and hyperconjugation effects of trifluoromethyl group. Due to the electron-deficient nature of acceptors, they are susceptible to nucleophilic attack, which limits the chemistry selection to produce themselves and couple them conjugatedly with the donor-bridge part of the chromophores.

5.2.1.4 Peripheral secondary groups

Two challenges still remaining today are optimization of poling efficiency and stability of the EO activity over the typical 25 year lifetime of devices. Chromophores with high hyperpolarizability are characterized by a large dipole moment. They tend to aggregate in anti-parallel order because of the presence of strong electrostatic interactions. High poling voltages and temperatures required to orient the chromophore in an acentric order can cause their degradation. Also, electrostatic interactions induce relaxation of the chromophores once the EO polymer is poled. To reduce electrostatic interactions, peripheral secondary groups (PSGs) as shown in Figure 5.2 can be introduced on the electron donor, bridge and/or electron acceptor group of chromophores [20]. Also, these PSGs can improve solubility, miscibility with host polymers by providing strong intermolecular forces with the polymer matrix, photochemical and thermal stability, and optical loss.

A very large number of examples show that functionalization of chromophores with alkyl chains improves solubility and reduces crystallinity but generally reduces the T_g of the material. Recently, more bulky and rigid structures such as triptycene [21] or trityl were reported and showed to increase tremendously the T_g of the chromophore. Also, they have good compatibility with amorphous polycarbonate (APC).

Poling efficiency can be improved in the presence of favorable interactions between PSGs. For example, the use of chromophores bearing both hydrogenated (HD) and fluorinated (FD) benzyl ether dendrons leads to materials with threefold increase in r_{33} value compared to other materials with unfunctionalized chromophores [22]. Moreover, pentafluorobenzyl ether dendrons have the advantage to reduce optical loss.

5.2.2 Current status of commercial technologies

Although there are more and more demonstrations of devices using EO polymer [23], GigOptix Inc. is the only company producing EO modulators using Telcordia-qualified EO polymer, so-called M3. M3 is based on a guest-host system composed of a chromophore with PSGs and a high T_g host polymer. This commercial material provides an r_{33} value of 91 pm/V after corrections and an index of refraction of 1.7010 at 1320 nm. The optical loss of this material is less than 1.2 dB/cm at 1550 nm. M3 has been used in 150 mm wafer production line to support optical modulators for 100 Gbps applications and beyond.

5.3 EO MATERIAL CHARACTERIZATION

To ensure the stable production of EO polymer based devices, it is critical to have accurate characterization and monitoring of materials at both the molecular thin film level. Specific instruments and measurement techniques are used for both initial evaluate of materials, and support ongoing monitoring of production to ensure stable and reliable devices to telecommunication quality standards.

5.3.1 Characterization methods for electro-optical polymer material

5.3.1.1 Properties

Chromophores are typically characterized at the fundamental material level by nuclear magnetic resonance (NMR), UV to visible spectrum (UV-Vis) spectroscopy, and Fourier transform infrared (FT-IR) spectroscopy. UV-Vis spectroscopy is a powerful tool that gives crucial information about the chromophore within the typical telecom operation wavelengths, and provides critical information related to extinction ratio and the solvochromic parameter. FT-IR is used to measure the intrinsic absorption at near IR other than regular structure characterization. Both glass transition temperature (T_g) and melting point can be measured by differential scanning calorimetry (DSC). Chromophore decomposition can be measured by DSC and thermal gravimetric analysis (TGA). High pressure liquid chromatography (HPLC) gives the purity of chromophores and percentage of each isomer if isomers exist. Some volatiles that cannot be detected by HPLC can be measured by TGA.

5.4 FUNDAMENTAL EO PERFORMANCE CHARACTERIZATION

To fully characterize the performance of an electro-optical (EO) polymer material, both the, electrical and optical properties must be considered independently as well as the integrated electro-optical properties. This section will focus on these three aspects and introduce the techniques used to characterize them.

5.4.1 Electrical properties

5.4.1.1 Conductivity

Material conductivity is critical for EO polymers not only in the final performance of the device, but is also plays a critical role in the poling efficiency of the material during device fabrication. EO polymers must be poled with a high voltage electric field (up to $\sim$150V/μm) at elevated temperature to align the chromophore acentrically and achieve bulk EO activity. In a typical device, the active EO polymer layer (so-called core layer where the majority of light resides) is sandwiched between two passive polymer layers (so-called cladding). In order to improve the poling efficiency on the core layer, the conductivity of each layer of the stack needs to be carefully engineered and monitored so that the majority of the poling electrical field will concentrate on the core to align the chromophore rather than being lost the cladding. In addition, the conductivity of the cladding layers can have a direct impact on the high frequency response, and must be considered when designing the integrated device.

The conductivity of each polymer layer is tested using ITO (indium tin oxide) coated glass as a substrate. A thin layer of polymer is spin-coated on an ITO with film thickness of around 2$\sim$3 μm. A thin layer of gold is deposited on the top of this film as the top electrode and the ITO film serves as the bottom electrode. Voltage is applied vertically across the film and LTC (leak through current) is measured as a function of temperature. The magnitude of the LTC at temperature is used to select the right combination of core and cladding to form the device stack.

5.4.2 Optical properties

There are two basic optical properties that should be considered when designing an EO polymer. They are refractive index and optical propagation loss.

5.4.2.1 Refractive index

Refractive index is important in a typical material system consisting of a waveguide structure with an active EO polymer as core and passive polymer as top and bottom cladding. In order to design an efficient waveguide to maximize the interaction between light and the chromophore, the refractive indices of each layer of the stack need to be carefully matched to ensure the best overlap between light mode and the active EO layer.

A simple prism-coupler refractometer is used to measure the refractive index of the polymer thin film. The thin film sample is in contact with the prism surface while a laser beam is shone onto the prism. The laser scans through a range of incidence angles until the critical angle is reached—this is where the light leaks into the material being measured and a minimum reflected light is detected through the prism. The critical angle for a series of optical modes is used to calculate the refractive index of the sample. Typically the refractive index of the EO polymer using a 1.5 μm polarized laser in the transverse magnetic (TM) incidence ranges from 1.6 to 1.8.

5.4.2.2 Optical loss

Optical propagation loss is a critical parameter when it comes to evaluating the performance of EO polymer. The main contributing factors to optical loss are intrinsic loss due to material absorption, scattering loss due to fabrication imperfections, and poling-induced loss. In order to reduce loss, the absorption spectrum of the EO polymer material is tuned at the optical wavelength of 1550 nm to be a minimum. This intrinsic loss is dependent on the structure of the chromophore and the functional groups inherent to the molecule. The instrument used to measure material absorption loss is SPA4000 Prism Coupler from Sairon Technology. The technique is based on high index liquid immersion of planar optical waveguides. Using a prism coupler identical to that found in the simple prism-coupler refractometer, laser light is focused into the EO polymer material at the critical angle of the fundamental optical mode. As the sample is immersed into a higher refractive index fluid, the light that is propagated through this waveguide leaks out at the point of immersion and this intensity is plotted against the distance the light travels through the material. An exponential fitting of the light intensity versus the distance traveled will give the propagation loss coefficient. Scattering loss due to imperfections relies heavily on sample preparation. This factor can be mitigated with material fabrication procedures to provide uniform and homogenous film quality.

Poling is also a factor in the optical loss of EO polymers and devices. Poling-induced loss also depends significantly on the leak through current (LTC) observed during poling. In general, optical loss increases with increasing poling field and increasing LTC. The loss observed with high LTC is most likely attributable to microfracture "treeing" that occurs just prior to dielectric breakdown [24] since LTC usually begins to spike under high poling fields just prior to dielectric breakdown of the polymer stack. Unfortunately, EO activity also peaks when the poling field is highest, which is in the treeing region just before dielectric breakdown. Thus, the interplay between LTC and EO activity is critical for optimizing the optical loss and V_μ of EO polymer devices.

5.4.3 Electro-optical properties

5.4.3.1 Poling and EO coefficient r_{33}

Poling is a crucial step that requires particular attention when manufacturing a modulator device. During poling, high voltage is applied across the polymer film at elevated temperature to align the chromophore and achieve optimal bulk EO activity. Most of the poling happens at, near, or above the glass transition temperature T_g when the chromophore is free to rotate and respond well to the electrical field. However, poling is a very complex process that requires a subtle balance between temperature and the electrical field applied. On one hand, the chromophore needs to exceed a certain temperature depending on the T_g of the choromophore polymer composite, but on the other hand, excessive temperature and thermal energy tend to randomize chromophore orientation therefore decreasing the effect of the electric field and poling efficiency. An ideal interplay between the temperature and voltage will generate the maximum poling efficiency for a given chromophore polymer system. With the help of an ultrafast laser system a procedure to in situ monitor the poling efficiency by detecting the second harmonic generation (SHG) signal produced by the EO polymer during poling is used. A laser beam is shined through the polymer film while the SHG signal generated is being recorded during poling. When the SHG signal reaches its peak, it indicates the poling efficiency is optimized.

To evaluate the final EO performance the chromophore, the EO coefficient r_{33} is measured after poling. A "Teng-Man" simple reflection technique is a quick screening tool for r_{33} measurement [25]. Using this method, there is a significant impact of film thickness on measured r_{33} due to multiple reflections at the film interfaces. Therefore, a procedure to measure r_{33} across different film thicknesses to correct out the effect due to thickness variation is required for accurate measurements. Typical thin film r_{33} measurements are obtained at 1310 nm using ITO coated glass as substrate.

5.4.4 Measurement results from prototype materials

A typical guest-host EO polymer system, "LPD-80," is a composite material of the chromophore shown in Figure 5.3 and amorphous polycarbonate (APC, poly[Bisphenol

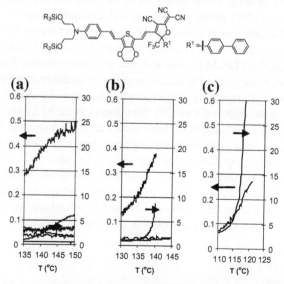

FIGURE 5.3 LPD-80 chromophore structure and SHG-LTC plots for (a) 30%, (b) 50%, and (c) 70% loading densities. In each graph, the left *y*-axis is SHG (arbitrary units) and the right *y*-axis is current (μA).

A carbonate]$_x$-co-[4,4'-(3,3,5-trimethylcyclohexylidene)diphenol carbonate]). As the guest host loading concentration is increased, the T_g decreases, and thus the poling efficiency decreases. From this study LPD-80 has an optimal T_g of 140°C, a thin film optical loss of 0.9 dB/cm, and a thin film r_{33} of 90 pm/V @ 1310 nm with 50% by weight chromophore loading [26].

5.5 DEVICE DESIGN

Compared to inorganic crystal materials such as LiNbO$_3$, EO polymers have the following advantages: (1) closely matched optical and millimeter wave refractive indices; (2) high EO activity r_{33}; (3) property tunable; (4) radiation resistance; and (5) ability to integrate with other photonic and electronic circuits. Recently, great progress has been made in proving that EO polymer modulators are meeting industry reliability requirements. EO polymer modulators have passed standard Telcordia stress tests (GR-Core 486) [27] and can operate at 85°C for 25 years [28]. With proper design, these advantages can be fully utilized to make EO polymer modulators superior to LiNbO$_3$ counterparts in performance.

5.5.1 Zero chirp single-ended EO modulators

Zero chirp EO modulators are desirable for next-generation communication systems. For LiNbO$_3$, zero chirp EO modulators use Z-cut dual electrode Mach-Zehnder modulators (DE-MZM), or X-cut single-ended MZM, or Z-cut single-ended MZM by using polarization-reversal (PR) technique [29,30] and Domain Inversion (DI) technique [31,33]. DE-MZM has to use complicated differential output drivers by adjusting phase and amplitude of electrical signal precisely to obtain zero chirp. X-cut MZM has larger driver voltage than Z-cut DE-MZM due to device structure features. Z-cut single-ended MZM modulators made by the PR technique suffer from either narrowband operation due to a quasi-velocity-matching [29] or chirp parameter depending on modulation frequencies due to the electrical loss [30]. Modulators made by DI technique have big half-wave voltage of 3~4 V and small bandwidth that is not more than 10 GHz [31,33]. EO polymer is especially suitable for making zero chirp single-ended modulators due to the easiness of achieving domain-inverted sections by push-pull poling process.

To design a photonic device, six aspects need to be considered, which are: (1) technology; (2) functionality; (3) fabrication capability; (4) packaging; (5) power consumption; and (6) cost. For zero chirp single-ended polymer EO modulator, these six aspects are considered by following specific parameters in the device design process including:

a. Half-wave voltage Vπ.
b. Optical insertion loss.
c. Extinction ratio.
d. EO bandwidth S21.
e. RF reflection S11.

Generally, a polymer EO modulator chip consists of three parts: optical waveguides, RF electrodes and bias circuits, which will be described individually in the following sections.

5.5.1.1 *Optical waveguide*

In a polymer EO modulator, polymer waveguide is located between the bottom electrode and the top RF electrodes and consists of three layers: bottom clad layer, core layer, and top clad layer. UV15LV is chosen as bottom clad which has an optical refractive index of 1.505 and an optical loss of ∼3 dB/cm at 1550 nm in TM mode. LP33ND is developed as the top clad which has an optical refractive index of 1.388 with an optical loss of 2 dB/cm at 1550 nm in TM mode. The core B74-APC has an optical refractive index of 1.68 at unpoled condition and an optical refractive index of 1.734 under poled condition with an optical loss of 1.2 dB/cm at 1550 nm in TM mode.

A reverted rib waveguide is used in the polymer EO modulator due to its fabrication advantage over normal rib waveguide [34]. Polymer rib waveguide should be designed as enlarged single mode [35] and have a total thickness as thin as possible to minimize the half-wave voltage without affecting the optical loss, and needs a trade-off between the $V\pi$ and insertion loss. Therefore, the waveguide rib (or trench) has a depth of 1 μm and a width of 3.7 μm with a slab thickness of 2.1 μm, which gives a single mode. Figure 5.4 shows the waveguide dimensions and the simulated fundamental mode is shown in Figure 5.5. The effective refractive index is 1.723 for the poled waveguide and 1.66 for the unpoled waveguide. The bottom clad thickness is chosen as 1.5 μm and the top clad is chosen as 1.3 μm. Table 5.1 shows the waveguide's propagation loss simulation results with software BeamPROP for different bottom and top clads (without considering the fabrication process induced loss).

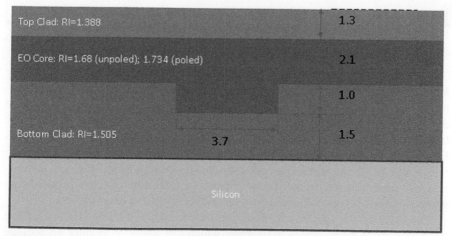

FIGURE 5.4 Schematic reverted rib waveguide (the unit for the dimensions is μm).

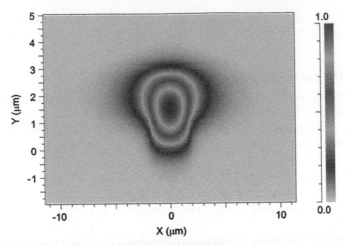

FIGURE 5.5 Fundamental mode of reverted rib waveguide.

Table 5.1 Waveguide propagation loss.

LP33ND Thickness (μm)	UV15LV Under-trench Thickness (μm)	Trench (μm)	Optical Propagation Loss (dB/cm)
1.3	1.5	1.0	1.73
1.2	1.4	1.0	1.83
1.4	1.6	1.0	1.67
1.3	1.5	1.2	1.75
1.2	1.4	1.2	1.85
1.4	1.6	1.2	1.73

Polymer EO modulator has a Mach-Zehnder interferometer architecture. The M3 core material B74-APC has a typical EO coefficient (r_{33}) of 65 pm/V after poling at 167°C 600V. For 9 mm active length, the half-wave voltage with the above designed waveguide after push-pull poling is 1.8 V, which is the same as that of the best $LiNbO_3$ 40G counterparts with dual drive [36]. In order to keep the insertion loss low, the chip should be designed as short as possible. With 9 mm active length, the total length of the polymer EO modulator is designed as 14.5 mm. The optical insertion loss simulation results are shown in Table 5.2 with different cladding thickness without considering the fabrication process induced loss and coupling loss.

For 40G zero chirp single-ended polymer EO modulator, the chip to fiber coupling has a free space micro optics architecture which can integrate monitor photodiode in the device package to keep the device dimensions small. However, since the side

Table 5.2 Polymer EO modulator insertion loss.

LP33ND Thickness (μm)	UV15LV Under Trench Thickness (μm)	Trench (μm)	MZ Optical Propagation Loss (dB)
1.3	1.5	1.0	2.993
1.2	1.4	1.0	3.107
1.3	1.5	1.2	3.028
1.2	1.4	1.2	3.125

mode of the output waveguide in the off state can be captured by the coupling micro lens due to the short chip, the extinction ratio will become smaller than 20 dB. To solve this issue, a side mode blocker is added on both sides of the waveguide near the facet. The blocker has a length of 300 μm along the waveguide and a width of more than 1000 μm. The distance between the two blockers is 16 μm (edge to edge) at the beginning and is 20 μm at the end 100 μm near the output waveguide facet. Figure 5.6 shows the top view of the side mode blocker, and Figure 5.7 shows the section view of the side mode blocker. The side mode blocker is made of Ti/Au and is 0.3 μm away from the surface of core B74-APC. Figure 5.8 shows the simulation results with side mode blocker at the off state with BeamPROP. With the side mode blocker, the extinction ratio of the EO modulator is more than 30 dB, and the extinction ratio of the monitor photodiode is more than 10 dB.

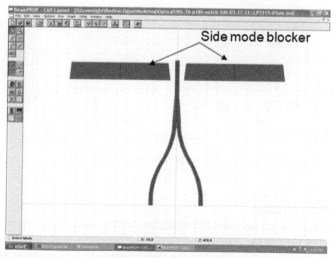

FIGURE 5.6 Top view of the side mode blocker.

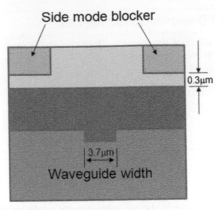

FIGURE 5.7 Section view of the side mode blocker.

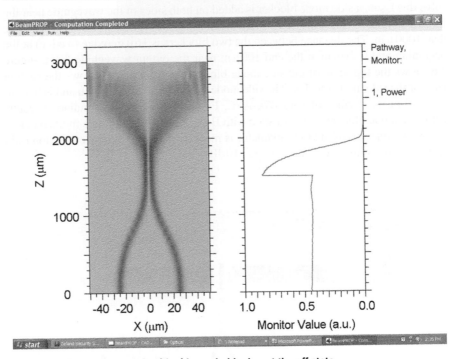

FIGURE 5.8 Simulation result with side mode blocker at the off state.

5.5.1.2 RF electrodes

Zero chirp single-ended polymer EO modulator can be realized by push-pull poling the two arms of the MZI to opposite align the chromophore with two microstrip line driving electrodes. With this structure the chirp parameter is frequency independent, and is better than the LiNbO$_3$ counterpart with PR technique whose chirp parameter is frequency dependent due to the RF loss difference at each frequency. However, for zero chirp single-ended polymer modulators it is very challenging to realize more than 40 GHz bandwidth (reference to 1 GHz) with S11 < −10 dB (from 1 GHz to 40 GHz) because of the impedance mismatch issue. In 2001, Zhang et al. reported 20 GHz bandwidth (reference to 2 GHz) polymer EO modulator by push-pull poling [37]. Since then, a decade has passed, but no progress had been reported in the zero chirp single-ended polymer EO modulator to reach 40 GHz bandwidth until May 2011 when Yu et al. reported 40 GHz zero chirp single-ended polymer EO modulators with low driving voltage [12]. Based on the work of 40 GHz zero chirp single-ended polymer EO modulators, small form factor polymer EO modulator products have been developed [39]. Now it becomes realistic that polymer EO modulators challenge LiNbO₃ EO modulators. In addition, EO polymers are uniquely suitable for making 100 Gb/s EO modulators that inorganic materials like LiNbO$_3$ are not capable of. Driven by the 100 Gb/s Ethernet (100 GbE) and the demand for increasing bandwidth, optical transmission with data-rates of more than 100 Gb/s have gain much interest recently. Serial 100 Gb/s transmission may also be a cost effective solution for some short reach applications, although multi-level modulation formats such as DQPSK and DP-QPSK are commonly used to achieve 100 Gb Ethernet. 100 Gb/s polymer EO modulator products have been reported [40–42], and zero chirp single-ended 100 Gb/s polymer EO modulators are commercially available now.

The issue of impedance mismatch has be resolved by the electrode design that ensures zero chirp single-ended polymer EO modulators to reach 40 GHz bandwidth and S11 < −10 dB while Vπ and insertion loss remain competitive to LiNbO$_3$ counterparts [38]. Typical zero chirp single-ended polymer EO modulator RF electrodes are schematically shown in Figure 5.9. In order to maintain Vπ competitive, the polymer waveguide stack is designed to be 6.0 μm and the active waveguide length is 9 mm. To match the impedance of RF source 50 Ω, the characteristic impedance of the microstrip lines on the two waveguide arms has to be 100 Ω, which results in the width of the microstrip lines being about 2 μm. Microstrip line of 2 μm has too much RF loss with which polymer EO modulators cannot reach 40 GHz bandwidth. On the other hand, if the microstrip line has a characteristic impedance of 50 Ω, the RF loss may not be good enough to reach 40 GHz bandwidth, and the S11 will be bigger than −10 dB in the entire 1–40 GHz range.

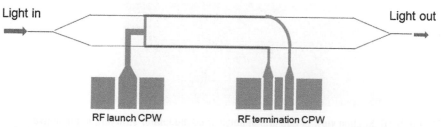

FIGURE 5.9 Schematic zero chirp single-ended polymer EO modulator RF electrodes.

In order to solve the dilemma, a transformer architecture has been used in the RF electrodes. The RF launch CPW and the RF termination CPWs have a characteristic impedance of 50 Ω. The RF electrode region from the launch CPW to the termination CPWs has been treated as a transformer with characteristic impedance varying from 50 Ω to 25 Ω. The thickness of the RF electrodes has been designed at 5.5 μm, and a layer 2.5 μm of polymer LM251 has been placed on the top of RF electrodes to match the RF and optical velocities. Figure 5.10 shows the section view of one arm of the polymer EO modulator structure in the active region.

The RF electrode's performance has been simulated with EM software IE3D. Parameters include RF losses, characteristic impedance, and RF effective dielectric constant. With 9 mm active length, the simulated EO response of the polymer modulator chips is shown in Figure 5.11. The zero chirp single-ended polymer EO modulator's 3 dBe bandwidth is simulated to be 40 GHz.

100G zero chirp single-ended polymer EO modulators can also use the same device architecture with active length of 5 mm. The RF electrodes have similar transformer design with different RF launch CPW and RF termination CPW structures to accommodate the device packaging house. The EO response simulation results for a 100G device are shown in Figure 5.12. The simulated 3dBe bandwidth is 74 GHz.

The designs for 40G and 100G have been implemented and devices have been fabricated. Figure 5.13 shows typical EO responses (S21) and Figure 5.14 shows

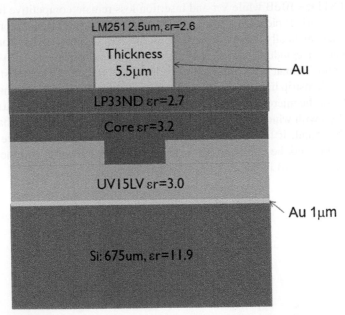

FIGURE 5.10 Section view of one arm of polymer EO modulator structure in the active region.

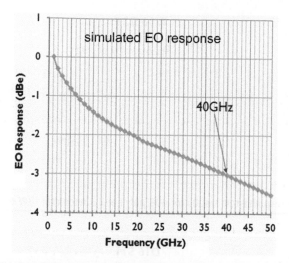

FIGURE 5.11 40G zero chirp single-ended polymer modulator EO response simulation results (BW 40 GHz).

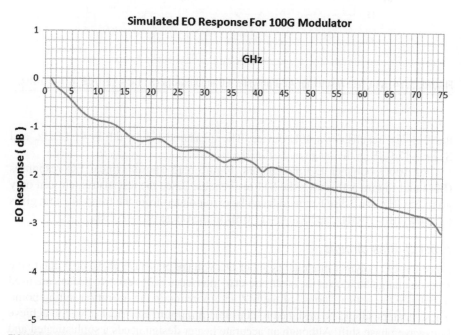

FIGURE 5.12 100G zero chirp single-ended polymer modulator EO response simulation results (BW 74 GHz).

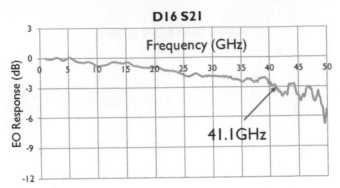

FIGURE 5.13 EO response test results for 40G polymer EO modulator (BW 41.1 GHz).

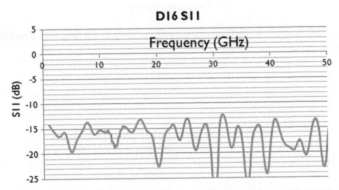

FIGURE 5.14 S11 test results for 40G polymer EO modulators (S11 < −12.0 dB in 1–40 GHz).

the S11 test results for a 40G device. The measured bandwidth is 41.1 GHz, and S11 < −12.0 dB in the range of 1–40 GHz.

For 100G polymer EO modulators, Figure 5.15 shows typical EO responses (S21) and Figure 5.16 shows the S11 test results. Since the test instrument can only go to 65 GHz, the 3 dBe bandwidth can be extrapolated from the trend line that gives a bandwidth > 70 GHz, and S11 < −11.0 dB in the range of 1–65 GHz.

5.5.1.3 Bias circuits

Since the EO polymer modulators under DC voltage bias have phase drift due to the free ion charge buildup at the electrode-clad interfaces [43], resistive heaters are used to balance the optical "arms" of the MZ and to maintain the operation setting point. These heaters should be designed to consume minimum electrical power to achieve half-wave phase shift. Although an accurate heater design needs a sophisticated and expensive software to do 3D simulations, a simple one-dimension model can be used to design the heater with a relatively high accuracy. The schematic cross-section of the heater structure is shown in Figure 5.17.

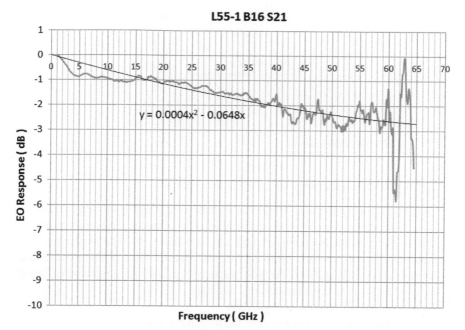

FIGURE 5.15 EO response test results for 100G polymer EO modulator (BW > 70 GHz).

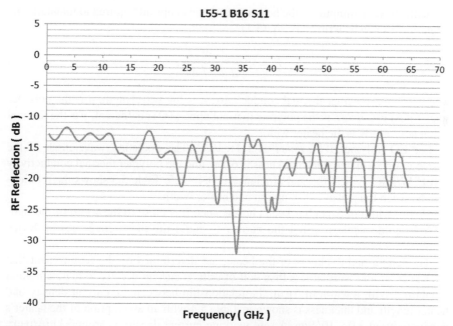

FIGURE 5.16 S11 test results for 100G polymer EO modulators (S11 < −11.7.0 dB in 1–65 GHz).

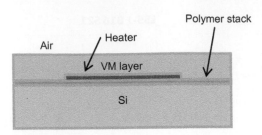

FIGURE 5.17 Device cross-section.

The following assumptions are used in the model:

1. Heater is working at steady state. Steady state means that heater dissipates power to keep heater temperature steady after reaching desired temperature. The silicon beneath the polymer stack acts as a heat sink since its thermal conductivity is 124 W/mK while the thermal conductivities of the polymer and air are 0.15 W/mK and 0.024 W/mK, respectively. The heat generated on heater is equal to the heat transferred to the substrate.
2. Heat transferred to the air is negligible.
3. Temperature is evenly distributed in the polymer stack.
4. Resistance of the heater is more than 10 times the resistance of two side leads of the heater, therefore, power wasted on two side leads is negligible.

With above assumptions, the heater power consumption required to maintain one π phase shift can be written as Eq. (5.1):

$$P = \kappa \frac{2\lambda W}{\gamma d\ dn/dT},\qquad(5.1)$$

where κ is the polymer thermal conductivity, λ is the wavelength, d is the polymer stack thickness, dn/dT is the polymer thermal optic (TO) coefficient, γ is the thermal efficiency, and W is the heater width.

For EO polymer B74, dn/dT is 1.4×10^{-4}. The operation wavelength of the device is 1.55 μm, the stack thickness is 6 μm, and the polymer thermal conductivity is 0.15 W/mK. Typical thermal efficiency is 0.4~0.5. Table 5.3 shows the relationship between the power consumption and heater width.

Heater width of 5 μm has been chosen since it is close to the width of the optical waveguide while it is not too narrow to cause misalignment issues in device fabrication.

The length and thickness of the heater are determined by the requirement that the heater should work at less than half of critical current density to maintain good reliability. With a width of 5 μm, the relationship between the current density and heater length and thickness is shown in Table 5.4 with an assumption of the heater's conductivity at 3.0×10^7 S/m. A safe working current density is around 1 mA/μm². So a length of 1600 μm has been chosen for the heaters.

Table 5.3 Power consumption as a function of heater width.

Heater Width (μm)	Power for Maintaining π Phase Shift (mW)
4	6.2
4.2	6.51
5	7.75
6	9.3
7	10.85
8	12.4
9	13.95
10	15.5
11	17.05
12	18.6

Table 5.4 Heater length and width as a function of current density.

Au Heater	Heater Current Density (mA/μm²)	
Heater Length (μm)	Heater Thickness 10 μm	Heater Thickness 5.5 μm
800	1.07	1.45
1000	0.96	1.29
1200	0.88	1.18
1600	0.76	1.02
2000	0.68	0.91

The width of the heater contact traces (or leads) must be designed such that its resistance has to be at least 10 times less than that of the heater. Figure 5.18 shows the top view of the contact trace and active heater.

Assuming the current is evenly distributed in the thin film traces, the resistance of each section can be estimated by square resistance. The electrode resistance of each section shown in Figure 5.18 is as follows:

R1 = 600/600 = 1 Rsquare.
R2 = 802.5/[(600+70)/2] = 2.40 Rsquare.
R3 = 5/70 = 0.07 Rsquare.
R4 = 1600/5 = 320 Rsquare.

Therefore, the resistance of the heater is 46 times that of trace.

Table 5.5 shows the current test results for a 1600 mm heater to bias the modulators at min, max, and quad states, from which the average pi phase shift power is 7.135 mW, very close to the simulated result in Table 5.3 (7.75 mW).

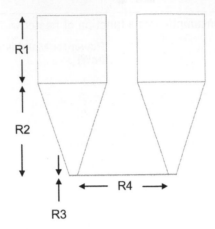

FIGURE 5.18 Top view of heater geometry.

Table 5.5 Heater bias test.

Modu-lator	Device	Current Required to Bias Stat, 1600 μm Heater (mA)						Average π Phase Shift Power
		Quad	Max	Quad	Min	Quad	Max	
2010	LX8401		0	42	61	76		7.135 mW
2006	LX8401				27	51	66	
2007	LX8401	37	57	73	85			
2009	LX8401				37	57	65	
2015	LX8401	41	58	71	82			
2016	LX8401				15	44	61	
2017	LX8401	40	60	75	86			
2018	LX8401	37	56	69	80			

5.6 WAFER FABRICATION

The typical Thin Film Polymer on silicon (TFPS) modulator device fabrication process consists of four sections. Each section contains several individual process modules. The first section is the substrate preparation, the second focuses on the optical stack buildup, and the third section completes the RF structures including the top electrode and velocity match. Finally, the fourth section includes wafer dicing and chip cleaning.

5.6.1 EO polymer processing fundamentals

5.6.1.1 Substrate cleaning

TFPS devices are typically fabricated on the surface of high resistivity ($>5000\,\Omega$-cm single crystal silicon) substrates. Several other suitable substrates can also be used such as single crystal sapphire, quartz, poly crystalline, or even polymer substrates. Standard Chemistry 1 (SC1) and Piranha are often used to clean substrates prior to metal or polymer being deposited on the substrates. These standard chemistries are very effective at removing particulate and organic material from the surface of the wafer, ensuring a clean, contaminate-free surface to deposit subsequent features on.

After metallic and polymer layers are deposited on the surface of the substrate, particular attention must be paid to the cleaning methods in an effort not to damage the TFPS device. Metal-Ion-Free developers are typically used for wafer cleaning, as well as hydrogen peroxide. Again, materials should always be tested for compatibility before a wafer cleaning is performed. Not only do the TFPS layers have to be compatible with the chemistries, but the interfaces between layers must also be robust to the cleaning method.

Standard immersion wet benches, dump rinsers, and spin rinse dryers are used to support most wet clean processes. Particular care should be taken to confirm that the rinse processes are long enough to adequately to remove any residual contaminants on the surface of the substrates. Standard $19\,\text{m}\,\Omega$ de-ionized water is used for rinse processes to prevent any particulate, organic, or ionic contamination of the device. Spray processing of developers is also utilized to clean surfaces due to the benefit of the physical aspect of the spray.

Care must be taken to develop photolithography and etch process that do not form unwanted polymer residues and/or defects. Typical photoresist strip chemistries are not compatible with TFPS materials. Either metal hard mask processes that are designed to have no residual photoresist after processing, or only soft baked photoresist, can be used. This allows the photoresist to be stripped with just a simple UV flood exposure and develop processes.

5.6.1.2 Metal sputtering

After initial substrate cleaning, a ground plane is deposited on the surface of the substrate. Adhesion to the substrate is critical, and in the case of depositing on the surface of the silicon, a thin metal adhesion layer is often used to enhance the adhesion of the highly conductive gold layer. Process parameters must be chosen to ensure a highly conductive, dense, smooth film is created. For the initial metal layer deposited on the Si substrate, standard "best practice" deposition methods can be used.

For metal films deposited directly on the surface of the thin film polymer layers, some care must be taken to ensure temperature, stress, and adhesion are optimized. In situ plasma treatment of polymer films prior to metal deposition is ideal to control the surface state of the substrate before the metal film is deposited. If this is not possible, hard limits must be set between plasma treatment and the onset of the metal deposition.

5.6.1.3 Spincoating

Standard methods of spincoating are typically used to deposit TFPS films. Solvents are used to adjust viscosities of materials to achieve target thicknesses. Variations in spin speed and time are adjusted to fine-tune processes for repeatability and uniformity. Particular care should be taken to control environmental factors such as exhaust flow, humidity, and temperature to ensure stable processing conditions.

5.6.2 Waveguide fabrication

A controlled, stable, consistent waveguide process is fundamental to ensure low optical loss in the final device. It is one of the most sensitive process steps in wafer process flow. Typical processing forms the waveguide in the bottom clad material using standard photolithography and oxygen plasma etching. A metal hard mask can be used to provide better control of sidewall roughness and dimension control.

5.6.3 Poling

5.6.3.1 Process requirements and background

One of the unique aspects of TFPS modulator technology is the ability to perform push-pull (PP) poling. This process allows alignment of the chromophores in the EO composite in opposite polarities in the two arms of a Mach-Zhender interferometer (MZI) as shown in Figure 5.19. The process requires sufficient isolation of each poling electrode arm to prevent arcing or shorting. Polyvinylpyrrolidone (PVP) is spun onto the surface of the poling electrode to prevent arcing through the air between the two oppositely biased arms of the poling electrode. This polymer was chosen due to the fact that it is both easily stripped and non-reactive to the polymer stack.

Poling is done by simultaneously applying opposite charged bias on each MZ arm at the material's (core) glass transition temperature while ground is connected on the bottom electrode. Typical poling results with variation of temperature and an voltage are shown in Table 5.1.

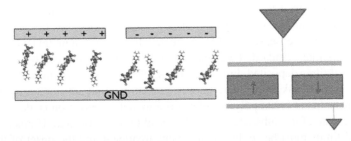

FIGURE 5.19 Push-pull poling schematic.

Table 5.6 $V\pi$ and IL of single MZI structures poled using various voltages and methods.

Format	Poling Type	Poling Voltage (v)	LTC Current (μA)	$V\pi$ (v)	IL (db)
DPSK	Standard	900	>300	3.6	−5.8
DPSK	Push-Pull	±700	<300	1.7	−6.5
DPSK	Push-Pull	±650	<100	1.8	−6.0
DPSK	Push-Pull	±600	<60	1.9	−5.4

5.6.3.2 R_{33} vs. optical loss

From Table 5.1, it can be seen that TFPS modulators poled at higher voltage demonstrated higher insertion losses. The current generated by push-pull poling process is observed to be less than standard poling, and thus it is expected that the high current is likely to cause higher insertion loss (IL). From the summary in Table 5.6, the optical poling conditions can be chosen to achieve the targeted $V\pi$ and IL.

The push-pull poling process was also applied to DQPSK devices. The resulting $V\pi$ was also less than 2.5 V. The resulting $V\pi$ of the DQPSK devices is roughly 1.9 V.

5.6.4 Side mode blocking structure

5.6.4.1 Process requirements and background

Due to the inverted quasi-rib structure of the TFPS device, a small amount of light does propagate in the EO layer slab mode of the device. These are commonly referred to as side modes. When the output waveguide is coupled with a free space lens, these side modes tend to scatter into the main mode of the device, and degrade the extinction ratio (ER). Near the output of the device, a shallow well is etched into the top clad to absorb the light in this area on either side of the waveguide.

5.6.4.2 Fabrication

Consideration is taken as to when the side mode blocking structure is formed on the wafer surface. It is challenging to ensure subsequent layers deposited after the side on the surface of the wafer will not be affected by the topography induced by this structure. Also, the thinning of top clad layer makes the core layer in this region more vulnerable to processing induced defects in subsequent steps. Less aggressive cleaning and baking cycles must be used after the notch has been formed in the device.

5.6.5 RF electrode fabrication

Traditional gold electroplating is used to form the RF electrode in TFPS devices. One of the challenges is selecting a low temperature cured mold material to mask the seed layer during plating. The mold material must be rigid enough to maintain vertical sidewalls during the plating process, but also be removable after the plating process has been completed. Also, the electroplated gold should be optimized to have the highest conductivity possible to ensure high EO bandwidth performance.

5.6.6 Velocity match

5.6.6.1 Process requirements and background

To achieve high bandwidth performance of a microstrip-driven TFPS modulator, a velocity match (VM) material is typically deposited on the surface of the TFPS stack and RF electrode using spin-coating technique. Typical VM materials have similar electrical properties compared to the cladding layers to provide optimal performance. It is also ideal to select a material with low stress and low thermal cure requirements.

5.6.6.2 Processing

Prior to deposition, the wafer is first prepared to ensure good adhesion of the VM layer to the top clad and RF electrode. This is achieved by applying an adhesion promoter. After deposition, certain areas in the device are cleared of the velocity match polymer to provide contact to the launch, termination, and bias connection pads. This is achieved by mask etching, and often a metal hard mask is used such as titanium due to the high selectivity of the mask to the polymer in a typical oxygen plasma etch.

5.6.7 Dicing

5.6.7.1 Process requirements and background

One of the challenges of TFPS device fabrication is wafer dicing. Particular care must be taken to ensure optical quality facets are consistently fabricated to guarantee the lowest coupling loss to the device. The highest level of success has been achieved with relatively thick ($>200\,\mu$m) overcoat materials deposited in a confined strip over the region to be diced. To enhance the rigidity of the overcoat, a $200\,\mu$m thick coverslip may also be attached to the dicing region using a low stress epoxy to hold it in place.

5.6.7.2 Facet coverslip attach

The device wafer is first prepared by applying a narrow glue bead to a $200\,\mu$m thick by $1000\,\mu$m wide D265 glass slide. The glue is a commercially available two-part epoxy called GV101. The glass slide is then placed on the surface of the wafer, centered over the optical facet. The epoxy is thermally cured at $60\,°$C in 100% nitrogen for 4 hours.

5.6.7.3 Dicing

Dicing and chip cleaning are the final steps done in the TFPS chip fabrication process prior to testing. Optical dicing conditions are selected to produce the smoothest optical

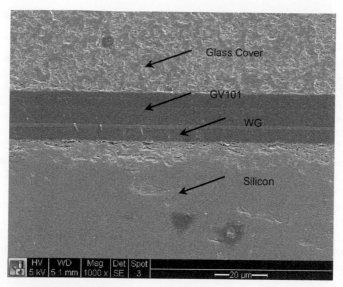

FIGURE 5.20 Diced facet SEM image.

facet possible without damaging the polymer layers or compromising the adhesion between various layers within the device stack. A balance of blade rotation speed, water flow, wafer feed speed, and blade grit size is tuned to support a repeatable process. The optical facet of the chip can be cleaned with a surfactant or solvent such as hexane, typically with a lint-free swab, to ensure no debris is left on the surface of the optical waveguide. A Scanning Electron Microscope (SEM) cross-section of the final device stack post dicing can be seen in Figure 5.20.

5.7 CONCLUSION

Recent advances in TFPS technology have provided the foundation to support commercial devices manufactured at production levels. A fundamental understanding of the material systems and fabrication techniques have been demonstrated, and will provide a stable platform for future developments to support next-generation optical communication applications.

References

[1] T.-D. Kim, J. Luo, J.-W. Ka, S. Hau, Y. Tian, Z. Shi, N.M. Tucker, S.-H. Jang, J.-W. Kang, A.K. Jen, Adv. Mater. 18 (2006) 3038–3042.

[2] (a) Y. Shi, C. Zhang, H. Zhang, J.H. Bechtel, L.R. Dalton, B.H. Robinson, W.H. Steier, Science 288 (2000) 119–122. (b) Y. Enami, C.T. Derose, D. Mathine, C. Loychik,

C. Greenlee, R.A. Norwood, T.D. Kim, J. Luo, Y. Tian, A.K.Y. Jen, N. Peyghambarian, Nat. Photon. 1 (2007) 423.

[3] (a) J. Luo, M. Haller, H. Ma, S. Liu, T.-D. Kim, Y. Tian, B. Chen, S.-H. Jang, L.R. Dalton, A.K.Y. Jen, J. Phys. Chem. B 108 (2004) 8523–8530. (b) D. Briers, G. Koeckelberghs, I. Picard, T. Verbiest, A. Persoons, C. Samyn, Macromol. Rapid. Comm. 24 (2003) 841–846. (c) S. Yang, Z. Peng, L. Yu, Macromolecules 27 (1994) 5858–5862.

[4] (a) P.A. Sullivan, B.C. Olbricht, A.J.P. Akelaitis, A.A. Mistry, Y. Liao, L.R. Dalton, J. Mater. Chem. 17 (2007) 2899–2903. (b) M.E. Van der Boom, Angew. Chem. Int. Ed. 41 (2002) 3363–3366.

[5] (a) B.H. Robinson, L.R. Dalton, J. Phys. Chem. A 104 (2000) 4785–4795. (b) P.A. Sullivan, H. Rommel, Y. Liao, B.C. Olbricht, A.J.P. Akelaitis, K.A. Firestone, J.-W. Kang, J. Luo, J.A. Davies, D.H. Choi, B.E. Eichinger, P.J. Reid, A. Chen, A.K.Y. Jen, B.H. Robinson, L.R. Dalton, J. Am. Chem. Soc. 129 (2007) 7523–7530.

[6] S. Song, S.J. Lee, B.R. Cho, D.H. Shin, K.H. Park, C.J. Lee, N. Kim, Chem. Mater. 11 (1999) 1406.

[7] (a) Y.J. Cheng, J. Luo, S. Hau, D.H. Bale, T.D. Kim, Z. Shi, D.B. Lao, N.M. Tucker, Y. Tian, L.R. Dalton, P.J. Reid, A.K.Y. Jen, Chem. Mater. 19 (2007) 1154. (b) S. Suresh, H. Zengin, B.K. Spraul, T. Sassa, T. Wada, J.D.W. Smith, Tetrahedron Lett. 46 (2005) 3913.

[8] J.A. Davies, A. Elangova, P.A. Sullivan, B.C. Olbricht, D.H. Bale, T.R. Ewy, C.M. Isborn, B.E. Eichinger, B.H. Robinson, P.J. Reid, X. Li, L.R. Dalton, J. Am. Chem. Soc. 130 (32) (2008) 10565–10575.

[9] X.-H. Zhou, J. Luo, J.A. Davies, S. Huang, A.K.Y. Jen, J. Mater. Chem. 22 (2012) 16390–16398.

[10] (a) Y. Bretonniere, C. Andraud, W. Buron, Novel nonlinear chromophores especially suited for use in electro-optical modulation, 2011 FR50087; PCT Int. Appl. 2011, WO 2011089351 A1. (b) P.A. Bouit, G. Wetzel, G. Berginc, B. Loiseaux, P. Feneyrou, Y. Bretonniere, K. Kamada, O. Maury, C. Andraud, Chem. Mater. 19 (2007) 5325–5335.

[11] C. Zhang, C. Wang, J. Yang, L.R. Dalton, G. Sun, H. Zhang, W. Steier, Macromolecules 34 (2001) 235.

[12] S. Thayumanavan, J. Mendez, S.R. Marder, J. Org. Chem. 64 (1999) 4289–4297.

[13] A.K.Y. Jen, Y.M. Cai, P.V. Bedworth, S.R. Marder, Adv. Mater. 9 (1997) 132–135.

[14] J. Roncali, P. Blanchard, P. Frere, J. Mater. Chem. 15 (2005) 1589.

[15] (a) J.M. Raimundo, P. Blanchard, N. Gallego-Planas, N. Mercier, I. Ledoux-Rak, R. Hierle, J. Roncali, J. Org. Chem. 67 (2002) 205–218. (b) J.M. Raimundo, P. Blanchard, I. Ledoux-Rak, R. Hierle, L. Michaux, J. Roncali, Chem. Comm. (2000) 1597–1598.

[16] (a) K. Schmidt, S. Barlow, A. Leclercq, E. Zojer, S.H. Jang, S.R. Marder, A.K.Y. Jen, J.L. Bredas, J. Mater. Chem. 17 (28) (2007) 2944–2949. (b) C. Cabanetos, W. Bentoumi, V. Silvestre, E. Blart, Y. Pellegrin, V. Montembault, A. Barsella, K. Dorkenoo, Y. Bretonnie, C. Andraud, L. Mager, L. Fontaine, F. Odobel, Chem. Mater. 24 (6) (2012) 1143.

[17] (a) S.H. Jang, J. Lou, N.M. Tucker, A. Leclerq, E. Zojer, M.A. Haller, T.-D. Kim, J.-W. Kang, K. firston, D. Bale, D. Lao, J.B. Benedict, D. Cohen, W. Kaminsky, B. Kahr, J.L. Bredas, P. Ried, L.R. Dalton, A.K.-Y. Jen, Chem. Mater. 18 (3) (2006) 2982. (b) M.J. Cho, S.K. Lee, J.-I. Jin, D.H. Choi, Macromol. Res. 14 (60) (2006) 603.

[18] C.R. Moylan, B.J. McNelis, L.C. Nathan, M.A. Marques, E.L. Hermstad, B.A. Brichler, J. Org. Chem. 69 (24) (2004) 8239.

[19] (a) D. Huang, D. Jin, D. Tolstedt, S. Condon, D. Ianakiev, H.W. Guan, S. Cong, E. Johnson, AS. Nishimoto, R. Dinu, Proc. SPIE 6020 (2005) 33. (b) S. Liu, M.A. Haller, H. Ma, L.R. Dalton, S.-H. Jang, K.-Y. Jen, Adv. Mater. 15 (603) (2003) 7–8. (c) M. He, T.M. Leslie, J.A. Sinicropi, S.A. Gamer, L.M. Reed, Chem. Mater. 14 (11) (2002) 4669. (d) Y. Cheng, J. Lou, S. Hau, D.H. Bale, T-D. Kim, Z. Shi, D.B. Lao, N.M. Tucker, Y. Tian, L.R. Dalton, P.J. Reid, A.K.Y. Jen, Chem. Mater. 19 (5) (2007) 1154.

[20] (a) S.R. Hammond, O. Clot, K.A. Firestone, D.H. Bale, D. Lao, M. Haller, G.D. Phelan, B. Carlson, A.K.Y. Jen, P.J. Reid, L.R. Dalton, Chem. Mater. 20 (10) (2008) 3425–3434. (b) T.D. Kim, J.W. Kang, J.D. Luo, S.H. Jang, J.W. Ka, N. Tucker, J.B. Benedict, L.R. Dalton, T. Gray, R.M. Overney, D.H. Park, W.N. Herman, A.K.Y. Jen, J. Am. Chem. Soc. 129 (2007) 488–489. (c) T.-D. Kim, J. Lou, A.K.Y. Jen, Bull. Korean Chem. Soc. 30 (4) (2009) 882.

[21] B.M. Polishak, S. Huang, J. Luo, Z. Shi, X.-H. Zhou, A. Hsu, A.K.Y. Jen, Macromolecules 44 (2011) 1261–1265.

[22] T.-D. Kim, J.-W. Kang, J. Luo, S.-H. Jang, J.-W. Ka, N. Tucker, J.B. Benedict, L.R. Dalton, T. Gray, R.M. Overney, D.H. Park, W.N. Herman, A.K.Y. Jen, J. Am. Chem. Soc. 129 (2007) 488–489.

[23] (a) Y. Enami, C.T. DeRose, C. Loychik, D. Mathine, R.A. Norwood, J. Luo, A.K.Y. Jen, N. Peyghambarian, Appl. Phys. Lett. 91 (1) (2007) (b) R.J. Michalak, Y.-H. Kuo, F. Nash, A. Szep, J. Caffey, P. Payson, F. Haas, B. McKeon, P. Cook, G. Brost, J.D. Luo, A.K.-Y. Jen, L.R. Dalton, W.H. Steier, IEEE Photon. Technol. Lett. 18 (11) (2006) 1207.

[24] R.M. Eichhorn, IEEE Trans. Elec. Ins. EI-12 (1) (1976)

[25] C.C. Teng, M.A. Mortazavi, G.K. Boudoughian, Appl. Phys. Lett. 66 (667) (1995)

[26] H. Chen, B. Chen, D. Huang, D. Jin, J.D. Luo, A.K.-Y. Jen, R. Dinu, Appl. Phys. Lett. 93 (043507) (2008)

[27] R. Dinu et al., Environmental stress testing of electro–optic polymer modulators, J. Lightwave Technol. 27 (2009) 1527–1532.

[28] D. Jin et al., EO Polymer modulators reliability study, in: Proceedings of the SPIE Organic Photonic Materials and Devices XII, vol. 7599, 2010, pp. 75990H–75990H-8.

[29] H. Murata et al., Novel guided-wave electrooptic single-sideband modulator by using periodically domain-inverted structure in a long wavelength operation, in: Optical Fiber Communication Conference, Atlanta, GA, 2003, Paper MF53.

[30] N. Courjal et al., LiNbO$_3$ Mach-Zehnder modulator with chirp adjusted by ferroelectric domain inversion, IEEE Photon. Technol. Lett. 14 (11) (2002) 1509–1511.

[31] S. Oikawa et al., Zero-chirp broadband Z-cut Ti:LiNbO$_3$ optical modulator using polarization reversal and branch electrode, J. Lightwave Technol. 23 (9) (2005) 2756–2760.

[32] N. Courjal et al., Modeling and optimization of low chirp LiNbO$_3$ Mach-Zehnder modulators with an inverted ferroelectric domain section, J. Lightwave Technol. 22 (2004) 1338–1343.

[33] F. Lucchi et al., Very low voltage single drive domain inverted LiNbO$_3$ integrated electro-optic modulator, Opt. Express 15 (17) (2007) 10739–10743.

[34] S.-K. Kim et al., Electrooptic polymer modulators with an inverted-rib waveguide structure, Photon. Technol. Lett. 15 (2003) 218–220.

[35] S. Pogossian et al., The single-mode condition for semiconductor rib waveguides with large cross section, J. Lightwave Technol. 16 (1998) 1851–1853.

[36] <http://jp.fujitsu.com/group/foc/en/services/optical-devices/40gln/>.

[37] H. Zhang et al., Push-pull electro-optic polymer modulators with low half-wave voltage and low loss at both 1310 and 1550 nm, Appl. Phys. Lett. 78 (20) (2001) 3136–3138.

[38] G. Yu et al., 40 GHz zero chirp single-ended EO polymer modulators with low half-wave voltage, in: CLEO 2011 Micro and Nano-photonics Modulators (CTuN), Baltimore, Maryland, May, 2011.

[39] R. Dinu et al., Small form factor thin film polymer modulators for telecom applications, in: OFC/NFOEC 2012 Modulators and Demodulators (OM3J), Los Angeles, California, March, 2012.

[40] S.R. Nuccio, R. Dinu, B. Shamee, D. Parekh, C. Chang-Hasnain, A. Willner, Modulation and chirp characterization of a 100-GHz EO polymer Mach-Zehnder modulator, in: National Fiber Optic Engineers Conference, Los Angeles, California, March 6, 2011, Poster Sessions II (JThA).

[41] J. Mallari et al., 100 Gbps EO polymer modulator product and its characterization using a real-time digitizer, in: OFC/NFOEC, 21–25 March 2010, Conference Publications (OThU2).

[42] V. Katopodis et al., Integrated transmitter for 100 Gb/s OOK connectivity based on polymer photonics and InP-DHBT electronics, in: European Conference and Exhibition on Optical Communication, Amsterdam, The Netherlands, June 16, 2012, Postdeadline Session II, (Th.3.B).

[43] R. Thapliya et al., Investigation of drift in electro-optic polymer waveguides, Appl. Phys. Lett. 93 (2008) 193309.

Nanophotonics for Low-Power Switches

Lars Thylen[a, b, c], Petter Holmström[a], Lech Wosinski[a, c],
Bozena Jaskorzynska[a], Makoto Naruse[d, e], Tadashi Kawazoe[e, f],
Motoichi Ohtsu[e, f], Min Yan[a], Marco Fiorentino[b], and Urban Westergren[a]

[a]Laboratory of Photonics and Microwave Engineering, Royal Institute of Technology (KTH),
SE-164 40 Kista, Sweden
[b]Hewlett-Packard Laboratories, Palo Alto, CA 94304, USA
[c]Joint Research Center of Photonics of the Royal Institute of Technology (KTH) and
Zhejiang University, Zhejiang University, Hangzhou 310058, China
[d]Photonic Network Research Institute, National Institute of Information and
Communications Technology, 4-2-1 Nukui-kita, Koganei, Tokyo 184-8795, Japan
[e]Nanophotonics Research Center, Graduate School of Engineering, The University of Tokyo,
2-11-16 Yayoi, Bunkyo-ku, Tokyo 113-8656, Japan
[f]Department of Electrical Engineering and Information Systems, Graduate School
of Engineering, The University of Tokyo, 2-11-16 Yayoi, Bunkyo-ku, Tokyo 113-8656, Japan

6.1 INTRODUCTION

This chapter will deal with photonic switches and the quest for the partly interlinked properties of low-power dissipation in operation and nanostructured photonics. In general, the nanofeatures will pertain to the device cross-sections, rather than to length.

Switches and modulators are key devices in ubiquitous applications of photonics: Telecom, measurement equipment, sensors, and the emerging field of optical interconnects in high performance computing systems. The latter could accomplish a breakthrough in offering a mass market for these switches.

A modulator can be said to be a 1×1 switch. Switches are generally built using amplitude and/or phase modulation of the optical wave. But it should be noted that access to a phase modulator enables the generation of virtually all basic types of modulation: amplitude, polarization, and frequency [1]. We will concentrate on the high-speed multi-GHz modulators and switches, which pose the greatest challenges, since modulators and in general devices with high reconfiguration rates require more power. At the same time it is important to bear in mind that, e.g., devices for cross connects do not in general require high-speed operation [2].

Optical Fiber Telecommunications VIA. http://dx.doi.org/10.1016/B978-0-12-396958-3.00006-8

The power dissipation we are dealing with here is joule heating, i.e. the high entropy dissipation that cannot easily be recovered. However, in, e.g., all-optical switches, treated below, "unused" photonic energy can, at least in principle, be recovered by using so-called photon recycling.

Why low power of operation? Low power of operation is of course a rather self-evident goal, given that costs are not impeding this. However, recent developments have made this issue more important, one example being the exploding power requirements in server parks necessitating placing them in colder regions of the world, such as Scandinavia (see Figure 6.1).

This development is to a large degree due to the fact that a large portion of power dissipation takes place in interconnects on different scales and to the deficient performance of the copper interconnects in currently used high-speed transmission. This has necessitated revisiting the old subject of optical interconnects, another example of a field not starting to boom until the time is ripe. Optical transmission is thus contemplated for transmission distances ranging from inter-cabinet to interchip or intrachip, so that we are eventually talking about transmission distances from multiple hundreds of kilometers down to submicron. But low power is generally favorable, such as for all battery-powered applications, e.g. for sensors. In general the power aspect did not receive prime attention in earlier years.

An aspect not so often touched upon in the scientific literature is cost. However, cost is an important issue, if mass deployment of switches, such as in interconnects, is contemplated.

Data Centers-Information Factories

FIGURE 6.1 Artist's view of a server park for supporting companies such as Google and Facebook as well as future cloud services. The footprint of the server park necessitates very high-speed transmission over distances ranging from sub-mm to kilometers. Partly from http://hightech.lbl.gov/htnews/htn-issue2.html.

The chapter will focus on electronically controlled switches, which in many important cases and using a simple model are operated by charging and discharging capacitors and thus changing absorption and/or refraction properties of the medium between the capacitor plates, see below. A phase change is as noted above more versatile since it in principle permits all modulation formats.

But switches come in other modes of operation such as all optically operated, where both the controlled and controlling signals are optical. This type of device has a long history, basically dating back to the development of the laser, and in general they are less power efficient and flexible than the electronically controlled ones, but can achieve very high speeds, such as THz, however, normally requiring more power and energy.

The chapter will try to give a rather speculative view on the subject, while presenting some results for present devices.

6.2 EXISTING AND EMERGING MATERIALS

Optical materials have a central role in enabling nanophotonic low-power switches. To give a simplified description, they have this central role in two ways:

1. By enabling an efficient conversion of applied stimulus (be it electric field or heat) into a change of the complex linear or nonlinear refractive index [1].
2. By making possible high confinement of the light fields, a property already treated in [3].

The first property determines the electric field or thermal energy required to achieve switching. This normally requires a phase change of order π, or an extinction of order 10 dB. The larger the efficiency of this conversion (i.e. the larger the electrooptic effect, quantum confinement Stark effect, nonlinear optical effects, etc. as treated below), the shorter the device length for, e.g., a given electric field.

The second property amplifies the first since it in general reduces the cross-section of the device and hence further diminishes the volume where energy exchange with the surroundings needs to take place. Here high index, low-optical loss materials such as silicon have acquired a prime role but so could also different versions of plasmonics-based switches with varying mixes of metals and dielectrics. In the latter case, confinement is in general traded against optical propagation losses, see, e.g., [4]. Both of these are treated in Section 6.3.1.2 below.

Table 6.1 summarizes some of the most important materials for nanophotonics low-power switches, embodying the features stated above, and describes physical mechanisms, operation mode, and characteristics (adapted from [1]).

Table 6.1 A compilation of different mechanisms to achieve the changes in complex refractive index required for modulation of optical radiation. Abbreviations used: EO: Electrooptic, SC: Semiconductor, LN: LiNbO3, KK: Kramers-Kronig relations [5].

Materials	Physical Mechanism	Operation Mode	Characteristics
• Electrooptic polymers • III–V SC • LN	Index change through Pockels effect (linear electrooptic effect) in materials lacking inversion symmetry	Voltage applied over dielectric material or reverse-biased pin-junction	Speed usually limited by microwave/optical wave walk-off or RC-constants. Nonresonant effect, low associated absorption. Index changes of order 0.1 achievable in EO polymers
Quantum well or in general low dimensional SC	Absorption change by Quantum Confined Stark Effect (QCSE)	Reverse-biased pin-junctions	RC- or transport-time limited response (fast, depends on number of wells and barrier heights). Strong λ dependence.
As above	Refractive index change through QCSE, related to the above by the KK relation	As above	See above; RC-limited response. $\Delta n/n = 0.001–0.01$ Strong λ dependence
Chalcogenides	Material phase change between crystalline and amorphous states, in general optical phase as well as amplitude change	Operation employing electric field, light, or thermal effects	Very large index changes achievable, >0.5, with material-dependent associated very large absorption. Memory function. Ms response times
Graphene	Absorption changes by voltage-mediated Fermi level change	Applied electric field	Absorption modulation
Chalcogenides and others	Third-order optical nonlinearity χ^3 phase change	Refractive index change mediated by light field (Kerr effect)	Large nonlinear effects and tight light confinement needed to venture into the low-power nanophotonics realm
Silicon	Plasma effect	Current injection, current depletion, or MOS effect	Changes in the real and imaginary parts of the index. Speed is limited by the recombination rate in the current injection. Drift is the limiting factor in depletion and MOS modulators

6.3 SWITCHES

6.3.1 Electronically controlled switches

6.3.1.1 Basic operation and power dissipation issues

The purpose of the electronic driver circuit in a photonic on/off (OOK) transmitter is to switch the output light power between a low-power ("zero") P_0 and a high-power ("one") P_1 at a given bitrate. A common circuit implementation is a driver that switches the current I through a resistor, in parallel with a modulator, between I_0 and I_1 corresponding to P_0 and P_1, respectively. The modulator is in general in this chapter in essence a capacitor. A system specification will include an average light power level and a minimum extinction ratio, so the switching has to be done very precisely.

The practical circuit design is usually balanced, especially at high bitrates, to reduce the risk of cross-talk and disturbances in the circuitry. In order to control the voltage across the modulator accurately to switch between the correct optical power levels, a differential pair of transistors, bipolar or field-effect, is used together with current sources which are constructed from current mirrors. Figure 6.2 shows a typical example of a driver circuit with bipolar transistors.

The power dissipation of this type of circuit stays close to constant in time due to the balanced circuit. The dissipation is determined by the larger current I_1 necessary to produce the voltage $V_1 = R_c I_1$ which gives the largest electric field in the capacitor. If the output impedance of the circuit is approximated by the load resistance R_c and the modulator is assumed to have a small capacitance C, the bandwidth of the transmitter is given by the charging time of the modulator, $R_c C$. A practically useful photonic transmitter is normally designed with a bandwidth so that $1/2\pi R_c C = B$ with the bitrate B in bits per second, i.e. the -3 dBe bandwidth in hertz is equal to the bit rate in b/s. With the drive circuit in close integration with the modulator, no impedance matching is necessary.

The dissipated energy per bit slot becomes approximately $C V_1^2$.

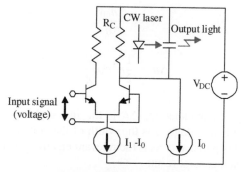

Minimum voltage ("1"): $R_c I_0$
Maximum voltage ("0"): $R_c I_1$

FIGURE 6.2 Modulator driver circuit.

6.3.1.2 Figures of merit for nanophotonics switches

There are several partly interrelated characteristics for switches. For space switches, these are cross-talk, insertion loss, drive power, RF operation bandwidth, and operation wavelength range to name the most important. However, in keeping with the topic of this chapter, we will instead focus on phase modulators, which, as noted, are ubiquitously usable, and in this section further focus on electrically controlled switches where operation is effected by refractive index changes which in turn change the phase of the light wave. This includes nearly all important types of switches, the operation of which basically relies on a π phase change:

$$\Delta n_{\mathrm{eff}} k_0 L = \pi, \tag{6.1}$$

where L is the device length, Δn_{eff} is the effective index change in the waveguide [3], and k_0 is the wave number in vacuum.

Current optical switches have transverse dimensions on the order of the wavelength in the medium and lengths 100 or 1000 times the wavelength, thus a very large aspect ratio. A nanophotonics switch could drastically shrink these dimensions, as we will see in some of the examples below and give a more appealing aspect ratio, amenable to integration. And as noted above, higher field confinement, enabling smaller *transverse* dimensions, would mean lower power dissipation. However, shrinking the *longitudinal* dimension would, according to Eq. (6.1), entail a larger required Δn_{eff} and hence, in the same configuration, higher voltage and larger power dissipation (Section 6.3.1.1). Below we analyze a slotline-type phase modulator (see, e.g., [6]), where the slot is filled with a low index electrooptic material surrounded by high index material such as silicon, also serving as electrodes.

Using Eqs. (6.1–6.4), where we assume an insertion loss less than or equal to 3 dB,

$$\alpha_m L \leq \ln(2), \tag{6.2}$$

$$\Gamma = \frac{\Delta n_{\mathrm{eff}}}{\Delta n}, \tag{6.3}$$

$$\Delta n = \frac{1}{2} n^3 r_{33} \frac{V}{g}, \tag{6.4}$$

and where α_m is spatial attenuation, Δn is the index change in the electrooptic medium, V is the applied voltage, g is the width of the slot or electrooptic material between the high index parts, and r_{33} is the relevant electrooptic tensor element (see, e.g., [7]), we arrive at a general result

$$\Gamma n^3 r_{33} \frac{V}{g} \frac{1}{\lambda_0 \alpha_m} \geq 1.44. \tag{6.5}$$

As can be seen, the attenuation per vacuum wavelength is important, as well as high Γ and index (and of course high r_{33}).

On the other hand, the energy per bit needed to charge the electrooptic material is

$$E = \varepsilon\varepsilon_0 V^2 \frac{LWg}{g^2} = \varepsilon\varepsilon_0 V_{\text{EO}} \left(\frac{V}{g}\right)^2, \tag{6.6}$$

where ε is the relative permittivity of the electrooptic material, W is its width, and V_{EO} is its volume. Using Eq. (6.5) for V/g we get a figure of merit (FOM) in terms of electrostatic energy per unit of electrooptic volume, a kind of specific energy:

$$\frac{E}{V_{\text{EO}}} \geq \varepsilon\varepsilon_0 \left(\frac{1.44\alpha_m\lambda_0}{\Gamma n^3 r_{33}}\right)^2. \tag{6.7}$$

The length is implicitly given by the maximum 3 dB loss length, Eq. (6.2). As an example, assuming the data of the Si/EOP/Si slotline Mach-Zehnder switch of Section 6.3.2.1 (and Table 6.4, case 3, EOP is electrooptic polymer) for $\lambda_0 = 1.55\,\mu\text{m}$, $\Gamma = 0.20$, $n = 1.7$, $r_{33} = 500\,\text{pm/V}$, $\alpha_m = 1.7\,\text{cm}^{-1}$, and $\varepsilon = 2.89$ we get 15 aJ/μm^3/bit and $L = 4.1\,\text{mm}$. However, for, e.g., the coupled-plasmonic-nanoarray switch of Section 6.3.2.4 and Table 6.4, case 2 below, the corresponding specific energy is 1 pJ/μm^3/bit for L = 200 nm, however, with a concomitantly higher loss.

A more relevant figure of merit would be EV_{EO}, since we want both to be as small as possible:

$$EV_{\text{EO}} \geq V_{\text{EO}}^2 \varepsilon\varepsilon_0 \left(\frac{1.44\alpha_m\lambda_0}{\Gamma n^3 r_{33}}\right)^2. \tag{6.8}$$

Here the data for the Si/EOP/Si slotline switch of Section 6.3.2.1 is 103 fJ/μm^3/bit and for the switch of Section 6.3.2.4 is 4×10^{-3} fJ/μm^3/bit, a huge difference, bought at the price of loss.

Equation (6.7) can be seen as a reasonably general equation, given that the applied field is in the EO medium (and not "wasted" such as in the buffer layer in LiNbO$_3$ devices) and if Γ is defined in a reasonable manner. It only involves material properties and waveguide properties in terms of Γ, but not waveguide dimensions. It tells what the material properties can do; to decrease total energy, one has to decrease the volume of the EO medium, so quite expectedly the nano and low-power features go hand in hand. Examples will be given in Section 6.4.

6.3.2 Some examples of electronically controlled switches

We will here focus on switches employing emerging electrooptic materials, such as electrooptic polymers, since they promise higher refractive index changes than the hitherto used LN and the quantum confined Stark effect (Table 6.1). We will not include electronically controlled chalcogenides in this treatment since they at present might have too high losses and perform in the ms time range. Operation in the ns range might be possible in future nanostructured materials [8]. Graphene amplitude modulation is summarily treated in Section 6.4.

6.3.2.1 Electrooptic polymers (EOPs) on a silicon platform

The use of electrooptic polymers for modulation, switching, and routing in optical communication systems is recently gaining once again more attention after several new developments with higher EO coefficient than in inorganic oxide crystals, such as lithium niobate, improved material, thermal, and photochemical stability as well as emerging new types of polymers that do not require a poling process.

Interest in this material dates to the mid-1980s together with general interest in nonlinear (NLO) polymers, initiated by the Air Force Office Scientific Research Program on NLO polymers that started in 1982, in parallel with academic research programs on several universities. It continued to late 1980s and the beginning of the 1990s, accompanied by a substantial industrial interest due to increased needs for ever-greater communication bandwidth. However, attempts to solve stability problems of this material were not successful and lithium niobate remained the dominant EO material and the interest in polymers decreased considerably. With the beginning of the new millennium some university research groups including Profs. Dalton, Jen, and Robinson's group at the University of Washington, Prof. Marks' group at Northwestern University, and Prof. Zyss' group at Ecole Normale Superieure, Cachan (France), showed considerable progress regarding EO polymer materials.

Novel types of NLO guest-host polymer systems have provided remedy to the problem of combining high performance and stability, like chromophores with twisted motifs or charge transfer chromophores in dendritic encapsulation. Poled guest-host polymers exhibit today very high electrooptic (EO) coefficient, approaching 500 pm/V, while keeping optical loss below 2 dB/cm [9,10]. Without long extents of conjugated double bonds, these chromophores exhibit excellent thermal and oxidative stability. Recent recurrence to these materials has been dictated by the growing interest in silicon photonics and photonics-electronics integration using CMOS technology, where polymers are matching very well since they are similar to photoresists, commonly used in CMOS technology. With considerably increased EO coefficient, polymers became also very attractive from a power budget point of view potentially allowing for large reduction of energy consumption. Finally mass production and, in most cases, ease of implementation using silicon platforms could lead to low-cost manufacturing of these in many applications relevant for active components.

Taking into account only the linear electrooptic effect with good approximation, the change of the material refractive index with applied electric field is given by Eq. (6.4) above, which can be rewritten as

$$\Delta n = \frac{1}{2}n^3 r_{33} E, \qquad (6.9)$$

where E is the electric RF field in the EOP, r_{33} is the EO coefficient, and n is the refractive index. Considering EO polymer with $r_{33} = 100$ pm/V, $n = 1.6$, material thickness 0.5 μm, and modulating voltage of 3 V ($E = 6 \times 10^6$ V/m), the change of refractive index will be $\Delta n = 1.23 \times 10^{-3}$. To introduce a π phase shift (equivalent to λ/2) for $\lambda = 1.55$ μm, an optical path length of $\lambda/(2 \times \Delta n)$ is necessary, equal to 0.6 mm. For a push-pull Mach-Zehnder [9] device, where each arm is driven in

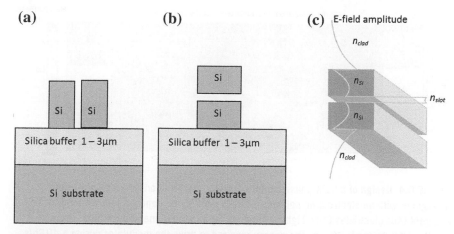

(a)   **(b)**   **(c)**

FIGURE 6.3 (a) Vertical and (b) horizontal configurations of silicon-based slotlines. (c) With light polarized perpendicular to the high-dielectric-constant silicon layers, light is heavily concentrated to the low dielectric constant intermediate region, the slot, which could be air or an electrooptic material.

antiphase it will be 0.3 mm. For a hybrid plasmonic structure with polymer thickness of 50 nm the length of this modulator will be 30 μm. Using improved polymer with $r_{33} = 300$ pm/V, modulator length would decrease to the value of 10 μm. For more accurate calculation, one should use the effective refractive index of light guided in the material, which will increase here with decreasing material thickness, but simultaneously the amount of light concentrated in the polymer can decrease if, for example, the silicon slot waveguide configurations shown in Figure 6.3 are used. The electrooptic coefficient for very thin polymer layers may also decrease from the value of bulk material. A more detailed design is given below.

One of the most important parameters of an optical modulator is its modulation speed. Here the polymer itself is not a limiting factor as the phase relaxation of its π-electrons counts in femtoseconds and so speeds of tens of THz should be possible. Instead the limitation lies rather in electronics, the configuration of electrodes, and the effects of RC time constants.

Presently EO modulators based on polymers are commercially available for 40 Gb/s and even for 100 Gb/s operation with 2.5–3.5 V modulation voltage, insertion loss of 7.5–9 dB, extinction ratio of 20 dB, and a form factor of about $100 \times 10 \times 10$ mm in packaged devices. Recently these modulators were environmentaly stress tested satisfying the stress specifications of the Telcordia GR-468 industry standard and showing excellent performances for thermal and photochemical stability [11]. Furthermore, the reliability study [12] showed <10% change in EO coefficient while operating at 85 °C for 25 years in the accelerated aging tests. However, this is not nanophotonics yet.

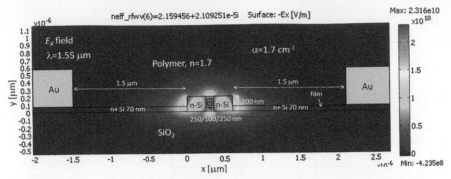

FIGURE 6.4 Design of a light-phase modulator based on a silicon-on-silica nanoslot waveguide with an electrooptic polymer (EOP) in the 100 nm nanoslot and with n-Si, n+-Si and gold (Au) electrodes [14]. Light is propagating into the plane of the figure. The active length of the device is 80 μm. The energy required to drive the modulator during a 40 Gb/s bitslot is 33 fJ, with $r_{33} = 500$ pm/V in the EOP and a 2 V drive for the π/2 phase shift required in a push-pull type modulator. This gives an average drive power requirement of 1.3 mW.

Currently instead of using large-core, several micrometers cross-section polymer waveguide structures, slot-waveguide architectures are usually designed, where most of the light is guided in a nanoscale slot filled with polymer in between material with much higher refractive index, usually silicon, Figures 6.3 and 6.4. This configuration greatly increases electric field inside the slot [13] allowing highly efficient interaction between the polymer material and the optical field. Figure 6.4 shows the design of a such light-phase modulator based on a silicon-on-silica nanoslot waveguide with an EOP in the 100 nm nanoslot and with n-Si, n+-Si, and gold electrodes [14]. A large overlap of the optical mode with the EOP layer is achieved. The large electro-optic effect (refractive index change) of the EOP then creates a very efficient phase modulator, see the caption of Figure 6.4.

6.3.2.2 Hybrid plasmonics

Due to the diffraction limit of light, the size of an optical mode propagating in a conventional dielectric waveguide cannot be smaller than approximately half of the wavelength in the material in which the light is propagating. To improve the integration density, one needs to break this limit. This can be done using surface plasmon polaritons (SPPs) that can have extremely short effective wavelength (and extremely high effective index) and high optical field confinement at a metal-dielectric interface, where the guiding relies on coupled plasmon-polaritons propagating as electromagnetic fields coupled to surface plasma oscillations of conduction electrons in the metal. By using metal as cladding and dielectric as core layer (metal-insulator-metal structure), it is possible to achieve a true subwavelength waveguide which could confine the light field to nanometer scales (see, e.g., [4,15]; however, losses in such structures realized with available metals are severely limiting the propagation length to the order of a few micrometers or below for confinement on the 10s of nanometer scale, depending on wavelength and metal.

Recently an alternative geometry of hybrid plasmonic waveguide has been proposed [16,17] where instead of one metal layer a silicon substitute is used. In this case the modal field becomes slightly larger, still keeping subwavelength confinement, whereas one can obtain a considerable decrease of losses. Simulations and experimental results [18] show that a propagation loss as low as 0.01 dB/μm (propagation length over 400 μm) can be achieved with a mode size in the lateral direction of 200 nm. This opens the possibility to realize functional photonic elements based on hybrid plasmonic waveguides that can be highly integrated keeping simultaneously relatively low loss. Some theoretical work and computer simulations have been done to investigate hybrid plasmonic modulators [15,19], see Figure 6.5. Due to the difficult fabrication of ultra-compact devices and not-straightforward characterization of the fabricated structures, no quantitative experimental work has been done so far.

As a particular version of a hybrid plasmonics polymer-based modulator, a slot waveguide layout can be used for high-speed (40 Gbit/s and beyond) modulators using EOPs as active materials on a silicon platform [20]. The obvious advantage of the design is low driving voltage and power consumption, and high extinction ratio. Recent progress in the research on polymeric and hybrid guest-host polymer

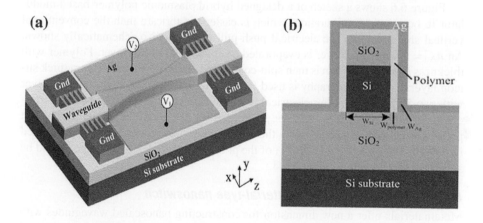

FIGURE 6.5 (a) Schematic perspective view of a proposed phase modulator based on a hybrid plasmonic waveguide. The plasmonic waveguide is connected to a silicon wire waveguide via an adiabatic taper. A half-wave voltage of $V_\pi = 2.5\,$V for an active waveguide length of 13 μm gives a switching energy of 9 fJ in the simulated device. (b) Cross-sectional view of the hybrid plasmonic waveguide. The silicon waveguide core is heavily n-type doped, $2 \times 10^{20}\,$cm^{-3}, to serve as the ground electrode when biasing the modulator. The lowest order even TE-polarized mode is employed for modulation. Reproduced with permission [19] © OSA 2011.

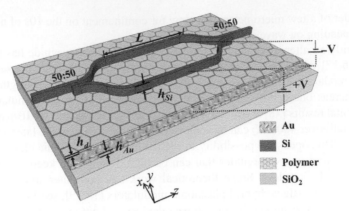

FIGURE 6.6 Sketch of the hybrid plasmonic polymer-based modulator in horizontal configuration. The electrical push-pull circuit is also schematically shown.

EO materials manifests advantages of these materials for developing high-speed and energy-efficient photonics components using silicon and hybrid plasmonics technology [10].

Figure 6.6 shows a sketch of a designed hybrid plasmonic polymer-based modulator in horizontal configuration, which is easier to fabricate than the conventional vertical slot structure. The electrical push-pull circuit is also schematically shown. An $h_{Au} = 100$ nm gold layer is evaporated on thick SiO_2 buffer layer. Polymer with thickness $h_d = 100–200$ nm is then spin-coated and covered by $h_{Si} = 400$ nm thick silicon layer. E-beam lithography is used for patterning of the MZ structure in Si layer with two 50:50 couplers (for electrical isolation between the arms) and two interferometer arms with length $L = \lambda/(4\Delta n_{eff})$ giving the π phase shift between the two light paths. L is calculated depending on the effective refractive index for the light guided in the polymer optical mode and r_{33} of the polymer material. Performance should be similar to the structure of Figure 6.4.

6.3.2.3 Modeling of a metamaterial-type nanoswitch

Metamaterials offer a new dimension for constructing nanoscaled waveguides with deep-subwavelength mode confinement. Here we first elucidate the possibility of realizing a composite material with an effective refractive index ($\bar{n}$) as high as $\sim$8 at 1.55 μm wavelength. The material configuration is schematically shown in the top inset in Figure 6.7a. It consists of alternating layers of silver and a dielectric (assumed to have an index of 3). When the layer thicknesses are significantly smaller than the operating wavelength, the overall structure appears to light as a homogenized medium with an effective refractive index, which can be calculated using the Maxwell-Garnett mixing formula [21]. In particular, when the metal filling ratio $f_m = 0.8$, the real and imaginary parts of $\bar{n}$ (for wave propagating in z direction, with

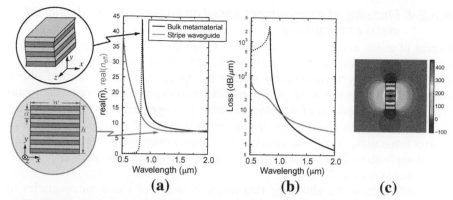

FIGURE 6.7 (a) Real part of the effective (mode) indices of a bulk metamaterial with metal filling ratio of 0.8 and the corresponding metamaterial-based stripe waveguide with air cladding. (b) Losses of the bulk metamaterial and the stripe waveguide. (c) H_x field of the fundamental mode of the stripe waveguide, with domain size at 300 × 300 nm². Reproduced with permission [24] © OSA 2011.

electric field polarized dominantly along y direction) as a function of wavelength are shown by the black curves in Figure 6.7a and b, respectively. A Drude model for the permittivity of silver has been used [22]. Very importantly, the medium has a rather large real ($\bar{n}$) of 7.92 at $\lambda_0 = 1.55\,\mu$m, while its loss is at (under these circumstances) a reasonably low value of 1.5 dB/μm.

The bottom inset in Figure 6.7a shows a stripe waveguide made of such a metamaterial. The metamaterial has a period $a = 20$ nm, total height $h = 5.5a$, and width $w = 40$ nm. Cladding is simply air. According to the red curves in Figure 6.7a and b, the stripe mode has an effective mode index (n_{eff}) as high as 7.35 at $\lambda_0 = 1.55\,\mu$m, with a loss of 3.3 dB/μm. The mode size (Figure 6.7c) is 0.01 μm², 20 times smaller than that for a silicon-wire waveguide.

More interestingly, a refractive index change in the dielectric material is amplified by a factor of roughly four in the effective index of the waveguided mode, giving a corresponding reduction in length or drive voltage when such a waveguide is employed for constructing a Mach-Zehnder interferometer. The metal layers serve as both driving electrodes and capacitors arranged in parallel for inducing necessary electric field change in the dielectric layers. At $\lambda_0 = 1.55\,\mu$m, given an index change in the dielectric layers of 0.1, we have a change in mode index $\Delta n_{eff} = 0.36$. This leads to a phase modulator as short as 2.2 μm. A drive voltage below 0.5 V at a device length of $\sim$2 μm would give a π phase shift of the optical field, as required in many modulators. The energy required to produce the π phase shift, e.g. to be used for on/off modulation, is 3 aJ, which is the energy required to charge the capacitor. However, with existing metals the insertion loss would be 7 dB, implying that the device would be used in a stand-alone fashion only, since concatenation would mean too high losses.

6.3.2.4 *Modeling of nanoswitches based on optical-near-field-coupled metal nanoparticles*

Chains of metal nanoparticles can transport electromagnetic energy via the excitation of their localized surface plasmons [24,25]. Such metal-nanoparticle arrays offer the possibility to considerably surpass the optical field confinement of silicon waveguides. The modal width of nanoparticle-array-waveguides at resonant operation is quite independent of wavelength, rather it is approximately three times the particle radius. This could allow denser lateral packing of waveguides as well as shorter resonators, filters, and switches. However, a main problem for many albeit not all applications is the optical loss associated with such high confinement metal-based metamaterials [24–26].

In this section the ability of two adjacent arrays of metal nanoparticles to achieve extremely compact directional couplers is analyzed theoretically [27]. Such couplers [26] form the basis of generic types of integrated photonics devices such as modulators and switches. Specifically, here, the coupling length l_c, filtering, and switching characteristics are investigated. The analysis is carried out for a hypothetical lossless silver, including only radiative losses, to demonstrate the potential of this type of circuit for applications in telecom and interconnects. This is in anticipation of a possible future breakthrough in developing metamaterials with at least a factor of 10 lower losses [29], or in achieving loss compensation [30], or else by employing stand-alone submicron devices. The arrays consist of 50-nm diameter metal particles spaced by $d=75$ nm center to center. A hybrid finite-element-method and multi-level fast-multipole-algorithm (FEM/MLFMA) is used to simulate the structure in the frequency domain with 3D resolution of the particles [26].

For the case of two identical arrays a spatially periodic sinusoidal-like coupling of the surface plasmon polariton excitations on the arrays is demonstrated numerically [28]. Extremely short coupling lengths, e.g. $l_c=490$ nm for $c=90$ nm, are demonstrated. For a center-to-center spacing $c=90$–130 nm of the two arrays, the coupling length can be very well fitted by the power law $l_c=l_0(c/d)^{5.6}$, where $l_0=170$ nm. A strong wavelength dependence of the coupling length allows for efficient filtering, e.g. corresponding to a 2.4-nm bandwidth (full width at half maximum of the cross-coupled intensity) for a 3-μm long coupler, indeed very good data for an integrated optics filter.

The impact of phase mismatch between the two nanoarray waveguides in order to effect optical switching is analyzed in the following. Such phase mismatch can be brought about by artificially changing the plasma frequency or by changing the host index or particle shape for one of the arrays, in order to get a propagation constant mismatch δk between the two waveguides. In Figure 6.8 we see that for an approximately 1.4 μm long coupler, i.e. one coupling length in the uppermost phase-matched case, a change of the plasma frequency by 0.32% ($\hbar\omega_{p,\text{lower}}=6.18 \to 6.16\,\text{eV}$) is enough for high-extinction-ratio switching, from a cross state to a bar state.

Another way of effecting switching is to change the host refractive index for one of the arrays, thereby changing the single-particle (Fröhlich) resonance frequency. Figure 6.9 shows the dispersion relation for longitudinally excited arrays of prolate spheroids.

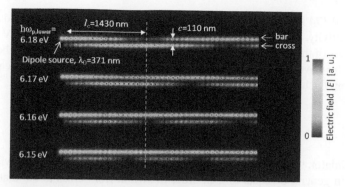

FIGURE 6.8 Optical switch based on a mismatched nanoarray waveguide directional coupler. The nanoparticles in the upper waveguide have the plasma frequency $\hbar\omega_{p,\text{upper}} = 6.18\,\text{eV}$, while those of the lower waveguide are shifted as indicated in the figure. Shown is a top view of the lossless Ag nanoparticle arrays indicating the *E*-field magnitude in a plane 3 nm above the particle surfaces with the particle positions indicated by black dots. The nanoparticles are 50 nm in diameter, separated by 75 nm center to center.

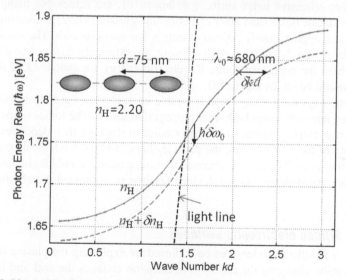

FIGURE 6.9 Dispersion relation in a longitudinally excited array of prolate spheroids of size $50 \times 25 \times 25\,\text{nm}^3$. Shifting the host refractive index by δn_H gives a shift δk of the waveguide *k* number at the operation point. The shift is exaggerated compared to the example in the text for clarity. Reproduced with permission [27] © OSA 2011.

The dispersion is obtained within a point-dipole model including the effects of retardation [30]. By using spheroids the Fröhlich resonance frequency

$$\omega_0 = \frac{\omega_p}{\sqrt{1 + \frac{1-N}{N}\varepsilon_H}}, \tag{6.10}$$

where $N=0.174$ is the depolarization factor [23] ($N=1/3$ for spheres) and ε_H is the host permittivity, is red-shifted, and further the bandwidth $\Delta E = 0.230\,\text{eV}$ between the extremes of the dispersion curve is smaller than for the spheres used above. This increases the wavevector shift δk that can be achieved in this longitudinal case. Now, by changing the host index $n_H = \sqrt{\varepsilon_H}$, we have from Eq. (6.10) that

$$\frac{\delta\omega_0}{\omega_0} \approx -\frac{\delta n_H}{n_H}. \tag{6.11}$$

For a modulator, the wavevector shift at a given frequency δk times the device length L should in general be equal to π in order to switch from cross to bar state or vice versa. A rough estimate can be made by writing

$$\delta k \cdot L = \pi \Rightarrow \delta\omega = \frac{\Delta E \cdot d}{L\hbar}. \tag{6.12}$$

Inserting $n_H = 2.20$ (as in $LiNbO_3$), we get $\delta n_H L \approx 2 \cdot 10^{-8}$ m. Thus, a 5 μm long device would require $\delta n_H = 4 \times 10^{-3}$ host index change, reachable with $LiNbO_3$. Much larger refractive index shifts, e.g. $\delta n_H = 0.1$, are achievable using gallium lanthanum sulfide (GLS) and other chalcogenide glasses [31,32], indicating switching lengths of approximately 200 nm or only a few nanoparticles. The small dimensions of even the 5-μm-long device imply an extremely low-power-dissipation device, where the necessary stored RF electric energy for switching in the above example would be on the order of fJ.

Possible ways of utilizing these favorable characteristics are elusive and yet to be explored due to the very high light propagation losses. The losses in metal nanoparticle arrays employing real silver were studied to elucidate their dependence on the permittivity of the host medium, the particle shape, and the mode polarization [25]. Losses just over 10 dB/μm were obtained for both transverse and longitudinal waves, still making stand-alone sub-μm devices feasible in optimized metal nanoparticle structures.

6.3.2.5 Silicon electrooptic switches

Modulation in silicon devices can be achieved by exploiting the plasma dispersion effect whereby changing the carrier concentration changes the real and imaginary parts of the index of refraction [32]. In a recent review article Reed and coworkers [34] have individuated three possible configurations for electrically driven silicon modulators. In a carrier injection modulator carriers are injected by forward biasing a p-i-n junction. This type of modulator is limited in speed by the recombination rate of the carriers and has relatively high power consumption. On the positive side carrier injection allows large changes in the carrier density and therefore high modulation depth. A second method of modulation is carrier depletion. In this case the modulation in the carrier density is obtained by depleting a p-n junction. Carrier depletion is fast and low power as little current flows in the junction during operation. Because of the need to overlap the p-n junction and the optical field, this method, in general, creates modulators that have large insertion losses and low modulation depth. A third modulation scheme is the charge accumulation. Here a MOS capacitor is used to accumulate charges in the optical waveguide. Like charge depletion this method leads to fast and low-power modulation. Because of the need to create an oxide barrier in the middle of the modulator waveguide, however, charge accumulation modulators are hard to fabricate.

Two main silicon modulator topologies have been used. Mach-Zehnder interferometers (MZIs) are the simplest and most robust type of modulators and for these reasons have been adopted in many silicon photonic products. An MZI modulator is relatively immune to thermal drifts and fabrication imperfections. MZIs are also relatively large (with typical areas larger or much larger than 1000s of μm^2), have high power consumption, and require complex RF electrode design to achieve high speeds. Green et al. [35] reported a 10 Gb/s injection-based MZI modulator with a total area of $\sim$100 $\times$ 200 μm^2 and a power consumption of 5 pJ/bit. Resonant micro-ring modulators have several advantages over MZIs. Because they are resonant devices micro-rings lend themselves to be used in wavelength multiplexing schemes, in addition they are much smaller than MZIs (with typical areas of 10s of μm^2) and consume much less power. On the negative side micro-rings are extremely sensitive to thermal shifts and fabrication imperfections. Li et al. [36] have demonstrated a depletion-based ring modulator with a 25 Gb/s modulation rate, a driving voltage of 1 V_{pp}, and 7 fJ/bit power consumption, see Figure 6.10. Chen et al. [37] demonstrated an injection-based modulator with a 3 Gb/s modulation rate, a driving voltage of 0.5 V_{pp}, and a power consumption of 86 fJ/bit.

6.3.2.6 Slow-wave switches

Slow light, i.e. light with an anomalously low group velocity, can be achieved in optical waveguides typically using photonic crystals or other grating or microresonator structures. A low group velocity $v_g = d\omega/dk$ corresponds to regions of the ω-k dispersion with a small slope. A shift of the dispersion in ω, which occurs in relation to the shift in material refractive index, thus gives an enhanced shift Δk, or equivalently an enhanced Δn_{eff}, in the slow-light waveguide, reducing the required switch length L for a π phase shift according to Eq. (6.1). Slow light may thus be used to enhance linear optical interactions such as the electrooptic effect and also gain/absorption in relation to the slowing of the group velocity; however, at the expense of a smaller bandwidth.

As an illustration of the slow-wave effect, very short 5-μm-long silicon directional couplers incorporating photonic crystal structures have been demonstrated, where the required refractive index change, 4.2×10^{-3} by the thermo-optic effect (switch energy $\sim$200 pJ), is 36 times smaller than in a conventional device, while the available optical bandwidth is sufficient to accommodate signals of the order of 100 GHz [38]. A low half-wave voltage of $V_\pi = 1$ V ($V_\pi L = 0.55$ V cm) has been

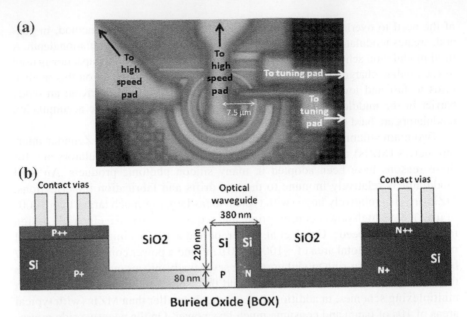

FIGURE 6.10 Silicon depletion-mode ring modulator with thermal tuning. A 25 Gbps modulation rate, extinction ratio >5 dB, has been demonstrated at a driving voltage of 1 V, corresponding to ~7 fJ/bit or ~0.18 mW switching power. The power consumption in the present device is dominated though by the tuning power of up to 66 mW to tune the whole 12.6-nm free spectral range. (a) Photograph of the ring modulator. The upper-right 25% of the ring is made as a Si resistor heater providing wavelength tuning, while 67% is a pn diode for electrooptic modulation. (b) Cross-section diagram of the ring waveguide electrooptic section. The doping profiles of the pn diode are graded by implant diffusion. Reproduced with permission [36] © OSA 2011.

demonstrated in a silicon MZI based on carrier depletion of a *p-n* junction [39]. In this structure a group index of 9.5 was achieved using a laterally corrugated waveguide in one arm of the switch. A substantially lower $V_\pi L = 0.056\,\text{V cm}$ was demonstrated by Nguyen et al. in a photonic crystal MZI using a Si *p-n* junction without a DC bias, where the electrooptic effect is predominant by carrier injection [40]. High-speed operation at 10 Gb/s with 7.9 dB extinction ratio was also achieved with a switching energy of 2.9 pJ/bit, somewhat improving the 5 pJ/bit power consumption obtained by Green et al. [35] not using slow light. The lowest half-wave voltage length product in a silicon photonic crystal MZI reported to date is $V_\pi L = 0.0464\,\text{V mm}$, $V_\pi = 0.58\,\text{V}$ for an active length of 80 μm, by carrier injection in a Si pin diode and a slow-light transmission bandwidth as wide as 18 nm; however, without confirming a high-speed limitation [41]. Employing electrooptic polymers, enabling potentially very high-speed operation due to the ultrafast material response time, a low $V_\pi L = 0.44\,\text{V mm}$ at a group index of 36 has been demonstrated in a photonic-crystal slot-waveguide MZI [42].

In general, rather impressive performance has been reported for photonic crystal-based devices, but in addition to bandwidth limitations, or small tolerances in operating wavelengths, there are also tolerance issues in fabrication. All optical versions of slow light switches are described in Section 6.3.3.1.

6.3.3 All-optical switches

6.3.3.1 All-optical switches based on waveguides and optical nonlinearities

An all-optical switch allows one optical beam to control another via nonlinear interactions between light and matter. The control (switching) beam alters the characteristics of the material which in turn affects the signal beam.

The possibility of controlling light with light has been claimed as a way to overcome the bandwidth limitations imposed by electronics, and be the "Holy Grail" of optical networking [43] and computing [44]. All-optical switches have been the subject of extensive research over several decades. While the bitrate transparency important for network upgradability is also provided by electronically controlled optical switches, all-optical operation offers ultra-high switching speeds currently not attainable by electronics.

Even though fully transparent networks are still a long way off, there is a growing demand for integrated, all-optical switches, in particular to relieve the nodal electronic bottleneck in metro and core networks [43], given that they, for the intended application, perform better than electronically controlled switches regarding speed, power, cost, size, cascadability, and general versatility. For the all-optical solution there is in general a difference between the power used for switching and that dissipated as heat, as also mentioned in Section 6.3.3.2.

Practical implementation of such switches has been hampered by higher switching energies than those of their electronically driven counterparts. This is a result of a relatively weak nonlinear response. It should, however, be noted that when all optical nonresonant nonlinearities are utilized, only a small part of the switching energy is actually dissipated locally; as noted above, dissipation is mainly due to the linear losses. The rest of the switching energy could in principle be recycled. This "photon recycling" could e.g., entail an opto-electronic-opto transformation or a direct reuse of the rest photonic beam energy, to make it useful in other parts of a network. Such employment of "photon recycling" does not seem to have attracted any interest so far, photon recycling is rather used in solar cells, LEDs etc. as is obvious from an abundant literature. Another drawback implied by the weak nonlinearity is that in order to assure the required phase shift at available control pulse energies, the device must be sufficiently long. Early demonstrated switches based on a passive optical fiber and the fast but very weak Kerr nonlinearity in silica had lengths of several hundred meters, not exactly nano [45].

Truly integrable all-optical switches with transverse dimensions of, e.g., $1-2\,\mu m$ and a length of the order of $0.5-2\,mm$ were only provided by active semiconductor optical amplifiers [46], which are still electrically pumped and hence neither free from electronics nor very energy efficient and still not nano photonics.

However, in the last few years we have witnessed great advances in material science and nanotechnology, which have changed the situation, and brought also passive switches closer to practical applications. Recent demonstrations showed that hundreds of GHz speeds [47], or femtojoule switching energies [48], are possible in compact nanodevices.

In the following we give examples of recent highlights indicating how the issues of speed, size, and switching energy have been addressed.

Most of the demonstrated all-optical switches rely on third-order nonlinear effects, in particular optical (quadratic) Kerr effect that induces an intensity-dependent change of the refractive index n:

$$\Delta n = n_2 I, \tag{6.13}$$

where I is the light intensity and n_2 is the nonlinear refractive index coefficient (also called nonlinear Kerr coefficient).

Apart from optical control, the switch geometries are similar to their electronically controlled counterparts described earlier. They use planar devices based on various types of cavities and waveguides capable of the wavelength or subwavelength light confinement at least in the transverse direction.

There are two main approaches taken to boost nonlinear response. One is to find a material in which nonlinear effects are strongest, as discussed in Section 6.2, and the other where one tries to find a structure whose geometrical properties optimize the nonlinear interaction. The structural approach takes advantage of field enhancement due to modification of local density of states, provided by, e.g., microcavities, photonic crystals [49], rough metal surfaces [50], which one can cover with nonlinear material [51], plasmonic metamaterials combined with carbon nanotubes [52], or quantum dots strongly coupled to waveguides [53], or nanocavities [54].

The nonlinear refractive index coefficient n_2, Eq. (6.13), is related to the real part of the third-order nonlinearity $\chi^{(3)}$ and is not only different for different materials but also depends on the wavelength. In accordance with the Kramers-Kronig relations n_2 is largely enhanced close to the material resonances, which is unfortunately associated with enhanced nonlinear loss, typically due to two-photon absorption (TPA), related to the imaginary part of $\chi^{(3)}$. Therefore, to compare nonlinear efficiency of materials, one uses the nonlinear figure of merit $NFOM = n_2/(\beta\lambda)$, where β is TPA efficiency (dimension m/W) and λ is the operation wavelength. For all-optical switching $NFOM > 1$ is required. Table 6.2 is a comparison of NFOM for some materials of interest at $\lambda = 1.55\,\mu m$ [55,56].

For wide bandgap bismuth oxide and silica TPA is negligible since two photons at $\lambda = 1.55\,\mu m$ cannot bridge the bandgaps. The nonlinear response in this nonresonant regime is almost instantaneous (tens of femtoseconds), suitable for ultrafast processing. Those materials, however, have too low refractive index and too small n_2 for nanophotonics devices. Silicon that has the smallest bandgap suffers from the strongest TPA and has $NFOM < 1$. TPA-generated free carriers are the dominant mechanism for the desirable refractive index change, but they also absorb photons and cause additional heat generation. Moreover, the carrier recombination time of

Table 6.2 Nonlinear performance of different materials, extracted from [55–57]. Note that the organic material has the largest n_2, and still adequate NFOM.

Material	$n_2 [\times 10^{-18} m^2 W^{-1}]$	$\beta_{TPA} [\times 10^{-12} mW^{-1}]$	NFOM
Chalcogenide glass (As_2S_3)	2.9	<0.01	>200
Chalcogenide glass (As_2Se_3)	11	2.5	~2
Bismuth oxide (Bi_2O_3)	1.1	Negligible	Large
Silicon (Si)	6	5	0.8
Silica (SiO_2)	0.022	Negligible	Large
$Al_{0.18}Ga_{0.82}As$	14.3	4	2.3
Twisted π-system chromophores (TMC-2)	18.7	7.5	3.2

few microseconds would not allow for high-speed switching, if it was not possible to mitigate those drawbacks by structural and material modification [58,59]. AlGaAs offers both acceptable NFOM and very strong nonlinearity. However, it also suffers from TPA-induced free carrier absorption (FCA) and speed limitations, possible to relieve in nanostructures [48]. Chalcogenide glasses are a very interesting alternative because they can have both acceptable NFOM and very high n_2. Even though, they may still experience a considerable TPA in the telecom band, the generation of free carriers and the implied speed limitations are insignificant [60]. Those glasses have also relatively high refractive index (between 2 and 3) allowing for much smaller devices [56] than silica, including photonic crystals [61]. Even though NFOM gives a good estimate of material nonlinear efficiency, one should also give particular attention to nanostructures, which can strongly enhance nonlinear efficiency.

To reduce switching energy one typically minimizes transverse light confinement or maximizes effective time of interaction between light and medium. Increasing the interaction time by circulating light in a cavity has been used since a long time. The effect is particularly pronounced for microcavities of high-quality factors (Q) and small modal volumes (V), such as, e.g., photonic crystal (PhC) cavities. The maximum light intensity inside the cavity is proportional to $1/V$, and the phase shift to obtain full switching is proportional to $1/Q$. The overall effect is a reduction of the operation power by a factor Q/V, basically the Purcell factor, for tailoring the non-linear performance [62]. It should be noted that while small cavity volume enhances the nonlinear efficiency without affecting the operation speed, Q ultimately limits the speed according to the proportionality of Q and cavity lifetime.

More recently, it was realized that in a properly designed PhC waveguide one can strongly reduce light group velocity v, and thereby enhance the light-matter

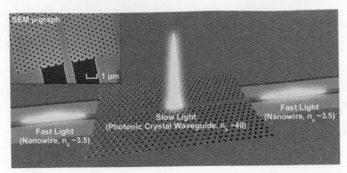

FIGURE 6.11 Schematic of slow-light propagation of pulses in a PhC waveguide [47]. Figure courtesy of the University of Sydney.

interaction time by $S = v_0/v_s = n_s/n_0$, where v_s is the reduced group velocity, v_0 is its "fast" reference value in a waveguide, and n_s, n_0 are the respective group indices.

This slow-light (SL) effect demonstrated in PhCs [63] is also associated with the large LDOS (local density of states), and makes a light pulse spatially compressed along the direction of propagation (see Figure 6.11) enhancing the peak power by S. Thus the overall nonlinear efficiency is boosted by $S^2 = (n_s/n_0)^2$ [64].However, the implied reduction of the operation power may be lower since linear loss per length is also enhanced [65]. The bandwidth available for SL operation scales inversely proportional with group index [66], and their product $n_s (\Delta\omega/\omega)$ (proposed as a more fair figure of merit than bandwidth-delay product [66]) is larger for higher index materials, e.g. 0.3 in Si compared to 0.25 in chalcogenides [61].

An advantage of the slow-light approach over the cavity approach is that SL waveguides can be dispersion engineered [65] to provide not only high energy concentration but also ultra-high bandwidth capability. Slow-light PhC waveguides, with $S = 20$, have also been demonstrated in lower index ($n = 2.6$) chalcogenides [61]. The reported linear loss was only 21 dB/cm, although the operation bandwidth was inevitably smaller (5 nm vs. 12 nm in Si).

A spectacular example of benefiting from a high Q cavity is the demonstration of 40 Gb/s switching with control pulse energies as low as $\sim$0.4 fJ [48]. This was realized in a photonic crystal nanocavity, based on Ref. [67], side-coupled to a waveguide patterned in an InGaAsP photonic crystal (Figure 6.12). The lattice-shifted cavity (H0) had modal volume of only $V \approx 0.025\,\mu m^3$ and very high Q ($\sim 1.8 \times 10^5$). Fast carrier outdiffusion due to the very small V, and the recombination enhanced by the large ratio of surface area to volume in PhC, reduced the lifetime of the free carriers to $\sim$20 ps. However, in this amplitude modulation scheme, control and signal pulses have to be synchronized, and the modulation is of return-to-zero (RZ) character.

This significant demonstration indicates that all-optical switches based on ultra-small, high Q cavities are capable of satisfying fJ power requirement for chip integration [68]. However, the carrier dynamics, even though strongly improved, restricts the achievable speeds to tens of Gb/s.

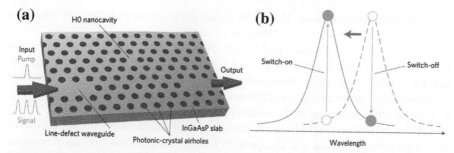

FIGURE 6.12 (a) Schematic of the switch of [48]. The H0 cavity is coupled to input and output PhC defect waveguides. (b) Operating principle: the pump light induces a wavelength shift in the resonant transmission spectrum. Switch-on or switch-off operations are selected by the initial setting of the signal wavelength. Reproduced with permission [48] © Macmillan Publishers Ltd: Nature 2010.

Efforts to overcome the speed and loss limitations imposed by free carriers, when switching relies on resonant $\chi^{(3)}$, have been focused on removing the carrier from the switching area as fast as possible. This was accomplished by structural enhancement of carrier surface recombination or outdiffusion, or by ion implantation [69]. By those means switching times in Si could be reduced to the picosecond range [48,59,70].

But there was also another approach proposed [71,72]—to suppress TPA-induced carrier generation and employ nonresonant Kerr effect instead. To make this possible, the control and signal wavelengths are chosen such that no combination of them can bridge the material band-gap (Figure 6.13). Thus only the "instantaneous," nonresonant $\chi^{(3)}$ is accessed, and THz switching limited only by the dynamics of the cavity ($\tau_{cav} = 0.3\,\text{ps}$, $Q = 320$) was demonstrated [72]. The proof-of-principle experiments were performed on a planar GaAs/AlAs micro-cavity resonant for $\lambda_{probe} = 1280\,\text{nm}$

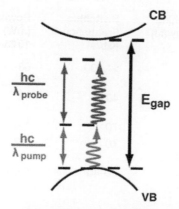

FIGURE 6.13 Schematic energy diagram for the switching concept of [71]. The estimated fluence of the control beam was $70\,\text{pJ/}(\mu\text{m})^2$ and the probe beam fluence was $0.18\,\text{pJ/}(\mu\text{m})^2$. Reproduced with permission [71] © AIP 2011.

in the original telecom band. The probe signal pulses were switched off when the cavity was detuned by the control beam at $\lambda = 2400\,\text{nm}$.

In this concept though, the speed benefit should be weighed against the restricted wavelength range and the drop of $\chi^{(3)}$ out of the resonance. The latter is less significant for resonances related to indirect band-gaps, like in silicon, where TPA vanishes at wavelengths approaching $\lambda = 2200\,\text{nm}$, while the nonlinear Kerr coefficient n_2 remains nearly the same as in the telecom bands [73].

III/V semiconductors and silicon devices attract the most attention because of their high refractive index and compatibility with CMOS. Their operating energies have now reached the femtojoule range required for on-chip integration, and hence became comparable with those of electronically controlled switches, see Table 6.3 as well as Table 6.4, depending upon material and switch length. This performance of the all-optical switches was achieved by clever design and utilizing the strongest resonant (carrier induced) $\chi^{(3)}$, but for the lowest switch energies involved, speed is limited to well below the picosecond regime.

The approach (Section 6.3.2.1) with an active nonlinear medium enclosed in between high index materials such as silicon gives power benefits for electronically as well as all-optically controlled switches, by virtue of good field overlap between controlled and controlling fields and increased power density.

The speed limit arising from slow carrier dynamics can even be overcome for the Kerr process, by using long control wavelengths, so that only nonresonant $\chi^{(3)}$ is accessed. Although this approach offers beyond-terahertz bandwidths, the speed benefit should be weighed against the restricted wavelength range and drop of $\chi^{(3)}$ far from the resonance.

Table 6.3 Comparison of the performance of various on-chip all-optical switches [48], illustrating the compromise between the switching energy and operation speed.

Device	Switching Energy (fJ)	Switching Frequency (Gb/s)	Power used (mW) at 10 Gb/s	Size (μm^3)
ZnSe/BeTe MQW ISBT WG [74]	6,800	>1,000	68	240
Si microring cavity [75]	1,000	40	10	4
PhC-QD MZI [76]	100	70	1.0	45
PhC cavity (GaAs) [77]	120	70	1.2	0.02
PhC cavity side-coupled to PhC WG (InGaAsP) [48]	0.42	20–40	0.0042	0.025

(WG—waveguide, MZI—Mach-Zehnder interferometer, ISBT—intersubband transition, MQW—multiple quantum well)

6.3.3.2 All-optical NOT gate based on near-field-coupled quantum dots

Energy transfer in nanostructures has been readily available in biology at least since the 1950s to determine proximity or shape changes of molecules [78], and energy transfer processes have been implemented in various material systems [79–85]. One thing that should be noted is that, in describing energy transfer, point-dipole models such as Förster resonant energy transfer do not allow optical transitions to dipole-forbidden energy sublevels, which is often the case with the experimental conditions of two closely located quantum dots (QDs) of slightly different size. Also, recent experimental observations in light harvesting antenna indicate the inaccuracy of dipole-based modeling [86,87]. On the other hand, the localized nature of optical near fields frees us from conventional optical selection rules, meaning that optical excitation could excite QDs to energy levels that are conventionally electric dipole forbidden [81]. The coherent interaction between two quantum dots via optical near-fields results in unidirectional optical energy transfer by an energy dissipation process occurring in the larger dot. It has been theoretically analyzed that the lower bound of this energy dissipation, or the intersublevel energy difference at the larger dot, when the excitation appearing in the larger dot originated from the excitation transfer via optical near-field interactions, could be 10^4 times more energy efficient compared with the bit flip energy of an electrically wired device [88]. Such energy transfer mediated by optical near-field interactions involving conventionally dipole-forbidden energy levels allows various functionalities such as switching operations [81,82], light concentrations [84,85], among others.

All-optical AND and NOT logic gates, each comprising two near-field-coupled InAs/GaAs epitaxial QDs, have been demonstrated experimentally [82]. A schematic figure of such a gate is shown in Figure 6.14a, including the two QDs in a mesa structure and an Au nanoparticle to enhance the output light emission. Figure 6.14b shows an S-TEM image of the demonstrated device. It would be interesting to perform a rough comparison of the energy dissipation to state-of-the-art CMOS gates of gate

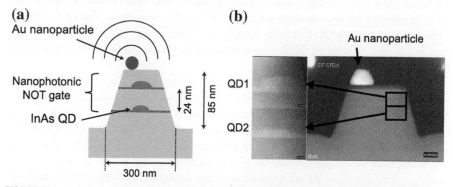

FIGURE 6.14 (a) Schematic figure of the nanophotonic NOT gate comprising two near-field-coupled QDs as well as an Au nanoparticle for the output interface. (b) Cross-sectional image of a mesa, taken with a scanning transmission electron microscope (S-TEM).

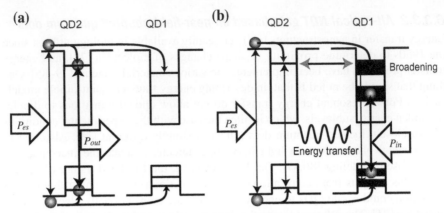

FIGURE 6.15 Energy diagram of the two QDs QD2 and QD1, constituting a NOT gate, based on near-field-mediated energy transfer of exciton polaritons. (a) "1" output case ($P_{in} = 0$), and (b) "0" output case ($P_{in} \neq 0$).

length around 22 nm [89]. It should be noted that the all-optical gates of Ref. [82] and references therein are quite different in concept from other attempts in the field [90]. We analyze specifically the NOT gate, which consists of a smaller and a larger QD, denoted respectively by QD2 and QD1, as schematically summarized in Figure 6.15. Note that the energy level in QD2 and the upper energy level of QD1 are slightly off-resonant. The output signal, P_{out}, is associated with the radiation from QD2.

The NOT gate has two input channels, one is a continuous energy source denoted by P_{es}, and the other is the input signal denoted by P_{in}. Here we assume that P_{es} excites an excited state in QD2 from where it rapidly relaxes to the ground state in QD2. When P_{in} is not present, Figure 6.15a, energy transfer from QD2 to QD1 does not occur (since the energy levels are off-resonant) and thus output radiation from QD2 is emitted by spontaneous emission. When P_{in} is present, Figure 6.15b, the exciton population in QD1 induces energy level broadening in QD1 allowing excitons in the ground state of QD2 to be transferred to QD1, and thus the output signal from QD2 decreases.

We assume that both, when on, obey Poisson photon statistics with the same average photon number n_{in} during one bit-slot, i.e. the photon number has the distribution $N_{in} \sim \text{Pois}(n_{in})$. The average time between photons in each pulse is taken to be τ_{sp}/F_P, i.e. the bare spontaneous emission time τ_{sp} reduced by the Purcell factor F_P associated with the nearby Au particle. This photon spacing should be suitable in order not to overwhelm the gate, while not giving an unnecessary delay. Here we assume that $\tau_{sp} = 1$ ns and $F_P = 10$, which is close to the experimentally reported values, for instance in Ref. [91]. With such parameters, the energy transfer time between the QDs, $\tau_{ET} = 50$ ps, is smaller than the spontaneous emission time and the average time between photons, $\tau_{sp}/F_P = 100$ ps.

To make a "first-order" estimate of the number of output photons required for a desired bit error rate, we consider only what should be the main error mechanism in the NOT gate, i.e. in the "0" output case, Figure 6.15b, that an exciton in the ground state of QD2 decays contributing to the output signal P_{out} before the intended near-field-mediated energy transfer from QD2 to QD1 is induced. The probability for this is $1 - p_{ET,0}$ where the probability for the intended transfer is approximately $p_{ET,0} = 1/\tau_{ET}/(1/\tau_{ET} + F_P/\tau_{sp}) = 2/3$. The photon numbers of the output pulses (as received by the detector) will then also be Poisson distributed for both logical "1" and "0" levels; that is, $N_{out,1} \sim \text{Pois}(n_{out,1})$ and $N_{out,0} \sim \text{Pois}(n_{out,0})$ with the average photon numbers $n_{out,1} = \eta_{out}n_{in}$ and $n_{out,0} = \eta_{out}n_{in}(1 - p_{ET,0}) = \eta_{out}n_{in}/3$. Here $\eta_{out} = 0.45$ is the output coupling efficiency [92]. Assuming equal rates of "1" and "0" bit occurrences, the bit error rate is determined by

$$p_e = 0.5\text{P}(0|1) + 0.5\text{P}(1|0) = 0.5\text{P}(N_{out,1} < n_d) + 0.5\text{P}(N_{out,0} > n_d), \quad (6.14)$$

where, e.g., $\text{P}(0|1)$ is the probability to have a "0" output, when "1" is intended, relative to the optimal decision level n_d. A bit error rate of $p_e < 10^{-9}$ demands that $n_{out,1} = 200$, $n_{out,0} = 67$ ($n_d = 121.5$ photons), and thus $n_{in} = 444$, photons.

Looking now at the energy dissipation of the NOT gate we distinguish between interband relaxation, where the energy is lost to radiation, and intraband relaxation causing local heating. We see that for a "1" output there is only intraband relaxation of the P_{es} signal from the excited to the ground state of QD2, while an intended "0" output entails interband dissipation of both input signals P_{es} and P_{in}, though excitons excited by P_{es} may decay emitting a photon from the lowest state of QD2 or QD1 depending on if an energy transfer has taken place. Again assuming equal rates of "1" and "0" bits the average switching energies per bit that is radiated due to interband relaxation or locally dissipated due to intraband relaxation respectively are then

$$E_{sw,rad} = n_{in}/2 \left[(1 - p_{ET,0})E_{out} + (1 + p_{ET,0})E_{in}\right], \quad (6.15)$$

and

$$E_{sw,heat} = n_{in} \left[E_{es} - E_{out} + p_{ET,0}/2(E_{out} - E_{in})\right], \quad (6.16)$$

where $E_{in} = 0.974\,\text{eV}$, $E_{es} = E_{out} + 0.1$ eV, and $E_{out} = 1.027\,\text{eV}$ are the photon energies of the input and output signals P_{in}, P_{es}, and P_{out} of the NOT gate. This gives $E_{sw,rad} = 70\,\text{aJ}$ and $E_{sw,heat} = 8\,\text{aJ}$ for the example above. In addition the input interface would dissipate $n_{in} (1/\eta_{in} - 1)(E_{es} + E_{in}/2) = 13\,\text{aJ}$ as heat, where $\eta_{in} = 0.9$ is the input coupling efficiency [92]. The energy dissipation of electronic wired devices, including interfacing, is estimated to be 6.3 MeV [93], indicating that the energy dissipation of nanophotonic devices discussed above is about 10^4 times more energy efficient than their electrical counterparts [92]. These data should also be compared to Table 6.3, considering that the NOT gate or modulator volume is less than $0.01\,\mu\text{m}^3$. Speed is, however, limited.

The comparison above considers energy dissipation including interfacing, which leads to higher energy efficiency with optical energy transfer. Superficially, however, the switching energy in state-of-the-art electronics and the total energy dissipation for optical energy transfer in logic gates are comparable as discussed below. For the electronics case, we assume a 22 nm (Intel state of the art) feature size [94]. This gives sub-ps gate delay time and ~20 aJ dissipated switch energy. Electronic gates typically have a considerably lower bit error rate than 10^{-9} used here for the optical gate, corresponding to perhaps a factor 2–3 times higher switch energy for the optical gate to be comparable. But a NOT gate consists of two transistors, and thus comparing as here to the switch energy of a single transistor should be fair. The locally dissipated switch energy $E_{sw,heat}$ of the nanophotonic NOT gate is thus somewhat smaller than for electronics. The radiated energy, $E_{sw,rad}$, does not cause local heating and is a unique mechanism inherent in optical energy transfer logic whereas dissipation in electronic devices contributes to local heat generation, which is the most severe issue in state-of-the-art electronics. We see that the CMOS gate is very fast per se, but when concatenated into circuits it is slowed down by, e.g., interconnects, which in the all-optical version would be somehow for free. It is further questionable that one should try to emulate the electronic system's architectures and operation. As an example, in the optical case, there is the feature of parallel processing which seems very interesting here.

6.4 SUMMARY AND CONCLUSIONS

This chapter has dealt with the issue of finding low power or switch energy nanosized optical switches.

As for the energy part, this is only one part of a increasingly higher level problem, since one usually has to assess the total power dissipation of the system at hand, especially in today's large information processing systems. In many cases, power dissipation is strongly dominated by electronics. But electronics is developing toward lower powers and it is a challenge in itself to lower power dissipation in highly integrated nanophotonics systems. However, from the summary in Table 6.4, we see that the most efficient nanophotonics devices seem to require plasmonics-type waveguides, where, e.g., micron length and submicron confinement as well as sub-fJ/bit switch energy can be achieved. But the very high loss coefficients involved here in essence only permit stand-alone devices for these structures, not the dense integration desired in some cases, e.g. for circuits for advanced modulation formats. And in fact, for electronically controlled devices and for emerging materials like the electrooptic polymers, the switch energy for say a 80 μm long device with 2 V drive ($V_\pi L = 160$ Vμm) will be 33 fJ (Table 6.4, case 3). This can be compared to reported data of an electrooptic-polymer-based switch designed for a low operation voltage giving $V_\pi = 0.25$ V for a 2 cm long device ($V_\pi L = 0.5$ Vcm) with a switch energy of 75 fJ [95]. Scaled for a length of 5 mm, this gives $V_\pi = 1$ V and a switch energy of 300 fJ, whereas the polymer switch of Table 6.4 for the same length of 5 mm would have a drive voltage of 32 mV and a switch energy of 0.53 fJ, clearly demonstrating the potential of electrooptical polymers with high r_{33} coefficients.

Table 6.4 Comparison of electronically controlled modulators (A: amplitude, P: phase), electrooptic polymer (EOP): 500 pm/V or chalcogenide with index change = 0.1; $V_\pi L$ for π phase shift, or >10 dB extinction ratio for case 4,5. Case 1-3 theory (V_π and L as examples), case 4,5 experiment.

Device and Wavelength λ_0	$V_\pi L$ [Vμm]	V_π [V]	L [μm]	IL [dB] (Attenuation) [dB/μm]	Confinement	Switch Energy [fJ/bit] (Capacitance) [fF]	Comments
1) P, Layered metal/chalcogenide waveguide, Section 6.3.2.3, [24] 1.55 μm	0.66	0.33	2	7 (3.5)	0.01 μm²	0.003 (0.01)	Chalcogenide thickness 4 nm, index change 0.1
2) P, Array of Ag nano-particles in EOP matrix, Section 6.3.2.4, [27] 0.680 μm	3	15	0.2	2.4 (12)	Appr. 0.01 μm²	(Very approximate) 2 (0.01)	200 nm electrode separation. Very rough approximation, real values probably much better. Trading lower voltage for length impeded by loss
3) P, Slotline Si/EOP/Si, Section 6.3.2.1, [14] 1.55 μm	160	2	80	0.1 (0.001)	Appr. 0.3 × 0.7 μm²	33 (8)	Doped Si serves as electrodes. 100 nm EOP
4) A, Silicon micro-ring modulator [36] 1.55 μm	41	1	41	~5	0.3 × 0.38 μm²	~50	Depletion mode modulator *Experiment*
5) A, III-V Electro absorption QCSE [99] 1.55 μm	400	2	200 active 500 total	3–5	4 μm²	300 (n/a)	Traveling-wave type EAM, 50 Ω transmission line *Experiment*

A recent contender is graphene, where experimental results were presented in, e.g., [96]: By electrically tuning the Fermi level of a graphene sheet in a waveguide configuration, absorption modulation with a modulation depth of 6 dB of the guided light was demonstrated at frequencies over 1 GHz, together with a broad operation spectrum from 1.35 to 1.6 μm. The device area was $25\,\mu m^2$ and switch energy for the stated capacitance of 0.22 pF is around 3 pJ with a drive voltage of around 4 V, taken from the reference. For the future it is predicted that the footprint would be around $1\,\mu m^2$, speed 50 Gb/s, switch energy 5 fJ/bit, with a drive voltage of 1 V [97]. In a recent theoretical calculation [98] a 3 dB modulation depth was reported with an 800-nm-long silicon waveguide, or with a 120-nm-long plasmonic waveguide, all based on three-dimensional numerical simulations. With a capacitance of 0.02 pF it is claimed that a switch energy of less than 0.2 pJ/bit is required.

A possibility to retain nanostructures as well as low power could be the use of near-field-coupled quantum dots, as in Section 6.3.3.2, with switch energies on the order of 100 aJ, of which only say 20% are dissipated, but with limited switching speed [92]. This brings us to the more traditional form of all-optical switching of Section 6.3.3.1, where the recent progress has reduced switching energies to sub-femtojoules in all-optical switching.

Thus, the performance in the last line of Table 6.3 is impressive [48]. However, in addition to the synchronization requirements for optical control signals in circuits containing multiple devices of this particular kind, there is a more fundamental difference in the application envelope of all-optical switches compared to electronically controlled devices: The modulating optical signal has to be generated somewhere, so in a way one moves the problem of generating this information carrying signal to somewhere else, e.g. to an electronically controlled optical switch. While Table 6.3 includes switching frequency, Table 6.4 states pertinent capacitances for estimation of reachable speeds, in general larger than several 100 GHz.

The International Technology Roadmap for Semiconductors (ITRS, [99]) predicts for NMOSFET technologies switch energies or power dissipation in their parlance of 0.07 fJ/μm for 6 nm gate length in year 2026, with an intrinsic delay of 0.26 ps. Even the speculative devices of Sections 6.3.2.3 and 6.3.2.4 would have difficulties to rival electronics. However, the QD-based devices of Section 6.3.3.2 come close in size and energy (depending on transistor gate width) but not in speed. The best all-optical switches can currently vie in power but not in speed.

On a more basic note, a comparison between electronics and photonics is difficult, since the "particles" involved obey different statistics. A more adequate comparison would be to compare photonics with microwave technology. Here the transistor played a very important role. Such a device in photonics is still elusive.

But one should not try to have photonics emulate electronics but rather focus on applications where photonics has a clear advantage. This is the case for all kinds of very high-speed transmission, with the corresponding transducers as well as devices for optical transmission network (re)configuration being key elements.

Above we have tried to elucidate the status of low-power nanophotonics switches; what can we expect for the future? One line of development is materials. Here we

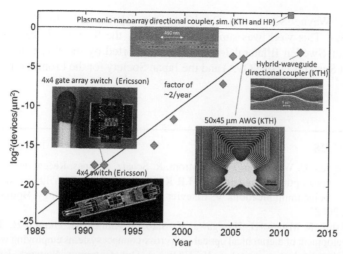

FIGURE 6.16 A Moore's law for integration density in terms of equivalent number of elements per square micron of integrated photonics devices, showing a growth faster than the IC Moore's law, adapted from [101]. The figure covers, in time order, a lithiumniobate 4×4 polarization independent switch array, an InP-based integrated gated amplifier switch array, an SOI arrayed-waveguide grating and a hybrid plasmonic (passive) directional coupler. All these are experimentally demonstrated. At the top is a simulation of two coupled nanoparticle arrays, forming a directional coupler (see Section 6.3.2.4), each array being a resonantly operated array of silver nanoparticles. If loss requirements of, e.g., 3 dB/cm were invoked, the two latter would occupy significantly lower places in the figure.

have novel electrooptic polymers as contenders. Chalcogenides are currently too slow (except for some applications), order ms and far too lossy; this type of material would otherwise be very interesting due to huge refractive index changes and due to memory functions. Further on the agenda is graphene with an at the moment somewhat unclear potential. Another track is novel network architectures, e.g. in the shape of 2D or 3D routing in QD arrays. The field is wide open to innovation in materials, device concepts, and applications and, furthermore, breakthroughs are required in order to pursue the exponential development in integration density depicted in Figure 6.16 (adapted from [100]). This development was to a large degree due to progress in material technology.

Acknowledgments

The authors gratefully acknowledge valuable discussions with Prof. Hans Ågren, Prof. Anders Hult, Assoc. Prof. Michael Malkoch, and Mr. Fei Lou, all with KTH, Prof. Min Qiu, KTH and Zhejiang University, Prof. Daoxin Dai and Prof. Sailing He of Zhejiang University, and Dr. Alexander Bratkovski, Dr. Mike Tan, Dr. Shih-Yuan

Wang, Dr. Wayne Sorin, and Dr. Zhang-Lin Zhou, all with Hewlett-Packard Laboratories. This work was supported in part by the Swedish Research Council, and a Japan–Sweden Bilateral Joint Project supported by the Swedish Agency for Innovation Systems (VINNOVA) and the Japan Society for the Promotion of Science (JSPS).

References

[1] L. Thylen, U. Westergren, P. Holmström, R. Schatz, P. Jänes, Recent developments in high-speed optical modulators, in: I.P. Kaminow, T. Li, A.E. Willner (Eds.), Optical Fiber Telecommunications V, Elsevier Science and Technology Books, Oxford, UK, 2008.

[2] K. Ishii, S. Mitsui, H. Hasegawa, K. Sato, S. Kamei, M. Okuno, H. Takahashi, Development of hierarchical optical path cross-connect systems employing wavelength/waveband selective switches, IEEE/OSA J. Opt. Commun. Network 3 (7) (2011) 559–567.

[3] T. Tamir, Integrated Optics, Topics in Applied Physics, Springer, 1975.

[4] E. Berglind, L. Thylen, L. Liu, Plasmonic/metallic passive waveguides and waveguide components for photonic dense integrated circuits: a feasibility study based on microwave engineering, IET Optoelectron. 4 (1) (2010) 1.

[5] A. Yariv, Quantum. Electronics. second ed., Wiley, New York, 1975 app. 1.

[6] V.R. Almeida, Q. Xu, C.A. Barrios, M. Lipson, Guiding and confining light in void nanostructure, Opt. Lett. 29 (11) (2004) 1209–1211.

[7] A. Yariv, Quantum. Electronics. second ed., Wiley, New York, 1975 (Chapter 14).

[8] Private Communication, Paul-Drude-Institut Group on Chalcogenides.

[9] L.R. Dalton, B. Robinson, A. Jen, P. Ried, B. Eichinger, P. Sullivan, A. Akelaitis, D. Bale, M. Haller, J. Luo et al., Electro-optic coefficients of 500 pm/V and beyond for organic materials, Proc. SPIE 5935 (2005) 5935061–5935113.

[10] L.R. Dalton, P.A. Sullivan, D.H. Bale, Electric field poled organic electro-optic materials: state of the art and future prospects, Chem. Rev. 110 (1) (2010) 25–55.

[11] R. Dinu, D. Jin, G. Yu, B. Chen, D. Huang, H. Chen, A. Barklund, E. Miller, C. Wei, J. Vemagiri, Environmental stress testing of electro-optic polymer modulators, J. Lightwave Technol. 27 (2009) 1527–1532.

[12] D. Jin, H. Chen, A. Barklund, J. Mallari, G. Yu, E. Miller, R. Dinu, EO polymer modulators reliability study, Proc. SPIE 7599 (2010) 75990H–75998H.

[13] V.R. Almeida, Q. Xu, C.A. Barrios, M. Lipson, Guiding and confining light in void nanostructure, Opt. Lett. 29 (11) (2004) 1209.

[14] P. Holmström, L. Thylén, Unpublished.

[15] S. Zhu, G.Q. Lo, D.L. Kwong, Theoretical investigation of silicon MOS-type plasmonic slot waveguide based MZI modulators, Opt. Express 18 (26) (2010) 27802–27819.

[16] R.F. Oulton, V.J. Sorger, D.A. Genov, D.F.P. Pile, X. Zhang, A hybrid plasmonic waveguide for subwavelength confinement and long-range propagation, Nat. Photon. 2 (2008) 496.

[17] D. Dai, S. He, A silicon-based hybrid plasmonic waveguide with a metal cap for a nano-scale light confinement, Opt. Express 17 (19) (2009) 16646–16653.

[18] Z. Wang, D. Dai, Y. Shi, G. Somesfalean, P. Holmstrom, L. Thylen., S. He, L. Wosinski, Experimental realization of a low-loss nano-scale Si hybrid plasmonic waveguide, paper JThA017, Optical Fiber Communication Conference (OFC), Los Angeles, USA, 6–10 March, 2011.

[19] X. Sun, L. Zhou, X. Li, Z. Hong, J. Chen, Design and analysis of a phase modulator based on a metal-polymer-silicon hybrid plasmonic waveguide, Appl. Opt. 50 (20) (2011) 3428.

[20] M. Hochberg, T. Baehr-Jones, G. Wang, J. Huang, P. Sullivan, L. Dalton, A. Scherer, Towards a millivolt optical modulator with nano-slot waveguides, Opt. Express 15 (13) (2007) 8401–8410.

[21] J.C.M. Garnett, Colors in metal glasses and in metal films, Philos. Trans. R. Soc. Lond. Ser. A 203 (1904) 385.

[22] P. Drude, Lehrbuch der Optik, Leipzig, 1906.

[23] M. Yan, L. Thylén, M. Qiu, Layered metal-dielectric waveguide: subwavelength guidance leveraged, modulation sensitivity in mode index, and reversed mode ordering, Opt. Express 19 (4) (2011) 3818–3824.

[24] S.A. Maier, Plasmonics: Fundamentals and Applications, Springer, New York, 2007.

[25] W.H. Weber, G.W. Ford, Propagation of optical excitations by dipolar interactions in metal nanoparticle chains, Phys. Rev. B 70 (2004) 125429.

[26] P. Holmström, J. Yuan, M. Qiu, L. Thylén, A.M. Bratkovsky, Passive and active plasmonic nanoarray devices, Proc. SPIE 8070 (2011) 80700T–1–80700T–6

[27] P. Holmström, J. Yuan, M. Qiu, L. Thylén, A.M. Bratkovsky, Theoretical study of nanophotonic directional couplers comprising near-field-coupled metal nanoparticles, Opt. Express 19 (2011) 7885–7893.

[28] T. Tamir, Guided-Wave Optoelectronics, Springer-Verlag, Berlin, 1988.

[29] J.B. Khurgin, G. Sun, In search of the elusive lossless metal, Appl. Phys. Lett. 96 (2010) 181102.

[30] P. Holmström, L. Thylen, A. Bratkovsky, Composite metal/quantum-dot nanoparticle-array waveguides with compensated loss, Appl. Phys. Lett. 97 (2010) 073110.

[31] Z.L. Sámson, S.-C. Yen, K.F. MacDonald, K. Knight, S. Li, D.W. Hewak, D.-P. Tsai, N.I. Zheludev, Chalcogenide glasses in active plasmonics, Phys. Stat. Sol. RRL 4 (10) (2010) 274–276.

[32] Y. Ikuma, Y. Shoji, M. Kuwahara, X. Wang, K. Kintaka, H. Kawashima, D. Tanaka, H. Tsuda, Small-sized optical gate switch using $Ge_2Sb_2Te_5$ phase-change material integrated with silicon waveguide, Electron. Lett. 46 (5) (2010) 368–369.

[33] R. Soref, B. Bennett, Electrooptical effects in silicon, IEEE J. Quant Electron. 23 (1987) 123–129.

[34] G.T. Reed, G. Mashanovich, F.Y. Gardes, D.J. Thompson, Silicon optical modulators, Nat. Photon. 4 (2010) 518–526.

[35] W.M. Green, M.J. Rooks, L. Sekaric, Y.A. Vlasov, Ultra-compact, low RF power, 10 Gb/s silicon Mach-Zehnder modulator, Opt. Express 15 (2007) 17106–17113.

[36] G. Li, X. Zheng, J. Yao, H. Thacker, I. Shubin, Y. Luo, K. Raj, J. Cunningham, A. Krishnamoorthy, 25 Gb/s 1 V-driving CMOS ring modulator with integrated thermal tuning, Opt. Express 19 (2011) 20435–20443.

[37] L. Chen, K. Preston, S. Manipatruni, M. Lipson, Integrated GHz silicon photonic interconnect with micrometer-scale modulators and detectors, Opt. Express 17 (2009) 15248–15256.

[38] D.M. Beggs, T.P. White, L. O'Faolain, T.F. Krauss, Ultracompact and low-power optical switch based on silicon photonic crystals, Opt. Lett. 33 (2) (2008) 147–149.

[39] A. Brimont, D.J. Thomson, J. Herrera, F.Y. Gardes, J.M. Fedeli, G.T. Reed, J. Martí, P. Sanchis, Slow light enhanced carrier depletion modulators with 1 V drive voltage, Proc. SPIE 8431 (2012) 84310K–1–84310K–7.

[40] H.C. Nguyen, Y. Sakai, M. Shinkawa, N. Ishikura, T. Baba, 10 Gb/s operation of photonic crystal silicon optical modulators, Opt. Express 19 (14) (2011) 13000–13007.

[41] A. Hosseini, X. Xu, H. Subbaraman, C.-Y. Lin, S. Rahimi, R.T. Chen, Large optical spectral range dispersion engineered silicon-based photonic crystal waveguide modulator, Opt. Express 20 (11) (2012) 12318–12325.

[42] X. Wang, C.-Y. Lin, S. Chakravarty, J. Luo, A.K.-Y. Jen, R.T. Chen, Effective in-device r_{33} of 735 pm/V on electro-optic polymer infiltrated silicon photonic crystal slot waveguides, Opt. Lett. 36 (5) (2011) 882–884.

[43] A.A.M. Saleh, J.M. Simmons, All-optical networking—evolution, benefits, challenges, and future vision, invited paper, Proc. IEEE 100 (5) (2012) 1105–1117.

[44] H.J. Caulfield, S. Dolev, Why future supercomputing requires optics, Nat. Photon. 4 (2010) 261–263.

[45] T. Yamamoto, E. Yoshida, M. Nakazawa, Ultrafast nonlinear optical loop mirror for demultiplexing 640 Gbit/s TDM signals, Electron. Lett. 34 (1998) 1013–1014.

[46] K.E. Stubkjaer, Semiconductor optical amplifier-based all-optical gates for high-speed optical processing, IEEE J. Sel. Top. Quant. Electron. 6 (2000) 1428–1435.

[47] B. Corcoran, C. Monat, M. Pelusi, C. Grillet, T.P. White, L. O'Faolain, T.F. Krauss, B.J. Eggleton, D.J. Moss, Optical signal processing on a silicon chip at 640 Gb/s using slow-light, Opt. Express 18 (8) (2010) 7770–7781.

[48] K. Nozaki, T. Tanabe, A. Shinya, S. Matsuo, T. Sato, H. Taniyama, M. Notomi, Sub-femtojoule all-optical switching using a photonic-crystal nanocavity, Nat. Photon. 4 (2010) 477–483.

[49] M. Soljacic, J.D. Joannopoulos, Enhancement of nonlinear effects using photonic crystals, Nat. Mater. 3 (2004) 211–219.

[50] P. Genevet, J.-P. Tetienne, E. Gatzogiannis, R. Blanchard, M.A. Kats, M.O. Scully, F. Capasso, Large enhancement of nonlinear optical phenomena by plasmonic nanocavity gratings, Nano Lett. 10 (2010) 4880.

[51] J. Renger, R. Quidant, L. Novotny, Enhanced nonlinear response from metal surfaces, Opt. Express 19 (2011) 1777.

[52] A.E. Nikolaenko, N. Papasimakis, A. Chipouline, F. De Angelis, E. Di Fabrizio, N.I. Zheludev, THz bandwidth optical switching with carbon nanotube metamaterial, Opt. Express 20 (6) (2012) 6068–6079.

[53] X. Ma, S. John, Quantum-dot all-optical logic in a structured vacuum, Phys. Rev. A 84 (2011) 013830.

[54] D. Englund, A. Majumdar, M. Bajcsy, A. Faraon, P. Petroff, J. Vučković, Ultrafast photon-photon interaction in a strongly coupled quantum dot-cavity system, Phys. Rev. Lett. 108 (2012) 093604.

[55] V.G. Ta'eed, N.J. Baker, L.B. Fu, K. Finsterbusch, M.R.E. Lamont, D.J. Moss, H.C. Nguyen, B.J. Eggleton, D.-Y Choi, S. Madden, B. Luther-Davies, Ultrafast all-optical chalcogenide glass photonic circuits, Opt. Express 15 (15) (2007) 9205–9221.

[56] B.J. Eggleton, T.D. Vo, R. Pant, J. Schroder, M.D. Pelusi, D.Y. Choi, S.J. Madden, B. Luther-Davies, Photonic chip based ultrafast optical processing based on high

nonlinearity dispersion engineered chalcogenide waveguides, Laser Photon. Rev. 6 (1) (2012) 97–114.

[57] G.S. He, J. Zhu, A. Baev, M. Samoc, D.L. Frattarelli, N. Watanabe, A. Facchetti, H. Ågren, T.J. Marks, P.N. Prasad, Twisted π-system chromophores for all-optical switching, J. Am. Chem. Soc. 133 (17) (2011) 6675–6680.

[58] J. Leuthold, C. Koos, W. Freude, Nonlinear silicon photonics, Nat. Photon. 4 (2010) 535–544.

[59] M.A. Foster, A.C. Turner, M. Lipson, A.L. Gaeta, Nonlinear optics in photonic nanowires, Opt. Express 16 (2) (2008) 1300–1320.

[60] A. Zakery, S.R. Elliott, Optical Nonlinearities in Chalcogenide Glasses and Their Applications, Springer Series in Optical Sciences, Springer, 2010.

[61] M. Spurny, L. O'Faolain, D.A.P. Bulla, B. Luther-Davies, T.F. Krauss, Fabrication of low loss dispersion engineered chalcogenide photonic crystals, Opt. Express 19 (3) (2011) 1991–1996.

[62] P. Bermel, A. Rodriguez, J.D. Joannopoulos, M. Soljacic, Tailoring optical nonlinearities via the Purcell effect, Phys. Rev. Lett. 99 (2007) 053601.

[63] M. Notomi, M.K. Yamada, A. Shinya, J. Takahashi, C. Takahashi, I. Yokohama, Extremely large group-velocity dispersion of line-defect waveguides in photonic crystal slabs, Phys. Rev. Lett. 87 (2001) 2539021.

[64] M. Soljacic, S.G. Johnson, S.H. Fan, M. Ibanescu, E. Ippen, J.D. Joannopoulos, Photonic-crystal slowlight enhancement of nonlinear phase sensitivity, J. Opt. Soc. Am. B 19 (9) (2002) 2052–2059.

[65] C. Monat, M. de Sterke, B.J. Eggleton, Slow light enhanced nonlinear optics in periodic structures, J. Opt. 12 (2010) 104003.

[66] S.A. Schulz, L. O'Faolain, D.M. Beggs, T.P. White, A. Melloni, T.F. Krauss, Dispersion engineered slow light in photonic crystals: a comparison, J. Opt. 12 (2010) 104004.

[67] Z.Y. Zhang, M. Qiu, Small-volume waveguide-section high-Q microcavities in 2D photonic crystal slabs, Opt. Express 12 (17) (2004) 3988–3995.

[68] D.A.B. Miller, Device requirements for optical interconnects to silicon chips, Proc. IEEE 97 (2009) 1166–1185.

[69] A. Chin, K.Y. Lee, B.C. Lin, S. Horng, Picosecond photoresponse of carriers in Si ion-implanted Si, Appl. Phys. Lett. 69 (1996) 653–655.

[70] M. Belotti, M. Galli, D. Gerace, L.C. Andreani, G. Guizzetti, A.R. Md Zain, N.P. Johnson, M. Sorel, R.M. De La Rue, All-optical switching in silicon-on-insulator photonic wire nano-cavities, Opt. Express 18 (2) (2010) 1450–1461.

[71] (a) G. Ctistis, E. Yüce, A. Hartsuiker, J. Claudon, M. Bazin, J.M. Gérard, W.L. Vos, Ultimate fast optical switching of a planar microcavity in the telecom wavelength range, Appl. Phys. Lett. 98 (2011) 161114-1–161114-3. (b) G. Ctistis, E. Yüce, A. Hartsuiker, J. Claudon, M. Bazin, J.M. Gérard, W.L. Vos, Addendum: "Ultimate fast optical switching of a planar microcavity in the telecom wavelength range" [Appl. Phys. Lett. 98, 161114, 2011], Appl. Phys. Lett. 99 (2011) 199901.

[72] X. Liu, R.M. Osgood, Jr., Y.A. Vlasov, W.M.J. Green, Mid-infrared optical parametric amplifier using silicon nanophotonic waveguides, Nat. Photon. 4 (2010) 557–560.

[73] A.D. Bristow, N. Rotenberg, H.M. van Driel, Two-photon absorption and Kerr coefficients of silicon for 850–2200 nm, Appl. Phys. Lett. 90 (2007) 191104.

[74] G.W. Cong, R. Akimoto, K. Akita, T. Hasama, H. Ishikawa, Low-saturation energy-driven ultrafast all-optical switching operation in (CdS/ZnSe)/BeTe intersubband transition, Opt. Express 15 (2007) 12123–12130.

[75] M. Waldow, T. Plötzing, M. Gottheil, M. Först, J. Bolten, T. Wahlbrink, H. Kurz, 25-ps all-optical switching in oxygen implanted silicon-on-insulator microring resonator, Opt. Express 16 (2008) 7693–7702.

[76] H. Nakamura, Y. Sugimoto, K. Kanamoto, N. Ikeda, Y. Tanaka, Y. Nakamura, S. Ohkouchi, Y. Watanabe, K. Inoue, H. Ishikawa, K. Asakawa, Ultrafast photonic crystal/quantum dot all-optical switch for future photonic networks, Opt. Express 12 (2004) 6606–6614.

[77] C. Husko, A. De Rossi, S. Combrié, Q.V. Tran, F. Raineri, C.W. Wong, Ultrafast all-optical modulation in GaAs photonic crystal cavities, Appl. Phys. Lett. 94 (2009) 021111.

[78] R.M. Clegg, The history of FRET, in: C.D. Geddes, J.R. Lakowicz (Eds.), Reviews in Fluorescence, vol. 3, Springer, New York, 2006., pp. 1–45.

[79] C.R. Kagan, C.B. Murray, M.G. Bawendi, Long-range resonance transfer of electronic excitations in close-packed CdSe quantum-dot solids, Phys. Rev. B 54 (12) (1996) 8633–8643.

[80] S.A. Crooker, J.A. Hollingsworth, S. Tretiak, V.I. Klimov, Spectrally resolved dynamics of energy transfer in quantum-dot assemblies: towards engineered energy flows in artificial materials, Phys Rev. Lett. 89 (18) (2002) 186802.

[81] M. Ohtsu, K. Kobayashi, T. Kawazoe, S. Sangu, T. Yatsui, Nanophotonics: design, fabrication, and operation of nanometric devices using optical near fields, IEEE J. Sel. Top. Quant. Electron. 8 (4) (2002) 839–862.

[82] T. Kawazoe, M. Ohtsu, S. Aso, Y. Sawado, Y. Hosoda, K. Yoshizawa, K. Akahane, N. Yamamoto, M. Naruse, Two-dimensional array of room-temperature nanophotonic logic gates using InAs quantum dots in mesa structures, Appl. Phys. B 103 (2011) 537–546.

[83] L. Medintz, A.R. Clapp, H. Mattoussi, E.R. Goldman, B. Fisher, J.M. Mauro, Self-assembled nanoscale biosensors based on quantum dot FRET donors, Nat. Mater. 2 (2003) 630–638.

[84] T. Franzl, T.A. Klar, S. Schietinger, A.L. Rogach, J. Feldmann, Exciton recycling in graded gap nanocrystal structures, Nano Lett. 4 (2004) 1599–1603.

[85] T. Kawazoe, K. Kobayashi, M. Ohtsu, Optical nanofountain: A biomimetic device that concentrates optical energy in a nanometric region, Appl. Phys. Lett. 86 (2005) 103102-1–103102-3.

[86] M. Kubo, Y. Mori, M. Otani, M. Murakami, Y. Ishibashi, M. Yasuda, K. Hosomizu, H. Miyasaka, H. Imahori, S. Nakashima, J. Phys. Chem. A 111 (24) (2007) 5136–5143.

[87] G.D. Scholes, G.R. Fleming, J. Phys. Chem. B 104 (8) (2000) 1854–1868.

[88] M. Naruse, H. Hori, K. Kobayashi, P. Holmström, L. Thylén, M. Ohtsu, Lower bound of energy dissipation in optical excitation transfer via optical near-field interactions, Opt. Express 18 (S4) (2010) A544–A553.

[89] M.T. Bohr, Nanotechnology goals and challenges for electronic applications, IEEE Trans. Nanotechnol. 1 (1) (2002) 56–62.

[90] D.A.B. Miller, Are optical transistors the logical next step? Nat. Photon. 4 (2010) 3–5.

[91] G. Sun, J.B. Khurgin, Plasmon enhancement of luminescence by metal nanoparticles, IEEE J. Sel. Top. Quant. Electron. 17 (2011) 110–118.

[92] M. Naruse, P. Holmström, T. Kawazoe, K. Akahane, N. Yamamoto, L. Thylén, M. Ohtsu, Energy dissipation in energy transfer mediated by optical near-field interactions and their interfaces with optical far-fields, Appl. Phys. Lett. 100 (2012) 241102-1–241102-4.

[93] F. Moll, M. Roca, E. Isern, Analysis of dissipation energy of switching digital CMOS gates with coupled outputs, Microelectron. J. 34 (9) (2003) 833–842.

[94] T. Baehr-Jones, B. Penkov, J. Huang, P. Sullivan, J. Davies, J. Takayesu, J. Luo, T.-D. Kim, L. Dalton, A. Jen, M. Hochberg, A. Scherer, Nonlinear polymer-clad silicon slot waveguide modulator with a half wave voltage of 0.25 V, Appl. Phys. Lett. 92 (2008) 163303.

[95] M. Liu, X. Yin, E. Ulin-Avila, B. Geng, T. Zentgraf, L. Ju, F. Wang, X. Zhang, A graphene-based broadband optical modulator, Nature 474 (2011) 64–67.

[96] Ming Liu, University of California at Berkeley, Private Communication.

[97] Z. Lu, W. Zhao, Nanoscale electro-optic modulators based on graphene-slot waveguides, J. Opt. Soc. Am. 29 (6) (2012) 1490–1496.

[98] M. Chacinski, U. Westergren, B. Stoltz, L. Thylén, Monolithically integrated DFB-EA for 100 Gb/s Ethernet, IEEE Electron. Dev. Lett. 29 (12) (2008) 1312–1315.

[99] <http://www.itrs.net/>.

[100] L. Thylén, S. He, L. Wosinski, D. Dai, Moore's law for photonic integrated circuits, J. Zhejiang Univ. Sci. 7 (12) (2006) 1961–1967.

Fibers for Short-Distance Applications

7

John Abbott, Scott Bickham, Paulo Dainese, and Ming-Jun Li

Corning Incorporated, Corning, NY, USA

7.1 INTRODUCTION

In the last 10 years multimode fiber and short-wave VCSELs have emerged as dominant technologies for short-reach high data rate networks [1]. Multimode glass optical fibers are used in networks, in office buildings, and in data centers where data rates and/or lengths are higher than can be met with copper networks, and the use of VCSELs makes the overall system cost low enough to enable widespread adoption.

The fast growth is driven by the demand for higher data rates for computer connections, data storage, and local communication including linking to Internet traffic. It is tied to a variety of standards enabling products targeted at or including specific short-distance objectives, including:

a. *Ethernet* is a standard supported by the IEEE. The IEEE 802.3 standards continue to expand and now include 10 Gb Ethernet with 10 Gb VCSEL sources, as well as 40 Gb and 100 Gb Ethernet with parallel fiber and array VCSELS [2,3]. The IEEE 803.3 standards group has recommended a project to develop at 4×25 objective using 25 Gb/s VCSELS [4]. Two recommended Ethernet references are the book by Cunningham and Lane [20] and a chapter in volume V B of this series [70].

b. *Fiber Channel* is a standard developed by the T11 Technical Committee of the International Committee for Information Technology Standards (NCITS). It arose in support of the supercomputer field but now has become a standard connection option for data centers and storage area networks (SANs). The multimode fiber is the same as in the IEEE standards; the implemented standards include 10GFC Serial, 10GFC Parallel, and 16GFC. A T11 project is currently working on 32GFC (28 Gb/s VCSELS) [6,5].

c. *Infiniband* [7] is a communications link standard supported by the Infiniband Trade Association [8] used in high-performance computing (HPC), with the goals of high throughput, low latency, and quality of service. DDR links are used in the Jaguar supercomputer at NCCS (Oak Ridge) [97]. Current QDR standards have a 10 Gb/s per-line data rate, and the roadmap includes FDR (14 Gb/s) and EDR (26 G/s) [9].

Optical Fiber Telecommunications VIA. http://dx.doi.org/10.1016/B978-0-12-396958-3.00007-X

Tables of the various data rates supported by these three standards are given in Tables A1–A3 at the end of the chapter.

The specification of the fiber itself is done by the ANSI-affiliated Telecommunications Industry Association (TIA) [10] and by the International Electrotechnical Commission (IEC) [11]. In Volume I of this series [12], the emergence of MMF with a 62.5 μm core and 2% Delta (maximum relative refractive index difference) for short-distance applications was discussed, and in Volume IV [1] it was noted that this was the majority of installed MMF. This fiber was not optimized for lasers, so for those applications, the new "laser-optimized" MMF with 50 μm core and 1% Delta developed specifically for 10 Gb/s VCSELs operating at 850 nm has become the norm. A common nomenclature for the multimode fibers (ISO/IEC 11801, ANSI/TIA-568-C.3) is now used and summarized in Table 7.1. OM1 is the 62.5 μm core MMF mentioned in the preceding volumes. OM2 is a 50 μm, 1% Delta fiber with a near-parabolic profile optimized at a wavelength between 850 nm and 1300 nm, meeting an overfilled bandwidth (OFL BW) specification of 500 MHz km at both wavelengths. The next two fiber types are laser-optimized 1% Delta fibers with 50 μm cores and higher bandwidth values than OM2. OM3 and OM4 fiber both meet a legacy OFL BW specification

Table 7.1 Multimode optical fiber nomenclature and link distances nomenclature reference: ISO/IEC 11801 and ANSI/TIA-568-C.3 link distances reference: IEEE802.3.

| Fiber Type | Core Diameter (μm) | Minimum Modal Bandwidth MHz km | | | IEEE 802.3 Link Distances | | |
| | | Overfilled Launch Bandwidth (OFL BW) | | Effective Laser Launch Bandwidth (EMB) | | | |
		850 nm	1300 nm	850 nm	1000BASE-SX	10GBASE-SX	40GBASE- and 100GBASE-SX
OM1	62.5[a]	200	500	not specified	275 m	33 m	not spec
OM2	50[a]	500	500	not specified	550 m	82 m	not spec
OM3	50	1500	500	2000	not spec	300 m	100 m
OM4	50	3500	500	4700	not spec	400 m	150 m

[a]*Common usage: IEC 11801 includes OM1/50 μm and OM2/62.5 μm (rarely seen).*

of 500 MHz km at 1300 nm, but are optimized for laser performance at 850 nm: OM3 has an 850 nm OFL BW of 1500 MHz km and an effective modal bandwidth (EMB) of 2000 MHz km. OM4 has an 850 nm OFL BW of 3500 MHz km and an EMB of 4700 MHz km.

The theory of light propagation in multimode fiber and the definition of bandwidth will be reviewed in Section 7.2. Newer references include the excellent chapter (in German) by Freude in the book edited by Volges and Petermann [13] and the JLT review by Freund [14]. Multimode fiber itself has now been around for some time; other recommended references are the books by Marcuse [15], Okoshi [16], and Murata [17]. The review article by Olshansky [18] covers many of the essential points. Finally, the book by Snyder and Love [19] is an outstanding reference on the theory of light propagation in optical fibers.

The actual performance of an MMF link (the bit error rate and intersymbol interference) depends on both the fiber and the laser. EMB, which includes both fiber and laser effects, will be discussed in Section 7.3, and the method of characterizing fiber with the differential-mode-delay measurement (DMD) and the laser with the encircled flux measurement (EF) will be summarized there as well. System link models and the measurements used to characterize high data rate MM fiber will be more fully discussed in Section 7.4. Gigabit Ethernet links are discussed in detail in the book by Lane and Cunningham [20]; the origin of the link model used today is summarized in the paper by Nowell et al. [21]. The modeling of OM3 links at 850 nm to develop the TIA standard supporting the IEEE 10 Gb Ethernet standard is summarized in the JLT paper by Pepeljugoski et al. [22].

The performance of MMF systems depends on both the laser and fiber used, and the limitations of each affect the system. The TIA and IEC determine the fiber specification, but it is the IEEE (and Fiber Channel, Infiniband, etc.) which determines the assumptions underlying a specific standard and the worst-case distance which can be supported with a given fiber type when all system assumptions are met. The distances for current IEEE 802.3 standards for short-distance, short-wavelength links with OM2, OM3, and OM4 fiber are given in Table 7.1 above as an example. Distances for Fiber Channel and Infiniband are different and set by those groups based on different assumptions, and individual vendors may guarantee different link lengths based on the specific link design including fiber, cable, and laser attributes.

A new development since 2002, when the section on multimode fiber in volume IV-B of this series [1] was written, is the introduction of bend-sensitive MMF. These fibers have been in production for several years [23] and will be discussed in Section 7.5.

A new area for multimode fiber is easy-connect short-distance applications for consumers or computer rooms. These very-short-distance-network fibers are just emerging and will be discussed in Section 7.6, along with MM fibers designed to minimize chromatic dispersion (DC-MM) and other new multimode fibers. The section closes with a detailed discussion of multicore fibers.

7.2 THEORY OF LIGHT PROPAGATION IN MULTIMODE FIBERS

This section reviews the basic ideas of light propagation in multimode fiber which is important for applications in data/computer centers. The "link performance" including the effect of launch conditions will be covered in the next section.

As noted in reference [1], the ideal refraction index profile $n(r)$ of a multimode fiber is optimized to maximize the modal bandwidth, which requires minimizing the spread in group delays. The work summarized in [1] and [22] led to the OM3 standard with a minimum effective modal bandwidth (EMB) of 2000 MHz km. Since then, an OM4 standard with a significantly higher EMB of 4700 MHz km has been introduced to meet the increased bandwidth demands in short-distance applications. As the refractive index profiles have become more perfect through improvements in the manufacturing processes, a variety of details become more important. We will try to highlight these as we quickly review the theory of light propagation in MM fibers.

Given the index profile $n(r)$ and the wavelength λ of interest (and the corresponding wave number $k = 2\pi\lambda$), the modes are usually solved using the scalar wave equation [19,15,16,18].

$$(\nabla^2 + k^2 n^2(r))\psi = \beta^2 \psi. \tag{7.1}$$

The guided modes take the form

$$\Psi_{l,m}(r, \theta) = \psi_{l,m}(r) \begin{Bmatrix} \cos l\theta \\ \sin l\theta \end{Bmatrix}. \tag{7.2}$$

When there is no angular dependence on the index profile, the *cosine* and *sine* "modes" are degenerate. Likewise each individual mode has two polarization states which are degenerate.

A variety of numerical methods have been proposed for solving the equation, which is a cylindrical version of the scalar wave equation seen in quantum mechanics. For the specific case of multimode optical fibers, these include [24–27]; there are also surveys [28,29] and a variety of papers focused on single-mode fibers or general eigenvalue problems which are also applicable [30–34]. Various applied mathematical techniques developed for quantum mechanics (for example, the WKB approximation [35]) have been applied to light propagation in optical fibers [18,36]. Commercial packages have also been developed which can numerically solve this governing scalar wave equation [29,37], as well as modeling of links including the effects of electronics and lasers [38]. Chapter 20 in volume V B of this series gives an overview of simulation tools including propagation in multimode fibers [71].

The scalar wave equation is itself an approximation to Maxwell's equations, appropriate when the maximum refractive index difference Δn is small and dn/dr is small. Maxwell's equations are a set of two nonlinear coupled equations for the electric field E and the magnetic field H. It is possible to calculate the polarization correction as a perturbation of the scalar wave solution result [19]. If the index profile has a small angular variation $n(r, \theta)$, this effect can also be included by perturbation techniques similar to what has been done for estimating PMD in single-mode fibers [39], following methods developed for more general perturbations [40]. There are examples in the literature of solving light propagation in MM fibers with a full solution to Maxwell's equations, for example to look at the effect of small scale index fluctuations in the manufacturing process [41].

Because glass is a dispersive medium and the refractive index varies with wavelength, the modal solutions to Eq. (7.1) are characterized by both the propagation parameter β_{lm} and the group delay τ_{lm} given by

$$\tau_{lm} = \frac{1}{c}\frac{d\beta_{lm}}{dk}. \tag{7.3}$$

When light is launched into a multimode fiber, the input pulse couples into the modes which propagate in the fiber, not at the phase velocity (the velocity of light in glass $= c/n_{\text{eff}} = ck/\beta$) but at the group velocity, with a relative delay given by Eq. (7.3). For multimode fibers with small dispersion and a small Δn where the scalar wave equation is applicable, the optimal profile is approximately a parabola. The profiles are further optimized to account for dispersion and Δn. These so-called "alpha profiles" ([19] Eq. (2.44)) [42,43] are described by the formulae:

$$n^2(r) = n_1^2\left\{1 - 2\Delta\left(\frac{r}{a}\right)^\alpha\right\}, \quad r \leqslant a,$$

$$n^2(r) = n_2^2, \quad r \geqslant a,$$

$$\Delta = \frac{(n_1^2 - n_2^2)}{2n_1^2},$$

$$\alpha_{\text{opt}} = 2 - C\Delta - 2p, \tag{7.4}$$

$$C \approx 2,$$

$$P = \frac{n_1}{N_1}\frac{\lambda}{\Delta}\frac{d\Delta}{d\lambda},$$

$$N_1 = n_1 - \lambda\frac{dn_1}{d\lambda}.$$

In the formula for the optimum alpha, $\alpha_{\text{opt}} = 2 - C\Delta - 2P$, the coefficient C depends on the assumed modal power distribution and exactly what metric is being minimized. For example, $C \sim 12/5$ [42] corresponds to minimizing the weighted RMS spread of the mode delays where the individual modes are equally weighted. P depends upon the material properties of the doped glass. The optimum alpha also varies with wavelength [17,44,22] and will be covered in more detail in the next section, with the effect of the source included.

In Eq. (7.1), modes which have the same "mode group number" [18] or principal mode number, $g = 2m + l + 1$, have the same propagation parameter β and propagate as a single-mode group rather than as individual modes. They are said to be in the same "degenerate mode group" [1] or "principal mode group" [14] or just "mode group" [18]. They have a common phase velocity and are strongly coupled. This effect is prevalent even for index profiles with perturbations, and this deserves emphasis. The number of individual modes in group "g" is equal to "g" (with a factor of 2 for the sine/cosine degeneracy for azimuthal modes with $l > 0$), and the delay of the "mode group" is the weighted average of the g individual mode delays.

The different mode delay (DMD) measurement [15,45] described in the next section verifies this interesting effect. For example, the components of mode groups 3 and 4 result in two delays (group 3 has two individual modes, and group 4 has three modes, but two of these have identical delays because of the sine/cosine degeneracy) corresponding to individual modes. However, in a real fiber measurement, the modes in each group always arrive at the same time, even if a careful measurement of the index profile and a numerical solution for the delays shows the individual delays are different. The strong coupling between the modes in each group results in the formation of a wave packet that propagates in the fiber with an average delay equal to the average delays of the constituent modes. It should also be noted that a special property of the "alpha profiles" in Eq. (7.4) is that the group delay τ_{lm} is the same for all individual modes in the group (with no coupling), but "non-alpha" errors (for example, a center-line dip in the index profile typical of early MM fibers [16] p. 83) verify the intra-group coupling and co-propagation of degenerate modes with different individual mode delays.

7.3 CHARACTERIZATION OF MM FIBER AND SOURCES FOR HIGH DATA RATE APPLICATIONS

The performance of telecommunication links with MM fiber depends on both the fiber and the source, particularly the mode group delays (a fiber property) and the modal power distribution (a property of the source and launch conditions). Chromatic dispersion (CD) is also important, because typical links are operating at 850 nm to take advantage of low-cost VCSEL transceivers; this effect will be covered in the next section, which discusses full system models. This section focuses on the measurements of the fiber and the source, which are used to ensure link performance. To motivate the discussion, we first summarize the "bandwidth" measurement which depends on both a specific fiber and a specific launch. We then review the characterization of the fiber (i.e. the HRDMD measurement) and the source (i.e. the "encircled flux," the launch power inside of a radius r), and tie the three together with the "EMBc" metric.

7.3.1 Bandwidth

The output pulse of a MM fiber, neglecting broadening due to chromatic dispersion, can be approximated as the sum of N_g pulses corresponding to the N_g mode groups, each with a modal power P_g corresponding to the total power in all modes in that group. Theoretically the pulse can be represented by a sum of delta functions each with a relative delay corresponding to the group

$$P(t) = \sum_g P_g \, \delta(t - \Delta\tau_g),$$

$$\sum_g P_g = 1,$$ (7.5)

$$\Delta\tau_g = \tau_g - \sum_g P_g \, \tau_g.$$

Again we note that the mode power distribution P_g is a function of the launch conditions and the delay $\Delta\tau g$ relative to the centroid is a function of the fiber; a subtle point is that the absolute arrival time of the pulse (that is, its centroid) is $\Sigma P_g \tau_g$, and this quantity depends on both modal delays and power distribution.

To achieve low bit error rate (BER) and low intersymbol interference (ISI) for high data rate systems largely limited by the transceiver and system penalties, it is important that little power resides outside the unit bit window.

If the input pulse and output pulse are measured in nsec, the 3 dB bandwidth of the fiber is defined as the frequency measured in GHz where the modulus of the transfer function drops to 50% of its peak value. For an ideal fiber with a "delta function" input and an output pulse given by Eq. (7.5), the transfer function $H(f)$ can be calculated with elementary properties of Fourier transforms:

$$H(f) = \sum_g P_g \exp(2\pi i f \Delta\tau_g)$$

$$|H(f)| = \sqrt{\left(\sum_g P_g \cos(2\pi f \Delta\tau_g)\right)^2 + \left(\sum_g P_g \sin(2\pi f \Delta\tau_g)\right)^2}.$$ (7.6)

We are interested in the low frequency part of this curve with $|H(f)| > 0.5$. If the transfer function is a Gaussian curve, the entire curve is represented by the 3 dB value and the power outside the bit window can be estimated for any bit rate and any link length. For short distance, high data rate systems where the fiber adds only an incremental penalty, the important metric is the "bit window penalty" at a lower frequency corresponding to the system. If the transfer function is non-Gaussian, some modification of the 3 dB estimate is used, including an extrapolation from the 1.5 dB value [46], Gaussian fitting [47], or another approach based on the bit rate/length and fraction of power outside the bit window [48,49]. The "effective modal bandwidth" or EMB of a fiber includes both the notion that P_g can be from an arbitrary source (not just the overfilled bandwidth (OFL BW) with P_m equal for each individual mode), and the notion that the literal 3 dB BW might need to be adjusted to take into account the non-Gaussian nature of the transfer function. If the transfer function and bandwidth are calculated using the spectral width of the transmitter, etc., it is referred to as the "effective bandwidth" (EB) [22].

Two important examples of these effects are: (a) If $|H(f)|$ is approximately a Gaussian function or (b) if $P(t)$ is approximately the sum of two equal pulses (a so-called dual Dirac pulse). In example (a) the bandwidth in GHz is approximately $0.187/\sigma$, where σ is the RMS pulse in nsec [50] (Appendix B). In example (b),

$$P(t) = .5\left\{\delta(t + \Delta\tau) + \delta(t - \Delta\tau)\right\},$$
$$\Delta\tau = \frac{\tau_2 - \tau_1}{2},$$
$$H(f) = \cos(2\pi f \Delta\tau),$$
$$f_{3dB} = \frac{1}{6\Delta\tau} = \frac{1}{3(\tau_2 - \tau_1)}. \qquad (7.7)$$

The measurements originally used to characterize MM fiber measured the bandwidth with a prescribed launch. The so-called overfilled bandwidth [51] used a mode conditioning patchcord to simulate a Lambertian source with equal power at all angles, putting equal power into all the individual modes. This approximated the launch of an LED which was the important source for 100 MHz systems. For 1 GHz systems the modified technique of a restricted mode launch (RML) [51] was used, using a nominally 50 μm 1% Delta MM fiber which was drawn down to a core diameter of ∼23 μm, and used this smaller core to simulate in a repeatable way the launch from 1 GHz VCSELs.

With the arrival of higher frequency 10Gb VCSELS around in the 2000–2001 timeframe, a different approach was needed. As summarized in [22] and [1], this led to the development of a high resolution version of the differential-mode-delay (DMD) measurement [45] to better estimate the fiber mode group delays.

7.3.2 Fiber characterization: HRDMD

The DMD measurement [46,45,15,1,22] is made by launching a carefully aligned small diameter spot (nearly diffraction-limited) into the MM fiber at different offset positions from the center. At each offset position x the output pulse $F_x(t)$ is recorded and the 2D array $F_x t$ is analyzed. Earlier versions of the DMD measurement used the centroid of the output pulse because they were oriented toward manufacturing and optimizing the index profile. "High resolution" DMDs (HRDMDs) use the entire pulse at each offset position to capture as much information as possible about the differential-mode delays in the MM fiber.

The power P_m going into each individual mode can be rigorously calculated from overlap integrals analogous to calculations in quantum mechanics [52–54].

Two parallel specifications of MM fiber using the HRDMD have been implemented in standards, the "minEMBc" approach and the "DMDmask" approach.

The "DMDmask" approach is summarized in [22,1] and specified in TIA/EIC 492AAAC/D [55]. Its original motivation lies in Eq. (7.7), which means in principle to determine the leading and trailing edge of the DMD pulse structure and make $\tau_2 - \tau_1$ small enough that the pulse is within the "bit window." In practice this is too stringent a requirement because it neglects realistic mode power distributions. To determine adequate and economical limits, an extensive Monte Carlo simulation was run [22] with a set of 5000 theoretical "fibers" and a set of 2000 theoretical "sources," determining τ_g and P_g. A disadvantage of the DMDmask approach is that the actual specification depends on the underlying assumptions for the fiber/lasers used in the Monte Carlo analysis, which is hidden on a day-to-day basis. This shortcoming is balanced by the advantage that it focuses on the DMD measurement, which is a fiber attribute that can be characterized by the manufacturer.

The "minEMBc" approach is also specified in TIA/EIC 492AAAC/D [55]. The original motivation [56] was to construct a weighted sum of the HRDMD pulses to simulate the output pulse of a VCSEL. Reference [1] gives an example in which the overfilled BW (OFL BW) is constructed from a weighted sum of the DMD pulses; a different weighting should therefore simulate different launch conditions [123]. Thus the RML BW idea used for 1 Gb systems could be extended to simulate as many lasers as needed. The "c" in "EMBc" indicates that it is calculated from the DMD data, not measured with an actual source; the "min" in "minEMBc" indicates that what will be reported is the minimum EMBc of a "test set" of sources. The minEMBc approach is different from the DMDmask approach in that it is quantifying the bandwidth, which reflects the system performance. It does not point directly at the index perturbation. The advantages are (a) it yields the bandwidth which needs to be guaranteed for system link models and (b) it is easier to modify the specification to encompass the new generations of lasers. EMBc also has a more direct link to bit error rate and ISI [48] and provides a method of predicting system performance for any particular fiber and particular laser [57–59]. It provides a logical way to include spectral information as well [60,61] to generate a "calculated" effective bandwidth (EB$_c$) [22] including chromatic dispersion. The two methods have been compared using actual fiber data and found to be equivalent [62].

Before discussing the EMBc metric in more detail we review the characterization of the source using encircled flux.

7.3.3 Source characterization (encircled flux)

In the TIA Monte Carlo simulation [22], each of the 2000 theoretical sources was generated with a consensus laser model producing laser modes; the coupling of these to the modes of a 50 μm 1% Delta fiber was calculated. The launch was characterized by P_m, the power going into the individual modes, and $EF_{in}(r)$, the integral of the modal power versus radius for all modes (weighted by P_m, remembering to count once for purely radial modes and twice for azimuthal modes):

$$EF_{in}(r) = \int_0^r \sum_m P_m \psi_m^2(r) r \, dr. \qquad (7.8)$$

$EF_{in}(r)$ is normalized so that it goes to 1.00 as r goes to infinity, and it corresponds to the integral of a measured near field profile on a short length of fiber. An example of

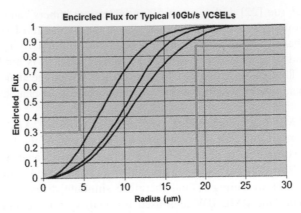

FIGURE 7.1 Encircled flux versus radius for three typical 10Gb/s VCSELS, with the inner and outer specifications marked.

measured $EF(r)$ is given in Figure 7.1. These measurements also show the specification developed for 10GbE, where $EF<0.30$ at a radius of $4.5\,\mu$m, and $EF>0.86$ at a radius of $19.0\,\mu$m. Joint fiber and laser specifications were needed for the laser (P_m or ultimately P_g) and the fiber (that is τ_g). The source specification resulted in an upper limit on $EF_{in}(r)$ at small radius and a lower limit on $EF_{in}(r)$ at large radius. A scatter plot of the fraction of power inside of $4.5\,\mu$m radius versus the radius in microns which encircles 86% of the power gives the plot shown in Figure 7.2. Figure 7.2 shows the distribution of theoretical sources used in the TIA 10GbE Monte Carlo modeling [22], as well as the ten theoretical 850nm sources used in developing the EMBc specification.

As noted above, there is strong coupling between the degenerate individual modes in a mode group, and the analysis for the TIA 10GbE work made the standard assumption [22,1] of full coupling within a group with no phase cancellations. The result of this ansatz is that the power is divided evenly between all the modes in the group, and there is a single power P_g and a single-mode delay τ_g for the group. Although this is an approximation, it agrees well with observations of DMD measurements.

When this intra-mode group coupling is considered, it results in an implicit encircled flux $EF_{fiber}(r)$ of interest, which corresponds to

$$EF_{fiber}(r) = \int_0^r \sum_g P_g \psi_g^2(r)r\,dr,$$

$$\psi_g^2(r) = \sum_{m\in g} \psi_m^2(r), \tag{7.9}$$

$$P_g = \sum_{m\in g} P_m.$$

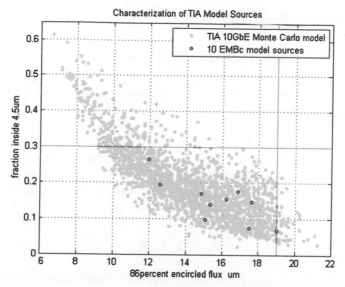

FIGURE 7.2 Distribution of sources used to develop the 10 GbE specification [22]. The red lines indicate the source specification which were developed. The red dots indicate ten of the 2000 sources used to develop the EMBc fiber specification (see text), and correspond to real lasers with the same measured encircled flux parameters. (For interpretation of the references to color in this figure legend, the reader is referred to the web version of this book.)

Thus the source determines the modal power weighting P_g, and the fiber determines the mode group delay τ_g. The impulse response is given by Eq. (7.5), and the bandwidth can be calculated as described above. Alternatively, the impulse response (or bandwidth) can be incorporated into a system model as described in the following section.

7.3.4 EMBc metric

We return to the discussion of the EMBc metric, which has been added to the standards after reference [1]. It had been found in the 10GbE development [22] that there was a functional relationship between the EMB calculated from P_g and τ_g and the ISI. That results in a complicated set of DMDmask parameters in order to guardband the modeled DMD to yield the desired ISI. It was tested and shown [56] that the required ISI for the set of 40,000 modeled links could be guaranteed if (a) there is an encircled flux spec on the source and (b) the EMB for a small selected set of fibers is high enough. This analysis led to the introduction of the 10 sources used in developing the TIA EMBc standard, as shown as red dots in Figure 7.2 [56].

The idea behind the EMB sources is to determine a weighting of the DMD pulses $P_x(t)$ at different offset positions x, which when summed together will give a single pulse $P(t)$ representing the correct mode power distribution P_g. The MPD P_g can be determined theoretically or by deconvolving an encircled flux measurement and solving the equation

$$I(r) = \sum_g P_g \psi_g^2(r).$$ (7.10)

Equation (7.10) can be set up as a non-negative least squares problem [63], although frequently a linear constrained least squares approach works equally well [46,64,65].

Once the modal powers P_g are known, the problem of approximating them with a weighted sum of DMD pulses involves a second least squares problem. In the DMD measurement, the power going into group g when the offset is at x can be written as an overlap integral [52–54] and reduced to a matrix P_{xg} or P_{gx} (depending on the mathematical algorithm). This matrix is used to solve for a weighting function W_x which satisfies

$$P_g = \sum_x P_{gx} W_x.$$ (7.11)

Equation (7.11) is well suited for a constrained least squares solution forcing some smoothness on W_x, with the caveat that W_x must be non-negative.

Once the weights are determined, the pulse simulating the laser with a given encircled flux (and hence P_g) is given by the weighted sum of the DMD pulses $P_x(t)$:

$$P(t) = \sum_x W_x P_x(t).$$ (7.12)

EMBc is thus determined from a calculated output pulse and a reference pulse measured on a short length. In the 10 GbE standard, fibers meeting the DMDmask were defined as meeting a bandwidth of 2000 MHz km, in the sense of having an average failure rate of 0.5% relative to the ISI target. To generate the same failure rate and hence an equivalent specification, the specification EMB of a fiber is defined as 1.13*minEMBc, where "minEMBc" is the minimum EMBc of the ten sources. The DMDmask and EMBc specifications are equivalent on average. For purposes of system modeling, we show in the next section that the EMBc value is more applicable (since the system performance of a particular fiber depends on the mode power distribution P_g and the mode delays τ_g). Neither the DMDmask nor minEMBc takes into account the particular laser, and hence neither will predict a change in system performance if the properties of the laser are significantly changed.

As noted in Section 7.2, the optimum "alpha" depends on wavelength. When the 850 nm and 1300 nm bandwidths of a series of alpha profiles with different alphas are calculated and plotted, a characteristic "triangular" curve results [17]. Figure 7.3 is the corresponding plot using the 5000 theoretical "fibers" in the TIA Monte Carlo study [22], using a single source of the 2000 "lasers." Here we use source 1946 which corresponds to source 5 of the 10 minEMBc sources, the red[1] dot furthest to the right

---

[1]For interpretation of color in Figure 7.2, the reader is referred to the web version of this book.

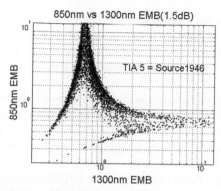

FIGURE 7.3 EMB(850 nm) versus EMB(1300 nm) in GHz km for the 5000 theoretical "fibers" in the Monte Carlo analysis of [22], using a fixed mode power distribution (source 1946 = "Source 5" of the 10 minEMBc sources). The 1.5 dB frequency has been extrapolated to 3 dB by multiplying by 1.414.

in Figure 7.2. The 1300 nm delays were calculated using the approximation in [22], and the 1300 nm mode power distribution is calculated by interpolating source 5 to the correct number of mode groups. The plot in Figure 7.3 represents the modal bandwidth without the chromatic dispersion of the laser, but it incorporates approximately the effect of the parameter P in Eq. (7.4). The high 850 nm BWs correspond to an alpha near 2.1, the high 1300 nm BWs correspond to an alpha near 2.0, and as alphas run from $\sim$1.8 to $\sim$2.3 the (x, y) plot traces out a triangular pattern. Non-alpha errors give points inside the triangle, and points which appear to lie outside the triangle typically have non-Gaussian transfer functions.

7.4 SYSTEM MODELS AND MEASUREMENTS FOR 1 GB AND 10 GB ETHERNET

This section will briefly review system and link models relating to short-distance multimode fibers; however, a detailed discussion is beyond the scope of the chapter.

Current multimode link models are based on the work by Nowell, Cunningham, and Hanson for 1 Gb/s systems [21]. This paper and the book by Cunningham and Lane [20] provide a clear and thorough introduction to modeling of MMF links, and current terminology has changed only when additional parameters or effects needed to be included.

The system modeling for 10 GbE systems is summarized in [22,66]. In developing the IEEE standard, a spreadsheet by Dawe et al. [67] incorporated the features of the 1 Gb/s model and extended them to 10 Gb/s (the spreadsheet also has tabs for single-mode link models). It is important to note that the 2003 references predate largescale manufacturing of the 850 nm 10 Gb lasers, and that the 10 GbE spreadsheet online is not the final version but a working draft. There is no "official" IEEE 10 GbE link spreadsheet, but rather "official" IEEE specifications which the link needs to meet. This basic spreadsheet has been updated by various authors in presentations at IEEE and Fiber Channel; a 25 Gb/s example is given in [68] (this also does not represent a final version).

A system link model following the ideas in the above references and spreadsheets must simultaneously deal with two issues: (a) the intersymbol interference (ISI) must lie below a targeted level and (b) the total system power must be above a center budgeted level. Depending on the assumptions for the link, the key constraint can be either (a) or (b), or in some cases both.

There are a number of simplifications in the 10 GbE spreadsheet model, and there has been a proliferation of proprietary models in Excel, Matlab, or actual model codes to address them. At higher bit rates, the method of accurately handling the jitter due to laser becomes more complicated. The link models generally approximate the fiber impulse response by a Gaussian model, which misses subtleties arising from multiple pulses. They also typically approximate the spectral character of the laser by an RMS spectral width; however, current 850 nm VCSELS have multiple modes, and hence multiple lines. The rise/fall parameters in the model are approximate values; in fact, each laser mode has its own rise/fall structure (as well as its own wavelength). Improving the link models is an area of current research in both IEEE [68] and Fiber Channel [69].

An image from the "IEEE spreadsheet model" [67] is given in Figure 7.4. The upper half of the spreadsheet allows entry of the many parameters needed for the system. In the middle the various penalties are calculated for a range of lengths, allowing one to determine what length will meet the criteria (a) and (b) mentioned above.

7.5 BEND-INSENSITIVE MM FIBER

The previous sections focus on the relationship of multimode fiber bandwidth to system performance. As systems have evolved toward longer lengths and faster data rates, multimode fiber manufacturers have had to keep pace by delivering better products. The development of the "laser-optimized" MMF is one such example that targets 10 Gb/s VCSELs operating at 850 nm. There are other problems to solve than just delivering higher bandwidth systems. Increasing line rates have put increasing pressure on the power budget for multimode fiber systems, and as a result, the impact of bend loss has become much more critical. This observation led to the development and commercial launch of bend-insensitive multimode fiber (BIMMF) using novel fiber designs that inhibit the leakage of the optical power into the cladding. These BIMMF have in turn enabled the development of smaller, more flexible cables that facilitate installation because they tolerate more convenient routing. Bend-insensitive multimode fibers have also been the keystone for smaller hardware components, such as connectors with shorter boots which permit easier finger access when installing or

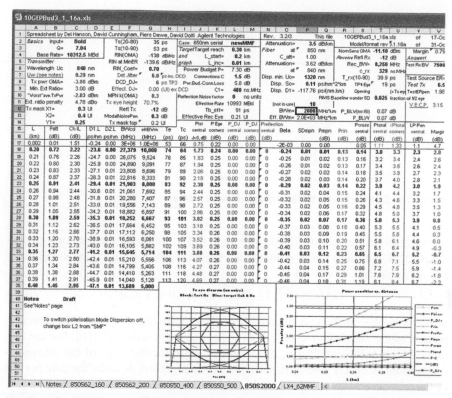

FIGURE 7.4 An example calculation with the spreadsheet from Ref. [67]. The parameters are set in the upper section, calculated link penalties versus distance are output in the middle section, and plots are produced at the bottom.

removing the connector [72]. Consequently, BIMMF have added value to the connector and the equipment rack by allowing tighter routing and higher densities.

In this section, we start with a discussion of conventional MMF (C-MMF) to introduce the nomenclature. We then introduce a trench into the cladding and show how that feature can compensate the "anomalous" dispersion exhibited by C-MMF [73]. We conclude with a discussion of the bend performance and illustrate how a simple empirical formula captures the length dependence for a given bend diameter.

Figure 7.5 shows a schematic for a BIMMF refractive index profile with a low index ring, or trench, in the cladding. The core has the same graded-index shape as in conventional $50\,\mu$m MMF, but the trench creates a barrier that improves the confinement of the propagating modes. It has been shown analytically and experimentally that careful control of the spacing between the core and the trench enables high bandwidth [74–76] while simultaneously yielding a significant improvement in macrobend performance [77].

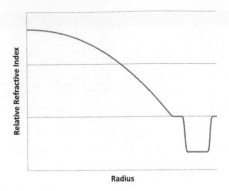

FIGURE 7.5 Schematic of a BIMMF refractive index profile.

As discussed in Section 7.2, the refractive index profile for the core of a graded-index MMF with core radius a, maximum refractive index n_1, and curvature parameter α (alpha) may be written

$$n^2(r) = n_1^2 \lfloor 1 - 2\Delta(r/a)^\alpha \rfloor, \quad 0 \leqslant r \leqslant a, \tag{7.13}$$

where $\Delta = \frac{n_1^2 - n_2^2}{2n_1^2}$ is the maximum relative refractive index, or delta, of the core with respect to the cladding, and the cladding has refractive index n2 [16]. For conventional 50 μm MMF, $a \approx 25$ μm, $\Delta \approx 0.01$, and the refractive index of the silica cladding is approximately 1.4525 at 850 nm. These parameters describe the core portion of the refractive index profile plotted in Figure 7.5.

It is straightforward to insert the refractive index profile given by Eq. (7.13) into the scalar wave equation for an optical waveguide [25] to numerically calculate the eigenvalues (i.e. the propagation constants, β) and eigenvectors (transverse electric fields) of the propagation modes. For the profile parameters given above, this procedure yields 100 LP_{nm} eigenvectors which correspond to 380 propagation modes which form 19 mode groups. While the mode coupling between mode groups is usually assumed to be vanishing slowly, the modes within a group are strongly coupled and are usually assumed to propagate through the fiber as a wave packet with a well-defined group velocity [78].

An important attribute of an MMF is the numerical aperture (NA), which is optically defined as the sine of the maximum angle (relative to the axis of the fiber) of the incident light that becomes completely confined in the fiber by total internal reflection. For an optical fiber with a given core diameter, a higher NA value means that the core is more strongly guiding and should have better bend performance. On the other hand, the NA for most commercial applications has upper and lower bounds (e.g. 0.185 and 0.215, respectively for OM2/3/4) which are specified by the TIA and IEC. These boundaries limit the degree to which the core parameters can be tuned to improve the bend performance, and hence fiber designers have focused on optimizing the properties of the trench in the cladding to balance bend performance with standards' compliance.

It can be shown that for optical fibers, the NA definition yields the relationship [16], $NA = \sqrt{n_1^2 - n_2^2}$, which can be transformed into

$$NA = n_1\sqrt{2\Delta} = n_2\sqrt{\frac{2\Delta}{1 - 2\Delta}}. \tag{7.14}$$

The number of guided modes in the multimode fiber is then given by

$$M = \frac{2\alpha}{\alpha + 2}\left(\frac{\pi a}{\lambda}\right)^2 (NA)^2. \tag{7.15}$$

For fiber core parameters $\Delta = 1.0\%$, $a = 25\,\mu m$, and $\alpha = 2.1$, M is equal to 376, which is very close to the value of 380 assumed for a conventional $50\,\mu m$ MMF. However, this value is really a theoretical maximum which reflects an ideal fiber which guides every mode group regardless of where the cutoff wavelengths are located with respect to the operating wavelength (or the proximity of the effective index to the index of the cladding). In a real fiber, at least two of the outer two mode groups have high losses and are often excluded from a modal propagation analysis [79]. For the above conventional MMF example, this reduces the effective number of mode groups from 19 to 17.

This assumption of dropping the outer two mode groups may not be applicable to a BIMMF. If the relative refractive index of the trench is sufficiently low, some of the outer modes that have high losses in the conventional MMF now propagate with much lower attenuation. The contributions of these modes to the fiber bandwidth can no longer be neglected in system lengths of 10–100 m. The number of mode groups that may be excluded from the modal analysis depends on the width and depth of the trench, since these dimensions determine how difficult it is for these modes to tunnel from the core region into the cladding.

The depth and location of the trench also impacts the delays of the outer few mode groups, as shown in Figure 7.6. When the trench is far away from the core,

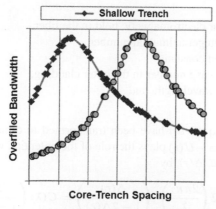

FIGURE 7.6 Illustration of the effect of the core-trench spacing on fiber bandwidth.

it still enhances the bend losses of the outer modes, but it is only a weak perturbation on the dispersive properties. In this regime, the outer modes exhibit "anomalous" delays [73], and the overfilled bandwidth is low because the outer modes travel faster than the other modes in the fiber. As the trench is moved closer to the core, the outer modes slow down and the bandwidth increases. There is an optimal spacing between the core and the trench at which the differential delays between the inner and outer mode groups in the fiber are minimized [74]. This optimal spacing depends on the core delta, the trench delta, and the trench width, but it is typically in the 1–4 μm range. For example, in Figure 7.6, the optimum location of the deep trench is approximately 0.4 μm farther from the core than the shallow trench. In both cases, if the trench is closer to the core than the optimum position, the outer modes slow down further and eventually travel more slowly than the other propagating modes. This again has a negative impact on the fiber bandwidth.

The previous paragraph gives a qualitative picture of how the characteristics of the trench impact the delays of a BIMMF. A more detailed analytical derivation shows that the delays of the outer modes are always larger when there is a trench in the cladding [76]. The trench "squeezes" the outer modes of the fiber, which eliminates the "anomalous" delays that occur in conventional MMF. In terms of power flow, the "anomalous" delays are a consequence of the outer modes having a large fraction of their energy propagating in the cladding. The trench suppresses this energy, eliminates the anomalous delay, and slows down the outer modes so that they can be equalized to those of the other modes propagating in the fiber.

We now focus in more detail on the macrobend performance of BIMMF compared to conventional MMF. We begin with the general two-dimensional scalar wave equation:

$$\left(\frac{1}{r}\frac{d}{dr}r\frac{d}{dr} + \frac{1}{r^2}\frac{\partial^2}{\partial\varphi^2} + \left(k^2(r) - k_2^2\right)\right)\Phi(r,\varphi) = \left(\beta^2 - k_2^2\right)\Phi(r,\varphi), \quad (7.16)$$

where:

$\varphi(r, \phi)$ is the scalar electromagnetic field in the fiber,
m is the positive integer azimuthal number,
β is the propagation constant,
$k_2 = 2\pi n_2/\lambda$ is the wave number in the outer cladding,
λ is the free space wavelength, and
$k(r) = 2\pi n(r)/\lambda$.

All variables in Eq. (7.16) have been transformed to dimensionless quantities. The term $k^2(r) - k_2^2 \equiv -U(r)$ plays the role of the potential, which is related to the refractive index profile, $\Delta(r)$, by

$$-U(r) = 2\left(\frac{2\pi n_2}{\lambda}\right)^2\left(\frac{\Delta(r)}{1 - 2\Delta(r)}\right) = C(\lambda)\left(\frac{\Delta(r)}{1 - 2\Delta(r)}\right). \quad (7.17)$$

The term $\beta^2 - k_2^2 \equiv -E$ in Eq. (7.16) plays the role of energy. This scalar wave equation can be transformed into the one-dimensional Schrödinger equation [77]

$$\frac{d^2\psi(r)}{dr^2} + \left(E - \frac{m^2 - \frac{1}{4}}{r^2} - U(r) \right)\psi = 0, \tag{7.18}$$

where m is the azimuthal quantum number obtained from

$$\frac{\partial^2 K(\varphi)}{\partial\varphi}^2 + m^2 K(\varphi) = 0, \tag{7.19}$$

and the effective potential is

$$V_{\text{eff}}(r) = \frac{m^2 - \frac{1}{4}}{r^2} + U(r). \tag{7.20}$$

We first consider the case of a conventional MMF with a graded index core, without a trench. Modes with small m values are strongly guided in the core of the MMF, and the effective potential forms a deep well with a minimum value below the zero energy level (at which the propagation constant is equal to the wave number in the outer cladding). As the value of the azimuthal quantum number increases, the minimum of $\beta^2 - k_2^2 \equiv -E$ increases and almost reaches the cladding level for the outermost mode groups.

We now add the bend-induced perturbation $Bx = Br\cos\phi$, where B is proportional to the curvature of the bend [80]:

$$\left(\frac{1}{r}\frac{\partial}{\partial r}r\frac{\partial}{\partial r} + \frac{1}{r^2}\frac{\partial^2}{\partial\varphi^2} + (E - U(r) + Br\cos\varphi) \right)\Phi(r,\varphi) = 0. \tag{7.21}$$

If the bend radius is large compared to the radius of the fiber, we can make the approximation $\cos\phi \approx 1 - \phi^2/2$ and rewrite Eq. (7.21) as

$$\left(\frac{1}{r}\frac{\partial}{\partial r}r\frac{\partial}{\partial r} + \frac{1}{r^2}\frac{\partial^2}{\partial\varphi^2} + \left(E - U(r) + Br - \frac{Br\varphi^2}{2} \right) \right)\Phi(r,\varphi) = 0. \tag{7.22}$$

The term $Br\phi^2$ breaks the radial symmetry of the effective potential "seen" by the modes propagating in the fiber. Comparing to the unperturbed fiber, we find that the effective potential in the cladding is now negative for all m values, but the shape is unchanged for small radii. The energy eigenvalues for the high azimuthal modes are now shifted very close to or even above the cladding level, which means they are weakly guided and will have high leakage losses [77].

If we now consider the case of a BIMMF with a trench in the cladding, the negative refractive index results in positive energy barrier in the cladding for the effective potential, $V_{\text{eff}}(r)$. This barrier inhibits tunneling losses, and if it is sufficiently high, the high azimuthal modes that were lossy with the bend perturbation are more tightly bound and will be more strongly guided. The reduction in the bend sensitivity of the outer modes will depend on the size of this tunneling barrier, as well as the strength of the bend perturbation B.

An example of the improvement in bend performance of a BIMMF is plotted in Figure 7.7, which compares the bend loss versus turns around a 10 mm diameter mandrel for two samples of conventional MMF with one sample of BIMMF. All three fibers have core diameter and NA values of approximately 50 μm and 0.2 μm, respectively. When there are only a few turns, the bend loss of the conventional MMF is high because several outer mode groups in fiber are simultaneously being attenuated. The rate of bend loss then decreases with more mandrel wraps as these outer modes are stripped out, and the total bend loss eventually plateaus. In contrast, the bend loss of the BIMMF is an order of magnitude lower and increases almost linearly with an increasing number of turns, indicating that only one or two of the outer mode groups is susceptible to the bend condition.

Both bend loss regimes can be approximated by the sigmoidal function [77],

$$B_m(dB) = 10 * \log_{10}\left[1 - \exp\left(-\frac{N}{\alpha(R, m)}\right)\right],\tag{7.23}$$

where N is the number of turns and $\alpha(R, m)$ is the bend loss parameter for mode group m for bend radius R. The bend loss parameters in Eq. (7.23) can be easily extracted from a few measurements for a given fiber design and bend radius, and this allows an accurate prediction of the bend loss for an arbitrary number of bends at that radius.

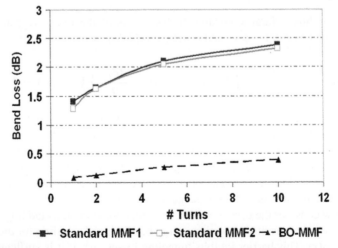

FIGURE 7.7 Measured bend loss versus turns around a 10 mm diameter mandrel.

7.6 CURRENT AND FUTURE DIRECTIONS FOR OPTICAL FIBERS FOR SHORT-REACH APPLICATIONS

7.6.1 Consumer applications and very short-distance networks

Historically, optical communications have penetrated into various segments of communications networks driven primarily by a need to increase transmission speeds, from long-haul and submarine into metropolitan networks, and more recently in the "last mile" connection to individual houses with Fiber to the Home (FTTH) networks [81]. It is becoming apparent that optical communication may find new opportunities in short-reach consumer electronics interconnects or home-networking [82]. Modern residential homes contain an ever-increasing number of consumer electronic (CE) devices, which must be interconnected with each other and to the outside service providers that deliver entertainment content, telecom services, and Internet access. Today, broadband residential services are delivered to homes by coaxial cable, twisted pair, optical fiber, or broadband wireless. In most cases, the signal is terminated in a modem, and a "backbone" in-home network is used to distribute the signal from the modem throughout the house. In addition to the backbone network, various interfaces exist today for a direct device-to-device communication. Examples are USB, HDMI, DisplayPort, FireWire, Thunderbolt, and others.

Figure 7.8 shows the evolution of transfer speed for these two types of networks in the home (backbone and device-to-device). A steady increase in data rates is observed in both, and it is interesting to note that device-to-device speeds are consistently significantly higher than the backbone speeds and therefore might be the first area where fiber is adopted in the future. Since 1995, the data rate for consumer protocols has increased by roughly two orders of magnitude reaching speeds around and above 10 Gb/s. USB 3.0, for example, was released in 2009 at 5 Gb/s, HDMI spans

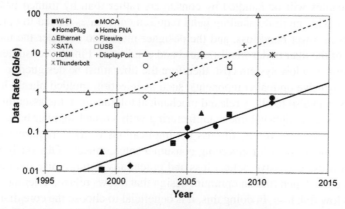

FIGURE 7.8 Data rate evolutions of various protocols commonly used in "backbone" home-networks (closed symbols) and for device-to-device direct communication (open symbol). The lines represent exponential fitting.

a range up to 10.2 Gb/s, and in 2011 Thunderbolt was introduced with two channels each at 10 Gb/s. With the increase in data rate, practical challenges arise for copper-based interconnects due to the higher cable losses at high frequency and crosstalk: maximum transmission distance is shortened, cables and connectors become bulkier due to the additional wires used for parallel multiplexing, and power consumption increases due to equalization and re-timing chips. For example, a USB 3.0 copper cable has two additional differential twisted pairs compared to USB 2.0, and yet its reach was reduced from 5 m to about 3 m [83]. Because of the additional wires, the cable diameter increased to 5–7 mm depending on the construction. The connector plug also became wider to accommodate the extra wires taking up more space on the size of small devices. Thunderbolt [84] electrical cables are also limited to 3 m in reach despite the use of equalization/re-timing chips embedded in the plug connectors (such implementation is referred to as active electrical cable). In comparison, a typical indoor two-fiber optical cable is approximately 3 mm in diameter and can reach much longer transmission distances. Optical fiber is then being considered as a potential alternative to overcome these challenges, either in the form of active optical cables with electro-optical conversion embedded in the connector plug, or in the form of optical ports with the electro-optical conversion embedded in the device.

From the point of view of fiber design, consumer interconnects represent a new application space that has different requirements from traditional applications such as in data centers. First, backwards compatibility with standard fibers is not a strong requirement since optical fiber is not yet widely used in this space: this opens up new options in the fiber design. Link lengths are shorter: typically a few meters for the most common connections between computers and peripherals, and up to 20–30 m for connections to devices such as a ceiling projector in a conference room. This alleviates the need for very high fiber bandwidth, although future upgradeability to higher data rates must be considered. The deployment conditions are more demanding in consumer interconnects: the fiber must withstand much smaller bend radii, since the cables will be handled by consumers rather than by trained professional technicians. This more demanding bend requirement creates a need to optimize the fiber for even lower bend loss, and the designer must also consider the mechanical reliability of the fiber under small-radius bend conditions. Finally, consumer applications require very low system cost, therefore the fiber must be designed to allow the use of inexpensive optical components, such as injection-molded plastic lenses and connectors, and also to allow relaxed mechanical tolerances in the assembly process to permit low-cost high-volume manufacturing with a broad supplier base.

In this section, we briefly discuss the implications of these requirements to the design of optical fibers for emerging consumer interconnects. Obviously misalignments can cause excessive optical loss and potentially compromise the link performance: one must then find an optimum design that allows relaxed tolerances but still maintains low link loss. In doing this, it is beneficial to choose the core diameter and NA of the fiber that minimize the total link loss in the presence of misalignments. In particular, a fiber with larger core and/or higher NA is less sensitive to misalignments at the transmitter and at in-line connectors, but an excessively large core or NA can impair the coupling efficiency at the receiver. For example, the use of Polymer Clad Silica (PCS) fibers with 200 μm core size and NA around 0.37 has been explored for low-cost links, however with relatively low-speed link. A group from IBM [85] presented an analysis using 200 μm core size fibers to enable optical subassemblies and ferrules fabricated with plastic molding technology. Reference [86] is an analysis of a low-cost optical link for automotive applications with simple opto-electronics module and plastic ferrules based on 200 μm core size PCS as well as 1000 μm core size polymer optical fiber.

A more comprehensive analysis was performed for a high-speed link operating at 10 Gb/s [87]. Modeling and experimentation were used to study the link efficiency and determine the optimum fiber core size and NA that minimize the sensitivity to misalignments in a low-cost consumer optimized optical link. The transmitter considered is an 850 nm (VCSEL) whose output is coupled into the fiber by a "photonic turn" element made of polymer material. A similar photonic turn element is used in the receiver with a photodiode of 60 μm diameter aperture capable of 10 Gb/s operation. The link contains three lengths of multimode fiber with identical core diameter and NA, and two Expanded-Beam (EB) connectors that produce an essentially collimated beam, approximately 300 μm in diameter. Compared to direct fiber-to-fiber connectors, the EB connectors greatly reduce sensitivity to lateral misalignment as well as sensitivity to dust contamination [88]. The result is shown in Figure 7.9 as a contour map showing the total link loss (including reflection losses) as a function of fiber core diameter and NA. A marked improvement of link loss occurs when the NA increases up to ∼0.3 and the core diameter increases up to ∼80 μm. Further increase

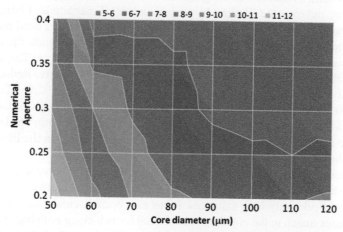

FIGURE 7.9 Simulated optical link loss under misaligned conditions as a function of fiber core diameter and numerical aperture using the set of in-plane misalignments [87]. The color scale indicates the total loss in dB, including the reflection losses. (For interpretation of the references to color in this figure legend, the reader is referred to the web version of this book.)

in the core diameter or NA provides only a marginal advantage because the coupling improvements at transmitter and in-line connectors become marginally better, while at the same time the coupling degradation when focusing the light onto the photodiode increases. The total link loss is approximately 6.2 dB for a fiber with 80 μm core diameter and NA = 0.3, compared to 11.5 dB for a standard multimode fiber with 50 μm core diameter and NA = 0.2. It is interesting to note that while in a perfectly aligned system it is sufficient to have the fiber numerical aperture approximately equal to the source numerical aperture (as transformed by the lenses in the photonic turn element), the presence of misalignments requires higher numerical aperture to maintain efficient coupling into the fiber. This can be understood considering that the fiber has a graded-index profile and as the incident beam is shifted towards the edge of the core due to misalignments; the local index contrast relative to the cladding is reduced compared to a launch at the center of the core.

As mentioned above, robust optical cables are necessary to withstand consumer handling, particularly in transient short-term tight bend conditions. For example, a temporary cable pinch is likely to occur sporadically—where a cable is completely folded onto itself for a short period of time (few seconds to minutes). In this condition, the optical fiber will experience a bend diameter comparable to the cable diameter itself (e.g. ~3 mm, depending on the cable design). Two implications to fiber design arise: (i) the fiber lifetime under bend must be evaluated to ensure survival in transient very small bend diameter, and (ii) the fiber design should have negligible bend loss to ensure an operating link even while the cable is bent. The lifetime of glass optical fiber is determined by *fatigue* growth of flaws present on the fiber surface under a certain level of *applied stress* [89,90]. In such small bend diameter regime (~3 mm), a reduction in the glass diameter reduces the applied stress in bend [91] and can increase the lifetime by several orders of magnitude. Change in the fatigue parameter has little impact on long-term reliability at larger bend radii, however, for fiber experiencing transient very small (~3 mm radius) bends, the increased fatigue resistance may substantially extend the lifetime of the fiber. The impact of different cable designs is also important, but is not discussed here. Reduced glass diameter fibers have been considered for small bend applications [92,93]. From [91], Figure 7.10 shows the lifetime for a reference fiber with 125 μm glass diameter as compared to a fiber with reduced glass diameter of 100 μm. At 3 mm bend diameter, we can see an increase of approximately 4 orders of magnitude in the lifetime, and therefore the fiber can withstand temporary very small bends.

The need for high data transfer rates requires the fiber to have a graded-index core, since a step-index core cannot achieve sufficiently high bandwidth and support >10 Gb/s over relatively long distances. It has also been discussed that the addition of a low-index trench in the cladding is beneficial for achieving both high bandwidth and low bend performance [76]. Figure 7.11 from [87] shows the bend performance for three profile designs: Design A represents a regular 50 μm core diameter with parabolic profile and ~1% delta, design B adds low-index trench at the cladding (delta around −0.4%), and in design C the core is further increased to 80 μm and delta to ~2%. By tuning the width and depth of the low-index trench in design C,

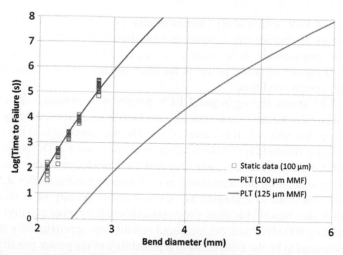

FIGURE 7.10 Solid lines represent lifetime predictions based on power law theory [94,95] for a fiber with reduced 100 μm glass diameter and a fiber with 125 μm diameter in a two-point bend deployment [91]. The points represent direct lifetime measurements [87].

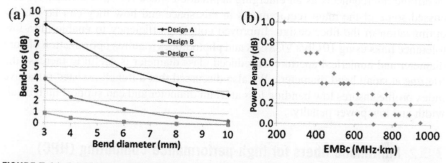

FIGURE 7.11 (a) Measured bend loss for three profile designs. Design A represents a regular 50 μm core diameter with parabolic profile and 1% delta, design B adds low index trench in the cladding and in design C we further increase the core to 80 μm and delta to ~2% while still maintaining the low index trench. The launch condition for these measurements is based on IEC 61280-4-1 (Table E.4); (b) Power penalty measured for various links with different bandwidths [87].

loss on the order of 1 dB in a 3 mm diameter bend was achieved (this example has a trench with about −0.5% delta and about 5 μm width). Fibers based on design C were manufactured and tested in a digital data transmission experiment at 10 Gb/s on a link containing the same components described in the beginning of this section. A Bit Error Rate Tester (BERT) provides the modulation current to the VCSEL and the power penalty introduced by the fiber under test is measured relative to a "back-to-back" BER with a short (approximately 1 m) length of fiber between the two connectors. By introducing a controlled misalignment between the VCSEL and the photonic turn element, it is possible to create different launch conditions, i.e. different distributions of modal power in the fiber, and measure the change of power penalty with launch condition.

Figure 7.11 shows the aggregated BER penalty data obtained on 15 distinct fibers (all with 50 m length), each tested in two launch conditions, namely with perfect alignment and with a 15 μm lateral offset. The horizontal axis is the calculated Effective Modal Bandwidth (EMBc), which is derived from a Differential-Mode-Delay (DMD) measurement on each fiber and the knowledge of the modal power distribution generated by the transmitter in each launch condition. For a 50 m link, a fiber with bandwidth of 232 MHz*km showed power penalty of at most ∼1 dB. Fibers with higher bandwidth show progressively smaller power penalty: for bandwidth exceeding 800 MHz*km, the measured penalty was approximately 0.1 0.1 dB, which is estimated to be the measurement repeatability of the power penalty with our experimental setup. A good correlation between the power penalty and the EMBc is observed, proving that the EMBc metric is a useful predictor of system-level performance.

To summarize, in this section we discussed the evolution of speeds in consumer electronic interconnects as an emerging application space for optical fiber. We discussed some of the main requirements in this space and how they can be met by optimization of the fiber design. Improved coupling efficiency to enable ultra-low tolerance links using 10 Gb/s VCSELs and photodiodes requires optimization of core diameter and numerical aperture. A reduced glass diameter is useful to increase the lifetime at small bend diameters. We also discussed that a trench-assisted refractive index profile ensures low bending loss at 3 mm diameter and can support high bandwidth and low power penalty.

7.6.2 Multimode fibers for high-performance computing (HPC)

K. Bergman reviewed optical interconnections in advanced computing systems in Chapter 19 of volume V B of this series [96], noting an increasing trend toward using multimode optical fibers with relatively low-cost and easy-to-use connectors. They are now prevalent within the top-ranked high-performance computing (HPC) systems [97]. They are also attractive for terabit switches and routers, digital cross-connect systems, and high-end servers. Optical interconnects are seeing use primarily as a way of meeting increased data rate requirements (expressed as MHz.km) and providing a path for improved energy efficiency. The cost of the connections is also an issue because of the vast number of connections, and multimode fibers are typically seen with the optical modules at higher data rates. For example, the peta-scale Jaguar computer at Oak Ridge [97] used a high-performance Infiniband DDR network (see Section 7.1) with over 3000 Infiniband ports.

The incorporation of optics into high-speed computing can be seen as covering multiple generations [98–100]:

a. Optical fibers at edge of the card (for short connections).
b. Optical fibers across the board connecting processors ("backplane").
c. Optical waveguides in or on the boards ("intra-card" or "intra-module").
d. Optical interconnects integrated with the processor ("intra-chip").

Multimode fiber is used in "steps" (a) and (b), with both the number and the data rate of the links increasing with time. The 1 PetaFlop/s Cray Jaguar (ORNL) and IBM Roadrunner (LLNL) systems used DDR links, with 40,000 optical links in the Roadrunner system. The "practical petascale" Blue Waters System is expected to have >2.5 million optical channels running at 10 Gb/s each [100,101].

For these applications current multimode fiber is cabled for parallel connections. Cables currently run 4/10 Gb/s or 10/10 Gb/s allowing 40 Gb/s or 100 Gb/s throughput. The IEEE 802.3bm project will address 4/25 Gb/s for 100 Gb/s. Examples of 15×15 Gb/s and 24×15 Gb/s push the aggregate rate to 225–360 Gb/s [97].

The industry continues to look at ways to increase the data rate capacity or "effective bandwidth" of multimode fiber. The TIA has maintained a summary document outlining various approaches [102]. Since the discussion in reference [1], two items in the tabulation [102] have been commercialized: the OM4 fiber has been developed enabling higher data rates, and parallel multimode fiber has become common as noted above. These two trends will continue. The performance of OM4 fiber is largely limited by the commercially available 850 nm VCSELs, both because 850 nm is not optimal for fiber attenuation or chromatic dispersion, and because it becomes difficult to make low-cost high-speed optical modules with tight tolerances.

One idea which has been proposed [102,103,60] is to modify the index profile of the multimode fiber slightly to compensate for the typical VCSEL's characteristic spectrum (3–4 narrow lines spaced a distance apart). The idea is that if the lines are launched into different radial locations, the contribution of the line spacing to chromatic dispersion can be reduced by adjusting the profile so that the optimum wavelength shifts slightly with radius. For a specific laser with known characteristics a significant improvement in "reach" can be obtained by this optimization; however, this makes the overall profile non-optimal for other lasers with different characteristics (in particular, it limits the significant improvement possible with a single-mode VCSEL [104]) because the coupling pattern is not a specified attribute.

Another opportunity is to use wavelength-division multiplexing (WDM), transmitting simultaneously at multiple wavelengths. It has been noted that a P_2O_5–GeO_2–SiO_2 composition "flattens" the bandwidth versus wavelength curve [105–107] with a single optimum alpha applied over a broader spectral range. An idea proposed by Olshansky [108] and recently revisited [109] further reduces the wavelength sensitivity of the parameter P in Eq. (7.4) and hence the variation of the optimum alpha with wavelength. Olshansky [108] noted that by using two different dopants with different dispersion characteristics, and making a profile with a "dual dopant dual alpha" profile, the intermodal dispersion could be minimized at two wavelengths, not just one, resulting in a broader wavelength capability for the fiber. Reference [109] applies the technique to the GeO_2–F–SiO_2 system. As noted in [101], the technique increases optical complexity (and system cost) as a tradeoff for enabling higher data rates.

Another opportunity, but one not covered in [101], is to use "spatial division multiplexing," sending multiple signals down the same fiber. This can involve launching into individual modes of a single fiber [110–112] or "multicore" fibers with multiple cores distributed in the same fiber [113,114]. The latter appear to offer significant advantages for high data rate, high-density HPC links of the future, and are discussed in detail in the next section.

Optical interconnects are being incorporated into data centers and the eventual trend will be to transfer short length HPC solutions into longer data center solutions, both for data rate and energy savings [115,116].

7.6.3 Multicore fiber for optical interconnect

Interconnect systems for box-to-box, rack-to-rack, board-to-board, and chip-to-chip include hundreds or even thousands of short length links ranging from several hundred meters to tens of meters. Installing and managing such a large number of links is very challenging. Current systems using individual fibers or fiber ribbons are costly, bulky, hard to manage, and not scalable. One of the promising solutions for high-density parallel optical data links is to use multicore fibers (MCFs). The use of a VCSEL array and multicore fiber to realize multichannel transmission was proposed in Ref. [117], where transmission over a 2×2 MCF using direct coupling with a linear VCSEL array at 1 Gb/s was demonstrated. Recently, transmission over a hexagonal 7 core multimode MCF using tapered multicore connectors and 850 nm VCSELs was reported [118]. Long-haul transmissions using hexagonal 7 core single-mode MCFs were also reported [119,120].

Figure 7.12 shows different MCF structure designs. Figure 7.12a is a hexagonal type of design. Figure 7.12b is a 1×4 linear array fiber, and Figure 7.12c is a 2×4 linear array fiber. The fibers in Figure 7.12a–c have a round fiber cladding. A ribbon-shaped MCF design can also be used as shown Figure 7.12d [121].

One most important aspect for MCF designs is the crosstalk among the cores. The crosstalk depends on core refractive index profile designs and the distance between two neighboring cores. One way to model the crosstalk is to use the coupled mode theory [122]. To start, consider two cores separated by a distance D. From the coupled mode theory, if light is launched into core 1, the powers P_1 and P_2 transmitted

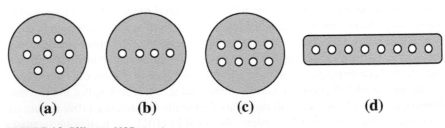

(a)   **(b)**   **(c)**   **(d)**

FIGURE 7.12 Different MCF structures.

in two cores will change sinusoidally. The power crosstalk can be calculated using the following equation:

$$X = 10\log\left(\frac{P_2}{P_1}\right) = 10\log\left(\frac{4\kappa^2}{4g^2\,\text{ctan}(gz) + (\Delta\beta)^2}\right), \tag{7.24}$$

where z is the propagation distance, κ is the coupling coefficient, $\Delta\beta$ is the mismatch in propagation constant between the modes in two cores when they are insulated, and g is a parameter depending on κ and $\Delta\beta$,

$$g^2 = \kappa^2 + \left(\frac{\Delta\beta}{2}\right)^2. \tag{7.25}$$

The crosstalk depends on the coupling coefficient κ that depends on the core design and distance between the two cores, and $\Delta\beta$ depends on the difference in refractive index profile between the two cores. This simple two-core model can provide good guidelines for designing MCFs although a more complicated model is needed to determine the crosstalk more accurately among all the cores in an MCF. To determine the coupling coefficient and the mismatch in propagation constant, we model a two-core waveguide structure using a finite element method by solving the full vectorial Maxwell's wave equations.

Figure 7.13 plots the crosstalk at 1550 nm as a function of distance between the two cores for three fiber designs. The fiber length used in the modeling is 100 m. We examine first crosstalk for two identical cores. Each core has a simple step-index profile, with core $\Delta = 0.34\%$ and core diameter $d = 8.4\,\mu\text{m}$. In this case, $\Delta\beta = 0$, i.e. the two cores are phase matched. The black line in Figure 7.13 shows modeled crosstalk results for two identical single-mode cores. It can be seen that the crosstalk decreases when the distance between the cores gets larger. To have a

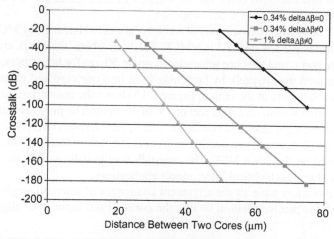

FIGURE 7.13 Crosstalk as a function of core separation.

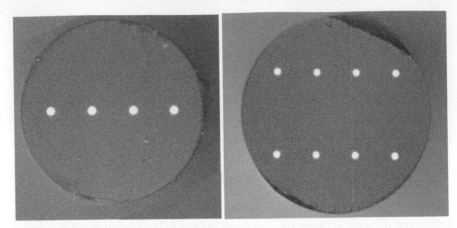

FIGURE 7.14 Pictures of 1 × 4 and 2 × 4 MCFs.

crosstalk less than −35 dB, the two cores have to be separated by more than 54 μm. However, if the two cores are slightly different, i.e. $\Delta\beta \neq 0$, the cores can be placed closer. For example, if we assume that the delta of the second core is changed by 1%, the crosstalk is much lower than the phase matched case. This is shown by the blue curve in Figure 7.13. It is apparent that a core spacing of 28 μm is enough to reach a crosstalk level of −35 dB. The core can be placed even closer if the coupling coefficient can be reduced further. To do this, the power in the core needs to be more confined to reduce the overlap between the cores. One way to increase the power confinement is to increase the core delta. This is shown in the green[2] line in Figure 7.13, where the fiber core profile is still a step, but the core delta is increased to 1%. The core diameter is reduced to 4.9 μm to have a single-mode operation at 1550 nm. In this case, the cores can be situated as close as 20 μm for a crosstalk level of −35 dB.

Figure 7.14 shows cross-sections of recently fabricated linear array MCFs, which we designate as a 1 × 4 and a 2 × 4 MCF. For both fibers, the cores are standard single-mode cores, and the core separation is about 50 μm for both fibers.

We measured the crosstalk by launching light into one core and measured optical power in each core at the output. Figure 7.15 shows the measured results from a 200 m 1 × 4 fiber wound on a fiber reel of 15 cm diameter. The plots are output power profiles with the light launched into cores 1, 2, 3, and 4, respectively. It can be seen that the crosstalk between two neighboring cores is below −45 dB. Similar crosstalk was measured on the 2 × 4 MCF. The crosstalk results indicate that the MCFs are suitable for short reach optical interconnect applications.

MCF connectors have all the alignment challenges of standard single core fiber connectors plus precise rotation so that the cores of the input fiber align with the

---

[2]For interpretation of color in Figure 7.13, the reader is referred to the web version of this book.

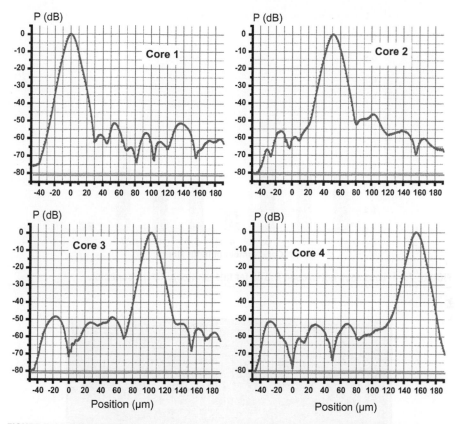

FIGURE 7.15 Measured crosstalk for the 1 × 4 MCF. The results were similar for the 2 × 4 MCF.

cores in the output fiber. In Figure 7.16, the loss for each of the cores was measured by first launching light into one core of the input MCF with the power measured before the connector. Then the connector was mated with the output fiber, and the output power was measured. Care was taken to strip optical power from the cladding by coiling the input and output fibers. The insertion loss is approximately 0.3 dB or less for all but one of the eight cores in this prototype MCF.

Another important attribute for MCF is the relative time required for signals to propagate in the multiple cores of the fiber. Keeping latency low in optical interconnects is increasingly important, especially for HPCs. Similarly for parallel communication links, the relative delay between parallel signals (skew) is also important. In current 10 Gb/s systems, skew below 750 ps is required, but this is expected to become more stringent at higher bit rates. The skew was measured for the 1 × 4 MCF shown in Figure 7.14, and the data shown in Figure 7.17 demonstrates that the maximum delay between cores 2 and 3 is less than 50 ps for a 200 m length of fiber.

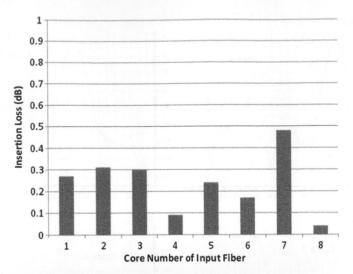

FIGURE 7.16 Loss does not exceed 0.5 dB for this multicore fiber connector.

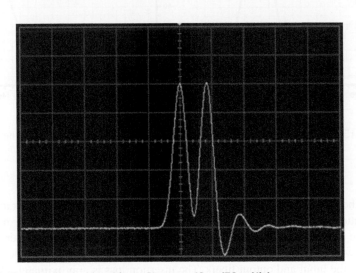

FIGURE 7.17 Skew for 1 × 4 multicore fiber was 42 ps (50 ps/div).

The results were similar for any two cores in this fiber, demonstrating that MCF is an attractive option for increasing the density of short length data networks.

Figure 7.17 plots the maximum measured skew between two of the cores in the 1 × 4 MCF. The measurement simultaneously launched 1550 nm pulses into each core of 200 m of uncoiled MCF. The maximum delay, seen between cores 2 and 3, was 42 ps.

APPENDIX A

See Tables A1–A3.

Table A1 Ethernet data rates over time.

Product	Through-put, Both Directions (MBps)	Line Rate (Gbaud)	IEEE Standard	IEEE 802.3 Revision	IEEE Spec Completed
Ethernet	2.5	0.01	10BASE-	802.3–1985	1985
Fast Ethernet	25	0.10	100BASE-	802.3u–1995	1995
Gigabit Ethernet	250	1.25	1000BASE-	802.3z–1998	1998
10 Gigabit Ethernet	2500	1 × 10.3	10GBBASE-	802.3ae–2002	2002
40 Gigabit Ethernet	10,000	4 × 10.3	40GBBASE-	802.3ba–2010	2010
100 Gigabit Ethernet	25,000	10 × 10.3	100GBASE-	802.3ba–2010	2010

Table A2 Fibre channel data rates: fiber channel roadmap reference: http://www.fibrechannel.org/roadmaps.

Product Naming	Throughput (MBps)	Line Rate (Gbaud)	T11 Spec Technically Completed	Market Availability (Year/est)
1GFC	200	1.0625	1996	1997
2GFC	400	2.125	2000	2001
4GFC	800	4.25	2003	2005
8GFC	1600	8.5	2006	2008
16GFC	3200	14.025	2009	2011
32GFC	6400	28.05	2012	2014
64GFC	12,800	TBD	2015	Market Demand
128GFC	25,600	TBD	2018	Market Demand
256GFC	12,800	TBD	2021	Market Demand
512GFC	25,600	TBD	2024	Market Demand

Table A3 Infiniband data rates: infiniband roadmap (reference: http://www. infinibanda.org/content/pages.php?pg=technology_overview).

Product				Line Rate Designation	Per 1× Line Rate Gb/s	Year
1×	4×	8×	12×			
	20-IB-DDR	40-IB-DDR	60-IB-DDR	DDR	5	2005
10G-IB-QDR	40G-IB-QDR	80GB-IB-QDR	120-IB-QDR	QDR	10	2008
14G-IB-FDR	56G-IB-FDR	112G-IB-FDR	168G-IB-FDR	FDR	14	2011
26G-IB-EDR	104G-IB-EDR	280G-IB-EDR	312G-IB-EDR	EDR	26	2013(est)
HDR	HDR	HDR	HDR	HDR	TBD	Market Demand
NDR	NDR	NDR	NDR	NDR	TBD	Market Demand

References

[1] G. DiGiovanni, S. Das, L. Blyler, W. White, R. Boncek, S.G. Golowich, Chapter 2: design of optical fibers for communications systems, in: I. Kaminow, T. Li (Eds.), Optical Fiber Telecommuncations IVB, Academic Press, New York, 2002.

[2] Index, IEEE 802.3ae 10 Gb/s Ethernet task force. <http://www.ieee802.org/3/ae/index.html>, 2012 (checked 29.07.12).

[3] Index, IEEE 802.3ba 40 G0b/s and 100 Gb/s Ethernet task force. <http://www.ieee802.org/3/ba/index.html>, 2012 (checked 29.07.12).

[4] Index, Next generation 40 Gb/s and 100 Gb/s optical Ethernet study group. <http://www.ieee802.org/3/100GNGOPTX/index.html>, 2012 (checked 29.07.12).

[5] Fibre Channel, Wikipedia summary and links. <http://en.wikipedia.org/wiki/Fibre_Channel>, 2012 (checked 29.07.12).

[6] Fiber Channel Speed Roadmap V13. <http://www.fibrechannel.org/roadmaps>, 2012 (link checked 29.07.12).

[7] Infiniband, Wikipedia summary and links. <http://en.wikipedia.org/wiki/InfiniBand>, 2012 (checked 29.07.12).

[8] Infiniband® Trade Association Home Page. <http://www.infinibandta.org/>, 2012 (chcked 29.07.12).

[9] InfiniBand® Roadmap. <http://www.infinibandta.org/content/pages.php?pg=technology_overview>, 2012 (checked 29.07.12).

[10] Telecommunications Industry Association (TIA) Home Page. <http://www.tiaonline.org/>, 2012 (checked 29.07.12).

[11] International Electrotechnical Commission (IEC) Home Page. <http://www.iec.ch/>, 2012 (checked 29.07.12).

[12] T. Li, Optical Fiber Communications: Volume 1: Fiber Fabrication, Academic Press, New York, 1985 Sections 1.5, 2.3, 3.7.

[13] W. Freude, Vielmodenfasern, in: E. Voges, K. Petermann (Eds.), Optische Kommunications-Technik [German], Springer, Berlin, 2002 (116 Ref).

[14] R. Freund, High-speed transmission in multimode fibers, J. Lightwave Technol. 28 (4) (2010) 569–586 (62 Ref).

[15] D. Marcuse, Principles of Optical Fiber Measurements, Academic Press, New York, 1981.

[16] T. Okoshi, Optical Fibers (English Translation), Academic Press, New York, 1982.

[17] H. Murata, Handbook of Optical Fibers and Cables. second ed., Marcel Dekker, New York, 1996.

[18] R. Olshansky, Propagation in glass optical waveguides, Rev. Mod. Phys. 51 (2) (1979) 341–367 (225 Refs).

[19] A.W. Snyder, J.D. Love, Optical Waveguide Theory, Chapman and Hall, London, 1983.

[20] D.G. Cunningham, W.G. Lane, Gigabit Ethernet Networking, Macmillan Technical Publishing, New York, 1999.

[21] M.C. Nowell, D.G. Cunningham, D.C. Hanson, L.G. Kazovsky, Evaluation of Gb/s laser based fibre LAN links: Review of the gigabit Ethernet model, Opt. Quant. Electron. 32 (2000) 169–192.

[22] P. Pepeljugoski, M.J. Hackert, J.S. Abbott, S.E. Swanson, S.E. Golowich, A.J. Ritger, P. Kolesar, Y.C. Chen, P. Pleunis, Development of system specification for laser-optimized, J. Lightwave Technol. 21 (5) (2003) 1256–1275.

[23] Corning ClearCurve Multimode Fiber Reaches 1 Million Kilometers Sold to Worldwide Customers. Corning Press Release 09 Feb 2012. <http://www.corning.com/news_center/news_releases/2012/2012020901.aspx>.

[24] W.L. Mammel, L.G. Cohen, Numerical prediction of transmission characteristics from arbitrary refractive-index profiles, Appl. Opt. 21 (4) (1982) 699–703.

[25] T.A. Lenahan, Calculation of modes in an optical fiber using the finite element method and EISPACK, BSTJ 62 (9, Part 1) (1983) 2663–2694.

[26] C. Bunge, Gigabit-transmission with multimode fiber, Doctoral Thesis, Tech. Univ. Berlin (Elektrotechnik und Informatik), 2003 (in German).

[27] R. Smink, Optical fibres-analysis, numerical modeling, and optimization, Doctoral Thesis, Tech. Univ. Eindhoven (Electrical Engineering), 2009.

[28] R. Scarmozzino, Numerical techniques for modeling guided-wave photonic devices, IEEE J. Sel. Top. Quant. Electron. 6 (1) (2000) 150–162.

[29] K.S. Chiang, Review of numerical and approximate methods for the modal analysis of general optical dielectric waveguides, Opt. Quant. Electron. 26 (1994) S113–S134.

[30] R.W. Davies et al., Singe-mode optical fiber with arbitrary refractive-index profile: Propagation solution by the numerov method, JLT LT-3 (3) (1985) 619–627.

[31] L. Kaufman, Eigenvalue problems in fiber optic design, SIAM J. Matrix Anal. Appl. 28 (1) (2006) 105–117.

[32] J-C. Baumert, J.A. Hoffnagle, Numerical method for the calculation of mode fields and propagation constants in optical waveguides, J. Lightwave Technol. LT-4 (11) (1986) 1626–1630.

[33] H. Etzkorn, T. Heun, Highly accurate numerical method for determination of propagation characteristics of dispersion-flattened fibres, OQE 18 (1986) 1–3.

[34] E. Kuester, D.C. Chang, Propagation, attenuation, and dispersion characteristics of inhomogeneous dielectric slab waveguides, IEEE Trans. Microwave Theory Tech. MTT-23 (1) (1975) 98–106.

[35] C. Bender, S. Orszag, Advanced Mathematical Methods for Scientists and Engineers, McGraw-Hill, New York, 1978.

[36] W. Streifer, C. Kurt, Scalar analysis of radially inhomogeneous guiding media, J. Opt. Soc. Am. 57 (6) (1967) 779–786.

[37] G. Shaulov, B. Whitlock, Multimode fiber communication system simulation, IEEE 802.3aq Plenary Meeting, Portland, Oregon, July 2004.

[38] A. Ghiasi, PAM-8 Optical Simulations, IEEE Next generation 40 Gb/s and 100 Gb/s optical Ethernet study group interim meeting, Minneapolis, Minnesota, May 2012.

[39] D. Chowdhury, D. Wilcox, Comparison between optical fiber birefringence induced by stress anisotropy and geometric deformation, IEEE J. Sel. Topics Quant. Electron. 6 (2) (2000) 227–232.

[40] R. Olshansky, Pulse broadening caused by deviations from the optimal index profile, Appl. Optics 5 (3) (Mar 1976) 782–788.

[41] A. Carnevale et al., Assessment of index layer structure effects on multimode transmission characteristics evaluated through modal analysis, in: Optical Fiber Communications Conference 1985, Paper TUQ6, San Diego, 1985.

[42] R. Olshansky, D.B. Keck, Pulse broadening in graded-index optical fibers, Appl. Opt. 15 (2) (1976) 483–491.

[43] D. Gloge, E.A.J. Marcatili, Multimode theory of graded-core fibers, BSTJ 52 (9) (1973) 1563–1578.

[44] J.W. Fleming, Dispersion in GeO2-SiO2 glasses, Appl. Opt. 23 (24) (1984) 4486–4493.

[45] R. Olshansky, S.M. Oaks, Differential mode delay measurement, in: Proceedings of the Fourth European Conference on Optical Communication, Genova, Italy, 1978, p. 128.

[46] Telecommunications Industry Association, FOTP-220: Differential mode delay measurement of multimode fiber in the time domain, *TIA-455-220-A*, January 2003.

[47] D. Schicketanz, Fitting of a weighted gaussian lowpass filter to the transfer function of graded-index fibres to reduce bandwidth ambiguities, Elec. Lett. 19 (1983) 651–652.

[48] J.S. Abbott, Characterization of multimode fiber for 10+ Gbps operation by predicting ISI from bandwidth measurement data, in: Proc. NIST Symposium on Optical Fiber Measurements, September 2004, pp. 175–177.

[49] K. Balemarthy et al., Fiber modeling resolution and assumptions: Analysis, data, and recommendations, IEEE 802.3aq10GBASE-LRM Task Force Plenary, November 2004, San Antonio, Texas.

[50] J.M. Senior, Optical Fiber Communications: Principles and Practice. Second ed., Prentice-Hall, New York, 1992.

[51] Telecommunications Industry Association, FOTP-204: Measurement of bandwidth on multimode fiber, *TIA/EIA-455-204*, December 2000.

[52] J. Saijonmaa et al., Selective excitation of parabolic-index optical fibers by Gaussian beams, Appl. Opt. 19 (14) (1980) 2442–2452.

[53] L. Jeunhomme, J.P. Pocholle, Selective mode excitation of graded index fibers, Appl. Opt. 17 (3) (1978) 463–468.

[54] G.K. Grau et al., Mode excitation in parabolic fibres by Gaussian beams, AEU Band 31 Heft 6, 1980, pp. 259–265.

[55] TIA/EIA-492AAAC, TIA/EIA-492AAAD detail specification.

[56] J. Abbott, N. Fontaine, M. Hackert, S. Swanson, P. Pepeljugoski, Improving the specification for 850 nm laser optimized 50 μm fiber, TIA FO-2.2.1 Working Group on Modal Dependence of Bandwidth, June 24, 2002 Meeting (Kiawah Island, SC).

[57] R.S. Freeland, Advanced BER analysis of a bend-insensitive fiber, in: Proceedings of the 58th IWCS Conference, 2009, pp. 468–475.

[58] A. Gholami et al., A complete physical model for gigabit Ethernet optical systems, in: Proceedings of the 57th IWCS Conference, 2008, pp. 289–294.

[59] R. Pimpinella et al., Correlation of BER performance to EMBc and DMD measurements for laser optimized multimode fiber, in: Proceedings of the 56th IWCS Conference, Lake Buena Vista, Florida, November 2007 (Paper 5–5).

[60] R. Pimpinella et al., Dispersion compensated multimode fiber, in: Proceedings of the 60th IWC Conference, 2011, pp. 410–418.

[61] D. Molin, et al., Chromatic dispersion compensated multimode fibers for data communications, in: Proceedings of the 60th IWCS Conference, 2011, pp. 419–423.

[62] A. Huth et al., Comparison of different methods for determination of 10 Gbit link length of laser optimized multimode fibers, in: Proceedings of the 55th IWCS Conference, 2006, pp. 258–260.

[63] C. Hanson, R. Hanson, Solving Least Squares Problems, SIAM, Philadelphia, 1974.

[64] J. Abbott, Modal excitation of optical fibers: estimating the modal power distribution, TIA 2.2tg Draft Notes, June 15, 1998. <http://www.ieee802.org/3/aq/public/upload/2004.html>.

[65] J. Abbott, Modal excitation of optical fibers: Initial results: Calculation of modal power distribution, TIA 2.2tg Draft Notes, June 25, 1998. <http://www.ieee802.org/3/aq/public/upload/2004.html>.

[66] P. Pepeljugoski et al., Modeling and simulation of next-generation multimode fiber links, J. Lightwave Technol. 21 (5) (May 2003) 1242–1255.

[67] D. Hanson, D. Cunningham, P. Dawe, D. Dolfi, 10 Gb/s link budget spreadsheet (version 3.1.16a), Excel spreadsheet on IEEE website. <http://www.ieee802.org/3/ae/public/index.html>.

[68] P. Pepeljugoski, 10GEPBud3_1_16a_25G with MPN changes for web, Excel spreadsheet on IEEE Website, Jan 10, 2012 MM Ad Hoc teleconference for Next Generation 40 Gb/s and 100 Gb/s Optical Ethernet Study Group. <http://www.ieee802.org/3/100GNGOPTX/public/mmfadhoc/meetings/index.html>.

[69] J. King, J. Tatum, 32G fibre channel modeling, Fibre Channel Presentation T11/11-241v0, 3 June 2011.

[70] C.F. Lam, W.I. Way, Optical Ethernet: Protocols, management, and 1-100G technologies, Optical Fiber Telecommunications V B: Systems and Networks, Elsevier, New York, 2008 (Chapter 9).

[71] R. Scarmozzino, Simulation tools for devices, systems, and networks, Optical Fiber Telecommunications V B: Systems and Networks, Elsevier, New York, 2008 (Chapter 20)

[72] W.C. Hurley, T.L. Cooke, Bend-insensitive multimode fibers enable advanced cable performance, Proceedings of the 58th International Wire and Cable Symposium, 2009, p. 450.

[73] G.A.E. Crone, J.M. Arnold, Anomalous group delay in optical fibres, Opt. Quant. Electron. 12 (1980) 511–517.

[74] J. Abbott et al., Bend resistant multimode optical fiber, US20090154888.

[75] D. Molin et al., Trench-assisted bend-resistant OM4 multi-mode fibers, Proceedings of the 59th International Wire and Cable, Symposium, 2010, p. 439.

[76] O. Kogan et al., Design and characterization of bend-insensitive multimode fiber, Proceedings of the 60th International Wire and Cable, Symposium, 2011, p. 154.

[77] S. Bickham et al., Theoretical and experimental studies of macrobend losses in multimode fibers, Proceedings of the 58th International Wire and Cable, Symposium, 2009, p. 458.

[78] P. Pepeljugoski et al., Modeling and simulation of next-generation multimode fiber links, J. Lightwave Technol. 21 (2003) 1242–1255.

[79] S.E. Golowich et al., A new modal power distribution measurement for high-speed short-reach optical systems, J. Lightwave Technol. 22 (2004) 457.

[80] M.B. Shemirani et al., Principal modes in graded-index multimode fiber in presence of spatial- and polarization-mode coupling, J. Lightwave Technol. 27 (2009) 1248–1261.

[81] R.E. Wagner et al., Fiber-based broadband-access deployment in the United States, J. Lightwave Technol. 24 (2006) 4526–4540.

[82] S. Ten, In home networking using optical fiber, in: National Fiber Optic Engineers Conference NFOEC, Optical Society of America, 2012, p. NTh1D. 4.

[83] B. Dunstan, USB 3.0 architecture overview, in: SuperSpeed USB Developers Conference, 2011.

[84] Intel Corporation, Thunderbolt technology—technology brief, http://www.intel.com/content/dam/doc/technology-brief/thunderbolt-technology-brief.pdf, 2012.

[85] J.M. Trewhella et al., Evolution of optical subassemblies in IBM data communication transceivers, IBM J. Res. Dev. 47 (2003) 251–258.

[86] T. Kibler et al., Optical data buses for automotive applications, J. Lightwave Technol. 22 (2004) 2184–2199.

[87] P. Dainese et al., Novel optical fiber design for low-cost optical interconnects in consumer applications, Opt. Express 20 (2012) 26528–26541.

[88] J.C. Baker, D.N. Payne, Expanded-beam connector design study, Appl. Opt. 20 (1981) 2861–2867.

[89] B.R. Lawn, T.R. Wilshaw, Fracture of Brittle Materials, Cambridge University Press, London, 1975.

[90] S.T. Gulati, Crack kinetics during static and dynamic loading, J. Non-Cryst. Solids 38–39 (1980) 475–480.

[91] M.J. Matthewson et al., Strength measurement of optical fibers by bending, J. Am. Ceram. Soc. 69 (1986) 815–821.

[92] R. Sugizaki et al., Small diameter fibers for optical interconnection and their reliability, in: Proceedings of the 57th International Wire & Cable, Symposium, 2008, pp. 377–381.

[93] M. Ohmura, K. Saito, High-density optical wiring technologies for optical backplane interconnection using downsized fibers and pre-installed fiber type multi optical connectors, in: Optical Fiber Communication Conference (OFC), Optical Society of America, 2006, p. OWI71.

[94] R.J. Charles, Static fatigue of glass: I, II, J. Appl. Phys. 29 (1958) 1657–1662.

[95] A.G. Evans, S.M. Wiederhorn, Proof testing of ceramic materials. An analytical basis for failure predictions, Int. J. Frac. 10 (1974) 379–392.

[96] K. Bergman, Optical interconnection networks in advanced computing systems, Optical Fiber Telecommunications V B: Systems and Networks, Elsevier, New York, 2008 (Chapter 19)

[97] F.E. Doany et al., Terabit/s-class optical PCB links incorporating 360 Gb/s bidirectional 850 nm parallel optical transceivers, J. Lightwave Technol. 30 (4) (2012) 560–571.

[98] National Center for Computational Sciences. Oak Ridge, TN Aug 2012 (Online). <http://ww.nccs.gov/jaguar/>.

[99] B.J. Offrein, Optical Interconnects for Computing Appications, Swisslasernet Workshop, IBM Zurich, October 2010.

[100] A.F. Benner, Optical Interconnects for HPC, OIDA Roadmapping Workshop, Stanford, CA, April 2011.

[101] C. Show, F. Doany, J. Kash, Get on the optical bus, IEEE Spectrum September 2010 (online). <http://spectrum.ieee.org/semiconductors/optoelectronics/get-on-the-optical-bus/0>.

[102] Telecommunications Industry Association (TIA), High date rate multimode transmission techniques, TIA Technical Services Bulletin TIA TSB-172, 2007.

[103] A. Gholami et al., Compensation of chromatic dispersion by modal dispersion in MMF- and VCSEL- based gigabit Ethernet transmissions, IEEE Photon. Technol. Lett. 21 (10) (2009) 645–647.

[104] A. Gholami, Physical modeling of 10 GbE optical communications systems, J. Lightwave Technol. 29 (1) (2011) 115–123.

[105] M.G. Blankenship et al., High phosphorus containing P205-GeO2-SiO2 optical waveguide, Proceedings of the OFC, 1979, Washington DC.

[106] E.E. Basch, Optical-Fiber Transmission, H.W. Sams, Indianapolis, 1987.

[107] I. Kaminow, D. Marcuse, H. Presby, Multimode fiber bandwidth: Theory and practice, Proc. IEEE 68 (10) (1980) 1209–1213.

[108] R. Olshansky, Multiple-α index profile, Appl. Opt. 18 (5) (1979) 683–689.

[109] P. Matthijsse et al., On the design of wide bandwidth window multimode fibers, Proceedings of the 54th IWCS, 2005, pp. 332–337.

[110] H. Bulow et al., Spatial mode multiplexers and MIMO processing, 17th OptoElectronics and Communications Conference (OECC 2012), Paper 5SE4-1, July 2012.

[111] J. Carpenter, T. Wilkinson, All optical mode-mixing using holography and multimode fiber couplers, J. Lightwave Technol. 30 (12) (2012) 1978–1984.

[112] B. Franz, et al., Mode group multiplexing over graded-index multimode fiber, Proceedings of the 2012 14th International Conference in Transparent Optical Networks (ICTON), Paper Th.A.1.3, 2012.

[113] M-J. Li et al., Multicore fiber for optical interconnect applications, 17th OptoElectronics and Communications Conference (OECC 2012), Paper 5SE4-2, July 2012.

[114] B.G. Lee et al., End-to-end multicore multimode fiber optic link operating up to 120 Gb/s, J. Lightwave Technol. 30 (6) (2012) 886–892.

[115] C. Kachris et al., The rise of optical interconnects in data centre networks, Proceedings of the 2012 14th International Conference in Transparent Optical Networks (ICTON), Paper We.C2, 2012.

[116] A. Benner, Optical interconnect opportunities in supercomputers and high end computing, OFC 2012 Tutorial OTuB.4.

[117] B. Rosinski, J.D. Chi, P. Grosso, J.L. Bihan, Multichannel transmission of a multicore fiber coupled with VCSEL, J. Lightwave Technol. 17 (5) (1999) 807–810.

[118] B. Zhu, T.F. Taunay, M.F. Yan, M. Fishteyn, G. Oulundsen, D. Vaidya, 70-Gb/s multicore multimode fiber transmissions for optical data links, IEEE Photon Technol. Lett. 22 (22) (2010) 1647–1649.

[119] J. Sakaguchi, Y. Awaji, N. Wada, A. Kanno, T. Kawanishi, T. Hayashi, T. Taru, T. Kobayashi, M. Watanabe, 109-Tb/s ($7 \times 97 \times 172$-Gb/s SDM/WDM/PDM) QPSK transmission through 16.8-km heterogeneous multi-core fiber, in: Proceedings of the OFC, Los Angeles, CA, USA, 2011, Paper PDPB6.

[120] B. Zhu, T. Taunay, M. Fishteyn, X. Liu, S. Chandrasekhar, M. Yan, J. Fini, E. Monberg, F. Dimarcello, Space-, wavelength-, polarization-division multiplexed transmission of 56-Tb/s over a 76.8 km seven core fiber, in: Proceedings of the OFC, Los Angeles, CA, USA, 2011, Paper PDPB7.

[121] Ming-Jun Li, Brett Hoover, Vladimir N. Nazarov, Douglas L. Butler, Multicore fiber for optical interconnect applications, in: 17th Opto-Electronics and Communications Conference (OECC2012) Technical Digest, Paper 5E4-2, Busan, Korea, July 2012.

[122] D. Marcuse, Theory of Dielectric Optical Waveguide. Second ed., Academic Press, 1991.

[123] J.S. Abbott, Characterization of multimode fiber for operation, in: NIST Symposium on Optical Fiber Measurements, 2000, pp. 29–34.

Few-Mode Fiber Technology for Spatial Multiplexing

David W. Peckham, Yi Sun, Alan McCurdy, and Robert Lingle Jr.

OFS, 2000 Northeast Expressway, Norcross, GA 30071, USA

8.1 MOTIVATION

8.1.1 Background

Recently, work on fibers designed to support space-division multiplexing (SDM) has been reported, for example see [1–4]. This work has generally focused either on fibers that contain multiple cores with weak coupling between the cores or on fibers with a single core that supports the propagation of a few modes. In addition though, work has also been reported on multicore fibers that have closely spaced cores with strong coupling between the cores [5–7]. This interest in SDM is due to the impending "capacity crunch," in which the fundamental, nonlinear Shannon limit to increasing the spectral efficiency of fiber optic transmission [8,9] will force carriers to deploy fiber cables at an accelerating rate, rather than simply deploying faster transmitters at decreasing marginal cost-per bit, thus destroying the economics of the backbone network. A rich new medium with $100\times$ to $1000\times$ times the capacity of standard single-mode fiber (SSMF) would be required. Few-mode fiber (FMF) technology combined with multicore fiber technology might create such a medium. This chapter will review the single-core FMF work, also briefly touching upon the concept of supermodes using strongly coupled multicore fiber.

Work on mode diversity multiplexed transmission on multimode fibers (MMF), e.g. 50 μm core diameter, 1% core relative delta fibers that support low loss propagation of 20 or more mode groups, has been studied for several years [10–13]. With mode diversity multiplexed transmission, independent transmission channels are established by selectively launching light into particular mode group(s) of a MMF and then subsequently detecting the data streams after transmission through the fiber. Some of the fiber related limitations of this SDM scheme are the relatively high loss of MMF, the strong mode coupling that results from the close spacing of the propagation constants of the mode groups, and the high differential mode attenuation (DMA) of the fibers. Transmission distance has been limited to less than 10 km so far.

Work on dual-mode fibers used to study transient effects of signal transmission in MMF [14] and for use as a large-core alternative to single-mode transmission fibers [15–17] was first reported as early as 1978. The dual-mode fibers were designed so that the operating wavelength was located between the theoretical cutoff wavelength

Optical Fiber Telecommunications VIA. http://dx.doi.org/10.1016/B978-0-12-396958-3.00008-1

and effective cutoff wavelengths of the third mode, i.e. the LP_{21}, so that low bending loss propagation of the first two modes was obtained. To produce high transmission bandwidth these fibers were designed to have low differential group delay (DGD) between the fundamental mode and the first higher-order mode. Obtaining low mode coupling was not an issue with the intended application. Fibers with step-index and graded-index core shapes were studied and fabricated.

8.1.2 Modern few-mode fiber design objectives

Recently, work has been reported on fibers that support the propagation of a few modes for use in SDM transmission. The effects of nonlinear propagation [18,19], mode coupling [20], and modal birefringence [21] in FMF have been studied. Example two-mode fibers using step-index core shapes [22,4] and graded-index (GRIN) core shapes [3] have been reported. For use in high capacity SDM transmission it is desirable that the waveguide:

- supports the low loss propagation of N unique modes, where N is at least 2 and possibly 10–20. Here low loss is considered to be that of conventional single-mode fiber;
- has low differential mode attenuation (DMA), for example less than about 0.02 dB/km. DMA is a fundamental, uncorrectable impairment that limits the capacity of transmission based on multiple-input, multiple-output (MIMO) signal processing [23,24];
- provides low DGD between all of the low loss modes so that the receiver design can be simplified. To support 1000 km transmission with ASIC technology that may be realizable on a 10-year timeframe, the accumulated DGD of a FMF transmission line probably needs to be equalizable with perhaps hundreds of T/2-spaced complex taps [25] for a time domain equalizer. This represents a technological, but not a fundamental, limitation. The relationship between fiber DGD and accumulated DGD will be discussed below. In a contrary view, large DGD may have the beneficial impact of reducing nonlinear crosstalk between modes;
- optimizes the strength of distributed mode coupling. It has been proposed that low mode coupling in the fiber will minimize the complexity of MIMO crosstalk mitigation hardware. In a contrary view, strong mode coupling has the benefit of minimizing the accumulation of DGD with distance as well as minimizing the impact of DMA in the fiber and mode-dependent gain in the amplifier (see Chapter 34 in this volume);
- provides a low level of transmission penalty caused by nonlinear propagation impairments, including maximizing the effective areas of the low loss modes;
- can be cost effectively realized with state-of-the-art fiber fabrication techniques.

It will be noted that alternative suggestions have been put forward as to the most beneficial properties for FMFs. In the following sections we will discuss FMF design strategies for step-index and graded-index fibers and consider the inevitable trade-offs that will be made in trying to achieve a design that meets any set of objectives.

8.2 MODAL STRUCTURE OF FIBER DESIGNS

8.2.1 Linearly polarized modes

We start with a discussion of the modes of step-index optical fibers. We will employ the weakly guiding approximation and consider the linearly polarized modes derived by Gloge [26] when considering the design of few-mode fibers. The following is a brief survey of some important topics in the theory of modal propagation in optical fibers. A full treatment of the theory of modal propagation in an optical fiber is beyond the scope of this chapter and the authors refer the reader to Gloge's paper and any of the textbooks that cover optical fibers (for example see [27]).

Consider a circularly symmetric optical fiber with cladding of infinite radial extent and radially varying index of refraction as shown in Figure 8.1. The index of the cladding is given by n_{clad} and the index of the core at $r=0$ is n_{core}. The index within the core, $n(r)$, at radial position r is given by

$$n(r) = n_{clad} + n_{core} * \left[1 - \left(\frac{r}{a} \right)^{\alpha} \right] \quad \text{for } r \leqslant a, \tag{8.1}$$

where a is the core radius, α is the core shape parameter. The ideal step-index core shape occurs when α becomes infinite.

It can be shown that the effective index, β/k, of a mode guided by this waveguide structure must satisfy the inequality

$$n_{clad} < \beta/k < n_{core}, \tag{8.2}$$

where β is the propagation constant of the mode and $k=2\pi/\lambda$ is the propagation constant of a plane wave in free space (for example see [28]). When the effective index is

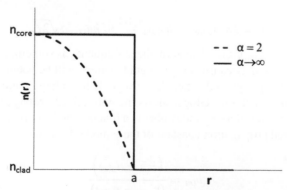

FIGURE 8.1 Index of refraction for step-index ($\alpha = \infty$) and parabolic ($\alpha = 2$) core shapes.

greater than the cladding index the solutions for the transverse fields in the cladding region are radially evanescent and therefore the modal energy is confined within the waveguide structure and the mode is referred to as a guided mode. A mode is said to be cut off when its effective index is equal to the cladding index since the solutions for the transverse fields in the cladding are oscillatory, rather than evanescent, and energy is carried away from the fiber axis. In general it is desirable for a mode to have effective index far above the cladding index since this results in rapid decay of the evanescent field in the cladding, and it being less susceptible to bending losses.

Gloge showed when the weakly guiding assumption holds, i.e. when $(n_{core} - n_{core}) \ll 1$, then the waveguide properties can be accurately approximated by linearly polarized modes that have no longitudinal field components, i.e. the polarization is in the plane transverse to the fiber axis. The fields and characteristic equation of the linearly polarized modes can be described by simple analytic formulas that simplify calculation of the waveguide properties. The properties of the LP modes are a good approximation of those of the real modes of weakly guiding fibers over a wide range of conditions. For these reasons, the LP mode analysis is often used when considering typical optical fibers used in optical communications systems.

The LP modes correspond to near-degenerate groups which may include the EH, HE, TE, and TM modes given by the more general analysis that does not make use the weakly guiding approximation. For the LP modes with no azimuthal variation of the fields, i.e. the azimuthal mode number is zero, the LP modes are comprised of two degenerate modes; the two polarizations of the HE_{1x} modes. For the LP modes with azimuthal variation of the fields, i.e. the azimuthal mode number is greater than zero, then the LP modes are comprised of four nearly degenerate modes; a set of HE, EH, TE, and TM modes. The LP_{lm} nomenclature is generally used to name the individual linearly polarized modes. Here the azimuthal and radial mode-numbers are given by l and m, respectively. The lowest order LP_{01} mode is often referred to as the fundamental mode and corresponds to the two polarizations of the HE_{11} mode. The first higher-order mode, the LP_{11} mode, is comprised of the two polarizations of the HE_{21} mode and the TM_{01} and TE_{01} modes, i.e. four nearly degenerate "real" modes.

The normalized frequency of a step-index fiber is defined as

$$V = ka(n_{core}^2 - n_{cladd}^2)^{1/2} \approx kn_{core}a\sqrt{2\Delta}, \tag{8.3}$$

where $\Delta \approx (n_{core} - n_{clad})/n_{clad}$. The normalized frequency is sometimes referred to as the waveguide strength because any given guided mode will be better confined to the core, i.e. more strongly guided, when the waveguide has a larger value of V.

Figure 8.2, modeled after Gloge, shows the normalized propagation constant of the guided LP_{lm} modes of a step-index fiber as a function of the normalized frequency, V. The normalized propagation constant of the ij mode, b_{ij}, is defined as

$$b_{lm} = \frac{\left(\frac{\beta_{lm}}{k} - n_{clad} \right)}{(n_{core} - n_{clad})} \tag{8.4}$$

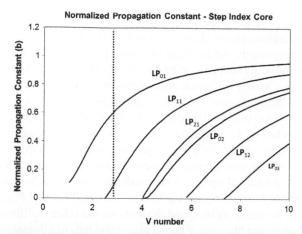

FIGURE 8.2 Normalized propagation constant, b_{lm}, as a function of V for LP_{lm} modes of a step-index fiber. Vertical (dashed) line is located at $V = 3.15$ where the DGD between the LP_{11} and LP_{02} modes is zero.

when V is less than 2.405, then only the fundamental LP_{01} satisfies the condition that the effective index is greater than n_{clad} and therefore the fiber is single moded. When V is greater than 2.405 then additional modes satisfy the propagation condition and the fiber supports the propagation of more than one LP mode.

When designing the index profile of a single-mode fiber it is usual to place the V value slightly greater than 2.405 at the shortest operating wavelength, say $V \sim 2.8$. Even though the fiber can theoretically support the propagation of the LP_{11} mode, the effective index of the LP_{11} mode is very low and the loosely bound LP_{11} mode is susceptible to excess loss caused by bending and waveguide imperfections. With a fiber of this design under practical deployment conditions, the LP_{11} is effectively cut off because of the excess losses that result from bending. This design trick of operating the waveguide at V-number slightly above the cutoff V-number results in a "stronger waveguide" and therefore the fundamental mode has better mode confinement and lower susceptibility to bending loss than would be otherwise possible. This same design approach can be used when designing FMFs.

8.2.2 Distributed mode coupling

As noted previously, it is desirable for FMFs to have low mode coupling between the modes that will be used for SDM multiplexing to minimize the crosstalk between the multiplexed data streams. An additional requirement is that the highest order mode used in the SDM scheme have low mode coupling to even higher order guided, leaky, or radiation modes, since energy coupled to these modes results in energy loss.

The field shapes of the guided modes of an *ideal* fiber satisfy an orthogonality condition and therefore energy does not couple between the modes. However, in a real fiber the orthogonality can be broken by imperfections in the fiber, e.g. inhomogeneities of the index of refraction or deformations of the fiber axis or core size, core noncircularity, etc., which can result in the coupling of energy between the modes. Imperfections in the transmission path or coupling points can cause optical modes to exchange power. This issue can be addressed with MIMO signal processing, but for a good understanding of the FMF properties, one must have a grasp of the potential and implications of mode coupling. For degenerate modes (such as the two polarizations of the LP_{01} which have identical phase constants) the mode coupling is usually strong; that is a substantial optical power will be transferred between the modes within a few tens of meters. In the case of other modes (LP_{11} to LP_{01}, for example) the coupling can be much weaker, and depends on the relative difference in phase constants. In such a case, the optical signal may travel tens of kilometers before there is significant coupling to another mode. Different FMF design strategies can result in either strong or weak mode coupling. Marcuse [29] and Olshansky [30] developed mode coupling theory for imperfect optical waveguides. This work found that energy will couple between two modes when the imperfections have a longitudinal spatial frequency component equal to the difference in the longitudinal propagation constants of the modes, $\Delta\beta$. The strength of the coupling between two modes is a strong function of $\Delta\beta$. Olshansky found that the coupling between modes of adjacent mode groups is proportional to

$$(\Delta\beta)^{-(4+2p)}, \tag{8.5}$$

where p characterizes the power spectrum of the perturbation and typically has values of 0, 1, or 2 depending on the nature of the external stresses, the fiber outer diameter, and coating properties. This result implies that to minimize mode coupling we must maximize the $\Delta\beta$ of the modes.

From Figure 8.2 one can see for a step-index two-mode fiber with V equal to about four that $\Delta\beta$ between the LP_{01} and LP_{11} modes and $\Delta\beta$ between the LP_{11} and the LP_{21} modes are simultaneously maximized. This condition results in low mode coupling between the LP_{01} and LP_{11} modes that are used for SDM and low mode coupling between the LP_{11} and the lossy LP_{21} mode. Similarly, for a four-mode step-index fiber that supports propagation of the LP_{01}, LP_{11}, LP_{21}, and LP_{02} modes, the mode coupling between the four modes will be minimum when the $V \sim 5.5$. However, the coupling between LP_{21} and LP_{02} will always be relatively much stronger than coupling between other pairs of modes.

8.2.3 Differential group delay (DGD)

When the group velocities of the modes that carry independent SDM data channels are different, then pulses that are simultaneously launched into the various modes of the fiber will arrive at the end of the fiber at different times. When mode coupling and DGD are both present then crosstalk between modes can spread across multiple bit periods. The MIMO signal processing electronics that address channel crosstalk in the SDM receiver hardware become more complex when the accumulated DGD between the modes grows and the crosstalk spreads over many bit periods [5]. Therefore for long distance SDM transmission it is desirable to minimize the DGD. Figure 8.3 following Gloge [26] plots the normalized group delay as a function of V for various modes of a step-index fiber. Figure 8.3 shows that the group delay curves of the LP_{01} and the LP_{11} modes cross and the DGD becomes zero when V is approximately equal to 3.15. Note that we found in the previous section that the $\Delta\beta$s of a two-mode step-index fiber are maximized when $V \sim 4$. So for two-mode step-index fibers it is not possible to simultaneously minimize both DGD and mode coupling. For a four-mode fiber, Figure 8.3 shows that a step-index design does not exist where the group delay between all of the lowest order four-modes is zero. Note that for step-index fibers there are values of V where the group delays of a subset of the guided modes are equalized. For example, the group delay of the LP_{02}, LP_{21}, and LP_{12} modes are approximately equal when V is equal to about 6.5. However, when $V \sim 6.5$ the fiber supports three more modes with quite different group delays.

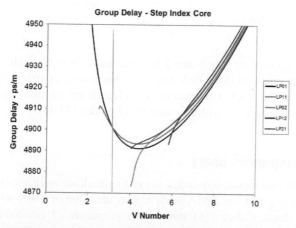

FIGURE 8.3 Group delay as a function of V for the first six LP_{lm} modes of a step-index fiber. The vertical red line is located at $V = 3.15$ where the DGD between the LP_{01} and LP_{11} modes is zero. However, when $V = 3.15$ the normalized propagation constant of the LP_{11} mode is very small and coupling between the LP_{11} mode and leaky modes will be large resulting in DMA. When $V \sim 4.5$, the normalized propagation constants and the difference between the propagation constants of the LP_{01} and LP_{11} are large which gives low sensitivity to mode coupling between the LP_{01} and LP_{11} modes and between the LP_{11} mode and leaky modes. Further, the propagation constant of the LP_{02} and LP_{21} modes are very small so that these modes will be very lossy and therefore only the two lowest order modes propagate with low loss. However, the LP_{01} and LP_{11} mode DGD is quite large when $V \sim 4.5$. (For interpretation of the references to color in this figure legend, the reader is referred to the web version of this book.)

8.2.4 Accumulation of DGD with propagation distance

In addition to the magnitude of the $\Delta\beta$ of the fiber profile design, mode coupling also depends on factors related to the deployment of the fiber. Here cabling and splicing effects need to be considered. Cabling stress will increase distributed mode coupling by providing an additional source of perturbations of the fiber. Splices and connectors provide points of discrete mode coupling.

When small and random mode coupling is considered, it can be shown that the DGD will grow linearly with length for distances much shorter than the correlation length and as the square root of length for long lengths [31]. The two-mode case is completely analogous to the results obtained for PMD [32]. If a short pulse is launched simultaneously in each mode then the variance in arrival times of portions of the pulse is given as a function of fiber length, L:

$$\langle (T - \langle T \rangle_{\text{av}})^2 \rangle_{\text{av}} = \frac{\text{DGD}^2 l_c L}{4} \left[1 - \frac{l_c}{2L}(1 - \exp(-2L/l_c)) \right], \qquad (8.6)$$

$$\lim_{L/l_c \to \infty} \langle (T - \langle T \rangle_{\text{av}})^2 \rangle_{\text{av}} = \frac{\text{DGD}^2 l_c L}{4}, \qquad (8.7)$$

where l_c is the correlation length and T is the time-of-flight through the fiber. Note from the second equation (long fiber limit) that the spread in arrival times scales as the square root of the product of the correlation length and the fiber length. A similar scaling law holds for guides with any number of modes [33,20].

8.2.5 Non step-index fibers

The inability of two-mode, step-index fibers to simultaneously provide low DGD, low mode coupling and low DMA leads to consideration of fibers with more complicated core shape. Cohen [15] studied the properties of two-mode fibers with step and graded-core profiles. It was pointed out in this work that when the core shape parameter α is 2.5 that the group delay curves of the LP_{01} and LP_{11} modes cross when V is ~5.5 and that the fiber is effectively two-moded. Figures 8.4 and 8.5 show curves of the normalized propagation constant and group delay, respectively, for the first six LP modes of a parabolic ($\alpha=2.00$) core fiber as a function V. For non step-index fibers, i.e. $\alpha \neq \infty$, we define V with Eq. (8.3) as previously defined for step-index fibers. When α is finite, the fiber is single moded when $V < 2.405 \cdot (1+2/\alpha)^2$ [34]. For example, cutoff occurs when $V=3.40$ for a parabolic profile. The core shape parameter α can be chosen to minimize DGD at a particular wavelength. Figure 8.4 illustrates that when $V\sim6$, the difference between the propagation constants of the LP_{01} and LP_{11} are large which gives low sensitivity to mode coupling between the LP_{01} and LP_{11} modes. Also when $V\sim6$ the

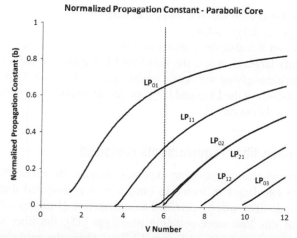

FIGURE 8.4 Normalized propagation constants for the first six modes of a parabolic ($\alpha = 2.00$) core fiber. When $V \sim 6$, the normalized propagation constants and the difference between the propagation constants of the LP_{01} and LP_{11} are large which give low sensitivity to mode coupling between the LP_{01} and LP_{11} modes and between the LP_{11} mode and leaky modes. Further, the propagation constant of the LP_{02} and LP_{21} modes is very small so that these modes will be very lossy and therefore only the two lowest order modes propagate with low loss. Also, when $V \sim 6$ and the difference between the LP_{01} and LP_{11} group delays is low.

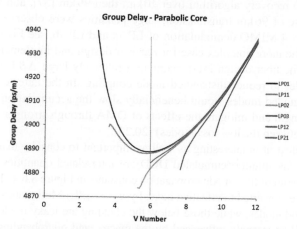

FIGURE 8.5 Group delay curves for six modes of a parabolic ($\alpha = 2.00$) core fiber. The difference between the group delay curves for the LP_{01} and LP_{11} modes is small for $V \sim 6$.

normalized propagation constant of the LP_{11} mode of the parabolic core fiber is quite large which minimizes the coupling of LP_{11} to leaky modes. Further, the propagation constant of the LP_{02} and LP_{21} modes is very small so that these modes will be very lossy and therefore only the two lowest order modes propagate with low loss.

Figure 8.5 show for a parabolic core shape that when $V \sim 6$ the difference between the LP_{01} and LP_{11} group delays is low.

Figures 8.4 and 8.5 also show that when the V of a parabolic core shape fiber has value slightly larger than six, the first four LP modes will have widely spaced propagation constants giving low mode coupling as well as low DGD. Further, the propagation constants of the LP_{02} and LP_{21} modes are maximized while higher order modes are effectively cut off.

8.2.6 Few-mode fiber requirements revisited

Recent fiber design and transmission experiments over few-mode fiber have been conducted with two limits in mind. In one case, it is assumed that mode coupling in an N-mode fiber will be confined to a subset of M modes where $M < N$. In one example of this low mode-coupling paradigm [35], the fiber supported five spatial modes (comprising LP_{01}, LP_{11}, and LP_{21}), where the only strong couplings were between $LP_{11}a$ and $LP_{11}b$ and then between $LP_{21}a$ and $LP_{21}b$. So it was only necessary to implement two 4×4 MIMO recovery algorithms instead of one 10×10 MIMO algorithm. In another paradigmatic case, it is assumed that all N fiber spatial modes mix sufficiently such that full $2N \times 2N$ MIMO recovery of the signals is necessary (where $2N$ accounts for two polarizations for each spatial mode). In Ref. [36] independent data streams were multiplexed onto three independent spatial modes (comprising LP_{01}, LP_{11}) and their x- and y-polarizations were then demultiplexed by a 6×6 MIMO recovery algorithm over 10 km, then 96 km [37], and then 1200 km [5]. In the case of 96 km transmission, large penalties were observed for reduced complexity 4×4 MIMO demodulation of $LP_{11}a$ and $LP_{11}b$, see Figure 13 of Ref. [37]. This is the more complex case for receiver design and implementation. In the former case, the fiber design must maintain a relatively large $\Delta\beta$ between nearest neighbor modes to reduce distributed mode coupling. In the latter case, a smaller $\Delta\beta$ between guided modes would beneficially slow the accumulation of DGD as a function of length and mitigate the effects of DMA through stronger mode-mixing (within and between the low loss modes) [20,23].

In this context, it is interesting as well as important to consider that DMA, mode coupling, and maximum accumulated DGD are interrelated quantities. Consider the schematic diagram of fiber mode propagation constants in Figure 8.6. (The propagation constant can be converted to effective index by $n_{eff} = \beta/k$.) Modes above the cladding index are bound modes, while those below the cladding are leaky modes. Attenuation and DMA will be strongly influenced by the macro- and microbending losses of the modes. Macrobending loss is generally minimized by maximizing β (or equivalently the effective index n_{eff}); keeping β_{min} above some minimum level is critical for minimizing loss in SSMF and DMA in FMF. SSMF is generally designed so that there is a highly lossy, though technically still bound, mode just above the cladding index; that mode is said to be effectively cut off through the high loss. The same principle will hold for step-index or graded-index FMF with no cladding structure: an optimized design would have a mode just above the cladding index as shown as a dashed line in Figure 8.6.

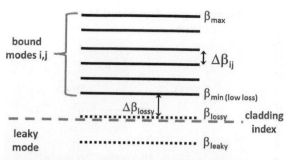

FIGURE 8.6 illustrative diagram of modal propagation constants for a hypothetical step-index or graded-index FMF with no cladding structure. In the case of a structured cladding, the lossy mode may drop below the cladding index and become a leaky mode. It is difficult in practice to engineer the modal spacings such that $\Delta\beta_{lossy} \gg \Delta\beta_{ij}$.

The strength of coupling between bound modes which carry signals and the lossy or leaky modes is governed by Eq. (8.5), leading to microbending loss in the presence of cable stress. Microbending loss for the lowest bound mode is a strong function of $\Delta\beta_{lossy}$, so maximizing $\Delta\beta_{lossy}$ is also a condition for minimizing DMA. In an ideal case for the strong mode-mixing paradigm, all $\Delta\beta_{ij}$ would be small, β_{min} would be relatively large, and $\Delta\beta_{lossy} \gg \Delta\beta_{ij}$. This would result in a fiber where modes couple strongly in pairwise fashion leading to (1) accumulation of DGD which is proportional to the square root of the fiber length (see Eq. (8.7)) even over shorter links of a few hundred km and (2) mitigation of the deleterious impact of DMA [23,38]. Furthermore macro- and microbending of the lowest guide mode would be small, leading to low DMA. In fact, these are difficult conditions to fulfill. It is typical that the spacing between adjacent modes does not vary strongly over a few modes, and there is typically no abrupt change in mode spacing across the cladding index. More commonly $\Delta\beta_{lossy} \approx \Delta\beta_{ij}$. In other words, *it is challenging to design a fiber to promote mixing between multiple low loss bound modes while simultaneously minimizing the loss of the lowest bound mode.*

Since low DMA is a fundamental requirement, we conclude that β_{min} must be kept greater than some threshold for low macrobending and $\Delta\beta_{lossy}$ (typically similar to $\Delta\beta_{ij}$) must be kept large enough to minimize microbending loss. Once these two criteria are fulfilled, there will typically be little flexibility to manipulate the magnitude of $\Delta\beta_{ij}$.

8.2.7 Impact of discrete mode coupling

While two regimes of strong and weak mode coupling for few-mode transmission have been contemplated, there is some doubt that a weak coupling regime will exist in a deployed transmission link. The picture can be clarified by considering that the strength of mode coupling in a FMF transmission line will depend on both distributed and discrete contributions. Figure 8.7 shows a four-quadrant chart illustrating the possibilities.

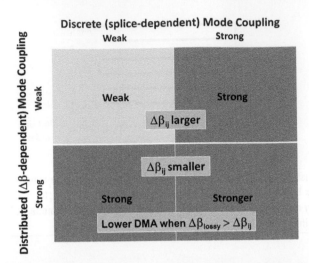

FIGURE 8.7 The strength of mode coupling in a deployed FMF transmission line will depend on both distributed and discrete mode coupling. Transmission in the weakly mode-coupled regime requires that both contributions be weak. If discrete mode coupling at splices, connections, and in-line components is sufficiently strong, then the mode spacings in the fiber $\Delta\beta$ will be of secondary importance for mode-mixing considerations but remain of primary importance for minimizing DMA.

In a deployed fiber cable splices will occur approximately every 5 km, so an 80 km amplified span will contain about 16 splices on average. Furthermore, other components such as wavelength selective switches and optical amplifiers will also be nodes for mode coupling. If the mode coupling at splices is sufficient such that the correlation length l_c is equal to 5–10 cable segments, then transmission will occur in a strongly mode-coupled regime regardless of the strength of distributed mode coupling in the fiber (i.e. regardless of $\Delta\beta_{ij}$). This will have the beneficial result that DGD will accumulate as $\sqrt{L}$ in the link, and mitigate the impact of DMA, but necessitate full $2N \times 2N$ MIMO processing in all cases. However, large $\Delta\beta$ will nonetheless generally give the lowest possible DMA and perhaps always be desirable for this fundamental reason.

At the time of this writing, the impact of splicing modern FMF on mode coupling has not yet been quantitatively determined, although early studies considered loss and mode-mixing at splices of traditional MMF [39,40]. If it be the case that splices, connectors, and components generally lead to the strong mode coupling regime, then the upper right quadrant of Figure 8.7 may prove to be the necessary approach to FMF design, yielding lowest possible DGD and DMA, though necessitating full $2N \times 2N$ MIMO processing at the receiver.

8.2.8 How many modes can a FMF support?

Table 8.1 shows the modal content of the LP modes in terms of the more fundamental HE, TE, and TM modes. The table illustrates that designing a FMF to support,

Table 8.1 Underlying content of LP modes that can be used for coherent communication; to calculate the total number of modes onto which data can be multiplexed, multiply by two to account for the two polarizations for each spatial mode pattern.

LP-Mode Designation	True Mode Content	Number of Degenerate Spatial Modes
LP_{01}	HE_{11}	1
LP_{11}	TE_{01}, TM_{01}, HE_{21}	2
LP_{21}	EH_{11}, HE_{31}	2
LP_{02}	HE_{12}	1
LP_{31}	EH_{21}, HE_{41}	2
LP_{12}	TE_{02}, TM_{02}, HE_{22}	2
LP_{41}	EH_{31}, HE_{51}	2
LP_{22}	EH_{12}, HE_{32}	2

e.g. 10 low loss, orthogonal spatial modes is equivalent to designing for the lowest six LP modes for transmission. Increasing the number of low loss modes requires increasing the V-number. If V is increased by raising the core diameter, then the modes will become more closely spaced, the mode A_{eff} will increase, and $\Delta\beta_{lossy}$ will become smaller leading to higher DMA. If V is increased by increasing the core Δ, then the mode A_{eff} will decrease, Rayleigh scattering losses will increase due to higher concentration of GeO_2, and the modal spacing will increase helping to minimize DMA. A judicious combination of adjusting core Δ and diameter, along with other degrees of freedom in the profile, will be necessary to guide 10–20 modes with low DMA and low attenuation losses.

8.3 FIBER DESIGNS OPTIMIZED FOR FEW-MODE TRANSMISSION

8.3.1 Early dual-mode optical fiber design

Precursors to the FMF designs being explored recently can be found in the early literature of fiber design. Dating back to the late 1970s, there was a need to reduce the splicing loss in long-haul transmission links. Dual-mode optical fiber having a larger core diameter than single-mode optical fiber, without sacrificing bandwidth, was proposed as an alternative to single-mode optical fiber. Unlike modern applications in spatial multiplexing, both the LP_{01} and LP_{11} modes were intended to carry the same signal. The inter-modal dispersion was reduced to increase the bandwidth. This is more or less similar to pursuing a high bandwidth and minimum differential modal delay in multimode optical fiber for modern VCSEL-based links.

The 1310 nm operation window, where the material dispersion is zero, was used to design fibers with zero dispersion difference between LP_{01} and LP_{11} modes.

Examples of the step-index profiles that achieved zero dispersion difference at one V-number (i.e. one wavelength) are shown by both simulations and fiber experiments in [41,42,15].

Gradient-index profiles were explored to increase the V-number where the dispersion difference between two guided modes was zero [15]. Suppression of the center dip in the index profile was found to improve the variation of the delay difference due to the V-number deviation [16]. The core shape parameter α is larger than 2.0 even for an index difference Δ as small as 0.2%. An optimization of α for zero delay difference as a function of Δ for maximum operational V-number suggested that α increases with an increase in Δ (Figure 2 in [16]).

At least one proposal was made to design a virtually dispersion-free, dual mode optical fiber at 1.55 μm [17]. However, the optimum index profile achieved through a simulation feedback routine was different from a simple alpha-parameter profile; instead, it was approximated by a complex analytical expression as shown by (Eq. (19) in [17]):

$$n^2(r) = (n_1^2 - n_2^2)\left\{1 - 4.6R^{3.44} - 0.1\left[e^{-\left(\frac{R}{0.2}\right)^2} - 0.018\right]\right\}, \quad R = \frac{r}{a} \leq 0.6,$$

$$n^2(r) = (n_1^2 - n_2^2)e^{\frac{(R-0.18)^2}{0.34}} - 0.0005, \quad 0.6 < R < 1,$$

$$n^2(r) = n_2^2, \quad R \geq 1,$$

where n_1 and n_2 are the core peak index and cladding index, respectively.

Differential modal attenuation (DMA) was not a concern for any of these designs. None of these designs used a trench structure.

8.3.2 Step-index style FMF for SDM

Several groups have investigated step-index FMF for spatial mode multiplexing. Li et al. [22] designed a two-mode step-index fiber with a core diameter of 11.9 μm, index difference of 5.4×10^{-3}, and loss of 0.26 dB/km. The effective index difference between LP_{01} mode and LP_{11} mode is 2.98×10^{-3}, which indicates a low inter-modal coupling. The DGD between these two guided modes is 3.0 ps/m at 1550 nm. 107-Gb/s dual-mode and dual-polarization coherent optical orthogonal frequency-division multiplexing (CO-OFDM)) over 4.5 km of such two-mode fiber was successfully demonstrated using a mechanically induced LP_{01}/LP_{11} mode converter as the mode-selective element.

Bigot Astrue and coworkers [4,43] studied weakly coupled step-index fibers with two, four, five, six, and seven LP modes intended for transmission in the low-complexity regime for MIMO processing. The design criteria were to set the effective index difference between adjacent LP modes larger than 0.5×10^{-3}, the effective area between 80 μm² and 160 μm², and macrobend losses less than 10 dB/km at 10 mm bend radius. The DGD versus LP_{01} mode was designed to be larger than 0.05 ps/m in order to limit inter-modal nonlinear effects. Table 8.2 from [4] shows the

Table 8.2 Characteristics at 1550 nm of the four LP modes of the FMF described in Ref. [4].

		LP01		LP11		LP21		LP02	
		Calc.	Meas.	Calc.	Meas.	Calc.	Meas.	Calc.	Meas.
$(n_{eff}-n_{cl})$	$\times 10^{-3}$	8.3	–	6	–	3.2	–	2.4	–
DGD versus Lp_{01}	ps/m	–	–	4.4	4.4	8.5	8.9	7.2	7.9
Chromatic Dispersion	ps/nm.km	21.9	21	24.9	26	20.9	19	7.8	8
A_{eff}	μm^2	124	124	118		133		127	
Loss	dB/km	<0.22	0.218	0.215		0.21		0.21	
Cable Cut-off Wavelength	nm	–	–	2980	>1700	1950	>1700	1800	>1700
Macro-bend loss$_{\{10\,mm\;bend\;radius\}}$	dB/turn	<0.005	0.002	<0.01		0.01		1.9	
Polarization Mode Dispersion	ps/km$^{0.5}$	<0.1	0.04						

characteristics of a four-LP mode fiber at 1550 nm. The effective index difference is 0.8×10^{-3} between LP_{21} and LP_{02} modes and larger than 2×10^{-3} between the remaining LP modes. The effective areas of the four LP modes are very similar, ranging from $118\,\mu m^2$ to $133\,\mu m^2$. The loss difference across all LP modes is less than 0.01 dB/km. The DGD target was easily achieved as shown in Table 8.1. 100 Gb/s data stream has been transported over a 40 km FMF using multiplexing on five spatial modes [35].

8.3.3 Variations on step-index fiber designs

DGD-managed waveguides consisting of both positive and negative DGD two-mode fiber have been proposed [44,45]. Maruyama et al. [44] achieved 700 ps/km net DGD over the entire C-band using a step-index two-mode fiber. Sakamoto et al. [45] designed a multi-step-index two-mode fiber shown in Figure 8.8. By adjusting the index difference between the two segments of core index, the DGD can be widely tuned from positive to negative. A net DGD < 111 ps/km was achieved over the entire C-band for a 20.4 km transmission span.

Figure 8.3 shows for a step-index fiber that the DGD between the LP_{01} and LP_{11} modes can be small only when the *V*-number is near 3.15, i.e. where the group delay curves cross. Therefore to obtain low DGD with a step-index profile requires tight tolerances on core delta and core radius to ensure that the *V*-number is close to 3.15.

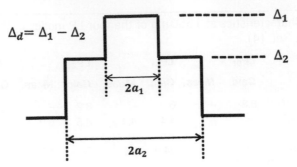

FIGURE 8.8 Index profile of a multi-step two-mode fiber [45].

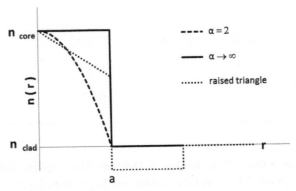

FIGURE 8.9 Raised-triangle, depressed cladding index profile that is optimized for low DGD and ease of manufacturing of two-mode fiber.

Figure 8.3 also shows that the group delay will remain low only over a narrow range of wavelengths. Figure 8.9 shows an index profile referred to as a raised-triangle, depressed cladding design that is optimized to provide two-mode operation with low DGD over a wider range of V than a step-index profile. Figure 8.10 shows the group delay curves calculated for a prototype raised-triangle, depressed cladding fiber fabricated using the VAD process. The group delay curves for the LP_{01} and LP_{11} modes fall very close to one another over a broad range of V-numbers. This behavior maintains low DGD, i.e. less than 100 ps/km over the entire C-band, while using standard fabrication techniques commonly employed for manufacturing single-mode transmission fibers and manufacturing tolerances typical for SSMF. The VAD and rod-in-tube manufacturing techniques were used to fabricate a few hundred kilometers of raised-triangle, depressed-cladding, two-mode optical fiber with low DGD, low DMA and good axial uniformity. The A_{eff} of the LP_{01} and LP11 modes were $155\,\mu m^2$ and $160\,\mu m^2$, respectively. The attenuation of the two-mode fiber was 0.2 dB/km. These prototype fibers were used in several SDM transmission experiments.

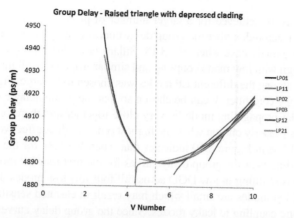

FIGURE 8.10 Low differential group delay between LP$_{01}$ and LP$_{11}$ modes over a wide range of V-numbers for the raised-triangle, depressed cladding profile shape. The DGD of a two-mode fiber design with this profile shape is insensitive to V-number. When V is ~5.31, the group delay curves of the first four modes cross resulting in four-mode operation with low DGD.

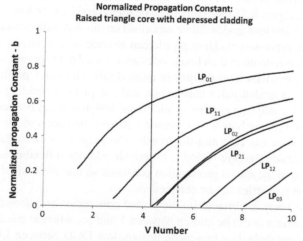

FIGURE 8.11 Normalized propagation constant as a function of V-number of a raised-triangle, depressed cladding design. When V~4.5 (solid vertical line), the waveguide supports the propagation of two modes with low DGD as indicated in Figure 8.10. When V~5.31 (dashed line), the waveguide supports the propagation of four modes with low DGD.

Ref. [37] shows 6×40 G QPSK transmission over 96 km with a penalty of < 1.2 dB using 6×6 feed-forward equalizers of 120 taps each (see Figure 8.11).

The previous example shows that zero DGD can be achieved for two-mode fibers with step-index, parabolic index, and raised-triangle core depressed cladding index shapes.

Can this property be achieved for a larger number of propagating modes? Figure 8.3 shows that for a step-index fiber the group delay curves of all propagating modes cross only for the two-mode case when $V=3.15$. Sillard et al. [4] report a four LP mode step-index design with low mode coupling and similar A_{eff} among all the modes. However, the DGD among the different LP modes was chosen to be large. Figure 8.5 shows that, for parabolic core fiber, V can be chosen so that the group delay curves of all but the highest order propagating mode lie very close together and therefore low DGD is achievable. V is properly chosen when the highest order mode is effectively cut off. The vertical dashed line in Figure 8.10 indicates that when $V \sim 5.31$ for the raised-triangle depressed cladding fiber the group delay curves for the first four LP modes simultaneously cross, thus resulting in low DGD across all four low loss modes. However, β for the LP_{02} and LP_{21} modes are small which may result in elevated sensitivity to macrobends and strong coupling to leaky modes. Since the group delay curves of the raised-triangle, depressed cladding fiber do not simultaneously cross for larger values of V, so this design does not provide low DGD for more than four LP modes.

8.3.4 Two-mode GRIN FMF for SDM

If it be assumed that deployed FMF waveguides will operate in the strong mode-mixing regime (see Figure 8.7), then there must be focus on achieving low DGD among a large number of low loss modes. Since variations on the step-index design such as the raised-triangle, depressed cladding profile can provide only a narrow design space for low DGD when no more than four modes are allowed to propagate, an alternative is to consider the GRIN fiber designs in more detail. The index profile considered here consists of a graded-index core region and a depressed cladding region (i.e. a "trench"). There could be a number of additional design features between the graded core and the trench, such as a shelf region between the core and the trench or an index step between the core and the trench. The purpose of these features to the index profile outside the core region is to provide additional flexibilities to manipulate the spacing of the modal propagation constants so that the desired combination of transmission properties can be obtained.

The simplest way to characterize the graded-core region was shown in Eq. (8.1). The alpha parameter α can be chosen between 1 and ∞, whereas $\alpha=2$ corresponds to an inverted parabola. For two-mode design, low DGD between LP_{01} and LP_{11} modes can be obtained with any α between 1 and ∞ combining proper values of other profile parameters such as n_{core}, r_{core}, trench depth, and position. However, for FMF design beyond two LP modes, α is preferentially chosen close to an inverted parabola shape to achieve low DGD among all LP modes. The trench feature has three functions. As shown in Figure 8.6, $\Delta\beta_{ij}$ should be as large as possible. A trench structure allows to push down β_{min} (LP_{11} mode in two mode) closer to the cladding index while maintaining low loss and push down β_{lossy} below the cladding index to become a leaky mode. The trench also promotes reduced bending loss and differential modal attenuation (DMA) of both LP_{01} and LP_{11} modes. The trench on the periphery of the raised index core also forms an index structure to manipulate DGD, especially of the high order mode(s). Two GRIN FMF designs were evaluated in fiber prototyping and measurements, as shown in detail in the following sections.

8.3.5 GRIN two-mode fiber example 1

Figure 8.12 shows the first example of a GRIN two-mode fiber design. The fiber has a graded-index core with a step down at the core/trench boundary. The relative Δ at the center of the core is about 1%. A trench with a confinement volume (defined by the index multiplied by area) of 0.75 is positioned adjacent to the core. Both LP_{01} mode and LP_{11} mode are well guided, as indicated by the effective indices marked as short bars in Figure 8.12. The effective index difference between the LP_{01} mode and the LP_{11} mode is 0.30%, indicating very low coupling between these two modes. By simulation, ideally, the DGD between LP_{01} mode and LP_{11} mode can be controlled less than $\pm0.001\,ps/m$ (or $\pm1\,ps/km$) within the C-band

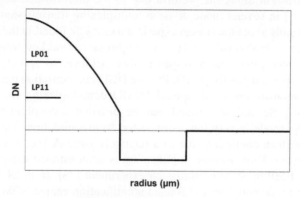

FIGURE 8.12 Index profile of a GRIN two-mode fiber.

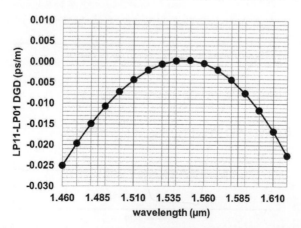

FIGURE 8.13 Differential group delays versus wavelength between LP_{11} mode and LP_{01} mode.

as shown in Figure 8.13. With deviations of the fiber parameters from the design targets stated above, the DGD curve shifts to left and right or up and down, and results in a positive or negative DGD across the C-band. The effective areas (A_{eff}) of the LP_{01} mode and LP_{11} mode are $59.6 \mu m^2$ and $59.8 \mu m^2$, respectively. The chromatic dispersion, calculated from the design index profile, is 18.6 ps/nm-km for both LP_{01} and LP_{11} modes.

The fiber preform was fabricated by using the modified chemical vapor deposition (MCVD) process. The preform is subsequently overclad with silica tubes and drawn into fiber of a diameter of 125 μm using a standard drawing process. The fiber index profile is scanned using an interferometric index profiler shown in Figure 8.14. The attenuation measured by OTDR using a launch from a super-large-effective-area fiber is 0.24 dB/km. The A_{eff} of LP_{01} mode and LP_{11} mode, calculated from the middle section of the preform scanning profile, are $64 \mu m^2$ and $67 \mu m^2$, respectively. The initial prototype yielded a variation of DGD in the order of ± 300 ps/km along the preform due to the preform non-uniformity. The fibers were used in several mode division multiplexing transmission experiments [46–48,25]. Details about the system experiments may be found in the references as well as Chapter xx in this volume. Here we highlight several key points relevant to the two-mode fiber. First, near-zero span DGD is achieved by splicing FMF pieces with both negative and positive DGD. Precise DGD compensation enabled a proof-of-concept demonstration of all-optical MIMO demultiplexing for transmission over 30 km [46]. Second, it is found that the distributed coupling between LP_{01} mode and LP_{11} mode is very low as expected from simulation. Third, this FMF has a higher Raman coefficient due to a relatively large Δ (i.e. a relatively large Ge-concentration). Thus, Raman amplification is demonstrated using tens of kilometers of this FMF in SDM transmission experiment [48]. Both DGD compensation, low inter-mode coupling and Raman amplification enable SDM transmission for long-haul distances. Ref. [19] demonstrated inter-modal nonlinear interactions for well-separated channels in this fiber.

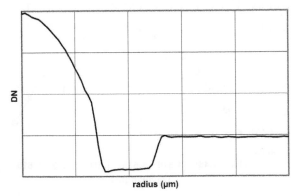

FIGURE 8.14 Index profile of a fabricated two-mode fiber of the design in GRIN example 1.

8.3.6 GRIN two-mode fiber example 2

Figure 8.15 shows another example of a GRIN two-mode fiber design. This design has a core-shelf-trench structure. The relative index difference at the center of the core is about half of the previous example. A trench with a confinement volume of 2.03 is positioned next to a narrow silica shelf adjacent to the core. Both LP_{01} mode and LP_{11} mode are well guided, as indicated by the effective indices marked as short bars in Figure 8.15. The effective index difference between LP_{01} mode and LP_{11} mode is 0.19%, indicating low coupling between these two modes. Theoretically, a low DGD between LP_{01} mode and LP_{11} mode can be achieved across an even wider wavelength range than the previous GRIN FMF example (<1 ps/km over 70 nm and <4 ps/km over 170 nm around the C-band as shown in Figure 8.16). Similar to the previous design, a deviation of the fiber parameters from optimized design values leads to a shift of the DGD curve to left and right or up and down, and results in a

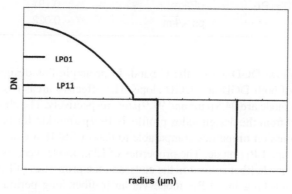

FIGURE 8.15 Index profile of a GRIN two-mode fiber.

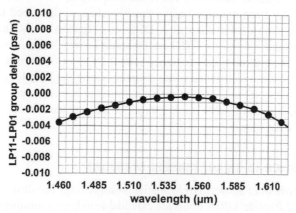

FIGURE 8.16 DGD between LP_{01} mode and LP_{11} mode.

Table 8.3 Typical properties at 1550 nm fabricated fiber Ref. [3].

Property	Unit	Value
Spool length	m	30,000
Distributed mode coupling LP_{01} to LP_{11}	dB	−25
DGD between LP_{11} and LP_{01}	ps/m	−0.076
Dispersion on LP_{01}	ps/(nm-km)	20.0
Dispersion slope LP_{01}	ps/(nm^2-km)	0.065
Effective area LP_{01}	μm^2	97
Dispersion LP_{11}	ps/(nm-km)	20.0
Dispersion slope LP_{11}	ps/(nm^2-km)	0.065
Effective area LP_{11}	μm^2	96
Attenuation OTDR LP_{01}	dB/km	0.198
Attenuation OTDR LP_{11}	dB/km	0.191
PMD LP_{01}	ps/$\sqrt{km}$	0.022

positive or negative DGD across the C-band. In principle this design allows good compensation of both DGD and DGD slope. The effective areas (A_{eff}) of the LP_{01} mode and LP_{11} mode are 94.8 μm^2 and 94.9 μm^2, respectively. The chromatic dispersion, calculated from the design index profile, is 19.9 ps/nm-km for both LP modes.

Fiber transmission properties comparable to that of SSMF are achieved for both the LP_{01} mode and LP_{11} mode. The properties of LP_{01} mode were measured with a standard single-mode fiber spliced to the ends, and the properties of LP_{11} mode were measured by launching into LP_{11} mode via an in-fiber long period grating mode converter. As shown in Table 8.3 of Ref. [3], the attenuation of LP_{01} mode and LP_{11} mode is 0.198 dB/km and 0.191 dB/km, respectively. The A_{eff} of LP_{01} is 97 μm^2, matching the simulation value well. The mode coupling as well as the DGD between LP_{01} and LP_{11} has been measured using the S^2 technique described in detail in the measurement section of this chapter. A DGD of 4.4 ps/km is obtained in a 10 km spool. 60 ps/km and −47 ps/km DGD are obtained in two 30 km spools, separately. By combining these three spools together, a total DGD of 6 ps/km is achieved over a 70 km span. Both the distributed mode coupling along the fiber length and the discrete mode coupling at the splicing points are found to be very small. The low mode coupling is confirmed by stable mode images of LP_{01} mode and LP_{11} mode on the output of the 70 km span.

8.3.7 Extension to higher number of modes

The two LP-mode fiber allows sixfold spatial multiplexing. The design principles for low DGD, two LP mode, GRIN fibers are extendable to higher numbers of LP modes. The solid curves in Figure 8.17 correspond to the $\alpha = 2.00$ GRIN fiber of Figure 8.4.

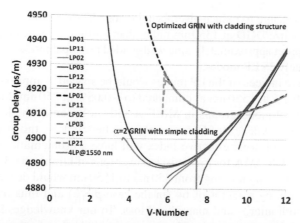

FIGURE 8.17 Group delays of LP$_{01}$, LP$_{11}$, LP$_{02}$, LP$_{21}$, LP$_{03}$, and LP$_{12}$ modes for a parabolic GRIN-FMF (lower curves) and a GRIN-FMF with a cladding structure, optimized for operation in the C-band (upper curves). The red vertical line near V-number equal to 7.5 corresponds to four mode operation. (For interpretation of the references to color in this figure legend, the reader is referred to the web version of this book.)

The dashed curves correspond to a GRIN design with an alpha profile core and a cladding structure which has been optimized for transmission in the C-band. The modal structure of this design remains similar to that of Figure 8.4. Four LP modes, specifically LP$_{01}$, LP$_{11}$, LP$_{02}$, and LP$_{21}$, are well guided at $V \approx 7.5$, and the cutoff wavelength of the next higher-order mode LP$_{12}$ is below 1550 nm. The V-number is chosen to achieve the large effective index difference between the lowest two guided LP modes and the cladding. The large spacing of the normalized propagation constant between different guided LP modes supports a large $\Delta\beta_{\text{lossy}}$, keeping DMA low. Figure 8.17 confirms the low DGD over a wide bandwidth.

8.3.8 How many modes can a FMF support? (Revisited)

It is interesting to consider how many low loss modes can be supported with a profile similar to those of Figures 8.12 and 8.15, as well as to ponder the pragmatic perspective for manufacturing. The fiber characteristics and manufacturability of high bandwidth, laser-optimized MMF, i.e. OM3 and OM4, may give some insight. OM4 is designed and manufactured to have low DGD (termed DMD or "differential mode delay" in standards) at 850 nm and thus high bandwidth when paired with high-speed VCSELs, for 100–500 m datacom links. In these applications the same signal is carried by each fiber mode, unlike the case of SDM. The so-called flat-mask DMD specification requires that the group delay between the fastest and slowest modes at 850 nm be less than or equal to 0.14 ps/m (OM4 grade) and 0.33 ps/m (OM3 grade) for the 18 or 19 principal mode groups (over 100 individual LP modes) of a near-parabolic core with radius 25 μm and $\Delta = 1\%$. The so-called bend-insensitive MMF adds a trench structure to reduce macrobending loss. The α profile parameter can be adjusted to equalize DGD at 1550 nm rather than 850 nm, in which case the 50 μm core would support approximately nine well-guided principal mode groups comprising approximately 25 LP modes.

The effective indices of the LP modes would be spaced at approximately equal intervals of ∼0.1% at 1550 nm as compared to ∼0.05% at 850 nm. Mode coupling between principal mode groups in MMF at 850 nm becomes the length scale of a kilometer. Since $\Delta\beta_{lossy}$ at 850 nm would also be ≈0.05%, this implies that the guided mode with lowest effective index will have a problematic microbending loss leading to high DMA for transmission over tens of kilometers. Doubling $\Delta\beta_{ij}$ from 0.05% to 0.1% by moving from 850 nm to 1550 nm would decrease the mode coupling strength by a factor of 64 (in the case $p = 6$) and thus improve DMA by reducing both micro- and macrobend loss. To our knowledge, DMA has not been measured for this scenario. However, it is likely that further improvement in DMA would be required by increasing $\Delta\beta_{ij}$ to 0.14% (additional ∼10× reduction in mode coupling at 1550 nm) or even 0.2% (additional 64× reduction). This would imply that limits on DMA may in turn limit the total number of low loss LP modes to ten or less in this family of profiles. It may be difficult to maintain the same effective area for all LP modes beyond four LP modes for the GRIN-FMF approach.

It is likely that the GRIN-FMF would be similarly challenging to manufacture as OM3 and OM4 MMF. (Thus GRIN-FMF will be inherently more expensive to manufacture than SSMF, and the cost to manufacture will be a strong function of DGD specifications.) For GRIN-MMF, the errors in DGD of the various modes occur randomly over the population of manufactured fibers with both positive and negative delays with respect to LP_{01}. If fibers are randomly cabled and spliced in ∼5 km segments in a deployment of total length L, then we expect DGD to accumulate as $\sqrt{L}$ based on these statistics as well as mode coupling effects already noted. If the manufacturing distribution of GRIN-FMF was ±70 ps/km (matching the limits of the OM4 specification), and if the correlation length l_c was 25 km, then the impulse response of a FMF channel would be about 11 ns for a 1000 km link. For a ∼30 Gbaud symbol rate, or ∼33 ps symbol period, this case would require about 670 T/2-spaced complex equalizer taps in the time domain (or equivalent frequency-domain equalizer capability). If the manufacturing distribution of the FMF matched the looser, but still demanding, OM3 specifications, then the impulse response rises to 26 ns, requiring approximately 1600 taps in the time domain, with 500 km requiring 1000 taps. It seems reasonable based on Moore's law that something between 100 and 1000 time domain taps (or its equivalent) would be feasible on a 10-year time frame at these baud rates.

8.3.9 Supermode fiber concept

Supermode fiber is a concept bridging between single-core FMF and multicore fiber (MCF). It is a strongly coupled multicore fiber. Figure 8.18 shows a cartoon of the

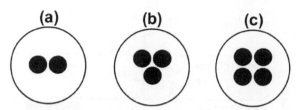

FIGURE 8.18 Cartoon of coupled-mode fiber structure: (a) two-core, (b) three-core, and (c) four-core.

strongly coupled multicore fiber structure including two cores, three cores, and four cores. The radius of the core is r, the distance between adjacent cores is d_1, and the distance between non-adjacent cores is d_2. One can view it as a few-mode fiber with supermodes as the eigenbasis of transmission modes.

Supermodes can be described as a linear combination of the LP modes of each individual core. The number of modes in a supermode fiber is equivalent to the number of the cores times the number of modes in each core. The mathematical treatment of the supermodes fiber is to use the coupled-mode equations. As the result of coupling, the effective index or propagation constant of the same LP mode for each individual core splits. We go through a simple exercise to find the effective indices (or propagation constants) of supermodes for coupled two-core, three-core, and four-core fiber examples shown in Figure 8.19. Following the analysis in [49], the coupled-mode equation is described by

$$\frac{d}{dz}A = -jMA,\tag{8.8}$$

where A_i ($i=$ the core number) refers to the complex amplitude of the electrical field of the ith core, and M is the coupling matrix. The coupling matrix M for two-core, three-core, and four-core fiber structure is written as following:

$$M = \begin{bmatrix} \beta_0 & c_1 \\ c_1 & \beta_0 \end{bmatrix} \quad \text{for two-core fiber structure,}$$

$$M = \begin{bmatrix} \beta_0 & c_1 & c_1 \\ c_1 & \beta_0 & c_1 \\ c_1 & c_1 & \beta_0 \end{bmatrix} \quad \text{for three-core fiber structure,}$$

$$M = \begin{bmatrix} \beta_0 & c_1 & c_2 & c_1 \\ c_1 & \beta_0 & c_1 & c_2 \\ c_2 & c_1 & \beta_0 & c_1 \\ c_1 & c_2 & c_1 & \beta_0 \end{bmatrix} \quad \text{for four-core fiber structure,}$$

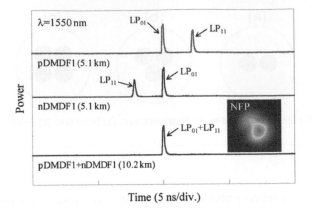

FIGURE 8.19 Time-of-flight measurements through two different FMF (red and blue) with equal and opposite DGD. Green shows measurement of the concatenated fibers with near-zero net DGD; an image of the near-filed pattern is also shown (from [45]). (For interpretation of the references to color in this figure legend, the reader is referred to the web version of this book.)

where β_0, c_1, and c_2 are the propagation constant of LP mode in each core, coupling coefficient between adjacent cores and non-adjacent cores. For a lossless system, M can be transformed into a diagonal matrix as following:

$$M^{\text{diag}} = \begin{bmatrix} \beta_0 - c_1 & 0 \\ 0 & \beta_0 + c_1 \end{bmatrix} \quad \text{for two-core fiber structure,}$$

$$M^{\text{diag}} = \begin{bmatrix} \beta_0 - c_1 & 0 & 0 \\ 0 & \beta_0 - c_1 & 0 \\ 0 & 0 & \beta_0 + 2c_1 \end{bmatrix} \quad \text{for three-core fiber structure,}$$

$$M^{\text{diag}} = \begin{bmatrix} \beta_0 - 2c_1 + c_2 & 0 & 0 & 0 \\ 0 & \beta_0 - c_2 & 0 & 0 \\ 0 & 0 & \beta_0 - c_2 & 0 \\ 0 & 0 & 0 & \beta_0 - 2c_1 + c_2 \end{bmatrix} \quad \text{for four-core fiber structure.}$$

The values of the matrix elements at the diagonal of the above matrixes are the propagation constants of the supermodes of the coupled-core structure. The coupled three-core fiber structure bears similarity to the two-mode FMF (including LP_{01} mode, and two degenerate LP_{11} modes), while the coupled two-core and four-core fiber structures have no equivalents in the FMF of a single solid core structure. So a coupled multicore fiber can break the constraint of the cylindrical symmetry in single solid core FMF.

Several groups have explored the design concept of supermodes experimentally and theoretically. Sasaki and coworkers [50] used a three-core fiber with a core diameter of 11.2 μm, both core pitch and core to fiber center distance of 22 μm and a relative index difference of 0.32% for each core. The fiber has an effective area of 104 μm^2 and attenuation of 0.177 dB/km. Randel and coworkers transmitted SDM 112-Gb/s PDM-QPSK signals over 24 km of such strongly coupled three-core fiber using MIMO-based signal processing. The same groups [5] demonstrated an aggregated single-wavelength capacity of 240 Gb/s over 1200 km of a coupled three-core fiber using 6 × 20 Gb/s QPSK modulation. This three-core fiber had homogeneous cores with a core diameter of 12.4 μm, core pitch 29.4 μm, and relative index difference of 0.27% for each core. The center of the core is 17 μm away from the center of the fiber. The effective area of the fiber was measured to be around 129 μm^2. The attenuation of all launch conditions was 0.181 dB/km. Li and coworkers [49] have explored the theory of three, four, and six coupled-core structures. Modeling results showing the use of supermodes leads to a larger mode effective area and higher mode density than the conventional multicore fiber does. There is also a comparison between MCF, coupled-MCF, and single-core FMF in Table 2 of Ref. [49]. Some statements there are not necessarily true for all FMF. For example, Ref. [3] demonstrated a FMF with low DGD, low modal crosstalk, low attenuation, and low modal dependent loss. However, the supermode concept with coupled cores may allow a different approach toward achieving large A_{eff}, low DGD, and low DMA. Imamura et al. [6] fabricated a design using seven weakly coupled cores. Selective super-mode excitation was demonstrated. By preferentially localizing the supermodes on particular cores, Ref. [6] shows the potential to achieve high spatial density using multicore connectors for the input and output. Ref. [51] further demonstrated wave division multiplexing in 6 × 5 × 40 Gb/s PDM-QSPK transmission over 4200 km of the same three-core fiber used in Ref. [5].

8.4 MEASUREMENT OF FEW-MODE FIBER

The important new transmission-related fiber parameters associated with FMF include DGD, DMA, and optical mode coupling. A challenge to finding these parameters is that their number increases as the square of the number of propagating modes. The actual number of measurements which must be made is somewhat less because of modal degeneracies. The ability to do mode-selective launch into the fiber and mode-selective receiving at the fiber output is a great asset when doing these measurements [52].

8.4.1 Differential group delay (DGD)

There are several means of carrying out measurements of DGD. In the case of a single LP_{01} mode, there is a substantial literature describing measurements of DGD between the two polarizations of the LP_{01} [53]. This problem is somewhat different from that of characterizing FMF used for spatial multiplexed transmission in that irregularities are required in the fiber cross-section for polarization-based DGD to be present.

In addition, the polarization mode DGD is typically very small (fs/km) and the mode coupling strong. By comparison, when measuring DGD between the LP_{01} and LP_{11} modes in a FMF, an approach as simple as launching a light pulse in each mode and detecting arrival time at the far end of the fiber is feasible. DGD values for FMF are typically in the tens of ps/km range (a thousand times larger than for polarization modes in SSMF). Also, FMF is often designed so that there is a small coupling between modal groups during light transmission through the fiber. This allows long sections of fiber to be measured with little concern for reduction of the DGD by mode coupling.

8.4.2 Time of flight (impulse response) measurements

As just mentioned, the DGD in FMF is often large enough to be measured by comparing the arrival time of light pulses launched simultaneously in the various modes. This is the method that has been used when measuring differential mode delay in multi-mode fibers [54]. A recent study in FMF [37] uses MZM intensity modulation of a continuous laser beam to create 100 ps pulses which are launched into the test fiber with a phase plate-based mode multiplexer (tuned to couple both to the LP_{01} and LP_{11} modes). The difference in arrival time of the pulses at the end of 96 km fiber allows determination of the DGD. In order to determine the sign of the DGD, it must be possible to determine which mode corresponds to which pulse at the receiving end. This can be done with mode demultiplexing at the receiver or by varying the launch properties (for example, the relative power in the two modes). An alternative is to use computer simulation of the modes of propagation for the given fiber index profile to determine which mode has the higher group velocity at the operating wavelength. A complicating factor in this test is the possible deterioration of the light pulse shape due to chromatic dispersion.

An example of such a measurement is shown in Figure 8.19 [45]. Here DGD measurements on two different fibers are shown: one fiber with LP_{01} speed greater than LP_{11}, "pDMDF1"; the other fiber with LP_{11} speed greater than LP_{01}, "nDMDF1"; along with a measurement of the DGD through a concatenation of the two fibers. The goal of this link design was to reduce the net DGD between the two modes to zero.

8.4.3 Spatially integrated interferometric measurements

Another method for determining modal DGD is to create an interferometer from the test fiber and measure the free spectral range of the resulting spectral beat signal. This multimode interference phenomenon was studied in the past in connection to MPI [55,56], high-order mode dispersion [57], and fiber sensors [58,59]. The method involves launching broadband optical power primarily into a reference mode with a small amount of power in one or more other modes of interest. For example, this can be done for the LP_{01}–LP_{11} system by using an offset launch from a SSMF pigtail. After traveling the length of the FMF, the light is collected by another offset SSMF and detected by an optical spectrum analyzer (OSA). The light in each mode of the

FMF will project onto the LP_{01} mode in the offset output fiber, creating interference because of the different group delays through the FMF. This interference is wavelength dependent because the relative modal phase shift through the FMF is $(\beta_{11} - \beta_{01})L = \Delta\beta L$, where the $\beta_i = 2\pi n_i/\lambda$ are the phase constants of the two modes, n_i is the (wavelength-dependent) effective index of mode "i," and L is the fiber length. To determine the periodicity in wavelength consider two nearby wavelengths which create the same interference. This will occur when the modal phase shifts at the two wavelengths differ by $2\pi k$ radians:

$$\Delta\beta\lambda_1 L - \Delta\beta\lambda_0 L = 2\pi L\{\Delta n(\lambda_1)/\lambda_1 - \Delta n(\lambda_0)/\lambda_0\} = 2\pi k,$$

where $\Delta n \equiv n_{11} - n_{10}$ is the difference in effective indices for the two modes and k is an integer. Assuming that Δn is a weak function of wavelength, it can be Taylor expanded about λ_0 and, with $k = 1$, an expression can be found for the wavelength periodicity, $\Delta\lambda$:

$$\Delta\lambda = \lambda_1 - \lambda_0 = -\lambda_0^2/(\Delta n_g(\lambda_0)L + \lambda_0,$$

where the group index difference has been introduced:

$$\Delta n_g = \Delta n(\lambda_0) - \lambda_0(\partial\Delta n/\partial\lambda|_{\lambda 0}.$$

Introducing the DGD coefficient (DGD per unit length), DGD_L, we have

$$DGD_L = \Delta n_g c \quad \text{and} \quad |DGD_L| = \lambda_0^2/(|\Delta\lambda|Lc). \tag{8.9}$$

This allows the magnitude of the DGD to be found from a spectral measurement of the two-mode interference.

Some measured results using this technique on a 10 m FMF are shown in Figure 8.20. The modal cutoff properties are shown by the red curve in Figure 8.20a. Using a multi-mode fiber reference and comparing to simulation, it is found that the LP_{01} and LP_{11} modes propagate above 1450 nm, with one set of higher order modes (HOM) appearing in the 1150–1450 nm range and another set experiencing cutoff at 1150 nm. The blue curve in Figure 8.20a is the transmitted power using offset SSMF fibers to launch and collect the signal through the FMF. The variable periodicity in the transmitted power is evident and is related to the DGD through above Eq. (8.9). It is found that the introduction of an appropriately sized mandrel can move the HOM cutoffs to shorter wavelength, so that two-mode operation can be obtained throughout the 1200–1600 nm region. Figure 8.20b shows the computed DGD obtained from measuring the spectral periodicity as a function of wavelength for different lengths of FMF. Note that the data falls on a continuous curve for all fiber lengths. A zero DGD point is apparent at ~1380 nm. Since $\Delta\lambda \rightarrow \infty$ in this regime, it is difficult to measure accurate values. A solution is to use a long fiber (980 m in this case) to allow more DGD to accumulate. Note that the finite resolution of the OSA prevents such a long fiber from providing useful DGD data away from the wavelength of zero DGD.

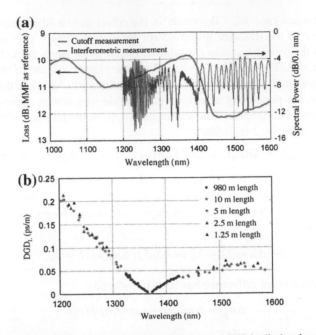

FIGURE 8.20 (a) Transmission measurement through 10 m of FMF (red) showing cutoff properties of some high order modes. Interferometric mode beating corresponding to these modes (blue). (b) DGD measurements derived from the spectral beat frequencies at different fiber lengths (from [37]). (For interpretation of the references to color in this figure legend, the reader is referred to the web version of this book.)

This method does have limitations. It is a challenge to measure the interference of more than two modes (the different wavelength periodicities must be isolated). Modal identification must be done by comparison to another measurement or simulation. Mode filtering can be used to alleviate this problem. Small DGD presents a challenge since the spectral range over which DL can be measured is limited (also, the variation of DGD with wavelength is obscured). Large DGD measurement is limited by the minimum resolution bandwidth capability of the OSA. Finally, like the time-of-flight method, this method is unable to distinguish variation in DGD along the length of the fiber under test.

8.4.4 Spatially resolved interferometric measurements (S^2)

A more sophisticated interference measurement has been developed over the last few years which incorporates information on the variation of the modal interference over the cross-section of the FMF fiber end face. This has been termed "spatial-spectral" or "s-squared (S^2)" imaging [60]. This method improves upon the spatially integrated method previously described in that interference at many spatial locations

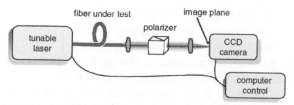

FIGURE 8.21 Schematic of the S^2 measurement setup (from [61]).

on the output fiber end face are recorded as a function of wavelength. From this data, relative power levels and phases of high order modes can be found relative to a reference mode. The analysis requires that the reference mode be large in power (roughly a factor of ten) compared to the other modes. The data capture is summarized in Figure 8.21. A tunable laser launches a signal (perhaps through an offset SSMF fiber into the fiber under test where a number of modes are excited. A SWIR camera images the end face of the fiber and captures the light intensity as a function of wavelength. Each cameral pixel then captures the optical power as a function of wavelength (similar to the oscillatory data found in Figure 8.20a) from a small portion of the fiber under test. Fourier transforming (FT) this data allows underlying periodicities to be found at that pixel. Low pass and band pass filtering of the FT data enables separation of the fundamental optical mode and the different HOM as well as determining the group delays. An example is shown in Figure 8.22 for a 20 m

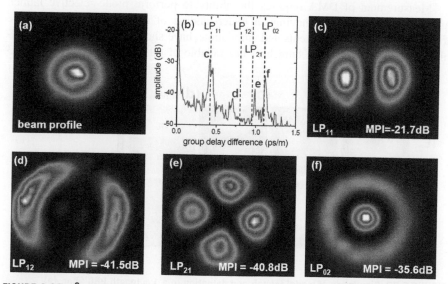

FIGURE 8.22 S^2 measurement on a 20 meter length of LMA fiber: (a) beam profile, (b) FT of the optical spectra with calculated group delays (dashed), (c)–(f) calculations of HOM modal patterns and MPI [62].

length of large mode area fiber. The "group delay spectra" is shown in Figure 8.22b where peaks are noted at each of the HOM group delays (relative to the fundamental mode). Based on the distribution of group delays, one can identify mode coupling points or regions. Squaring and integrating (over the band pass regions) the HOM FT data give the intensity of each HOM at the given pixel position. Summing the HOM intensities over all the pixels yields the power in each HOM relative to the fundamental (reference) mode which is equivalent to the multipath interference (MPI). In addition, mode patterns and relative phases can be found which allow each mode to be identified. Figure 8.22c–f shows examples of the resulting mode patterns.

8.4.5 Differential mode attenuation (DMA)

An important feature of FMF fibers is how different modes are attenuated as they propagate. Of particular concern is the fact that MIMO processing is unable to compensate for DMA. In one theory [24], the degradation is dependent on the DMA over the entire end-to-end optical link. A 20 dB DMA results in loss of system capacity of ~20%. This 20 dB is equivalent to 0.02 dB/km of DMA for a 1000 km link, 0.05 dB/km for a 400 km link, 0.1 dB/km for a 200 km link, and 0.2 dB/km over a 100 km link. Therefore it seems important to limit DMA to <0.02 dB/km to support backbone transmission over traditional distances. A DMA<0.01 dB/km would sacrifice only 5% of the system capacity over 1000 km. Hence it is generally true that FMF designs aim to minimize DMA as previously discussed. This can be accomplished by minimizing microbending and macrobending as well as by frequent mode coupling.

Measurement of DMA requires the ability to perform mode-selective launch and reception as well as having an understanding of the mode coupling in the transmission path. Experiments and simulations have shown that spatial light modulators and mode selective launchers are capable of this test [35,52]. Figure 8.23

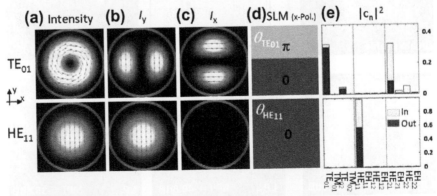

FIGURE 8.23 Simulation of modal launch and detection in a 1 m length PBG fiber. (a) Two modes are launched into the fiber (TE01 and HE11), (b) intensities polarized in x and (c) y directions, (d) SLM phase modulation applied to x-polarization, (e) resulting modal intensities at input and output of fiber (from [63]).

shows a simulation demonstrating a method for launching and recovering modes in a PBG fiber. An indirect method of assessing DMA effects is to optimize channel performance by optimizing the power launch into each mode. Some portion of the difference in required power is due to DMA.

8.4.6 Mode coupling

Measurement of mode coupling can be approached in several different ways. One method is to look for deviations from linear growth of the DGD with fiber length. Measuring the DGD as a function of fiber length, with a means of randomizing the mode coupling perturbations, can allow computation of a linear DGD and a correlation length. Another method involves mode-selective launch and detection. One mode is launched and the power in each mode is detected at the fiber output. This measurement can be done as a function of length to isolate imperfections in the launch and reception apparatus from the coupling in the FMF. This information can also be obtained from equalizer tap positions in coherent MIMO transmission experiments [25]. A final method utilizes the S^2 interference technique described earlier. Mode coupling comes out in the MPI calculations performed here. In addition, information on where the coupling occurs, and whether it is discrete or distributed, can be obtained.

8.5 FUTURE PERSPECTIVE

Research in few-mode fiber, multicore fiber (coupled and uncoupled), and relevant space-division multiplexing transmission technologies have progressed at an impressive speed. To achieve the $100-1000\times$ capacity increase needed to justify massive deployment costs for a new technology, the fiber optics community will need the benefit of multiple concepts and techniques. It is likely that transmission over hundreds of kilometers will always occur in the strong mixing regime due to discrete mode coupling points throughout the deployed transmission line. This will slow the accumulation of DGD with distance and lessen the impact of DMA as well. It will nevertheless be challenging to develop practical, cost-effective fibers with low DMA and low DGD supporting 10–20 orthogonal spatial modes. A graded-index core shape with cladding structure shows promise for low DGD over wide bandwidths as well as low DMA. Nevertheless, FMF meeting tight DGD tolerances to support MIMO signal recovery will be significantly more costly to manufacture than SSMF.

Acknowledgments

The authors acknowledge valuable conversations and collaboration with Lars Gruner-Nielsen, Jeff Nicholson, Jinkee Kim, Kasyapa Balemarthy, and Benyuan Zhu of OFS; Peter Winzer, Roland Ryf, Rene Essiambre, and Sebastian Randel of Alcatel-Lucent; and Prof. Joseph Kahn of Stanford University.

References

[1] B. Zhu, T.F. Taunay, M.F. Yan, J.M. Fini, M. Fishteyn, E.M. Monberg, F.V. Dimarcello, Seven-core multicore fiber transmissions for passive optical network, Opt. Express 18 (11) (2010) 11117–11122.

[2] J.M. Fini, T.F. Taunay, B. Zhu, M.F. Yan, Low cross-talk design of multi-core fibers, in: CLEO Paper CTuAA3, 2010.

[3] L. Gruner-Nielsen, Y. Sun, J.W. Nicholson, D. Jakobsen, R.J. Lingle, B. Palsdottir, Few mode transmission fiber with low DGD, low mode coupling and low loss, in: OFC/NFOEC Paper PDP5A.1, Los Angeles, California, 2012.

[4] M. Bigot-Astruc, D. Boivin, P. Sillard, Design and fabrication of weakly-coupled few-modes fibers, in: IEEE Summer Topical Meeting Paper TuC1.1, Seattle, Washington, 2012.

[5] R. Ryf, A. Sierra, R.J. Essiambre, A. Gnauck, S. Randel, M. Esmaeelpour, S. Mumtaz, P.J. Winzer, R. Delbue, P. Pupalaikis, A. Sureka, T. Hayashi, T. Taru, T. Sasaki, Coherent 1200-km 6 × 6 MIMO mode-multiplexed transmission over 3-core microstructured fiber, in: ECOC PD. Paper Th.13.C.1., Geneva, 2011.

[6] K. Imamura, H. Inaba, K. Mukasa, R. Sugizaki, Weakly coupled multi core fibers with optimum design to realize selectively excitation of individual super-modes, in: OFC/NFOEC Paper OM2D.6, Los Angeles, 2012.

[7] Y. Kokubun, T. Komo, K. Takenaga, S. Tanigawa, S. Matsuo, Quantitative mode discrimination and bending crosstalk of four-core homogeneous coupled multi-core fiber, in: OFC/NEOFC Paper OM2D.7, Los Angeles, 2012.

[8] A.R. Chraplyvy, The coming capacity crunch, in: Proc. European Conf. on Opt. Commun. (ECOC) Plenary Talk, 2009.

[9] R.-J. Essiambre, G. Kramer, P. Winzer, G. Foschini, B. Goebel, Capacity limits of optical fiber networks, J. Lightwave Technol. 28 (2010) 662–701.

[10] H. Stuart, Dispersive multiplexing in multimode optical fiber, Science 289 (5477) (2000) 281–283.

[11] T. Koonen, H.V.D. Boom, I.T. Monroy, G.-D. Khoe, High capacity multi-service in-house networks using mode group diversity multiplexing, in: OFC/NFOEC Paper FG4, 2004.

[12] S. Schoellmann, C. Xia, W. Rosenkranz, Experimental investigations of mode group diversity multiplexing on multimode fibre, in: OFC/NEOFC Paper OWR3, Anaheim, California, 2006.

[13] B. Franz, D. Suikat, R. Dischler, F. Buchali, H. Buelow, High speed OFDM data transmission over 5 km GI-multimode fiber using spatial multiplexing with 2 × 4 MIMO processing, in: ECOC Paper Tu.3.C.4, Torino, Italy, 2010.

[14] S. Kawakami, Transmission characteristics of a two-mode optical waveguide, IEEE J. Quant. Electron. QE-14 (8) (1978) 608–614.

[15] L. Cohen, Propagation characteristics of double-mode fibers, Bell Syst. Tech. J. 59 (6) (1980) 1061–1071.

[16] K.I. Kitayama, Y. Kato, S. Seikai, N. Uchida, Structural optimization for two-mode fiber: theory and experiment, IEEE J. Quant. Electron. QE-17 (6) (1981) 1057–1063.

[17] M.M. Cvijetic, G. Lukatela, Design considerations of dispersion-free dual-mode optical fibers: 1.55 μm wavelength operation, IEEE J. Quant. Electron. QE-23 (5) (1987) 469–472.

[18] C. Koebele, M. Salsi, G. Charlet, S. Bigo, Nonlinear effects in long-haul transmission over bimodal optical fibre, in: ECOC Paper MO.2.C.6, Torino, 2010.

[19] R.-J. Essiambre, R. Ryf, M. Mestre, A. Gnauck, R. Tkach, A. Chraplyvy, S. Randel, Y. Sun, X. Jiang, R.J. Lingle, Inter-modal nonlinear interactions between well separated channels in spatially-multiplexed fiber transmission, in: ECOC, Amsterdam, 2012.

[20] K.P. Ho, J.M. Kahn, Statistics of group delays in multimode fiber with strong mode coupling, J. Lightwave Technol. 29 (21) (2011) 3119–3128.

[21] H. Kogelnik, P.J. Winzer, Modal birefringence in weakly guiding fibers, J. Lightwave Technol. 30 (14) (2012) 2240–2245.

[22] A. Li, A.A. Amin, X. Chen, W. Shieh, Transmission of 107-Gb/s mode and polarization multiplexed CO-OFDM signal over a two-mode fiber, Opt. Express 19 (9) (2011) 8808–8814.

[23] K.-P. Ho, J. Kahn, Mode-dependent loss and gain: statistics and effect on mode-division multiplexing, Opt. Express 19 (17) (2011) 16612–16635.

[24] P.J. Winzer, G.J. Foschini, MIMO capacities and outrage probabilities in spatially multiplexed optical transport systems, Opt. Express 19 (17) (2011) 16680–16696.

[25] S. Randel, R. Ryf, A. Gnauck, M.A. Mestre, C. Schmidt, R. Essiambre, P. Winzer, R. Delbue, P. Pupalaikis, A. Sureka, Y. Sun, X. Jiang, R.J. Lingle, Mode-multiplexed 6×20-GBd QPSK transmission over 1200-km DGD-compensated few-mode fiber, in OFC/NFOEC Paper PDP5C.5, Los Angeles, California, 2012.

[26] D. Gloge, Weakly guiding fibers, Appl. Opt. 10 (10) (1971) 2252–2258.

[27] J. Buck, Fundamentals of Optical Fibers, Wiley, 1995.

[28] L. Jeunhomme, Single-mode Fiber Optics, Mercel Dekker, New York, 1983, p. 3.

[29] D. Marcuse, Theory of Dielectric Optical Waveguides, Academic press, New York, 1974.

[30] R. Olshansky, Mode coupling effects in graded-index optical fibers, Appl. Opt. 14 (4) (1975) 935–945.

[31] S. Personick, Time dispersion in dielectric waveguides, Bell Syst. Tech. J. 50 (3) (1971) 843–859.

[32] C. Poole, Statistical treatment of polarization dispersion in single-mode fiber, Opt. Lett. 13 (8) (1988) 687–689.

[33] D. Marcuse, Pulse propagation in multimode dielectric waveguides, Bell Syst. Tech. J. 51 (6) (1972) 1199–1232.

[34] K. Okamoto, T. Okoshi, Analysis of wave propagation in optical fibers having core with a-power refractive-index distribution and uniform cladding, IEEE Trans. Microwave Theory Tech. MTT-24 (1976) 416–421.

[35] C. Koebele, M. Salsi, L. Milord, R. Ryf, C. Bolle, P. Sillard, 40 km transmission of five mode division multiplexed data streams at 100 Gb/s with low MIMO-DSP complexity, in: ECOC Postdeadline Papers Th. 13, C.3, 2011.

[36] R. Ryf, S. Randel, A. Gnauck, C. Bolle, R.-J. Essiambre, P. Winzer, D. Peckham, A. McCurdy, R.J. Lingle, Space-division multiplexing over 10 km of three-mode fiber using coherent 6×6 MIMO processing, in: OFC/NEOFC Paper PDPB10, Los Angeles, California, 2011.

[37] R. Ryf, S. Randel, A. Gnauck, C. Bolle, A. Sierra, S. Mumtaz, M. Esmaeelpour, E. Burrows, R.-J. Essiambre, P.-J. Winzer, D. Peckham, A. McCurdy, R.J. Lingle, Mode-division multiplexing over 96 km of few-mode fiber using coherent 6×6 MIMO processing, J. Lightwave Technol. 30 (4) (2012) 521–531.

[38] K.-P. Ho, J.M. Kahn, Frequency diversity in mode-division multiplexing systems, J. Lightwave Technol. 29 (24) (2011) 3719–3726.

[39] M. Ohashi, K. Kitayama, S. Seikai, Mode coupling at arc-fusion splices in graded index fibers, IEEE J. Quant. Electron. 18 (2) (1982) 274–277.

[40] I.A. White, S.C. Mettler, Modal analysis of loss and mode mixing in multimode parabolic index splices, Bell Syst. Tech. J. 62 (5) (1983) 1189–1207.

[41] K. Kitayama, Y. Yato, S. Seikai, N. Uchida, M. Akiyama, O. Fukuda, Transmission characteristic measurement of two-mode optical fiber with a nearly optimum index-profile, IEEE Trans. Microwave Theory Tech. MTT-28 (1988) 604–608.

[42] K. Kitayama, Y. Kato, S. Seikai, N. Uchida, M. Ikeda, Experimental verification of modal dispersion free characteristics in a two-mode optical fiber, IEEE J. Quant. Electron. QE-15 (1979) 6–7.

[43] P. Sillard, M. Bigot-Astruc, D. Boivin, H. Maerten, L. Provost, Few-mode fiber for uncoupled mode-division multiplexing transmissions, in: ECOC Tech. Digest Paper Tu.5.LeCervin.7, 2011.

[44] R. Maruyama, N. Kuwaki, S. Matsuo, K. Sato, M. Ohashi, Mode dispersion compensating optical transmission line composed of two-mode optical fibers, in: OFC/NFOEC Tech. Digest Paper JW2A.13, 2012.

[45] T. Sakamoto, T. Mori, T. Yamamoto, S. Tomita, Differential mode delay managed transmission line for wide-band WDM-MIMO system, in: OFC/NFOEC Tech. Digest Paper OM2D.1, 2012.

[46] N.K. Fontaine, C.R. Doerr, M.A. Mestre, R. Ryf, P. Winzer, L. Buhl, Y. Sun, X. Jiang, R.J. Lingle, Space-division multiplexing and all-optical MIMO demultiplexing using a photonic integrated circuit, in: OFC/NFOEC Paper PDP5B.1, Los Angeles, California, 2012.

[47] R. Ryf, M. A. Mester, A. Gnauck, S. Randel, C. Schmidt, R. Essiambre, P. Winzer, R. Delbue, P. Pupalaikis, A. Sureka, Y. Sun, X. Jiang, D. Peckham, A.H. McCurdy, R.J. Lingle, Low-loss mode coupler for mode-multiplexed transmission in few-mode fiber, in: OFC/NFOEC Paper PDP5B.5, Los Angeles, California, 2012.

[48] R. Ryf, M.A. Mestre, S. Randel, C. Schmidt, A.H. Gnauck, R.-J. Essiambre, P.J. Winzer, R. Delbue, P. Pupalaikis, A. Sureka, Y. Sun, X. Jiang, D.W. Peckham, A. McCurdy, R.J. Lingle, Mode-multiplexed transmission over a 184-km DGD-compensated few-mode fiber span, IEEE Summer Topicals, Seattle, Washington, 2012.

[49] C. Xia, N. Bai, I. Ozdur, X. Zhou, G. Li, Supermodes for optical transmission, Opt. Express 19 (17) (2011) 16653–16664.

[50] S. Randel, M. Magarini, R. Ryf, R.-J. Essiambre, A. Gnauck, P.-J. Winzer, T. Hayashi, T. Taru, T. Sasaki, MIMO-based signal processing of spatially multiplexed 112-Gb/s PDM-QPSK signals using strongly-coupled 3-core fiber, in: ECOC Paper Tu.5.B1, 2011.

[51] R. Ryf, R.-J. Essiambre, A.-H. Gnauck, S. Randel, M. Mestre, C. Schmidt, P. Winzer, R. Delbue, P. Pupalaikis, A. Sureka, T. Hayashi, T. Taru, T. Sasaki, Space-division multiplexed transmission over 4200-km 3-core microstructured fiber, in: OFC/NEOFC Paper PDP5C.2, Los Angeles, 2012.

[52] R. Ryf, C. Bolle, J.V. Hoyningen-Huene, Optical coupling components for spatial multiplexing in multi-mode fibers, in: ECOC Tech. Digest Paper Th.12.B.1, 2011.

[53] H. Kogelnik, R. Jopson, L. Nelson, Polarization-mode Dispersion in Optical Fiber Telecommunications IVB: Systems and Impairments, Academic Press, 2002.

[54] R. Olshansky, S.M. Oaks, Differential mode delay measurements, in: Proc. Fourth Eur. Conf. Opt. Commun., Genoa, 1978, p. 128.

[55] D.G. Duff, F.T. Stone, J. Wu, Measurements of modal noise in single-mode lightwave systems, in: Proc. OFC, Paper TUO1, 1985.

[56] S. Ramachandran, J.W. Nicholson, S. Ghalmi, M.F. Yan, Measurement of multipath interference in the coherent crosstalk regime, IEEE Photon. Technol. Lett. 15 (8) (2003) 1171–1173.

[57] D. Menashe, M. Tur, Y. Danziger, Interferometric techniques for measuring dispersion of high order modes in optical fibers, Electron. Lett. 37 (24) (2001) 1439–1440.

[58] A. Kumar, N. Goel, R. Varshney, Studies on a few-mode fiber-optic sensor based on LP01-LP02 mode interference, J. Lightwave Technol. 19 (3) (2001) 358–362.

[59] K.A. Murphy, M.S. Miller, A.M. Vengsarkar, R.O. Claus, Elliptical-core two-mode optical-fiber sensor implementation method, J. Lightwave Technol. 8 (11) (1990) 1688–1696.

[60] J.W. Nicholson, A.D. Yablon, J.M. Fini, M.D. Mermelstein, Measuring the modal content of large-mode-area fibers, IEEE J. Sel. Top. 15 (1) (2009) 61–69.

[61] A.M. DeSantolo, D.J. DiGiovanni, F.V. DiMarcello, J.M. Fini, M. Hassan, L. Meng, E.M. Monberg, J.W. Nicholson, R.M. Ortiz, R.S. Windeler, High resolution S2 mode imaging of photonic bandgap fiber, in: OSA/CLEO Paper CFM4, 2011.

[62] J.W. Nicholson, A.D. Yablon, S. Ramachandran, S. Ghalmi, Spatially and spectrally resolved imaging of modal content in large-mode-are fibers, Opt. Express 16 (10) (2008) 7233–7243.

[63] O. Shapira, A.F. Abouraddy, Q. Hu, D. Shemuly, J.D. Joannopoulos, Y. Fink, Enabling coherent superpositions of iso-frequency optical states in multimode fibers, Opt. Express 18 (12) (2010) 12622–12629.

Multi-Core Optical Fibers

9

Tetsuya Hayashi

Sumitomo Electric Industries, Yokohama, Japan

9.1 INTRODUCTION

Recent single-core transmission systems have achieved capacities up to about 100 Tb/s per fiber by employing time-, wavelength-, polarization-division multiplexing, and multilevel modulations [1–3]. However, the transmission capacity of the single-core fiber is rapidly approaching its fundamental limit, and the current trends of traffic growth—increasing by a factor of 10 in 5 years—and system capacity growth will result in capacity crunch in the near future [4,5]. In such a situation, spatial division multiplexing (SDM) is an attractive technology for further enlargement of the fiber capacity or spatial capacity—the capacity per cross-sectional area of the fiber— [5,6] (see the chapter titled "Fiber Nonlinearity and Capacity: Single-Mode and Multimode Fibers" for more details of the fiber capacity).

Concepts of SDM optical fiber transmission are not so new [7–9]. The MCF for high-density transmission was first proposed in the late 1970s [7,8]. This MCF was drawn from one preform including multiple cores, and therefore its cladding was circular shaped and similar to recent MCFs (see Figure 9.1a). However, since there was a difficulty in aligning and splicing the multiple cores in circular MCFs, another type of MCF, the so-called bunch fiber, was proposed [10]. The bunch fiber is drawn from multiple preforms and each preform includes one respective core, and thus has non-circular cladding (see Figure 9.1b and c). The bunch fiber had been developed for use in subscriber lines [11–13]. However, the passive optical network (PON) [14] was able to provide cost- and space-efficient networks for the subscriber lines, and the MCF became unnecessary for the density improvement and has not been commercialized for SDM transmission.

In the late 2000s, since the future capacity crunch became a reality, the MCF has come to attract lots of attention again [15–18]. Various types of MCFs have been proposed, and characteristics of the MCF have been intensively investigated. This research on the MCFs is not only for high-capacity and long-reach applications [19–21], but also for high-density and short-reach applications [15,22,23], and high subscriber-count PON systems [24].

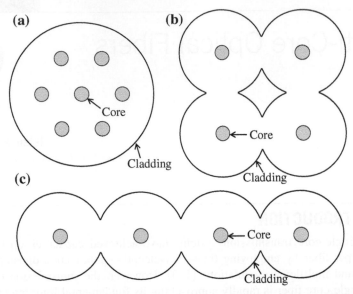

FIGURE 9.1 Schematic cross-sections of (a) a normal MCF and (b,c) bunched MCFs.

In this chapter, the author describes an MCF for SDM transmission, wherein each core is used as an individual waveguide, which is the so-called uncoupled MCF, weakly coupled MCF, etc. Another type of MCF, the so-called coupled MCF, strongly coupled MCF, etc., is covered in the chapter titled "Few-Mode Fiber Technology for Spatial Multiplexing," since the strongly coupled cores can be regarded as one microstructured few-mode waveguide. Details of the SDM methods and other types of fibers for SDM are covered in other related chapters. The MCF also has been studied in non-communication fields such as fiber optic sensing [25], but it is out of the scope of this chapter. The uncoupled MCF is just referred to as MCF in this chapter.

9.2 INTER-CORE CROSSTALK

To improve the spatial capacity of the MCF, it is important to increase the core density—core count per fiber cross-sectional area—while preserving the capacity of each core, or avoiding characteristics degradations. Shortening of the core-to-core distance is a very efficient way to improve the core density. When shortening the core-to-core distance, suppression of the inter-core crosstalk is crucial to send data independently over the individual cores in the MCF. To suppress the crosstalk in a short core-to-core distance, the modes of the cores should be confined strongly into the cores in order to decrease the overlap of the modes. Accordingly, there is a tradeoff between crosstalk, core-to-core distance, effective area, and cutoff wavelength—the latter two and core structure govern the mode confinement. Whereas such a tradeoff can be understood from the simple coupled-mode theory, it has been revealed that the crosstalk in the MCF does not obey the simple unperturbed coupled-mode equation, and proper adjustments on the perturbations can suppress the crosstalk and relax the tradeoff. This section describes such crosstalk characteristics of MCF in detail.

9.2.1 Coupled-mode equation with perturbed propagation constants and inter-core crosstalk in multi-core fiber

The coupled-mode equation describing the crosstalk in MCF is represented as

$$\frac{d}{dz}\boldsymbol{E} = -j\left(\boldsymbol{\beta}_{\text{eq}} + \boldsymbol{\kappa}\right)\boldsymbol{E}, \tag{9.1}$$

$$\boldsymbol{\beta}_{\text{eq}} = \boldsymbol{\beta}_{\text{c}} + \boldsymbol{\beta}_{\text{b}} + \boldsymbol{\beta}_{\text{s}}, \tag{9.2}$$

where $\boldsymbol{E} = [E_1, E_2, \ldots, E_N]^T$ is the vector of the electric fields E_n of each core, $\boldsymbol{\beta}_{\text{c}} = \text{diag}(\beta_{\text{c},1}, \beta_{\text{c},2}, \ldots, \beta_{\text{c},N})$ is the $N \times N$ diagonal matrix, which includes unperturbed *constant* propagation constants $\beta_{\text{c},n} = kn_{\text{eff},n}, k = 2\pi/\lambda$ is the wave number in a vacuum, λ is the wavelength in a vacuum, n_{eff} is the effective refractive index, $\boldsymbol{\beta}_{\text{b}} = \text{diag}(\beta_{\text{b},1}, \beta_{\text{b},2}, \ldots, \beta_{\text{b},N})$ includes bend-induced perturbation $\beta_{\text{b},n}$ of the propagation constants, $\boldsymbol{\beta}_{\text{s}} = \text{diag}(\beta_{\text{s},1}, \beta_{\text{s},2}, \ldots, \beta_{\text{s},N})$ includes structural fluctuation induced perturbation $\beta_{\text{s},n}$ of the propagation constants, and $\boldsymbol{\kappa}$ is the $N \times N$ matrix, including the mode coupling coefficients κ_{nm}. Since a bent fiber with the refractive index profile $n(r, \theta)$ can be described as a corresponding straight fiber with the equivalent refractive index [26]:

$$n_{\text{eq}}(r, \theta) \approx n(r, \theta)\left(1 + \gamma\frac{r\cos\theta}{R_{\text{b}}}\right), \tag{9.3}$$

where (r, θ) is the local polar coordinate in the cross-section of the fiber, $\theta = 0$ in the radial direction of the bend, and R_{b} is the bending radius of the MCF. $\gamma \approx 1$ can include stress corrections. Thus, the equivalent effective index $n_{\text{eqeff},n}$ of Core n can be represented as (see Figure 9.2)

$$n_{\text{eqeff},n} \approx n_{\text{eff},n}\left(1 + \gamma\frac{r_n\cos\theta_n}{R_{\text{b}}}\right), \tag{9.4}$$

and $\beta_{\text{b},n}$ can be represented as

$$\beta_{\text{b},n} \approx \beta_{\text{c},n}\gamma\frac{r_n\cos\theta_n}{R_{\text{b}}}. \tag{9.5}$$

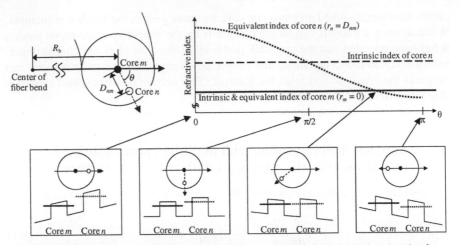

FIGURE 9.2 Schematics of equivalent refractive index variation induced by bend and twist. The center of Core *m* is taken as the origin of the local coordinate for simple description.

Since E can be expressed as

$$E = \exp\left(-j \int_0^z \beta_{eq} dz\right) A, \tag{9.6}$$

where $A = [A_1, A_2, \ldots, A_N]^T$ is the vector the complex amplitude A_n of E_n, Eq. (9.1) can be rewritten as

$$\frac{d}{dz} A = -j \exp\left(j \int_0^z \beta_{eq} dz\right) \kappa \exp\left(-j \int_0^z \beta_{eq} dz\right) A. \tag{9.7}$$

A component of Eq. (9.7) is represented as

$$\frac{d}{dz} A_n = \sum_{m \neq n} -j \kappa_{nm} \exp\left[-j \int_0^z \left(\beta_{eq,m} - \beta_{eq,n}\right) dz\right] A_m. \tag{9.8}$$

When $\beta_{eq} = \beta_c$ is constant, a well-known form of the coupled-mode equation is obtained as

$$\frac{d}{dz} A_n = \sum_{m \neq n} -j \kappa_{nm} \exp\left[-j \left(\beta_{c,m} - \beta_{c,n}\right) z\right] A_m. \tag{9.9}$$

If a simple two core case (Cores m and n) is considered with Eq. (9.9), and if only Core m is excited, the powers of the two cores are given as

$$|A_m(z)|^2 = 1 - |A_n(z)|^2, \quad |A_n(z)|^2 = F_{nm} \sin^2(q_{nm}z), \tag{9.10}$$

where

$$F_{nm} = \left[1 + \left(\frac{\beta_{c,n} - \beta_{c,m}}{2\kappa_{nm}}\right)^2\right]^{-1}, \quad q_{nm} = \left[\kappa_{nm}^2 + \left(\frac{\beta_{c,n} - \beta_{c,m}}{2}\right)^2\right]^{\frac{1}{2}}. \quad (9.11)$$

In this ideal case, a slight difference in effective refractive indices n_{eff} induces a large difference in $\beta_c = 2\pi n_{eff}/\lambda$ since the wavelength $\lambda \sim 10^{-6}$ is very small, and therefore decreases the power conversion efficiency F considerably when κ is small [18,27]. However, it was found that β_b and β_s of the actual MCF are not negligible and fluctuated along the longitudinal direction of the MCF, and the crosstalk in the actual MCF cannot be predicted using the simple unperturbed coupled-mode equation in Eq. (9.9) [28–32].

For example, Figure 9.3 shows an example of the longitudinal evolution of the coupled power or $|A_n|^2$ in an MCF, simulated using Eqs. (9.2), (9.5), and (9.8). The MCF was bent at the constant R_b, and twisted continuously at a constant twist rate of 2 turns/m. Discrete-like dominant changes (bend-induced resonant couplings) were observed at every phase-matching point where the difference Δn_{eqeff} in the equivalent effective indices n_{eqeff} between the cores equals zero, and the crosstalk changes in the other positions can be regarded just as local fluctuations. Even if the propagation constants are different between the cores, such resonant coupling can occur in small bending radii [30–32]. By taking the origin of the local coordinate at the midpoint of Cores m and n, the threshold bending radius R_{PM}—that is, the maximal R_b where the phase matching can induced by the bend—is obtained from Eq. (9.4) as

$$R_{PM,nm} \approx \gamma D_{nm} \frac{(n_{eff,n} + n_{eff,m})/2}{|n_{eff,n} - n_{eff,m}|} = \gamma D_{nm} \frac{(\beta_{c,n} + \beta_{c,m})/2}{|\beta_{c,n} - \beta_{c,m}|}, \quad (9.12)$$

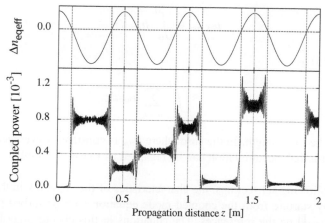

FIGURE 9.3 An example of longitudinal evolution of coupled power in a bent and twisted MCF. (Replotted from the data in Ref. [32].)

where D_{nm} is the core-to-core distance between Cores m and n, or the distance between the centers of the cores. The dominant changes appear random, because the phase differences between Cores m and n are different for every phase-matching point. The phase differences can easily fluctuate in practice by slight variations of the perturbations. Therefore, the crosstalk can be understood as a practically stochastic parameter.

Since the perturbations β_b and β_s behave as random processes, coupling in the MCF is described using power-coupled equations [28,29]:

$$\frac{dP_n}{dz} = \sum_{m \neq n} h_{nm} [P_m - P_n], \tag{9.13}$$

where P_n is the average power in Core n, h_{nm} is the power coupling coefficient from Core m to Core n, and z is the longitudinal position of the MCF. The crosstalk— that is, average/mean crosstalk, or statistical average/mean μ_X of crosstalk X, to be exact—can be defined in many ways, but one of the widely used definitions of the average crosstalk from Core m to Core n is

$$\mu_{X,nm} \equiv P_{nm} / P_{mm}, \tag{9.14}$$

where P_{nm} is the output power of Core n when only Core m is excited. The average crosstalk to Core n also can be defined as

$$\mu_{X,n} \equiv \sum_{m \neq n} P_{nm} / P_{nn}, \tag{9.15}$$

which may correspond to the crosstalk-to-signal ratio. If μ_X is small enough, or $P_m - P_n \approx P_m|_{z=0}$ can be assumed in Eq. (9.13), Eq. (9.14) can be approximated as

$$\mu_{X,nm} \approx \int_0^L h_{nm} dz, \tag{9.16}$$

and Eq. (9.15) as

$$\mu_{X,n} \approx \int_0^L \sum_{m \neq n} h_{nm} dz, \tag{9.17}$$

where L is the fiber length. In the case where h_{nm} is constant or averaged along the MCF, the crosstalk accumulation can be regarded as linear accumulation (+10 dB/ decade).

In the following subsections, the characteristics of the crosstalk in the MCF are described by starting from the coupled-mode equation with perturbed propagation constants, based on the references. The equations in this chapter may be different from those in the references because of the differences of the definitions of symbols and functions.

9.2.2 Redefinition of mode coupling coefficient

Formulations in Section 9.2.1 are based on the conventional coupled-mode theory (CMT) [33], which is called orthogonal CMT, since orthogonality of the modes is assumed. In the orthogonal CMT, the mode coupling coefficient is defined as

$$\kappa_{nm} \equiv \frac{\omega \varepsilon_0 \iint \left(n^2 - n_m^2\right) e_n^* \cdot e_m \, dx dy}{\iint \hat{z} \cdot \left(e_n^* \times h_n + e_n \times h_n^*\right) dx dy}, \tag{9.18}$$

where $\omega = 2\pi c/\lambda$ is the angular frequency, c is the speed of light in vacuum, n is the actual index profile, n_m is the index profile of Core m in the absence of the other cores, e and h are the normalized vector core modes of the electric and magnetic fields, respectively and the superscript * indicates the complex conjugate. Although the mode coupling coefficients should be symmetric as $\kappa_{nm} = \kappa_{mn}^*$ for power conservation in loss-less systems, they are asymmetric between non-identical cores, therefore the orthogonal CMT with non-identical cores is not self-consistent [34]. To cope with such problems, the nonorthogonal CMT was developed [35–37], which was formulated using the cross power term:

$$C_{nm} \equiv \frac{\iint \hat{z} \cdot \left(e_n^* \times h_m + e_m \times h_n^*\right) dx dy}{\iint \hat{z} \cdot \left(e_n^* \times h_n + e_n \times h_n^*\right) dx dy}. \tag{9.19}$$

However, when an overlap of the modes of different cores is small enough and the cross power term is negligible, the nonorthgonal CMT can be reduced to a self-consistent orthogonal CMT with the redefined mode coupling coefficient [38,39]:

$$\bar{\kappa}_{nm} = \bar{\kappa}_{mn} \equiv (\kappa_{nm} + \kappa_{mn})/2. \tag{9.20}$$

In the case of the low crosstalk MCF, the overlap of the modes can be assumed to be well suppressed, therefore the orthogonal CMT with the redefined mode coupling coefficient can be employed for theoretical considerations. In this chapter, the coupling coefficients κ in the equations are not barred explicitly, but the redefined mode coupling coefficient in Eq. (9.20) should be used.

9.2.3 General expression of the power coupling coefficient under random perturbations

To obtain the power coupling coefficient of the MCF, the coupling within a small segment $[z_1, z_2]$ is discussed, wherein the longitudinal perturbations can be considered as a stationary process [31,40]. From Eq. (9.7), the evolution of A in the small segment can be expressed as

$$A(z_2) = T|_{z_1}^{z_2} A(z_1), \tag{9.21}$$

$$T|_{z_1}^{z_2} = \exp\left[-j \int_{z_1}^{z_2} \exp\left(j \int_{z_1}^{z} \beta_{eq} dz'\right) \kappa \exp\left(-j \int_{z_1}^{z} \beta_{eq} dz'\right) dz\right]$$

$$\approx \mathbf{I} - j \int_{z_1}^{z_2} \exp\left(j \int_{z_1}^{z} \beta_{eq} dz'\right) \kappa \exp\left(-j \int_{z_1}^{z} \beta_{eq} dz'\right) dz, \quad (9.22)$$

where T is the transfer matrix of A, $\mathbf{I}$ is the identity matrix, and a first-order approximation was used by assuming that κ is small enough. The mode coupling within the small segment can be represented by non-diagonal elements of $T|_{z_1}^{z_2}$. Elements of $T|_{z_1}^{z_2}$ are expressed as:

$$T_{nm} \approx \delta_{nm} - j \int_{z_1}^{z_2} \kappa_{nm} \exp\left[-j \int_{z_1}^{z} (\beta_{eq,m} - \beta_{eq,n}) \, dz'\right] dz. \quad (9.23)$$

By assuming that κ are constant over the small segment, non-diagonal elements of $T|_{z_1}^{z_2}$, which contribute mode coupling, can be rewritten as

$$T_{nm} \approx -j\kappa_{nm} \int_{z_1}^{z_2} \exp\left[j\Delta\beta_{c,nm} (z - z_1)\right] \exp\left(j \int_{z_1}^{z} \Delta\beta_{b+s,nm} dz'\right) dz, \quad (9.24)$$

where $\Delta\beta_{c,nm} = \beta_{c,n} - \beta_{c,m}$ is the unperturbed propagation-constant difference, and $\Delta\beta_{b+s,nm} = (\beta_{b,n} + \beta_{s,n}) - (\beta_{b,m} + \beta_{s,m})$ is the difference of the longitudinal perturbations on the propagation constants. By defining $f(z)$ as the accumulated phase difference induced by the perturbations:

$$f(z) \equiv \exp\left(j \int_{z_1}^{z} \Delta\beta_{b+s,nm} dz'\right), \quad (9.25)$$

the average power coupled within the segment can be expressed as

$$\langle |T_{nm}|^2 \rangle = \langle T_{nm} T_{nm}^* \rangle$$

$$\approx \kappa_{nm}^2 \int_{z_1}^{z_2} \int_{z_1}^{z_2} \exp\left[j\Delta\beta_{c,nm} (z' - z)\right] \langle f(z')f^*(z)\rangle dz'dz, \quad (9.26)$$

where $\langle\cdot\rangle$ represents the ensemble average. Since the autocorrelation function (ACF) of $f(z)$ is represented as

$$R_{ff}(z' - z) = \langle f(z') f^*(z)\rangle. \quad (9.27)$$

Equation (9.26) can be rewritten as

$$\langle |T_{nm}|^2 \rangle \approx \kappa_{nm}^2 \int_{z_1}^{z_2} \int_{z_1-z}^{z_2-z} R_{ff}(\zeta) \exp\left[j\Delta\beta_{c,nm}\zeta\right] d\zeta dz, \quad (9.28)$$

by substituting $\zeta = z' - z$. In the case where the correlation length of $R_{ff}(\zeta)$ is adequately smaller than the segment length $z_2 - z_1$, boundary terms of the integral with respect to ζ can be neglected, and the integral can be regarded as a Fourier transform.

Based on the Wiener–Khinchin theorem, the power spectrum density (PSD) is the Fourier transform of the ACF:

$$S_{ff}^{(\tilde{v})}(\tilde{v}) = \int_{-\infty}^{\infty} R_{ff}(\zeta) \exp\left(j2\pi\hat{v}\zeta\right) d\zeta, \tag{9.29}$$

where $\tilde{v} = n_{\text{eff}}/\lambda = \beta/(2\pi)$ represents the (not angular) wave number in the medium. Note that plus sign is taken as sign of the exponent in the spatial Fourier transform by considering the sign of the propagation constant in Eqs. (9.1) and (9.6). From the Parseval's theorem, the average power of $f(z)$, or expected value of $|f(z)|^2$, is equivalent to the integral of the PSD over whole $\tilde{v}$, and the following equation holds between $f(z)$ and the PSDs of $f(z)$:

$$\int_{-\infty}^{\infty} S_{ff}^{(\tilde{v})}(\tilde{v})d\tilde{v} = \int_{-\infty}^{\infty} S_{ff}^{(\tilde{v})}(\tilde{v})\frac{d\tilde{v}}{d\beta}d\beta = \text{E}\left[|f(z)|^2\right] = 1. \tag{9.30}$$

Thus, in this chapter, the PSD $S_{ff}^{(\beta)}(\beta)$ with the scale of the propagation constant β (the angular wave number in the medium) is defined as:

$$S_{ff}^{(\beta)}(\beta) \equiv S_{ff}^{(\tilde{v})}(\tilde{v})\frac{d\tilde{v}}{d\beta} = \frac{1}{2\pi}\int_{-\infty}^{\infty} R_{ff}(\zeta) \exp\left(j\beta\zeta\right) d\zeta. \tag{9.31}$$

Then, Eqs. (9.28), (9.29), and (9.31) give the power coupling coefficient $h_{nm}^{(z_1,z_2)}$ in the segment (z_1, z_2) [31,40]:

$$h_{nm}^{(z_1,z_2)} = \frac{\langle|T_{nm}|^2\rangle}{z_2 - z_1} \approx \kappa_{nm}^2 S_{ff}^{(\tilde{v})}\left(\frac{\Delta n_{\text{eff},nm}}{\lambda}\right) = 2\pi\kappa_{nm}^2 S_{ff}^{(\beta)}\left(\Delta\beta_{c,mn}\right). \tag{9.32}$$

Because the unit of the PSD with the scale of the angular wave number is different, the expression of Eq. (9.32) is a bit different from those in Refs. [31,40]; however, they are essentially the same.

9.2.4 Effects of bend and twist

Fini et al. investigated crosstalk behavior under the perturbations induced by random bend and twist using Eq. (9.32) [31]. Figure 9.4a and b shows two sets of examples of the curvature and twist (bend orientation) functions. In Ref. [31], the standard deviations of the curvature and the twist rate were assumed as 20% of its nominal value and 2.5 rad/m, respectively.

Figure 9.4c shows the calculated average crosstalk plotted as a function of effective index mismatch Δn_{eff} for several nominal bending radii R_{b0}. The results indicate that the average crosstalk can be significantly decreased by the appropriate design of Δn_{eff}, taking account of bending conditions of actual MCFs—for example, $\Delta n_{\text{eff}} > 1 \times 10^{-4}$ at $R_{b0} = 1$ for the calculated MCF. Given the standard deviation of 20% of R_{b0}, $D = 42$ μm, and $n_{\text{eff}} = 1.444$, R_{PM} of $\sim 6 \times 10^{-1}$ m at $\Delta n_{\text{eff}} = 1 \times 10^{-4}$ is consistent with the results in Figure 9.4c. Hayashi et al. also

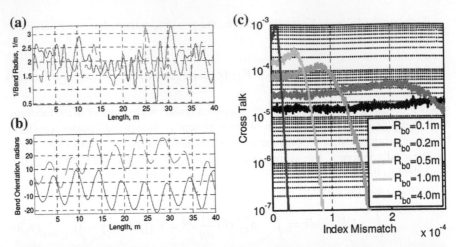

FIGURE 9.4 Examples of random profiles of (a) curvature and (b) bend oriatation, and (c) average crosstalk as a function of effective index mismatch for several nominal bending radii, in the case of $\lambda = 1550$ nm, $\kappa/k \approx 4.6 \times 10^{-9}$ ($\kappa \approx 1.9 \times 10^{-2}$), $D = 42$ µm, $L = 40$ m. (Reproduced from Ref. [31]. ©2010 OSA).

investigated the effect of the bend on the crosstalk by evaluating the average cross-talk between two dissimilar neighboring cores in an MCF from calculations using Eqs. (9.2), (9.5), and (9.8), and from measurements [32]. From both the calculations and the measurements, they observed large crosstalk increase—more than 20 dB—at bending radii less than R_{PM}.

9.2.5 Effects of the random structural fluctuations

Though $f(z)$ in Eq. (9.25) can include the structural fluctuation-induced perturbation β_s, it is a bit difficult to assume a proper perturbation β_s in the form of Eq. (9.25). The effect of the perturbation β_s can be evaluated from the equation obtained by formulating the effects of the perturbations β_b and β_s separately as follows [39,41].

By assuming that β_b is constant and β_s is variable over the small segment $[z_1, z_2]$, Eq. (9.23) can be rewritten as

$$T_{nm} \approx -j\kappa_{nm} \int_{z_1}^{z_2} \exp\left[j\Delta\beta_{c+b,nm}(z-z_1)\right] \exp\left(j\int_{z_1}^{z} \Delta\beta_{s,nm}dz'\right) dz, \quad (9.33)$$

where $\Delta\beta_{c+b,nm} = (\beta_{c,n} + \beta_{b,n}) - (\beta_{c,m} + \beta_{b,m})$, and $\Delta\beta_{s,nm} = \beta_{s,n} - \beta_{s,m}$. By defining the accumulated phase difference $g(z)$ induced by the structural fluctuation in the equation:

$$g(z) \equiv \exp\left(j\int_{z_1}^{z} \Delta\beta_{s,nm}dz'\right), \quad (9.34)$$

the average power coupled within the segment can be expressed as

$$\langle |T_{nm}|^2 \rangle \approx \kappa_{nm}^2 \int_{z_1}^{z_2} \int_{z_1-z}^{z_2-z} R_{gg}\left(\zeta\right) \exp\left[j\Delta\beta_{c+b,nm}\zeta\right] d\zeta\,dz. \qquad (9.35)$$

In a similar way of deriving Eq. (9.32), by assuming that the correlation length l_c of R_{gg} is short enough than the small segment $[z_1, z_2]$, the power coupling coefficient can be obtained as [39,41]:

$$h_{nm}(z) \approx \kappa_{nm}^2 S_{gg}^{(\tilde{\nu})}\left(\frac{\Delta n_{\text{eqeff},nm}}{\lambda}\right) = 2\pi\kappa_{nm}^2 S_{gg}^{(\beta)}\left[\Delta\beta_{c+b,nm}(z)\right]. \qquad (9.36)$$

Equation (9.36) holds if the change of $\Delta\beta_b$—that is, changes of R_b and θ—are gradual enough compared to l_c. Note that the expression of Eq. (9.36) is also different from that in Refs. [39,41], because of the difference of the definitions of the PSD as is the case with Eq. (9.32).

Koshiba et al. investigated the effects of correlation length l_c and shape of the ACF R_{gg} on the average crosstalk μ_X, and agreement between measured μ_X and μ_X obtained using Eq. (9.36) [39]. In Ref. [39], some types of R_{gg} were assumed, such as exponential ACF (EAF):

$$R_{gg}(\zeta) = \exp\left(-|\zeta|/l_c\right) \qquad (9.37)$$

and Gaussian ACF (GAF):

$$R_{gg}(\zeta) = \exp\left[-(\zeta/l_c)^2\right]. \qquad (9.38)$$

The EAF and the GAF have been introduced to microbending loss analysis [42,43]. Since the PSD of these ACFs are the Lorentzian distribution for the EAF, and Gaussian distribution for the GAF; closed forms of the power coupling coefficient $h_{nm}(z)$ corresponding to Eqs. (9.37) and (9.38) were obtained as [39,41]:

$$h_{nm}(z) = \kappa_{nm}^2 \frac{2l_c}{1 + \left[\Delta\beta_{c+b,nm}(z)\,l_c\right]^2}, \quad \text{(for EAF)}, \qquad (9.39)$$

$$h_{nm}(z) = \kappa_{nm}^2 \sqrt{\pi} l_c \exp\left\{-\left[\frac{\Delta\beta_{c+b,nm}(z)\,l_c}{2}\right]^2\right\}, \quad \text{(for GAF)}. \qquad (9.40)$$

Figure 9.5 shows the relationships between the bending diameter $2R_b$ and the average crosstalk μ_X. Measured and calculated values are plotted in the graphs. In the phase-matching region where the bending diameter is smaller than about 600 mm, the average crosstalk is hardly affected by either the correlation length l_c or the shape of the ACF R_{gg}, because the bend-induced perturbation is dominant.

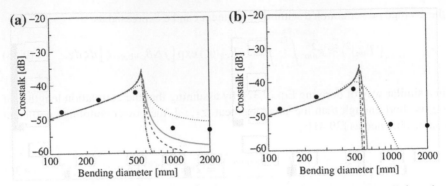

FIGURE 9.5 Relationships between the bending diameter and the average crosstalk (noted as "Crosstalk" in the graph). Closed circles: measured average crosstalk, lines: average crosstalk calculated using (a) h_{nm} for EAF and (b) h_{nm} for GAF. (dotted lines: $l_c = 0.01$ m, solid lines: $l_c = 0.05$ m, dashed lines: $l_c = 0.1$ m, dash-dotted lines: $l_c = 0.5$ m). (Reproduced from Ref. [39]. ©2011 OSA.)

From the crosstalk behavior in the non-phase-matching regions where the bending diameter is larger than about 600 mm, the ACF R_{gg} proper for the evaluated MCF can be recognized as the EAF, and the correlation length l_c of the R_{gg} can be considered to be around 5 cm [39].

9.2.6 Average crosstalk under gradual and random fiber rotation

Let $p_{\theta_{nm}}(\theta_{nm})$ and $p_{R_b}(R_b)$ be the probability density functions of θ_{nm} and of R_b, respectively, along the MCF; by assuming that $p_{\theta_{nm}}(\theta_{nm})$ and $p_{R_b}(R_b)$ are statistically independent, and the twist of the MCF is gradual enough; from Eqs. (9.16) and (9.36), the average crosstalk $\mu_{X,nm}$ can be expressed as

$$\mu_{X,nm} \approx LE(h_{nm})$$

$$\approx L \int_0^\infty p_{R_b}(R_b) \int_0^{2\pi} p_{\theta_{nm}}(\theta_{nm}) 2\pi \kappa_{nm}^2 S_{gg}^{(\beta)}\left[\Delta\beta_{c+b,nm}(R_b,\theta_{nm})\right] d\theta_{nm} dR_b,$$

$$\tag{9.41}$$

where $E(\cdot)$ represents the expected value, and

$$\Delta\beta_{c+b,nm}(R_b,\theta_{nm}) = \Delta\beta_{c,nm} + \Delta\beta_{b,nm}^{\text{dev}}(R_b)\cos\theta_{nm}, \tag{9.42}$$

$$\Delta\beta_{b,nm}^{\text{dev}}(R_b) = \gamma\beta_{c,n}\frac{D_{nm}}{R_b}. \tag{9.43}$$

By assuming that the twist of the MCF is random enough and the MCF is adequately long, $p_{\theta_{nm}}(\theta_{nm})$ can be assumed to be constant ($=1/(2\pi)$) over all θ_{nm}; therefore, by substituting $b = -\Delta\beta_{b,nm}^{\text{dev}}(R_b)\cos\theta_{nm}$ and using $\sin(\arccos x) = \sqrt{1-x^2}$, Eq. (9.41) can be rewritten as

$$\mu_{X,nm} \approx L \int_0^\infty p_{R_b}(R_b) \int_0^\pi 2\kappa_{nm}^2 S_{gg}^{(\beta)}\left[\Delta\beta_{c,nm} + \Delta\beta_{b,nm}^{\text{dev}}(R_b)\cos\theta_{nm}\right] d\theta_{nm} dR_b$$

$$\approx L \int_0^\infty p_{R_b}(R_b) \int_{-\Delta\beta_{b,nm}^{\text{dev}}}^{\Delta\beta_{b,nm}^{\text{dev}}} \frac{2\kappa_{nm}^2}{\sqrt{\left[\Delta\beta_{b,nm}^{\text{dev}}(R_b)\right]^2 - b^2}} S_{gg}^{(\beta)}(\Delta\beta_{c,nm} - b)\, db\, dR_b$$

$$\approx L \int_0^\infty p_{R_b}(R_b)\, \tilde{h}_{nm}(\Delta\beta_{c,nm}, R_b)\, dR_b, \tag{9.44}$$

where

$$\tilde{h}_{nm}(\Delta\beta_{c,nm}, R_b) = 2\pi\kappa_{nm}^2 \left[p_{\Delta\beta_b}(\Delta\beta_{c,nm}) * S_{gg}^{(\beta)}(\Delta\beta_{c,nm})\right] \tag{9.45}$$

is the power coupling coefficient as a function of R_b averaged over θ_{nm}, and the arc-sine distribution:

$$p_{\Delta\beta_b}(\Delta\beta_{c,nm}) = \begin{cases} \left[\pi\sqrt{\left(\Delta\beta_{b,nm}^{\text{dev}}\right)^2 - \Delta\beta_{c,nm}^2}\right]^{-1}, & |\Delta\beta_{c,nm}| \leqslant \Delta\beta_{b,nm}^{\text{dev}}, \\ 0, & \text{otherwise,} \end{cases} \tag{9.46}$$

is the probability distribution of $\Delta\beta_b$, and the operator $*$ denotes the convolution.

Especially in the case where $|\Delta\beta_{c,nm}|$ and the bandwidth of $S_{gg}^{(\beta)}$ are adequately smaller than $\Delta\beta_{b,nm}^{\text{dev}}$, $S_{gg}^{(\beta)}$ becomes a narrow delta-function-like distribution and the convolution contains only a gradually varying part of $p_{\Delta\beta_b}(\Delta\beta_{c,nm})$; therefore, Eq. (9.45) can be approximated as

$$\tilde{h}_{nm}(\Delta\beta_{c,nm}, R_b) \approx 2\pi\kappa_{nm}^2 p_{\Delta\beta_b}(\Delta\beta_{c,nm})$$

$$\approx \frac{2\kappa_{nm}^2}{\sqrt{\left(\gamma\beta_{c,n}D_{nm}/R_b\right)^2 - \Delta\beta_{c,nm}^2}}, \tag{9.47}$$

which is also obtained from Eq. (9.32) by approximating the PSD $S_{ff}^{(\beta)}$ as the probability distribution of $\Delta\beta_b$ in Eq. (9.46) with constant R_b [44]. In the case of homogeneous MCFs ($\Delta\beta_{c,nm} = 0$), Eq. (9.47) is reduced to

$$\hat{h}_{nm}(R_b) \approx \frac{2\kappa_{nm}^2 R_b}{\gamma\beta_{c,n}D_{nm}}. \tag{9.48}$$

9.2.7 Statistical distribution of the crosstalk

So far, the average/mean value μ_X of the crosstalk X has been discussed. In this section, how the crosstalk X is statistically distributed is discussed, especially in the phase-matching case that $R_b < R_{pk}$. In the phase-matching case, e.g. the case of Figure 9.3, longitudinal evolution of the crosstalk X_{nm} from Core m to Core n can be approximated by approximating the coupled-mode equation as the discrete changes [45,46]:

$$A_n (N_{PM}) \approx A_n (N_{PM} - 1) - j K_{nm} (N_{PM}) \exp\left[-j\phi_{rnd} (N_{PM})\right] A_m (N_{PM} - 1)$$

$$\approx A_n (0) - j \sum_{l=1}^{N_{PM}} K_{nm} (l) \exp\left[-j\phi_{rnd} (l)\right] A_m (l - 1), \qquad (9.49)$$

where $A_n(N_{PM})$ represents the complex amplitude of Core n after N_{PM}th phase-matching point, $\phi_{rnd}(N_{PM})$ is the phase difference between Cores m and n at N_{PM}th phase-matching point, and K_{nm} is the coefficient for the discrete changes caused by the coupling from Core m to Core n. ϕ_{rnd} can be regarded as a random sequence, and ϕ_{rnd} significantly varies with slight variations of the bend, twist, and other perturbations. Here, the adequately low crosstalk ($|A_n(N_{PM})| \ll 1$) is assumed for the approximations that $A_m(0) \approx 1$ and the crosstalk $X_{nm} \approx |A_n(N_{PM})|^2$. Since $\Re[K_{nm} \exp(j\phi_{rnd})]$ and $\Im[K_{nm} \exp(j\phi_{rnd})]$ have the variance $\sigma_{2df,nm}^2$ of $|K_{nm}|^2/2$, probability density functions (pdf) of $\Re A_n(N)$ and $\Im A_n(N)$ converge to Gaussian distributions whose variance $\sigma_{2df,nm}^2$ is $N_{PM}|K_{nm}|^2/2$ if N_{PM} is adequately large, based on the central limit theorem. When assuming random polarization mode coupling, coupled power can be equally distributed to two polarization modes statistically. Therefore pdf's of $\Re A_n(N_{PM})$'s and $\Im A_n(N_{PM})$'s of two polarization modes converge to the Gaussian distribution with the variance $\sigma_{4df,nm}^2$ of $N_{PM}|K_{nm}|^2/4$. In this case, the crosstalk X can be represented as a sum of the powers of $\Re A_n(N_{PM})$'s and $\Im A_n(N_{PM})$'s of two polarization modes. Since the sum χ^2 of the powers of four normally distributed random numbers divided by their own variances is chi-squared distributed with 4 degrees of freedom (df):

$$f_{\chi^2,4df} (x) = \frac{x}{4} \exp\left(-\frac{x}{2}\right), \qquad (9.50)$$

$X_{nm}/\sigma_{4df,nm}^2$ can be chi-square distributed with 4df, and its statistical average $\mu_{X,nm}/\sigma_{4df,nm}^2$ is 4. Therefore, the pdf and the statistical average $\mu_{X,nm}$ of X_{nm} can be obtained as

$$f_{X,4df} (X_{nm}) = f_{\chi^2,4df}\left(\frac{X_{nm}}{\sigma_{4df,nm}^2}\right) \frac{d}{dX}\left(\frac{X_{nm}}{\sigma_{4df,nm}^2}\right)$$

$$= \frac{X_{nm}}{4\sigma_{4df,nm}^4} \exp\left(-\frac{X_{nm}}{2\sigma_{4df,nm}^2}\right), \qquad (9.51)$$

$$\mu_{X,nm} = 4\sigma_{4df,nm}^2 = N_{PM} |K_{nm}|^2. \qquad (9.52)$$

In the case of the homogeneous MCF ($\Delta\beta_{c,nm} = 0$), by assuming that the MCF is bent at a constant R_b and twisted at a constant twist rate, the average crosstalk from Core m to Core n can be derived as [46]:

$$\mu_{X,nm} \approx \frac{2\kappa_{nm}^2 R_b L}{\gamma\beta_{c,n} D_{nm}}, \tag{9.53}$$

which corresponds to Eq. (9.48). In the case that R_b varies along the MCF, Eq. (9.53) may hold by substituting the average bending radius for R_b, based on Eqs. (9.44) and (9.48).

Hayashi et al. observed the statistical distribution of the crosstalk in the actual MCF by utilizing wavelength sweeping for randomizing ϕ_{rnd} [47,48]. Figure 9.6

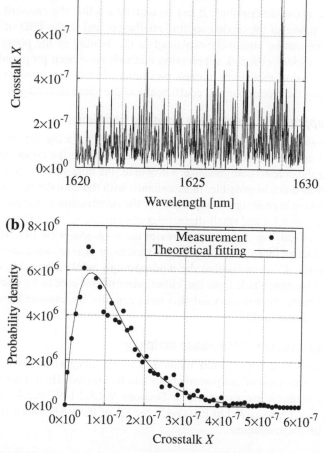

FIGURE 9.6 Examples of (a) a crosstalk spectrum of a 17-km-long homogeneous MCF and (b) a probability distribution of the crosstalk obtained from the spectrum. (Replotted from the data in Refs. [47,48].)

shows examples of a spectrum of the crosstalk of the actual MCF and the probability distribution of the crosstalk obtained from the spectrum. The spectrum was measured using a tunable laser whose linewidth is specified at 100 kHz, which was adequately narrow to avoid averaging of the crosstalk over the wavelength. The measured crosstalk distribution is well fitted by the theoretical fitting line of Eq. (9.51). So, indeed, the crosstalk in the measured MCF is considered to be normally distributed on the complex amplitude planes of the two polarizations. They also reported that the measured average crosstalk obtained from the spectrum was in good agreement with the average crosstalk calculated using Eq. (9.53) with $\gamma = 1$, and proportional to the bending radius as shown in Eq. (9.53).

9.2.8 Crosstalk suppression

Based on the above descriptions, it can be understood that the crosstalk is proportional to the power of the mode coupling coefficient and to the PSD of the perturbations, which can be intuitively explained as the amount of the phase matching. Accordingly, various crosstalk suppression methods have been proposed and demonstrated by ways of reductions of these parameters. In this subsection, the methods for suppressing the mode coupling coefficient and the phase matching are described.

9.2.8.1 Suppression of the mode coupling coefficient

The mode coupling coefficient can be suppressed by reducing the overlap of the modes, so it can be suppressed by confining the modes to the cores strongly and/or enlarging the core-to-core distance. Of course, the core-to-core distance should be shortened as much as possible, in accordance with the core density. Accordingly, high confinement core design is important for the suppression of the mode coupling coefficient. High-index and small-diameter core structure is one of the options, but it degrades the effective area A_{eff} and increases the nonlinear noise [49,50]. Specially designed refractive index profiles—such as trench- or hole-assisted core structures, shown in Figure 9.7—can confine the mode strongly while preserving a large A_{eff} [45,51–54]. One trench/hole layer can either surround one core or be shared between neighboring cores. Photonic-crystal structures can also be leveraged for the strong mode confinement [15–17].

9.2.8.2 Suppression of the phase matching

The phase matching suppression methods can be categorized into some types according to what kind of the perturbations can be utilized. Here, four types of the suppression methods are explained in Sections 9.2.8.2.1–9.2.8.2.4. A schematic example of $\tilde{h}_{nm}\left(\Delta\beta_{c,nm}, R_b\right)$ in Eq. (9.45) for $S_{gg}^{(\beta)}$ of the EAF shown in Figure 9.8 will help with understanding.

9.2.8.2.1 Utilization of the propagation constant mismatch

To suppress the phase matching, one of the methods is utilizing the propagation constant mismatch $\Delta\beta_c(= k\Delta n_{\text{eff}})$ [13,18,30,31]. The propagation constant mismatch

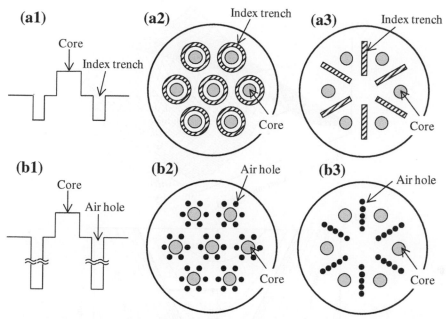

FIGURE 9.7 Schematic examples of (1) refractive-index profiles and (2,3) cross-sections of (a) trench- and (b) hole-assisted MCFs.

$\Delta\beta_c$, which is larger than the maximum bend-induced perturbation $\Delta\beta_b^{dev}$, can prevent the bend-induced phase-matching between the dissimilar cores (the non-phase-matching regions in Figure 9.8). In other words, the bending radius of the MCF has to be managed to be *adequately* larger than R_{PM} in Eq. (9.12)—some margin is needed for avoiding the phase matching induced by the spectral broadening of S_{gg} due to the structural fluctuations. In the heterogeneous MCFs, it is preferred if the correlation length l_c of the structural fluctuation can be elongated [39], because the spectral broadening of S_{gg} can be narrowed and the PSD leakage in the non-phase-matching region can be suppressed, as shown in Figures 9.5 and 9.8.

If most part of an MCF is deployed in gentle-bend conditions, a slight difference in propagation constants or effective indices may be enough for the phase matching suppression, as already described in Section 9.2.4. Matsuo et al. reported that the crosstalk suppression was observed in the bending radii of 1 m or larger with a difference of a few percent in core diameter between cores whose relative refractive index difference Δ_{core} to cladding was ~0.4% and mode field diameters were around 9.6 µm at 1550 nm [55].

If MCFs are deployed in more bend-challenged conditions, the large Δn_{eff} is required for avoiding the bend-induced phase matching. For example, Hayashi et al. reported that a large difference of ~15% in core diameter around 9 µm results in $R_{PM} \sim 7$ cm in the case of $\Delta_{core} = \sim 0.38\%$ [32]. To induce large Δn_{eff} without

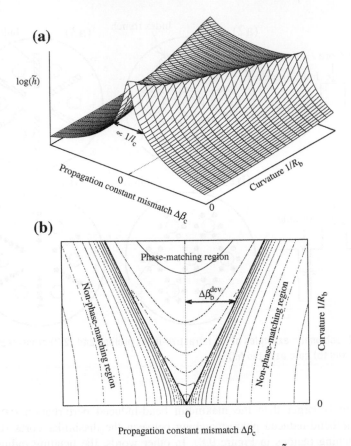

FIGURE 9.8 A schematic example of the power coupling coefficient $\tilde{h}$ averaged over the rotation direction, as a function of the propagation constant mismatch $\Delta\beta_c$ and the curvature $1/R_b$, in the case where the twist of an MCF is gradual and random enough. (a) A 3-dimensional plot and (b) a contour map of $\log(\tilde{h})$. Thick solid lines in (b) are the thresholds between the phase-matching region and the non-phase-matching region.

large variations in propagation characteristics, Kokubun et al. proposed the double cladding structure where cores are surrounded by respective first claddings and the first claddings are surrounded by a second cladding; large index differences can be induced while suppressing variations in characteristics of core modes, by changing the refractive indices of the cores and first claddings [56]. Saitoh et al. reported that, for step-index cores, up to two kinds of dissimilar cores can be designed while achieving A_{eff} of around 80 μm^2 at 1550 nm, cable cutoff wavelength λ_{cc} less than 1530 nm, low macrobending loss, and Δn_{eff} larger than 1×10^{-3}—corresponding to $R_{PM} \sim 5$ cm at $D = 35$–40 μm [57]. Yao et al. proposed to leverage the large difference in core structure for both inducing large Δn_{eff} and adding different functions to dissimilar cores by arranging normal single-mode cores and dispersion compensating cores alternately [54].

It should be noted that most of the crosstalk measurements and transmission experiments reported in papers are considered to have been conducted in the phase-matching region in the case where the cores are designed to have *slightly* different structures, since the typical winding radii of fiber spools are around 10 cm, and R_{PM} less than 10 cm requires a very large differences in core structure as mentioned above.

9.2.8.2.2 Utilization of the bend-induced perturbation

The bend can also be utilized for the phase matching suppression [45–47]. As shown in Figure 9.8, enlargement of the bend-induced perturbation—caused by the increase of the curvature or the decrease of the bending radius—can spread the PSD and suppress the crosstalk even in the case of a homogeneous MCF ($\Delta\beta_c = 0$); identical core structure is rather desirable for suppressing the PSD. The PSD changes gradually with the bend radius, and there is no drastic PSD increase like that around R_{PM} in case of heterogeneous MCFs, since the PSD is suppressed in the phase-matching region. As described in Sections 9.2.6 and 9.2.7, the average crosstalk of a homogeneous MCF is proportional to the average bending radius. Therefore, if the average bending radius of the MCF is managed to be smaller than a certain value, or if the MCF is deployed in bend-challenged conditions, low crosstalk can be achieved with identical cores.

9.2.8.2.3 Utilization of the longitudinal structural fluctuation

As shown in Figure 9.8, the power spectrum of the perturbations is broadened by the longitudinal structural fluctuations. If an MCF has a very short correlation length l_c, the power spectrum spreads broadly over the propagation constant mismatch $\Delta\beta_c$, and thus the PSD may be suppressed even in the case of an unbent homogeneous MCF ($\Delta\beta_c = 0, 1/R_b = 0$). A homogeneous MCF utilizing the longitudinal structural fluctuations was conceptually proposed by Takenaga et al. as "quasi-homogeneous MCF" in Refs. [28,29]. To the author's knowledge, the crosstalk suppression by the structural fluctuation has not been actually observed yet, because the bend-induced perturbations are much larger than the fluctuation induced perturbations in the measurement conditions. However, the structural fluctuation may work when the MCF is cabled and installed in very-gently-bent conditions.

9.2.8.2.4 Utilization of the power spectrum sampling induced by short- and constant-period spin

When the MCF is bent at a constant radius and spun at a constant twist rate, the accumulated phase difference $f(z)$ induced by the perturbations in Eq. (9.25) can be recognized as a frequency-modulated signal whose baseband signal is a simple sinusoid. Since a power spectrum of a periodically modulated signal is sampled at an interval of the frequency of the baseband signal, the power spectrum of $f(z)$ is sampled at the interval of the spin rate f_{spin} (turns per unit length), or $\omega_{spin} = 2\pi f_{spin}$ (radians per unit length), which means that the peak spacing in the spectrum with respect to Δn_{eff} is λf_{spin}. In this case, the power coupling coefficient in Eq. (9.32) can be suppressed except for the sampled peaks of the spectrum, and the power spectrum may become very different from that shown in Figure 9.8. Fini et al. investigated the effects of deterministic constant fiber spinning and other random perturbations on the spectrum [44]. Though there are some challenges, such that high f_{spin} is required to separate the sampled peaks in the spectrum—f_{spin} of 100 turns/m corresponds to the peak separation of 1.55×10^{-4} in Δn_{eff} at $\lambda = 1550$ nm, and that the discrete peaks are broadened by the random perturbations—the short-period spin has a potential to suppress the crosstalk like in the non-phase-matching regions even with a slight Δn_{eff} and a very small bending radius, as pointed out in Ref. [44].

9.2.9 Target level of crosstalk suppression

Though various approaches for crosstalk suppression have been proposed, the appropriate suppression level of the crosstalk is still under investigation. In studies on MCFs themselves, the average crosstalk of -30 dB after 100-km or intended-length of propagation is widely adopted as a target level [18,46,57–60]. Some transmission experiments on MCFs imply the effects of the crosstalk on the transmission performance [61–63]. For example, Winzer et al. reported that 1 dB of Q-penalty was induced at a bit-error ratio (BER) of 1×10^{-3} by a crosstalk of -18 dB, -24 dB, and -32 dB for quadrature phase-shift keying (QPSK), 16-ary quadrature amplitude modulation (QAM), and 64-QAM, respectively; from calculations and experiments at 21.4 Gbaud based on non-stochastic crosstalk realization [61]. Simulations taking account of the crosstalk stochasticity and transmission experiments over actual MCFs will reveal further details of the relationships between the crosstalk and the transmission quality.

9.3 CUTOFF WAVELENGTH VARIATION DUE TO EFFECTS OF SURROUNDING CORES

Other than the crosstalk, another important issue for realizing the short core pitch is the effect of the surrounding cores on the cutoff wavelength of each core. Though trench- or hole-assisted core refractive index profiles are often employed for confining the power strongly into the core, the cutoff wavelength of a certain core in an MCF can be affected by trenches or holes of cores surrounding the certain core [52,64]. If the trenches or holes of the surrounding cores are arranged too near to the surrounded core, they confine not only the modes of the surrounding cores but also the mode of the surrounded core. Figure 9.9 shows an example of a relationship between the cutoff wavelength of the center core of a trench-assisted seven-core fiber and the core pitch, which was reported by Takenaga et al. [52]. Though index profiles of cores in Fiber A and Fiber B were reported to be almost the same, the elongation of the cutoff wavelength clearly shows the effect of the surrounding cores. This effect

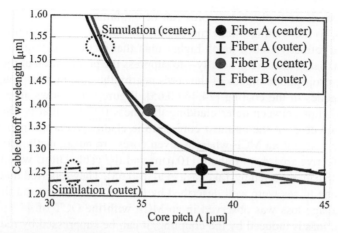

FIGURE 9.9 Examples of dependences of the cutoff wavelength on the core pitch. The MCFs are hexagonally-arranged seven-core fiber with trench-assisted structure. The cutoff wavelength becomes longer when the core pitch is shorter, because higher order modes of the center core are confined by trenches of the outer cores. (Reproduced from Ref. [52]. ©2010 OSA.)

of the surrounding cores on the cutoff wavelength of the surrounded core also have to be taken into consideration in the design of MCF.

To improve the core density without characteristics degradation, the core-to-core distance has to be shortened while suppressing both the crosstalk and the cutoff wavelength variation by designing high-confinement cores, inducing proper perturbations, and finding the best tradeoff between the core-to-core distance, the crosstalk, the cutoff wavelength, and the effective area.

9.4 EFFICIENT UTILIZATION OF FIBER CROSS-SECTIONAL AREA

Once the minimum core-to-core distance that is acceptable in respect of the characteristics, is determined it is desired to improve the ratio of the core-arrangeable area to the fiber cross-sectional area, and core density in the core-arrangeable area, in order to pack as many cores into the cladding as possible.

9.4.1 Outer cladding thickness and excess loss in outer cores

One of the challenges in improving the ratio of the core-arrangeable area is the excess loss in outer cores. When cores in an MCF are arranged too near the cladding–coating interface, loss characteristics of the outer cores of MCFs can be degraded depending on the fiber design [24,46,58,60,65].

The excess loss in the outer cores can be explained as leakage loss, or tunneling loss, to the coating. Since the coating of optical transmission fibers generally has a refractive index much higher than that of the cladding—even much higher than that of the core generally—to suppress the propagation of the cladding modes, the propagation modes of the cores nearby the coating may couple to the radiation modes in the coating [24,46,60,65]. Figure 9.10 shows an example of the relationships between outer cladding thickness (OCT)—that is, the distance between the center of the outmost core and the cladding/coating interface—and the leakage loss, for an MCF wherein seven cores are hexagonally arranged and have an effective area A_{eff} of about $110\,\mu\text{m}^2$ and the cable cutoff wavelength λ_{cc} of 1.33–$1.38\,\mu\text{m}$, reported by Takenaga et al. [60,65]. They reported that the measured leakage losses were well described by the simulated ones, and that no significant leakage loss was observed in an MCF with the OCT of $47.7\,\mu\text{m}$. Since the leakage loss is induced by the coupling, it can be suppressed by reducing the overlap of the core mode and the coating modes. Therefore, the strong mode confinement is effective not only for the crosstalk suppression but also for the leakage loss suppression and thinning of the OCT. For example, Hayashi et al. reported that the leakage loss at $1625\,\text{nm}$ was calculated to be lower than $0.001\,\text{dB/km}$ at an OCT of about $30\,\mu\text{m}$ with high confining trench-assisted cores where A_{eff} is $80\,\mu\text{m}^2$ and λ_{cc} is around $1.50\,\mu\text{m}$, and that no significant leakage loss was observed from the measurements [46].

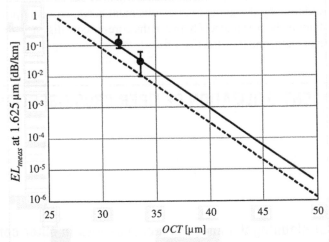

FIGURE 9.10 An example of relationship between the outer cladding thickness (OCT) and the leakage loss (*EL_meas* in the graph) of an outer core. Closed circles: measured values, solid line: a regression line of the measured values, dashed line: simulated value. (Reproduced from Ref. [60]. ©2011 OSA.)

9.4.2 Cladding diameter and mechanical reliability

Another challenge in improving the ratio of the core-arrangeable area is the limitation of the cladding diameter (CD). Since some margin layers—the outer cladding layer and the coating layer—are required for preserving the characteristics of the MCF, enlargement of the CD can reduce the ratio of the margin layers area where there is a constant margin thickness. However, the large CD increases the failure probability of the fiber [66]. Imamura et al. and Matsuo et al. reported that a failure probability of a fiber with a CD of about 200–225 μm can be kept to be that of the standard 125-μm fiber by relaxing the minimum bending radius from 15 mm to 30 mm [67,68]. Imamura et al. reported that a failure probability of a fabricated MCF with a CD of 210 μm and a proof stress of 1.7% was equivalent to that of the standard 125-μm fiber with a proof stress of 1% based on the calculation using the measurement results of the fabricated MCF [69].

9.4.3 Core arrangement

To improve the packing density of cores in a core-arrangeable area, how the cores are arranged is important.

To increase the core counts and improve the core density of heterogeneous MCFs, Koshiba et al. proposed offset arrangement of dissimilar kinds of homogeneous-core lattices [18]. Figure 9.11 shows the schematic cross-sections of heterogeneous MCFs with offset arrangements of homogeneous-core lattices. A homogeneous-core lattice consists of only identical cores, and the identical cores are adequately separated to suppress the crosstalk. Since crosstalk between dissimilar cores is lower than that between identical cores in the case where R_b is adequately larger than R_{PM}, the offset arrangement of homogeneous-core lattices whose cores are dissimilar with other lattices may improve the core density with a negligible increase of crosstalk. Tomozawa and Kokubun investigated the relationships between the core count and a number of

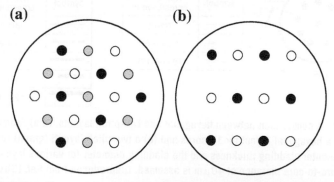

FIGURE 9.11 Schematic cross-sections of heterogeneous MCFs with (a) hexagonal and (b) rectangular core lattices. A difference of core colors represents the difference of core types.

types of cores in the heterogeneous MCFs, and their results indicate that the core count is approximately proportional to the number of types of cores [70].

For homogeneous MCFs, since the hexagonal packing of congruent circles is the densest on an infinite plane, one might assume that a proper hexagonal core layout can provide optimal packing; however, there are cases when non-hexagonal layouts can increase the core count, more so than hexagonal layouts can, depending on the core-arrangeable area and the minimum core-to-core distance. Such non-hexagonal layouts may decrease the core density in core-*arranged* area compared to the hexagonal layouts, but it may be preferable to leverage the core-arrangeable area as much as possible so as to decrease the ratio of the area of the margin layers and improve the core density of the MCF. Matsuo et al. proposed the two-pitch core layout shown in Figure 9.12(b) [68], in order to leverage the core-arrangeable

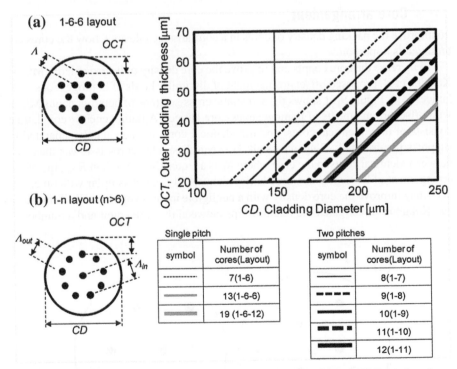

FIGURE 9.12 A comparison between hexagonal and two-pitch layouts. (a,b) Examples of layouts: (a) a hexagonal layout for 13 cores and (b) a two-pitch layout; (graph) relationships between the outer cladding thickness and the cladding diameter for various layouts when the minimum core-to-core distance of 40 μm is assumed. (Reproduced from Ref. [68]. ©2011 OSA.)

area, which is not enough for a hexagonal layout for 13 cores (Figure 9.12a), yet too much for the 7-core hexagonal layout. The graph in Figure 9.12 shows the relationships between the OCT and the CD for various hexagonal and two-pitch layouts when the minimum core-to-core distance of 40 μm is assumed. It can be seen that small changes of the OCT and the CD can be leveraged for increasing the core count by employing the two-pitch layouts. The two-pitch layout also has the effect of suppression of the elongation of the cutoff wavelength and of the crosstalk to the center core—which are the worst in the case of single-pitch, due to the number of surrounding cores.

9.5 CONCLUSION

Recent research on MCFs has revealed various important characteristics of MCFs for SDM transmission, such as inter-core crosstalk, loss degradation in outer cores, and a cutoff wavelength variation due to surrounding cores. The research has well-elucidated mechanisms of such characteristics specific to the MCF; therefore, an MCF can be designed with intended characteristics. Many groups are trying to further improve the characteristics of MCFs and find the best tradeoff between the parameters, such as crosstalk, core-to-core distance, effective area, and cutoff wavelength. For your information, actual characteristics of fabricated MCFs are organized from the references and shown in Table 9.1, so that rough images of the tradeoff between the parameters and the recent status of MCF developments can be grasped.

Though the fundamentals of the MCF have been elucidated considerably, there are many challenges left for commercialization of the MCF. For long-reach application, the development of various peripheral technologies are required, such as power-/space-efficient multi-core amplifiers, pump combiners, low-loss and robust coupling techniques, e.g. splicing and connecting. How to combine the SDM technique using the uncoupled multi-core with other SDM techniques, e.g. few-mode multi-core fibers [71–73], can also be an issue for further capacity improvement. There are no so many issues in short-reach application compared with long-reach application. Simple replacement of single-core devices (fibers, connectors, couplers, etc.) by multi-core devices may improve the core density and is very effective for high-density transmission. Of course, for both long- and short-reach application, further development of the MCFs themselves, along with the developments of SDM transmission systems, are desired so that the performances of the systems can be improved.

Table 9.1 List of characteristics of typical fabricated MCFs in the references.

Ref.	Core Count	Wave-guide Structure[a]	A_{eff}[b] (μm^2)	λ_{cc} (μm)	D_{min} (μm)	L (km)	$\mu_{X,dB}$[c] (dB)	h[c] (km)	R_b (cm)	OCT (μm)	Loss[b] (dB/km) Center Core	Loss[b] (dB/km) Outmost Core	CD (μm)
[28]	7	MC	40	1.12	35.4	1.97	−36	1×10^{-4}	n/a	27.6	0.43	0.44–0.47	125.9
[28]	7	MC	43	1.27	40.4	2.46	−63	2×10^{-7}	n/a	30.8	0.37	0.37–0.39	142.4
[52]	7	MC	72	1.26	39.4	5	−27	4×10^{-4}	10.5	29.4	n/a	n/a	137.5
[52]	7	TA	71	1.26	38.3	3.01	−50	3×10^{-6}	10.5	30.2	0.24	n/a	136.9
[52]	7	TA	77	1.26–1.39	35.4	2.63	−42	2×10^{-5}	10.5	27.3	0.21	n/a	125.4
[19,74]	19	TA	72	1.35–1.46	35	10.4	−32	6×10^{-5}	9	30	0.225	0.225	200
[62]	7	MC	n/a[d]	1.44	46.8	23.5	−45	1×10^{-6}	9	46.5	0.23	0.26	186.5
[46,47]	7	TA	80	1.50	45	17.4	−79.5	6×10^{-10}	14	30	0.176	0.175–0.181	150
[58,59]	7	MC	103–110[e]	1.34–1.37[e]	40	2	−18.5	7×10^{-3}	14	30.5	0.254	0.398–0.444	141
[58,59]	7	MC	99–103[e]	1.36–1.38[e]	40	2	−19.5	6×10^{-3}	14	67.5	0.21	0.270–0.273	215
[59]	7	MC	99–110[e]	1.41–1.49[e]	46	3.5	−38	5×10^{-5}	14	62.5	0.213	0.237	217

(Continued)

Table 9.1 Continued.

Ref.	Core Count	Wave-guide Structure[a]	A_{eff}[b] (μm^2)	λ_{cc} (μm)	D_{min} (μm)	L (km)	$\mu_{X,dB}$[c] (dB)	h[c] (km)	R_b (cm)	OCT (μm)	Loss[b] (dB/km) Center Core	Outmost Core	CD (μm)
[63]	7	TA	110	<1.40	49	75	−65	4×10^{-9}	n/a	48.5	0.190–0.199		195
[60,65]	7	TA	111	1.33	40.7	2.77	−45	1×10^{-5}	10.5	31.6	0.202	n/a[f]	144.6
[60,65]	7	TA	114	1.38	42.6	1.77	−55	2×10^{-6}	10.5	33.6	0.198	n/a[f]	152.4
[60,65]	7	TA	112	1.37	43	1.91	−56	1×10^{-6}	10.5	47.7	0.198	n/a[f]	181.3
[68]	10[g]	TA	116	1.28	40.5	3.96	−40	3×10^{-5}	10.5	43.0	0.242	0.22–0.24	204.4
[54]	6[h]	DC w/ HW	133	1.49	40.1	5	−48.5 ±5.0[i]	$<9 \times 10^{-6i}$	10.5	54.8	n/a	0.211–0.218	189.7

[a]MC: matched cladding structure, TA: trench-assisted structure, DC w/ HW: depressed cladding with hole-walled structure like Figure 9.7(b3).
[b]Values at 1550nm.
[c]Values between nearest neighboring two cores (D=D_{min}) at 1550nm; h is calculated from μ_X/L.
[d]The mode field diameter at 1550nm is reported to be 9.6 μm.
[e]The MCFs are designed as heterogeneous MCFs with slightly dissimilar cores. [f]Excess losses at 1625nm are reviewed in Section 9.4.1.
[g]Cores are arranged in a two-pitch layout.
[h]Cores are arranged in a concentric pattern.
[i]Values of crosstalk X (not average crosstalk μX); h is calculated from $10^{-43.5/10}/L$ by assuming that μ_{X,dB} is lower than −43.5dB

References

[1] A. Sano, T. Kobayashi, S. Yamanaka, A. Matsuura, H. Kawakami, Y. Miyamoto, K. Ishihara, H. Masuda, 102.3-Tb/s (224 × 548-Gb/s) C- and extended L-band all-Raman transmission over 240 km using PDM-64QAM single carrier FDM with digital pilot tone, in: Opt. Fiber Commun. Conf. (OFC), 2012, p. PDP5C.3.

[2] S. Zhang, M.-F. Huang, F. Yaman, E. Mateo, D. Qian, Y. Zhang, L. Xu, Y. Shao, I. Djordjevic, T. Wang, Y. Inada, T. Inoue, Y. Ogata, Y. Aoki, 40×117.6 Gb/s PDM-16QAM OFDM transmission over 10,181 km with soft-decision LDPC coding and nonlinearity compensation, in: Opt. Fiber Commun. Conf. (OFC), 2012, p. PDP5C.4.

[3] J.-X. Cai, Y. Cai, C. Davidson, A. Lucero, H. Zhang, D. Foursa, O. Sinkin, W. Patterson, A. Pilipetskii, G. Mohs, N. Bergano, 20 Tbit/s capacity transmission over 6,860 km, in: Opt. Fiber Commun. Conf. (OFC), 2011, p. PDPB4.

[4] R.-J. Essiambre, G. Kramer, P.J. Winzer, G.J. Foschini, B. Goebel, Capacity limits of optical fiber networks, J. Lightwave Technol. 28 (4) (2010) 662–701.

[5] R.-J. Essiambre, R.W. Tkach, Capacity Trends and Limits of Optical Communication Networks, Proc. IEEE 100 (5) (2012) 1035–1055.

[6] T. Morioka, New generation optical infrastructure technologies: "EXAT initiative" towards 2020 and beyond, in: OptoElectron. Commun. Conf. (OECC), Hong Kong, 2009, p. FT4.

[7] S. Inao, T. Sato, S. Sentsui, T. Kuroha, Y. Nishimura, Multicore optical fiber, in: Opt. Fiber Commun. Conf. (OFC), 1979, p. WB1.

[8] S. Inao, T. Sato, H. Hondo, M. Ogai, S. Sentsui, A. Otake, K. Yoshizaki, K. Ishihara, N. Uchida, High density multicore-fiber cable, in: Int. Wire Cable Symp. (IWCS), 1979, 370–384.

[9] S. Berdagué, P. Facq, Mode division multiplexing in optical fibers, Appl. Opt. 21 (11) (1982) 1950–1955.

[10] N. Kashima, E. Maekawa, F. Nihei, New type of multicore fiber, in: Opt. Fiber Commun. Conf. (OFC), 1982, p. ThAA5.

[11] S. Sumida, E. Maekawa, H. Murata, Design of bunched optical-fiber parameters for 1.3-μm wavelength subscriber line use, J. Lightwave Technol. 4 (8) (1986) 1010–1015.

[12] F. Nihei, Y. Yamamoto, N. Kojima, Optical subscriber cable technologies in Japan, J. Lightwave Technol. 5 (6) (1987) 809–821.

[13] G. Le Noane, D. Boscher, P. Grosso, J.C. Bizeul, C. Botton, Ultra high density cables using a new concept of bunched multicore monomode fibers: a key for the future FTTH networks, in: Int. Wire Cable Symp. (IWCS), 1994.

[14] J.R. Stern, J.W. Ballance, D.W. Fraulkner, S. Hornung, D.B. Payne, K. Oakley, Passive optical local networks for telephony applications and beyond, Electron. Lett. 23 (24) (1987) 1255–1257.

[15] D.M. Taylor, C.R. Bennett, T.J. Shepherd, L.F. Michaille, M.D. Nielsen, H.R. Simonsen, Demonstration of multi-core photonic crystal fibre in an optical interconnect, Electron. Lett. 42 (6) (2006) 331–332.

[16] K. Imamura, K. Mukasa, R. Sugizaki, Y. Mimura, T. Yagi, Multi-core holey fibers for ultra large capacity wide-band transmission, in: Eur. Conf. Opt. Commun. (ECOC), 2008, p. P.1.17.

[17] K. Imamura, K. Mukasa, Y. Mimura, T. Yagi, Multi-core holey fibers for the long-distance >100 km) ultra large capacity transmission, in: Opt. Fiber Commun. Conf. (OFC), 2009, p. OTuC3.

[18] M. Koshiba, K. Saitoh, Y. Kokubun, Heterogeneous multi-core fibers: proposal and design principle, IEICE Electron. Express 6 (2) (2009) 98–103.

[19] J. Sakaguchi, B.J. Puttnam, W. Klaus, Y. Awaji, N. Wada, A. Kanno, T. Kawanishi, K. Imamura, H. Inaba, K. Mukasa, R. Sugizaki, T. Kobayashi, M. Watanabe, 19-core fiber transmission of $19 \times 100 \times 172$-Gb/s SDM-WDM-PDM-QPSK signals at 305Tb/s, in: Opt. Fiber Commun. Conf. (OFC), 2012, p. PDP5C.1.

[20] R. Ryf, R. Essiambre, A. Gnauck, S. Randel, M.A. Mestre, C. Schmidt, P. Winzer, R. Delbue, P. Pupalaikis, A. Sureka, T. Hayashi, T. Taru, T. Sasaki, Space-division multiplexed transmission over 4200 km 3-core microstructured fiber, in: Opt. Fiber Commun. Conf. (OFC), 2012, p. PDP5C.2.

[21] X. Liu, S. Chandrasekhar, X. Chen, P.J. Winzer, Y. Pan, B. Zhu, T.F. Taunay, M. Fishteyn, M.F. Yan, J.M. Fini, E.M. Monberg, F.V. Dimarcello, 1.12-Tb/s 32-QAM-OFDM superchannel with 8.6-b/s/Hz intrachannel spectral efficiency and space-division multiplexing with 60-b/s/Hz aggregate spectral efficiency, in: Eur. Conf. Opt. Commun. (ECOC), 2011, p. Th.13.B.1.

[22] B.G. Lee, D.M. Kuchta, F.E. Doany, C.L. Schow, C. Baks, R. John, P. Pepeljugoski, T.F. Taunay, B. Zhu, M.F. Yan, G.E. Oulundsen, D.S. Vaidya, W. Luo, N. Li, 120-Gb/s 100-m transmission in a single multicore multimode fiber containing six cores interfaced with a matching VCSEL array, in: IEEE Photonics Society Summer Topical Meeting Series, 2010, p. TuD4.4.

[23] B. Zhu, T.F. Taunay, M.F. Yan, M. Fishteyn, G. Oulundsen, D. Vaidya, 7×10-Gb/s multicore multimode fiber transmissions for parallel optical data links, in: Eur. Conf. Opt. Commun. (ECOC), 2010, p. We.6.B.3.

[24] B. Zhu, T.F. Taunay, M.F. Yan, J.M. Fini, M. Fishteyn, E.M. Monberg, Seven-core multicore fiber transmissions for passive optical network, Opt. Express 18 (11) (2010) 11117–11122.

[25] M.J. Gander, E.A.C. Galliot, R. McBride, J.D.C. Jones, Bend measurement using multicore optical fiber, in: Int'l Conf. Opt. Fiber Sensors, Williambsurg, 1997, p. OWC6.

[26] D. Marcuse, Influence of curvature on the losses of doubly clad fibers, Appl. Opt. 21 (23) (1982) 4208–4213.

[27] S. Kumar, U.H. Manyam, V. Srikant, Optical fibers having cores with different propagation constants, and methods of manufacturing same, US Patent 6611648, 26 August 2003.

[28] K. Takenaga, S. Tanigawa, N. Guan, S. Matsuo, K. Saitoh, M. Koshiba, Reduction of crosstalk by quasi-homogeneous solid multi-core fiber, in: Opt. Fiber Commun. Conf. (OFC), 2010, p. OWK7.

[29] K. Takenaga, Y. Arakawa, S. Tanigawa, N. Guan, S. Matsuo, K. Saitoh, M. Koshiba, An investigation on crosstalk in multi-core fibers by introducing random fluctuation along longitudinal direction, IEICE Trans. Commun. E94.B (2) (2011) 409–416.

[30] J.M. Fini, T.F. Taunay, B. Zhu, M.F. Yan, Low cross-talk design of multi-core fibers, in: Conf. Lasers and Electro-Opt. (CLEO). 2010, p. CTuAA3.

[31] J.M. Fini, B. Zhu, T.F. Taunay, M.F. Yan, Statistics of crosstalk in bent multicore fibers, Opt. Express 18 (14) (2010) 15122–15129.

[32] T. Hayashi, T. Nagashima, O. Shimakawa, T. Sasaki, E. Sasaoka, Crosstalk variation of multi-core fibre due to fibre bend, in: Eur. Conf. Opt. Commun. (ECOC), 2010, p. We.8.F.

[33] A.W. Snyder, Coupled-mode theory for optical fibers, J. Opt. Soc. Am. 62 (11) (1972) 1267–1277.

[34] A. Hardy, W. Streifer, Coupled mode theory of parallel waveguides, J. Lightwave Technol. 3 (5) (1985) 1135–1146.

[35] W. Streifer, M. Osinski, A. Hardy, Reformulation of the coupled-mode theory of multiwaveguide systems, J. Lightwave Technol. 5 (1) (1987) 1–4.

[36] S.-L. Chuang, A coupled mode formulation by reciprocity and a variational principle, J. Lightwave Technol. 5 (1) (1987) 5–15.

[37] H.A. Haus, W.P. Huang, S. Kawakami, N.A. Whitaker, Coupled-mode theory of optical waveguides, J. Lightwave Technol. 5 (1) (1987) 16–23.

[38] W.-P. Huang, Coupled-mode theory for optical waveguides: an overview, J. Opt. Soc. Am. A 11 (3) (1994) 963–983.

[39] M. Koshiba, K. Saitoh, K. Takenaga, S. Matsuo, Multi-core fiber design and analysis: coupled-mode theory and coupled-power theory, Opt. Express 19 (26) (2011) B102–111.

[40] D. Marcuse, Coupled power theory, in: theory of dielectric optical waveguides. second ed., Academic Press, San Diego, 1991.

[41] M. Koshiba, K. Saitoh, K. Takenaga, S. Matsuo, Recent progress in multi-core fiber design and analysis, in: OptoElectron. Commun. Conf. (OECC), Busan, Korea, 2012, p. 5E3-1.

[42] K. Petermann, Microbending loss in monomode fibers, Electron. Lett. 12 (4) (1976) 107–109.

[43] D. Marcuse, Microdeformation losses of single-mode fibers, Appl. Opt. 23 (7) (1984) 1082–1091.

[44] J.M. Fini, B. Zhu, T.F. Taunay, M.F. Yan, K.S. Abedin, Crosstalk in multicore fibers with randomness: gradual drift vs. short-length variations, Opt. Express 20 (2) (2012) 949–959.

[45] T. Hayashi, T. Taru, O. Shimakawa, T. Sasaki, E. Sasaoka, Low-crosstalk and low-loss multi-core fiber utilizing fiber bend, in: Opt. Fiber Commun. Conf. (OFC), 2011, p. OWJ3.

[46] T. Hayashi, T. Taru, O. Shimakawa, T. Sasaki, E. Sasaoka, Design and fabrication of ultra-low crosstalk and low-loss multi-core fiber, Opt. Express 19 (17) (2011) 16576–16592.

[47] T. Hayashi, T. Taru, O. Shimakawa, T. Sasaki, E. Sasaoka, Ultra-low-crosstalk multi-core fiber feasible to ultra-long-haul transmission, in: Opt. Fiber Commun. Conf. (OFC), 2011, p. PDPC2.

[48] T. Hayashi, T. Taru, O. Shimakawa, T. Sasaki, E. Sasaoka, Characterization of crosstalk in ultra-low-crosstalk multi-core fiber, J. Lightwave Technol. 30 (4) (2012) 583–589.

[49] G. Bosco, P. Poggiolini, A. Carena, V. Curri, F. Forghieri, Analytical results on channel capacity in uncompensated optical links with coherent detection, Opt. Express 19 (29) (2011) B438–B449.

[50] A. Mecozzi, R.-J. Essiambre, Nonlinear Shannon limit in pseudolinear coherent systems, J. Lightwave Technol. 30 (12) (2012) 2011–2024.

[51] K. Saitoh, T. Matsui, T. Sakamoto, M. Koshiba, S. Tomita, Multi-core hole-assisted fibers for high core density space division multiplexing, in: OptoElectron. Commun. Conf. (OECC), Sapporo, Japan, 2010, p. 7C2-1.

[52] K. Takenaga, Y. Arakawa, S. Tanigawa, N. Guan, S. Matsuo, K. Saitoh, M. Koshiba, Reduction of crosstalk by trench-assisted multi-core fiber, in: Opt. Fiber Commun. Conf. (OFC), 2011, p. OWJ4.

[53] M. Tanaka, T. Yamamoto, T. Kinoshita, S. Kusunoki, H. Taniguchi, A study on low crosstalk multi-core fiber (in Japanese with English abstract), IEICE Tech. Rep. OFT2011-58 111 (381) (2012) 15–18.

[54] B. Yao, K. Ohsono, N. Shiina, F. Koji, A. Hongo, E.H. Sekiya, K. Saito, Reduction of crosstalk by hole-walled multi-core fibers, in: Opt. Fiber Commun. Conf. (OFC), 2012, p. OM2D.5.

[55] S. Matsuo, K. Takenaga, Y. Arakawa, Y. Sasaki, S. Tanigawa, K. Saitoh, M. Koshiba, Crosstalk behavior of cores in multi-core fiber under bent condition, IEICE Electron. Express 8 (6) (2011) 385–390.

[56] Y. Kokubun, T. Watanabe, Dense heterogeneous uncoupled multi-core fiber using 9 types of cores with double cladding structure, in: Microopics Conference (MOC), 17th, Sendai, Japan, 2011, p. K-5.

[57] K. Saitoh, M. Koshiba, K. Takenaga, S. Matsuo, Low-crosstalk multi-core fibers for long-haul transmission, Proc. SPIE 8284 (2012) 82840I.

[58] K. Imamura, K. Mukasa, T. Yagi, Investigation on multi-core fibers with large Aeff and low micro bending loss, in: Opt. Fiber Commun. Conf. (OFC), 2010, p. OWK6.

[59] K. Imamura, Y. Tsuchida, K. Mukasa, R. Sugizaki, K. Saitoh, M. Koshiba, Investigation on multi-core fibers with large Aeff and low micro bending loss, Opt. Express 19 (11) (2011) 10595–10603.

[60] K. Takenaga, Y. Arakawa, Y. Sasaki, S. Tanigawa, S. Matsuo, K. Saitoh, M. Koshiba, A large effective area multi-core fiber with an optimized cladding thickness, Opt. Express 19 (26) (2011) B543–B550.

[61] P. Winzer, A. Gnauck, A. Konczykowska, F. Jorge, J.Y. Dupuy, Penalties from in-band crosstalk for advanced optical modulation formats, in: Eur. Conf. Opt. Commun. (ECOC), 2011, p. Tu.5.B.7.

[62] B. Zhu, J.M. Fini, M.F. Yan, X. Liu, S. Chandrasekhar, T.F. Taunay, M. Fishteyn, E.M. Monberg, F.V. Dimarcello, High capacity space-division-multiplexed DWDM transmissions using multicore fiber, J. Lightwave Technol. 30 (4) (2012) 486–492.

[63] H. Takara, H. Ono, Y. Abe, H. Masuda, K. Takenaga, S. Matsuo, H. Kubota, K. Shibahara, T. Kobayashi, Y. Miyamoto, 1000-km 7-core fiber transmission of 10 × 96-Gb/s PDM-16QAM using Raman amplification with 6.5 W per fiber, Opt. Express 20 (9) (2012) 10100–10105.

[64] Y. Arakawa, K. Takenaga, S. Tanigawa, Y. Sasaki, S. Matsuo, K. Saitoh, M. Koshiba, Length dependence of cutoff wavelength of trench-assisted multi-core fiber, in: OptoElectron. Commun. Conf. (OECC), Kaohsiung, Taiwan, 2011, p. 6C25.

[65] K. Takenaga, Y. Arakawa, Y. Sasaki, S. Tanigawa, S. Matsuo, K. Saitoh, M. Koshiba, A large effective area multi-core fibre with an optimised cladding thickness, in: Eur. Conf. Opt. Commun. (ECOC), 2011, p. Mo.1.LeCervin.2.

[66] IEC/TR 62048 ed2.0, Optical Fibres—Reliability—Power Law Theory, International Electrotechnical Commission, May 2011.

[67] K. Imamura, K. Mukasa, T. Yagi, Effective space division multiplexing by multi-core fibers, in: Eur. Conf. Opt. Commun. (ECOC), 2010, p. P1.09.

[68] S. Matsuo, K. Takenaga, Y. Arakawa, Y. Sasaki, S. Tanigawa, K. Saitoh, M. Koshiba, Large-effective-area ten-core fiber with cladding diameter of about 200 μm, Opt. Lett. 36 (23) (2011) 4626–4628.

[69] K. Imamura, I. Shimotakahara, K. Mukasa, N. Oyama, R. Sugizaki, A study on reliability for large diameter multi-core fibers, in: Int. Wire Cable Symp. (IWCS), 2011, pp. 284–288, P-2.

[70] K. Tomozawa, Y. Kokubun, Maximum core capacity of heterogeneous uncoupled multi-core fibers, in: OptoElectron. Commun. Conf. (OECC), Sapporo, Japan, 2010, p. 7C2-4.

[71] K. Mukasa, K. Imamura, R. Sugizaki, 7-core 2-mode fibers with large Aeff to simultaneously realize "3M", in: OptoElectron. Commun. Conf. (OECC), Busan, Korea, 2012, p. 5C1-1.

[72] K. Takenaga, Y. Sasaki, N. Guan, S. Matsuo, M. Kasahara, K. Saitoh, M. Koshiba, A large effective area few-mode multi-core fiber, in: IEEE Photonics Society Summer Topical Meeting Series, Seattle, 2012, p. TuC1.2.

[73] C. Xia, R. Amezcua-Correa, N. Bai, E. Antonio-Lopez, D. May-Arriojo, A. Schulzgen, M. Richardson, J. Liñares, C. Montero, E. Mateo, X. Zhou, G. Li, Hole-assisted few-mode multi-core fiber for high-density space-division multiplexing, in: IEEE Photonics Society Summer Topical Meeting Series, Seattle, 2012, p. TuC4.2.

[74] K. Imamura, H. Inaba, K. Mukasa, R. Sugizaki, 19-core multi core fiber to realize high density space division multiplexing transmission, in: IEEE Photonics Society Summer Topical Meeting Series, Seattle, 2012, p. TuC4.3.

Plastic Optical Fibers and Gb/s Data Links

10

Yasuhiro Koike[a] and Roberto Gaudino[b]

[a]Faculty of Science and Technology, Keio University,
[b]Politecnico di Torino, Dipartimento di Elettronica e Telecomunicazioni (DET)

10.1 INTRODUCTION

The peripheral component of the communication networks toward the final user, referred to as "the last 1 mile," is estimated to account for approximately 95% of the overall network. Electrical wiring such as unshielded twisted pair (UTP) and coaxial cables has been mainly adopted in local area networks (LANs) for the connection to the user terminals. However, its bandwidth and transmission distance are severely limited, and UTPs have hardships to realize high-speed data transmissions of the order of Gb/s over 100 m. On the other hand, silica-based single mode fibers used in backbone systems can achieve extremely high data rate and long distance communications. However, precise and time-consuming techniques are demanded for termination, connection, and branching because of its quite small diameter of less than 10 μm, which induces high costs for LAN systems with huge numbers of connections and junctions. The same considerations hold true for home networking, whenever a wired solution is preferred inside the house to a wireless connection. In contrast, plastic optic fibers (POFs), consisting of plastic core and cladding, can have much larger core diameters (up to 1000 μm) than silica-based optical fibers because of the inherent flexibility, although POFs exhibit relatively high attenuations. POFs, comparable to UTPs, provide no fear to stick into human skin, and to be broken by bending and physical impact. Highly accurate alignments are not required for POF connections because of the large core. These characteristics enable easy and low cost installation and safe handling.

The demand for high-speed communications over private intranets and the Internet is growing explosively with increasing available data volume in personal devices. POFs are attracting a great deal of attention, because electrical wiring induces critical problems such as electromagnetic interference, high power consumption, bandwidth limitations, high attenuation, and heat generation, even in their simpler step index (SI POF) version [1]. As high-speed data processing and communication systems are required, graded index (GI) POFs become promising candidates for optical interconnects as well as optical networking in LANs at homes, offices, hospitals, and buildings, because of their high bandwidth in addition to the advantages for consumer use

Optical Fiber Telecommunications VIA. http://dx.doi.org/10.1016/B978-0-12-396958-3.00010-X

such as high tolerances to misalignments and bendings, high mechanical strength, and long-term reliabilities [2–4].

Although several types of POFs, such as single mode, step index, multi step index, multicore, graded index, and microstructured, were reported [5, 6], this chapter mainly explains representative SI and GI POFs, since they appear as the only ones that so far had an actual commercial exploitation.

This chapter is organized as follows: POF structures and fabrication methods are reviewed in Section 10.2. The most important characteristics for optical fibers, attenuation and bandwidth, are described in Sections 10.3 and 10.4. Section 10.5 introduces applications of SI and GI POFs.

10.2 STRUCTURE AND FABRICATION OF PLASTIC OPTICAL FIBER

The principle of light propagation through optical fiber is simply explained as follows [7,8]: an optical fiber generally consists of two coaxial layers in cylindrical form. One is a core in the central part of the fiber. The other is a cladding in the peripheral part, which completely surrounds the core. Although the cladding is not required for light propagation in principle, it plays important roles in practical use, such as protection of core surface from imperfection and refractive index change caused by physical contact or absorbing contaminants, and enhancement of mechanical strength. The core has a refractive index slightly higher than the cladding. Therefore, when the incident angle of the light input to the core is greater than the critical angle determined by Snell's law, the input light is confined to the core region and propagates through the fiber, because the light is repeatedly reflected back into the core region due to the total internal reflection at the interface between the core and cladding. The propagation of light along the fiber can be described in terms of electromagnetic waves called "modes" which are patterns of electromagnetic field distributions. The fiber can guide a certain discrete number of modes that must satisfy the electric and magnetic field boundary conditions at the core-cladding interface according to its material and structure, and light wavelength.

Optical fibers can be commonly classified into two types: single mode and multimode fibers [9]. As the names suggest, the single-mode fiber (SMF) allows only one propagating mode, while the multimode fiber (MMF) guides a large number of modes. The POFs are typically designed to be highly multimodal, due to the requirements of obtaining a very large diameter and, usually, a large numerical aperture (NA). Both the SMF and MMF are again divided into two classes: step index (SI) and graded index (GI) fibers. Figure 10.1 conceptually illustrates the refractive index profiles and ray trajectories in SI and GI MMFs. This section describes the structures, characteristics, and fabrication methods of SI and GI POFs.

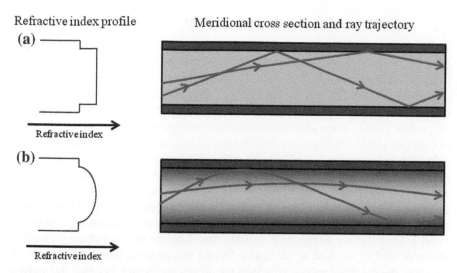

Refractive index profile   Meridional cross section and ray trajectory

(a)

Refractive index

(b)

Refractive index

FIGURE 10.1 Refractive index profiles and ray trajectories of (a) SI and (b) GI MMFs.

10.2.1 Step index plastic optical fiber

SI POF has a very simple radial profile for the refractive index, as shown in Figure 10.1a: the fiber core has a constant refraction index, and it is surrounded by a cladding material with slightly lower (and approximately constant) index. As explained in [62, Section 2.8] SI POF can be fabricated by heat drawing a preform usually made of poly(methyl methacrylate) (PMMA). Anyway, due to the much larger core area compared, for instance, to single-mode glass optical fibers, the efficiency of the preform process is quite low, simply because a much large material volume is required for the same length of fiber.

Thus, SI POF large volume production is usually obtained by continuous extrusion processes in order to obtain a high production rate. This method has several variants, but its main principles can be summarized as follows [63]: a purified monomer (e.g. MMA) along with an initiator and a chain transfer agent are fed into the polymerization reactor where polymerization reaction takes place. The polymer, which is for the core of the SI-POF, is then fed into an extruder by a gear pump. The extruder is capable of devolatilization to remove monomer residues and to return them back to the polymerization reactor. The core material then proceeds into a coextrusion die (or a spinning block) where the concentric core-cladding structure of an SI POF is formed along with a cladding material fed by a separate extruder.

Although the description of this process may appear to be simple, numerous variables including the purity of the monomer, the reaction temperature, the degree of polymerization, amount of the initiator and the chain transfer agent, extrusion conditions play an important role in determining the optical and physical properties of the fiber, mostly in terms of extrinsic attenuation factors, as will be explained in Section 10.3.3.

10.2.2 Graded index plastic optical fiber

The early GI POF was fabricated by heat drawing a preform with GI profile. The preform has a cylindrically symmetric refractive index profile that gradually decreases from the center axis to the periphery. The first GI preform was prepared by copolymerization of, at least, two monomers with different refractive indices and monomer reactivity ratios [2]. There are various reports about fabrication methods of GI profiles [10,11], some representative methods are introduced here. An interfacial gel polymerization (IGP) technique was developed to improve the attenuation and the refractive index profile [12,13]. In this method, a polymer tube is first prepared by polymerizing with rotation. The polymer tube filled with mixture of monomer and dopant with higher refractive index than polymer matrix is heated from the surroundings to induce polymerization. The inner wall of the tube is slightly swollen by the mixture to form the polymer gel phase. The reaction rate of the polymerization is generally faster in the gel phase due to the gel effect. Therefore, the polymer phase grows from the inner wall of the tube to the center. During this process, the monomer can diffuse into the gel phase more easily than the dopant molecules, because the molecular volume of dopant, which typically contains benzene rings, is larger than that of the monomer. Thus, the dopant molecules are gradually concentrated in the center region to form a nearly quadratic refractive index profile corresponding to the dopant concentration distribution. The polymerization reaction rate plays a crucial role to control the refractive index profile, because it strongly affects the diffusion process of monomer and dopant molecules into the polymer gel phase. The IGP technique can provide an optimum refractive index profile well fitted to the power law form, explained in Section 10.4.3, of the PMMA-based GI POF as shown in Figure 10.2a.

Another formation method of the refractive index profile in the preform, a direct diffusion (DD) method, was adopted for perfluorinated (PF) polymer-based GI POF. This GI POF was commercialized from Asahi Glass Co., Ltd (AGC) with a trade name of "Lucina." In the DD method, the dopant material is directly diffused into the molten polymer [14]. However, the refractive index profile formed by the DD method was not necessarily controlled to the optimum power law profile, because

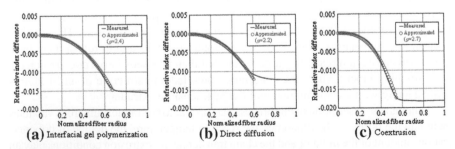

(a) Interfacial gel polymerization **(b)** Direct diffusion **(c)** Coextrusion

FIGURE 10.2 Measured refractive index profiles of GI POFs fabricated by various processes, compared with the power law approximations.

the dopant diffusion into the polymer is basically governed by Fick's diffusion theory [15]. The measured refractive index profile formed by the DD process had a large discrepancy from the power law approximation, particularly at the core-cladding boundary as shown in Figure 10.2b. The tailing part at the core-cladding boundary is the typical profile by Fick's diffusion with a constant diffusion coefficient. The large tailing induces the large modal dispersion, and restricts the bandwidth.

An innovative fabrication method, coextrusion process, was developed for mass production of GI POFs [16–18]. In this process, the GI POF is directly fabricated from polymers, not from the GI preform. The polymers without dopants for the cladding and polymers containing low molecular weight dopants for the core are separately prepared. The two polymers are melted by heat and fed separately into a coextrusion die that has a concentric core-cladding circular structure. The extruded polymer with double layers obtains the dopant concentration distribution due to dopant diffusion from the core to the cladding region during passing through a heat diffusion zone. Thus, the refractive index distribution of the core is originally the SI type and changes to the GI profile by the diffusion process. The dopant concentration distribution (i.e. the GI profile) can be controlled by changing the diffusion temperature and time, and by adjusting the molecular weights and the glass transition temperatures (T_gs) of the core and cladding polymers. The refractive index profile formed by the coextrusion process was well fitted to power law form as shown in Figure 10.2c. This is because the diffusion coefficient depends on the dopant concentration in the coextrusion process. The increase in the dopant concentration decreases the T_g due to plasticization, and thus increases the diffusion coefficient. The T_g is an important factor to determine the diffusion coefficient. This continuous fabrication process is currently utilized in commercially available GI POFs.

10.3 ATTENUATION OF PLASTIC OPTICAL FIBER

Attenuation of fiber mainly determines the maximum transmission distance of optical communication systems without amplifiers or repeaters, as well as the maximum output power from the light source and the minimum receiver sensitivity. The attenuation is basically caused by absorption, scattering, and radiation of optical power. Attenuation, or transmission loss, is defined as the ratio of input and output powers, and expressed by linearly adding the losses due to the individual components in units of dB [19,20]. Optical power decreases exponentially with distance, as light travels along a fiber. Therefore, attenuation is discussed for given fiber length and is typically described in dB/km

$$\alpha = -\frac{10}{L} \log_{10} \left(\frac{P_{out}}{P_{in}} \right), \tag{10.1}$$

$$\alpha = \alpha_a + \alpha_s + \alpha_r, \tag{10.2}$$

where α is the total attenuation, P_{in} is the input optical power into the fiber, P_{out} is the output optical power from the fiber, L is the fiber length, α_a, α_s, and α_r are the losses caused by absorption, scattering, and radiation, respectively. The total attenuation can be easily determined by the cutback method [21]. The optical power is first measured at the output end of the long fiber, and then the fiber is cut off a few meters from the input end, and the output power is again measured at this output end of the short fiber. The launching condition must be identically maintained throughout the measurement.

The extrinsic factors of fiber attenuation such as contaminations, refractive index perturbations, chemical impurities, structural imperfections, and geometrical fluctuations are not discussed here, because the extrinsic attenuation can be dramatically affected by the fabrication method. Thus, the intrinsic absorption and scattering are explained in this section.

10.3.1 Absorption loss

All materials absorb light at wavelengths corresponding to resonance frequencies in the molecules. Intrinsic absorption losses in optical fibers originate from electronic transitions in the ultraviolet wavelength region and molecular vibrations in the near-infrared wavelength region.

Absorption due to electronic transition between the different energy levels within the materials is known as electronic transition absorption. The light absorption occurs when a photon excites an electron to a higher energy level. Electronic transition absorption peaks typically appear in the ultraviolet wavelength region. The optical loss caused by electronic transition absorption α_{et} obeys Urbach's empirical rule [22]

$$\alpha_{et} = A \exp\left(\frac{B}{\lambda}\right), \tag{10.3}$$

where λ is the wavelength of the light, A and B are inherent constants of materials. These parameters were empirically determined in PMMA, polystyrene (PS), and polycarbonate (PC), which are general polymer matrices for POFs [23,24]. Figure 10.3 shows the calculated electronic transition absorption losses of PMMA, PS, and PC. The electronic transition absorption loss of PMMA was remarkably small in the visible wavelength region, and negligible in POF applications, because the loss decreases exponentially with increasing wavelength. In contrast, PS and PC exhibited the large electronic transition absorption losses, because of the transition from π to π^* in the benzene rings [25].

The absorption in the wavelength from visible to near-IR region is mainly caused by the harmonics and their couplings of vibrations of chemical bonds. Vibrational absorptions of carbon–hydrogen (CH) bonds strongly affect the transmission losses of POFs in this wavelength region. The molecular vibrational absorption loss α_{mv} due to the overtone vibration of the CX bonds, based on Morse potential and Lambert-Beer law, is described by

$$\alpha_{mv} = 3.2 \times 10^8 \frac{\rho}{M} N^{CX} \left(\frac{E_v^{CX}}{E_1^{CH}}\right), \tag{10.4}$$

where ρ is the density, M is the molecular weight of the monomer, N^{CX} is the number of CX bonds in monomer unit, v is the vibrational quantum number, and E_v^{CX}/E_1^{CH}

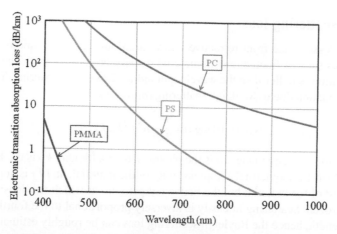

FIGURE 10.3 Calculated electronic transition absorption losses of PMMA, PS, and PC.

denotes the energy ratio of the vth vibration of the CX bond to the fundamental vibration of the CH bond [26]. Figure 10.4 shows the calculated absorption losses caused by the overtones of molecular vibrations of various CX bonds; here, the typical physical constants of PMMA were adopted, and these parameters were assumed to be not affected by substituting deuterium (D), fluorine (F), and chlorine (Cl) for the hydrogen. The absorption intensities decrease by one order with increasing the v value by 1. In addition, the fundamental vibration wavelengths shifted to much longer wavelengths by substituting heavier atoms for the hydrogen. Therefore, the effect of the vibrational absorption can be dramatically reduced and negligible in the visible and the infrared regions by replacing the hydrogen with heavier atoms.

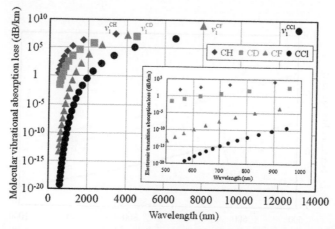

FIGURE 10.4 Calculated absorption losses due to overtones of molecular vibrations of CH, CD, CF, and CCl bonds. Inset shows magnified wavelength range of interest.

10.3.2 Scattering loss

Scattering losses arise from refractive index variations caused by microscopic heterogeneities such as fluctuations in material density or composition. The refractive index variations smaller than the wavelength induce Rayleigh scattering. The optical power attenuation due to Rayleigh scattering α_{Rs} is described by

$$\alpha_{Rs} = 10 \log(e) \frac{8\pi^3}{3\lambda^4} n^8 p^2 k_B T_f \beta_T, \tag{10.5}$$

where n is the refractive index, p is the photoelastic coefficient, k_B is the Boltzmann constant, β_T is the isothermal compressibility of the material, and T_f is not the actual temperature of the material but the fictive temperature at which the density fluctuations are frozen. Scattering intensity is inversely proportional to the fourth power of the wavelength, hence the Rayleigh scattering loss can be roughly estimated by [5]:

$$\alpha_{Rs}(\lambda) = \alpha_{Rs}(\lambda_0) \left(\frac{\lambda_0}{\lambda}\right)^4. \tag{10.6}$$

Thus the scattering loss decreases dramatically with increasing wavelength. In addition, lower refractive index or lower isothermal compressibility of material also leads to lower scattering loss. The inevitable scattering loss was experimentally analyzed, based on thermal fluctuation theory derived by Einstein [27], in PMMA, PS, and PC [24,25,28]. Figure 10.5 shows the scattering losses of PMMA, PS, and PC calculated by Eq. (10.6). The scattering losses of PS and PC were much higher than those of PMMA, because of the optical anisotropy of the benzene rings. The intrinsic scattering loss of PMMA has only a slight influence on the total attenuation, although extrinsic scattering strongly affects the transmission loss in practice.

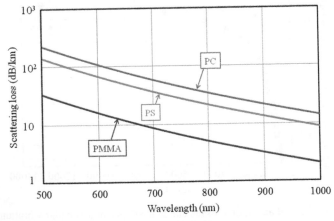

FIGURE 10.5 Calculated scattering losses of PMMA, PS, and PC.

10.3.3 Low loss plastic optical fiber

The attenuations of the first reported SI and GI POFs were greater than 1000 dB/km. The various analyses on the intrinsic factors of fiber attenuation described above revealed the theoretical limit of transmission loss, and clarified that such high attenuations were not strongly affected by the intrinsic factors, but mainly induced by the extrinsic factors, e.g. contaminations and structural imperfections. PS- and PMMA-based POFs with low loss were obtained by improvements of fabricating procedures to reduce the extrinsic loss factors [2,29]. There are various reports about further reductions in attenuation by substituting heavier atoms for the hydrogen [30–33]. Figure 10.6 shows the measured attenuation spectra of the GI POFs based on PMMA, perdeuterated PMMA (PMMA-d8), and PF polymer that is commercially available from AGC with a trade name of "CYTOP." The significant decreases in the transmission losses of GI POFs because of the substitution of the hydrogen were experimentally observed, all the GI POFs shown in Figure 10.6 were fabricated by the IGP technique. The CYTOP-based GI POF exhibited extremely low attenuation because of the low absorption loss due to the absence of the CH bond and the low scattering loss resulted from the low refractive index as mentioned above, although the extrinsic factors were still not eliminated [34].

AGC commercialized the first CYTOP-based GI POF with a trade name of "Lucina" which was fabricated by the DD process. Several types of the low loss CYTOP-based GI POFs are commercially available in recent years from AGC with a trade name of "Fontex," from Sekisui Chemical Co., Ltd with a trade name of "Ginover," and from Chromis Fiberoptics with a trade name of "GigaPOF." These GI POFs are manufactured by the coextrusion processes, which are mass production methods.

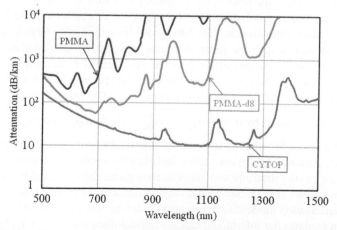

FIGURE 10.6 Measured attenuation spectra of PMMA, PMMA-d8, and CYTOP-based GI POFs.

10.4 BANDWIDTH OF PLASTIC OPTICAL FIBER

Bandwidth of fiber determines the maximum transmission data rate or the maximum transmission distance. Most common POF transmission systems adopt on-off keying by direct modulation of the optical source (laser or LED). If an input pulse waveform can be detected without distortion at the other end of the fiber, except for the decrease in optical power, the maximum link length is limited by the fiber attenuation. However, in addition to the optical power attenuation, the output pulse is generally broader in time than the input pulse. The pulse broadening restricts the transmission capacity, namely bandwidth of the fiber. The bandwidth is determined by the impulse response as follows [20]. Optical fibers are usually considered as quasilinear systems, thus the output pulse is described by

$$p_{out}(t) = h(t) * p_{in}(t). \tag{10.7}$$

The output pulse $p_{out}(t)$ from the fiber can be calculated in the time domain through the convolution (denoted by *) of the input pulse $p_{in}(t)$ and the impulse response function $h(t)$ of the fiber. Fourier transformation of Eq. (10.7) provides the simple expression as the product in the frequency domain

$$P_{out}(f) = H(f)P_{in}(f), \tag{10.8}$$

where $H(f)$ is the power transfer function of the fiber at the baseband frequency f. The power transfer function defines the bandwidth of the optical fiber as the lowest frequency at which $H(f)$ is reduced to the half value of that at DC. The power transfer function is easily calculated from the Fourier transform of the experimentally measured input and output pulses in the time domain, or from the measurement of the output power from the fiber in the frequency domain from DC to the bandwidth frequency. The higher bandwidth means the less pulse broadening and enables the higher speed data transmission. The bandwidth limitation also largely determines the maximum link length for given data rate in some MMF systems. The pulse broadening, theoretically proportional to the fiber length in the absence of mode coupling, is mainly caused by two mechanisms of dispersion in POFs. One is intermodal dispersion, the other is intramodal dispersion. The rms width of the impulse response σ is calculated by $\sigma = \sqrt{\sigma_{inter}^2 + \sigma_{intra}^2}$, where σ_{inter} and σ_{intra} are the rms widths of the pulse broadening induced by intermodal and intramodal dispersions, respectively. Another dispersion, called polarization mode dispersion, arises from nonuniformity of the structure and material throughout the fiber, and two orthogonal polarization modes are affected by the slightly different refractive indices [35]. The effect of polarization mode dispersion usually can be ignored in MMFs. Therefore, this section explains the intermodal and intramodal dispersions, and the bandwidths of POFs.

10.4.1 **Intermodal dispersion**

Intermodal dispersion can be briefly explained as follows. When an optical pulse is input into a MMF, the optical power of the pulse is generally distributed to huge number of modes of fiber. Different modes travel at different propagation speeds along the fiber, which means that the different modes launched at the same time reach the output end of the fiber at different times. Therefore, the optical pulse is broadened in time as it travels along the MMF.

This pulse broadening effect, well known as modal dispersion, is significantly observed in SI MMF. As shown in Figure 10.1, different rays travel along their paths with different lengths, here each distinct ray can be thought of as a mode for simple interpretation. The lights travel at the same velocity along their optical paths because of the constant refractive index throughout the core region in SI MMF. Consequently, the same velocity and the different path lengths result in the different propagation speeds along the fiber, thus leading to the pulse spread in time. The pulse broadening caused by the modal dispersion seriously restricts the transmission speeds of MMF, because overlaps of the broadened pulses induce intersymbol interferences, and disturb correct signal detections, and hence increase bit error rate [36].

The modal dispersion is generally a dominant factor of the pulse broadening in MMFs, and in particular on SI fibers. As shown in [62, Section 2.1.5.8] for ideal step index fibers, the difference in propagation delay between the fastest and slowest mode is given by:

$$\Delta t_{mod} = \frac{L}{2cn_2}NA^2,\tag{10.9}$$

where L is the fiber length, n_1 and n_2 are the core and cladding refraction indices, and $NA = $ numerical aperture $= \sqrt{n_1^2 - n_2^2}$. For instance, for a typical $NA = 0.5$, one would get after 50 m of SI POF a $\Delta t_{mod} = 14.8$ ns, and thus an electrical-to-electrical available bandwidth that can be estimated as $B \cong 0.44/\Delta t_{mod} \approx 30$ MHz under worst case conditions. The actual bandwidth depends on several other parameters, mainly related to the numerical aperture of the source, to the level of mode mixing inside the POF, and to the fact that higher order modes (that travel slower) usually see higher attenuation, and thus give a smaller contribution to the overall transfer function (see again [62, Section 2.1.5.8]) compared to lower order modes. As a result, the actual available bandwidths for SI POF are higher than those given by the theoretical treatment above, but still limited to less than approximately 100 MHz over 50 m.

However, the modal dispersion can be dramatically reduced by forming a near-parabolic refractive index profile in the core region of MMF, which allows much higher bandwidth (i.e. higher speed data transmission) [37]. The typical GI MMF has a cylindrically symmetric refractive index profile that gradually decreases from the core axis to the core-cladding interface. The ray confined to near the core axis, corresponding to a lower order mode, travels shorter geometrical length at slower light velocity along the path because of the higher refractive index. The sinusoidal ray passing through near the core-cladding boundary, considered as a higher order mode, travels longer geometrical length at faster velocity along the path, particularly in the lower refractive index region far from the core axis. As a result, the output times from the fiber end of lights through the shorter geometrical length at the slower velocity and through the longer geometrical length at the faster velocity can be almost the same by the optimum refractive index profile. Thus, perturbation of the refractive index distribution strongly affects the bandwidth of MMF. The optimization of the refractive index profile in GI MMF is an important issue to reduce the modal dispersion. A power law index profile approximation is a well-known method to analyze the optimum profile of GI MMF [38]. In the power law profile approximation, the refractive index distribution of a GI MMF is described by:

$$n(r) = \begin{cases} n_1 \left[1 - 2\Delta \left(\dfrac{r}{a} \right)^g \right]^{\frac{1}{2}} & \text{for } 0 \leq r \leq a, \\ n_2 & \text{for } r > a, \end{cases} \tag{10.10}$$

where $n(r)$ is the refractive index as a function of radial distance r from the core center, n_1 and n_2 are the refractive indices of the core center and the cladding, respectively, a is the core radius. The profile exponent g determines the shape of the refractive index profile and Δ is the relative index difference given by

$$\Delta = \frac{n_1^2 - n_2^2}{2n_1^2}. \tag{10.11}$$

Equation (10.10) includes the SI profile when $g = \infty$.

An optimum profile exponent g_{opt}, which minimizes the modal dispersion and the difference in delay of all the modes and maximizes the bandwidth, is expressed based on the analysis of the Maxwell's equation as follows:

$$g_{opt} = 2 + \varepsilon - \Delta \frac{(4 + \varepsilon)(3 + \varepsilon)}{5 + 2\varepsilon}, \tag{10.12}$$

$$\varepsilon = -\frac{2n_1}{N_1} \frac{\lambda}{\Delta} \frac{d\Delta}{d\lambda}, \quad N_1 = n_1 - \lambda \frac{dn_1}{d\lambda}. \tag{10.13}$$

Equation (10.12), in the absence of wavelength dependence of refractive index of material, becomes the simple expression:

$$g_{opt} = 2 - \frac{12}{5}\Delta. \tag{10.14}$$

High bandwidth can be typically achieved when the profile exponent g is approximately equal to 2.0. However, the refractive indices of materials generally depend on the wavelength, which induces profile dispersion.

Profile dispersion is caused by the wavelength dependence of the refractive index [39]. The profile dispersion p is given by

$$p = \frac{\lambda}{\Delta} \frac{d\Delta}{d\lambda}. \tag{10.15}$$

The g_{opt} value depends on the relative index difference Δ, which is a function of the refractive indices of the core and the cladding. These refractive indices are determined by the wavelength and the dopant characteristics. If polymer matrix and dopant have identical wavelength dependence of refractive index, the g_{opt} value obeys Eq. (10.14). However, the wavelength dependence of refractive index of the dopant is generally different from that of the polymer matrix, hence the shape of the refractive index profile depends on the wavelength. Therefore, even if the optimum refractive index profile is provided at a certain wavelength, this refractive index profile is different from the optimum profile at another wavelength. The profile dispersion also depends on the wavelength of the light signal. The effect of the profile dispersion can be compensated by the refractive index profile by taking it into account, which is easily explained by Eqs. (10.12), (10.13), and (10.15). Thus, the g_{opt} value is shifted by the profile dispersion. Consequently the intermodal dispersion can be minimized by the optimum refractive index profile with taking the profile dispersion into account. Then, intramodal dispersion becomes important to achieve high bandwidth. Additionally, SMFs exhibit even higher bandwidths than GI MMFs, because the modal dispersion does not exist in principle, and thus intramodal dispersion seriously restricts the bandwidth of SMF [40].

10.4.2 Intramodal dispersion

Intramodal dispersion or chromatic dispersion is the pulse widening caused by the finite spectral width of the light source. Intramodal dispersion comprises material and waveguide dispersions.

Material dispersion is induced by the wavelength dependence of the refractive index of the core material [41]. The group velocity of a given mode depends on the wavelength, and thus the output pulse is broadened in time even when optical signals with different wavelengths travel through the same path. This effect is generally much smaller than the modal dispersion in MMFs; however, it is no longer negligible when the modal dispersion is suppressed enough.

Waveguide dispersion arises from the wavelength dependence of the optical power distribution of a mode between the core and the cladding. Light at shorter wavelength is more completely confined to the core region, light at longer wavelength is more distributed in the cladding. The light at longer wavelength has the greater portion in the cladding, and thus travels at higher propagation speed, because the refractive index of the cladding is lower than that of the core.

10.4.3 High bandwidth plastic optical fiber

A huge number of modes, typically more than tens of thousands, can propagate in POFs because of the large cores. Therefore, reduction of modal dispersion has been a key issue in such multimode POFs. The measured output pulses from PMMA-based SI, multi SI, and GI POFs under overfilled mode launch condition are shown in Figure 10.7, compared with the input pulse. The bandwidth of the SI POF was

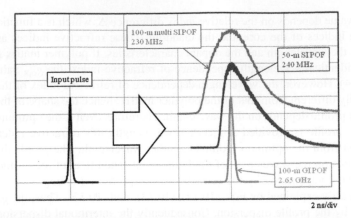

2 ns/div

FIGURE 10.7 Measured output pulse waveforms through 50 m SI, 100 m multi SI and 100 m GI POFs under overfilled mode launch condition.

seriously restricted. The impulse responses and the bandwidths of the SI and multi SI POFs were almost the same, despite the fiber length of the multi SI POF being twice as long as that of the SI POF. The multi SI POF can realize the communication at a data rate twice as high as the SI POF; however, the bandwidth was still limited to be only hundreds of MHz for 100 m. The GI POF was proposed to improve the bandwidth limitation induced by the large modal dispersion of the SI POF [2], and the refractive index profile can be precisely controlled to be the optimum one. The bandwidth of GI POF was dramatically enhanced to be several GHz for 100 m by the precise control of the refractive index profile. This is the effect of the decrease in the modal dispersion mainly caused by the GI profile.

Figure 10.8 shows the dispersion effect on the bandwidth characteristics of the PMMA-based GI POFs. Here, the bandwidths were estimated by the following equations [42,43], the Gaussian pulses were assumed, and the rms spectral width was set to 1 nm

$$\sigma_{\text{inter}} = \frac{LN_1\Delta}{2c} \frac{g}{g+1} \left(\frac{g+2}{3g+2}\right)^{1/2} \left[C_1^2 + \frac{4C_1C_2\Delta\,(g+1)}{2g+1} + \frac{4\Delta^2 C_2^2\,(2g+2)^2}{(5g+2)\,(3g+2)} \right]^{1/2},$$

(10.16)

$$\sigma_{\text{intra}} = \frac{L\sigma_\lambda}{c\lambda} \left[\left(-\lambda^2 n_1''\right)^2 - 2\lambda^2 n_1''\,(N_1\Delta)\left(\frac{g-2-\varepsilon}{g+2}\right)\left(\frac{2g}{2g+2}\right) \right.$$
$$\left. + (N_1\Delta)^2 \left(\frac{g-2-\varepsilon}{g+2}\right)^2 \frac{4g^2}{(g+2)\,(3g+2)} \right],$$

(10.17)

$$C_1 = \frac{g-2-\varepsilon}{g+2}, \quad C_2 = \frac{3g-2-2\varepsilon}{2\,(g+2)}.$$

(10.18)

The maximum bandwidth, near 100 GHz for 100 m, was theoretically obtained when g was almost 2.0, if only the modal dispersion was taken into consideration. However, large disagreements were observed between the calculated and measured bandwidths. In contrast, the relation between the bandwidth and the profile exponent of PMMA-based GI POF was accurately estimated by the Wentzel-Kramers-Brillouin (WKB) method, when all the dispersion factors were taken into account.

On the other hand, the g_{opt} value which represents the highest bandwidth shifted to around 2.4 because of the profile dispersion as discussed above, where the dopant used in the GI POFs shown in Figure 10.8 was diphenyl sulfide. This result indicates that the bandwidth characteristics are strongly influenced by the profile dispersion, even if the refractive index profile is largely deviated from the ideal one.

The calculated maximum bandwidth at g_{opt} was drastically reduced due to the large material dispersion of PMMA, when all the dispersions were considered. Significant decrease in bandwidth was mainly observed in the range of the profile exponent near g_{opt} value. When the profile exponent deviates far from the g_{opt}, the material dispersion has little effect on the bandwidth because of the dominantly large modal dispersion.

The bandwidth of the PMMA-based GI POF can be optimized by the accurate control of the refractive index profile [44]. Further bandwidth improvement has been investigated by using PF polymers [45]. PF polymers have valuable characteristics of low material dispersions as well as the low transmission losses. Figure 10.9 shows material dispersions of PMMA, CYTOP, and silica calculated from their wavelength dependences of the refractive indices. CYTOP has the material dispersion much smaller than PMMA and even lower than silica, particularly from the visible to near-infrared region, which means that CYTOP-based GI POF with an optimum refractive index profile can realize higher bandwidth than silica-based GI MMF. The material

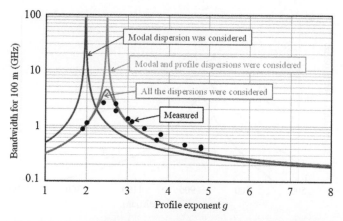

FIGURE 10.8 Calculated bandwidth of 100 m PMMA-based GI POF as a function of profile exponent g at a wavelength of 650 nm with consideration of individual dispersions, compared with the measured bandwidths under overfilled mode launch condition. The rms spectral width was assumed to be 1 nm.

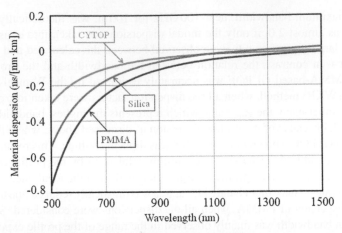

FIGURE 10.9 Material dispersions of PMMA, CYTOP, and silica.

dispersion curve of CYTOP was insensitive to the wavelength, compared with silica and PMMA. Hence the bandwidth of CYTOP-based GI POF is also expected to be insensitive to the wavelength.

The wavelength dependences of the bandwidths of 100 m PMMA- and CYTOP-based GI POFs and silica-based GI MMF were theoretically estimated from the wavelength dependences of their refractive indices as shown in Figure 10.10, where the refractive index profiles of all the fibers were optimized at a wavelength of 850 nm, and the rms spectral width of the light source was assumed to be 1.0 nm.

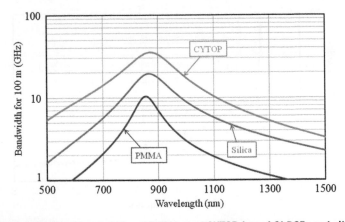

FIGURE 10.10 Calculated bandwidths of PMMA- and CYTOP-based GI POFs and silica-based GI MMF as a function of wavelength. Their refractive index profiles were optimized at a wavelength of 850 nm, and the rms spectral width of the light source was assumed to be 1 nm.

The bandwidth at 850 nm of the PMMA-based GI POF, almost 10 GHz, was higher than that at 650 nm because of the lower material dispersion, although light at 850 nm cannot travel long distance in PMMA because of the large transmission loss. The wavelength dependence of the optimum profile was small in CYTOP-based GI POF, as mentioned above, because CYTOP had low material and profile dispersions. Therefore, the CYTOP-based GI POF can maintain the high bandwidth characteristics over a wide wavelength range, compared with silica-based GI MMFs. Consequently, CYTOP-based GI POF systems can utilize various light sources with large variety of wavelengths. Thus, CYTOP-based GI POFs are predicted to have a higher bandwidth than silica-based GI MMFs. The PF GI POFs, indeed, demonstrated 40 Gb/s and even higher data transmissions over 100 m [46,47].

10.5 APPLICATION AND FUTURE PROSPECT OF PLASTIC OPTICAL FIBER

POFs have been used extensively in short distance datacom applications, such as digital audio interface. POFs are used for data transmission equipment, control signal transmission for numerical control machine tools and railway rolling stocks, etc. In the late 1990s, POF also had become to be used for optical data bus in automobiles. This section introduces recent practical use and future prospects of POF applications.

10.5.1 Step index large-core plastic optical fiber

In this section, we present the current state-of-the-art in the applications of step index large-core POF, from now on simply indicated as SI POF, as already introduced in previous Section 10.2.1. Under this classification, we will consider here only large-core POF (with core diameter typically around 1 mm) and, unless otherwise noted, made of PMMA material. SI POF has long been used for short-reach optical communications [1], mainly due to:

- Its extreme ease of connection, thanks to the very large-core diameter.
- Its high mechanical resilience, including a very low bending loss.
  Overall, the success of SI POF is closely related to its ease of installation and maintenance, making it the lower cost solutions compared to any other type of optical fibers for short distance links (tens of meters). These are clearly key points to keep in mind to understand the success of SI POF, despite its relatively low transmission properties compared to all other types of both glass fibers and GI POF, as it is evident from the comparison introduced in previous sections. In order to understand physical properties of the most commonly used SI POF, we briefly review the specification introduced in 2008 by IEC to standardize a "datacom grade" SI POF, thanks to the A4a.2 SI POF classification that was included in a revision of the document IEC 60793-2-40 [52]. The new SI POF class A4a.2 introduces significant upgrades to the previous version A4a.1,

defining an SI POF that must have attenuation smaller than 18 dB/100 m for all launching conditions at 650 nm, and a minimum modal bandwidth at 650 nm equal to at least 40 MHz over 100 m. As of today (2012), some fiber producers already certify their SI POF products according to this new sub-class A4a.2. The first fiber specified as an A4a.2 POF was the HMKU-1000 from Asahi Chemical in 2010 [53], then other producers followed, such as Mitsubishi with its product GH-4000. In Figure 10.11, the measured loss vs. wavelength of these two fibers are compared to the A4a.2 specifications.

From Figure 10.11 we can derive what are the typical wavelengths used for SI POF transmission. Attenuation is so high at the infrared wavelengths used for glass fibers (such as 980, 1300, and 1550 nm) that only optical sources in the visible region can be used for SI POF. The most common one is the red window around 650 nm, mostly due to the large availability of red lasers, LED and RC-LED. Other interesting windows where the attenuation is even lower are the green and blue wavelength window.

Due to the reasons explained in previous Section 10.4, the available bandwidth is low, even though the A4a.2 specification (at least 40 MHz over 100 m) is quite conservative in this respect, since usually slightly higher bandwidths are available in practical installations. In fact, there is a strong dependence on the launching conditions and on the used type of source: for a given length of SI POF, the use of low numerical aperture (NA) edge-emitting lasers tends to give higher available bandwidth compared to larger NA sources, since a lower number of modes are excited at the transmitter end. In many experimental

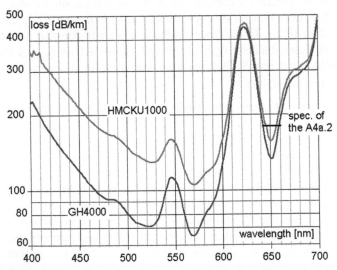

FIGURE 10.11 Loss of two commercial SI-POF produced according to the A4a.2 specification (data taken from [53]).

situations, we measured an available electrical-to-electrical bandwidth of the order of 80 MHz over 50 m.

For all these reasons, today commercial SI POF applications are related to the following areas of short-range optical transmission:

*   Industrial automation, to cover distances from 50 to 100 m and bit rates up to 100 Mbit/s.
*   Automotive, mainly thanks to the MOST (Media Oriented Systems Transport) consortium that uses SI POF for short-reach interconnection inside cars for the internal infotainment network. Originally implementing a digital connection over SI POF at 25 Mbit/s (MOST25) this de facto standard has now evolved to 150 Mbit/s (MOST150) with new products hitting the market from 2011.
*   Patch cords for digital audio equipments, as originally introduced by Toshiba with TOSLINK™.

These are the three mainstream SI POF applications where they have reached mass production. Another very recent and interesting development is the use of SI POF for home networking, as an alternative to more traditional wired solutions using either copper UTP cables or powerline solutions. In this specific scenario, the use optical fibers is interesting since they can be installed in already existing electrical ducts, where the use of copper UTP cable is critical due to both potential electromagnetic interferences and, in some countries, due to safety regulation. Anyway, glass optical fibers are usually seen as too expensive in the home scenario, while SI POF is attractive for the aforementioned reasons (ease of installation, ruggedness, and overall low cost) and due to the fact that the target distance is relatively low (usually set at 50 m). Today, some home-networking equipment vendors have started to include an SI POF port inside their residential gateways, usually running at Fast Ethernet rates (100 Mbit/s). Many research groups have been active on this area starting around 2005, and in particular in Europe, where several EU-funded projects (such as POF-ALL [54], POF-PLUS [55], and ALPHA [56]) have tried to find solutions to upgrade the bit rate over a typical span of 50 m of SI POF from the today commercially available 100 Mbit/s to at least 1 Gbit/s. The rationale behind these projects is that in order for SI POF to be competitive in the long term inside the "technologically crowded" home-networking arena, the available bit rate over the target 50 m length should be significantly increased above the current 100 Mbit/s, but this imposes significant physical layer challenges, mainly due to the low SI POF bandwidth. Several solutions have been proposed by different groups:

*   At the component level, a massive amount of work has been done to improve the electrical bandwidth of SI POF transmitters, particularly to achieve 1 Gbit/s using resonant cavity LED (RC-LED [57]), or red lasers for multi Gbit/s [55], by a proper optimization of the optical source and its electrical driver.
*   At the physical layer, it was clearly shown that the traditional NRZ OOK modulation with direct detection and threshold decision is impossible at 1 Gbit/s and above, since it would generate a completely closed eye diagram

due to insufficient channel bandwidth, so that a combination of one or more of the following digital transmission techniques is required:

- Electronic equalization at the receiver, performed either digitally with FFE-DFE equalizer structures [58] or through ad hoc analog circuitry [59].
- Advanced modulation formats at the transmitter end, including M-PAM, DMT, and OFDM [60].

All these efforts have shown the technical feasibility of at least 1 Gbit/s over 50 m of SI POF, even using inexpensive RC-LED as transmitters, and more than 1 Gbit/s using VCSELs and edge-emitting lasers, even though all of them require some form of digital signal processing at the transmitter and/or receiver. Starting from 2010, there has also been a standardization initiative inside VDE (German Commission for Electrical, Electronic & Information Technologies) toward the definition of a new physical layer for gigabit transmission [61]. Today, we are confident that the transmission solutions and optoelectronic devices technology for Gigabit Ethernet over SI POF for home networking are ready, even though its practical mass production will depend on market requirements.

In the POF-ALL project [54], we also demonstrate lower rate transmission (Fast Ethernet at 100 Mbit/s) but over an extended reach of 200+ meters made possible by the use of green LEDs at the transmitter and electronic equalization at the receiver, a scientific result that may find application in some specific sectors, such as industrial automation.

10.5.2 Graded index plastic optical fiber

Recently, demand for optical fiber communications substituting electrical wiring is rapidly increasing not only in long haul networks but also in short reach networks such as LANs and even interconnections. A bit rate of tens of Gb/s will be required in these networks. The GI POFs are attracting a great deal of attention in intrabuilding and interconnection networks, because the GI POF has high bandwidth, low bending loss, high flexibility, and large core, which allows easy handling and rough connections, and thus enables easy installation of a high-speed network at a low cost [2,3,47–49]. The GI POF is expected to be used even in automotive or aircraft networks, because the GI POF, in principle, has no electromagnetic interference problems, and because the GI POF is lighter in weight than a metal cable, which leads to higher energy efficiency [1,50]. Typical copper cable networks adopted in conventional buildings are dispersed network systems, where servers and switches are dispersed on each floor, which are connected by copper cables. On the other hand, a novel centralized network has been proposed using high-speed GI POFs mentioned above, where there is only one main server, and GI POFs are directly distributed to any outlets and terminals from the main server, without any floor switches or servers in the intermediate [51]. Therefore, quite simple processes of maintenance and troubleshooting for the networks can be employed. As a result, the maintenance fee of the centralized network is one fifth of that of a conventional dispersed network. Such a Gb/s hospital system with the centralized network using GI POF has been realized in a cardiac hospital with 320 beds. The total length of GI POFs installed in this hospital is amazingly as long as 230 km, which is likened to much longer branched veins compared with main veins of the human body. The GI POF network systems have been adopted in various LANs at hospitals, campuses, and condominiums, for example.

The data rate required to handle images and motion pictures is easily expected to dramatically increase in the near future. Although various efforts to achieve a highly networked information society have produced a steady stream of successful results, current broadband services are still far from the true sense of such society. Nevertheless, distances among people can be dramatically reduced, if face-to-face communication systems are more easily realized. In such systems, high-speed, real time, and interactive communication based on optical fibers and high resolution and large sized displays will become major technologies. In order to further promote broadband systems, programing research and development based on real-life situations should be required. These studies should be undertaken in collaboration with hardware fields including peripheral elements of communications, home electronics, audiovisual equipment, etc. as well as technologies for optical fiber system installation, and should also involve simultaneous connection with software fields such as data communication systems, Internet services, contents. A new standard of high resolution display interface (i.e. high definition multimedia interface) is requiring a data rate as high as 10 Gb/s, which can be realized by GI POF. Thus, telemedicine and distance learning based on face-to-face and real-time communication will be realized. Low loss and high bandwidth GI POFs will accelerate a remarkable paradigm shift in network architecture.

References

[1] P. Polishuk, Plastic optical fibers branch out, IEEE Commun. Mag. 44 (9) (2006) 140–148.
[2] Y. Koike, High-bandwidth graded-index polymer optical fibre, Polymer 32 (10) (1991) 1737–1745.
[3] K. Makino, T. Ishigure, Y. Koike, Waveguide parameter design of graded-index plastic optical fibers for bending-loss reduction, J. Lightwave Technol. 24 (5) (2006) 2108–2114.
[4] Y. Koike, K. Koike, Polymer optical fibers, Encyclopedia of Polymer Science and Technology, John Wiley & Sons, Inc., 2002.
[5] J. Zubia, J. Arrue, Plastic optical fibers: an introduction to their technological processes and applications, Opt. Fiber Technol. 7 (2) (2001) 101–140.
[6] M.A. van Eijkelenborg, M.C.J. Large, A. Argyros, J. Zagari, S. Manos, N.A. Issa, I. Bassett, S. Fleming, R.C. McPhedran, C.M. de Sterke, N.A.P. Nicorovici, Microstructured polymer optical fibre, Opt. Express 9 (7) (2001) 319–327.
[7] M. Born, E. Wolf, Principles of Optics: Electromagnetic Theory of Propagation, Interference and Diffraction of Light, Cambridge University Press, 1999.
[8] E. Hecht, Optics, fourth ed., Addison-Wesley, 2002.
[9] J. Hecht, L. Long, Understanding Fiber Optics, Prentice Hall, Columbus, 2002.

[10] B.G. Shin, J.H. Park, J.J. Kim, Low-loss, high-bandwidth graded-index plastic optical fiber fabricated by the centrifugal deposition method, Appl. Phys. Lett. 82 (26) (2003) 4645–4647.

[11] A.G. Villegas, M.A. Ocampo, G. Luna-Bárcenas, E. Saldivar-Guerra, Obtainment of graded index preforms by combined frontal co-polymerization of MMA and BzMA, in: Proc. Macromol. Symp., 2009, pp. 336–341.

[12] Y. Koike, Optical resin materials with distributed refractive index process for producing the materials, and optical conductors using the materials, US Patent 5 541 247, JP Patent 3 332 922, EU Patent 0 566 744, KR Patent 170 358, CA Patent 2 098 604, originally filed in 1991.

[13] Y. Koike, T. Ishigure, E. Nihei, High-bandwidth graded-index polymer optical fiber, J. Lightwave Technol. 13 (7) (1995) 1475–1489.

[14] M. Naritomi, H. Murofushi, N. Nakashima, Dopants for a perfluorinated graded index polymer optical fiber, Bull. Chem. Soc. Jpn. 77 (11) (2004) 2121–2127.

[15] A. Fick, Ueber diffusion, Ann. Phys. 170 (1) (1855) 59–86.

[16] B.C. Ho, J.H. Chen, W.C. Chen, Y.H. Chang, S.Y. Yang, J.J. Chen, T.W. Tseng, Gradient-index polymer fibres prepared by extrusion, Polym. J. 27 (3) (1995) 310–313.

[17] I.S. Sohn, C.W. Park, Diffusion-assisted coextrusion process for the fabrication of graded-index plastic optical fibers, Ind. Eng. Chem. Res. 40 (17) (2001) 3740–3748.

[18] R. Hirose, M. Asai, A. Kondo, Y. Koike, Graded-index plastic optical fiber prepared by the coextrusion process, Appl. Opt. 47 (22) (2008) 4177–4185.

[19] T. Okoshi, Optical Fibers, Academic Press, 1982.

[20] G. Keiser, Optical Fiber Communications, Wiley Encyclopedia of Telecommunications, John Wiley & Sons, Inc., 2010.

[21] D. Marcuse, Principles of Optical Fiber Measurements, Academic Press, 1981.

[22] F. Urbach, The long-wavelength edge of photographic sensitivity and of the electronic absorption of solids, Physical Review 92 (1953) 1324.

[23] T. Kaino, Absorption losses of low loss plastic optical fibers, Jpn. J. Appl. Phys. 24 (12) (1985) 1661–1665.

[24] T. Yamashita, K. Kamada, Intrinsic transmission loss of polycarbonate core optical fiber, Jpn. J. Appl. Phys. 32 (1993) 2681–2686.

[25] T. Kaino, M. Fujiki, S. Nara, Low-loss polystyrene core-optical fibers, J. Appl. Phys. 52 (12) (1981) 7061–7063.

[26] W. Groh, Overtone absorption in macromolecules for polymer optical fibers, Makromol. Chem. 189 (12) (1988) 2861–2874.

[27] A. Einstein, Theory of opalescence of homogenous liquids and liquid mixtures near critical conditions, Ann. Phys. 33 (16) (1910) 1275–1298.

[28] Y. Koike, N. Tanio, Y. Ohtsuka, Light scattering and heterogeneities in low-loss poly(methyl methacrylate) glasses, Macromolecules 22 (3) (1989) 1367–1373.

[29] T. Kaino, M. Fujiki, K. Jinguji, Preparation of plastic optical fibers, Rev. Elec. Commun. Lab. 32 (3) (1984) 478–488.

[30] T. Kaino, K. Jinguji, S. Nara, Low loss poly(methyl methacrylate-d8) core optical fibers, Appl. Phys. Lett. 42 (7) (1983) 567–569.

[31] R. Nakao, A. Kondo, Y. Koike, Fabrication of high glass transition temperature graded-index plastic optical fiber: Part 1—material preparation and characterizations, J. Lightwave Technol. 30 (2) (2012) 247–251.

[32] R. Nakao, A. Kondo, Y. Koike, Fabrication of high glass transition temperature graded-index plastic optical fiber: Part 2—fiber fabrication and characterizations, J. Lightwave Technol. 30 (7) (2012) 969–973.

[33] E. Nihei, T. Ishigure, Y. Koike, High-bandwidth, graded-index polymer optical fiber for near-infrared use, Appl. Opt. 35 (36) (1996) 7085–7090.

[34] N. Tanio, Y. Koike, What is the most transparent polymer? Polym. J. 32 (1) (2000) 43–50.

[35] C. Poole, R. Wagner, Phenomenological approach to polarisation dispersion in long single-mode fibres, Electron. Lett. 22 (19) (1986) 1029–1030.

[36] M.C. Nowell, D.G. Cunningham, D.C. Hanson, L.G. Kazovsky, Evaluation of Gb/s laser based fibre LAN links: review of the Gigabit Ethernet model, Opt. Quant. Electron. 32 (2) (2000) 169–192.

[37] D. Gloge, E.A.J. Marcatili, Multimode theory of graded-core fibers, Bell Syst. Tech. J. 52 (9) (1973) 1563–1578.

[38] R. Olshansky, Propagation in glass optical waveguides, Rev. Mod. Phys. 51 (2) (1979) 341–367.

[39] E. Marcatili, Modal dispersion in optical fibers with arbitrary numerical aperture and profile dispersion, Bell Syst. Tech. J. 56 (1977) 49–63.

[40] T.D. Croft, J.E. Ritter, V.A. Bhagavatula, Low loss dispersion-shifted single-mode fiber manufactured by the OVD process, J. Lightwave Technol. 3 (5) (1985) 931–934.

[41] M. DiDomenico Jr., Material dispersion in optical fiber waveguides, Appl. Opt. 11 (3) (1972) 652–654.

[42] R. Olshansky, D.B. Keck, Pulse broadening in graded-index optical fibers, Appl. Opt. 15 (2) (1976) 483–491.

[43] M. Soudagar, A. Wali, Pulse broadening in graded-index optical fibers: errata, Appl. Opt. 32 (1993) 6678.

[44] Y. Koike, T. Ishigure, High-bandwidth plastic optical fiber for fiber to the display, J. Lightwave Technol. 24 (12) (2006) 4541–4553.

[45] T. Ishigure, Y. Koike, J.W. Fleming, Optimum index profile of the perfluorinated polymer-based GI polymer optical fiber and its dispersion properties, J. Lightwave Technol. 18 (2) (2000) 178–184.

[46] H. Yang, S.C.J. Lee, E. Tangdiongga, C. Okonkwo, H.P.A. van den Boom, F. Breyer, S. Randel, A.M.J. Koonen, 47.4 Gb/s transmission over 100 m graded-index plastic optical fiber based on rate-adaptive discrete multitone modulation, J. Lightwave Technol. 28 (4) (2010) 352–359.

[47] P.J. Decker, A. Polley, J.H. Kim, S.E. Ralph, Statistical study of graded-index perfluorinated plastic optical fiber, J. Lightwave Technol. 29 (3) (2011) 305–315.

[48] T. Ishigure, K. Makino, S. Tanaka, Y. Koike, High-bandwidth graded-index polymer optical fiber enabling power penalty-free gigabit data transmission, J. Lightwave Technol. 21 (11) (2003) 2923–2930.

[49] K. Makino, T. Nakamura, T. Ishigure, Y. Koike, Analysis of graded-index polymer optical fiber link performance under fiber bending, J. Lightwave Technol. 23 (6) (2005) 2062–2072.

[50] T. Kibler, S. Poferl, G. Bock, H.P. Huber, E. Zeeb, Optical data buses for automotive applications, J. Lightwave Technol. 22 (9) (2004) 2184–2199.

[51] N. Ohtsu, K. Uehara, T. Ishigure, Y. Koike, Construction of the all GI-POF network in the hospital, in: Proc. Int. Conf. POF, 2004, pp. 492–495.

[52] IEC Specification IEC 60793-2-40 ed3.0, Optical fibres—Part 2-40: Product specifications—sectional specification for category A4 multimode fibres.

[53] POF-PLUS Final Report. <http://www.ict-pof-plus.eu/>.

[54] I. Mollers, D. Jager, R. Gaudino, A. Nocivelli, H. Kragl, O. Ziemann, N. Weber, T. Koonen, C. Lezzi, A. Bluschke, S. Randel, Plastic optical fiber technology for

reliable home networking: overview and results of the EU project POF-ALL, IEEE Commun. Mag. 47 (8) (2009) 58–68, doi:10.1109/MCOM.2009.5181893.

[55] C.M. Okonkwo, E. Tangdiongga, H. Yang, D. Visani, S. Loquai, R. Kruglov, B. Charbonnier, M. Ouzzif, I. Greiss, O. Ziemann, R. Gaudino, A.M.J. Koonen, Recent results from the EU POF-PLUS project: multi-gigabit transmission over 1 mm core diameter plastic optical fibers, IEEE J. Lightwave Technol. 29 (2) (2011) 186–193, doi:10.1109/JLT.2010.2096199.

[56] M. Popov, Recent progress in optical access and home networks: results from the ALPHA project, in: 37th European Conference and Exhibition on Optical Communication (ECOC 2011), 18–22 September 2011.

[57] B. Charbonnier, P. Urvoas, M. Ouzzif, J. Le Masson, J.D. Lambkin, M. O'Gorman, R. Gaudino, EU project POF-PLUS: gigabit transmission over 50 m of step-index plastic optical fibre for home networking, in: Optical Fiber Communication Conference, OFC 2009, March 2009, pp. 22–26.

[58] A. Nespola, S. Straullu, P. Savio, D. Zeolla, J.C.R. Molina, S. Abrate, R. Gaudino, A new physical layer capable of record gigabit transmission over 1 mm step index polymer optical fiber, IEEE J. Lightwave Technol. 28 (20) (2010) 2944–2950.

[59] M. Atef, R. Swoboda, H. Zimmermann, 1.25 Gbit/s over 50 m step-index plastic optical fiber using a fully integrated optical receiver with an integrated equalizer, IEEE J. Lightwave Technol. 30 (1) (2012) 118–122, doi:10.1109/JLT.2011.2179520.

[60] S.C.J. Lee, F. Breyer, S. Randel, R. Gaudino, G. Bosco, A. Bluschke, M. Matthews, P. Rietzsch, R. Steglich, H. van den Boom, A. Koonen, Discrete multitone modulation for maximizing transmission rate in step-index plastic optical fibers, IEEE J. Lightwave Technol. 27 (11) (2009) 1503–1513, doi:10.1109/JLT.2009.2013480.

[61] VDE 0885-763-1 Physical Layer Parameters and Specification for High Speed Operation Over Plastic Optical Fibres, to be published by DKE-AK 412.7.1/DKE-UK 412.7.

[62] Olaf Ziemann, Jürgen Krauser, Peter E. Zamzow, Werner Daum, POF Handbook: Optical Short Range Transmission Systems, second ed., Springer-Verlag, Berlin, 2008.

[63] Chang-Won Park, Manufacture of POF, in: 12th International Conference on POF (ICPOF 2003), Seattle, USA, September 2003.

Integrated and Hybrid Photonics for High-Performance Interconnects

Nikos Bamiedakis[a], Kevin A. Williams[b], Richard V. Penty[a], and Ian H. White[a]

[a]*Centre of Advanced Photonics and Electronics, Electrical Engineering Division, Department of Engineering, University of Cambridge, 9 JJ Thomson Avenue, Cambridge, CB3 0FA, UK,*
[b]*COBRA Research Institute, Eindhoven University of Technology, Postbus 513, 5600 MB, Eindhoven, The Netherlands*

11.1 INTRODUCTION

Significant progress has been made in the deployment of optical links within high-performance computing systems. Fiber-based interconnection schemes now enable high-capacity and high-density rack-to-rack interconnection within supercomputers and data centers. Optical links with data rates up to 10 Gb/s/channel are installed and significant progress is being made toward the deployment of 100 Gb/s optical interconnections (4×25 Gb/s). Intra-rack optical backplanes are expected to appear shortly. Polymer waveguide-based optical backplanes and board-level interconnects have also attracted significant research and development. They can be embedded within electronic printed circuit boards (PCBs) to offer cost-effective, high-aggregate data capacities at even shorter distances. State-of-the-art systems currently deploy point-to-point links based on arrays of VCSEL sources and multimode waveguides. Future generation systems may be expected to deploy three-dimensional interconnects. Wavelength multiplexing schemes are expected to be deployed in order to achieve even higher aggregate data capacities and density interconnections within the board itself and conceivably between and within modules and chips. The trend to implement shorter and shorter links using optical techniques is driving research into new technologies and techniques. Integrated photonic circuits and systems are being devised to facilitate networking even at the chip level. As distances get shorter new design rules come into play. Demands on physical size, interconnectivity, energy use, thermal loading, bandwidth density, and latency and perhaps most importantly cost now dominate short-reach interconnect design.

In this chapter we review the requirements in high-performance interconnects, identifying techniques and technologies for addressing connectivity, size, bandwidth, latency, energy and cost. State-of-the-art technologies for waveguide interconnects are compared and contrasted. We review the design constraints which are apparent today, and the potential for emerging optical technologies. Components for routing in both multimode and single-mode optical systems are discussed in Section 11.2. Fixed waveguide optical routing and optically switched architectures are then described in Section 11.3, reviewing state-of-the-art performance and identifying the opportunities and challenges as concepts are further developed. The prospects for deployment of optical interconnects ever closer to the electronic chip and, ultimately, the processor are discussed in Section 11.4. Finally, design choices and considerations are summarized to understand the likely roles for the broad range of technologies and architectures.

11.1.1 Short-reach optical interconnects

Photonics is the dominant technology for long-distance and high-capacity telecommunications systems. For the highest performance systems, highly-sophisticated, spectrally-efficient modulation schemes are deployed with precision low-line width laser sources and temperature stabilization in gold-plated packages. Massive capacity has been achieved by aligning wavelength multiplexed data to standardized grids. Data-communication systems, being closer to the end users, have a completely different cost model, and many of the control techniques and packaging concepts simply cannot be used. Here, electronics is co-packaged with optical components to form compact modules with only digital inputs and outputs and a fiber connection at the package. Energy consuming temperature control is avoided and wavelengths inefficiently roam the optical spectrum.

Computer communications are yet different again. This requires the bandwidth performance of telecommunications but across the operating temperature ranges of data communications, at a price point which until recently has been unthinkable in photonics. The specifications are expected to be different for each order-of-magnitude reduction in transmission distance. While shorter distances will relax some performance metrics, formidable challenges arise as distances decrease. To reflect the wide variation in requirements, we categorize short-reach interconnects into three ranges: rack-to-rack, card-to-card, and chip-to-chip, depending on the application space and the required link lengths (Figure 11.1).

11.1.1.1 Rack-to-rack (1–10 m)

Active optical cables have been introduced into rack-to-rack communications in supercomputers and large data centers [2–4] over the last decade. Here electro-optic conversions are built into the fiber connector shell to release PCB real estate. Prior to 2005, practically all supercomputer interconnects were implemented with electrical cables. In 2002, the top-performing supercomputer, the NEC Earth Simulator, contained no optics. In 2005, a combination of electrical and optical cabling was first implemented in the ASCI Purple supercomputer ($\sim$100 Tflop/s): while copper cabling was deployed for links shorter than 10 m, optical cables were used for 20–40 m cabling. Approximately 3000 parallel optical links were used in this system

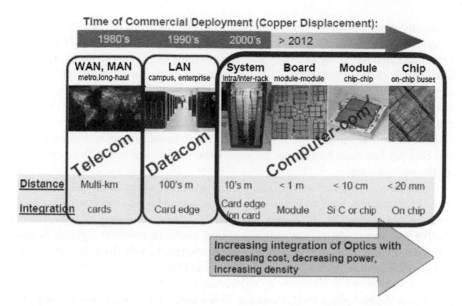

FIGURE 11.1 Introduction of optical links into shorter reach communication systems [1].

with each link comprising $12 + 12$ channels (bidirectional) with a bit rate of 2 Gb/s per channel. In 2008, the IBM Roadrunner broke the 1 petaflop/sec milestone. This system comprised approximately 40,000 optical links for rack-to-rack communications with 55 miles of active optical cables and operating data rates of 5 Gb/s per channel. Currently, Gigabit Ethernet, 10 GbE and optical Infiniband connections comprise the larger percentage of currently installed systems within high-performance supercomputers [5,6] and the standardization of 100 GbE is expected to drive further the development of larger capacity interconnection systems. Research is now focused on multiplexing techniques such as wavelength (multiple optical channels per fiber) and space division multiplexing (multiple fibers per cable) to enable larger data capacities and densities. A simple scaling in optical wiring is, however, unsustainable, and switch technologies are required to provide interconnectivity. In the short term, the combination of electronic switch fabrics and point-to-point optical links will likely be sufficient, but researchers are actively exploring optical switching technologies to avoid the latency, size, and energy overhead associated with electrical to optical conversions and electronic signal processing [7].

11.1.1.2 Card-to-card (0.1 m–1 m)

Intra-rack connections have so far been implemented with electrical cables. However, the new generation of supercomputers planned for 2015 is expected to operate at over 10 petaflop/sec and these are planned to deploy optics even within the rack for board-to-board communications [8]. The P7-IH system based on VCSEL transmitters and multimode ribbon fibers is a prominent example [2,4,9]. Even greater interconnection densities and reduced power consumption are required for exascale supercomputers ($\sim$1 exaflop/s). Optics may penetrate into lower levels of the system interconnection hierarchy and increasingly push their use in shorter communication links within such systems [10–13]. Requirements for the successful deployment of any optical technology are cost-effectiveness and compatibility with existing system materials, architectures, and manufacturing processes. As communication distances decrease, the required interconnection density in each interconnection level of the system, as well as the total number of links, is increased. The power consumption and cost per link must be reduced to achieve the power- and cost-efficiency targets for the entire system.

Fiber-based, polymer waveguide, hollow metallic waveguide, and free-space optical technologies have all been considered for use in board-level optical interconnects up to 1 m of length. Proposed interconnection schemes based on optical fibers involve the integration of conventional optical fibers in low-cost rigid or flexible substrates to form hybrid PCBs [14–16]. Such schemes benefit from the deployment of the well-established fiber-based technologies and their respective connectorization schemes and provide a relatively straightforward way to tackle the interconnection problem at the backplane level. However, fiber-based solutions are of relatively high cost and are complex to assemble.

Polymer waveguide-based solutions exploit the relatively low cost of the materials with properties better suited for integration within conventional electronic PCBs. In recent years, a new class of polymer materials with suitable optical, thermal, and mechanical properties has been developed [17–20]. These materials are flexible, allowing integration onto both rigid and flexible substrates, exhibit low losses (<0.02–0.05 dB/cm) at the near-infrared wavelengths of interest, possess high thermal stability, and withstand the high-temperature soldering and lamination processes for standard PCBs. Moreover, these materials exhibit the required environmental stability to ensure long lifetimes and reliable operation in the typical operating conditions of current electronic systems. They also allow patterning with a wide range of techniques suitable for large-scale manufacturing such as printing and stamping. The development of cost-effective integration methods for the formation of polymer-based optical networks within standard PCBs is a particularly active research area in both academia and industry.

Metallic hollow waveguides offer a low-loss low-latency optical medium which can be easily fabricated to form optical backplanes [21,22]. The main drawback of this technology is, however, the requirement for high accuracy in the source alignment at the waveguide inputs. Angular deviations exceeding 0.5° can lead to high insertion losses due to power coupling to higher-order waveguide modes which exhibit higher propagation losses. Therefore, the cost-effective assembly and packaging of such systems remains a challenge.

Free-space optics constitute an interesting approach to implementing optical interconnects for card-to-card communications. These involve the use of free-space optical elements such as micro-lenses, holograms, and mirror structures to route efficiently free-space optical beams over the backplane area [23–25]. Such systems can offer high-density board-level interconnection enabling a large number of optical links within small area, but typically have strict fabrication requirements to achieve efficient operation while their assembly and packaging within electronic systems is not straightforward.

11.1.1.3 Chip-to-chip (<0.1 m)

Communications between modules and chips are expected to involve close-proximity, intimate, or even monolithic integration of optoelectronic components and circuits with electronics. There is considerable and extensive current research into silicon-on-insulator (SOI)-based optical circuits, and hybrid attached III–V optoelectronic integrated circuits. The current developmental effort is focusing primarily on co-packaging and flip-chip assembly of III–V lasers within modules. Compatibility of photonic integration with CMOS foundry processes is an ongoing area of research and development [26–28]. As the field of silicon photonics has evolved, the initial challenges have focused on building block demonstration with clear successes for the recent demonstrations for modest waveguide loss nanowires [26], compact Mach-Zehnder interferometers, low-energy resonant optical modulators, long-wavelength photodetection using SiGe, and wavelength selective switches [29–33]. With the exception of laser sources, which are most efficiently implemented in III–V materials, these devices offer a comprehensive range of components for forming on-chip photonic networks. However, formidable circuit level challenges remain, with many of the most efficient device concepts being critically sensitive to polarization, wavelength, temperature, and nanoscale fabrication variation [34]. While die-level assembly has been most actively explored so far, the combination of III–V membranes with a combination of optoelectronic and high-density integrated optics components on a silicon die offers additional intriguing possibilities [35] and challenges.

11.1.2 Bandwidth, connectivity, and latency

The emergence of chip multi-processors and the increasing demand for new user applications drive the need for higher bandwidth interconnection networks at all levels of communication. A diverse range of specifications are required for on-chip, chip-to-chip, board-to-board, and rack-to-rack interconnection. Required memory communication bandwidths are predicted to be in the realm of 100 Gbyte/second in the next few years. In many applications, up to 1 byte/second bandwidth requirement is used as a rule of thumb for on-die cache for each floating point operation [36]. These bandwidth requirements scale by a factor of ten for each step up the memory hierarchy, from Gbytes/second at the rack-to-rack level through tens of Gbytes/second to bulk storage, hundreds of Gbytes/second to bulk memory, and Tbytes between cores [37]. Teraflop processors now face the challenge of transferring Tbytes of data per second between tens to possibly hundreds of cores [38,39].

The bandwidth density for optics can be phenomenally high. Demonstrations of 12.5 Tb/s per fiber core are feasible with optical multiplexing [40]. Ultrahigh density parallel optics at 640 Gb/s within 80 μm of a PCB edge has been demonstrated [41].

Increasingly high-density transceivers are also being implemented to enhance the bandwidth density of optoelectronic conversion. Chip-scale interconnection with 4×12 matrices of VCSELS has been operated at >20 Gb/s. This has allowed an aggregate data rate of 960 Gb/s from a chip area of only 1.4 mm $\times$ 3.75 mm [42].

The transmission of large data volumes with small latencies constitutes an important technological challenge. This impedes performance enhancement with increases in system level scaling and is usually referred to as "the interconnection bottleneck." The latency can be low for optics because there is no gate delay and speed of optical transfer is purely time-of-flight. Delays of the order of 50–140 ps/cm are observed depending on whether a low-refractive index medium such as silica, or a higher refractive index material such as silicon, is used. However, additional delays are incurred by the digital logic within transmitters and receivers, and through any ingress and egress buffering, motivating interest in optical switched networks with minimum optoelectronic conversions.

11.1.3 Energy and power

Computer system design is now energy limited at all levels in the network hierarchy. Data transport is *the* energy problem with estimates of up to 200 times more energy required to transport a bit from the nearest neighbor than to perform the logical operation itself [36]. As shorter optical links displace copper in higher numbers and densities—a consequence of Rent's rule [43]—the energy dissipation limits per link are reduced radically. As energy is primarily dissipated through the charging and discharging of capacitance in electronic interconnects [44], comparisons and targets are commonly made in terms of energy per bit. At distances of several tens of meters, active optical cables are being deployed, with energy efficiency at and below 100 pJ/bit. Reports from the laboratory indicate total link energy of the order of 5 pJ/bit for 10 Gb/s optical transceivers operating on multimode fiber links [45]. State-of-the-art 90 nm CMOS driving circuits for VCSEL-based MMF links exhibit full-link transceiver power consumptions of 1.37 pJ/bit at 15 Gb/s and 3.6 pJ/bit at 25 Gb/s [46]. Electronic transceivers operating at low voltage and with equalization have been demonstrated to operate with 2.8–6.5 pJ/bit total energy for board or backplane interconnects [47]. Chip-to-chip lines are demonstrated at the level 2 pJ/bit [48] and 18 Gb/s optical receivers are demonstrated at 0.36 pJ/bit power efficiencies [49]. Energy targets of 1 pJ/bit and 100 fJ/bit have been identified for backplane and on-chip global wiring respectively [11] and technology extrapolations from Miller show that competitive off-chip optical interconnects must operate at less than 1 pJ/bit and scale down to less than 100 fJ/bit by 2022 [10]. Electronic transceivers for 10-mm on-chip differential wires are already consuming as little as 0.37–0.63 pJ/bit for 4–6 Gb/s/channel [50]. Co-integration of optical receivers on 45-nm SOI has allowed 52 fJ/bit [51] showing a path to even lower consumption. The energy budget is therefore squeezed rapidly as distances reduce. As a result, optical interconnects are identified as a key technology even down to the global on-chip wiring level for addressing future energy and power challenges.

Historically the optical transceiver has been engineered for overcoming transmission loss in long-distance telecommunication links rather than energy reduction in ultra-short links. Energy is expended at a number of points in an optical link. If comparisons are to be made with the displaced copper interconnect, the energy use of the optical link must be described in terms of full electronic energy consumption. This budget includes the energy required to overcome any optical energy loss. Table 11.1 sets out the key power consuming operations for a generalized optical link.

The energy expended within the link is dissipated primarily within the electronic components. A typical data-communication link operating with 1 mW mean power per channel will have dissipated most of this optical power before the receiver, but state-of-the-art laser drivers dissipate tens of milliwatts for 20 Gb/s [52] and a receiver front end may operate at 10 mW for 10 Gb/s [53]. While optical transmission losses may not appear to be a primary concern, the square law scaling between electrical and optical power leads in practice to considerable electrical energy consumption. Optimization of impedance mismatches at the transceivers remains an important area of active study. Possibilities to remove serialization and de-serialization through bit-parallel multi-wavelength transmission have been proposed.

The pJ/bit energy metric will continue to be valuable in comparing technologies, but it varies strongly with optical line rate and the specific physical deployment for optical interconnects. Important components such as lasers require a constant DC power which is only weakly dependent on bit rate and some classes of optical switch and modulator operate with power levels which are largely independent of data bandwidth. This has led some researchers to benchmark electronics and optics at a fixed data rate [54]. While this form of analysis provides an important reality check, in the longer term, as transmission rates far outstrip electronic signaling rates, these comparisons may do a disservice to certain classes of optical interconnects.

Large interconnection networks are most efficiently implemented with switches to avoid overprovisioning of optical wiring, transceivers, and the associated energy

Table 11.1 Energy dissipation in an optical interconnect.

De/serializer	Data from parallel buses to high rate bit streams
Encoders/decoders	Protect against transmission errors
Optical power generation	DC power to the laser source. Higher powers are able to compensate higher losses and operate in more sophisticated photonic networks
Electrical amplifiers	Lasers and photodiodes are current sinks and sources respectively, requiring careful impedance matching to CMOS
Optical amplifiers	DC power to pump a broadband amplifier. Widely used in telecom, but not used data networks
Clock re-acquisition	Data packets originating from different transmitters may not share the same clock or phase
Optical wavelength registration	Precision temperature control used in telecoms but no registration is used in data communications

overheads. Switched networks allow for efficiency savings through statistical multiplexing. Electronic switch complexity is advancing rapidly—a Gennum GX3146 offers 146×146 connectivity for 3.5 Gb/s connections at an energy efficiency of the order of 35 pJ/bit [55]. The transceiver complexity and energy overhead to interoperate such switches with point-to-point optical wiring is considerable, and the ultimate scaling in line rate and connectivity will be limited by package technologies and therefore cost.

11.1.4 Cost and integration

Cost is a parameter of paramount importance for the introduction of optical links into computer communications. A commonly accepted cost target for the adoption of optical links within real-world computing systems is considered to be $1 per Gb/s of interconnection, roughly an order of magnitude less than the typical market pricing/capability of GbE data-communication transceivers [13,56]. Such a low figure indicates that the on-board integration process of optics requires minimum capital expenditure for infrastructure and material costs. As performance improves, cost pressures will continue, and the Exaflop supercomputer of 2020 is expected to require 8×10^8 optical links operating at 10 Gbit/s each at a cost of $0.025/Gb/s [12]. Many of the optical solutions promoted for computer communications, such as dense wavelength division multiplexing, already exploit concepts from the highest cost photonic components. To meet bandwidth and energy efficiency, challenges emerge in integration, packaging, and systems level assembly.

11.1.4.1 Optoelectronic integration

The monolithic integration of multiple photonic components is a critical route to cost reduction in high-performance photonics [57]. VCSEL and photodiode arrays are already integrated in array and matrix form for implementation with multimode-fiber, space-domain-multiplexed systems. Wavelength domain multiplexing is now implemented with photonic integrated circuits [58]. This reduces the number of optical connections for a given bandwidth providing a route to simplified assembly for multi-channel, high-functionality optoelectronics. In a manner comparable to scaling in electronics, the maturing of photonic integration technology means that the cost overheads are not impacted directly by the number of components, but by the area of the chip and the complexity in assembly [59]. The emergence of foundries for III–V optoelectronics [60] and silicon photonics [61] set to consolidate these recent integration successes.

11.1.4.2 Electronic-Optoelectronic integration

Optoelectronic transmitters and receivers are ultimately connected to digital logic through sophisticated and specialist amplifiers. Despite the considerable thermal power dissipated in the drivers, and the high-temperature sensitivity of many optoelectronic parts, there is a strong motivation toward more intimate levels of integration. This is particularly the case in the rapidly developing field of silicon photonics. Monolithic integration has been studied, but here optical waveguides require feature sizes of the order of 500 nm but electronics will hit the 10 nm scale in 5 years. This 50:1 scaling problem may warrant a separate optical layer. The absence of efficient light sources in silicon may also motivate the use of III–V membranes as self-contained chip-scale optoelectronic networks or the 3D stacking of separate electronic and optoelectronic chips. Such a System in Package (SiP) concept may even be extended to include passive optics, micro-electro-mechanical systems (MEMS), and other packages and devices. Each active component can be optimized and tested separately then integrated using wafer or chip level die stacking with Through Silicon Vias (TSVs). It will have both electrical and optical I/Os with mechanical features which allow the passive alignment of the SiP to the optical media.

11.1.4.3 Optoelectronic to optical connections

The development of cost-effective packaging is currently one of the biggest challenges for photonics. The complexity of attaching micro- and fiber-optics to the chips is high. Even for data-communication parts where automated alignment and attachment of optics is routine, there is still a need to use sophisticated fiber-optic connectors. Large alignment tolerances and relaxed assembly specifications are desired in the fabrication of the optical waveguides. Various passive alignment schemes have been developed to enable direct interface of the active devices with the optical waveguides. These typically rely on the use of alignment features formed on the substrate/PCB or submounts/interposers that ensure accurate positioning of the active devices with respect to the optical waveguides. Moreover, free-space elements such as double micro-lens structures, mode transformation optics, and tapered waveguide structures have been proposed and have shown improvement in alignment tolerances for system assembly.

11.1.4.4 Assembly

The deployment of automated assembly processes such as pick-and-place is highly desirable as they can provide significant cost benefits. State-of-the-art pick-and-place machines exhibit 3σ placement accuracies of better than $\pm 10\,\mu m$ which are comparable to those required for the alignment of active devices with PCB-integrated multimode waveguides [62]. This is within tolerance for the electronic connections made to optoelectronic chips, but places considerable demand on even the most relaxed-tolerance optical alignment processes. Indeed, many optical connection schemes are still implemented on a per-chip basis, with high-value products receiving individual technician inspection and adjustment.

Passive optical alignment has been an ongoing goal for the photonics community, and particularly for single-mode components which can require submicrometer alignment precision in multiple dimensions. Tolerances will need to be relaxed through improved integrated mode-expansion optics and innovations in pick-and-place technology.

11.2 COMPONENTS

Embedded optical wiring and monolithic circuits require a means to route light in a predetermined manner. We review the broad range of waveguiding technologies which have been proposed for data routing. We introduce the key optical components which may facilitate added functional value including the optical splitter and combiner, and examples of optical waveguide-based components such as waveguide crossings, bends, and out-of-plane couplers.

11.2.1 Waveguides

A high-refractive index core surrounded by a low-refractive index cladding allows waveguiding in a broad range of materials and dimensions. Table 11.2 summarizes the current status of the key technologies in hybrid and integrated photonics. Core and cladding dimensions are listed with bend radii, modal properties, and optical power loss.

   In the short-reach environment, waveguide losses are now sufficiently low to enable in-depth research and development for all distances. The correlations of waveguide cross-section with minimum bend radius and optical losses are determined by the refractive index contrast between the core and cladding materials. The smallest circuits will require the smallest waveguides, but may tolerate higher transmission losses. Possibilities now arise to explore more sophisticated circuits for photonic interconnection and these low-loss connections offer new opportunities with respect to electronics. For example, separating switch elements out by a hundred microns does not degrade the bandwidth and only incurs a delay of the order of 10 ps. Broadcast can be implemented with negligible delay. Multimode and single-mode components are now reviewed separately.

11.2.2 Multimode components

Multimode waveguide components are actively considered for use in board-level interconnects with distances within the range 0.1–1 m as they allow relaxed alignment tolerances and therefore reduced alignment and packaging costs. Waveguides with cross-sections of $\sim 50 \times 50\,\mu m^2$ have been typicaly used as they exhibit 1 dB alignment tolerances of the order of 10–15 μm which can be achieved with conventional automated pick-and-place machines. Such waveguide sizes are also compatible with standard MMF patchcords and allow high coupling efficiencies with commercially available low-cost 10 Gb/s VCSEL and PIN PD components. Future generations operating at higher data rates ≥ 25 Gb/s will most likely have reduced core sizes of $\sim 30\,\mu$m to match the smaller apertures of higher bandwidth photodiodes. For the link lengths of interest in board-level communications (<1 m) multimode dispersion is sufficiently low to allow 40 Gb/s data rates [77,78]. Bandwidth length products of typical multimode polymer waveguides ($50 \times 50\,\mu m^2$, index step of $\Delta n = 0.02$) range from ~ 10 GHz m for the worst case and highly unlikely uniform input mode power distribution, to ~ 30 GHz m for a 50 μm MMF launch, and ~ 150 GHz m for a restricted launch input using either a single-mode fiber or a free-space optical input.

Table 11.2 Comparison of optical waveguide technologies.

Waveguide Technology	Waveguide Cross-Section (μm) Core	Cladding	Minimum Bend Radius	Modal Properties	Polarization Independence	Optical Power Loss[b]
Hybrid plasmon polariton [63]	0.05 × 0.06	-	-	Single mode	No	0.3dB/μm
Silicon on Silica	0.2 × 0.5	2	1μm	Single mode	No	2.0dB/cm
InP						
Membrane	0.2 × 0.4		5μm	Single mode	-	7.0dB/cm
Rib waveguide	0.3 × 1.5	2 × 10	100μm		Feasible	6.0dB/cm
Ridge waveguide	0.3 × 2.0	2 × 10	500μm		Feasible	4.0dB/cm
Silicon oxynitride						
High aspect Si_3N_4						
LPCVD [64]	0.1 × 2.8	10	2mm	Single mode	No	3.0dB/m
Thermal [65]	0.04 × 2.2	15	5mm		No	0.1dB/m
Symmetric [66]	3.0 × 3.0	12	2mm		Yes	5.0dB/m
Silica PLCs						
Ge doped core	3.0 × 3.1	12	2mm	Single mode	Yes	0.3dB/m
P doped [67]	5.0 × 5.0		30mm	Single mode	Yes	0.9dB/m
Polymer materials [68–70]						
Siloxanes [71,72]	5 to 70	depends on core dimensions	3–10mm	Single mode or Multimode	Yes	1–5dB/m
Acrylates [73]						
Perfluorinated [74,75]						
Silica fiber[a]						
Nonlinear [76]	<9.0	60	7.5mm	Single mode	Yes	
SMF-28	9.0	125		Single mode	Yes	0.2dB/km
OM3	50	125		Multimode	Yes	0.2dB/km

[a]circular core and cladding for fibers; [b]loss specified for 1.5 μm optical wavelength with the exception of polymer materials which are specified for 0.85 μm

A wide range of polymeric components have been reported in the literature and their use in complex on-board topologies has been widely demonstrated [79–84]. The optical transmission characteristics of such multimode waveguide components depend on the input excitation as different waveguide modes experience different optical attenuation as they propagate along the structures. Overfilled launch conditions achieved with multimode fiber inputs and fiber-shakers provide a worst-case and highly unlikely input scenario. At the other extreme, well-aligned restricted input launches such as with SMF inputs and low-NA free-space optics yield low insertion loss values which might not match the component losses obtained when multiple waveguide components are concatenated to form complex optical paths. Standardization of the optical characterization method for such highly multimoded components (similar to that prescribed for the use of MMFs in GbE and 10 GbE standards) is therefore required in order to establish reliable design rules in the formation of on-board optical waveguide networks. Here we review the key enabling components for sophisticated on-board optical networks.

11.2.2.1 Waveguide crossings

Waveguide crossings are of particular interest as there is no direct equivalent to electrical multi-plane wiring. Their use can maximize the usable board surface and eliminate the need for multiple on-board optical layers. The induced crossing losses depend on the crossing angle and the launch conditions at the waveguide input [81,85,86]. Higher-order waveguide modes exhibit higher losses as they propagate through the intersection. Ninety degree polymeric waveguide crossings (cross-section $50 \times 20 \, \mu m^2$) exhibiting losses of order 0.01 dB/crossing for a MMF input and crosstalk values below −60 dB in intersecting waveguides have been reported. This shows a potential tolerance to large numbers of crossings on the board [87].

11.2.2.2 Waveguide bends

The use of low-loss waveguide bends within boards allows the interconnection of non-colinear ports and eliminates the need for in-plane 45° mirror structures which require additional fabrication steps and are susceptible to non-ideal sidewall verticality and surface roughness. Polymeric 90° waveguide bends typically require bending radii of the order of 10 mm due to the large waveguide dimensions and the index step of ∼0.02 typically used in such MMF-compliant systems [80,82]. Tighter waveguide bends can be achieved by increasing the waveguide index step and reducing the waveguide dimensions. The minimum bending radius is an important design constraint for the implementation of complex interconnection architectures and depends on the available board area. S-bend waveguide structures are additionally useful in the creation of waveguide components such as Y-splitters and waveguide couplers where a gradual offset between parallel guides is required [80,88]. Raised-cosine S-bends with bending losses below 1 dB for a minimum bending radius of 5 mm have been demonstrated [80]. Sufficiently long S-bend structures provide a gradual reduction of the radius of curvature minimizing transition and bending losses along the waveguide path.

11.2.2.3 Power splitters/combiners

Optical power splitters and combiners enable the interconnection of multiple ports to a single optical input or output port and the resulting advanced on-board interconnection architectures [81,89]. Large polymeric multimode splitters up to 1×32 have been reported in the literature, achieving good splitting uniformity and low excess losses [75,90,91]. The main concern with such multimode structures is the dependence of the splitting ratio on the launch condition and input position. Overfilled input launches (such as with MMF inputs) yield a uniform mode power distribution at the splitter input and therefore achieve a uniform power splitting at the splitter outputs. Relatively high output uniformity with maximum output imbalance below 1 dB can be achieved in such cases at the expense of higher excess losses at the apex point of Y-branch splitters [80,92]. Such devices are not suitable for use, however, with restricted input launches as they result in a large variation in splitting uniformity for the different input positions. Mode mixers need to be introduced at the splitter input to ensure a relatively uniform mode power distribution at the device input and therefore achieve uniform splitting.

Multimode combiners achieve signal combining when excited with an underfilled launch without the 3 dB loss per junction obtained in single-mode combiners. This is due to the fact that lower-order modes at the combiner inputs can couple to higher modes at the merging sections, which are still guided at the output arm of the device. As a result, a different behavior is observed when these devices are employed as power splitters and combiners. A difference of 5 dB in insertion loss has been demonstrated for a 1×8 Y-splitter/combiner when a SMF input has been used at the combiner inputs [80].

Figure 11.2 shows images of fabricated polymer multimode components (90° crossings, 90° bends, and Y-splitters/combiners) and their respective optical transmission characteristics under different launch conditions.

11.2.3 Single-mode components

Optical interference allows a particularly rich set of photonic operations but the methods operate most effectively for single-mode waveguides where the core-cladding refractive index profile imposes precise control of the phase front. High confinement structures with submicron-square cross-sectional areas also allow the most efficient optoelectronic processes for lasing, amplification, detection as well as the highest density optical integration. Key components required for signal routing in the single-mode regime include mirrors, waveguide micro-bends, splitters, and crossings.

11.2.3.1 Total internal reflection micro-mirrors

High-density photonic circuits require the routing of light within a small area. Some of the smallest proposed schemes have implemented photonic band-gap defects [93,94], resonant corners [95], and total internal reflection mirrors [96,97]. Total internal reflection mirrors have perhaps been most widely used for beam redirection in integrated optoelectronic switching circuits, with the first monolithic

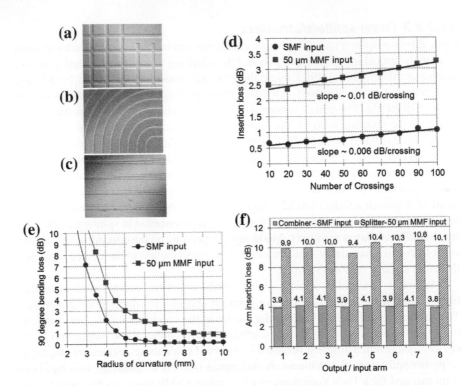

FIGURE 11.2 Polymer multimode waveguide components: (a) 90°crossings, (b) 90° bends, (c) Y-splitters/combiners, and their optical transmission characteristics: (d) insertion loss of 90°crossings as a function of the number of crossings, (e) bending losses for 90° bends, and (f) insertion losses of all arms of a 1 × 8 Y-splitter/combiner when used as a splitter with a 50 μm MMF launch (shaded bars) and as a combiner with a SMF input (solid bars).

optoelectronic 16 × 16 switch [98] being one of the most recent and sophisticated examples. In these schemes, mode-expanding input guides are used in the waveguide plane to reduce diffractive losses. Such mirrors can lead to excess losses which are sensitive to small fluctuations in fabrication and this has motivated studies into relaxed-tolerance integrated parabolic mirrors [99] and multimode interference reflector mirrors [100].

11.2.3.2 Waveguide micro-bends

Curved waveguides are more widely used in integrated photonics as they offer arbitrary waveguide rotation and modest loss with a regular process technology. However, the size depends on the refractive index contrast in the materials used. Bend radii of order several hundreds of microns require only modest refractive index contrast of order 0.01 [101,102]. Lower bend radii of order tens of microns require considerably higher refractive index contrast within the plane of the micro-bend to minimize radiation losses but this can be readily achieved with deeply etched waveguides with index steps exceeding 1.0 [103–105]. Waveguiding performance is, however, very sensitive to waveguide bend modal properties, fabrication tolerance, and polarization rotation. The latter effect has been so pronounced that it has led to a body of research on polarization converters [106,107]. A whispering gallery micro-bend design has been proposed [108] to cover radii of a few microns to a few tens of microns range which is expected to be particularly robust to fabrication variations. To reduce bend radii even further toward the micron level, researchers have considered high-refractive index contrast in both waveguide planes to minimize substrate losses, primarily using ultrahigh contrast silicon-on-insulator (SOI) and InP membranes. The ever-reducing radii place increasing constraints on fabrication methods, however, with single-mode waveguide bends becoming increasingly sensitive to nanoscale dimensional variations, sidewall tilt, surface roughness, and substrate leakage [109,110].

11.2.3.3 Splitters and combiners

The splitting and combining of light allows the implementation of bandwidth independent replication, fan-out and fan-in of data. In the case of single-mode splitters, the combiner may be considered to be a reciprocal process. No significant delay is incurred, and energy dissipation is low unless copies of data are intentionally discarded. Splitters using single-mode waveguide inputs and outputs have been demonstrated by means of spatial windowing and interference.

Spatial splitters have been implemented in integrated optics by means of symmetric T-splitters [111] and asymmetric arrangement of waveguides [112]. Both approaches rely on the combination of adiabatic mode expansion by a low-divergence ($\sim 1°$) tapered waveguide and a total internal reflection mirror implemented by a deep-etched waveguide. The quality and absolute positioning of the mirror directly influences excess losses, splitting ratio, and the forward propagating modal integrity.

Interference-based splitters exploit the phenomena of self-imaging when a single-mode waveguide is appropriately connected to a multimode waveguide [113]. This is shown through simulation in Figure 11.3 for the case of a 1×2 and a 2×2 multimode interference device. The displaced boundary of the waveguide leads to higher-order modes which constructively and destructively interfere to create replicas of the input along the multimode waveguide. Depending on width and length, the input can be either reproduced or replicated. Techniques to ensure tolerance to fabrication variations are now quite robust for 1×2 and 2×2 splitting and combining operations [114], and geometries to ensure power uniformity in 1×8 splitters have also been demonstrated for switch circuits [115]. This may be generalized to $N \times N$ for use in integrated star couplers which are a key component for the arrayed waveguide grating used in wavelength multiplexers [116]. While several hundreds of

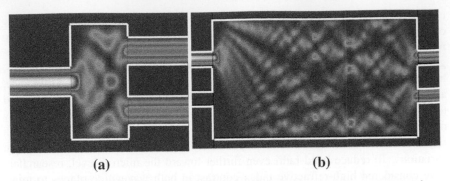

(a) **(b)**

FIGURE 11.3 Power splitters in integrated optoelectronics (not to scale): (a) a symmetric power splitter in a 1 × 2 multimode interference device and (b) two input two output multimode showing image formation along the device length.

microns of multimode waveguide are required for the more advanced splitting and combining functions, simple splitters may also be implemented within a length of a few tens of microns for optoelectronic circuits, and a few microns for optoelectronic membranes and SOI technology.

11.2.3.4 Compact waveguide crossings

In contrast to electronic integrated circuits, optical integrated circuits are largely implemented in one single plane and therefore the mesh interconnection of components mandates waveguide crossings. Light can be guided over short distances without the use of a physical waveguide but any signal leakage between intersecting waveguides through scattering and diffraction can lead to some data corruption through crosstalk. A broad body of work has assessed and optimized waveguide intersects in terms of miniaturization, and optical loss.

Early research indicated that the primary source of loss will be diffractive losses from a small aperture, but maintaining an angle of intersection between 45 and 135° allows for minimum and near-invariant excess loss. A near-exponential dependence of loss on waveguide width at the intersection motivates wider waveguide intersections using tapered waveguides. This allows losses as low as 0.02 dB per intersection in ridge waveguide III–V of the waveguide crossings [117]. Linear adiabatic tapering allows the connection of these wide aperture waveguides to high density wiring and optoelectronic parts, but required waveguide angles of order <1° can lead to relatively long overall crossing footprints. This motivates work on the more compact parabolic horn concept as used in the field of radio antenna. These have been miniaturized to micron levels for high confinement SOI with theoretical and experimentally measured losses of order 0.14 dB and 0.24 dB, respectively. The ultrahigh confinement in a rib waveguide structure can, however, be ameliorated by engineering the waveguide dimensions in three dimensions. Multi-step etch technologies [118] are increasingly used, and allow the formation of low-divergence ridge waveguides which may then be connected to high confinement waveguides for circuit miniaturization. Multimode

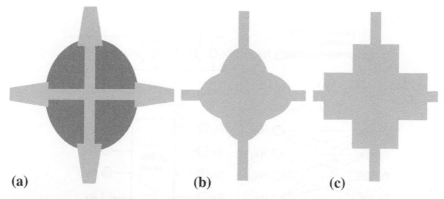

(a)   (b)   (c)

FIGURE 11.4 Single-mode optical waveguide crossings: (a) tapered waveguide with shallow ridge waveguide at the intersection, (b) parabolic taper for reduced footprint, (c) multimode interference crossing.

interference also offers an elegant route to reducing crosstalk at intersections [119]. Here the self-imaging properties of a center-launched mode in a wide, appropriate-length waveguide allow the optical signal to be self-focused at the center-point and output of the multimode crossing. Figure 11.4 shows schematic diagrams for inter-secting waveguides using the multi-step etch, parabolic horn, and multimode inter-ference structures respectively.

Optical power leaking into the intersected waveguide can in principle lead to signal corruption on a second circuit path. Experiments where crosstalk has been deliberately incurred in an photonic integrated circuit lead to the observation that the optical power penalty (a metric for signal degradation) increases by 1 dB as the levels of crosstalk increase from −15 dB to −5 dB [120], consistent with a number of other observations on incoherent crosstalk [121]. Coherent crosstalk is observed for con-siderably lower levels of cross-coupling, and has even been identified in re-circulated switch experiments where shallow orthogonal single-mode waveguide crossings are implemented [122]. This higher level of sensitivity may increase the crossing specifi-cations for highly advanced switching architectures which implement optical domain feedback including some forms of delay-line buffering [123].

11.3 **ARCHITECTURES**

Passive interconnection architectures include point-to-point links, meshed wave-guide architectures, and optical buses. Point-to-point links and passive shuffle net-works constitute non-blocking interconnection schemes that allow high aggregate data capacities but require the use of a large number of links that scale as N^2 with the number N of interconnected modules. On the other hand, broadcast architectures such as a shared bus interconnection scheme require fewer optical links within the

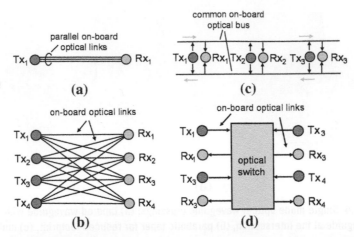

FIGURE 11.5 Illustration of main types of demonstrated board-level optical interconnection systems: (a) high-capacity point-to-point on-board optical links, (b) passive shuffle networks, (c) optical bus architectures, and (d) optical switch architectures.

system but rely on efficient control protocols to avoid and resolve transmission collisions. The majority of the optical backplane systems demonstrated so far utilize point-to-point parallel links, as they are straightforward to implement with arrays of waveguides. A few examples of demonstrated optical backplane systems for card-to-card communication (<1 m) are provided in the next section. Figure 11.5 illustrates the three main types of passive on-board optical interconnection systems that can be found in the literature and switched interconnection. The first group of demonstrated systems utilizes point-to-point parallel links between different electronic modules to achieve high aggregate interconnection capacities. The second group exploits optical waveguide shuffle networks, while the third group includes systems based on optical bus architectures. The use of fixed routing eliminates the need for optical switches and their respective control systems but necessitates a larger number of optical interconnects within the system. Switched routing schemes (the fourth and final group shown) minimize the number of interconnects required but rely on the use of high-speed, low-loss, and low-power optical switching elements and efficient control protocols.

11.3.1 Point-to-point on-board optical links

Significant work has been carried out by leading multinational companies, such as IBM, Fujitsu, and Siemens, on the development of board-level optical interconnects. The majority of the work focuses on the board-level integration of a large number of parallel optical links (mainly waveguide based for board-to-board communications) and the development of cost-effective packaging and assembly methods. Different

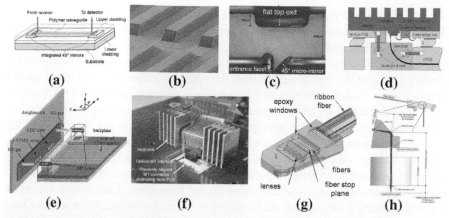

FIGURE 11.6 Optical coupling schemes: (a) principle of out-of-plane coupling [155], (b) integrated polymeric 45° micro-mirrors [156], (c) discrete 45° micro-mirror optical component [129], (d) proposed coupling scheme using bent optical [147], (e) end-fired coupling schematic [146], (f) demonstrated end-fired coupling scheme [145], (g) schematic of the MICRO-POD connector [9] fibers and (h) its principle of operation [9].

system designs and connectorization schemes have been proposed. Some important considerations regarding the design of a hybrid optoelectronic system and the integration of optical components on standard printed circuit boards are discussed below and some examples of the different demonstrated systems with respect to optical coupling are shown in Figure 11.6.

11.3.1.1 Electronic-photonic interface

Two different approaches have been taken concerning the position of the electronic-photonic interface, and specifically, the active optical components (LDs, PDs) and their driving circuits [4]. The first approach involves positioning the electro-optic conversion units close to the card/module edge in order to facilitate the interface of the card/module with the rest of the system. In optical backplanes, for example, such a scheme allows direct optical coupling of the laser diodes and photodiodes with the input/outputs of the optical waveguide structures on the backplane [124]. Such a scheme, however, requires suitable electrical interconnects to route the high-speed signals from the I/O ports of the high-performance computing chips to the optical modules imposing additional requirements in the system design. The second approach favors locating the active optical modules in close proximity to the high-performance chips facilitating the interface between the electrical and optical layers of the system. Various types of submounts that facilitate both the electrical connection of the active optical modules to the electronic chips and the coupling of the light beams into and out of the optical backplanes have been demonstrated [41,125–127].

11.3.1.2 Chip-to-waveguide coupling

Two coupling schemes are considered for use in polymer-based optical interconnects: out-of-plane and end-fired coupling.

Out-of-plane coupling schemes rely on the use of micro-optical 90° beam-turning elements to route the optical signals from the optical waveguides to the board surface and vice versa. Such a beam-turning functionality is typically achieved with 45° micromirror structures that are either formed in the polymer layers or integrated into micro-optical structures fitted within appropriately positioned slots in the board [128–130]. Examples of such systems are shown in Figure 11.6. The use of out-of-plane coupling schemes facilitates the positioning of the active devices on the board surface and their electrical connection to their respective driving circuits. Nevertheless, they require additional fabrication and assembly steps for the formation of the mirror structures. Various fabrication methods for the formation of integrated polymer mirrors have been proposed such as laser ablation [131,132], embossing [133], laser direct writing [134], and dicing [135,136]. Additional mirror metallization processes are typically required to achieve high mirror reflectivity. Polymer integrated mirrors with losses below 0.7 dB have been reported [137–139]. The use of micro-lenses is often required in such coupling schemes in order to achieve high coupling efficiency by mitigating diffractive losses in the relatively long free-space beam paths. The use of polymer-based optical structures integrated in the optical layers such as optical rods and cubes has also been proposed to optimize the coupling efficiency in such schemes [140–143].

End-fired optical coupling schemes are potentially simpler as they do not require beam-turning elements to interface the LDs and PDs with the optical waveguides (Figure 11.6e and f). However, they do require efficient 90° routing of high-speed electrical data signals from the board surface to the active devices. Therefore, such coupling schemes have been proposed for use mainly at the board edge where direct interface between the active optical modules and the backplane is straightforward [144–146].

Additional optical coupling schemes of interest involve the use of bent multimode fibers (Figure 11.6d) to interface the active optical components located on the board surface with the polymer waveguides embedded in the common substrate [147], multimode fiber-based evanescent couplers where the coupling can be controlled by mechanically adjusting the separation between the on-board waveguides and ribbon fibers [148] and on-board grating couplers [149,150].

11.3.1.3 Optical plug-n-play

Particular attention has been paid to the development of low-cost connectorization schemes allowing plug-and-play functionality in the connection of hybrid optoelectrical cards/modules onto optical backplanes. The majority of the proposed schemes rely on the use of MT-based connectors to make use of well-established low-cost technologies [151–153]. Such pin-based connection schemes offer mechanical robustness and component alignment within the required tolerances $<10\,\mu m$ to achieve high coupling efficiency while allowing two-dimensional arrays of input/output connections. Nevertheless, specific modifications in the connectors are in general required to match the design of each proposed backplane system, and alignment pins can only be

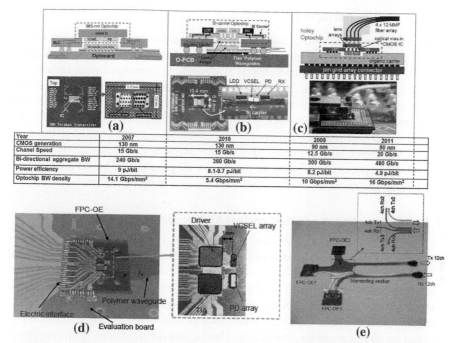

Year	2007	2010	2009	2011
CMOS generation	130 nm	130 nm	90 nm	90 nm
Chanel Speed	15 Gb/s	15 Gb/s	12.5 Gb/s	20 Gb/s
Bi-directional aggregate BW	240 Gb/s	360 Gb/s	300 Gb/s	480 Gb/s
Power efficiency	9 pJ/bit	8.1-9.7 pJ/bit	8.2 pJ/bit	4.9 pJ/bit
Optochip BW density	14.1 Gbps/mm²	5.4 Gbps/mm²	10 Gbps/mm²	16 Gbps/mm²

FIGURE 11.7 Examples of demonstrated board-level optical interconnection systems by (a-c) IBM [160] and (d-e) Fujitsu [161] with some details noted: (a) 985 nm Optochip, (b) Si-carrier 850 nm Optochip, (c) holey 850 nm Optochip interfacing with multimode ribbon fiber, (d) a 4-channel OE-PCB module with polymer waveguides on flexible substrate, and (e) a 12-channel opto-electronic system employing three of these OE modules.

made space efficient if used at the board edge. For mid-board connections, schemes employing optical modules with integrated mirror structures and lenses, such as the Avago MICRO-POD connectors (Figure 11.6g and h), appear to be more attractive solutions owing to their smaller footprint and the freedom they offer in positioning the interface anywhere on the backplane [9,130,154]. Such connectors are typically injection-molded or precision micro-machined and require the formation of appropriate alignment features on the board to ensure efficient mounting. They can be interfaced with either micro-mirror structures integrated within the optical layers of the boards enabling out-of-plane optical coupling or by fitting through-board slots to achieve end-fired coupling into/out of the optical waveguides [144].

On-board optical interconnection systems have been demonstrated for a range of polymer materials, coupling schemes, optoelectronic board designs, packaging, and assembly processes [45,136,157,158]. To date, the highest demonstrated data capacity fully integrated on-board link implemented with board-level polymer waveguides is 225 Gb/s (15 on-board polymer waveguides, each operating at 15 Gb/s) [159]. Figure 11.7 shows some recent examples of demonstrated board-level optical interconnections systems by IBM and Fujitsu.

11.3.2 Shuffle networks

Passive shuffle networks make use of spatially re-sequenced point-to-point optical links to enable high-speed optical interconnection between many electrical cards/ modules. The architecture allows strictly non-blocking communication between all connected entities but requires the implementation of a large number of on-board optical links to achieve full connectivity. The required number of links is N^2, where N is the number of connected entities, and offers an aggregate data capacity of $R \times N^2$, where R is the data rate of each optical link. Such an interconnection architecture relies on the use of large numbers of low-cost optical active components and passive on-board optical components to implement all possible on-board connections. Examples of demonstrated optical shuffle backplanes are presented below.

11.3.2.1 Edge-board coupled optical backplane

The powerful combination of polymer waveguide arrays and waveguide components (crossings, bends) can offer on-board interconnection between different electrical cards/modules. A strictly non-blocking waveguide-based scalable interconnection architecture has been proposed for the formation of an optical backplane [162]. All connections are made to the board edge and the optical backplane is expected to interface with the cards via either multimode ribbon fibers or VCSEL/PD arrays mounted on MT-compatible connectors. The Tx and Rx interfaces comprise arrays of N optical waveguides (where N is the number of connected cards/modules) and are positioned on opposite board edges to minimize any crosstalk due to non-ideal optical coupling at the waveguide inputs and light leakage at the waveguide bends. A single point-to-point link for each pair of transmitter (Tx) and receiver (Rx) in the system is formed on the backplane allowing non-blocking operation of each communication link in the system. Each link is implemented in the optical layer with a waveguide path which comprises one 90° waveguide bend and a number of waveguide crossings which varies for the different on-board interconnection paths (up to 90 for a 10-card backplane). A schematic of the proposed scalable shuffle interconnection architecture is shown in Figure 11.8a, while Figure 11.8b shows the waveguide layout for a 10-card optical backplane [162]. The 10-card backplane is fabricated on an FR4 substrate and measures 10 cm × 10 cm. The optical layer comprises a total of one hundred 90° bends and approximately 1800 waveguide crossings. An image of the fabricated sample is shown in Figure 11.8c, with red light illuminating the longest on-board optical path. Insertion losses for 50 μm MMF inputs are below 8 dB for all on-board optical paths, while worst-case crosstalk values are found to be −25 dB. Error-free 10 Gb/s data transmission over the optical backplane with low-cost 850 nm VCSEL sources has been demonstrated. The aggregate data capacity of this 10-card optical backplane is 1 Tb/s (10 Gb/s × 100 links).

11.3.2.2 Mid-board coupled optical backplane

A shuffle optical backplane has been proposed by Xyratex and UCL to allow interconnection between electrical cards in large-scale storage systems [124]. The

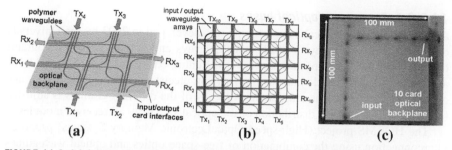

(a)   **(b)**   **(c)**

FIGURE 11.8 (a) Schematic of the proposed scalable meshed waveguide architecture, (b) waveguide layout for a 10-card optical backplane, and (c) photograph of the fabricated backplane on an FR4 substrate with longest waveguide path illuminated with red light (For interpretation of the references to color in this figure legend, the reader is referred to the web version of this book.)

backplane relies on the use of polymer waveguides integrated onto a PCB and mid-board slots to allow the connection of the electrical cards. Special optical connectors have been developed to enable mechanically robust and high-accuracy positioning of the electrical cards onto the backplane [146]. These connectors use end-fire optical coupling into and out of the polymer waveguides, while deploying a double micro-lens structure to achieve high coupling efficiency. A proof-of-principle demonstrator accommodating four cards has been developed and is shown in Figure 11.9. Each on-board optical link achieves 10.7 Gb/s operation, providing an aggregate data capacity of 86 Gb/s for four cards.

11.3.3 Optical buses

Optical bus interconnection schemes are attractive for use in applications where communication between a variable number of users with short bursts but high through-puts is required. Approaches to implement on-board optical buses include free-space optics, metallic hollow waveguides, and on-board polymer waveguides.

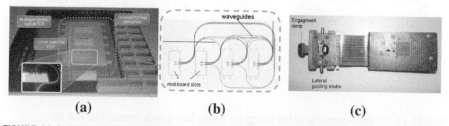

(a)   **(b)**   **(c)**

FIGURE 11.9 (a) Fabricated optical backplane with various features noted, (b) corresponding shuffle waveguide layout interconnecting four electrical cards, and (c) compatible MT-based optoelectronic connector enabling end-fired optical coupling [124].

11.3.3.1 Free-space optical buses

An early free-space optical bus prototype was developed in 2003 by the University of Texas [163]. It employs VCSEL and PD arrays and volume grating holograms to route the free-space optical beams over the entire backplane area (Figure 11.10a and b). The demonstrated system comprises four electrical cards and one arbiter and operates at 1.25 Gb/s. The diffraction efficiency of the different gratings is adjusted to ensure uniform power distribution of the optical signals to all receiver modules.

The HOLMS project (High-speed OptoeLectronic Memory Systems) proposed interconnection using the combination of free-space optics and optical waveguides integrated on PCBs in 2006 [164,165]. The system design targets the interconnection of processors with multiple memories in a multi-processor system. It uses ribbon fibers terminated with MT connectors, PCB-integrated polymer waveguides, and a 3D free-space optical interface module to interconnect all the different entities of the system (Figure 11.10c). The demonstrated optical interface module has the capacity to interface 192 optical channels and exhibits an average optical path loss of 8.4 dB for the various system I/O interconnections.

Finally, a free-space optical interconnection system was proposed by Hewlett Packard in 2009 for communication between electrical cards [25]. Two types of free-space links were implemented for short-reach card-to-card communication, one based on the use of a double telecentric lens with unity magnification and another on magnetically coupled proximity free-space modules. The first scheme allows

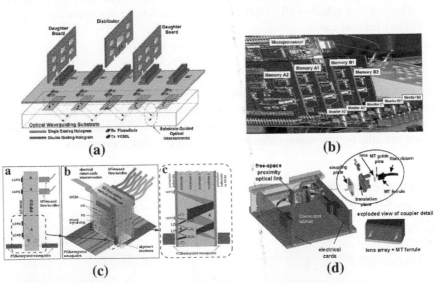

FIGURE 11.10 (a) Schematic of the hologram-based free-space optical bus and (b) photograph of demonstrated based system [163], (c) details of free-space optical interface module in HOLMS system [164], and (d) schematic proximity magnetically coupled card-to-card free-space optical link [25].

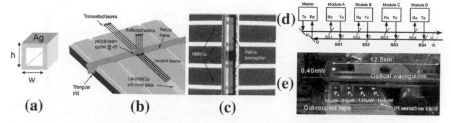

FIGURE 11.11 (a) Schematic of the metallic hollow waveguide (w, $h = 150\,\mu$m), (b) illustration of an optical tap using the pellicle beam splitter, and (c) photograph of fabricated module with features noted, (d) schematic of the optical bus interconnection scheme, and (e) photograph of fabricated optical system with 4 outputs illuminated [22].

interconnection with large alignment tolerances up to ± 2 mm and is resilient to mechanical vibrations in the system, while the second one allows direct 5 Gb/s interconnection between closely placed electrical cards, avoiding the use of the backplane.

11.3.3.2 Metallic hollow waveguide bus

An optical bus using hollow metallic waveguides has been designed and fabricated onto a silicon substrate [22]. The metallic hollow waveguides have a cross-section of $150 \times 150\,\mu$m^2 and 12.5 cm long silver-coated (Ag) sidewalls. The backplane comprises four parallel waveguides (optical channels) and 1×4 VCSEL and PD arrays each operating at 5 Gb/s and can accommodate up to eight optical taps. The signal distribution at the cards is achieved with pellicle beam splitters which are inserted into the substrate and allow partial beam reflection at each receiver interface (Figure 11.11). The waveguides exhibit optical losses of 0.05 dB/cm while the excess losses of the optical taps are 0.15 dB. A relatively large 3 dB variation in received optical power is observed for the different optical taps and this may be due to the high sensitivity of the reflectance of the beam splitters to the light state of polarization.

11.3.3.3 Polymeric optical bus

A multi-channel scalable regenerative optical bus architecture based on the use of on-board polymer waveguides has been proposed [166]. The architecture makes use of optical signal splitters and combiners to perform the signal "add" and "drop" functions at each card interface and utilizes electrical 3R regenerators (re-shape, re-time, re-transmit) to enable bus extension with multiple bus segments and therefore, connection of arbitrary number of cards onto the bus (Figure 11.12). As a proof-of-principle, a 4-channel 3-card polymeric optical bus module is designed and fabricated onto FR4 substrates [167]. The principle of operation of the regenerative architecture is demonstrated using two polymeric bus modules and a prototype 3R regenerator comprising back-to-back transceivers and clock recovery electronic chips [166]. Error-free (bit-error-ratio below 10^{-12}) 10 Gb/s data transmission between card interfaces located on the different bus segments is achieved through the 3R regenerator unit.

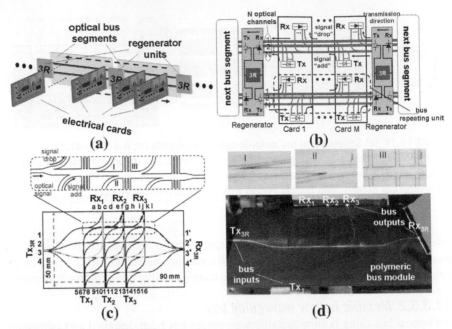

FIGURE 11.12 (a) Illustration of the proposed regenerative optical bus architecture, (b) schematic of a bus segment with optical and electrical parts noted, (c) waveguide layout of the designed 4-channel 3-card bus repeating unit with inset schematic of main optical bus. Photographs of (d) waveguide components of the optical bus: (I) splitter, (II) combiner, and (III) 90° crossings and (e) the fabricated bus sample with the paths noted in (c) illuminated with red and green light.

11.3.4 Optical switch architectures

Optical switching is a promising solution to reduce massively optical wiring without unduly compromising connectivity. The waveguide bandwidth becomes decoupled from transceiver bandwidth through space-domain switching and wavelength routing. Switches may be compared in terms of speed of path reconfiguration. This allows technologies allocated in three classes:

(i) provisioning and protection: connections are made and broken on the millisecond time scale and are widely deployed in the form of reconfigurable optical add drop multiplexers (ROADMs) in telecommunications backbones,

(ii) packet-switch compliance: connections are made and broken on the nanosecond time scale and may react to changes in data traffic allowing a statistically multiplexed gain for the interconnect,

(iii) bit-level processing: operations are performed on the picosecond time scale using all-optical processing techniques. This may offer exciting optical signal processing functionalities in the future as energy requirements are reduced.

While each class of switches is an active area of research in its own right, the focus for the coming decade is very likely to be in the area of packet-level routing using integrated circuits. We therefore focus exclusively on switch technologies which enable nanosecond time scale and direct-electronically-configured routing. Switch elements which may enable cross point switches are reviewed before considering high-radix elements with the potential added value of broadcast functionality, and finally multi-stage circuits for enhanced connectivity.

11.3.4.1 Crosspoint switch elements

Optical crosspoint switch elements require 2×2 elements which can be replicated across one chip in a bus or matrix format. The optical ring resonator is a conceptually attractive example for narrow-band optical signals. The combination of paired rings at waveguide intersects offers a potential route data by coupling signals between orthogonal bus waveguides. High-order resonators with multiple carefully coupled resonators offer the prospect of broadened pass-bands without compromising switch extinction [168], but challenges remain in creating fabrication tolerant circuits [34,169] which are both aligned to transceiver wavelengths and operate with modest electronic power. The physical mechanism exploited to tune the switch in and out of resonant coupling defines the overall switch properties and to date only all-optical [30] and thermo-optic tuning [170] have been demonstrated with relaxed wavelength tolerance.

Mach-Zehnder interferometers offer broadband operation and low inherent loss. The 2×2 form for the interferometer relies on precision fabrication in multimode interference couplers to perfectly match signals in phase and amplitude. While switch extinctions of order $-20\,dB$ can be achieved, good crosstalk suppression of the order of $-40\,dB$ and beyond has required multi-stage elements, and providing robust electronic control signals remains an ongoing challenge. Until recently, Mach-Zehnder interferometers have been relatively long due to the limited $V_\pi L$ product. The required voltage to achieve a π phase shift has scaled inversely with phase modulator size. However, recent work using SOI at IBM has demonstrated relatively compact elements of the order of a few hundred microns. The two-arm Mach-Zehnder interferometer concept may also be further generalized with multiple waveguides to create the phase array switch. Here, carefully calibrated analog signals are applied to tens of phase modulators to interferometrically image signals to one of sixteen [171] or even one hundred [172] output ports. In the latter case, a combination of two stages of phase array switches and one array of Semiconductor Optical Amplifier (SOA) gates was additionally used to ensure good switch extinction.

Excellent switch contrast can be achieved while retaining the low-loss operation by combining interference effects with the gain from an SOA gate. The combination of interferometric and gated switching in the same element [173,174] provides an interesting route to avoiding excess energy loss while retaining the advantages or excellent signal extinction, low operating voltage, and wavelength independence. The vertically coupled SOA gates which have lead to signal extinction levels of the order of $-60\,dB$ and the use of frustrated mode interference in MMI-based switches, both offer a crossbar compliant configuration, although device implementations can be challenging. A broad range of concepts for SOA-based switches have been reviewed in [123]. Figure 11.13 summarizes the waveguide and electrode arrangements for the paired ring resonant switch, the high-order resonant switch, the Mach-Zehnder switch, and the SOA gate-based switch elements which have been most extensively reported over the last 5 years.

11.3.4.2 High-radix switches

Optical components may be integrated to create switches with more than two inputs and two outputs while using only one active element per path. The use of one element per path becomes important when the switch elements themselves lead to some level of data corruption. In contrast to CMOS electronics, optical, and optoelectronic elements operate in an analog mode, leading to small amounts of signal leakage and corruption in off-state and on-state switches respectively. This is true for the most prominent example of single-stage, high-radix switches using SOA gates. While gates operate with high gain and can allow extinction ratios of up to even −70 dB [115], a power penalty is often incurred, and the optical input power levels must be controlled. Due to the specialist nature of making photonic integrated circuits, relatively few high-radix optical switches have made it to the level of rigorous system level assessment. However, extensive literature exists on switching systems using discretely packaged SOA gates. This has in part been due to the commercial availability of discrete SOAs over the last two decades but there are also additional important advantages in terms of relative temperature insensitivity, massive bandwidth operation to the point they can be considered wavelength insensitive, low voltage operation, ease of integration, and high signal extinction. An important advantage

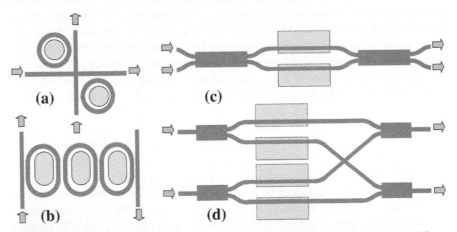

FIGURE 11.13 Optical switch elements: (a) paired ring resonators to allow 2 input and 2 output connections, (b) broadband high-order resonant switch, (c) broadband symmetric Mach-Zehnder interferometer, and (d) broadband broadcast and select switch using SOA gates. One electrode is shown for clarity.

with respect to interferometric switches is the possibility to broadcast and multicast signals without incurring any additional delay. The broadcast and select architecture is particularly well suited to the SOA gate as losses incurred in the optical fan-out and fan-in may be compensated in the gate itself. While most published examples are implemented monolithically up to 4×4 connectivity, multiple integrated circuits have been assembled to create 8×8 connectivity subsystems [115]. SOA gate architectures can be highly desirable for facilitating broadcast and multicast, although this can lead to energy loss when such operation is not required.

Figure 11.14 shows the example for a 4×4 monolithically integrated broadcast and select switch implemented on InGaAsP/InP. The circuit may be simply controlled with digital logic if round-robin path allocation [175] is acceptable, or with field programmable logic [176] if a more sophisticated form of arbitration is required. The compact nature of the circuit, with a footprint of $1 \, \text{mm} \times 4 \, \text{mm}$, means that the gate states may be synchronously updated with relatively simple electronics.

11.3.4.3 Multi-stage switches

Increases in interconnectivity in switched networks are achieved through the interconnection of small- or moderate-radix switches in multiple stages of switches. Extensive literature has been developed for electronic switching networks [177], and a number of these architectures have been re-implemented with optical circuits. There are, however, physical layer constraints which may influence optimal choices for interconnection. Signal impairments can arise from a number of sources: wavelength registration tolerance in wavelength-specific switch elements or build-up of noise in amplified links. Theoretical studies have, however, been performed for physical layer performance for some of the important classes of switch elements including

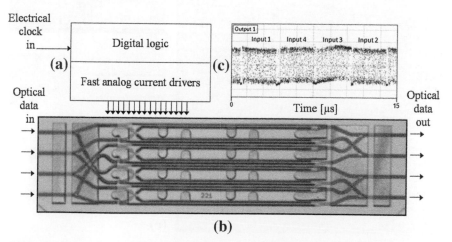

FIGURE 11.14 Integrated optical switch fabric operated with electronic control plane: (a) electronic control plane, (b) optical switch matrix with four inputs and four outputs, and (c) oscilloscope time trace (15 μs span) showing dynamic routing from four inputs.

SOA gates [178] and microring circuits [179]. While line rates are constrained to 40 Gb/s with critical wavelength alignment for the ring resonant switches so far demonstrated, 320 Gb/s has now been demonstrated in monolithic circuits with four stages of SOA-based crossbar switches [180]. Monolithic multi-stage circuits have also been demonstrated with 16 inputs and 16 outputs (16×16) [98]. This first such demonstration was made in 2009 and even allowed net gain on the chip while routing 10 Gb/s data [181]. Figure 11.15 shows the monolithic implementation of a Clos network using three stages of 4×4 elements to create an overall connectivity of 16×16. Identical switch stages are used to create a uniform topology and improve wafer level uniformity. The splitters and mirrors highlighted in Section 11.2 are shown as insets. Each path has three gating SOAs and goes through eight shuffle sections with a mean path length of 9 mm. The switch contains 192 SOAs, 210 waveguide crossings, 288 splitters, 424 etched corner mirrors and has dimensions of 6.3 mm $\times$ 6.5 mm.

11.3.4.4 Wavelength selective switching

The co-integration of wavelength filters with optical switches allows an additional route to enhanced connectivity. The combination of space- and wavelength-selective routing allows the number of connections to be defined by the product of physical input and the wavelength channel numbers. Thus order-of-magnitude scaling is possible with respect to both space-switch-only and wavelength routed circuits. Opto-electronic integrated circuits capable of fast space- and wavelength-selective routing have primarily focused on one WDM input, one WDM output ROADM concept where wavelengths are switched to and from four colored connections [182–184].

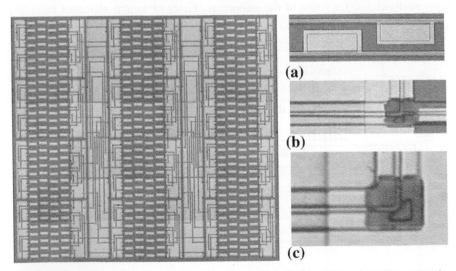

FIGURE 11.15 Sixteen input 16 output switch implemented with a three stage Clos network. Insets show: (a) SOA gates, (b) an example power splitter, and (c) an example of a total internal reflection mirror.

[8] A.F. Benner et al., Optics for high-performance servers and supercomputers, in: Conference on Optical Fiber Communication and National Fiber Optic Engineers Conference (OFC/NFOEC), 2010, pp. 1–3.

[9] D. Childers et al., Miniature detachable photonic turn connector for optical module interface, in: 61st IEEE Electronic Components and Technology Conference (ECTC), 2011, pp. 1922–1927.

[10] D. Miller, Device requirements for optical interconnects to silicon chips, Proc. IEEE 97 (2009) 1166–1185.

[11] D.A.B. Miller, Optical interconnects to electronic chips, Appl. Opt. 49 (2010) F59–F70.

[12] A.F. Benner, Cost-effective optics: enabling the exascale roadmap, in: 17th IEEE Symposium on High Performance Interconnects (HOTI), 2009, pp. 133–137.

[13] J.A. Kash et al., Optical interconnects in exascale supercomputers, in: 23rd Annual Meeting of the IEEE Photonics Society, 2010, pp. 483–484.

[14] M. Schneider, T. Kuhner, T. Alajoki, A. Tanskanen, M. Karppinen, Multi channel in-plane and out-of-plane couplers for optical printed circuit boards and optical backplanes, in: 59th Electronic Components and Technology Conference (ECTC), 2009, pp. 1942–1947.

[15] M. Ohmura, K. Salto, High-density optical wiring technologies for optical backplane interconnection using downsized fibers and pre-installed fiber type multi optical connectors, in: Conference on Optical Fiber Communication and National Fiber Optic Engineers Conference (OFC/NFOEC), 2006, p. 3.

[16] I.-K. Cho, J.-H. Ryu, M.-Y. Jeong, Interchip link system using an optical wiring method, Opt. Lett. 33 (2008) 1881–1883.

[17] J.V. DeGrootJr , Cost-effective optical waveguide components for printed circuit applications, Passive Components and Fiber-based Devices, vol. IV, SPIE, Wuhan, China, 2007 pp. 678116

[18] L. Eldada, Advances in Polymer-based Dynamic Photonic Components, Modules, and Subsystems, Passive Components and Fiber-Based Devices III, Parts 1 and 2, SPIE, vol. 6351, 2006, pp. 244–253.

[19] M.P. Immonen, M. Karppinen, O.K. Kivilahti, Investigation of environmental reliability of optical polymer waveguides embedded on printed circuit boards, Circ. World 33 (2007) 9–19.

[20] S. Kopetz, D. Cai, E. Rabe, A. Neyer, PDMS-based optical waveguide layer for integration in electrical-optical circuit boards, AEU—Int. J. Electron. Commun. 61 (2007) 163–167.

[21] M.R.T. Tan et al., A high-speed optical multidrop bus for computer interconnections, IEEE Micro 29 (2009) 62–73.

[22] M. Tan et al., A high-speed optical multi-drop bus for computer interconnections, Appl. Phys. A: Mater. Sci. Process. 95 (2009) 945–953.

[23] M.W. Haney, M.P. Christensen, Performance scaling comparison for free-space optical and electrical interconnection approaches, Appl. Opt. 37 (1998) 2886–2894.

[24] A.C. Walker et al., Operation of an optoelectronic crossbar switch containing a terabit-per-second free-space optical interconnect, IEEE J. Quant. Electron. 41 (2005) 1024–1036.

[25] H. Kuo et al., Free-space optical links for board-to-board interconnects, Appl. Phys. A: Mater. Sci. Process. 95 (2009) 955–965.

[26] B.G. Lee et al., Ultrahigh-bandwidth silicon photonic nanowire waveguides for on-chip networks, IEEE Photon. Technol. Lett. 20 (2008) 398–400.

[27] M. Haurylau et al., On-chip optical interconnect roadmap: challenges and critical directions, IEEE J. Sel. Top. Quant. Electron. 12 (2006) 1699–1705.

[28] A. Barkai et al., Integrated silicon photonics for optical networks [Invited], J. Opt. Netw. 6 (2007) 25–47.

[29] W.M. Green, M.J. Rooks, L. Sekaric, Y.A. Vlasov, Ultra-compact, low RF power, 10 Gb/s silicon Mach-Zehnder modulator, Opt. Express 15 (2007) 17106–17113.

[30] Y. Vlasov, W.M.J. Green, F. Xia, High-throughput silicon nanophotonic wavelength-insensitive switch for on-chip optical networks, Nat. Photon. 2 (2008) 242–246.

[31] M.A. Popovic et al., Hitless-reconfigurable and bandwidth-scalable silicon photonic circuits for telecom and interconnect applications, in: Conference on Optical Fiber Communication and National Fiber Optic Engineers Conference (OFC/NFOEC), 2008, pp. 1–3.

[32] A. Shacham, K. Bergman, L.P. Carloni, Photonic networks-on-chip for future generations of chip multi-processors, IEEE Trans. Comput. 57 (2008) 1246–1260.

[33] R.T. Chen, Hybrid and monolithic integration of planar lightwave circuits (PLCs), Integrated Optics: Devices, Materials, and Technologies, vol. XII, SPIE, San Jose, CA, USA, 2008., pp. 689611–689614.

[34] T. Baehr-Jones et al., Myths and rumours of silicon photonics, Nat. Photon. 6 (2012) 206–208.

[35] J. van der Tol et al., Photonic integration in indium-phosphide membranes on silicon, IET Optoelectron. 5 (2011) 218–225.

[36] P. Kogge, The tops in flops, IEEE Spectrum 48 (2011) 48–54.

[37] J.R. Bautista, Interconnect challenges in a many core compute environment, in: 17th IEEE Symposium on High Performance Interconnects (HOTI), 2009, pp. 148–148.

[38] S. Borkar, Thousand core chips: a technology perspective, Proceedings of the 44th Annual Design Automation Conference, ACM, San Diego, California, 2007, pp. 746–749.

[39] S. Vangal et al., An 80-Tile 1.28TFLOPS network-on-chip in 65 nm CMOS, in: IEEE International Solid-State Circuits Conference (ISSCC), 2007, pp. 98–589.

[40] J.X. Cai et al., 112 × 112 Gb/s transmission over 9,360 km with channel spacing set to the baud rate (360% spectral efficiency), in: 36th European Conference and Exhibition on Optical Communication (ECOC), 2010, pp. 1–3.

[41] B.G. Lee et al., 20-µm-pitch eight-channel monolithic fiber array coupling 160 Gb/s/channel to silicon nanophotonic chip, Conference on Optical Fibre Communication and National Fiber Optic Engineers Conference (OFC/NFOEC), Optical Society of America, 2010, pp. PDPA4.

[42] L. Chao-Kun, A. Tandon, K. Djordjev, S.W. Corzine, M.R.T. Tan, High-speed 985 nm bottom-emitting VCSEL arrays for chip-to-chip parallel optical interconnects, IEEE J. Sel. Top. Quant. Electron. 13 (2007) 1332–1339.

[43] B.S. Landman, R.L. Russo, On a pin versus block relationship for partitions of logic graphs, IEEE Trans. Comput. C-20 (1971) 1469–1479.

[44] M.R. Feldman, S.C. Esener, C.C. Guest, S.H. Lee, Comparison between optical and electrical interconnects based on power and speed considerations, Appl. Opt. 27 (1988) 1742–1751.

[45] L. Schares et al., Terabus: terabit/second-class card-level optical interconnect technologies, IEEE J. Sel. Top. Quant. Electron. 12 (2006) 1032–1044.

[46] J. Proesel, C. Schow, A. Rylyakov, 25 Gb/s 3.6 pJ/b and 15 Gb/s 1.37 pJ/b VCSEL-based optical links in 90 nm CMOS, in: IEEE International Solid-State Circuits Conference (ISSCC), 2012, pp. 418–420.

[47] G. Balamurugan et al., A scalable 5–15 Gbps, 14–75 mW low-power I/O transceiver in 65 nm CMOS, IEEE J. Solid-State Circ. 43 (2008) 1010–1019.

[48] J. Poulton et al., A 14-mW 6.25-Gb/s transceiver in 90-nm CMOS, IEEE J. Solid-State Circ. 42 (2007) 2745–2757.

[49] M.H. Nazari, A. Emami-Neyestanak, An 18.6 Gb/s double-sampling receiver in 65 nm CMOS for ultra-low-power optical communication, in: IEEE International Solid-State Circuits Conference (ISSCC), 2012, pp. 130–131.

[50] K. Byungsub, V. Stojanovic, An energy-efficient equalized transceiver for rc-dominant channels, IEEE J. Solid-State Circ. 45 (2010) 1186–1197.

[51] M. Georgas, J. Orcutt, R. J. Ram, V. Stojanovic, A monolithically-integrated optical receiver in standard 45-nm SOI, in: Proceedings of the 37th European Solid-State Circuits Conference (ESSCIRC), 2011, pp. 407–410.

[52] D. Kucharski et al., A 20 Gb/s VCSEL driver with pre-emphasis and regulated output impedance in 0.13 μm CMOS, in: IEEE International Solid-State Circuits Conference (ISSCC), vol. 221, 2005, pp. 222–594.

[53] D. Guckenberger, J.D. Schaub, D. Kucharski, K.T. Kornegay, 1 V, 10 mW, 10 Gb/s CMOS optical receiver front-end, in: IEEE Radio Frequency Integrated Circuits (RFIC), Symposium, 2005, pp. 309–312.

[54] R.S. Tucker, Scalability and energy consumption of optical and electronic packet switching, J. Lightwave Technol. 29 (2011) 2410–2421.

[55] <http://www.gennum.com>.

[56] J. Bautista, The potential benefits of photonics in the computing platform, Optoelectronic Integrated Circuits, vol. VII, SPIE, San Jose, CA, USA, 2005., pp. 1–8.

[57] C. Doerr, Planar lightwave devices for WDM, in: I.P. Kaminow, T. Li (Eds.) Optical Fiber Telecommunications IV, A Components, Academic Press, 2002, pp. 402–477.

[58] R. Nagarajan et al., InP photonic integrated circuits, IEEE J. Sel. Top. Quant. Electron. 16 (2010) 1113–1125.

[59] M.K. Smit, J. van der Tol, M. Hill, Moore's law in photonics, Laser Photon. Rev. 6 (2012) 1–13.

[60] M.K. Smit et al., Generic foundry model for InP-based photonics, IET Optoelectron. 5 (2011) 187–194.

[61] P. Dumon, W. Bogaerts, R. Baets, J.M. Fedeli, L. Fulbert, Towards foundry approach for silicon photonics: silicon photonics platform ePIXfab, Electron. Lett. 45 (2009) 581–582.

[62] Finetech, FINEPLACER® micro hvr. <http://eu.finetech.de/>.

[63] V.J. Sorger et al., Experimental demonstration of low-loss optical waveguiding at deep sub-wavelength scales, Nat. Commun. 2 (2011) 1–5.

[64] J.F. Bauters et al., Ultra-low-loss high-aspect-ratio Si3N4 waveguides, Opt. Express 19 (2011) 3163–3174.

[65] J.F. Bauters et al., Planar waveguides with less than 0.1 dB/m propagation loss fabricated with wafer bonding, Opt. Express 19 (2011) 24090–24101.

[66] T. Larsen, A. Bjarklev, D. Hermann, J. Broeng, Optical devices based on liquid crystal photonic bandgap fibres, Opt. Express 11 (2003) 2589–2596.

[67] R. Adar, M.R. Serbin, V. Mizrahi, Less than 1 dB per meter propagation loss of silica waveguides measured using a ring resonator, J. Lightwave Technol. 12 (1994) 1369–1372.

[68] H. Ma, A.K.Y. Jen, L.R. Dalton, Polymer-based optical waveguides: materials, processing, and devices, Adv. Mater. 14 (2002) 1339–1365.

[69] M. Zhou, Low-loss polymeric materials for passive waveguide components in fiber optical telecommunication, Opt. Eng. 41 (2002) 1631–1643.

[70] L. Eldada, L.W. Shacklette, Advances in polymer integrated optics, IEEE J. Sel. Top. Quant. Electron. 6 (2000) 54–68.

[71] J.V. DeGroot, Jr A. Norris, S.O. Glover, T.V. Clapp, Highly transparent silicone materials, Linear and Nonlinear Optics of Organic Materials, vol. IV, SPIE, Denver, CO, USA, 2004., pp. 116–123.

[72] D.K. Cai, A. Neyer, R. Kuckuk, H.M. Heise, Optical absorption in transparent PDMS materials applied for multimode waveguides fabrication, Opt. Mater. 30 (2008) 1157–1161.

[73] L.W. Shacklette et al., Ultra-low-loss acrylate polymers for planar light circuits, Adv. Funct. Mater. 13 (2003) 453–462.

[74] Y. Kuwana, S. Takenobu, K. Takayama, S. Yokotsuka, S. Kodama, Low loss and highly reliable polymer optical waveguides with perfluorinated dopant-free core, in: Optical Fiber Communication Conference and National Fiber Optic Engineers Conference (OFC/NFOEC), 2006, p. 1–3.

[75] S. Takenobu, Y. Kuwana, K. Takayama, and Y. Morizawa, Ultra-wide-band low loss and PDL 1 × 32 splitter polymer optical waveguide chip and module, in: Conference on Optical Fiber Communication and the National Fiber Optic Engineers Conference (OFC/NFOEC), 2007, pp. 1–3.

[76] M. Takahashi et al., Investigation of a downsized silica highly nonlinear fiber, J. Lightwave Technol. 25 (2007) 2103–2107.

[77] W. Xiaolong, W. Li, J. Wei, R.T. Chen, Hard-molded 51 cm long waveguide array with a 150 GHz bandwidth for board-level optical interconnects, Opt. Lett. 32 (2007) 677–679.

[78] F.E. Doany, P.K. Pepeljugoski, A.C. Lehman, J.A. Kash, R. Dangel, Measurement of optical dispersion in multimode polymer waveguides, in: Digest of the LEOS Summer Topical Meetings: Biophotonics/Optical Interconnects and VLSI Photonics/WGM Microcavities, MB4.4, 2004, pp. 31–32.

[79] G.L. Bona et al., Characterization of Parallel Optical-Interconnect Waveguides Integrated on a Printed Circuit Board, Micro-Optics, VCSELs, and Photonic Interconnects, SPIE, vol. 5453, 2004, pp. 134–141.

[80] N. Bamiedakis et al., Cost-effective multimode polymer waveguides for high-speed on-board optical interconnects, IEEE J. Quant. Electron. 45 (2009) 415–424.

[81] A. Hashim, N. Bamiedakis, R.V. Penty, I.H. White, Multimode 90°-crossings, combiners and splitters for a polymer-based on-board optical bus, in: Conference on Lasers and Electro-Optics (CLEO), 2012, pp. 1–2.

[82] I. Papakonstantinou, K. Wang, D.R. Selviah, F.A. Ferandez, Transition, radiation and propagation loss in polymer multimode waveguide bends, Opt. Express 15 (2007) 669–679.

[83] D.R. Selviah et al., Integrated optical and electronic interconnect printed circuit board manufacturing, Circ. World 34 (2008) 21–26.

[84] H. Sung Hwan et al., Two-dimensional optical interconnection based on two-layered optical printed circuit board, IEEE Photon. Technol. Lett. 19 (2007) 411–413.

[85] D.A. Zauner, A.M. Jorgensen, T.A. Anhoj, J. Hubner, High-density multimode integrated polymer optics, J. Opt. A—Pure Appl. Opt. 7 (2005) 445–450.

[86] T. Sakamoto et al., Optical interconnection using VCSELs and polymeric waveguide circuits, J. Lightwave Technol. 18 (2000) 1487–1492.

[87] N. Bamiedakis et al., Low loss and low crosstalk multimode polymer waveguide crossings for high-speed optical interconnects, in: Conference on Lasers and Electro-Optics (CLEO), 2007, pp. 1–2.

[88] N. Bamiedakis, R.V. Penty, I.H. White, J.V. DeGroot, T.V. Clapp, Cost-effective polymer multimode directional couplers for high-speed on-board optical interconnects, in: 14th European Conference on Integrated Optics (ECIO), Eindhoven, The Netherlands, WeB.5, 2008, pp. 45-48.

[89] X. Dou, A.X. Wang, X. Lin, H. Haiyu, R.T. Chen, Optical bus waveguide metallic hard mold fabrication with opposite 45° micro-mirrors, Proc. SPIE 7607 (2010) 76070P.

[90] H. Sung Hwan et al., Bendable and splitter-integrated optical subassembly based on a flexible optical board, IEEE Photon. Technol. Lett. 22 (2010) 167–169.

[91] C. Choon-Gi, H. Sang-Pil, K. Byeong Cheol, A. Seung-Ho, J. Myung-Yung, Fabrication of large-core 1×16 optical power splitters in polymers using hot-embossing process, IEEE Photon. Technol. Lett. 15 (2003) 825–827.

[92] N. Bamiedakis et al., Low-loss, high-uniformity 1×2, 1×4 and 1×8 polymer multimode Y-splitters enabling radio-over-fibre multicasting applications, in: 13th European Conference on Integrated Optics (ECIO), Conpehagen, Denmark, 2007.

[93] A. Chutinan, S. Noda, Waveguides and waveguide bends in two-dimensional photonic crystal slabs, Phys. Rev. B 62 (2000) 4488–4492.

[94] T. Baba, N. Fukaya, J. Yonekura, Observation of light propagation in photonic crystal optical waveguides with bends, Electron. Lett. 35 (1999) 654–655.

[95] C. Manolatou et al., High-density integrated optics, J. Lightwave Technol. 17 (1999) 1682–1692.

[96] P. Buchmann, H. Kaufmann, GaAs single-mode rib waveguides with reactive ion-etched totally reflecting corner mirrors, J. Lightwave Technol. 3 (1985) 785–788.

[97] S. De-Gui et al., Modeling and numerical analysis for silicon-on-insulator rib waveguide corners, J. Lightwave Technol. 27 (2009) 4610–4618.

[98] H. Wang, A. Wonfor, K.A. Williams, R.V. Penty, I.H. White, Demonstration of a lossless monolithic 16×16 QW SOA switch, in: 35th European Conference on Optical Communication (ECOC), 2009, pp. 1–2.

[99] W. Zhuoran et al., Integrated small-sized semiconductor ring laser with novel retro-reflector cavity, IEEE Photon. Technol. Lett. 20 (2008) 99–101.

[100] L. Xu et al., MMI-reflector: a novel on-chip reflector for photonic integrated circuits, in: 35th European Conference on Optical Communication (ECOC), 2009, pp. 1–2.

[101] M.K. Smit, E.C.M. Pennings, H. Blok, A normalized approach to the design of low-loss optical waveguide bends, J. Lightwave Technol. 11 (1993) 1737–1742.

[102] B.M.A. Rahman, D.M.H. Leung, S.S.A. Obayya, K.T.V. Grattan, Numerical analysis of bent waveguides: bending loss, transmission loss, mode coupling, and polarization coupling, Appl. Opt. 47 (2008) 2961–2970.

[103] L.H. Spiekman et al., Ultrasmall waveguide bends: the corner mirrors of the future? IEEE Proc.—Optoelectron. 142 (1995) 61–65.

[104] M. Popovic et al., Air trenches for sharp silica waveguide bends, J. Lightwave Technol. 20 (2002) 1762–1772.

[105] A. Sakai, G. Hara, T. Baba, Propagation characteristics of ultrahigh-Δ optical waveguide on silicon-on-insulator substrate: optics and quantum electronics, Jpn. J. Appl. Phys. Lett. 40 (2001) 383–385.

[106] C. van Dam et al., Novel compact polarization converters based on ultra short bends, IEEE Photon. Technol. Lett. 8 (1996) 1346–1348.

[107] S.S.A. Obayya, B.M.A. Rahman, K.T.V. Grattan, H.A. El-Mikati, Improved design of a polarization converter based on semiconductor optical waveguide bends, Appl. Opt. 40 (2001) 5395–5401.

[108] R. Stabile, K.A. Williams, Relaxed dimensional tolerance whispering gallery microbends, J. Lightwave Technol. 29 (2011) 1892–1898.

[109] A. Himeno, H. Terui, M. Kobayashi, Loss measurement and analysis of high-silica reflection bending optical waveguides, J. Lightwave Technol. 6 (1988) 41–46.

[110] A. Sakai, T. Fukazawa, T. Baba, Estimation of polarization crosstalk at a micro-bend in Si-photonic wire waveguide, J. Lightwave Technol. 22 (2004) 520–525.

[111] R.U. Ahmad et al., Ultracompact corner-mirrors and T-branches in silicon-on-insulator, IEEE Photon. Technol. Lett. 14 (2002) 65–67.

[112] G. Sherlock et al., Integrated 2 × 2 optical switch with gain, Electron. Lett. 30 (1994) 137–138.

[113] L.B. Soldano, E.C.M. Pennings, Optical multimode interference devices based on self-imaging: principles and applications, J. Lightwave Technol. 13 (1995) 615–627.

[114] M.T. Hill, X.J.M. Leijtens, G.D. Khoe, M.K. Smit, Optimizing imbalance and loss in 2 × 2 3-dB multimode interference couplers via access waveguide width, J. Lightwave Technol. 21 (2003) 2305–2313.

[115] Y. Kai et al., A compact and lossless 8 × 8 SOA gate switch subsystem for WDM optical packet interconnections, in: 34th European Conference on Optical Communication (ECOC) 2008, 2008, pp. 1–2.

[116] M.K. Smit, C. Van Dam, PHASAR-based WDM-devices: principles, design and applications, IEEE J. Sel. Top. Quant. Electron. 2 (1996) 236–250.

[117] H.G. Bukkems, C.G.P. Herben, M.K. Smit, F.H. Groen, I. Moerman, Minimization of the loss of intersecting waveguides in InP-based photonic integrated circuits, IEEE Photon. Technol. Lett. 11 (1999) 1420–1422.

[118] W. Bogaerts, P. Dumon, D.V. Thourhout, R. Baets, Low-loss, low-cross-talk crossings for silicon-on-insulator nanophotonic waveguides, Opt. Lett. 32 (2007) 2801–2803.

[119] C. Hui, A.W. Poon, Low-loss multimode-interference-based crossings for silicon wire waveguides, IEEE Photon. Technol. Lett. 18 (2006) 2260–2262.

[120] A. Albores-Mejia et al., Low-penalty monolithically-cascaded 1550 nm-wavelength quantum-dot crossbar switches, Conference on optical fiber communication and national fiber optic engineers conference (OFC/NFOEC), Optical Society of America, 2009., pp. JWA29.

[121] R. Stabile, K.A. Williams, Photonic integrated semiconductor optical amplifier switch circuits, in: P. Urquhart (Ed.), Advances in Optical Amplifiers, Intech Open, 2011.

[122] A. Rohit, A. Albores-Mejia, J. Bolk, X. Leijtens, K.A. Williams, Multi-hop dynamic routing in an integrated 4 × 4 space and wavelength select cross-connect, Conference on Optical Fiber Communication and National Fiber Engineers Conference (OFC/NFOEC), Optical Society of America, 2012, pp. OTh3D.5.

[123] K.A. Williams, Integrated semiconductor-optical-amplifier-based switch fabrics for high-capacity interconnects [Invited], J. Opt. Netw. 6 (2007) 189–199.

[124] R.C.A. Pitwon et al., Design and implementation of an electro-optical backplane with pluggable in-plane connectors, Optoelectronic Interconnects and Component Integration, vol. IX, SPIE, San Francisco, California, USA, 2010, pp. 76070J.

[125] F.E. Doany et al., Multichannel high-bandwidth coupling of ultradense silicon photonic waveguide array to standard-pitch fiber array, J. Lightwave Technol. 29 (2011) 475–482.

[126] C.S. Patel et al., Silicon carrier with deep through vias, fine pitch wiring and through cavity for parallel optical transceiver, in: Proceedings of 55th Electronic Components and Technology Conference, 2005, pp. 1318–1324.

[127] A.L. Glebov, M.G. Lee, K. Yokouchi, Backplane photonic interconnect modules with optical jumpers, Photon. Packag. Integrat. 5731 (2005) 63–71.

[128] H. Schroder et al., Waveguide and packaging technology for optical backplanes and hybrid electrical-optical circuit boards. in: Proceedings of the SPIE 6124 Optoelectronic Integrated Circuits VIII, 2006, pp. 612407.

[129] J. Van Erps, N. Hendrickx, C. Debaes, P. Van Daele, H. Thienpont, Discrete out-of-plane coupling components for printed circuit board-level optical interconnections, IEEE Photon. Technol. Lett. 19 (2007) 1753–1755.

[130] S. Hiramatsu, K. Miura, K. Hirao, Optical backplane connectors using three-dimensional waveguide arrays, J. Lightwave Technol. 25 (2007) 2776–2782.

[131] G. Van Steenberge et al., Laser ablation of parallel optical interconnect waveguides, IEEE Photon. Technol. Lett. 18 (2006) 1106–1108.

[132] N. Hendrickx, J. Van Erps, G. Van Steenberge, H. Thienpont, P. Van Daele, Laser ablated micromirrors for printed circuit board integrated optical interconnections, IEEE Photon. Technol. Lett. 19 (2007) 822–824.

[133] I.-K. Cho, W.-J. Lee, M.-Y. Jeong, H.-H. Park, Optical module using polymer waveguide with integrated reflector mirrors, IEEE Photon. Technol. Lett. 20 (2008) 410–412.

[134] A. McCarthy, H. Suyal, A.C. Walker, Fabrication and characterisation of direct laser-written multimode polymer waveguides with out-of-plane turning mirrors, Conference on Lasers and Electro-Optics Europe (CLEO Europe), 2005, pp. 477.

[135] Y. Ishii, S. Koike, Y. Arai, Y. Ando, SMT-compatible large-tolerance "OptoBump" interface for interchip optical interconnections, IEEE Trans. Adv. Packag. 26 (2003) 122–127.

[136] S. Nakagawa et al., High-density optical interconnect exploiting build-up waveguide-on-SLC board, in: Proceedings of the 58th Electronic Components and Technology Conference (ECTC), 2008, pp. 256–260.

[137] A.L. Glebov, J. Roman, M.G. Lee, K. Yokouchi, Optical interconnect modules with fully integrated reflector mirrors, IEEE Photon. Technol. Lett. 17 (2005) 1540–1542.

[138] M. Immonen, M. Karppinen, J.K. Kivilahti, Fabrication and characterization of polymer optical waveguides with integrated micromirrors for three-dimensional board-level optical interconnects, IEEE Trans. Electron. Packag. Manuf. 28 (2005) 304–311.

[139] S. Hiramatsu, T. Mikawa, Optical design of active interposer for high-speed chip level optical interconnects, J. Lightwave Technol. 24 (2006) 927–934.

[140] Y. Takagi et al., Low-loss chip-to-chip optical interconnection using multichip optoelectronic package with 40-Gb/s optical I/O for computer applications, J. Lightwave Technol. 28 (2010) 2956–2963.

[141] R. Byung Sup et al., PCB-compatible optical interconnection using 45°-ended connection rods and via-holed waveguides, J. Lightwave Technol. 22 (2004) 2128–2134.

[142] M.S. Bakir, J.D. Meindl, Sea of polymer pillars electrical and optical chip I/O interconnections for gigascale integration, IEEE Trans. Electron Dev. 51 (2004) 1069–1077.

[143] M. Shishikura, Y. Matsuoka, T. Ban, T. Shibata, A. Takahashi, A high-coupling-efficiency multilayer optical printed wiring board with a cube-core structure for high-density optical interconnections, in: Proceedings of the 57th Electronic Components and Technology Conference (ECTC), 2007, pp. 1275–1280.

[144] N. Bamiedakis, J. Beals, IV A.H. Hashim, R.V. Penty, I.H. White, Optical transceiver integrated on PCB using electro-optic connectors compatible with pick-and-place assembly technology, Optoelectronic Interconnects and Component Integration IX, SPIE, San Francisco, California, USA, 2010 76070O.

[145] L. Dellmann et al., 120 Gb/s optical card-to-card interconnect link demonstrator with embedded waveguides, in: Proceedings of the 57th Electronic Components & Technology Conference, 2007, pp. 1288–1293.

[146] I. Papakonstantinou, D.R. Selviah, R. Pitwon, D. Milward, Low-cost, precision, self-alignment technique for coupling laser and photodiode arrays to polymer waveguide arrays on multilayer PCBs, IEEE Trans. Adv. Packag. 31 (2008) 502–511.

[147] H. Sung Hwan et al., Passively assembled optical interconnection system based on an optical printed-circuit board, IEEE Photon. Technol. Lett. 18 (2006) 652–654.

[148] J.J. Yang, A.S. Flores, M.R. Wang, Array waveguide evanescent ribbon coupler for card-to-backplane optical interconnects, Opt. Lett. 32 (2007) 14–16.

[149] R.T. Chen et al., Fully embedded board-level guided-wave optoelectronic interconnects, Proc. IEEE 88 (2000) 780–793.

[150] C. Gee-Kung et al., Chip-to-chip optoelectronics SOP on organic boards or packages, IEEE Trans. Advanced Packaging 27 (2004) 386–397.

[151] G. Van Steenberge et al., MT-compatible laser-ablated interconnections for optical printed circuit boards, J. Lightwave Technol. 22 (2004) 2083–2090.

[152] I. Papakonstantinou et al., Optical 8-channel, 10 Gb/s MT pluggable connector alignment technology for precision coupling of laser and photodiode arrays to polymer waveguide arrays for optical board-to-board interconnects, in: Proceedings of the 58th Electronic Components and Technology Conference (ECTC), 2008, pp. 1769–1775.

[153] H. Uemura et al., Extremely-compact and high-performance (160 Gbps = 20 GB/s) optical semiconductor module using lead frame embedded optoelectronic ferrule, in: 58th Electronic Components and Technology Conference (ECTC), 2008, pp. 1936–1940.

[154] J. Van Erps et al., Low-cost micro-optics for PCB-level photonic interconnects—art. no. 64760L, in: Photonics Packaging, Integration, and Interconnects VII, vol. 6478, 2007, pp. XI–XXIV.

[155] A.L. Glebov, M.G. Lee, K. Yokouchi, Integration technologies for pluggable backplane optical interconnect systems, Opt. Eng. 46 (2007) 015403–015410.

[156] A. Neyer, S. Kopetz, E. Rabe, W.J. Kang, S. Tombrink, Electrical optical circuit board using polysiloxane optical waveguide layer, in: Proceedings of the 55th Electronic Components and Technology Conference (ECTC), 2005, pp. 246–250.

[157] X. Wang, J. Wei, W. Li, B. Hai, R.T. Chen, Fully embedded board-level optical interconnects from waveguide fabrication to device integration, J. Lightwave Technol. 26 (2008) 243–250.

[158] H.-H. Hsu, Y. Hirobe, T. Ishigure, Fabrication and inter-channel crosstalk analysis of polymer optical waveguides with W-shaped index profile for high-density optical interconnections, Opt. Express 19 (2011) 14018–14030.

[159] C.L. Schow et al., 225 Gb/s bi-directional integrated optical PCB link, in: Optical Fiber Communication Conference and Exposition and the National Fiber Optic Engineers Conference (OFC/NFOEC), 2011, pp. 1–3.

[160] C. Schow, Power-efficient transceivers for high-bandwidth, short-reach interconnects, Conference on Optical Fiber Communication and National Optical Engineers Conference (OFC/NFOEC), Optical Society of America, 2012., pp. OTh1E.4.

[161] K. Tanaka et al., Cost-effective transceiver technologies for high-bandwidth optical interconnection in high-end server systems, Conference on Optical Fiber Communication and National Optical Engineers Conference (OFC/NFOEC), Optical Society of America, 2012, pp. OW4I.2.

[162] J. Beals et al., A terabit capacity passive polymer optical backplane based on a novel meshed waveguide architecture, Appl. Phys. A: Mater. Sci. Process. 95 (2009) 983–988.

[163] H. Xuliang, G. Kim, G.J. Lipovski, R.T. Chen, An optical centralized shared-bus architecture demonstrator for microprocessor-to-memory interconnects, IEEE J. Sel. Top. Quant. Electron. 9 (2003) 512–517.

[164] M. Jarczynski, T. Seiler, J. Jahns, Integrated three-dimensional optical multilayer using free-space optics, Appl. Opt. 45 (2006) 6335–6341.

[165] P. Lukowicz et al., Optoelectronic interconnection technology in the HOLMS system, IEEE J. Sel. Top. Quant. Electron. 9 (2003) 624–635.

[166] N. Bamiedakis, A. Hashim, R.V. Penty, I.H. White, Regenerative polymeric bus architecture for board-level optical interconnects, Opt. Express 20 (2012) 11625–11636.

[167] N. Bamiedakis, A. Hashim, R. Penty, I. White, 4-Channel polymeric optical bus module for board-level optical interconnections, Conference on Lasers and Electro-Optics (CLEO), Optical Society of America, 2012., pp. 6.

[168] K.A. Williams, A. Rohit, M. Glick, Resilience in optical ring-resonant switches, Opt. Express 19 (2011) 17232–17243.

[169] S.K. Selvaraja, W. Bogaerts, P. Dumon, D. Van Thourhout, R. Baets, Subnanometer linewidth uniformity in silicon nanophotonic waveguide devices using CMOS fabrication technology, IEEE J. Sel. Top. Quant. Electron. 16 (2010) 316–324.

[170] A. Rohit, R. Stabile, K.A. Williams, Dynamic routing in a fifth order resonant switch array, in: European Conference on Optical Communications (ECOC), 2012, p. Tu.1.E.1.

[171] I.M. Soganci et al., Monolithically integrated InP 1 × 16 optical switch with wavelength-insensitive operation, IEEE Photon. Technol. Lett. 22 (2010) 143–145.

[172] I.M. Soganci et al., Monolithic InP 100-port photonic switch, in: 36th European Conference and Exhibition on Optical Communication (ECOC), 2010, pp. 1–3.

[173] R. Varrazza, I.B. Djordjevic, S. Yu, Active vertical-coupler-based optical crosspoint switch matrix for optical packet-switching applications, J. Lightwave Technol. 22 (2004) 2034–2042.

[174] G.A. Fish, B. Mason, L.A. Coldren, S.P. DenBaars, Compact, 4×4 InGaAsP-InP optical crossconnect with a scaleable architecture, IEEE Photon. Technol. Lett. 10 (1998) 1256–1258.

[175] R. Stabile et al., Multipath routing in a fully scheduled integrated optical switch fabric, in: 36th European Conference and Exhibition on Optical Communication (ECOC), 2010, pp. 1–3.

[176] R. Stabile, M. Zal, and K.A. Williams, Integrated optical switch circuit operating under FPGA control, in: Proceedings of the 16th Annual Symposium of the IEEE Photonics Benelux Chapter, Ghent, Belgium, 2011, pp. 85–88.

[177] W.J. Dally, B. Towles, Principles and practices of interconnection networks, Elsevier Science, 2004.

[178] I. White et al., Scalable optical switches for computing applications [Invited], J. Opt. Netw. 8 (2009) 215–224.

[179] A. Bianco et al., Optical interconnection networks based on microring resonators, in: IEEE International Conference on Communications (ICC), 2010, pp. 1–5.

[180] A. Albores-Mejia et al., 320 Gb/s data routing in a monolithic multistage semiconductor optical amplifier switching circuit, in: 36th European Conference and Exhibition on Optical Communication (ECOC), 2010, pp. 1–4.

[181] A. Wonfor, H. Wang, R.V. Penty, I.H. White, Large port count high-speed optical switch fabric for use within datacenters [Invited], J. Opt. Commun. Netw. 3 (2011) A32–A39.

[182] C.G.P. Herben et al., Polarization independent dilated WDM cross-connect on InP, IEEE Photon. Technol. Lett. 11 (1999) 1599–1601.

[183] K.T. Shiu, S.S. Agashe, S.R. Forrest, An InP-based monolithically integrated reconfigurable optical add–drop multiplexer, IEEE Photon. Technol. Lett. 19 (2007) 1445–1447.

[184] C.G.M. Vreeburg et al., First InP-based reconfigurable integrated add-drop multiplexer, IEEE Photon. Technol. Lett. 9 (1997) 188–190.

[185] A. Rohit, A. Albores-Mejia, J. Bolk, X.J.M. Leijtens, K.A. Williams, Multi-path routing at 40 Gb/s in an integrated space and wavelength selective switch, in: IEEE Photonics Conference (PHO), 2011, pp. 342–343.

[186] A. Rohit, A. Albores-Mejia, J. Bolk, X. Leijtens, K. Williams, Multi-path routing in an monolithically integrated 4×4 broadcast and select WDM cross-connect, 37th European Conference and Exposition on Optical Communications (ECOC), Optical Society of America, 2011, pp. 1–3

[187] D.G. Cunningham, I.H. White, Does multimode fibre have a future in data-communications? Electron. Lett. 43 (2007) 63–65.

CMOS Photonics for High Performance Interconnects

12

Jason Orcutt, Rajeev Ram, and Vladimir Stojanović

Massachusetts Institute of Technology.

12.1 ON-CHIP INTERCONNECTS AND POWER—A SYSTEM ARCHITECT'S VIEW

Over the last few years, there has been a dramatic shift in microprocessor architecture. Since 2002, uniprocessor performance has increased far more slowly than over the preceding 20 years, due to increasing power consumption, wire delay, and memory latency, as well as diminishing returns from further exploitation of instruction-level parallelism (ILP). Already, in 2006, uniprocessor performance was three times slower than historic trends would have predicted in 2002.

Manufacturers have turned away from ever larger and ever less efficient superscalar processors and instead are placing multiple cores on a die. Figure 12.1 illustrates this trend as well as the approximate bandwidth requirements of near-future manycore systems.

All multicore designs place a small amount of memory close to each core, as either a scratchpad RAM or an L1 cache. Communication between cores either occurs via messages sent over an on-chip network or through a shared banked L2 cache. The ratio of on-chip versus off-chip communication bandwidth varies depending on the application. Communication between cores can exploit locality to reduce latency, and on-chip wiring provides high-bandwidth connectivity.

Regardless of application domain, however, large multicore chips will be primarily constrained by *the latency, bandwidth, and capacity of the external memory system.* Therefore, the most significant near-term challenge is how to effectively connect on-chip processors to off-chip DRAM.

Our vision for such architecture is one that puts a large chunk of today's super-computer on a single printed circuit board, with a few manycore sockets and hundreds of DRAM chips providing thousands of memory banks on the same board, as illustrated in Figure 12.2.

12.1.1 The evolution of microprocessor architecture

From a high level, the general purpose, multicore microprocessor has several subsystems of interest for the current discussion: computation, clock distribution, core-to-core communication on die, socket-to-socket communication, and

Optical Fiber Telecommunications VIA. http://dx.doi.org/10.1016/B978-0-12-396958-3.00012-3

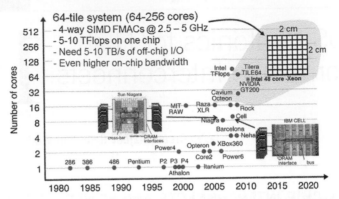

FIGURE 12.1 Processor core-count scaling trends and projected system bandwidth requirements.

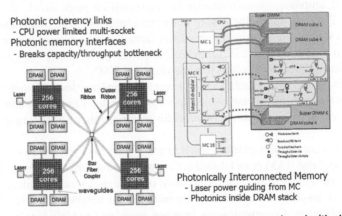

FIGURE 12.2 Envisioned four-socket 10 TFLOPs/s super-computer on-board with photonic networks for coherency and for DRAM access [1,2].

core-to-memory communication. These subsystems must share on-chip power, silicon area, backend metallization wiring, and off-chip I/O pins. The chief constraint that has traditionally driven trade-offs between each of these subsystems is how much power may be allocated for a given function. Standard total power dissipation for thermally limited microprocessors has stabilized to be approximately 100 W [3,4]. Joint optimization of various microprocessor subsystems has driven a coarse breakdown of power as follows [5,6–10]: 60% is allocated to computation internal to the various processing cores; 20% is allocated for all on-chip communication including clock distribution and core-to-core links; and 20% is allocated for off-chip communication including socket-to-socket and core-to-memory links.

Simultaneously, the performances of the various subsystems are linked together to enable optimal system function. Fundamentally, the communication bandwidth should support sufficient data exchange on core-to-core and core-to-memory links such that the operands are always available at each processing core to enable high utilization of the computing hardware. This means that bandwidth provisioning should not be scaled arbitrarily just to meet the total power dissipation specification. Specifically in the context of the core-to-memory bandwidth of a system, the relationship between floating point operations (FLOPs) in the computing cores and the data fetched from main memory is typically desired to be 1 byte of data per FLOP [11–16].

Although this high-level breakdown is not fundamental for all application types, deviation from this linked scaling forced by technical obstacles can be seen to result in sub-optimal systems. The byte per FLOP desired bandwidth target has not been met in general purpose CPUs and instead is typically well below 0.5 B/FLOP in Intel's Xeon line of processors [17–20]. The primary challenges are the number of I/O pins and total power dissipation that may be devoted to the memory interface [18]. The performance impact of this change has been reduced for many applications by the introduction of a large on-chip memory cache [21]. Since the memory bandwidth available to fetch operands from main memory can no longer be sufficiently provisioned, the size of on-chip caches has greatly increased to hide the bottleneck from impacting several applications. The cache sizes have greatly increased as a fraction of die area. If the memory size of data that is frequently used for the current application is sufficiently small to fit into the on-chip cache, the application execution avoids main-memory access bottlenecks and proceeds at a computationally limited pace. This circumstance is unfortunately not the case for most high-performance computing applications and many consumer-level tasks. Additionally, the die area used for the caches may not be used for additional processing cores. In the current memory-limited environment, this limit does not present a significant trade-off since the increased computation power on chip would not have sufficient memory bandwidth to run at peak speed for most applications. The elimination of the memory interconnect bottleneck therefore offers a coupled scaling of processor performance; increased bandwidth increases the efficiency of processing cores while simultaneously increasing the number of cores that may fit on a single chip.

For end-system comparisons, it is helpful to start with some assumptions on the computational performance of the logic cores to set the interconnect need. As the saturation of logic clock rates has existed for much of the last decade with no signs of change, it is assumed that the core clock rate will remain between 2 and 3 GHz. In contrast, the continuing advances in reducing computation energy cost will be set in line with the active DARPA Ultra High Performance Computing (UHPC) program to be 10 pJ/FLOP [22]. The total processor power considered will be 100 W as the heat dissipation limited average of the last decade of processors [3,4]. The budgeted computation energy allows for 6 TFLOP in the target processor. To maintain the 1 B/FLOP target, the read and write data paths each require 48 Tbit/s link bandwidth. Since the off-chip communication must be shared between system connectivity, e.g. socket-to-socket CPU links and storage-area-network links, and the memory interface, roughly half of the off-chip communication power budget is available for

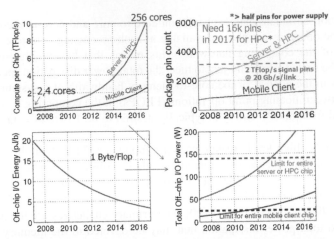

FIGURE 12.3 Energy and bandwidth-density projections for electrical links for manycore processors (packaging trends—Source: ITRS).

the memory interface. In summary, these system assumptions require an off-chip memory bandwidth of 96 Tbit/s within an energy budget of 10 W.

In Figure 12.3, we couple these requirements with the core-count and computational capacity scaling, as well as the link energy and bandwidth density limitations due to packaging.

Looking at Figure 12.3, it is clear that many core processors using electrical links face significant barriers to operate at 1 B/FLOP of memory bandwidth, and are at best confined to 0.1 B/FLOP due to inability of the electrical links to simultaneously satisfy conflicting energy cost and bandwidth density requirements. To increase bandwidth density with limited number of package pins, links must increase the data rate. At higher data rates, however, the electrical link energy-efficiency decreases.

The core of these issues is, as we see, the bandwidth-density and energy-efficiency scaling of electrical links.

12.1.2 Future scaling of electrical core-to-memory links

The fundamental challenge in off-chip communications is that as the size and power dissipation of the computation and memory devices have shrunk with semiconductor scaling, the macroscopic electrical connections have largely stayed the same. As the system is scaled to fit more computational and memory devices within a fixed area and power budget, these macroscopic electrical links connecting them become a bottleneck for system performance. Although it is generally a foolish bet to place a limit on achievable electronic circuit performance, some general targets need to be set as will be outlined in the following three sections.

The above-mentioned inability of electrical links to simultaneously increase the bandwidth density and energy efficiency can best be illustrated in Figure 12.4, which shows the performance of best recent electrical links (Source: ISSCC).

It is clear from Figure 12.4 that even nominally lower-energy chip-to-chip links do more than quadruple the energy cost for doubling of the data rate. Best 20 Gb/s links have energy costs above 10 pJ/bit, and as seen in Figure 12.3, even these links will not be able to meet the bandwidth-density requirements from slow package pin-scaling. Links at rates of at least 60–80 Gb/s are needed in order to meet these requirements, but looking at the Figure 12.4 data, it is unlikely that they would be built with energy cost under 10 pJ/bit in any foreseeable process node. This creates an opportunity for photonic technology that we will focus on in the rest of this chapter.

To put these scaling trends in practical perspective, the figure below illustrates the memory requirements for the 4-socket 10 TFLOPs/s system illustrated in Figure 12.2, supported by various existing electrical memory interfaces as well as projected capabilities of photonics technology (see Figure 12.5).

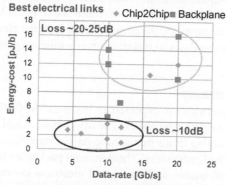

FIGURE 12.4 Energy and data rate of best recent electrical links (Source: ISSCC).

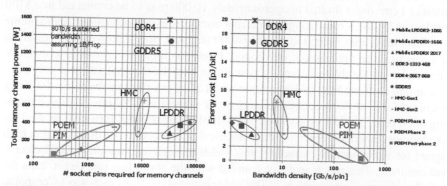

FIGURE 12.5 Memory interface scaling trends and requirements for a 10 TFLOPs/s system (Source: Intel Developer Forum).

This figure further supports the observations that low data-rate electrical interfaces can in fact achieve low link energies, but have prohibitively poor bandwidth densities and are impractical for use in future processor memory systems. The photonic memory interface, on the other hand, offers simultaneous scaling of both bandwidth density and energy efficiency. The reasons behind this capability of the photonic technology are not just inherent to the photonic transport mechanism, but are enabled by the overall photonic link system optimization as will be described in this chapter.

12.1.2.1 Channel bandwidth

Although the limited bandwidth of the off-chip electrical channel is often given as a motivation for the transition to photonic interconnect, it is optimal to operate the photonic interconnect at a higher signaling rate than the electrical interconnect that is being replaced. The reason for this is that the practical clock rate of the interconnect to operate with high energy efficiency is limited to a low multiple of the computation clock rate [23]. To push the clock rate higher, complicated clocking circuits must be added to the serialization/deserialization (SERDES) circuits of the link and would dramatically increase the power dissipation above target levels. By setting the maximum serialization rate to $4\times$ the core clock for both the electronic and photonic links considered in this analysis to maintain energy efficiency, the data line rate will be set at 8–12 Gbit/s.

12.1.2.2 Channel density

The fixed channel bandwidth is expected to be more than offset by a substantial increase in parallelism enabled by optics. The off-chip electrical channel density currently faces hard packaging limits that are the largest impediment toward increasing the off-chip bandwidth of processors. The dominant packaging technology for CPUs, flip-chip bonding, will enable scaling-limited electrical connections at a pitch of approximately $100\,\mu m$ in a 2D array. This density has been supported by low-cost packaging technology at approximately \$0.01 per pin [24,25]. As a result, a yield-limited $1\,cm^2$ die is limited to approximately 10,000 pins to be connected in a \$100 package [26–28]. Considering that current microprocessor packages in high volume products have been limited to roughly 4000 pins and slowly growing, this estimate appears to be an optimistic upper bound. From this limit, it is possible to estimate the total available channel count by first eliminating the required power supply pins. Assuming a current limit of 100 mA per pin in a fine pin-pitch package [6], the 100 A required for a 100 W processor running at 1 V consumes 2000 pins. Of the remaining 8000 pins, it is further assumed that half of the total signal pins may be used for the memory interface. Since off-chip electrical channels must be differential at high speed for signal integrity concerns, this yields a rough assumption of 2000 electrical channels. This rough analysis, coupled with the above line rate target, caps the off-chip electrical bandwidth at 20 Tbit/s bidirectional (10 Tbit/s each way). Compared to the system target of 96 Tbit/s, this is seen to be well below a quarter of the target bandwidth with very optimistic assumptions.

12.1.2.3 Energy efficiency

Energy-efficiency of future electrical off-chip links is likely the most difficult parameter to estimate. It may seem that since this has been a critical parameter for many years and that the materials and techniques used to make the electrical channel have remained constant that therefore performance would be saturating at a minimum value. Instead, designers have scaled the capacity of individual memory channels, continually improved drive circuit topologies as well as replaced analog equalization technology with lower power digital techniques. Complicating the matter further, results presented by academic researchers often do not address the complexities of a real system that would result in increased power dissipation. The pace of progress has been roughly a $2\times$ improvement in energy efficiency every 5 years. A reasonable target for near-term commercial systems is therefore on the order of 5–10 pJ/bit.

To set the target for chip-to-chip photonic links, we consider the absolute minimum power that an electrical chip-to-chip link has to dissipate (provided that it is using the current-mode signaling at the transmitter, Figure 12.9). This minimum power comes from the fact that an electrical link has to drive the $50\,\Omega$ off-chip trace impedance. Transmit voltage swings of less than 200 mV are very hard to detect at the receiver, due to channel losses, noise and receiver offsets [29]. We need at least 4 mA of current at the transmitter driver, Figure 12.9, to generate a 200 mV differential voltage swing at the transmitter. With a 1 V supply, this requires 4 mW of power. With link rates limited to below 10 Gb/s by the channel quality, the energy cost per bit is greater than 400 fJ/bit. Additional cost comes from the energy needed to drive the transmitter driver capacitance (device size that provides 4 mA of current). This is often dynamic power and depends on the link data rate, unlike the transmitter power above. The receiver samplers also dissipate at least some data-rate-dependent dynamic power. Note that a picture is a bit more complicated in reality, since low-swing voltage mode schemes can further reduce the transmit power dissipation at the expense of increased dynamic circuit power in the transmitter [29].

Academic papers have routinely reported power per bit numbers in the low pJs for several years [23,30–35]. Recent work has reported power levels approaching 0.2 pJ/bit for specific channel assumptions at lower data rates [36–38]. However, the best academic results would have to be significantly modified to handle the poor electrical channel present in a massively parallel microprocessor package. Given this outlook, it may be reasonable to assume as an upper limit that deployed systems may achieve the one to two orders of magnitude improvement to reach 0.5 pJ/bit within the next 5 years. Perhaps a more realistic limit would place 10 Gbps link performance in the 1–2 pJ/bit range for a high capacity system where the electrical channel is significantly degraded. To meet the total system target bidirectional memory bandwidth of 96 Tbit/s, these links would burn 24 W for the most optimistic 0.5 pJ/bit case. The optimistic estimated electrical link performance still falls over a factor of 2 short of the desired system targets. Again as was seen for the packaging density case, realistic limits place the electrical links at least a factor of 4 away from targets required for system enablement.

12.1.3 On-chip electrical interconnects

State-of-the art on-chip busses [39] and crossbars [40] exclusively use inverter-based repeaters, although they are known for their increased power consumption and inherent trade-off between bandwidth and latency [41]. Other electrical on-chip communication techniques (such as equalized point-to-point links [42,43]) are likely to be used in future, due to increased energy-efficiency and short latency. These techniques essentially decouple the latency-bandwidth trade-off at the expense of trading of the signal-to-noise ratio with power, data rate, and latency. In this aspect, these links are much closer to proposed photonic links, which are by default designed with some SNR constraints. The latency of equalized electrical wires can be considered near-speed of light for clock frequencies up to 10 GHz, and reticle limited 20 mm × 20 mm die sizes, since a signal can reach any part of the chip in a single cycle over this type of interconnect (unlike the wire with repeaters). For data rate density of 1 Gb/s/μm, the repeated wire power density is nearly 10 times larger than that of an equalized wire. The energy efficiency of optimized repeated wire is roughly 2 pJ/bit while equalized wire has 0.2 pJ/bit (10 times better energy efficiency). With constant frequency and supply voltage, we expect these energy-costs per bit to scale sub-linearly with process technology. This data sets the upper bound on density and energy-cost per bit of a photonic link alternative. For a photonic on-chip link, energy-cost per bit would have to be better than 100 fJ/bit to exceed the efficiency of the equalized on-chip electrical link at the 45 nm process node.

12.2 PHOTONIC NETWORK ARCHITECTURE

A variety of WDM network schemes exist [44] for flexible interconnections among large numbers of nodes. To prevent complexity explosion for large numbers of nodes, most schemes use either tunable transmitters or tunable receivers or both, and rely on sensing schemes or fixed slot allocation for collision avoidance. Existing on-chip WDM network proposals also involve tunable cross-point resonators [45,46], which unfortunately do not scale well with number of ports and have low energy efficiency and slow tuning. In our previous work [Micro08, NOCS09] [47,92], that with current state of the photonic components, the most efficient and scalable on-chip networks can be constructed with photonics used for efficient transport and electronics used for arbitration and switching. For example, Clos networks are very attractive from latency and throughput standpoints, since any node in the network is only a few hops away. However, electrical implementation has prohibitive energy cost due to long electrical links required between router stages and available density of global on-chip wires. Our studies have shown that just replacing the electrical links with simple point-to-point WDM photonic links presents an attractive option even with the current state of photonic technology [47].

Although photonic switching can be useful in some instances as we will illustrate later in the chapter, more complex photonic switching is still immature and does not directly improve the performance of on-chip networks. In order to both understand the sources of potential benefits as well as the limitations of the photonic link technology, we will focus next on an example photonic memory interface that utilizes the photonic WDM bus as the main building block.

12.2.1 Optical link design and components development

Optical links described above may consist of link-to-link connections at fixed wavelength as well as fully tunable hitless switchable networks. The key optical and optoelectronic devices include: fixed and/or tunable filters that can access/add a given wavelength from/to an optical waveguide, ultra-low power electro-optic modulators and high-sensitivity low-noise detectors to approach fundamental limits in power dissipated per Gb/s transmission speed.

To maintain a simple overall system architecture, we will consider the *simplest* case of photonic I/O in which point-to-point electrical links to the DRAM memory chips are replaced by photonic interconnect. Each memory photonic access point is integrated within the standard electronic memory controllers (MCs) that are distributed across the processor die as shown in Figure 12.6. The photonic access point consists of three fibers one to deliver the CW wavelength comb to the memory module, one to transmit upstream data, and one to receive downstream data. This three-fiber data bus is then connected to a DRAM memory module that can support

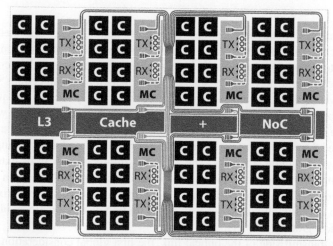

FIGURE 12.6 Example of a physical floorplan for the integrated photonic link infrastructure. Single wavelength optical power supply interfaces are located in the middle of the die and passively split for distribution as a wavelength-division-multiplexed optical power supply. The center region of the die is coarsely designated for the L3 cache and network-on-chip NoC in addition to all-shared control and other electrical I/O.

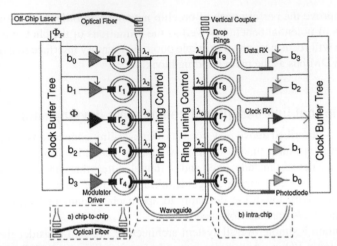

FIGURE 12.7 A WDM photonic bus. A continuous-wave (CW) multi-λ laser is coupled onto the chip through a vertical-coupling grating structure. Once on-chip, frequency selective ring-resonant modulators encode digital bitstreams onto their resonant wavelengths. Each wavelength propagates along the waveguide (and possibly off-chip) until it is routed through a matching drop ring to an integrated photodiode (PD). An optical receiver forms a bit decision based upon the PD photocurrent. Clock signals are routed both optically along the waveguide and electrically through local H-trees. Ring tuning circuits are used to tune the resonance of the modulator and drop rings [72].

high sustained I/O bandwidth such as the recently announced hybrid memory cube (HMC) architecture shown in Figure 12.7. The example chip floorplan has been simplified to include only eight wavelengths and eight memory module connections implemented as four horizontally stacked eight-fiber ribbon connections. The numbers of wavelength and memory channels do not necessarily need to be equal as they happen to be in this example. The passive power splitting can deliver the wavelength comb to an arbitrary number of waveguides for memory access. A unique feature of the proposed floorplan is that the entire on-chip optical network may be implemented without any waveguide crossings. The zero-crossing topology can be extended to a nearly arbitrary size network. Placing the input optical coupling ports in the center of the processor die minimizes the on-chip routing distance. The zero-crossing constraint then only trivially increases waveguide routing distance by local obstructions of the vertical grating couplers.

For the transmit side of the optical link, the first global consideration is that the laser sources are off-chip and coupled on-chip through vertical couplers. Since each laser must be split into various waveguides and each output waveguide needs components of all input laser sources, the bulk of the multiplexing and splitting operations can be achieved in one or two star-couplers [48–50] as shown in Figure 12.6 or any other suitable passive splitter.

After the light supply is distributed to all the memory controllers, the focus is on the information transport mechanism—i.e. a WDM bus. A full WDM optical link is illustrated in Figure 12.7.

The major considerations that will be addressed in the following sections then include the link-level trade-offs and interactions between the electronic and photonic components, primarily the transmit and receive optical components, and clocking/serialization and tuning control.

The performance of the ring resonator devices sets the maximum wavelength channel count that may be demonstrated in a given technology platform. The resonant frequency of each ring resonator must be precisely controlled with thermal feedback. The power required to heat the rings must also be included into the overall link energy budget. The ring modulators are narrow-band devices that can be electronically driven to either block or allow the light for a single wavelength channel to pass. The receive side of the link then requires the aggregated wavelength channels to be demultiplexed in a similar ring resonator filter bank before the light may be absorbed with integrated detectors. For both transmit and receive sides of the photonic link, the energy dissipated for the driver and receiver circuits will be considered in the context of achievable device capacitance.

12.2.2 Wavelength division multiplexing

The high optical bandwidth of the optical channel enables the chief density advantage of optics: wavelength division multiplexing. In this approach, multiple carrier wavelengths, each carrying separate data signals, are aggregated onto one waveguide to reduce the density of off-chip interfaces. It is important to first explain the relevant technologies required to examine the number of available channels as well as the optical loss and physical area utilization of the required components. The target specification that governs how closely channels may be packed together is the allowed optical cross-talk between adjacent channels. This number is typically set to be an allowed cross-talk ceiling of 20 dB. In the context of relatively power-equalized channels utilizing low extinction modulators, this represents a conservative bound.

12.2.2.1 Provisionable optical bandwidth

The provisionable optical bandwidth is set by the free spectral range (FSR) of the ring resonator filters used to demonstrate the wavelength division multiplexing filter banks. The FSR is proportional to the wavelength squared divided by the group index and ring radius. Using the nominal design dimensions (a 7 μm radius ring formed by an 80 nm × 500 nm waveguide) for our target SOI-CMOS process, the bend losses as a function of radius for various wavelengths are shown in Figure 12.8. Fitting the minimum radius as a function of wavelength and using the correct group indices, the maximum achievable FSR as a function of wavelength is shown in Figure 12.9. As a point of reference, 4 THz at a center wavelength of 1250 nm corresponds to a 20.8 nm wavelength range. The number of wavelength division multiplexing filter channels that may fit within this span may then be determined.

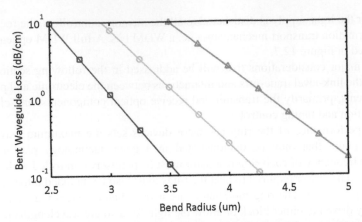

FIGURE 12.8 Bend loss simulation detail for a 500 nm width waveguide at wavelengths of 1200 nm (blue squares), 1250 nm (green circles), and 1300 nm (red triangles). (For interpretation of the references to color in this figure legend, the reader is referred to the web version of this book.)

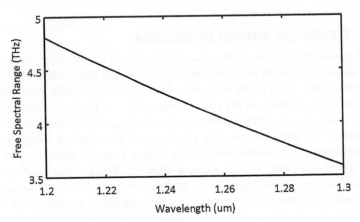

FIGURE 12.9 Maximum free spectral range (FSR) as a function of wavelength for a 1 dB/cm bend loss limit for a 500 nm width waveguide.

12.2.2.2 First-order filter realizations

First-order (single ring) microring resonator filters represent the simplest and smallest optical component available for wavelength division multiplexing. We may now analyze how well such a filter may perform for the wavelength division multiplexing application within the system. To target operating wavelength ranges between 1200 nm and 1250 nm, the nominal bend radius for the filters will be set to 4 μm—for Si waveguide cross-section of 80 nm × 500 nm—enabling roughly 4 THz of usable optical bandwidth. We will set the optical bandwidth to minimize any bit error rate penalties for 10 Gbit NRZ data transmission. A generic constraint for optical data transmission is typically given by specifying a filter full width at half maximum (FWHM) bandwidth that is 1.5 times the symbol rate or 15 GHz in this case. This general metric, however, does not consider the filter shape to truly assess the resulting data impairment.

To analyze the required bandwidth of a first-order ring resonator, we will instead consider the 90% bandwidth of the filters. Since the optical data bandwidth is well confined within 7 GHz for such a data stream, we will set the constraint that the 90% transmission bandwidth should be greater than 7 GHz. As shown in Figure 12.10, a 1.8% power coupling coefficient to the bus waveguides produces a ring filter with a 7.2 GHz wide 90% pass band and a FWHM of 22 GHz. Taking into account a typical propagation loss of 3 dB/cm, the drop loss for such a filter is 0.4 dB with an off-resonant insertion loss of 4×10^{-4} dB.

The packing density of the filter bank is then set by the allowable level of adjacent channel cross-talk. In standard telecommunication systems, most optical components are required to meet strict requirements for adjacent channel cross-talk suppression of 50 dB or more. This requirement is set by the possibility of large optical power differences between adjacent wavelength channels. In the chip-to-chip interconnect scheme proposed here all wavelengths on a given waveguide are transmitted or received at a single location for all channels and traverse identical optical paths. This system design feature ensures equal power levels across all wavelength channels. The extreme constraints of the telecommunication systems may therefore be significantly relaxed. Instead a proper specification can be determined by considering the impact of cross-talk from an aggressor channel transmitting a "1" in a channel receiving a "0." The important metric is then the relative strength of the cross-talk to the extinction ratio of

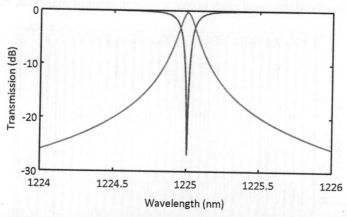

FIGURE 12.10 First-order ring resonator through (green) and drop (blue) transmission functions for a 4 μm bend radius, with 3 dB/cm waveguide and a 1.8% power coupling coefficient to the through and drop bus waveguides. (For interpretation of the references to color in this figure legend, the reader is referred to the web version of this book.)

the modulation. If we consider modulation extinction ratios of 10 dB, a 20 dB cross-talk suppression specification ensures that the aggressor "1" is an order of magnitude suppressed relative to the received "0" level. Although this is the constraint that we will consider in the following analysis, even this limit is relatively arbitrary. For a given system implementation the resulting signal-to-noise ratio from the combination of receiver technology, modulator extinction ratio, modulator cross-talk, filter channel bandwidth, and filter channel cross-talk should be considered jointly to optimize performance.

Examining the transmission characteristics of the target filter design, the cross-talk suppression of 20 dB may be achieved for an adjacent channel spacing of 110 GHz. Within the provisionable optical bandwidth of a single FSR, the 32-channel filter bank may be fabricated by stepping the radius of adjacent filters by 3 nm as shown in Figure 12.11. The transmission matrix model including the full lossy dispersion relation of the waveguide illustrates many relevant non-idealities that a fabricated filter bank would possess. The simulated on-resonance insertion loss of 0.44 dB is increased by 0.04 dB relative to an isolated filter in agreement with the target cross-talk specification of −20 dB. Second, for a fixed radius offset, the adjacent channel spacing changes by approximately 2% between the highest and lowest wavelength channels due to waveguide dispersion. More importantly, since the group index of the waveguide changes significantly over this wavelength range, the effective free spectral range also changes. As a result the adjacent channel separation between the highest and lowest wavelength channels out of band filter replicas changes by over 60% as is clearly evident in Figure 12.11. The 32-channel filter bank target still yields desirable system transmission performance while including all of these non-ideal characteristics.

The factor of 4 analysis in filter bank channel count relative to what has been demonstrated in this thesis reduces the minimum number of connecting optical fibers to the CPU die to 250. This is still a large number of fibers, but there is sufficient real estate on-die at the target pitch to enable over an order of magnitude more fiber

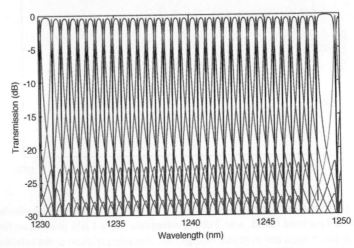

FIGURE 12.11 Thirty two channel filter bank simulated with 3 nm radius step per channel.

connections. Additionally, the fiber count is comparable to the state of the art in high-density single-mode fiber connectors.

12.2.2.3 Higher-order filter realizations

Similar to higher-order electronic filters, sharper transitions from the pass-band to the stop-band are achievable in coupled resonator, or higher-order, filter designs. The trade-off is primarily between the filter complexity and the number of optical fibers required to achieve a specified off-chip bandwidth. The difficulty in assessing the merits of a given design is therefore split between the device performance in terms of complexity, tuning power, and drop-loss, and the cost of the packaging. Further, from a system perspective, each optical fiber is expected to be connected to a single DRAM stack. The DRAM chip will have to provide high utilization of the provisioned per-fiber bandwidth. If 32 wavelength channels are utilized, each DRAM stack will be provisioned for 320 Gbit/s of read and 320 Gbit/s of write bandwidth. For comparison, a DDR3-1600 module has a data bandwidth of up to 100 Gbit/s today. Proposals for future stacked memory solutions such as the hybrid memory cube have predicted provisioned bandwidths up to 1 Tbit/s. Significant computer architecture considerations are required to address such trade-offs. In light of this uncertainty, first-order resonators will be the focus of this discussion.

12.2.3 Channel locking and thermal stabilization

A single ring filter for a wavelength division multiplexing filter bank may not be deterministically designed to be resonant at a specific absolute frequency. Instead, the fluctuations in silicon layer thickness and patterned waveguide width alter the exact group index and therefore the resonance frequency of the fabricated device. The variation in these parameters, while large in absolute uncertainty, may be very small relative to other nearby fabricated devices on the same wafer. For this reason, it is possible to match or offset filter resonant frequencies with high precision. Local frequency relative variations are below 50 GHz on average. The ideal modeled filterbank of Figure 12.12 would then have channels randomly shifted relative to

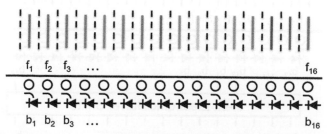

FIGURE 12.12 Ideal matching of filter bank resonance frequency to a receiver demultiplexing filter bank to convert incoming frequencies f_1 through f_{16} into parallel data bits b_1 through b_{16}.

each other to a small degree and a large uncertainty of what filter channel resonance is closest to a specific wavelength. However, if the filter channels are designed to completely fill the free spectral range, the resonance aliasing will distribute a channel to be nearby each desired wavelength.

The total data link can greatly reduce the burden of correcting each wavelength channel if nearest-neighbor wavelength grid locking is allowed for the filter channels. This may be achieved without loss of performance in point-to-point links where all wavelength channels are dropped at a single physical location. To understand this principle, first consider the ideal receiver filter bank shown in Figure 12.12 that is fabricated by offsetting adjacent channel radii. Each filter bank output channel will receive bits for the corresponding wavelength channel analogous to a multi-wire, parallel electrical bus. The good relative and poor absolute dimensional control will result in the filter channels aliasing over the free spectral range as shown in Figure 12.13. If no further provisioning is added to the system, the bits at each filter bank output channel will not correspond to the data from the intended wavelength channel. This simple, deterministic channel offset can then be corrected with digital electronics [72]. The barrel shifter pictured in Figure 12.13 can then correctly realign the data bits for the expected channel ordering. The control signal for this barrel shifter must then be initially configured during system initialization. The relative bit ordering between the nearest-neighbor wavelength-locked transmitter and receiver must be established at system initialization to set the barrel shifter offset.

The locking of the fabricated filters to this nearest-neighbor wavelength grid can then be accomplished through thermal tuning. The impact of temperature on the resonant frequency of the resonators is dominated by the thermo-optic coefficient of the silicon core material. Since the refractive index of the material increases with temperature primarily due to a reduction in the semiconductor's bandgap, the resonance frequency reduces or "red shifts." The measured effective thermo-optic coefficient has varied from 7.9 GHz/°C to 9.6 GHz/°C. The measured value for the SOI-CMOS platform was 9.6 GHz/°C and will be the value used for all further tuning efficiency

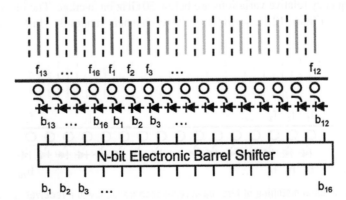

FIGURE 12.13 Filter bank receiver with aliased channels due to fabrication variation. Channel offsets can be corrected by the use of the electronic barrel shifter.

analysis. If efficient or directly integrated heaters are added to the filter bank, the net electrical tuning efficiency will be the product of the effective thermo-optic coefficient and the thermal impedance. The best measured heater thermal impedance achieved in our previous work is 44 °C/mW to yield a tuning power of 2.4 μW/GHz that is comparable to record results from the literature. For the sake of coarse estimation, the average tuning power per ring filter can be calculated using the channel spacing and process variation. Since the thermal tuning can only operate in one direction, i.e. red shift the filter, twice the stochastic variation estimate should be used. Additionally, since the filters must align to a predetermined wavelength grid, an average tuning range of one filter channel spacing frequency should be considered. The resulting total frequency shift is then 210 GHz to yield an average power estimate of 0.5 mW per filter. The control electronics required to achieve the locking is then assumed to be insignificant in comparison since the control loop can run roughly two orders of magnitude slower than the data link. Since only one half of the optical link will be located on the processor die, a 50 fJ/bit overhead for thermal tuning energy must be added to the on-chip link energy calculation for the 10 Gbps data rate.

12.2.4 Modulators

The most obvious contribution to on-chip electrical power dissipation for the interconnect is then the optical modulator itself. For the depletion device that offers the lowest possible power dissipation, the modulator acts as a capacitive load on the electrical driver. It is first necessary to specify an exact target device to analyze performance. Since depletion modulators have not yet been fabricated as part of this thesis, a target device based upon the measured technology parameters will be considered. The 3 dB/cm waveguide loss of the integrated CMOS platform enables narrow optical resonances. The optical bandwidth of the resonator can then be set to a 3 dB bandwidth equal to 70% of the data rate, or 7 GHz, with appropriate choices of the bus through and drop coupling coefficients.

At such narrow line widths, a small electro-optic phase shift is required to achieve sufficient modulation. Relatively low doping concentrations may then be used to form the pn-junction region. The n-type and p-type regions are then chosen to equal the existing transistor well-doping concentrations of 2×10^{-17} cm^{-3} that are available in the standard CMOS process. A maximum scaled-CMOS compatible voltage of 1.5 V in reverse bias results in a depletion width of 175 nm. This maximum depletion region width forms one logical state of the modulator. The other state is achieved by forward biasing the device below the diode turn on voltage that would result in considerable current flow. If this safe forward bias voltage is taken to be 0.7 V, the minimum depletion region width is then 47 nm. The depletion width variation of 128 nm then represents 43% variation of the effective mode width that is approximately 300 nm wide. Finally assuming a 70% silicon core confinement factor, the on and off state of the designed depletion-mode modulator is shown in Figure 12.15. The calculated optical performance characteristics are then a 17.6 dB extinction ratio with a 0.5 dB insertion loss. The narrow linewidth relative to the 22 GHz FWHM filter bank ensures that the modulator acts only on one wavelength channel. The total span of Figure 12.14 corresponds to one wavelength grid channel spacing of 110 GHz to demonstrate the isolation between neighboring wavelength channels.

The power dissipation of the modulator may then be determined by the total capacitance of the junction and all associated parasitic capacitances. If a planar junction model is used, the calculated capacitance for the forward, zero, and reverse bias states are 4.4 fF, 1.9 fF, and 1.2 fF, respectively. This simple model for a parallel plate capacitor, however, neglects the fringing field lines that result from the thin junction. Rigorously calculating the capacitance for the two states including the first metal level using the Raphael parasitic modeling tool from Synopsys, forward, zero, and reverse bias capacitances of 8.0 fF, 5.2 fF, and 4.1 fF, respectively, were obtained. Additionally, the wire length between the driving circuit and the modulator must be considered. If the photonic and electronic devices must be separated by approximately 10 μm for the localized substrate removal post-processing, an extra 2 fF of wiring capacitance must be added. For simplicity, we will assume average capacitances of 8.6 fF and 6.7 fF for the forward and reverse states, respectively. Since the modulator will only dissipate power when its state is toggled, the energy dissipation is multiplied by an estimated switching activity factor of 50%. The energy dissipation per bit for the modulator itself for a NRZ data stream is then given by the average of the forward and reverse states:

$$\text{Energy}_{\text{mod}} = \frac{1}{8}\bar{C}_{\text{forward}}V_{\text{forward}}^2 + \frac{1}{8}\bar{C}_{\text{reverse}}V_{\text{reverse}}^2. \tag{12.1}$$

For the stated parameters and assumptions, the energy dissipation for the modulator itself is then calculated to be 2.4 fJ/bit.

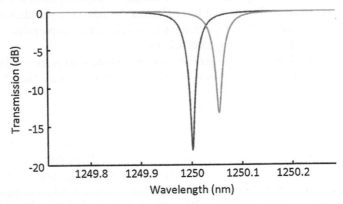

FIGURE 12.14 On (red line) and off (blue line) state transmissions for depletion-mode modulator. The total wavelength span matches the 110 GHz designed channel spacing for the 32 channel filter bank. Power coupling coefficients are 0.8% and 0.2% for the through and drop ports, respectively. Design input laser wavelength is 1250 nm for peak modulation. (For interpretation of the references to color in this figure legend, the reader is referred to the web version of this book.)

Two additional components are then required to interface the modulator to the logic level output of the latched data of the microprocessor. First, the data rate of the optical interconnect is expected to be up to $4\times$ the clock rate. Although the serialization/deserialization operation can in general require significant energy consumption due to the requirement for local clock generation, low multiplexing ratios can be achieved with simple circuits. First, doubling the data rate only requires triggering on both the rising and falling clock edges. The final data rate doubling can then be achieved by delaying both edges through an inverter delay chain by half the duty cycle. Since this delay has the potential to introduce clock jitter as the supply voltage fluctuates, it is difficult to extend this technique to higher multiplexing ratios. Since all of these multiplexing operations may be performed with less than 100 minimum width static logic transistors in the deeply scaled logic process, the power dissipation of this multiplexing is approximately 30 fJ/bit when the full parasitic capacitance of transistor gates, drains, and interconnecting wires are considered.

The $\sim$9 fF load of the modulator represents a significantly larger capacitance than the $\sim$90 aF input capacitance of a minimally sized CMOS logic gate. This factor of 100 capacitance ratio is significantly larger than factor-of-4 fan-out ratio (FO4) that is typically used to construct CMOS logic. For the standard FO4 logic design, the electrical stage delay is $\sim$10 ps in our technology. For the 10 Gbps data operation, it is desirable to have rising and falling edges of approximately 10 ps. The driving buffer chain can then be constructed with a FO4-loaded modulator driver. If three FO4 buffers are required to return the drive level to minimum width inverter strength, the total buffer chain therefore represents approximately 20 fF of additional switching capacitance. Since the switching activity factor is again 50% for the NRZ data stream, the pre-buffer and driver only contribute 5 fJ/bit to the total modulator power dissipation.

The summary is that the actual modulation operation of the optical interconnect link can be expected to consume $\sim$38 fJ/bit for such a depletion mode device designed in the thin-SOI technology platform developed in this thesis. This small number is roughly an order of magnitude less than the electrical link target. It is apparent that other aspects of the photonic link can be expected to dominate the total energy consumption. Already, it has been seen that the thermal control of the wavelength division multiplexing filter banks requires more power than the fundamental transmission operation.

As a point of comparison, the predicted $\sim$38 fJ/bit performance is significantly lower than the best transmitter efficiency of 135 fJ/bit achieved through hybrid integration [51]. The largest difference between this previous work and the predicted performance of a monolithically integrated modulator is the excess parasitic capacitance of 200 fF in the hybrid integration work from the photonic side metallization-to-substrate capacitance of 150 fF, wiring parasitic of 10 fF on each side and 30 fF resulting from the actual hybrid integration bond interface [52]. If a similar low-capacitance depletion modulator was used in a similar hybrid integration platform, the expected energy efficiency of the transmitter would rise to 109 fJ/bit by performing the same set of calculations with the additional capacitive parasitic. If the metallization-to-substrate parasitic capacitance was replaced with a 100 fF electro-static discharge protection diode for high-volume manufacturability, a realistic link efficiency of 94 fJ/bit would then be predicted. Although the transmit energy would still be low in comparison to an electrical link, over a factor of 2 reduction in energy efficiency may be expected for the transmitter in a hybrid-integrated photonic inter-connect platform.

12.2.5 Detectors

The receive operation then draws electrical power to convert the photocurrent signal that results from the incident optical signal into a logic-level data signal for use in the circuit. In terms of the on-chip electrical power dissipation, the photodetector efficiency is an irrelevant quantity for the moment. As such, the required energy is primarily a function of circuit design. A key parameter for the circuit performance, however, is the capacitance of the photodiode. Unlike in the modulator case where a simple nearly linear relationship exists between efficiency and capacitance, the mag-nitude of the device capacitance dictates available circuit topologies [53]. The most efficient designs are only practical for extremely small device capacitances.

For traditional telecommunication receivers, the optical module is constructed of several discrete modules such as the photodetector (typically III–V), transimpedance amplifier (typically SiGe), and clock and data recovery chip (typically CMOS). The power dissipation for such modules has traditionally been orders of magnitude larger than is necessary for the integrated photonic interconnect link. Recently, progress in integrated receivers has reduced the power to 2.8 pJ/bit [54]. However, even in this work, a discrete photodetector that must be wirebonded to the receiver limits scal-ability. Using photodetectors fabricated in a standard CMOS process for vertically incident 850 nm light, previous works have achieved fully integrated receivers [55–58]. Due to the pF-scale device capacitance and low bandwidths of the integrated detectors, the lowest energy consumptions for full receivers are 4.7 pJ/bit at 8.5 Gbps [56] and 6.6 pJ/bit at 10 Gbps [55].

Instead of relying on large mode area detectors intended for optical fiber, the small mode size of silicon photonic waveguides may be interfaced to small integrated detectors with a small capacitance. Germanium-on-silicon detectors have demon-strated capacitances as low as 1.2 fF [59]. Capacitances in this range should enable low-power receiver topologies such as sense amplifiers [53]. The lowest reported monolithically integrated device capacitance 20 fF [60], however, was combined with an inefficient transimpedance amplifier topology where the total power dissipa-tion was measured to be 1.5 pJ/bit [61]. Other workers have achieved a low-device capacitance of 10 fF [62] without integrated circuits and therefore relied on hybrid packaging to demonstrate a receiver. After wire bonding to a 90 nm transimpedance amplifier, total energy efficiencies of 4.7 pJ/bit at 15 Gbps [63] and 7.7 pJ/bit at 20 Gbps [64] were measured due to packaging parasitics.

The previous state of the art for an energy-efficient receiver was achieved through advanced micro-bump packaging a separately fabricated germanium detector with a 40 nm CMOS receiver [51]. Although the fundamental diode capacitance is only

10 fF, the bond pads add 30 fF with an extra wiring capacitance of 10 fF [65]. At 10 Gbps data rate, the received energy efficiency was measured to be 395 fJ/bit [51]. It is unclear that this result can scale to lower values in the future in a hybrid integration technology. Additionally, traditional electrostatic discharge (ESD) protection, which would significantly increase the interface capacitance, was eliminated in this research result. Manufacturing reliability may require the addition of ESD protection diodes and increase the interface parasitic capacitance by up to 100 fF.

In this work, integrated silicon-germanium photodetectors are connected to novel sense-amplifier-based receivers through the low-metal levels of the CMOS to enable record low energy efficiency by eliminating parasitic capacitances. A similar circuit topology was originally developed for a flip-chip bonded photodetector, but the interface parasitic prevented the demonstration of ultra-low power [66]. In our platform, the receiver energy efficiency has been measured to be 60 fJ/bit at 3.5 Gbps as fabricated in a 45 nm SOI CMOS process through an integrated electrical test circuit as discussed in [53] with a photodiode capacitance estimated to be 5 fF. It is expected that this number may decrease as the technology becomes more refined. The achieved data rate is limited by the timing integrity of the digital backend interface electronics. Since the efficiency increases with data rate as sources of static power dissipation are shared by an increasing data volume, further analysis will assume a 50 fJ/bit achievable performance for a 10 Gbps data rate. Given the current capacitance load of the integrated photodiode, the required received photocurrent modulation is ~1.2 μA for a 3.5 Gbps data rate. Since the sensitivity is expected to degrade with data rate, a receiver sensitivity of 5 μA in the SOI-CMOS platform will be used for further analysis. Since the DRAM process transistors typically lag CMOS device performance by approximately one technology generation, it may not be possible to achieve the same sensitivity in the memory process. As such, a factor of two degradation may be expected to result in a 10 μA receiver sensitivity in the DRAM process. From a CPU optical power sourced system perspective, this increased minimum receiver sensitivity on the DRAM side is offset by a lower optical power loss budget for the CPU write path. The more sensitive receivers that may be demonstrated on the CPU die detect light that must complete the round-trip communication to the DRAM module.

The photodetector efficiency for the integrated photonic system that may be ultimately achievable is the hardest parameter to estimate at the current state of platform development. Over the course of this work, a maximum external quantum efficiency of 1% has been demonstrated at 1240 nm in the SOI-CMOS platform for a first-generation device. Since the silicon germanium fabrication module within the memory process is still under development, no detector device results are available for that platform. Proof-of-concept photodetectors in a fully integrated CMOS process have demonstrated 0.85 A/W responsivity that may be operated with 99% of peak efficiency under short-circuit biasing as required for use in the sense-amplifier-based receiver [67]. The process flexibility of the memory integration process should enable similar results to be achieved. Therefore, 80% photodetector efficiency for the memory links will be assumed for optical link budget analysis. Since the zero-change integration method for the thin-SOI-CMOS process forbids process modification, the higher background doping of the silicon-germanium photodiode may limit device efficiency. As a result of this uncertainty, 50% photodiode efficiency will be assumed for the SOI-CMOS process for optical link budget analysis.

12.2.6 On-chip electrical link energy

The total on-chip energy consumption is a combination of thermally tuning the ring resonator filters, the modulator driver circuit and the receiver circuits. Since the 96 Tbps total off-chip bandwidth is a combination of transmit and receive links, the on-chip power dissipation is equivalent to 48 Tbps full photonic links. The total link energy budget from the previous section totals is then 188 fJ/bit including 100 fJ/bit for thermal tuning, 38 fJ/bit to drive the modulator, and 50 fJ/bit to receive the data. The total on-chip power dissipation for all of the links is then 9 W to fall within the required system limit of 10 W.

It is also worth comparing the monolithic photonic integration platform to a hybrid integration photonic platform. The modulator power dissipation has already been estimated to be limited to 94 fJ/bit under optimistic assumptions. The best available receiver energy literature value of 395 fJ/bit may be able to be further reduced, but it is difficult to estimate without significantly more detailed analysis. For the moment, we will assume that this can be reduced by 200 fJ/bit. If the same thermal tuning efficiency is then assumed, a total link energy of ~400 fJ/bit is estimated. The total power dissipation would then be estimated to equal 19 W. Although this represents a significant improvement over the most optimistic electrical scaling assumption, it still does not directly enable the desired 6 TFLOP processor system.

The final aspect of the link energy budget that has not yet been considered is the circuitry required to recover the clock from the received data [68]. In traditional fiber optic data links, the clock is typically recovered from the data during the receive operation. The circuitry to accomplish this, however, is complex and dissipates significant power. Typical low-power reports for discrete CMOS chips to accomplish this task at 10 Gbps are approximately ~50 mW or 5 pJ/bit [69,70]. Since the wavelength division multiplexed data channels traverse the same physical channel, it is possible to use a single recovered clock to receive all of the data. If this 5 pJ/bit overhead was shared between 32 wavelength channels, the effective clock recovery energy overhead would be reduced to ~150 fJ/bit. This overhead would double the total link energy.

Instead, the approach that has been taken in this work is to implement source-clock-forwarded links in which one wavelength channel is used to transmit and recover a clock [72]. The clock recovery circuit can then be implemented as an injection-locked oscillator to improve power efficiency. An injection-locked receiver has been designed and fabricated by collaborators on this project that has been measured to dissipate 47 μW [71]. The total system bandwidth may then remain unchanged by marginally increasing the data rate on the other 31 wavelength channels from 10 Gbps to 10.32 Gbps. The total effective energy efficiency for the clock-forwarded link would then be increased by only 2 fJ/bit for the rest of the data channels including the modulation energy. The total effective link energy of 190 fJ/bit would still meet the projected system targets.

While these calculations allow us to understand the breakdown of different energy components, it is interesting to explore the overall link power optimization, when the total link energy is taken into account—including the thermal tuning requirements and laser wall-plug energy costs. This kind of optimization can help us figure out the right degree of wavelength multiplexing for a given throughput requirement as well as the right integration strategy.

12.2.7 Link integration considerations

To facilitate a more detailed exploration of device and circuit interactions, here we utilize our models and results from [72].

12.2.7.1 Single channel link trade-offs

The interactions between the modulator and receiver and the impact on wall-plug laser power are exposed through the power optimization across modulator insertion loss, extinction ratio, and receiver topologies for different link data rates.

Figure 12.15 shows the energy-per-bit breakdowns for four integration scenarios.

It is immediately obvious from the plots that the laser power is the dominant energy consumer, increasing quickly with data rate as aggressive modulation rates force a relaxation of modulator insertion loss and extinction ratio. We can see that the laser power is highly sensitive to C_P, as the laser power for $C_P = 25$ fF is roughly $5\times$ that of $C_P = 5$ fF. Though the modulator tries to offset the laser cost by increasing its extinction ratio and decreasing insertion loss, it inevitably reaches a limit on its capabilities. The higher loss simply amplifies the laser power component, resulting in a $3\times$ laser power difference between the 15 dB and 10 dB loss cases. Matching previous analysis, the optimization performed on the underlying circuit models always chose the integrating receiver over the transimpedance amplifier (TIA) + sense-amplifier combination as the optimal receiver in all scenarios. It is worth noting here that in both cases, the receivers are clocked (using the clock-forwarding concept used frequently in electrical links), unlike typical telecom optical links which use clockless receivers with TIA + limit-amplifier combination. Such clockless receivers would be prohibitively expensive at this scale of photonic link integration and should be avoided at all cost.

Though our results present a grim outlook for the $C_P = 25$ fF (optical die with TSV) scenario, we note that lower losses may be achievable with a dedicated optical die, allowing TSV integration to remain competitive. Link and technology parameters used in this exploration are shown in Table 12.1.

12.2.7.2 Full WDM link evaluation

Expanding upon our analysis for a single-channel data link, we explore the impact of backend components required in a high-speed multi-channel WDM link. We incorporate a model of optical clock distribution and source-synchronous clocking as well

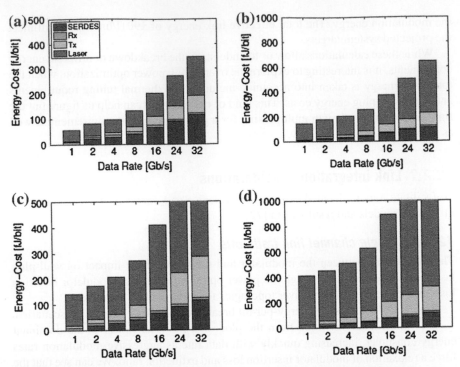

FIGURE 12.15 Single link in trade-offs for various integration scenarios [72]. Total capacitance at the receiver (including the photodetector, interconnect parasitics as well as receiver input capacitance) of $C_P=5$ fF represents monolithic integration, while $C_P=25$ fF is expected for a TSV connection to an optical die. Channel losses of 10–15 dB correspond to on-chip and chip-to-chip links, respectively. (a) Loss $=10$ dB, $C_p=5$ fF; (b) Loss $=15$ dB, $C_P=5$ fF; (c) Loss $=10$ dB, $C_p=25$ fF; (d) Loss $=15$ dB, $C_p=25$ fF.

Table 12.1 Link evaluation parameters.

Parameter	Value
Process node	32 nm Bulk CMOS
V_{DD}	1.0 V
Device to circuit parasitic cap C_P	5–25 fF
Wavelength band λ_0	1300 nm
Photodiode responsivity	1.1 A/W
Wall-plug laser efficiency P_{laser}/P_{elec}	0.3
Channel loss	10–15 dB
Insertion loss IL_{dB}(optimized)	0.05–5.0 dB
Extinction ratio ER_{dB}(optimized)	0.01–10 dB
Bit rrror rate (BER)	10^{-15}
Core frequency	1 GHz
SERDES topology	Mux/Demux Tree

as the overhead of the tuning control and compensation circuits described previously and in more detail in [72].

As we already mentioned, the integrated photonic links must utilize the forwarded clock scheme in order to be able to utilize the energy-efficient sense-amplifier based receiver topologies. At the differential receiver, the signal is regenerated, buffered, and used to clock the other data receivers, amortizing the cost of this clock receiver. Since the clock-TX and data-TX share the same clock fabric, relative jitter between the sent clock and data is minimal. On top of previously discussed benefits of optical links, optical clock signaling does not suffer from power rail injected noise or cross-talk, so no jitter is added in the channel. This obviates the need for an RX phase or delay-lock loop, greatly reducing the power and area overhead. The low latency of optical waveguides also means that all data receivers can operate on the same clock phase.

In our full-link evaluation, we explore links with four different aggregate through-put design points, 64 Gb/s, 256 Gb/s, 512 Gb/s, and 1024 Gb/s, corresponding to min-imum, medium, high, and maximum bandwidth scenarios.

Figure 12.16 shows that tuning power dominates at lower data rates (since there are more channels given fixed throughput) and decreases with data rate. Modulator, laser, SERDES, and receiver energies increase with data rate and dominate at high rates. At all throughput scenarios, an optimal energy balance is achieved at around 4–8 Gb/s. An overall energy optimal point occurs at less than 200 fJ/bit for a link with 256 Gb/s of aggregate throughput and 4 Gb/s data rate. It is interesting to note—and an indicator of a well-balanced system—that at the energy-optimal point, the energy consumption is roughly an even three-way split between tuning, laser, and mod/rx/SERDES.

As tuning power is now mostly dominated by the backend electrical components, this energy will scale favorably with technology and can be optimized using custom design. A full electrical tuning backend is also unnecessary on both modulate- and receive-side—barrel-shifts and bit-reordering only need to be performed once—meaning backend power can be cut by another 50%. Refinement of photodetector responsivity and parasitic capacitances as well as lower-loss optical devices with improved electrical laser efficiencies can bring about further reductions in wall-plug laser power. It can be expected that energy/bit will drop to sub-100 fJ with device development, process scaling, and overall link component refinement.

Projecting back to the overall system requirements, in Figure 12.17 we illustrate the bandwidth density of the link compared to die and package-limited electrical link bandwidth densities [29]. While it is clear that optimizing the photonic link energy leads to reduced optical bandwidth density, this energy-optimal point still has 1–2 orders of magnitude higher bandwidth density than the electrical die and package limited links, respectively.

12.2.8 Optical loss budget

Although the required optical power does not figure directly into the on-chip ther-mally limited power budget, it may be undesirable or impossible to implement an

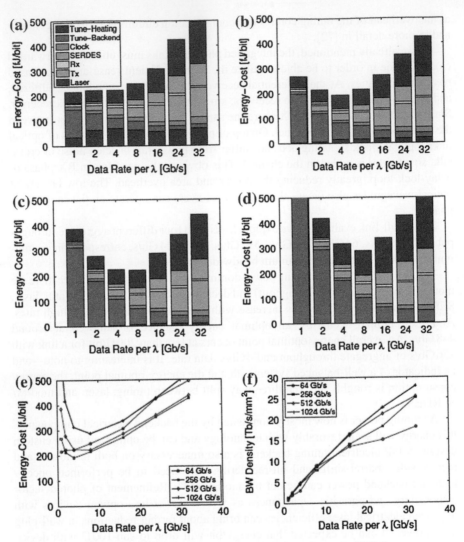

FIGURE 12.16 Full link evaluation for various throughput requirements [72], for Loss = 10 dB, C_P = 5 fF. For tuning, we assume a bit-reshuffler backend and electrically assisted tuning with local variation $\sigma_T L$ = 40 GHz and systematic variation $\sigma_T S$ = 200 GHz. Note that the number of WDM channels changes with data rate (Channels = Throughput/Data Rate). (a) 64 Gb/s (min); (b) 256 Gb/s (medium); (c) 512 Gb/s (high); (d) 1024 Gb/s (max); (e) throughput summary; (f) bandwidth density.

optical system if the required laser power is unreasonably high. The energy to generate the photons must still be considered if the actual end goal of energy-efficient computing is to be achieved in terms of power drawn from the system wall-plug per FLOP. The required optical power can then be calculated by working backwards from the receiver sensitivity targets and considering the sources of optical loss.

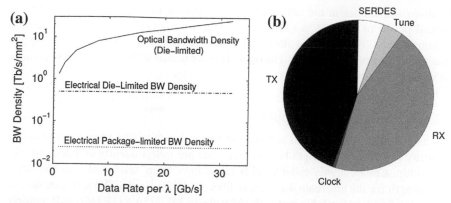

FIGURE 12.17 WDM link area and bandwidth density for a total throughput of 256 Gb/s and a $C_P = 5$ fF. (a) link bandwidth density; (b) WDM link area breakdown for a channel rate of 4 Gb/s [29].

All of the relevant insertion loss metrics have been presented in the previous sections and the total link optical losses for both link directions are shown in Table 12.2 to arrive at an average link loss of 8.9 dB. Combining the link-loss information with the estimated receive sensitivities, the required optical power per link is calculated in Table 12.3. In terms of the system wall-plug efficiency, ~29 fJ/bit of power should be added to the on-chip 158 fJ/bit calculation if the lasers used are 30% efficient as will

Table 12.2 Integrated photonic link optical loss budgets for upstream (data from cpu to memory "write" path) and downstream (optical power from cpu to memory module, data from memory module to cpu "read" path).

	Coupler Insertion Loss (dB)	Splitting Excess Loss (dB)	Modulator Insertion Loss (dB)	Waveguide Propagation Loss (dB)	Receiver Demux Loss (dB)	Total Link Loss (dB)
Upstream	3	1	0.5	3	0.4	7.9
Downstream	5	1	0.5	3	0.4	9.9

Table 12.3 Total optical power required per link. energy per bit for wall-plug analysis is included by assuming a laser efficiency of 30%.

	Photodiode Efficiency	Receiver Sensitivity (μA)	Link Optical Power (μW)	Energy at 10 Gbps (fJ/bit)
Upstream	0.8	10	77.07	25.69
Downstream	0.5	5	97.72	32.57

be discussed further in the next section. Considering the total system, the average optical power of $87.4\,\mu$W per link, each single wavelength source laser is required to output only 26.2 mW. This allows the power in each waveguide to stay below the nonlinear absorption threshold. The optical power is also well within the power range of widely available laser sources.

12.2.9 Optical power supply

As discussed at the start of this section, the nominal architecture proposal is to utilize an array of single wavelength laser sources that are independently coupled onto the CPU chip. Cheap, highly reliable DFB lasers have been developed near the target wavelength for the upstream data link in fiber-to-the-home (FTTH) applications. An example of widely available part is the Mitsubishi ML725AA11F InGaAsP multiple quantum well (MQW) laser diode [96]. It is designed to be operated in an uncooled environment between $-40\,°$C and $85\,°$C for an output power of 10 mW. Each laser is guaranteed to operate at a single wavelength, but the exact wavelength is not specified within a 40 nm range due to fabrication variance. The wavelength comb source may then be constructed by binning the fabricated DFB lasers and constructing a comb of lasers spaced by approximately 110 GHz each. Since no global constraints on absolute wavelength within this 40 nm range are required by the construction of the system, the comb laser source should be able to utilize all fabricated lasers. The lasers may then be allowed to drift with temperature at a rate of approximately 0.1 nm/°C as long as the system packaging maintains similar temperatures for the laser diode bars within approximately 1 °C.

The laser efficiency is an important metric when considering the link wall-plug efficiency. As seen in the previous section, 30% wall-plug efficiency yields sufficient link energy efficiency to impact the total link energy by less than 20%. For the uncooled MQW lasers such as the Mitsubishi DFB, a total efficiency of approximately 35% is achieved at 25 °C [73]. Since the laser is uncooled, the efficiency significantly degrades at higher temperature due to the limited energy separation in the quantum well-active region and increased non-radiative recombination pathways. The same MQW laser efficiency at 85 °C is only 16% [74]. This degradation of laser efficiency with increasing temperature can be greatly reduced by using quantum dot gain media that have been demonstrated to work well at the wavelength range of interest [72].

In systems with more complicated optical networks and therefore higher optical powers, the laser efficiency may become a more important parameter for total system energy efficiency. The highest wall-plug semiconductor laser efficiencies of greater than 70% today are found in the high-power lasers above 50 W output powers [77–81]. The wavelength range of these high efficiency sources are typically focused between 700 nm and 980 nm for use as amplifier pump sources or materials processing. The same material systems are equally applicable for achieving high efficiencies in the $1.2\,\mu$m range [82]. For smaller output powers, up to a 63% wall-plug efficiency has also been achieved in a single transverse mode design suitable for coupling to a single-mode fiber [83].

Recently, significant concern has been raised within the supercomputer community regarding the reliability of the VCSELs that are being deployed for rack-to-rack system connectivity. Although each VCSEL is a highly reliable component, the vast number of devices required has made the aggregate failure rate a legitimate system concern. Briefly, failure rates of most semiconductor components are specified by two related metrics: failures in time (FIT) and mean time between failures (MTBF). The FIT rate is calculated by measuring the number of failures observed relative to the total number of device-hours tested relative to 1 billion device-hours:

$$\text{FIT} = \frac{\#\text{Failures}}{\#\text{Devices}} \cdot \frac{1 \times 10^9 \text{ Hours}}{\text{Hours of Operation}}. \tag{12.1}$$

The MBTF is then expressed as the lifetime for a single device by the FIT rate:

$$\text{MTBF} = \frac{1 \times 10^9 \text{ Hours}}{\text{FIT}}. \tag{12.2}$$

The FIT rate for a 10 Gbps VCSEL has been measured by Finisar to be approximately 2.3 [84]. At first, this specification sounds very reliable when compared to the 19–23 year MTBF, which corresponds to a ~5000 FIT rate, of an Intel server motherboard [85,86]. The total failure rate, however, scales linearly with the number of devices. To enable the 96 Tbps total bandwidth per motherboard, 9600 VCSELs are required to yield a MTBF of only 2.3 years. The problem for the supercomputing system can be seen by considering the case of IBM's recently canceled Blue Waters supercomputer that was scheduled to use 1 million total VCSELs. The expected MBTF for VCSELs in the entire system would only be ~18 days. It is also important to point out that the FIT rate of VCSELs scales super-linearly with data rate [87]. It is therefore impossible to solve the reliability problem by reducing the number of VCSELs for the same total data rate.

The expected reliability for the laser sources required for the proposed silicon photonic interconnect architecture can then be analyzed in a similar context. As an example reliability specification for uncooled 1310 nm edge-emitting lasers, CyOptics reports a reliability of less than 15 FIT as measured in 200 billion field service hours [88]. The measured reliability includes directly modulated lasers that are expected to be less reliable than the CW operation required for the silicon photonic source lasers. Even using the 15 FIT estimate, the system reliability is dramatically improved relative to the VCSEL case. The 96 Tbps system bandwidth is enabled by only requiring one laser for each of the 32 wavelengths. The aggregate MTBF for the system is then 237 years and an order of magnitude longer than the published Intel server motherboard MTBF.

There are several other possible implementations for the multi-wavelength optical source. One interesting option is to use the multiple longitudinal modes of a spatially hole-burned Fabry-Perot (FP) laser. Since the active-layer diffusion is suppressed in quantum dot (QD) gain media, excellent comb-laser characteristics have been obtained using QD-FPs [74]. Alternatively, a photonic integrated circuit (PIC) source may be used to combine multiple on-chip lasers together into a single output fiber [89]. The total power out of one off-chip source may then be split to feed the various data channels. Yet another potential solution is to generate an on-chip wavelength comb through non-linear processes from a high-power off-chip single-wavelength laser source [90]. Many technological considerations may influence the effectiveness of these approaches including efficiency, reliability, and non-linear absorption in on-chip waveguides.

12.3 FUTURE CORE-TO-DRAM PHOTONIC NETWORKS

The photonic interconnect scheme presented to this point simplifies the total computer architecture in several important aspects. First, careful attention has not been paid to the memory side of the link. Second, the on-chip network has not been considered in conjunction with the photonic network. Finally, no attention has been paid as to how the network can be expanded to a multi-socket system. We briefly consider each of these issues, referring to published literature for more detailed analysis where available.

The memory side of the photonic link in the presented architecture only consists of three fiber connections. The input CW optical light is modulated with the upstream data and the downstream data is received all at a centrally located point on the DRAM module. This scheme does not leverage the photonic interconnect to distribute the data across the DRAM module. Since the primary energy dissipation in the photonic network is contained at the transmit and receive points, energy-efficient data transport across the large DRAM die may allow for further system energy minimization [91]. Even with the simple point-to-point interconnect proposal presented here, however, significant memory system benefit may be gained from the photonic integration. In the case of the hybrid memory cube DRAM proposal shown in Figure 12.3, the off-chip electrical interface layer requires a separate logic chip to be added to the DRAM memory stack. For a fixed layer count, this constraint reduces the memory capacity of a single module. Since the small footprint of the photonic interface module allows for integration with the DRAM modules themselves, the photonic interconnect integration may significantly increase the memory capacity of the total stacked chip. Additionally, the removal of the logic chip from the proposal may simplify fabrication by eliminating the requirements associated with the heterogeneous bonding of semiconductor layers fabricated using different wafer processes from different manufacturing vendors.

12.3.1 Memory module architecture

Little attention has been paid to the manner in which the photonic access network interfaces with the on-chip electrical network for the CPU side of the link. To first order, this matches a direct replacement of the point-to-point off-chip electrical network that it replaces. Additional capabilities of the photonic network include low-energy on-chip routing and independent wavelength routing. As discussed for the memory case, cross-chip routing can enable energy savings by eliminating receive and retransmission "hops." The photonic network distribution of memory access points across the die to be proximate to different cores may be a suitable architectural implementation. This does then place the burden of accessing distant memory controllers from cores not in the local area entirely on the on-electrical on-chip network. It is then only assumed that the 20% of total on-chip power budgeted for the network-on-chip (NoC) would be sufficient to provide suitable communication without creating a bottleneck for total system performance.

12.3.2 Wavelength-routed architecture

Instead of a single photonic access point being integrated with a memory controller and having full control of an attached DRAM module, each photonic access point controlled only a fraction of the wavelengths that would be aggregated onto the output fiber connected to a given DRAM module. The total wavelength space for each DRAM module connection is then partitioned between "groups" of cores on the chip that are locally connected together with a small-scale electrical network. To gain access to any block of data in main memory, any core must only communicate electrically within its own group before accessing the off-chip photonic link. The number of such groups then divides the effective radix of the on-chip electrical network to a manageable size for a high number of cores. As shown in Figure 12.18 for a 16 core

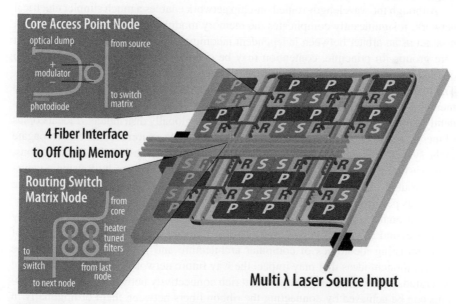

FIGURE 12.18 Wavelength routed photonic network to divide access to a global memory space into four on-chip groups with independent electrical mesh networks.

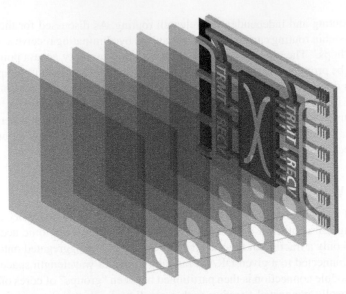

FIGURE 12.19 Stacked photonic memory module in which independent access requests must be arbited between and distributed to the relevant DRAM chip.

chip, the cores are partitioned into four separate electrical networks while wavelength routing still enables access to the complete memory space (see Figure 12.19).

Although the wavelength-routed on-chip network enables a much simpler electrical network, it significantly complicates the memory module. The memory module must now act as an arbiter between independent incoming memory requests from different core groups. In principle, contention may be simply resolved by denying access to one request, e.g. sending a NACK signal, whenever memory access requests conflict. The difficult nature of the problem is maintaining cache coherency for locally stored memory. Since the electrical networks updating a shared memory operate independently, updates to a piece of shared memory are not globally distributed. If local copies of memory stored in caches are not known to be obsolete, incorrect data can propagate in the system without additional architectural provisioning. Despite the complications, a full system analysis into the trade-offs of a multi-socket system built using a wavelength-routed photonic network such as the 256 total core system shown in Figure 12.20 was performed to a cycle-accurate level and published in the literature [92].

Additionally, many interesting ideas for future networks arose from this early work. Beyond a host of interesting exchanges that emerged from the intersection of the contrasting ideologies of computer architecture and device engineering, some new component ideas that may impact the way future networks are constructed were generated. One important concept is that rich connectivity between independent systems can be achieved by connecting the ribbon fibers between them orthogonally in passive off-chip connectors. A simple example of such a connector is shown as the cross-plane connector in Figure 12.20. In general, this notional connector, informally

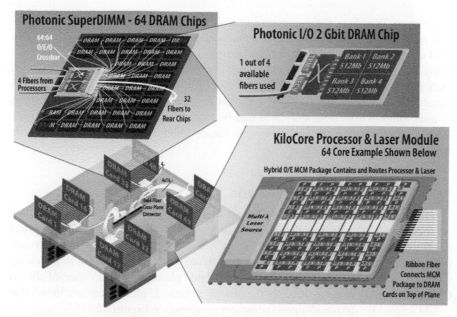

FIGURE 12.20 Full system implementation of a multi-socket system using a wavelength routed photonic access platform.

labeled as a "star-coupler" has proven especially helpful in connecting the socket-to-socket coherency network of more recent system proposals [1].

12.3.3 Survey of current and future architecture work

Recent computer architecture has focused on advanced photonic memory access networks that include explicit socket-to-socket coherency networks [91]. This topology is currently the focus of the newly launched DARPA POEM. The coherency connectivity is closely modeled after Intel's quick path interconnect (QPI) proposal [94]. Interested readers can find the details of this proposal in the published literature [91]. For comparison, several other research groups have put forward alternative network proposals [95–97].

In this section, a basic network architecture has been used to motivate the device work. From a simplified analysis of the electronic problem, competing density and energy-efficiency constraints appear to prevent realization of the desired computing systems by at least a factor of 4 in each constraint. Photonic interconnect has the capability to enable high density off-chip communication interfaces to memory for scaled microprocessor systems within the desired system energy budget. To realize the opportunity, integrated photonic devices must achieve certain performance metrics. The waveguide loss of 3 dB/cm enables not only low-loss cross-chip routing, but also low wavelength multiplexing filter insertion losses and highly resonant modulators. Wavelength division multiplexing filter banks must be fabricated with high precision to minimize the thermal tuning requirements that comprise almost one third of the total on-chip electrical link energy. The multiple fiber-to-chip interfaces of the proposed optical network encourage significant development to minimize the vertical coupler insertion loss. Integrated photodiode efficiency has a linear proportionality to the total optical power required for the links. The architecture targets assume that 0.5 A/W photodetectors can be fabricated in the logic process with a 0.8 A/W photodetector integrated through the insertion of a new silicon germanium module into the memory target.

12.3.3.1 Existing state of the art

The key optical and optoelectronic devices include (Figure 12.2): optical filters that can access/add a given wavelength from/to an optical waveguide, ultra-low power electro-optic modulators, and high-sensitivity low-noise detectors to approach fundamental limits in power dissipated per Gb/sec. After early success in providing a proof-of-concept for the feasibility of integrated photonics within state-of-the-art CMOS, our goal shifted to utilizing the devices to provide photonic interconnect that would impact the design of future computer systems (see Figure 12.21).

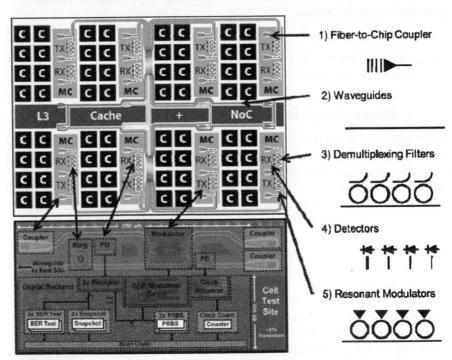

FIGURE 12.21 Identification of key photonic devices on the multicore processor and photonic interconnect architecture [99].

To assess the performance of the photonic devices, we analyze the achievable total interconnect system metrics for a reference the 96 Tb/s aggregate core-to-memory bandwidth compute system. Using our recently demonstrated, modeled, and predicted photonic device performance characteristics, total interconnect system performance metrics are shown in Figure 12.3 (see Table 12.4).

The current state of photonic devices is competitive with existing and next-generation on-chip power dissipation numbers. The modeled system improves significantly. Comparing the 2012–2013 timeline of these photonic devices to the latest hybrid memory cube (HMC) architecture, the estimated on-chip power dissipation is over an order of magnitude smaller than the electrical projection. Additionally, the physical system becomes significantly more realizable with an 832 fiber count and 1.8 mW of optical power per link. Finally, device targets based on several design iterations of improvement to target the 2013–2015 timeline still beat research electrical link projections by a factor of 5 while providing physically realizable fiber and input optical power requirements.

Table 12.4 Status of achieved, modeled, and predicted photonic device performance relative to various generations of electrical link technologies.

Component	Current	Modeled	Future
	DDR3	**HMC**	**Research**
	DDR3	HMC	Research
Thermal tuning	4.3 pJ (1.3 mW)	200 fJ (0.5 mW)	35 fJ (0.2 mW)
Modulator driver	2 pJ	500 fJ	38 fJ
Receiver energy	53 fJ	50 fJ	50 fJ
Receiver sensitivity	5 µA	5 µA	3 µA
Total shotonic	6.6 pJ	755 fJ	123 fJ
Electrical	64.7 pJ	10.8 pJ	1 pJ
	2011	**12–13**	**13–15**
Coupler	3.5 dB	3 dB	1.5 dB
Waveguide	3 dB	2 dB	2 dB
Mod. insert.	1 dB	1 dB	0.5 dB
Mod. rate	600 Mb/s	5 Gb/s	10 Gb/s
Filter insert.	0.5 dB	0.5 dB	0.5 dB
WDM count	8	32	128
Detector eff.	0.008	0.1	0.5
Laser power per link	44 mW (244 pJ)	1.8 mW (1.2 pJ)	48 µW (16 fJ)
# Fibers	26,674	832	240

References

[1] S. Beamer, K. Asanović, C. Batten, A. Joshi, V. Stojanović, Designing multi-socket systems using silicon photonics, in: Proceedings of the 23rd International Conference on Supercomputing, Yorktown Heights, NY, June 2009, pp. 521–522.

[2] S. Beamer, C. Sun, Y.-J. Kwon, A. Joshi, C. Batten, V. Stojanović, K. Asanović, Re-architecting DRAM with monolithically integrated silicon photonics, in: 37th International Symposium on Computer Architecture (ISCA-37), Saint-Malo, France, June 2010, pp. 129–140.

[3] H. Wei, R.S. Mircea, S. Kevin, S. Karthik, G. Shougata, V. Sivakumar, Compact thermal modeling for temperature-aware design, Proceedings of the 41st Annual Design Automation Conference, ACM, San Diego, CA, USA, 2004.

[4] Z. Xin, A.Q. Huang, A novel distributed control and its tolerance analysis for microprocessor power management, in: Applied Power Electronics Conference and Exposition, 2006, APEC '06. 21st Annual IEEE, 2006, p. 7.

[5] J. Bravo-Abad, E.P. Ippen, M. Soljačić, Ultrafast photodetection in an all-silicon chip enabled by two-photon absorption, Appl. Phys. Lett. 94 (2009) 241103.

[6] D. Brooks, R.P. Dick, R. Joseph, L. Shang, Power, thermal, and reliability modeling in nanometer-scale microprocessors, IEEE Micro 27 (2007) 49–62.

[7] . N. Magen, A. Kolodny, U. Weiser, N. Shamir, Interconnect-power dissipation in a microprocessor. in: Proceedings of the 2004 International Workshop on System Level Interconnect Prediction (SLIP '04), ACM, New York, NY, USA, pp. 7-13.

[8] D. Brooks, V. Tiwari, M. Martonosi, Wattch, ACM SIGARCH Comput. Architect. News 28 (2000) 83–94.

[9] P.E. Gronowski, W.J. Bowhill, R.P. Preston, M.K. Gowan, R.L. Allmon, High-performance microprocessor design, IEEE J. Solid-State Circ. 33 (1998) 676–686.

[10] V. Tiwari, D. Singh, S. Rajgopal, G. Mehta, R. Patel, F. Baez, Reducing power in high-performance microprocessors, in: Proceedings of the ACM/IEEE Design Automation Conference, 1998, pp. 732–737.

[11] T.H. Dunigan, Jr J.S. Vetter, J.B. White, III P.H. Worley, Performance evaluation of the Cray X1 distributed shared-memory architecture, IEEE Micro 25 (2005) 30–40.

[12] R. Espasa, F. Ardanaz, J. Emer, S. Felix, J. Gago, R. Gramunt, I. Hernandez, T. Juan, G. Lowney, M. Mattina, A. Seznec, Tarantula: a vector extension to the alpha architecture, in: Proceedings of the 29th Annual International Symposium on Computer Architecture 2002, 2002, pp. 281–292.

[13] S.R. Alam, J.S. Vetter, An analysis of system balance requirements for scientific applications, in: The 2006 International Conference on Parallel Processing (ICPP 2006), 2006, pp. 229–236.

[14] P.M. Kogge, An exploration of the technology space for multi-core memory/logic chips for highly scalable parallel systems, in: Innovative Architecture for Future Generation High-Performance Processors and Systems, 2005, p. 10.

[15] L. Christophe, S. Jack, F. Jean, J. Norm, The potential energy efficiency of vector acceleration, in: Proceedings of the ACM/IEEE Conference on SC 2006, 2006, p. 1.

[16] X. Li, J. Palma, P. Amestoy, M. Dayd, M. Mattoso, J.O. Lopes, Evaluation of Sparse LU Factorization and Triangular Solution on Multicore Platforms High Performance Computing for Computational Science—VECPAR 2008, vol. 5336, Springer, Berlin/Heidelberg, 2008, pp. 287–300.

[17] J. Chang, S.-L. Chen, W. Chen, S. Chiu, R. Faber, R. Ganesan, M. Grgek, V. Lukka, W.W. Mar, J. Vash, S. Rusu, K. Zhang, A 45 nm 24 MB on-die L3 cache for the 8-core multi-threaded Xeon® Processor, in: Symposium on VLSI Circuits 2009, 2009, pp. 152–153.

[18] S. Rusu, S. Tam, H. Muljono, J. Stinson, D. Ayers, J. Chang, R. Varada, M. Ratta, S. Kottapalli, S. Vora, A 45 nm 8-core enterprise Xeon® processor, IEEE J. Solid-State Circ. 45 (2010) 7–14.

[19] P. Gepner, D. Fraser, M. Kowalik, R. Wyrzykowski, J. Dongarra, K. Karczewski, J. Wasniewski, Evaluating Performance of New Quad-Core Intel XeonÂ® 5500 Family Processors for HPC Parallel Processing and Applied Mathematics, vol. 6067, Springer, Berlin/Heidelberg, 2010, pp. 1–10.

[20] S. Sawant, U. Desai, G. Shamanna, L. Sharma, M. Ranade, A. Agarwal, S. Dakshinamurthy, R. Narayanan, A 32 nm westmere-EX Xeon enterprise processor, in: Digest of Technical Papers—2011 IEEE International Solid-State Circuits Conference (ISSCC), 2011, pp. 74–75.

[21] R. Kuppuswamy, S.R. Sawant, S. Balasubramanian, P. Kaushik, N. Natarajan, J.D. Gilbert, Over one million TPCC with a 45 nm 6-core Xeon® CPU, in: Digest of Technical Papers—2009 IEEE International Solid-State Circuits Conference ISSC, 2009, pp. 70–71,71a.

[22] Defense Advanced Research Project Agency, vol. 2011, 2011. <https://www.fbo.gov/index?s=opportunity&mode=form&id=cbc05c86eb555a334708b570564dddca&tab=core&_cview=0>.

[23] F. O'Mahony, G. Balamurugan, J.E. Jaussi, J. Kennedy, M. Mansuri, S. Shekhar, B. Casper, The future of electrical I/O for microprocessors (2009) 31–34.

[24] J.H. Lau, Low Cost Flip Chip Technologies: For DCA, WLCSP, and PBGA Assemblies, McGraw-Hill, New York, 2000.

[25] M. Datta, T. Osaka, Y. Shacham-Diamand, Microelectronic Packaging Trends and the Role of Nanotechnology Electrochemical Nanotechnologies, Springer, New York, 2010.

[26] T. Tekin, Review of packaging of optoelectronic, photonic, and MEMS components, IEEE J. Sel. Top. Quant. Electron. 17 (2011) 704–719.

[27] P. Harvey, Y. Zhou, G. Yamada, D. Questad, G. Lafontant, R. Mandrekar, S. Suminaga, Y. Yamaji, H. Noma, T. Nishio, H. Mori, T. Tamura, K. Yazawa, Takiguchi, T. Ohde, R. White, A. Malhotra, J. Audet, J. Wakil, W. Sauter, E. Hosomi, Chip/Package design and technology trade-offs in the 65 nm cell broadband engine, in: Proceedings of the 57th Electronic Components and Technology Conference, ECTC '07, 2007, pp. 27–34.

[28] K. Gilleo, Area Array Package Design: Techniques in High-Density Electronics, McGraw-Hill, New York, 2004.

[29] M. Georgas, J.S. Orcutt, R.J. Ram, V. Stojanovic, A monolithically-integrated optical receiver in standard 45-nm SOI, IEEE J. Solid-State Circ. 47 (2012) 1693–1702.

[30] K. Chang, L. Haechang, W. Ting, K. Kaviani, W. Prabhu, W. Beyene, N. Chan, C. Chen, T.J. Chin, A. Gupta, C. Madden, Mahabaleshwara, L. Raghavan, S. Jie, S. Xudong, An 8 Gb/s/link, 6.5 mW/Gb/s memory interface with bimodal request bus, in: IEEE Asian Solid-State Circuits Conference, A-SSCC 2009, 2009, pp. 21–24.

[31] B. Leibowitz, R. Palmer, J. Poulton, Y. Frans, S. Li, J. Wilson, M. Bucher, A.M. Fuller, J. Eyles, M. Aleksic, T. Greer, N.M. Nguyen, A 4.3 GB/s mobile memory interface with power-efficient bandwidth scaling, IEEE J. Solid-State Circ. 45 (2010) 889–898.

[32] J. Poulton, R. Palmer, A.M. Fuller, T. Greer, J. Eyles, W.J. Dally, M. Horowitz, A 14-mW 6.25-Gb/s transceiver in 90-nm CMOS, IEEE J. Solid-State Circ. 42 (2007) 2745–2757.

[33] M. Kossel, C. Menolfi, J. Weiss, P. Buchmann, G. von Bueren, L. Rodoni, T. Morf, T. Toifl, M. Schmatz, A T-coil-enhanced 8.5 Gb/s high-swing SST transmitter in 65 nm bulk CMOS With >16 dB return loss over 10 GHz bandwidth, IEEE J. Solid-State Circ. 43 (2008) 2905–2920.

[34] R. Palmer, J. Poulton, A. Fuller, J. Chen, J. Zerbe, Design considerations for low-power high-performance mobile logic and memory interfaces, in: Proc. IEEE ASSCC, 2008, pp. 205–208.

[35] K. Hu, T. Jiang, J. Wang, F. O'Mahony, P.Y. Chiang, A 0.6 mW/Gb/s, 6.4–7.2 Gb/s serial link receiver using local injection-locked ring oscillators in 90 nm CMOS, IEEE J. Solid-State Circ. 45 (2010) 899–908.

[36] K. Hu, T. Jiang, J. Wang, F. O'Mahony, P.Y. Chiang, A 0.6 mW/Gbps, 6.4–8.0 Gbps serial link receiver using local injection-locked ring oscillators in 90 nm CMOS, in: Symposium on VLSI Circuits 2009, 2009, pp. 46–47.

[37] H. Kangmin, J. Tao, W. Jingguang, F. O'Mahony, P.Y. Chiang, A 0.6 mW/Gb/s, 6.4–7.2 Gb/s serial link receiver using local injection-locked ring oscillators in 90 nm CMOS, IEEE J. Solid-State Circ. 45 (2010) 899–908.

[38] H. Kangmin, J. Tao, S. Palermo, P.Y. Chiang, Low-power 8 Gb/s near-threshold serial link receivers using super-harmonic injection locking in 65 nm CMOS, in: 2011 IEEE Custom Integrated Circuits Conference (CICC), 2011, pp. 1–4.

[39] D.C. Pham, T. Aipperspach, D. Boerstler, M. Bolliger, R. Chaudhry, D. Cox, P. Harvey, P.M. Harvey, H.P. Hofstee, C. Johns, J. Kahle, A. Kameyama, J. Keaty, Y. Masubuchi, M. Pham, J. Pille, S. Posluszny, M. Riley, D.L. Stasiak, Overview of the architecture circuit design and physical implementation of a first-generation cell processor, IEEE J. Solid-State Circ. 41 (1) (2006) 179–196.

[40] P. Kongetira, K. Aingaran, K Olukotun Niagara, A 32-way multithreaded SPARC processor, IEEE Micro (2005) 21–29.

[41] R. Ho, K.W. Mai, M.A. Horowitz, The future of wires, Proc. IEEE 89 (4) (2001) 490–504.

[42] R. Ho, K. Mai, M. Horowitz, Efficient on-chip global interconnects, in: Symposium on VLSI Circuits, Digest of Technical Papers, 2003, pp. 271–274.

[43] D. Schinkel, E. Mensink, E.A. Klumperink, E. van Tuijl, B. Nauta, A 3-Gb/s/ch transceiver for 10-mm uninterrupted RC-limited global on-chip interconnects, IEEE J. Solid-State Circ. 41 (2006) 297–306.

[44] B. Mukherjee, WDM-based local lightwave networks. I. Single-hop systems, Netw. IEEE 6 (3) (1992) 12–27.

[45] B.E. Little, H.A. Haus, J.S. Foresi, L.C. Kimerling, E.P. Ippen, D.J. Ripin, Wavelength switching and routing using absorption and resonance, IEEE Photon. Technol. Lett. 10 (6) (1998) 816–818.

[46] S.J. Emelett, R. Soref, Design and simulation of silicon microring optical routing switches, J. Lightwave Technol. 23 (4) (2005) 1800–1807.

[47] A. Joshi, C. Batten, Y.-J. Kwon, S. Beamer, K. Asanović, V. Stojanović, Silicon-photonic clos networks for global on-chip communication, in: 3rd ACM/IEEE International Symposium on Networks-on-Chip, San Diego, CA, May 2009, pp. 124–133.

[48] C. Dragone, C.H. Henry, I.P. Kaminow, R.C. Kistler, Efficient multichannel integrated optics star coupler on silicon, IEEE Photon. Technol. Lett. 1 (1989) 241–243.

[49] P.D. Trinh, S. Yegnanarayanan, B. Jalali, 5 × 9 integrated optical star coupler in silicon-on-insulator technology, IEEE Photon. Technol. Lett. 8 (1996) 794–796.

[50] K. Okamoto, H. Takahashi, M. Yasu, Y. Hibino, Fabrication of wavelength-insensitive 8 × 8 star coupler, IEEE Photon. Technol. Lett. 4 (1992) 61–63.

[51] F. Liu, D. Patil, J. Lexau, P. Amberg, M. Dayringer, J. Gainsley, H.F. Moghadam, Z. Xuezhe, J.E. Cunningham, A.V. Krishnamoorthy, E. Alon, R. Ho, 10 Gbps, 530 fJ/b optical transceiver circuits in 40 nm CMOS, in: Symposium on VLSI Circuits (VLSIC) 2011, 2011, pp. 290–291.

[52] X. Zheng, D. Patil, J. Lexau, F. Liu, G. Li, H. Thacker, Y. Luo, I. Shubin, J. Li, J. Yao, P. Dong, D. Feng, M. Asghari, T. Pinguet, A. Mekis, P. Amberg, M. Dayringer, J. Gainsley, H.F. Moghadam, E. Alon, K. Raj, R. Ho, J.E. Cunningham, A.V. Krishnamoorthy, Ultra-efficient 10 Gb/s hybrid integrated silicon photonic transmitter and receiver, Opt. Express 19 (2011) 5172–5186.

[53] M. Georgas, J. Orcutt, R.J. Ram, V. Stojanovic, A monolithically-integrated optical receiver in standard 45-nm SOI, in: Proceedings of the ESSCIRC (ESSCIRC) 2011, 2011, pp. 407–410.

[54] T. Takemoto, F. Yuuki, H. Yamashita, T. Ban, M. Kono, Y. Lee, T. Saito, S. Tsuji, S. Nishimura, A 25-Gb/s, 2.8-mW/Gb/s low power CMOS optical receiver for 100-Gb/s Ethernet solution, in: 35th European Conference on Optical Communication ECOC '09, 2009, pp. 1–2.

[55] J. Youn, M. Lee, K. Park, W. Choi, A 10-Gb/s 850-nm CMOS OEIC Receiver with a Silicon Avalanche Photodetector, IEEE J. Quant. Electron. 48 (2) (2012) 229–236.

[56] L. Dongmyung, H. Jungwon, H. Gunhee, P. Sung Min, An 8.5-Gb/s fully integrated CMOS optoelectronic receiver using slope-detection adaptive equalizer, IEEE J. Solid-State Circ. 45 (2010) 2861–2873.

[57] T.S.C. Kao, F.A. Musa, A.C. Carusone, A 5-Gbit/s CMOS optical receiver with integrated spatially modulated light detector and equalization, IEEE Trans. Circuits Syst I: Reg. Papers 57 (2010) 2844–2857.

[58] H. Shih-Hao, C. Wei-Zen, C. Yu-Wei, H. Yang-Tung, A 10-Gb/s OEIC with meshed spatially-modulated photo detector in 0.18-μm CMOS technology, IEEE J. Solid-State Circ. 46 (2011) 1158–1169.

[59] C.T. DeRose, D.C. Trotter, W.A. Zortman, A.L. Starbuck, M. Fisher, M.R. Watts, P.S. Davids, Ultra compact 45 GHz CMOS compatible germanium waveguide photodiode with low dark current, Opt. Express 19 (2011) 24897–24904.

[60] M. Gianlorenzo, C. Giovanni, W. Jeremy, G. Cary, A four-channel, 10 Gbps monolithic optical receiver in 130 nm CMOS with integrated Ge waveguide photodetectors, in: Optical Fiber Communication Conference and Exposition and The National Fiber Optic Engineers Conference, 2007, p. PDP31.

[61] D. Kucharski, D. Guckenberger, G. Masini, S. Abdalla, J. Witzens, S. Sahni, 10 Gb/s 15 mW optical receiver with integrated germanium photodetector and hybrid inductor peaking in 0.13 μm SOI CMOS technology, in: Digest of Technical Papers—2010 IEEE International Solid-State Circuits Conference (ISSCC), 2010, pp. 360–361.

[62] S. Assefa, F. Xia, S.W. Bedell, Y. Zhang, T. Topuria, P.M. Rice, Y.A. Vlasov, CMOS-integrated high-speed MSM germanium waveguide photodetector, Opt. Express 18 (2010) 4986–4999.

[63] G.L. Benjamin, A. Solomon, S. Clint, M.G. William, R. Alexander, A.J. Richard, A.K. Jeffrey, A.V. Yurii, Hybrid-integrated germanium photodetector and CMOS receiver operating at 15 Gb/s, in: CLEO:2011—Laser Applications to Photonic Applications, 2011, p. CFB4.

[64] A. Solomon, G.L. Benjamin, S. Clint, M.G. William, R. Alexander, A.J. Richard, A.V. Yurii, 20 Gbps receiver based on germanium photodetector hybrid-integrated with 90 nm CMOS amplifier, in: CLEO:2011—Laser Applications to Photonic Applications, 2011, p. PDPB11.

[65] X. Zheng, J. Lexau, Y. Luo, H. Thacker, T. Pinguet, A. Mekis, G. Li, J. Shi, P. Amberg, N. Pinckney, K. Raj, R. Ho, J.E. Cunningham, A.V. Krishnamoorthy, Ultra-low-energy all-CMOS modulator integrated with driver, Opt. Express 18 (2010) 3059–3070.

[66] A. Emami-Neyestanak, D. Liu, G. Keeler, N. Helman, M. Horowitz, A 1.6 Gb/s, 3 mW CMOS receiver for optical communication, in: Symposium on VLSI Circuits Digest of Technical Papers 2002, 2002, pp. 84–87.

[67] J. Witzens, G. Masini, S. Sahni, B. Analui, C. Gunn, G. Capellini, 10 Gbit/s transceiver on silicon, in: Silicon Photonics and Photonic Integrated Circuits, Strasbourg, France, 2008, p. 699610.

[68] B. Casper, F. O'Mahony, Clocking analysis, implementation and measurement techniques for high-speed data links—a tutorial, IEEE Trans. Circuits Syst. I: Reg. Papers 56 (2009) 17–39.

[69] F.-T. Chen, M.-S. Kao, Y.-H. Hsu, C.-H. Lin, J.-M. Wu, C.-T. Chiu, and S.-H. Hsu, A 10 to 11.5 GHz rotational phase and frequency detector for clock recovery circuit, in: The IEEE International Symposium on Circuits and Systems (ISCAS) 2011, 2011, pp. 185–188.

[70] J. Savoj, B. Razavi, A 10-Gb/s CMOS clock and data recovery circuit with a half-rate binary phase/frequency detector, IEEE J. Solid-State Circ. 38 (2003) 13–21.

[71] J. Leu, A 9 GHz injection locked loop optical clock receiver in 32-nm CMOS, in: Electrical Engineering and Computer Science. vol. M.S. Cambridge, MA: Massachusetts Institute of Technology, 2010.

[72] M. Georgas, J. Leu, B. Moss, S. Chen, V. Stojanovic, Addressing link-level design tradeoffs for integrated photonic interconnects, in: IEEE Custom Integrated Circuits Conference (CICC) 2011, pp. 1–8.

[73] Mitsubishi, ML7XX11 Series, vol. 2011, 2004.

[74] A. Gubenko, I. Krestnikov, D. Livshtis, S. Mikhrin, A. Kovsh, L. West, C. Bornholdt, N. Grote, A. Zhukov, Error-free 10 Gbit/s transmission using individual Fabry-Perot modes of low-noise quantum-dot laser, Electron. Lett. 43 (2007) 1430.

[75] M. Kanskar, T. Earles, T.J. Goodnough, E. Stiers, D. Botez, L.J. Mawst, 73% CW power conversion efficiency at 50 W from 970 nm diode laser bars, Electron. Lett. 41 (2005) 245–247.

[76] A. Knigge, G. Erbert, J. Jonsson, W. Pittroff, R. Staske, B. Sumpf, M. Weyers, G. Trankle, Passively cooled 940 nm laser bars with 73% wall-plug efficiency at 70 W and 25 °C, Electron. Lett. 41 (2005) 250–251.

[77] L. Lin, L. Guojun, L. Zhanguo, L. Mei, L. Hui, W. Xiaohua, W. Chunming, High-efficiency 808-nm InGaAlA/AlGaAs double-quantum-well semiconductor lasers with asymmetric waveguide structures, IEEE Photon. Technol. Lett. 20 (2008) 566–568.

[78] P. Crump, H. Wenzel, G. Erbert, P. Ressel, M. Zorn, F. Bugge, S. Einfeldt, R. Staske, U. Zeimer, A. Pietrzak, G. Trankle, Passively cooled TM polarized 808-nm laser bars With 70% power conversion at 80-W and 55-W peak power per 100-μm stripe width, IEEE Photon. Technol. Lett. 20 (2008) 1378–1380.

[79] L. Hanxuan, T. Towe, I. Chyr, D. Brown, N. Touyen, F. Reinhardt, J. Xu, R. Srinivasan, M. Berube, T. Truchan, R. Bullock, J. Harrison, Near 1 kW of continuous-wave power from a single high-efficiency diode-laser bar, IEEE Photon. Technol. Lett. 19 (2007) 960–962.

[80] P. Crump, W. Dong, M. Grimshaw, J. Wang, S. Patterson, D. Wise, M. DeFranza, S. Elim, S. Zhang, M. Bougher, J. Patterson, S. Das, J. Bell, J. Farmer, M. DeVito, R. Martinsen, 100-W+ diode laser bars show >71% power conversion from 790-nm to 1000-nm and have clear route to >85%, Proc. SPIE 6456 (2007) 64560M.

[81] D. Botez, L.J. Mawst, A. Bhattacharya, J. Lopez, J. Li, T.F. Kuech, V.P. Iakovlev, G.I. Suruceanu, A. Caliman, A.V. Syrbu, 66% CW wallplug efficiency from Al-free 0.98 μm-emitting diode lasers, Electron. Lett. 32 (1996) 2012–2013.

[82] J. Heerlein, S. Gruber, M. Grabherr, R. Jager, P. Unger, High-efficiency laterally-oxidized InGaAs-AlGaAs single-mode lasers, in: IEEE 16th International Semiconductor Laser Conference, ISLC NARA 1998, 1998, pp. 219–220.

[83] Finisar, 10 Gb Oxide Isolated VCSEL Reliability Report, vol. 2011, 2007.

[84] Intel, Intel Server Board S3420GP Calculated MTBF Estimates, vol. 2011, 2009.

[85] Intel, Intel Server Board S1200BTL Calculated MTBF Estimates, vol. 2011, 2011.

[86] B.M. Hawkins, R.A. Hawthorne III, J.K. Guenter, J.A. Tatum, J.R. Biard, Reliability of various size oxide aperture VCSELs, in: Proceedings of the 52nd Electronic Components and Technology Conference 2002, 2002, pp. 540–550.

[87] CyOptics, Laser & detector chip & TO-can products, vol. 2011, 2011.

[88] C.H. Joyner, J.L. Pleumeekers, A. Mathur, P.W. Evans, D.J.H. Lambert, S. Murthy, S.K. Mathis, F.H. Peters, J. Baeck, M.J. Missey, A.G. Dentai, R.A. Salvatore, R.P. Schneider, M. Ziari, M. Kato, R. Nagarajan, J.S. Bostak, T. Butrie, V.G. Dominic, M. Kauffman, R.H. Miles, M.L. Mitchell, A.C. Nilsson, S.C. Pennypacker, R. Schlenker, R.B. Taylor, T. Huan-Shang, M.F. Van Leeuwen, J. Webjorn, D. Perkins, J. Singh, S.G. Grubb, M. Reffle, D.G. Mehuys, F.A. Kish, D.F. Welch, Large-scale DWDM photonic integrated circuits: a manufacturable and scalable integration platform, in: The 18th Annual Meeting of the IEEE Lasers and Electro-Optics Society, LEOS 2005, 2005, pp. 344–345.

[89] J.S. Levy, A. Gondarenko, M.A. Foster, A.C. Turner-Foster, A.L. Gaeta, M. Lipson, CMOS-compatible multiple-wavelength oscillator for on-chip optical interconnects, Nat. Photon. 4 (2010) 37–40.

[90] S. Beamer, C. Sun, Y.-J. Kwon, A. Joshi, C. Batten, V. Stojanovic, K. Asanovic, Re-architecting DRAM memory systems with monolithically integrated silicon photonics, SIGARCH Comput. Archit. News 38 (2010) 129–140.

[91] C. Batten, A. Joshi, J. Orcutt, A. Khilo, B. Moss, C. Holzwarth, M. Popovic, H. Li, H. Smith, J. Hoyt, F. Kaertner, R. Ram, V. Stojanović, K. Asanovic, Building manycore processor to DRAM networks with monolithic CMOS silicon photonics [Invited], IEEE Micro 29 (4) (2009) 8–21.

[92] R. Kumar, G. Hinton, A family of 45 nm IA processors, in: Digest of Technical Papers—2009 IEEE International Solid-State Circuits Conference ISSCC, 2009, pp. 58–59.

[93] A. Shacham, K. Bergman, L.P. Carloni, On the design of a photonic network-on-chip, Proceedings of the First International Symposium on Networks-on-Chip, IEEE Computer Society, 2007 53–64

[94] P. Koka, M.O. McCracken, H. Schwetman, X. Zheng, R. Ho, A.V. Krishnamoorthy, Silicon-photonic network architectures for scalable, power-efficient multi-chip systems, SIGARCH Comput. Archit. News 38 (2010) 117–128.

[95] W. Howard, M. Petracca, A. Biberman, B.G. Lee, L.P. Carloni, K. Bergman, Nanophotonic optical interconnection network architecture for on-chip and off-chip communications, in: Optical Fiber communication/National Fiber Optic Engineers Conference, OFC/NFOEC 2008, 2008, pp. 1–3.

[96] A.K. Kodi, R. Morris, A. Louri, Z. Xiang, On-chip photonic interconnects for scalable multi-core architectures, in: 3rd ACM/IEEE International Symposium on Networks-on-Chip 2009, NoCS 2009, 2009, p. 90.

[97] J.S. Orcutt, B. Moss, C. Sun, J. Leu, M. Georgas, J. Shainline, E. Zgraggen, H. Li, J. Sun, M. Weaver, S. Urosevic, M. Popovic, R.J. Ramand, V. Stojanovic, An open foundry platform for high-performance electronic-photonic integration, Opt. Express 20 (11) (2012) 12222–12232.

Hybrid Silicon Lasers

13

Brian R. Koch[a], Sudharsanan Srinivasan[b], and John E. Bowers[b]

[a]Aurrion, Inc., Goleta, CA, 93117 USA,
[b]University of California, Santa Barbara, Department of Electrical and Computer Engineering, Santa Barbara, CA, 93106 USA

13.1 INTRODUCTION TO HYBRID SILICON LASERS

Silicon photonics has long been viewed by many researchers as a way of making inexpensive photonic components by leveraging the mature infrastructure of CMOS fabrication [1–13]. The envisioned levels of integration range from simply making high-quality passive components on silicon to more complex schemes combining silicon passives, actives, and CMOS electronics all on the same chip. There are many factors that determine the value of silicon photonics, including the proposed applications, expected volumes of products, the manufacturing technology, and the integration schemes. At least in some cases the value is now apparent and silicon photonics is gaining attention outside the academic realm. Its use in products could soon become widespread.

One of the most difficult parts of silicon photonics has consistently been realizing the light source. A significant amount of high-quality research has been put into generating light from silicon [13–17], doped silicon materials [18–24], and silicon germanium [25–28]. Unfortunately to date none of these have succeeded in generating an electrically pumped laser with performance comparable to III–V materials. In a world where power efficiency is becoming more and more important, the benefit of a potentially less expensive all-silicon solution is outweighed by the significant performance improvements offered by III–V materials. The solution in making a silicon photonics transmitter then becomes essentially a packaging and/or integration problem. Various methods have been proposed and some clever methods are even in use in products today [29–33]. These methods typically involve using pre-fabricated lasers or gain blocks made out of III–V materials and somehow placing them onto separately fabricated silicon photonics chips. This method can work well but is not ideal because these blocks must be carefully placed, and usually one at a time. Any misalignment from nominal placement results in excess loss (or worse) at the interface between the laser and silicon chip. Thus this alignment step usually requires expensive equipment that is also relatively slow, particularly for high volume manufacturing. Ultimately a wafer-scale processing technique is necessary.

Optical Fiber Telecommunications VIA. http://dx.doi.org/10.1016/B978-0-12-396958-3.00013-5

This is where the integration methods of hybrid silicon lasers offer an advantage. These lasers are made by placing III–V chips onto pre-patterned silicon wafers with rough alignment. Subsequently, the alignment marks on Si are used to precisely align the III–V processes to the pattern on Si, using stepper lithography. This allows for a wafer-scale process for all lasers in this platform. The intimate optical contact at the silicon/III–V interface and the precise alignment can allow for very low losses, offering superior performance to other laser sources. In this chapter we will discuss the design, important fabrication methods, and several specific examples of demonstrated hybrid silicon lasers.

13.2 DESIGN OF HYBRID SILICON LASERS

In this section we will discuss some very basic laser fundamentals and general concepts associated with designing hybrid silicon lasers. Specific cavity designs along with the associated experimental results will be covered in a later section.

A laser consists of a gain medium and an optical cavity that provides feedback into the gain section. Light generated by the gain medium resonates inside the cavity and lasing occurs when the amount of gain becomes equal to the loss (light emitted out of the cavity plus any losses inside the cavity). In a hybrid silicon laser the gain medium is a semiconductor material (typically III–V compounds although II–VI and other materials are possible) that is placed in intimate contact with a light guiding waveguide in silicon. The light is generated in the semiconductor material and coupled to the silicon waveguide via a hybrid optical mode that exists in both the III–V and silicon. Essential to the hybrid silicon laser is the taper region in which the light is effectively transferred from the III–V or hybrid mode completely into the silicon waveguide. This allows for transfer of light from the gain section to the mirrors or waveguide loop that forms the optical cavity in the silicon. In general the concepts associated with hybrid silicon laser theory have been developed from previous works: theory of optical gain, laser cavities, optical gratings or other cavity mirrors, and coupled or twin waveguides. Here we will cover some of the more unique aspects of these principles as applied to hybrid silicon lasers.

13.2.1 Hybrid modes

A critical concept in hybrid silicon lasers is the hybrid mode, which is an optical mode that exists partially and simultaneously in different materials. In fact, this occurs in many semiconductor lasers including more conventional semiconductor quantum well lasers which can consist of many different layers of different materials. The optical mode is confined in the transverse direction within several different layers simultaneously as determined by the indices of refraction in the different vertical layers. The lateral photon confinement is achieved by patterned etching and deposition. The optical mode tends to exist in regions of higher refractive index. Therefore, designing a laser requires forming an area of high refractive index, with controlled optical gain, surrounded on all sides by areas of lower refractive index.

Thus to form a hybrid silicon laser we must design the III–V epi structure, silicon waveguide, and possibly etched regions such that the optical mode can exist simultaneously in the III–V material and silicon as shown in Figure 13.1. In practice the exact shape and position of the mode can be highly dependent on the indices of refraction for the different materials. Simulations of the mode profiles can be carried out using the film mode matching (FMM) technique. Because of the mode shape's high sensitivity to the indices of refraction, it is also possible to dramatically change the laser properties by simply changing the underlying silicon waveguide width [35–38] as shown in Figure 13.2. Essential laser properties such

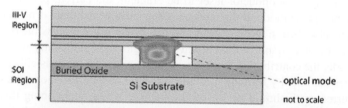

FIGURE 13.1 A hybrid optical mode in a hybrid silicon laser structure [34]. © Fang, 2008.

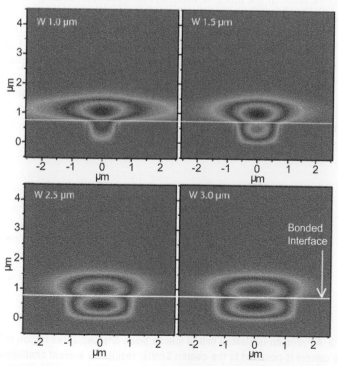

FIGURE 13.2 Hybrid optical modes for different widths of silicon waveguide from 1 to 3 μm, while all other aspects of the structure are kept fixed [34]. © Fang, 2008.

as the confinement factor in the quantum wells (and therefore optical gain), optical loss, and injection efficiency (as determined by the optical mode's overlap with the current path) can all be altered from these changes in the silicon waveguide width. This can be valuable in integrated devices such as hybrid silicon amplifiers, photodetectors, and modulators each with ideal confinement factors that are very different from the laser.

13.2.2 Current confinement and flow

In order for any semiconductor laser to be efficient, the current path must overlap well with the optical mode. If the mode exists in areas where there is no current flow, light will be absorbed in those regions and will make lasing impossible or very inefficient. On the other hand, if the current flow region is significantly wider than the optical mode, the contribution from the current outside the mode is either wasted or destabilizes the laser by forming multiple lateral modes. Furthermore, the resistance for the current path should be as low as possible while maintaining this efficient overlap. Defining the current path can be done in different ways, including etching a laser's sidewalls to remove material where the mode does not exist or implanting hydrogen to the sides of the mode in order to make the surrounding material electrically insulating, which is shown in Figure 13.3.

The hybrid lasers demonstrated to date have all used p-type material at the top of the III–V structure and a relatively thin n-type layer at the bottom. Current flows down from a p-contact positioned above the optical mode, through quantum wells and then out to the sides through the n-layer to the n-contact(s) which are positioned to the side of the laser. This structure has been used to minimize the thickness of the lateral current conducting layer resulting in the highest optical confinement factor and minimum interaction of the mode with p-type InP material, which has higher optical loss than its n-doped counterpart.

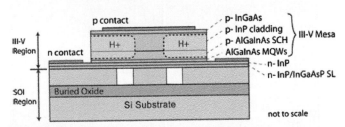

FIGURE 13.3 A hydrogen implanted hybrid silicon laser. The implanted region is highly resistive so current is confined to the center. Similar results for current confinement can be obtained by simply removing the material, but this has other important implications in the laser design [34]. © Fang, 2008.

13.2.3 **Hybrid laser tapers**

In order to take advantage of the benefits offered by silicon photonics, it is essential to transfer the laser light from the hybrid laser mode completely into the silicon waveguide layer. In fact for most hybrid laser designs, the laser light is transferred to the silicon waveguide (and back) within the laser cavity, so any loss from this transition leads to both higher threshold current and lower differential quantum efficiency. Therefore, it is extremely important to minimize any losses associated with this transfer. A common way of doing this is by including a taper. (A transition statement to the next paragraph is necessary here).

To create efficient tapers, we can use some of the general principles discussed in the previous section along with other techniques. For example, changing the width of the silicon waveguide below the III–V material can change the effective index and draw more of the optical mode into the silicon [35–38]. Further transfer can be provided by subsequently or simultaneously tapering the width of III–V laser section as well. It is also possible to use a grating-based coupler to couple between the III–V and silicon [39–41]. More specific explanations are provided below.

In some structures the silicon waveguide and III–V can be treated as coupled waveguides and the correct coupling length can be determined to allow for efficient transfer from the III–V waveguide to the silicon waveguide [38,42,43]. Although it is possible to achieve very short coupling lengths using resonant coupling, such a taper is a very sensitive structure so normally adiabatic coupling is used to transfer power from one waveguide to the other [38].

Figure 13.4 is a schematic showing the coupling between the III–V material and the silicon waveguide. Two modes of the uncoupled waveguides have phase mismatch δ and amplitude spatial overlap k which are given by [38,42]:

$$\delta = \frac{(\beta_b + M_b) - (\beta_a + M_a)}{2}$$

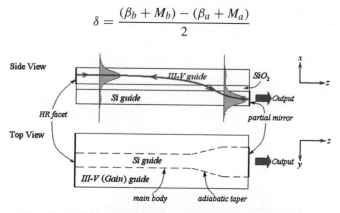

FIGURE 13.4 Schematic of the power transfer between the III–V material and silicon waveguide in a hybrid silicon supermode laser. The structure consists of a III–V guide, and bonding layer (SiO$_2$ in this case), and a silicon guide. As the vertical and lateral dimensions of these elements change the refractive index profiles change and thus the distribution of energy changes [38]. © OSA 2008.

and

$$k = (k_{ab}k_{ba})^{1/2},$$

where

$$k_{ab} = \frac{\omega \varepsilon_o}{4} \iint_\infty \left[n_c^2(x,y) - n_b^2(x,y) \right] \varepsilon^{(a)}(x,y) \varepsilon^{(b)}(x,y) dx \, dy$$

and

$$k_{ba} = \frac{\omega \varepsilon_o}{4} \iint_\infty \left[n_c^2(x,y) - n_a^2(x,y) \right] \varepsilon^{(a)}(x,y) \varepsilon^{(b)}(x,y) dx \, dy.$$

β_a and β_b are propagation constants of the modes in the uncoupled waveguides, M_a and M_b are small corrections in β_a and β_b caused by the presence of the other waveguide. $\varepsilon^{(a)}(x,y)$ and $\varepsilon^{(b)}(x,y)$ are the power normalized modal distributions of the uncoupled waveguides. $n_a(x,y)$, $n_b(x,y)$, and $n_c(x,y)$ are the refractive index profiles of the uncoupled III-V and Si waveguides "a" and "b" and coupled waveguide system "c," respectively. Thus we can see that by adjusting the refractive index profiles, we can control δ and k, which controls the distribution of energy in the two waveguides.

The results for the simulation of a hybrid laser taper are shown in Figure 13.5. In this simulation the silicon waveguide varies from 0.51 to 0.89 μm over the length of the taper (which is varied). For very short taper lengths a resonant coupling is achieved at around 20 μm, but as we mentioned earlier this is very sensitive so it is

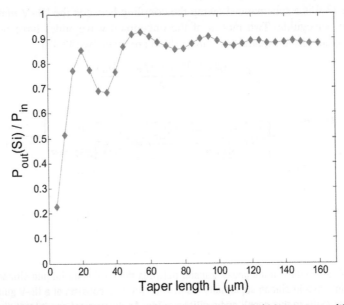

FIGURE 13.5 Power coupling ratio versus taper length for a hybrid laser taper with a silicon waveguide width varied from 0.51 to 0.89 μm exponentially over the length of the taper [38]. © OSA 2008.

risky to rely on this being repeatable. For longer lengths of ~50 μm the adiabatic coupling is reached with 90% efficiency. Using longer lengths is not advantageous in this particular design. A later theoretical study [43] showed that the shortest adiabatic taper for a given system can be formed using a taper with shape given by:

$$\tan\left[\arcsin\left(2k\epsilon^{\frac{1}{2}}(z - z_0)\right)\right],$$

where z is the distance in the direction of propagation and z_0 is the phase matching point corresponding to $\delta k = 0$. Figure 13.6 shows a comparison of different taper functions, showing that the above-mentioned design is the shortest.

13.2.4 Other important concepts

One potential advantage of fabricating lasers on silicon substrates is that the thermal conductivity of silicon is significantly higher than most compound semiconductor laser materials. However, hybrid silicon lasers are made using a buried silicon diox-ide layer beneath the relatively thin silicon epi layer. Since the buried oxide layer has relatively poor thermal conductivity, utilizing the Si substrate for heat dissipation becomes challenging and hence compromises the device thermal performance. One approach is to make thermal "shunt" paths through this buried oxide layer using metal or polysilicon shunts. Performance improvements in long lasers have been minimal; however, improvements in small lasers, for example microring lasers, were signifi-cant [44,45].

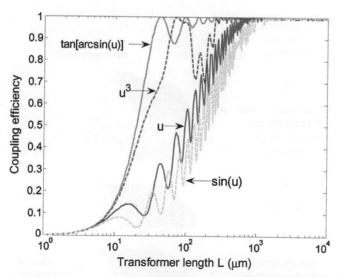

FIGURE 13.6 Coupling efficiency versus length for several different taper design functions [43]. © OSA 2009.

13.3 WAFER BONDING TECHNIQUES AND FABRICATION

Perhaps the largest technological challenge associated with making hybrid silicon lasers is the ability to place the gain material in intimate contact with the silicon waveguides. In this section we will discuss the different methods that are used to bond III–V materials to silicon, including direct bonding and adhesive bonding.

13.3.1 Direct wafer bonding

Although high-quality direct wafer bonding is a technology that has been proven highly manufacturable and has been commercially available since the early 2000s, there are considerable challenges associated with bonding highly dissimilar materials such as compound III–V semiconductors and silicon. There is a large thermal expansion coefficient mismatch that limits the temperature that can be used in bonding [46].

More recently an O_2 plasma-assisted wafer bonding technique has been developed to avoid this problem [47,48]. The process typically used to make III–V/silicon hybrid devices is shown in Figure 13.7. The process involves rigorous sample

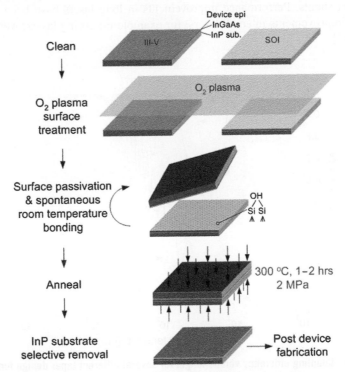

FIGURE 13.7 The O_2 plasma-assisted direct bonding procedure [49]. © Hindawi 2008.

cleaning, native oxide removal using HF and NH_4OH solutions, followed by an O_2 plasma step that converts the hydrophobic sample surfaces to hydrophilic via the growth of a thin native oxide on the surfaces. This step results in a high density of -OH groups after contacting H_2O based solutions. The two surfaces are placed in contact and held by Van der Waals forces or hydrogen bonds [50], and then the chemical reactions given in the equations below form covalent bonds [51]:

$$Si - OH + M - OH \rightarrow Si - O - M + H_2O(g), \, Si + 2H_2O \rightarrow SiO_2 + 2H_2(g).$$

In these equations M stands for metal with high electronegativity such as the group III and V elements, and (g) stands for gas. This process is accelerated using annealing at elevated temperature (typically ~300 °C). The bonding process results in a ~5–15 nm thick oxide layer between the final bonded III–V and silicon materials.

The gas byproducts H_2O and H_2 can accumulate and cause voids at the bonding surface as shown in Figure 13.8a and b, and these can even cause wafer delamination. Removing these gases is therefore a critical step. One method to reduce the accumulation of bonding outgasses is to utilize a layer of amorphous dielectric at the interface to absorb the gases. However, this causes tradeoffs between void formation and optical/ thermal performance [53], so the layer must be kept thin—typically 30 nm of SiO_2 on each side for a total thickness of 60 nm. The results of this bonding method are quite promising. A statistical study showed that the average void density is less than $5\,cm^{-2}$, which is actually believed to be due to surface defects in the III–V epilayer rather than outgassing. In contrast to the images in Figure 13.8, the images in Figure 13.9 show that the bonding voids have been effectively eliminated. When bond strength is measured using a conventional crack-opening method, failure occurs within the bulk III–V consistently, indicating that the bond strength is larger than the InP fracture energy of $0.63\,J/m^2$ [54]. When bonded samples are diced with a 10,000 rpm dicing blade, the III–V layer fragments no more than 15 μm from the edge, which is another indication of strong and uniform bonding.

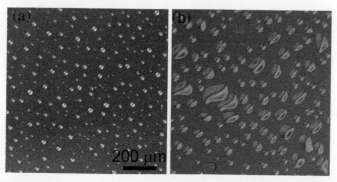

FIGURE 13.8 Nomarski mode microscope images of InP thin epitaxial layers directly bonded to SOI after a 2 h anneal at 300 °C, showing a large number of voids. (a) and (b) are from different SOI and III–V vendors, with the same scale in both images [52]. © Wiley 2010.

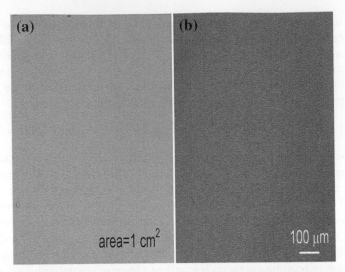

FIGURE 13.9 (a) Photograph of a 1 cm^2 area of III–V thin epi material transferred to SOI. (b) Closeup of (a) showing that the interface is void free at the microscopic scale as well [52]. © Wiley 2010.

Another method to mitigate the effects of outgassing is to place arrays of vertical outgassing channels (VOCs) on the SOI material before bonding [55]. The operation principle of these VOCs is shown in Figure 13.10. These holes can range in size and spacing but are typically a few microns in diameter with a spacing of around $50 \mu m$. The holes are etched through the silicon waveguide layer so that the buried oxide (BOX) layer is exposed. Any gases generated during bonding can flow to these regions and then diffuse into the buried oxide and eventually out of the structure. Bonded samples using this technique show similar performance to the previously mentioned method.

Similar bonding methods utilizing direct bonding have been used that allow current to flow across the III–V/silicon interface [50]. This expands the existing technology to design novel electrically pumped lasers [56].

13.3.2 Adhesive bonding

Another low temperature bonding technique involves using an adhesive material as a thin layer to effectively bond the two materials of interest together [57]. Several materials have been used as the adhesive layer, such as epoxies, spin-on-glasses, and polyimides. With due considerations of optical and thermal properties and bond strength, the most promising material was determined to be divinyl-tetramethyldisiloxane-benzocyclobutene (DVS-BCB) [52].

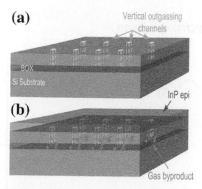

(a)

Vertical outgassing channels

BOX
Si Substrate

InP epi

(b)

Gas byproduct

FIGURE 13.10 Operation principle of vertical outgassing channels (VOCs) for direct wafer bonding. (a) The holes are etched through the silicon epi layer down to the buried oxide before bonding. (b) During and after bonding, the gas byproducts can diffuse into the holes and out through the buried oxide [52]. © Wiley 2010.

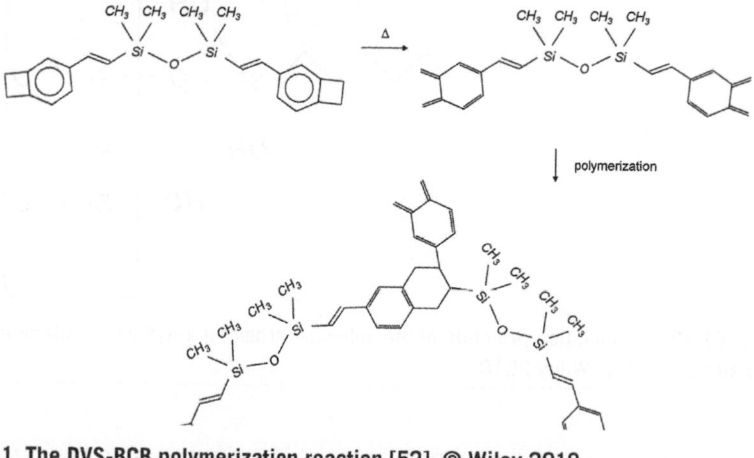

FIGURE 13.11 The DVS-BCB polymerization reaction [52]. © Wiley 2010.

The DVS-BCB monomer polymerization reaction is shown in Figure 13.11. The molecule is symmetrical consisting of a silicon backbone with two benzocyclobutene rings. The monomer is partially cured to form an oligomer by adding mesitylene. This solvent also affects the thickness of the DVS-BCB layer (to be discussed more later). Upon curing, the benzocylobutene ring opens to form o-quinodimethane that is highly reactive and undergoes a Diels-Alder reaction with an available vinylsiloxane group to form the structure shown in Figure 13.12 [58,52]. No byproducts are formed during this reaction. In addition to this, adhesive promoters such as vinyltrihydroxysilane interact with free hydroxyl groups on an oxidized silicon surface, as shown in Figure 13.12.

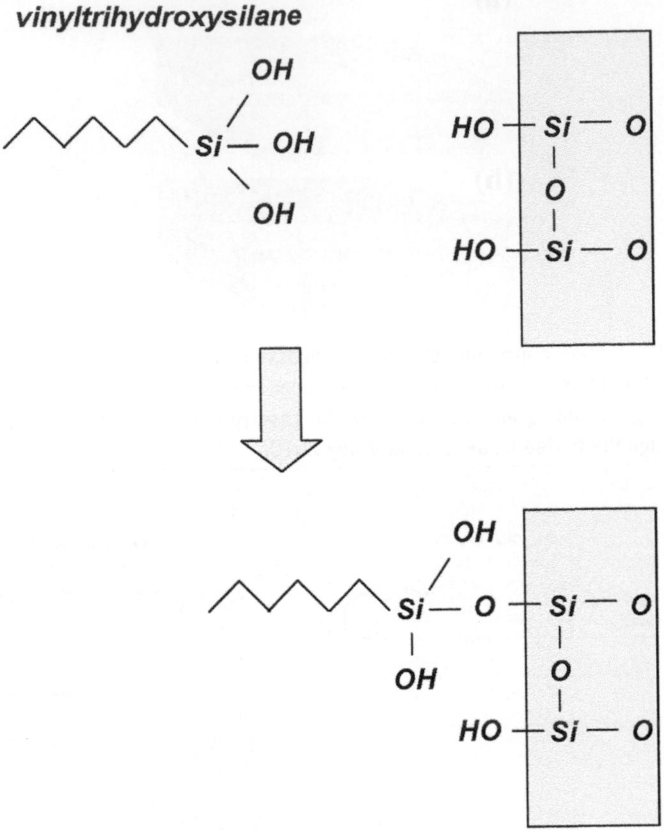

FIGURE 13.12 Operation principle of the adhesion promoter vinylhydroxysilane for DVS-BCB on silicon [52]. © Wiley 2010.

For adhesive bonding with BCB, the samples must be cleaned in order to remove particles from the surface and to remove hydrocarbon contamination. Hydrocarbons and organic materials can be removed from SOI using a Piranha solution (warm $3H_2SO_4:H_2O_2$). To clean the III–V surface, a pair of sacrificial InP/InGaAs layers are removed via selective wet etching with $3HCl:H_2O$ and $H_2SO_4:3H_2O_2:H_2O$. Next adhesive promoter AP3000 is applied to the SOI surface by spin coating, and then the DVS-BCB is deposed on the SOI surface also by spin coating. A 150°C anneal for 1 min is used to evaporate the remaining mesitylene solvent (used for thinning, discussed later), and to reflow the DVS-BCB for planarization. Next the III–V is placed on the SOI by tweezers or potentially by a commercial wafer bonder in a clean room environment, and then the DVS-BCB film is cured in an atmosphere of less than 100 ppm oxygen (in a nitrogen chamber) to prevent oxidation of the DVS-BCB.

The thickness of the DVS-BCB layer can be controlled by varying the amount of mesitylene that is added to the adhesive before bonding. Figure 13.13(a) shows the layer thickness as a function of added mesitylene and (b) shows images of the results for two different bonding thicknesses [52].

13.3.3 Fabrication

Prior to bonding of hybrid silicon lasers, typically all-silicon processing is completed. This may include silicon etches to form the rib waveguides and etches all the way down to the buried oxide. In the more complex photonic integrated circuits, complete silicon modulators and multiplexers may also be formed prior to bonding [59–61]. This can include many more silicon steps, such as implantations for doping the silicon and metallization. It is important to note that in any region where the III–V material is to be bonded, the surface must be kept perfectly planar.

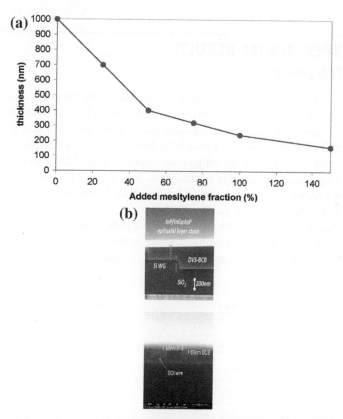

FIGURE 13.13 (a) Influence of DVS-BCB dilution with mesitylene on the bonding layer thickness. (b) Cross-sectional device images of structures having different bonding layer thicknesses achieved by DVS-BCB bonding [52]. © Wiley 2010.

Bonding can be done using whole III–V wafers to whole silicon wafers, or pieces of III–V onto whole silicon wafers using the techniques described previously in this section. The latter can be preferable due to more efficient usage of III–V material, but this depends on the devices being fabricated. (That is, PICs with large silicon-only areas do not require a lot of III–V material.) During the bonding step, only rough alignment of the silicon and III–V is required. This is because the marks on the silicon die can be used for reference for all of the III–V process steps that occur after the bonding step. Thus the only penalty for misalignment is wasted III–V material.

After the bonding is completed, the III–V substrate is removed, and after that point the III–V processing steps can be largely the same as those used for conventional III–V lasers. These include wet and dry etches of the laser cladding material, QW material, and etches to the contacts, metal contact deposition, hydrogen implantation for current channel formation, and probe metal deposition. Some of the devices require more specialized procedures, which can be found in the references.

13.4 EXPERIMENTAL RESULTS

13.4.1 Fabry-Perot lasers

The first demonstrations of hybrid silicon lasers had Fabry-Perot cavities, first optically pumped [35,62]; providing a proof of concept that a hybrid III–V/silicon mode could lase. Soon after electrically pumped lasers were demonstrated [63,64]. The simplest cavities were formed completely by hybrid optical modes formed by bonding III–V material onto straight silicon waveguides on an SOI wafer. The mirrors are formed by polishing the edge of the chips. A schematic cross-sectional view of the facet of a hybrid silicon electrically pumped laser is shown in Figure 13.14. The lasers emitted light at approximately 1570 nm.

Since the first demonstration of Fabry-Perot hybrid silicon lasers, many more demonstrations have been made using different methods. A 1310 nm laser was demonstrated with a threshold current of 30 mA, maximum output power of 5.5 mW, and

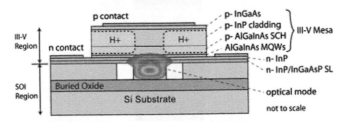

FIGURE 13.14 Facet cross-section of an electrically pumped hybrid silicon Fabry-Perot laser. This is also what the hybrid waveguide section of many other hybrid laser structures looks like [64]. © OSA 2006.

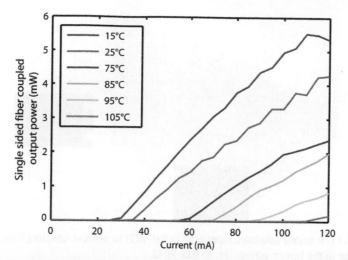

FIGURE 13.15 Continuous-wave single-sided output power at different temperatures for a 1310 nm hybrid silicon Fabry-Perot laser [65]. © OSA 2007.

was capable of continuous-wave (CW) lasing up to 105 °C [65]. Results are shown in Figure 13.15. More recently, 1310 nm Fabry-Perot lasers were demonstrated on the DVS-BCB bonding platform [66]. Continuous-wave lasing results of these lasers are shown in Figure 13.16.

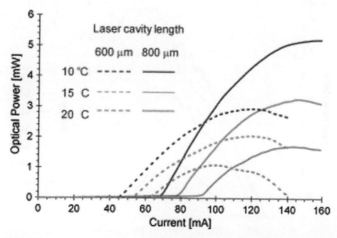

FIGURE 13.16 CW lasing results from a 1310 nm FP laser fabricated using adhesive bonding [66]. © IEEE 2011.

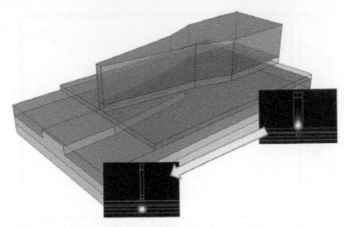

FIGURE 13.17 A double adiabatic taper used in Ref. [67] to achieve coupling from the III–V gain section to the silicon waveguide. © IEEE 2012.

Another type of Fabry-Perot laser that includes tapers to the silicon waveguide was developed in Refs. [63,67]. In this structure the light is transferred from the laser gain section to the silicon waveguides via a taper section and then the reflection occurs at the silicon only facet. The taper design schematic from Ref. [67] is shown in Figure 13.17. It consists of a simultaneous tapering of the silicon waveguide and the III–V waveguide above, resulting in a simulated coupling efficiency of 80%.

This laser had a lasing wavelength of ~1590 nm and a threshold of 30 mA at 20 °C. The light output and voltage versus current (LIV curve) are shown in Figure 13.18.

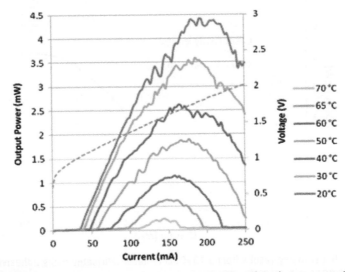

FIGURE 13.18 LIV curve for a Fabry-Perot laser with double adiabatic tapers used to transfer from the III–V gain section to the silicon waveguide within the cavity [67]. © IEEE 2012.

13.4.2 Ring lasers

Unfortunately the use of Fabry-Perot lasers in silicon photonics is severely limited by the fact that the laser mirrors are formed by the edge of the chip. Therefore the light emitted by the laser cannot be sent into other components on the silicon chip. One way to make a laser that can be integrated with other silicon photonic devices is to use a ring or racetrack cavity as shown in Figure 13.19. The cavity consists of a bus waveguide and a racetrack loop waveguide that are coupled together by a directional coupler. Depending on the length of the coupler and gap between the waveguides, a certain amount of light is coupled from one waveguide to the other and vice versa. The loop contains III–V gain material that is pumped electrically and some of the light generated is allowed to exit the loop through the directional coupler. In this cavity design the directional coupler controls the mirror loss-less coupling between the ring and bus waveguides keeps more power in the laser but means less light is emitted leading to lower threshold currents but lower output powers as well.

The results for these lasers varied significantly depending on the ring radius for the waveguide bends and the length of the directional coupler. One of the best designs was a laser with $200\,\mu m$ radius bends and total cavity length of $2656\,\mu m$. This device had a threshold current of $175\,mA$, and differential quantum efficiency of 17%. Figure 13.20 shows the continuous-wave lasing results for this device at different temperatures. In this first demonstration, the entire area shown in Figure 13.19 was covered with III–V material that had to be pumped (except for the photodetectors which were electrically isolated and reverse biased to absorb light). With the use of tapers from the III–V to silicon, shorter gain sections could be utilized to enable improved efficiency.

One problem with ring cavity lasers is that lasing tends to occur in both the clockwise and counter-clockwise directions, with unpredictable switching in

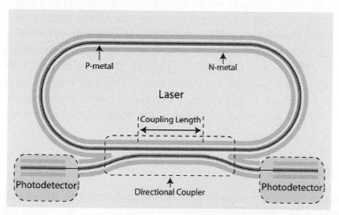

FIGURE 13.19 Layout of a racetrack laser with integrated photodetectors [68]. © OSA 2007.

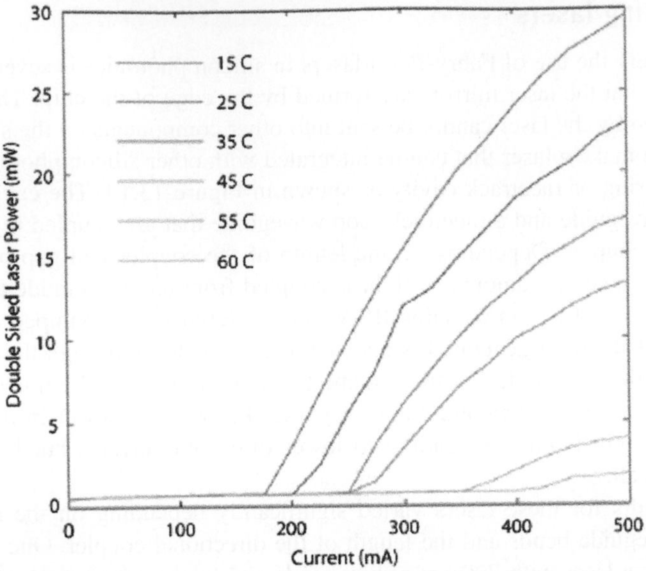

FIGURE 13.20 Continuous-wave lasing results of a racetrack hybrid silicon laser for different temperatures [68]. © OSA 2007.

the power distribution between the two. This means that the light emitted from one side is both variable over time and inherently inefficient in use of the electrical driving power. This can be avoided by injecting a small amount of light into one side to induce lasing out of the other side, but is undesirable due to the excess power required. Another disadvantage is that multiple wavelengths of light are emitted simultaneously in a somewhat uncontrolled fashion. As with the Fabry-Perot laser, the only wavelength selective element in a ring laser is the III–V gain medium (with the exception of the directional coupler which usually has some broad wavelength selectivity). Since the gain medium has a relatively wide spectrum compared to the cavity mode spacing, many wavelengths are usually emitted. (This is not the case for very small rings that have wide cavity mode spacing—these will be discussed later.) This can lead to problems with propagation in fiber due to dispersion, noise due to mode competition in the laser, and this also limits the use of these lasers for wavelength division multiplexing applications.

13.4.3 DBR lasers

One of the most common ways to achieve single wavelength lasing is to use gratings to form the mirrors of the laser cavity. This is called a distributed Bragg reflector (DBR) laser. The gratings can be designed to have an appropriate reflectivity spectrum so that a single cavity mode is selected to lase. Figure 13.21 shows the structure of a hybrid silicon DBR laser [69]. The laser consists of a gain region, two taper regions, a back DBR mirror, a front DBR mirror, and passive output waveguides. The magnitude of a DBR mirror's reflectivity (R, in power) is determined by the length (L) of the grating and the grating strength (κ) via the equation $R = \tanh^2(\kappa * L)$. The wavelength selectivity (optical bandwidth) of the grating is also determined by these factors. By changing the grating strength and length it is possible to achieve any combination of grating bandwidth and reflectivity, although some grating strengths and lengths are impractical or impossible to achieve in practice.

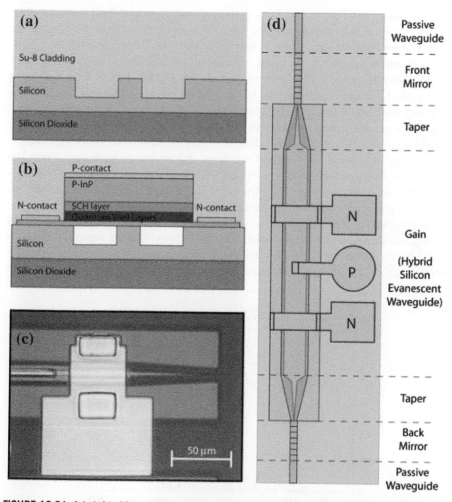

FIGURE 13.21 A hybrid silicon DBR laser [69]. (a) Cross-section of the device in the silicon only section. (b) Cross-section of the device in the hybrid gain section. (c) Top-down microscope image of the gain and taper region. (d) Top-down schematic of the device. © IEEE 2008.

For this laser the grating strength was $80\,\mathrm{cm}^{-1}$ as determined by the optical mode's interaction with the grating, which was etched 25 nm into the top surface of the silicon waveguide with a duty cycle of 75%. The grating strength can be estimated from the following parameters: the grating period (Λ), refractive index difference (Δn) and the averaged refractive index ($\langle n \rangle$) between the etched and unetched regions, the grating order (m), and the duty cycle (D) of the grating:

$$\kappa = \frac{1}{\Lambda}\frac{\Delta n}{\langle n \rangle}\sin(m\pi D).$$

The reflectivity of a 300 μm long back mirror and a 100 μm long front mirror are 97% and 44%, respectively. This means that 96% of the output power from the laser can be emitted from the front side, which is another feature that makes the DBR laser more practical than ring and FP designs (and conventional DFB designs to be discussed later).

The DBR laser had a 440 μm long gain section and two 80 μm long tapers that are formed by linearly narrowing the III–V mesa region, and additional tapers for the quantum wells and n-layer. These tapers were designed to provide an adiabatic transition between the hybrid and passive waveguide regions. These tapers had a measured loss of $\sim 1.2\,\mathrm{dB}$ at lasing wavelength. Figure 13.22 is a plot of the light output and voltage versus current for the DBR laser in continuous-wave operation. The device had an output power of 11 mW at 15 °C, with a threshold current of 65 mA and differential efficiency of 15%. The kinks in the light versus current curves are due to longitudinal mode hops in the laser. In fact, these mode hops also cause the kinks observed in the voltage versus current curve. This can be eliminated by the appropriate design of the DBR mirror spectrum to provide more smooth transitions as the cavity modes move in the wavelength domain relative to the grating reflectivity spectra. The side mode suppression ratio of this laser is $\sim 50\,\mathrm{dB}$ as shown in

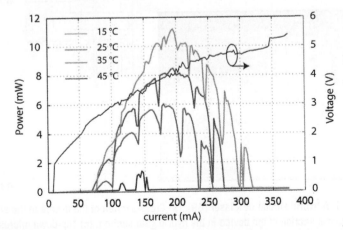

FIGURE 13.22 Light and voltage versus current for the DBR hybrid laser [69]. © IEEE 2008.

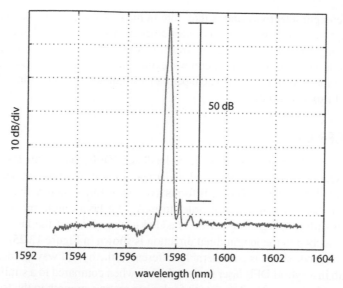

FIGURE 13.23 Optical spectrum of the single mode DBR laser, with approximately 50 dB of side mode suppression [69]. © IEEE 2008.

Figure 13.23. Current modulation on these lasers showed open eye diagrams for data rates up to 4 GHz.

Another DBR laser was demonstrated using a deeply etched DBR grating design in the silicon waveguide. The back mirror was designed for 90% reflectivity and a single slot was etched out to form a 50% reflective front mirror [70]. Since the gratings were short and deeply etched, the laser (shown in Figure 13.24) has much

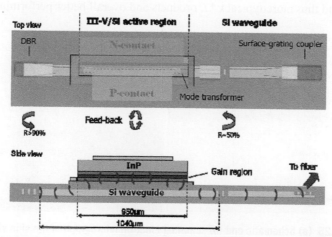

FIGURE 13.24 A hybrid silicon laser DBR with deeply etched gratings [70]. © OSA 2011.

lower wavelength selectivity and is therefore in many ways similar to a Fabry-Perot laser, except that it can be integrated with other components. This laser also had a surface grating coupler placed after the front mirror to measure the output. In order to transfer the laser power into the silicon waveguide, the silicon waveguide width was increased at the ends of the laser. The taper had a simulated 90% coupling efficiency, inclusive of process variations.

13.4.4 DFB lasers

Similar to the DBR laser is the distributed feedback (DFB) laser. This laser consists of a gain section with gratings placed underneath, rather than a gain section with gratings on either side of it. The DFB forms a resonant gain structure that can enable much lower threshold currents. In order to make the laser single mode, a quarter wavelength shift in the grating period is added to the center of the laser. The final DFB cross-section in the longitudinal direction is shown in Figure 13.25.

Since the mode shape is considerably different in the hybrid waveguide, the grating strength in a hybrid DFB laser is much higher when compared to a similar grating in a silicon only waveguide. The reason for higher grating strength in the former case is because the mode overlap with the grating is high, whereas only the evanescent tail of the mode interacts with the grating in the latter case. Therefore the same ~25 nm etch in the silicon waveguide creates a grating strength of $247\,cm^{-1}$ for the hybrid mode, compared to $80\,cm^{-1}$ obtained in the latter case.

The DFB laser in Ref. [71] consists of a $200\,\mu m$ long gain region with the grating underneath, and two $80\,\mu m$ long adiabatic tapers to the silicon only waveguide, which in this case are outside of the laser cavity (although still connected to the same laser gain section). These lasers had a $\kappa * L$ product of ~8. Ideally the product would be closer to 1, but in order to achieve this for such a high κ value the laser must be very short, resulting in thermal limitations. Shallower grating etches would allow for lower κ, and thus more typical $\kappa * L$ products and overall better performing lasers.

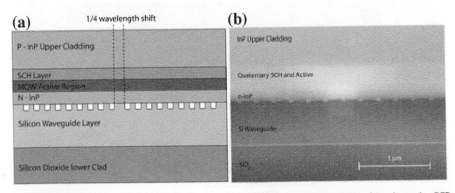

FIGURE 13.25 (a) Schematic and (b) scanning electron micrograph of the side view of a DFB cross-section with the waveguide extending from left to right showing the gratings below the III–V material [71]. © OSA 2008.

Still, the performance of these lasers is reasonable, with threshold currents of ~30 mA at room temperature and a maximum double-sided output power of 5.4 mW. The plots for light output and voltage versus current are shown in Figure 13.26 as a function of temperature. The optical spectrum is shown in Figure 13.27, again

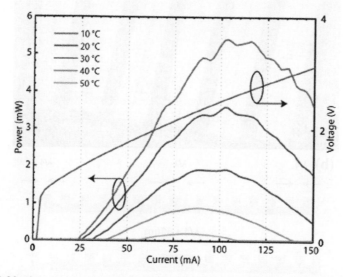

FIGURE 13.26 Light power and voltage versus current for a hybrid silicon DFB laser [71]. © OSA 2008.

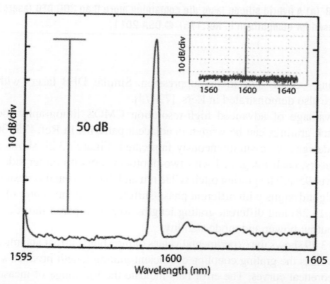

FIGURE 13.27 Optical spectrum of a hybrid silicon DFB laser [71]. © OSA 2008.

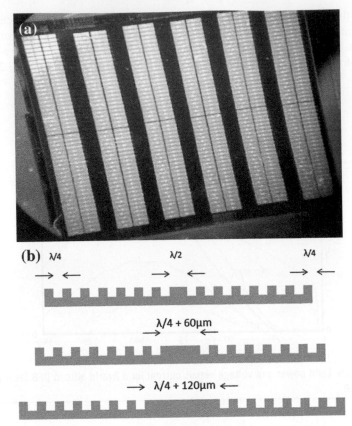

FIGURE 13.28 (a) A hybrid silicon laser die containing more than 300 DFB lasers [75]. (b) Designs used for comparison in Ref. [75]. © OSA 2011.

demonstrating ~50 dB side mode suppression. Similar DFB lasers with different designs were also demonstrated in Refs. [72–74].

One advantage of advanced high-resolution CMOS lithography tools is that different pitch gratings can be written in adjacent patterns. In Ref. [75] 36 different DFB laser designs were simultaneously fabricated. Figure 13.28 shows a die with 300 DFB lasers, each integrated with two photodetectors on either side. The laser yield was over 95%. The grating pitch is 238 nm and the stop band is designed around 1600 nm. Eight designs with different phase-shift lengths at the center of the grating (see Figure 13.28) and different grating lengths were compared in order to find the effects on output power and mode stability.

Figure 13.29 shows the experimental values of threshold current and maximum output power versus the grating coupling coefficient-grating length product, $\kappa * L$, alongside the theoretical curves. The error bars indicate the full range of measured values

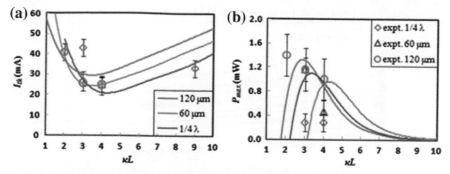

FIGURE 13.29 Threshold current (a) and maximum output power (b) versus $\kappa * L$ for three phase-shift lengths: ¼ wavelength (solid line and diamonds), 60 μm (dashed line and triangles), and 120 μm (dot-dashed line and circles) [75]. © OSA 2011.

from various positions on the chip. Both theoretical and experimental data show that the threshold decreases to a minimum value and then grows with reducing device length, L_{tot}, due to changes in mirror loss and average internal model loss. Similarly, the output power increases at the optimum value and then falls off. As the length of the phase-shift region increases, the minimum threshold current in Figure 13.29a moves to higher value and lower $\kappa * L$. The reason is twofold: First, threshold current density remains constant, so longer phase-shift lengths result in a longer cavity, and subsequently higher threshold current. Second, for a smaller $\kappa * L$, shorter device length leads to higher thermal impedance and hence a higher threshold current, as is confirmed by the experimental data. The threshold and output power is expected to vary as described by

$$I_{th} = I_{th0} \, e^{\left(Z_T \left(\frac{P_D - P_{out} + T}{T_0}\right)\right)}$$

and

$$P_{out} = \eta_i \left(\frac{\alpha_m}{\langle \alpha_i \rangle + \alpha_m}\right) \left(\frac{hv}{q}\right) e^{\left(\frac{-(Z_T(P_D - P_{out} + T))}{T_1}\right)} (I - I_{th}),$$

where h, v, q, $\langle \alpha_i \rangle$, α_m, n_i, and Z_T are Planck's constant, photon frequency, elementary charge, average internal modal loss, mirror loss, injection efficiency, and thermal impedance respectively. $T_0 = 51\,K$, $T_1 = 100\,K$, $Z_L = 3.6975\,(K\,cm)/W$, and $Z_T = Z_L/L_{tot}$ are obtained experimentally [76]. L_{tot} is the length of the gain region which is the sum of grating length and phase-shift length in our case. The values of internal quantum efficiency (η_i) and internal loss ($\langle \alpha_i \rangle$) extracted from fitting the experimental data to theory are 0.39 and 11 cm^{-1}, respectively. We assume a logarithmic dependence of gain on carrier density and the material gain and transparency carrier density used in the relation are $g_0 = 966\,cm^{-1}$ and $N_{tr} = 1.86 \times 10^{18}\,cm^{-3}$.

The graphs in Figure 13.29 suggest that we can achieve high power extraction and differential quantum efficiency by having large phase-shift lengths and lower

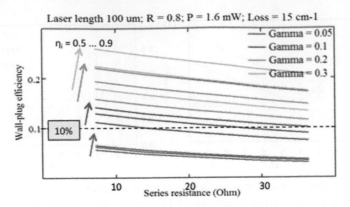

FIGURE 13.30 Wall plug efficiency versus series resistance for various quantum well confinement factors (Gamma) and injection efficiencies (η_i).

$\kappa * L$ products. The mode stability and side mode suppression ratio (SMSR) of these lasers, which are essential features specified in real-world applications, are discussed in detail in Ref. [75].

High wall plug efficiency devices can be targeted by increasing the confinement factor and injection efficiency while reducing the device series resistance as illustrated in Figure 13.30. With these modifications and advancements, wall plug efficiency of 20% is possible.

13.4.5 Microring and microdisk lasers

While the lasers mentioned above have clear potential in optical links between chips or in other short- to long-reach applications, smaller lasers with lower power requirements are desirable for some short-reach interconnects. One promising approach to this is the microring or microdisk laser, which can have very low threshold currents and electrical drive powers.

Microring lasers using direct wafer bonding were developed and presented in Refs. [77,78,52]. An image of the structure with integrated photodetectors is shown in Figure 13.31, highlights of the fabrication process are shown in Figure 13.32, and some experimental lasing results are shown in Figure 13.33. The lasers consist of a silicon waveguide disk coupled to a silicon bus waveguide, and an etched III–V microring waveguide placed above the outside edge of the SOI microdisk. These lasers can include an additional quantum well undercut for enhanced performance as shown in Figure 13.32d.

Design curves for ring lasers of different diameters are shown in Figure 13.34. Achieving thresholds below 1 mA should be possible by using smaller rings and larger coupling gap, but this will be at an expense of the output optical power. The intended optical link will determine which designs are most desirable.

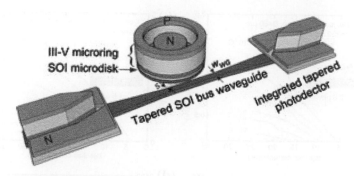

FIGURE 13.31 A hybrid silicon micro-ring laser with integrated tapered photodetectors to measure its characteristics [77]. © IEEE 2011.

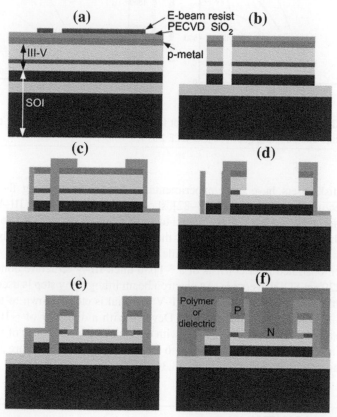

FIGURE 13.32 Fabrication process overview for a hybrid silicon mircoring laser [77], including e-beam resist deposition and definition (a), etching of the III–V material to form the ring diameter (b), etching to the *n*-contact layer (c), undercutting of the quantum wells (d), and metal deposition (e–f). © IEEE 2011.

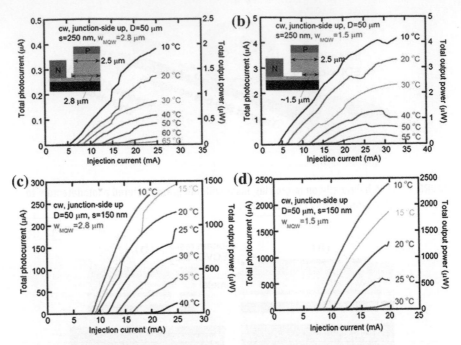

FIGURE 13.33 Experimental light output power for microring lasers with different quantum well undercuts and coupling gaps [77]. © IEEE 2011.

Microdisk lasers have been experimentally demonstrated using the adhesive bonding technique. In this design [78–82], shown in Figure 13.35, the III–V material forms a microdisk and is directly coupled to an SOI bus waveguide. When current is injected, the whispering gallery mode of the disk shown in Figure 13.35b lases and is partially coupled into the SOI waveguide.

The demonstrated laser consists of a ~1 μm thick III–V structure bonded above a 500 by 220 nm SOI waveguide. An electron beam lithography step is used to define the circumference of the disk and the III–V material is etched down to the n-layer of the III–V around the circumference. Devices with a diameter of ~10 μm typically have threshold currents near 1 mA with output powers in the tens of microwatts range. An array of four disk lasers was also fabricated [82]. The structure and spectra for the four lasers are shown in Figure 13.36.

13.5 RELIABILITY

As we have seen, hybrid silicon lasers can have good performance. One remaining question is whether those hybrid devices are able to show similar reliability as conventional pure III–V counterparts. A reliability study on fabricated hybrid Si

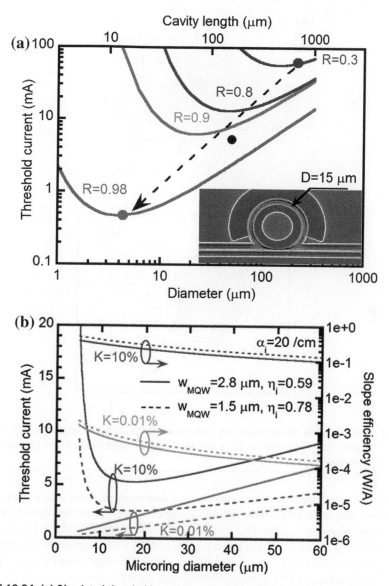

FIGURE 13.34 (a) Simulated threshold currents versus ring diameter with experimental data points. (b) Simulated threshold and slope efficiency for different quantum well widths and ring coupling ratios versus ring diameter [77]. © IEEE 2011.

distributed feedback (DFB) lasers with a quarter-wave shifted cavity design was conducted [83] and the influence on reliability of a superlattice between the lasing region and the bonded interface was investigated. Fourteen devices of each design

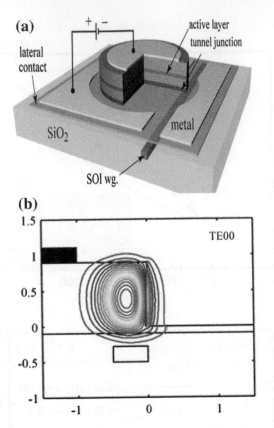

**FIGURE 13.35 A hybrid silicon microdisk laser made on the adhesive bonding platform.
(a) A schematic view of the device. (b) A simulated spatial mode shape of the fundamental
mode [52]. © Wiley 2010.**

(no superlattice, lattice matched superlattice, and strained superlattice) were characterized. Regardless of the design and aging condition, most devices showed no significant degradation over a 5000 h test. A typical result is shown in Figure 13.37 for different types of III–V epi materials [83]. With 50% degradation in threshold as the failure criterion, mean time to failure (MTTF) values from two devices that showed degradation at 70 °C were 48,000 h and 40,000 h for structures with lattice matched superlattices and strained superlattices, respectively. These numbers are comparable to those of InP substrate devices characterized in Ref. [84].

13.6 SPECIALIZED HYBRID LASERS AND SYSTEM DEMONSTRATIONS

One of the key attributes of hybrid silicon lasers is the ability to incorporate them with long silicon waveguides, other passive elements, and integration with other

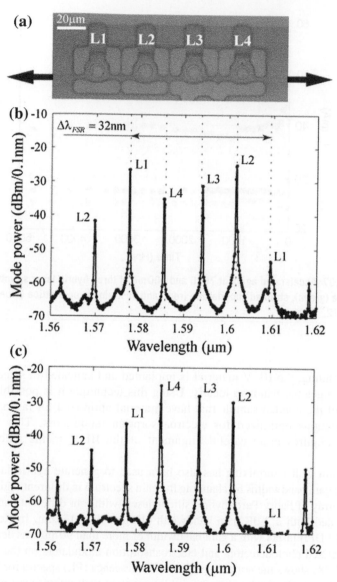

FIGURE 13.36 (a) Photograph of an array of four microdisk lasers and the associated lasing spectra from the left (b) and right (c) side grating couplers [52]. © Wiley 2010.

silicon photonics components such as modulators for externally modulated lasers (EMLs). In this section, a few examples of more complex laser cavities, integrated components, and system demonstrations are presented.

Hybrid silicon sampled grating lasers integrated with electroabsorption modulators were demonstrated in Refs. [85–89] using quantum well intermixing.

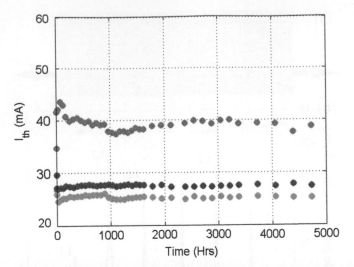

FIGURE 13.37 Accelerated aging at 70 °C and 100 mA of three hybrid laser structures: no superlattice (green), strained superlattice (red), and unstrained superlattice (blue) [83]. © IEEE 2012.

Before bonding, the III–V material is implanted and annealed in selected areas, which changes the bandgap locally. Using this technique it is possible to make sections of the bonded sample that have material optimized for lasers and other nearby sections optimized for electroabsorption modulators. This technique, however, requires more careful alignment of the III–V to the silicon before bonding.

Quantum well intermixing has also been used to generate hybrid lasers arrays with wide gain bandwidths by changing the gain spectrum in different regions locally [90]. An array of Fabry-Perot hybrid silicon lasers with four different bandgaps was demonstrated with a 3-dB gain bandwidth of more than 150 nm. In Ref. [87], an array of 13 DFB lasers with a gain bandwidth of more than 90 nm was demonstrated with integrated photodetectors and electroabsorption modulators on the same chip. Figure 13.38a shows the normalized photoluminescence (PL) spectra for four different bandgaps on the same chip achieved by quantum well intermixing in Ref. [88]. Figure 13.38b shows an array of six DFB lasers with six modulators and six photodetectors, and a separate array of 15 DFB lasers.

Mode locked lasers have also been demonstrated on the hybrid silicon laser platform, by electrically isolating a section of the laser cavity to form a saturable absorber. Fabry-Perot cavities with 10 GHz and 40 GHz repetition rates were demonstrated [91], and ring cavities with 30 GHz and 1 GHz have also been demonstrated [92–94]. The ability to use long low-loss cavities on this platform as was done to generate the 1 GHz laser opens interesting new possibilities for mode locked lasers.

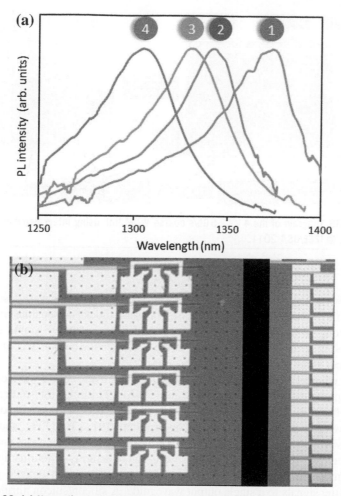

FIGURE 13.38 (a) Normalized photoluminescence spectra for four bandgaps created from a single starting quantum well material using quantum well intermixing. (b) An array of six DFB lasers integrated with six modulators and six photodetectors (left), and an array of 15 DFB lasers integrated with 15 photodetectors (right) [87]. © OSA 2011.

While the typical use of a mode locked laser is for its ability to generate short pulses, it is also possible to use the corresponding comb of continuous wavelengths that are simultaneously generated. The longitudinal modes of a mode locked laser are phase locked to each other and can therefore be locked to a WDM wavelength grid if the repetition rate of the laser is correct. This property makes them potentially useful as a WDM wavelength source from a single laser. Power efficiency per wavelength channel can also be increased in this manner. In Ref. [95] more than 100 lasing modes were generated from a single mode locked hybrid silicon laser.

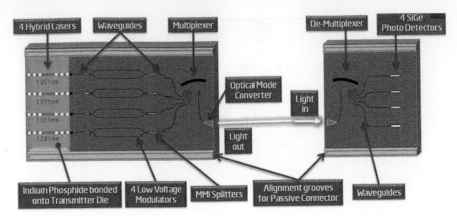

FIGURE 13.39 Diagram of the 4 × 12.5 Gb/s coarse WDM link using integrated hybrid silicon lasers [60]. © IEEE/OSA 2011.

Multi-wavelength arrayed waveguide grating (AWG) lasers were also demonstrated on the hybrid silicon platform [96]. Four gain blocks on separate waveguides were all coupled to the same arrayed waveguide grating with a single combined output. Lasing occurred with the cavity formed between the output facet of the AWG and the facet at the ends of the SOA array.

A four channel by 12.5 Gb/s silicon photonics link using coarse wavelength division multiplexing was demonstrated in Refs. [59–61]. The link's transmitter included an array of four hybrid DBR lasers utilizing two different epi materials bonded side by side (two lasers per epi). The lasers operated at 1291, 1311, 1331, and 1351 nm with the wavelengths selected by the silicon DBR mirrors at the front and back of the laser cavity. The outputs of the lasers were sent to an array of four parallel silicon Mach-Zehnder modulators operating at 12.5 Gb/s each, and an Echelle grating to multiplex the signals to a single output waveguide. A silicon/silicon nitride inverted taper was used to expand the optical mode at the output before fiber coupling. The full transmitter was packaged and tested with a silicon photonic receiver. A schematic of the entire optical link is shown in Figure 13.39. All four channels of the link had bit error rates (BERs) better than 1e−12 at 10 Gb/s when operated simultaneously. At 12.5 Gb/s three of the four channels had BER better than 1e−12 while the fourth was 3e−10.

13.7 CONCLUSIONS

Hybrid silicon laser technology has the potential to enable high volume manufacturable, and high performance integrated transmitters on a silicon platform. The performance of hybrid silicon lasers has continually improved since their first demonstration, with numerous experimental demonstrations of various types of lasers. Individual demonstrations to date have included thresholds below 2 mA, output powers above 30 mW, high temperature continuous lasing up to 105 °C, and fully integrated silicon photonics links at 40 Gb/s with bit error rates below 1e−12. The reliability of these lasers appears to be comparable to standard InP lasers. As laser designs become optimized and processing capabilities improve, we expect to see continued improvements in hybrid silicon laser performance and eventual implementation in deployed photonic links.

Acknowledgments

The authors would like to thank the many researchers at UCSB, Aurrion, Intel, and HP with whom we have worked and who have significantly contributed to hybrid silicon laser technology.

References

[1] R.A. Soref, Silicon-based optoelectronics, Proc. of the IEEE 81 (12) (1993) 1687–1706.

[2] R. Soref, J. Lorenzo, All-silicon active and passive guided-wave components for λ = 1.3 and 1.6 μm, J. Quant. Electron. 22 (6) (1986) 873–879.

[3] R.A. Soref, B.R. Bennett, Kramers-Kronig analysis of E-O switching in silicon, Proc. SPIE Integr. Opt. Circuit Eng. 704 (1986) 32–37.

[4] R.A. Soref, J. Schmidtchen, K. Petermann, Large single-mode rib waveguides in GeSi and Si-on-SiO$_2$, J. Quant. Electron. 27 (8) (1991) 1971–1974.

[5] P.D. Trinh, S. Yegnanarayanan, B. Jalali, Integrated optical directional couplers in silicon-on-insulator, Electron. Lett. 31 (24) (1995) 2097–2098.

[6] U. Fischer, T. Zinke, K. Petermann, Integrated optical waveguide switches in SOI, Proc. IEEE Int. SOI Conf. (1995) 141–142.

[7] C.Z. Zhao, G.Z. Li, E.K. Liu, Y. Gao, X.D. Liu, Silicon on insulator Mach-Zehnder waveguide interferometers operating at 1.3 μm, Appl. Phys. Lett. 67 (17) (1995) 2448–2449.

[8] B. Jalali, S. Yegnanarayanan, T. Yoon, T. Yoshimoto, I. Rendina, F. Coppinger, Advances in silicon-on-insulator optoelectronics, J. Sel. Topics Quant. Electron. 4 (6) (1998) 938–947.

[9] D.A.B. Miller, Optical interconnects to silicon, J. Sel. Topics Quant. Electron. 6 (6) (2000) 1312–1317.

[10] A. Liu, R. Jones, L. Liao, D. Samara-Rubio, D. Rubin, O. Cohen, R. Nicolaescu, M. Paniccia, A high-speed silicon optical modulator based on a metal-oxide-semiconductor capacitor, Nature 427 (2004) 615–618.

[11] L. Liao, D. Samara-Rubio, M. Morse, A. Liu, H. Hodge, D. Rubin, U.D. Keil, T. Franck, High-speed silicon Mach-Zehnder modulator, Opt. Express 13 (8) (2005) 3129–3135.

[12] L. Liao, A. Liu, D. Rubin, J. Basak, Y. Chetrit, H. Nguyen, R. Cohen, N. Izhaky, M. Paniccia, 40 Gbit/s silicon optical modulator for high-speed applications, Electron. Lett. 43 (22) (2007) 1196–1197.

[13] H. Rong, R. Jones, A. Liu, O. Cohen, D. Hak, A. Fang, M. Paniccia, A continuous-wave Raman silicon laser, Nature 433 (2005) 725–728.

[14] L. Dal Negro, M. Cazzanelli, L. Pavesi, S. Ossicini, D. Pacifici, G. Franzò, F. Priolo, F. Iacona, Dynamics of stimulated emission in silicon nanocrystals, Appl. Phys. Lett. 82 (26) (2003) 4636–4639.

[15] L. Ferraioli, M. Wang, G. Pucker, D. Navarro-Urrios, N. Daldosso, C. Kompocholis, L. Pavesi, Photoluminescence of silicon nanocrystals in silicon oxide, J. Nanomater. (2007) ID 43491.

[16] Y. Gong, S. Ishikawa, S. Cheng, M. Gunji, Y. Nishi, J. Vuckovic, Photoluminescence from silicon dioxide photonic crystal cavities with embedded silicon nanocrystals, Phys. Rev. B 81 (2010) 235317.

[17] S.S. Walavalkar, A.P. Homyk, C.E. Hofmann, M.D. Henry, C. Shin, H.A. Atwater, A. Scherer, Size tunable visible and near-infrared photoluminescence from vertically etched silicon quantum dots, Appl. Phys. Lett. 98 (2011) 153114.

[18] Y. Miwa, H. Sun, K. Imakita, M. Fujii, Y. Teng, J. Qiu, Y. Sakka, S. Hayashi, Sensitized broadband near-infrared luminescence from bismuth-doped silicon-rich silica films, Opt. Lett. 36 (21) (2011) 4221–4223.

[19] A. Pitanti, D. Navarro-Urrios, N. Prtljaga, N. Daldosso, F. Gourbilleau, R. Rizk, B. Garrido, L. Pavesi, Energy transfer mechanism and Auger effect in Er^{3+} coupled silicon nanoparticle samples, J. Appl. Phys. 108 (2010) 053518.

[20] M. Wodjak, M. Klik, M. Forcales, O.B. Gusev, T. Gregorkiewicz, D. Pacifici, G. Franzo, F. Priolo, F. Iacona, Sensitization of Er luminescence by Si nanoclusters, Phys. Rev. B 69 (2004) 233315.

[21] I. Izeddin, A.S. Moskalenko, I.N. Yassievich, M. Fujii, T. Gregorkiewicz, Nanosecond dynamics of the near-infrared photoluminescence of Er-doped SiO_2 sensitized with Si nanocrystals, Phys. Rev. Lett. 97 (2006) 207401.

[22] G.M. Miller, R.M. Briggs, H. Atwater, Achieving optical gain in waveguide-confined nanocluster-sensitized erbium by pulsed excitation, J. Appl. Phys. 108 (2010) 063109.

[23] J.M. Ramirez, F. Ferrarese Lupi, O. Jambois, Y. Berencen, D. Navarro-Urrios, A. Anopchenko, A. Marconi, N. Prtljaga, A. Tengattini, L. Pavesi, J.P. Colonna, J.M. Fedeli, B. Garrido, Erbium emission in MOS light emitting devices: from energy transfer to direct impact excitation, Nanotechnology 23 (2012) 125203.

[24] Y. Gong, M. Makarova, S. Yerci, R. Li, M.J. Stevens, B. Baek, S.W. Nam, R.H. Hadfield, S.N. Dorenbos, V. Zwiller, J. Vuckovic, L. dal Negro, Linewidth narrowing and Purcell enhancement in photonic crystal cavities on an Er-doped silicon nitride platform, Opt. Express 18 (2010) 2601–2612.

[25] S. Das, R.K. Singha, A. Dhar, S.K. Ray, A. Anopchenko, N. Daldosso, L. Pavesi, Electroluminescence and charge storage characteristics of quantum confined germanium nanocrystals, J. Appl. Phys. 110 (2011) 024310.

[26] X. Sun, J. Liu, L.C. Kimerling, J. Michel, Toward a germanium laser for integrated silicon photonics, J. Sel. Top. Quant. Electron. 16 (1) (2010) 124–131.

[27] J. Michel, J. Liu, L. Kimerling, R. Camacho-Aguilera, J. Bessette, Y. Cai, A germanium-on-silicon laser for on-chip applications, CLEO 2011, Paper CFL3, 2011.

[28] S. Cheng, G. Shambat, J. Lu, H. Yu, K. Saraswat, T.I. Kamins, J. Vuckovic, Y. Nishi, Cavity-enhanced direct band electroluminescence near 1550 nm from germanium microdisk resonator diode on silicon, Appl. Phys. Lett. 98 (2011) 211101.

[29] M. Graeme, Low-cost hybrid photonic integrated circuits using passive alignment techniques, LEOS 2006 (2006) 98–99.

[30] G.D. Maxwell, Hybrid integration technology for high speed optical processing devices, Optical Internet (COIN), 2008.

[31] G.D. Maxwell, Hybrid integration of InP devices, in: Conference on Indium Phosphide and Related Materials, 2011, pp. 22–26.

[32] A. Narasimha et al., A 40-Gb/s QSFP optoelectronic transceiver in a 0.13 μm CMOS silicon-on-insulator technology, in: Optical Fiber Communications Conference (OFC), 2008.

[33] A. Narasimha et al., An ultra low power CMOS photonics technology platform for H/S optoelectronic transceivers at less than $1 per Gbps, in: Optical Fiber Communications Conference (OFC), 2010.

[34] A.W. Fang, Silicon evanescent lasers, Ph.D. dissertation, University of California, Santa Barbara, 2008.

[35] A.W. Fang, H. Park, R. Jones, O. Cohen, M.J. Paniccia, J.E. Bowers, A continuous-wave hybrid AlGaInAs-silicon evanescent laser, Photon. Technol. Lett. 18 (10) (2006) 1143–1145.

[36] A.W. Fang, H. Park, Y. Kuo, R. Jones, O. Cohen, D. Liang, O. Raday, M.N. Sysak, M.J. Paniccia, J.E. Bowers, Hybrid silicon evanescent devices, Mater. Today 10 (7) (2007) 28–35.

[37] A. Yariv, X.K. Sun, Supermode Si/III–V hybrid lasers, optical amplifiers and modulators: a proposal and analysis, Opt. Express 15 (2007) 9147–9151.

[38] X.K. Sun, A. Yariv, Engineering supermode silicon/III–V hybrid waveguides for laser oscillation, J. Opt. Soc. Am. B 25 (2008) 923–926.

[39] T. Dupont, L. Grenouillet, A. Chelnokov, P. Viktorovitch, Contradirectional coupling between III–V stacks and silicon-on-insulator corrugated waveguides for laser emission by distributed feedback effect, Photon. Technol. Lett. 22 (19) (2010) 1413–1415.

[40] T. Dupont, L. Grenouillet, A. Chelnokov, S. Messaoudene, J. Harduin, D. Bordel, J.-M. Fedeli, C. Seassal, P. Regreny, P. Viktorovitch, A III–V on silicon distributed-feedback laser based on exchange-Bragg coupling, in: Group IV Photonics (GFP), 2010, pp. 19–21.

[41] Y. De Koninck, G. Roelkens, R. Baets, Cavity enhanced reflector based hybrid silicon laser, in: IEEE Photonics Society Annual Meeting, 2010, pp. 469–470.

[42] A. Yariv, Optical Electronics in Modern Communications, fifth ed., Oxford University Press, 1997 pp. 526–531.

[43] X.K. Sun, H.-C. Liu, A. Yariv, Adiabaticity criterion and the shortest adiabatic mode transformer in a coupled-waveguide system, Opt. Lett. 34 (3) (2009) 280–282.

[44] M. Sysak, H. Park, A. Fang, O. Raday, J. Bowers, R. Jones, Reduction of hybrid silicon laser thermal impedance using Poly Si thermal shunts, in: Optical Fiber Communications Conference (OFC), 2011, Paper OWZ6.

[45] M.N. Sysak, D. Liang, R. Jones, G. Kurczveil, M. Piels, M. Fiorentino, R.G. Beausoleil, J.E. Bowers, Hybrid silicon laser technology: A thermal perspective, J. Sel. Topics Quant. Electron. 17 (6) (2011) 1490–1498.

[46] Q.-Y. Tong, G. Cha, R. Gafiteanu, U. Gosele, Low temperature wafer direct bonding, J. Microelectromech. Syst. 3 (1994) 29–35.

[47] D. Pasquariello, K. Hjort, Plasma-assisted InP-to-Si low temperature wafer bonding, J. Sel. Top. Quant. Electron. 8 (2002) 118–131.

[48] K. Schjolberg-Henriksen, S. Moe, M.M.V. Taklo, P. Storas, J.H. Ulvensoen, Low-temperature plasma activated bonding for a variable optical attenuator, Sensor Actuat. A—Phys. 142 (2008) 413–420.

[49] H. Park, A.W. Fang, D. Liang, Y.-H. Kuo, H.-H. Chang, B.R. Koch, H.-W. Chen, M.N. Sysak, R. Jones, J.E. Bowers, Photonic integration on hybrid silicon evanescent device platform, Adv. Opti. Technol, 2008, Article ID 682978.

[50] A. Black, A.R. Hawkins, N.M. Margalit, D.I. Babic, A.L. Holmes, Y.-L. Chang, Jr. P. Abraham, J.E. Bowers, E.L. Hu, Wafer fusion: Materials issues and device results, J. Sel. Top. Quant. Electron. 3 (3) (1997) 943–951.

[51] Y.L. Chao, Q.Y. Tong, T.H. Lee, M. Reiche, R. Scholz, J.C.S. Woo, U. Gosele, Ammonium hydroxide effect on low-temperature wafer bonding energy enhancement, Electrochem. Solid-State. Lett. 8 (2005) G74.

[52] G. Roelkens, L. Liu, D. Liang, R. Jones, A. Fang, B. Koch, J. Bowers, III–V/silicon photonics for on-chip and inter-chip optical interconnects, Laser Photon. Rev. 4 (6) (2010) 751–779.

[53] D. Liang, A.W. Fang, H. Park, T.E. Reynolds, K. Warner, D.C. Oakley, J.E. Bowers, Low-temperature, strong SiO_2-SiO_2 covalent wafer bonding for III–V compound semiconductors-to-silicon photonic integrated circuits, J. Electron. Mater. 37 (2008) 1552–1559.

[54] W.P. Maszara, Silicon-on-insulator by wafer bonding: a review, J. Electrochem. Soc. 138 (1991) 341–347.

[55] D. Liang, J.E. Bowers, Highly efficient vertical outgassing channels for low-temperature InP-to-silicon direct wafer bonding on the silicon-on-insulator substrate, J. Vac. Sci. Technol. B 26 (2008) 1560–1567.

[56] K. Tanabe, K. Watanabe, Y. Arakawa, III–V/Si hybrid photonic devices by direct fusion bonding, Nature 2 (2012) 349.

[57] F. Niklaus, G. Stemme, J.Q. Lu, R.J. Gutmann, Adhesive wafer bonding, J. Appl. Phys. 99 (2006) 031101.

[58] P. Garrou, R. Heistand, M. Dibbs, T. Mainal, C. Mohler, T. Stokich, P. Townsend, G. Adema, M. Berry, I. Turlik, Rapid thermal curing of BCB dielectric, IEEE Trans. Compon. Hybrids Manuf. Technol. 16 (1993) 46–52.

[59] A. Alduino et al., Demonstration of a high speed 4-channel integrated silicon photonics WDM link with hybrid silicon lasers, in: Integrated Photonics Research Conference (IPR) 2010, Paper PDIWI5, 2010.

[60] B. Koch et al., A 4×12.5 Gb/s CWDM Si photonics link using integrated hybrid silicon lasers, in: Conference on Lasers and Electro-optics (CLEO), 2011.

[61] H. Park, M.N. Sysak, H. Chen, A.W. Fang, Di Liang, L. Liao, B.R. Koch, J. Bovington, Y. Tang, K. Wong, M. Jacob-Mitos, R. Jones, J.E. Bowers, Device and integration technology for silicon photonic transmitters, J. Sel. Top. Quant. Electron. 17 (3) (2011) 671–688.

[62] H. Park, A.W. Fang, S. Kodama, J.E. Bowers, Hybrid silicon evanescent laser fabricated with a silicon waveguide and III–V offset quantum wells, Opt. Express 13 (23) (2005) 9460–9464.

[63] G. Roelkens, D. Van Thourhout, R. Baets, R. Notzel, M. Smit, Laser emission and photodetection in an InP/InGaAsP layer integrated on and coupled to a Silicon-on-Insulator waveguide circuit, Opt. Express 14 (18) (2006) 8154–8159.

[64] A.W. Fang, H. Park, O. Cohen, R. Jones, M.J. Paniccia, J.E. Bowers, Electrically pumped hybrid AlGaInAs-silicon evanescent laser, Opt. Express 14 (2006) 9203–9210.

[65] H. Chang, A.W. Fang, M.N. Sysak, H. Park, R. Jones, O. Cohen, O. Raday, M.J. Paniccia, J.E. Bowers, 1310 nm silicon evanescent laser, Opt. Express 15 (2007) 11466–11471.

[66] S. Stankovic, R. Jones, M.N. Sysak, J.M. Heck, G. Roelkens, D. Van Thourhout, 1310-nm hybrid III–V/Si Fabry-Pérot laser based on adhesive bonding, Photon. Technol. Lett. 23 (23) (2011) 1781–1783.

[67] M. Lamponi, S. Keyvaninia, C. Jany, F. Poingt, F. Lelarge, G. de Valicourt, G. Roelkens, D. Van Thourhout, S. Messaoudene, J.-M. Fedeli, G.H. Duan, Low-threshold heterogeneously integrated InP/SOI lasers with a double adiabatic taper coupler, Photon. Technol. Lett. 24 (1) (2012) 76–78.

[68] A.W. Fang, R. Jones, H. Park, O. Cohen, O. Raday, M.J. Paniccia, J.E. Bowers, Integrated AlGaInAs-silicon evanescent race track laser and photodetector, Opt. Express 15 (2007) 2315–2322.

[69] A.W. Fang, B.R. Koch, R. Jones, E. Lively, D. Liang, Y. Kuo, J.E. Bowers, A distributed Bragg reflector silicon evanescent laser, Photon. Technol. Lett. 20 (20) (2008) 1667–1669.

[70] B.B. Bakir, A. Descos, N. Olivier, D. Bordel, P. Grosse, E. Augendre, L. Fulbert, J.M. Fedeli, Electrically driven hybrid Si/III–V Fabry-Pérot lasers based on adiabatic mode transformers, Opt. Express 19 (2011) 10317–10325.

[71] A.W. Fang, E. Lively, Y.-H. Kuo, D. Liang, J.E. Bowers, A distributed feedback silicon evanescent laser, Opt. Express 16 (2008) 4413–4419.

[72] T. Okumura, T. Maruyama, H. Yonezawa, N. Nishiyama, S. Arai, Injection-type GaInAsP/InP/Si distributed-feedback laser directly bonded on silicon-on-insulator substrate, Photon. Technol. Lett. 21 (5) (2009) 283–285.

[73] T. Dupont, L. Grenouillet, A. Chelnokov, S. Messaoudene, J. Harduin, D. Bordel, J.-M. Fedeli, C. Seassal, P. Regreny, P. Viktorovitch, A III–V on silicon distributed-feedback laser based on exchange-Bragg coupling, in: Group IV Photonics (GFP) 2010, 2010, pp. 19–21.

[74] T. Dupont, L. Grenouillet, A. Chelnokov, P. Viktorovitch, Contradirectional coupling between III–V stacks and silicon-on-insulator corrugated waveguides for laser emission by distributed feedback effect, Photon. Technol. Lett. 22 (19) (2010) 1413–1415.

[75] S. Srinivasan, A.W. Fang, D. Liang, J. Peters, B. Kaye, J.E. Bowers, Design of phase-shifted hybrid silicon distributed feedback lasers, Opt. Express 19 (2011) 9255–9261.

[76] M.N. Sysak, H. Park, A.W. Fang, J.E. Bowers, R. Jones, O. Cohen, O. Raday, M.J. Paniccia, Experimental and theoretical thermal analysis of a hybrid silicon evanescent Laser, Opt. Express 15 (23) (2007) 15041–15046.

[77] Di Liang, M. Fiorentino, S. Srinivasan, J.E. Bowers, R.G. Beausoleil, Low threshold electrically-pumped hybrid silicon microring lasers, J. Sel. Top. Quant. Electron. 17 (6) (2011) 1528–1533.

[78] Di Liang, M. Fiorentino, T. Okumura, H.-H. Chang, D.T. Spencer, Y.-H. Kuo, A.W. Fang, D. Dai, R.G. Beausoleil, J.E. Bowers, Electrically-pumped compact hybrid silicon microring lasers for optical interconnects, Opt. Express 17 (22) (2009) 20355–20364.

[79] J. Van Campenhout, P. Rojo-Romeo, D. Van Thourhout, C. Seassal, P. Regreny, L. Di Cioccio, J.M. Fedeli, C. Lagahe, R. Baets, Electrically pumped InP-based microdisk lasers integrated with a nanophotonic silicon-on-insulator waveguide circuit, Opt. Express 15 (2007) 6744–6749.

[80] J. Van Campenhout, P. Rojo-Romeo, D. Van Thourhout, C. Seassal, P. Regreny, L. Di Cioccio, J.M. Fedeli, R. Baets, Design and optimization of electrically injected InP-based microdisk lasers integrated on and coupled to a SOI waveguide circuit, J. Lightwave Technol. 26 (2008) 52–63.

[81] J. Van Campenhout, P. Rojo-Romeo, D. Van Thourhout, C. Seassal, P. Regreny, L. Di Cioccio, J.M. Fedeli, R. Baets, Thermal characterization of electrically injected thin-film InGaAsP microdisk lasers on Si, J. Lightwave Technol. 25 (2007) 1543–1548.

[82] J. Van Campenhout, L. Liu, P. Rojo Romeo, D. Van Thourhout, C. Seassal, P. Regreny, L. Di Cioccio, J.-M. Fedeli, R. Baets, Nanophotonic devices for optical interconnect, Photon. Technol. Lett. 16 (2008) 1363–1375.

[83] S. Srinivasan, J.E. Bowers, Reliability of hybrid III–V on Si distributed feedback lasers, in: International Semiconductor Laser Conference 2012, pp. 10–11.

[84] J.S. Huang, Temperature and current dependences of reliability degradation of buried heterostructure semiconductor lasers, IEEE. T. Device. Mat. Re. 5 (1) (2005) 150–154.

[85] M.N. Sysak, J.O. Anthes, J.E. Bowers, O. Raday, R. Jones, Integration of hybrid silicon lasers and electroabsorption modulators, Opt. Express 16 (2008) 12478–12486.

[86] M.N. Sysak, J.O. Anthes, Di Liang, J.E. Bowers, O. Raday, R. Jones, A hybrid silicon sampled grating DBR tunable laser, in: Group IV Photonics (GFP), 2008, pp. 55–57.

[87] S.R. Jain, M.N. Sysak, G. Kurczveil, J.E. Bowers, Integrated hybrid silicon DFB laser-EAM array using quantum well intermixing, Opt. Express 19 (2011) 13692–13699.

[88] S.R. Jain, Y. Tang, H. Chen, M.N. Sysak, J.E. Bowers, Integrated hybrid silicon transmitter, J. Lightwave Technol. 30 (5) (2012) 671–678.

[89] S.R. Jain, M.N. Sysak, G. Kurczveil, J.E. Bowers, Integrated broadband hybrid silicon DFB laser array using quantum well intermixing, in: International Semiconductor Laser Conference (ISLC), 2010, pp. 139–140.

[90] B.R. Koch, M.N. Sysak, R. Jones, Gain measurements of quantum well intermixed hybrid silicon evanescent lasers, in: Group IV Photonics (GFP) 2009, 2009, pp. 211–213.

[91] B.R. Koch, A.W. Fang, O. Cohen, J.E. Bowers, Mode-locked silicon evanescent lasers, Opt. Express 15 (2007) 11225–11233.

[92] A.W. Fang, B.R. Koch, K. Gan, H. Park, R. Jones, O. Cohen, M.J. Paniccia, D.J. Blumenthal, J.E. Bowers, A racetrack mode-locked silicon evanescent laser, Opt. Express 16 (2008) 1393–1398.

[93] B.R. Koch, A.W. Fang, H.N. Poulsen, H. Park, D.J. Blumenthal, J.E. Bowers, R. Jones, M.J. Paniccia, O. Cohen, All-optical clock recovery with retiming and reshaping using a silicon evanescent mode locked ring laser, in: Optical Fiber Communication Conference (OFC), 2008.

[94] M. Heck, M.L. Davenport, H. Park, D.J. Blumenthal, J.E. Bowers, Ultra-long cavity hybrid silicon mode-locked laser diode operating at 930 MHz, in: Optical Fiber Communication Conference (OFC), 2010.

[95] B.R. Koch, A.W. Fang, O. Cohen, M. Paniccia, D.J. Blumenthal, J.E. Bowers, Multiple wavelength generation from a mode locked silicon evanescent laser, in: International Semiconductor Laser Conference (ISLC), 2008, pp. 175–176.

[96] G. Kurczveil, M.J. Heck, J.D. Peters, J.M. Garcia, D. Spencer, J.E. Bowers, An integrated hybrid silicon multiwavelength AWG laser, J. Sel. Top. Quant. Electron. 17 (6) (2011) 1521–1527.

VCSEL-Based Data Links

14

Julie Sheridan Eng and Chris Kocot

Finisar Corporation, Sunnyvale CA, USA

14.1 INTRODUCTION

Fiber optic systems were first deployed for long-haul telecommunications networks for link distances up to 1000 km. As technology advanced and costs decreased, fiber optics began to replace copper connections in datacenter applications at distances of approximately 300 m. One of the technologies that made this migration possible was the Vertical-Cavity Surface-Emitting Laser or VCSEL. GaAs shortwave (850 nm) VCSEL-based data links were first introduced into the market in 1996, and the first Ethernet and Fibre Channel standards for VCSEL-based links were ratified in 1997. Today, VCSEL-based links support Ethernet, Fibre Channel, and Infiniband standards for data communications to 14 Gb/s and 300 m, with standards in development to support up to 28 Gb/s. Seventy-five percent of all Ethernet and Fibre Channel optical data links are VCSEL-based, making the VCSEL-based data link one of the highest volume optical links in the fiber optic network today. More than 20M VCSEL-based transceivers were shipped for commercial deployment in 2011 [1].

VCSEL-based links have taken the majority of the volume due to their cost-competitiveness, low power consumption, and ability to support links up to 300 m distance, which is adequate for datacenter applications. VCSEL-based links are lower cost than other optical device options such as single-mode long wavelength lasers or external modulators primarily because they are multi-mode and are coupled into multi-mode fiber, the core of which is either 50 or 62.5 μm in diameter. In contrast, single-mode optical devices emit a beam with a smaller spot size which is coupled into single-mode fiber, the core of which is 8 μm in diameter. This delta in optical requirements makes a significant difference in optical alignment cost in manufacturing, resulting in lower costs for the multimode link. In addition, due to the more relaxed optical system, VCSEL-based data links are able to utilize plastic optical lenses made from materials such as Ultem, rather than more expensive glass optical lenses required for single-mode coupling. Furthermore, due to their surface normal light output, VCSELs can be optically tested at wafer level, rather than after bar singulation for in-plane lasers, which further reduces the cost. And, because short wavelength multimode VCSELs can be driven at lower currents to achieve the necessary output optical power, the resulting optical transceiver is lower power consumption than other options.

Optical Fiber Telecommunications VIA. http://dx.doi.org/10.1016/B978-0-12-396958-3.00014-7

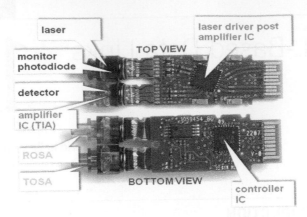

FIGURE 14.1 Internal photo of VCSEL-based fiber optic transceiver.

The interior of a typical short wavelength (850 nm) VCSEL-based optical transceiver is shown in Figure 14.1. The VCSEL is hermetically sealed in a Transistor Outline (TO) can, and plastic molded lensing is used to couple the light from the VCSEL into the multimode fiber (not shown). On the receive side, the light is focused via plastic optics onto a GaAs PIN photodetector mounted in a TO can. The resulting electrical signal is amplified using a transimpedance amplifier (TIA) collocated with the PIN in the TO can. The transmitter and receiver optical packages are referred to as the TOSA (Transmitter Optical SubAssembly) and the ROSA (Receiver Optical SubAssembly). Flex circuits connect the TOSA and ROSA to a printed circuit board where electronic circuitry including a laser driver, a post-amplifier, a microcontroller, and passive circuitry is located. VCSEL-based optical transceivers for 1–4 Gb/s are shipped in a mechanical outline referred to as the Small Form Pluggable, or SFP, form factor. This mechanical form factor has been enhanced to support higher data rates including 8–14 Gb/s, and the higher data rate version is referred to as the SFP+. SFP and SFP + optical transceivers support 1G and 10G Ethernet, and 4X, 8X, and 16X Fibre Channel (Storage Area Network) applications. Typical uses for such systems include 48 SFP + across a 14 in. line card, or 96 or 144 SFP + in a system. In 2010, storage area network (SAN) industry volume crossed over so that the volume of optical transceivers shipped supporting data rates of 8 Gb/s and above are now greater than transceivers supporting 4 Gb/s, as shown in Figure 14.2.

In order to achieve higher aggregate bandwidth and higher bandwidth density, VCSEL-based parallel optics transceivers have been introduced. In a parallel optics module, an array of shortwave VCSELs is coupled into an array lens, which guides the light into ribbon multimode fiber; receiving light is coupled into a PIN array; integrated circuits in array form are used. Parallel optics form factors include: QSFP, which supports four channels of 10 Gb/s or 40 Gb/s; CXP, which supports 10 or 12 channels of 10–12 Gb/s or 100 Gb/s; and CFP, which supports 4 × 10 Gb/s, 40 Gb/s,

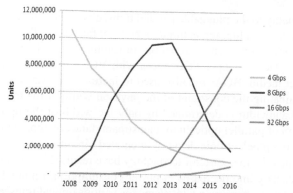

FIGURE 14.2 Units shipped of VCSEL-based transceivers for Storage Area Network Fibre Channel Applications. As indicated, the volume of 8 Gb/s optical transceivers surpassed the volume of 4 Gb/s transceivers in 2010. Up to 2011 is actual volume; 2012 and beyond is projected.

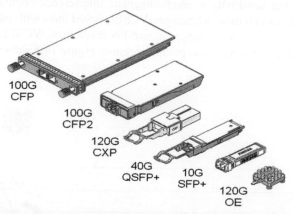

FIGURE 14.3 Recent commercial VCSEL-based transceiver form factors. SFP + is a single-channel 10G form factor. The others are all multichannel form factors supporting parallel optics VCSEL-based solutions. The OE is an optical engine that is mid-board mount, rather than face plate pluggable.

10×10 Gb/s, or 100 Gb/s. These form factors are shown in Figure 14.3. Most of these form factors are currently undergoing changes to support 25–28 Gb/s.

All of these form factors are face-plate pluggable, which refers to the ability to plug the transceiver into a line card or chassis through the front panel, without requiring disassembly of the chassis. The majority of optical transceiver volume that ships today is face-plate pluggable. Also shown in Figure 14.3 is a board-mounted optical assembly or BOA, also referred to as an "Optical Engine." This transceiver has similar functionality to the pluggable parallel transceivers, but mounts on the interior of a chassis rather than plugging in through the front face plate. This type of VCSEL-based transceiver is used as an interconnect for routing signals between high density switch chips, replacing copper PCB traces with VCSELs and multimode fiber as the interconnect media. BOAs are also used to increase the bandwidth density of a chassis by eliminating the constraint of the physical width of the transceiver so that the aggregate bandwidth is determined by only the width of the optical connector. The introduction of parallel optics into mainstream datacom applications, combined with advances in optoelectronic devices, integrated circuits, optical packaging, and printed circuit board and connector technology have resulted in significant increases in bandwidth density and decreases in power consumption. Figure 14.4 shows the increase in bandwidth density in Gbit/s/cm^3 for various optical transceiver form factors introduced since 2000. Also indicated is the decrease in total power consumption in watts per Gbit. Bandwidth density has increased and power consumption has decreased 50–70× over the past decade.

In addition to providing extremely high bandwidth density, the optical link offers advantages such as lower power consumption, lower weight, more flexible routing, and lower sensitivity to electromagnetic interference compared to copper links. However, with all those advantages, VCSEL-based links will only be deployed where they can compete with copper on cost. For this reason, VCSEL-based optical transceivers are under tremendous pricing pressure. Figure 14.5 shows industry data

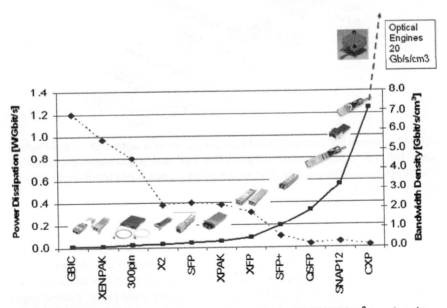

FIGURE 14.4 Power dissipation per Gb/s and bandwidth density in Gb/s/cm^3 as a function of form factor for form factors introduced since 2000. Bandwidth density has increased and power consumption has decreased 50–70× over the past 10 years.

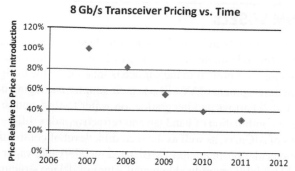

FIGURE 14.5 **Actual Fibre Channel 8 Gb/s transceiver pricing as a function of time relative to price at introduction in 2007. 8 Gb/s optical transceiver pricing has dropped by over 3× in the 4 years since introduction.**

of the relative average selling price of 8 Gb/s optical transceivers as a function of time [1]. Over the 4 years since introduction, the price of an 8 Gb/s optical transceiver dropped to less than one third the price at introduction.

For very low-cost markets where optical links are replacing copper, Active Optical Cables or AOCs have been introduced. An AOC is a length of multi-mode fiber with a VCSEL-based transceiver permanently attached at each fiber end. This implementation reduces the total cost of the product compared to the implementation of two independent VCSEL-based transceivers and a separate multi-mode fiber by eliminating costly optical specs such as those referenced by the IEEE 802.3 for a 10 Gb Ethernet VCSEL link, called 10G-BASE SR. These optical specifications are required in order to provide interoperable links across many vendors and a variety of fiber plant installations. However, for a short (less than 100 m) link, where one vendor owns both ends of the link, this complex and costly optical testing can be eliminated by producing and selling an AOC that simply works error-free from electrical connector to electrical connector, with no guaranteed specifications for the optics. In addition, integration of the transceiver with the fiber allows elimination of industry-standard optical connectors and ferrules, which also decreases the cost of the resultant product. AOCs are a future trend for VCSEL-based links in applications where they are replacing copper in computing and Ethernet systems.

This chapter will introduce and review VCSEL technology and design principles relative to the requirements of datacom optical links. In particular, shortwave multi-mode VCSELs and long wavelength single-mode VCSELs will be reviewed. Challenges to achieving VCSEL performance at higher speeds such as 25 Gb/s and beyond will be discussed. In addition, competing device technologies such as low-power Distributed Bragg Reflector (DBR) lasers and silicon photonics will be reviewed and compared to VCSELs for datacom link applications. The migration of VCSEL-based optical links from datacenter 300 m links into computing interconnect links as short as a few meters will also be discussed.

14.2 850 NM VCSELs

The concept of the vertical-cavity surface-emitting laser (VCSEL) was proposed in 1977 by Iga at the Tokyo Institute of Technology [2]. It is unique among semiconductor lasers in that emission normal to the surface is inherent to the design. The reason for the explosive development of VCSELs is their simpler packaging design, smaller size, fabrication of 2D arrays and integration, on wafer testing, and manufacturability. The particular combination of bandgap and refractive index differences in GaAs, AlAs, and alloys of AlGaAs, as well as the incredible level of control that was developed for both molecular beam epitaxy (MBE) and metalorganic vapor phase epitaxy (MOVPE), enables highly reproducible manufacture of device structures. The small volume of the resonator and the active region enable low-threshold current and efficient high-speed modulation at low currents. Properly designed VCSELs are able to operate over a wide range of temperatures with minimal change in performance.

Data communication was the first considerable driver for the development of the VCSEL technology [3], and led to significant improvements of VCSEL performance in terms of efficiency, speed, and reliability. VCSELs are now well established as cost-effective and power-efficient optical sources in transmitters for short-reach, high-capacity optical interconnects. A great deal of effort has consequently been undertaken to improve the high-speed performance of VCSELs. To reach high modulation bandwidths and bit rates, efforts were primarily focused on maximizing the resonance frequency through improved differential gain ($\partial g/\partial n$) [4], minimizing the electrical parasitics [5], minimizing self-heating [6], and optimizing the photon lifetime [7].

Recently, 850 nm VCSELs with a 3 dB bandwidth of 23 GHz at room temperature were reported by Chalmers University [8]. In addition, Finisar has demonstrated 850 nm VCSELs with a 3 dB bandwidth of 24 GHz at room temperature and 15 GHz at 95 °C (see Figure 14.15). Furthermore, error-free transmission at bit rates exceeding 40 Gb/s using VCSELs emitting at 850 nm, 980 nm, and 1.1 μm was reported by multiple groups [9–11]. A VCSEL-based link operating error-free at a record-high speed of 56 Gbps was recently demonstrated by Finisar and IBM [47,99].

Today, the majority of Ethernet and Fibre Channel deployed links operate at less than 10 Gbps. Looking forward, standards organizations (Infiniband, IEEE 802.3, and ANSI X3.T11) are working on, and in some cases have already published, 850 nm specifications that will operate at aggregate speeds up to 100 Gbps. Table 14.1 summarizes the standards that are currently under development.

A majority of the recent VCSEL development effort was centered on the new Fibre Channel standard operating at 14 Gbps. Significant departure from previous standards definitions was required to support the market requirements for distance and fiber support. Some of the additional complexities included in the module specifications are the use of clock and data recovery (CDR) circuits, and the move from 8B/10B to 64B/66B bit data encoding to keep the signaling rate lower. The next section of this chapter will describe the 14 Gpbs VCSELs that were developed to support this data rate [12].

Table 14.1 Recent fiber link standard and status.

Standard	Speed (Gbps)	Status
IEEE 802.3ae	10	Complete
ANSI X3.T11 FC PI-4	10	Complete
ANSI X3.T11 FC PI-5	14	Complete
IEEE 802.3ba	$4, 10 \times 10$	Complete
Infiniband QDR	$1, 4, 12 \times 8$	Complete
Infiniband FDR	$1, 4, 12 \times 14$	Complete
Infiniband EDR	$1, 4, 12 \times 25$	2013
ANSI X3.T11 FC PI-6	28	2013
IEEE 802.3	$4, 10 \times 25$	2013

The challenge currently facing the industry is in the specification of optical components for use at the next data rate, which is forming around 25–28 Gbps. As with previous standards development activity, tradeoffs exist between power consumption, link distance, module cost, and component complexity. These issues have become more acute as data rates have increased, and are further complicated by a strong desire from the marketing arm of Fibre Channel to maintain a link distance of 100 m on multimode optical fiber at 850 nm. This may require the inclusion of some form of electronic dispersion compensation (EDC) and aggressive transmitter pre-emphasis to compensate for the bandwidth limitations of the VCSEL, detector, and optical fiber. Other difficulties are driven by the electrical channel from Media Access Controller (MAC) IC to the module (typically up to 9 in. in FR4 printed circuit board material), which may require electrical compensation and module CDR circuits. While these are common features of electrical transceivers, they are not traditionally included in optical modules, and represent a significant increase in power dissipation, possibly driving the maximum power specification from 1 W to 1.5 W. This change in power consumption would drive significant change in the thermal management requirements of the end switch, ultimately reducing module face plate density on the switch [3]. Some of the tradeoffs most relevant to an 850 nm 28 Gbps VCSEL-based optical link on multimode fiber are summarized in Table 14.2. The values summarized do not include the benefit of electrical compensation.

Subsequent material presented to Fibre Channel [100] calculated the benefits of equalization, which is essentially required when transmitter dispersion penalty (TDP) is larger than 3.8 dB [4]. In that presentation, a continuous time fractionally spaced equalizer (CTFSE) with either six feed forward (CTFSE (6,0)) or four feed forward and two feedback taps (CTFSE (4,2)) could improve the link budget by more than 4 dB for a 70 m link. More importantly, it would significantly relax rise and fall time requirements for a given link budget by as much as 10 ps, as compared to Table 14.2, which brings 28 Gbps links to a more attainable regime of link distance, RIN, and spectral bandwidth. While it is a laudable goal to have a VCSEL

Table 14.2 From reference [101], waveform requirements, T_R/T_F, RIN, and RMS bandwidth $\Delta\lambda$ for 850 nm links over OM3 and OM4 fiber for various link distances.

	Fiber	Distance			
		25 m	50 m	75 m	100 m
T_R/T_F	OM3	16	14	11	6
	OM4	16	16	15	14
RIN	OM3	−131	−131	−132	−135
	OM4	−131	−132	−132	−132
$\Delta\lambda$	OM3	0.6	0.5	0.4	0.3
	OM4	0.6	0.5	0.4	0.35

with very high edge speeds, it may not be the best overall tradeoff in terms of reliability, EMI, and ultimately, cost of the complete link. One of the next sections of this chapter details some of the VCSEL characteristics and tradeoffs in driver pre-/post-emphasis and EDC that are pertinent for creating robust optical links operating at 28 Gbps.

14.2.1 Basics of VCSEL operation

The basics of VCSEL operation and the epitaxial structures are extensively described in the literature, e.g. [13]. Epitaxial design of a typical 850 nm GaAs QW VCSEL contains an output p-DBR mirror with 27–30 alternating pairs of Al0.2Ga0.8As-Al0.9Ga0.1As quarter wavelength layers, an active region containing typically three 75–80A GaAs MQWs, and a high-reflectivity n-type DBR mirror. Transverse optical and current confinement is achieved by various means, depending on the material system. GaAs-based VCSELs employ oxide confinement, where a high aluminum AlGaAs layer (~99%) located in a null position of the standing wave above the active region is oxidized in a high temperature wet oxidation process to form the oxide aperture with a typical diameter of 6–10 μm. InP-based VCSELs and GaSb-based VCSELs commonly employ a buried tunnel junction (BTJ) [20].

The modulation speed of a VCSEL is limited by the intrinsic damping of the resonant carrier-photon interaction, and by effects of self-heating and electrical parasitics. Guidelines for design are derived from a single-mode rate equation analysis, which results in the following (intrinsic) transfer function, representing the small-signal modulation response at frequency f [14]:

$$H(f) = \frac{f_r^2}{f_r^2 - f^2 + j\,\gamma_d\,\frac{f}{2\pi}}. \tag{14.1}$$

The resonant carrier-photon interaction is characterized by the resonance frequency f_r and the damping rate γ. The rate at which the resonance frequency increases with current (I) above threshold I_{th} is quantified by the D-factor

$$f_r = D \cdot \sqrt{I - I_{th}},$$

(14.2)

$$D = \frac{1}{2\pi} \cdot \sqrt{\frac{\eta_i \Gamma v_g}{q V_a} \cdot \frac{dg_{mat}/dn}{\chi}},$$

(14.3)

where η_i is the internal quantum efficiency, Γ is the optical confinement factor, v_g is the group velocity, q is the elementary charge, V_a is the volume of the active (gain) region, dg/dn is the differential gain, and χ is the transport factor. The rate at which the damping rate increases with resonance frequency is quantified by the K-factor

$$K = 4\pi^2 \left(\tau_p + \frac{\varepsilon \chi}{v_g dg/dn} \right) \quad \gamma_d = \gamma + K f_r^2,$$

(14.4)

where τ_p is the photon lifetime and ε is the gain compression factor. The damping offset γ is inversely proportional to the differential carrier lifetime. The intrinsic modulation bandwidth is set by the rate at which damping increases with resonance frequency [14]. According to calculations, the intrinsic bandwidth benefits from a small photon lifetime, small gain compression, and large differential gain, which results in a small K-factor and less damping at a given resonance frequency. However, these fundamental laser parameters cannot be independently optimized since a smaller photon lifetime leads to higher cavity loss, a higher threshold carrier density, and a lower differential gain. This calls for tradeoffs in the design.

Since VCSELs have a relatively high series resistance, thermal effects caused by current-induced self-heating have more pronounced impacts on speed than edge-emitting lasers. Increasing temperature contributes to a reduction of differential gain, gain and internal quantum efficiency, and an increase of the threshold current. This eventually causes the photon density, output power, and resonance frequency to saturate at a certain current [15]. This is referred to as the thermal rollover current. Therefore, the thermally limited bandwidth benefits from a large D-factor, which enables the VCSEL to reach a high resonance frequency before saturation. Again, this emphasizes the importance of large differential gain [14]. In addition, smaller aperture VCSELs can reach a higher resonance frequency since the D-factor is inversely proportional to the square root of the active region volume [16]. However, there is an optimum aperture size, since an aperture that is too small leads to very high resistance, and therefore excessive self-heating. The optimum aperture size depends strongly on resistance and thermal impedance, which scale with aperture size. For GaAs-based oxide-confined VCSELs, the highest resonance frequency is typically reached for an aperture size in the range 3–7 μm.

Current-induced self-heating also affects damping (the K-factor) through the temperature dependencies of photon lifetime, differential gain, and gain compression [10]. In a properly designed high-speed VCSEL with a sufficiently small photon lifetime, the K-factor is small at low currents and may increase at higher currents, primarily because of the reduction of differential gain with temperature [14,15]. This results in a super-linear increase of damping with resonance frequency squared, and therefore excessive damping at high currents. In addition, the escape rate of carriers out of the quantum wells (QWs) may approach the capture rate at very high active region temperatures, high ambient temperatures, and high currents, causing the transport factor (χ) to exceed unity [14]. This implies that effects of transport may further reduce the bandwidth through reduced resonance frequency and increased damping [see Eqs. (14.3) and (14.4)].

With the strong effects of self-heating on the dynamics of VCSELs (particularly the resonance frequency), it is clear that thermal effects impose severe limitations on the achievable modulation bandwidth. Therefore, high-speed VCSELs should be designed for low heat generation (low resistance [18], low non-radiative recombination rates, low internal optical loss, etc.) and low thermal impedance (DBRs with high thermal conductivity [19] and mounting for efficient heat sinking).

The contributions to the VCSEL capacitance depend strongly on the design and technique used for current and optical confinement. Capacitances associated with the bond pad and the p-n junction (diffusion and depletion capacitances) are common to all designs. In an oxide-confined VCSEL, there is an additional contribution from charge stored over the oxide layer (the "oxide capacitance" [22]), while in a BTJ VCSEL, there is an additional contribution from charge stored over the depletion region of the reversed biased (blocking) p-n junction surrounding the BTJ [20].

14.2.2 **VCSEL detailed design**

Important aspects of the active-region design for 850 nm are carefully analyzed in an article by Healy and O'Reilly [4]. As previously outlined in this chapter, strained quantum wells (QWs) can provide a higher differential gain due to effects of both quantum confinement and strain [21]. In Ref. [4], the authors analyzed the effects of QW thickness and In concentration (strain) on differential gain using the 8-band k.p theory.

When adding In to the QWs for improving differential gain, the QW thickness must be reduced, and the Al concentration in the barriers increased, in order to maintain the gain peak at 850 nm. The accumulated strain must also be considered when using multiple InGaAs/AlGaAs QWs, as the critical thickness should not be exceeded. In addition, a significant population of X-band states in the conduction band will occur if the concentration of Al in the barriers is too high, which may slow down the trapping of electrons in the QW ground state and affect the high-speed performance. Therefore, the Al concentration in the barriers is limited to 37%, where the occupation of X-related states is negligible.

The QW designs considered in [4] have In concentrations varying from 0% to 15%. To maintain the gain peak close to 850 nm, the Al concentration in the InGaAs/AlGaAs QWs is increased to 37%, and the QW thickness is reduced, with increasing In concentration. Constant optical confinement factor (~3%) was also maintained for the VCSELs with InGaAs/AlGaAs QWs, and the number of QWs was increased from 3 to 6, with increasing In concentration and decreasing QW thickness. The calculations show that the addition of In to the QW increases the heavy hole (HH) to light hole (LH) splitting, which is partly due to the strain-induced splitting of the HH and LH band edge states, and partly to the reduction in QW width that is required to maintain a fixed emission wavelength. The incorporation of 10% In in the QW (0.71% strain) increases the HH-LH splitting to 76 meV, while 12% In in the QW layer increases it further to 84 meV, with 15% In giving a calculated splitting of 95 meV.

The calculations also show that the addition of In significantly reduces the density of states near the valence band maximum, which is due to the increased separation and reduced band mixing between the highest HH and LH valence bands. This reduction in the valence band DOS near the band edge reduces the carrier density per unit area required to reach transparency, which also contributes to an improved differential gain in the In-containing QW structures [21]. Furthermore, the combination of the increased In concentration in the QW and Al concentration in the barrier pushes the barrier valence band edge further away from the highest valence state in the QW, with this energy separation increasing from 130 meV in the GaAs/AlGaAs to over 200 meV in the InGaAs/AlGaAs QW structures. Consequently, the modal and the material gain of the structure with 6 QWs and 15% In is higher (by a factor of 2) than the corresponding gain of the 3 QW GaAs structure. As a result, the oscillation frequency Fr will also increase according to Eqs. (14.1) and (14.2).

At Finisar, we have adapted a commercial 3-D finite-element device simulator that addresses optical, electrical, and thermal effects in our VCSEL design. This allows us to gain a deeper understanding of the underlying device physics affecting the VCSEL's high-speed performance, especially the influence of various material systems and carrier dynamics.

We employed the simulator to explore the effect of the In composition in the QWs on the oscillation frequency (Fr). A basic diagram of the VCSEL structure used in the simulations is shown in Figure 14.6. The wavelength was kept constant at 850 nm by reducing the well thickness with increasing In concentration. The active regions in the simulated structures have 3, 4, or 5 QWs. The well widths were in the range between 80 and 40 A, depending on the In composition. We also varied the barrier widths in the range between 80 and 35 A. The simulations show that the oscillation frequency increases as we increase the In composition and reduce the barrier thickness (smaller active volume). The DBR reflectivity was optimized to achieve the appropriate photon lifetime. The parasitic resistance and capacitance was also minimized in these structures. For high In compositions in the range of 17–19%, the oscillation frequency approached 20 GHz at room temperature and 15 GHz at 95 °C.

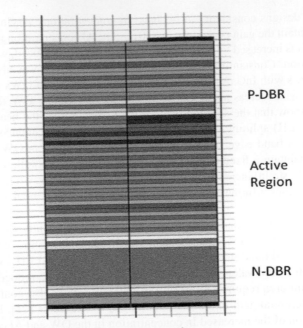

FIGURE 14.6 VCSEL layer structure used for simulations.

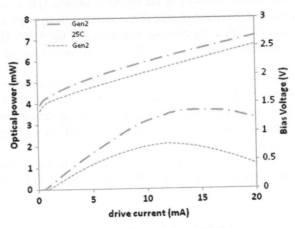

FIGURE 14.7 Typical 14G VCSEL LIV curves at 25 °C and 95 °C.

Design optimization was first applied to 14 Gb/s VCSELs [12]. This new design has now been introduced in Finisar products.

Figure 14.7 shows typical LIV characteristics for the 14 Gb/s devices. For the nominal 8 μm aperture size, the series resistance is 58 Ω. The relatively low-resistance specification was needed for parallel applications, where VCSEL driver chips are directly coupled and typically have no back termination, causing device resistance to affect rise and fall times. The slope efficiency is 0.3 mW/mA at 25 °C and 0.2 mW/mA at 95 °C. This value of slope and power output helps facilitate a low 0.5 mA threshold current. The combination of these parameters allows the use of these devices at module-ambient temperatures of 85 °C.

Figure 14.8 shows measured frequency response of a typical VCSEL at 25 °C and 95 °C. Three decibel bandwidth at 25 °C and at 95 °C saturates at 16 GHz and 14 GHz, respectively, for this design. The simulated S21 curves at 25 °C shown in Figure 14.9 very closely match the experimental data.

This frequency response meets the requirements of the 14 Gb/s products. The damping rate is an equally important parameter that has to be optimized independently. As mentioned earlier in this chapter, a lower damping rate (lower K) leads to a higher bandwidth. According to Eq. (14.4) the K-factor can be decreased by reducing the photon lifetime in the cavity. Several research groups used this method to achieve a high modulation bandwidth of their VCSELs [6,25]. However, the ultimate goal in data link applications is a high mask margin at the lowest possible operating current. A low bias current makes it easier to meet the reliability requirements. If the damping factor K is too low, the bias current must be relatively high in order to sufficiently dampen the oscillations to achieve a substantial mask margin. On the other hand, if

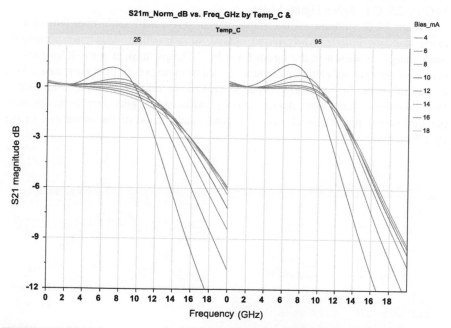

FIGURE 14.8 Normalized S21 versus frequency at 25 °C and 95 °C for the 14G Finisar VCSEL.

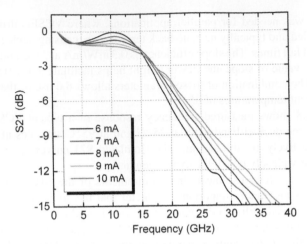

FIGURE 14.9 Simulated S21 versus frequency at 25 °C and various bias currents for the 14G Finisar VCSEL.

damping is too high, the 3 dB bandwidth is reduced, which also adversely affects the eye mask margin. A comparison between simulated and measured eye diagrams at 14 Gb/s, 25 °C is shown Figure 14.10.

Figure 14.11 summarizes eye mask margin levels achievable over temperature using standard Finisar TO-46-based optical subassemblies. Maximum attainable mask margin is shown in the figure for a FC 14 Gbps mask at a mask hit ratio of 0.1%. A conventional pattern generator is used with no pre- or post-emphasis added to further shape the waveform, which is typical for many modern VCSEL driver chips. As shown here, these devices show good mask margins over the entire temperature range, potentially allowing for further expansion of the lower end of the operation range.

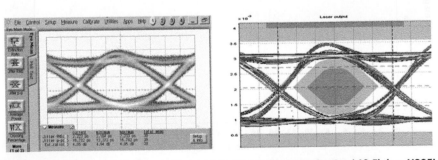

FIGURE 14.10 Measured (left) and simulated (right) eye diagrams for the 14G Finisar VCSEL (14 Gb/s, 25 °C).

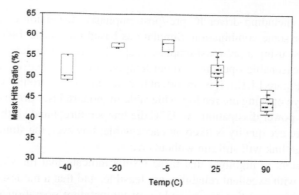

FIGURE 14.11 Eye mask performance for Finisar 14G VCSEL over temperature with 0.1% mask hit ratio.

14.2.3　14 Gb/s reliability

Many design and growth iterations have focused on stabilizing the wear-out reliability of the 14 Gb/s VCSEL. At this point, Finisar's most recent design/growth iteration of these VCSELs has completed more than 2000 h of wear-out evaluation in a full matrix of test conditions—ranging from temperatures of 150–85 °C—and drive currents of 9–15 mA. FIT testing is also under way. Based on the wear-out data, we currently achieve an activation energy of 1.09 eV and an acceleration factor of 3.27 for the most recent round of growths. For comparison, typical 14 Gbps devices with 8 μm apertures show an activation energy of 0.7 eV and a current acceleration factor of 2.0. Table 14.3 shows expected times to 1% failure based on the reliability model at its use conditions. This material shows an excellent wear-out time, despite its high current acceleration factor, largely because higher bandwidth and damping frequency allow for operation at lower drive currents.

14.2.4　28 Gb/s VCSEL design

The current focus of 850 nm VCSEL development activity in the communications industry is making data links operate up to 28 Gbps. As the speed increases standards are likely to incorporate some form of electronic compensation into the optical

Table 14.3 Expected times to 1% failure.

Module	Die Temperature (°C)	Current (mA)	1% Fail 2 dB (years)
14G	55	6	902
	95	6	17

specifications, including driver IC pre-/post-emphasis, dispersion compensation in the receiver, or some combination. Simulations based on Finisar 14G VCSEL data were conducted using a pre-/post-emphasis capable driver chip design. An eye diagram with a reasonable opening at room temperature was achieved after applying equalization (Figure 14.12). It is important to note that the electronic "tricks" used to generate the eye opening are readily achievable in modern laser drivers with reasonable electrical power dissipation. At 95 °C die temperature, bandwidth is reduced to the point where eye quality is likely not acceptable; however, the simulation reveals that the average link will still run without errors.

It is important to highlight that these results indicate that the viability of 28G VCSEL links with excellent reliability are feasible, and that a modest improvement in operating characteristics can provide over-temperature operation with adequate eye quality.

Research teams from Chalmers University and TU Berlin have recently demonstrated datacom VCSELs at 850 nm with a modulation bandwidth of 23 GHz. They have also reported 40 Gb/s error-free transmission at this wavelength [7]. The Chalmers VCSEL design is shown in Figure 14.13 [24]. The active region employs

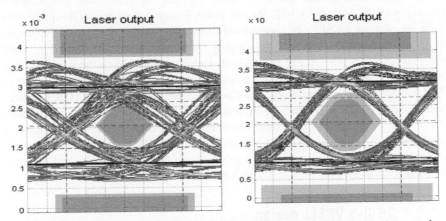

FIGURE 14.12 Simulated eye diagrams for the Finisar 14G VCSEL at 25 °C. The data rate is 28 Gb/s. The eye on the right was equalized (5 taps + jitter equalization).

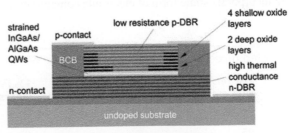

FIGURE 14.13 Chalmers University VCSEL structure [24].

five strained InGaAs/AlGaAs QWs for improved differential gain and a separate confinement heterostructure designed for fast carrier capture and reduced gain compression [4]. For low resistance, they use graded interfaces and modulation doping in the AlGaAs-based DBRs, and a binary compound (AlAs) in most of the bottom DBR to lower the thermal impedance [19]. Low capacitance is achieved by means of a multi-oxide layer design [22,23] with two deep oxide layers defining the aperture for current and optical confinement, and an additional four shallow oxide layers for capacitance reduction [24]. Transport simulations show that this causes a negligible increase of resistance, while reducing the oxide capacitance by 50%. The bond pad capacitance is kept at a minimum by a thick layer of benzocyclobutene under the pad and a small pad size. The high VCSEL bandwidth was achieved by reducing the photon lifetime in the cavity. This was accomplished by shallow etching of the topmost layer in the top DBR, which controls the reflectivity of the top DBR by the phase of the reflection at the surface [6].

The small signal modulation response (S21) of the Chalmers VCSEL is shown in Figure 14.14. The maximum modulation bandwidth is 23 GHz at a 7.8 mA bias current [6].

Finisar also demonstrated an 850 nm VCSEL with a 24 GHz modulation bandwidth (as shown in Figure 14.15). This is the highest bandwidth ever reported for 850 nm datacom VCSELs. The high bandwidth of the Finisar VCSEL was achieved by maximizing the differential gain rather than by reducing the photon lifetime. As a result, the oscillation frequency Fr is very high for this VCSEL, as can be seen in Figure 14.17. As a reference, the oscillation frequencies of the more standard VCSEL designs are also shown in this figure (GaAs- and InGaAs-based). The K-factor (damping) of the Finisar VCSEL is relatively high and the S21 curve shows almost no overshoot at an 8 mA bias current (see Figure 14.15). The simulated S21 agree very well with the experimental data (see Figure 14.16).

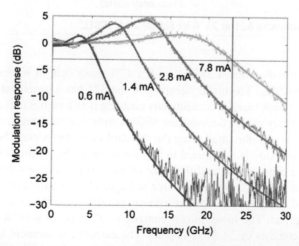

FIGURE 14.14 Normalized S21 at various bias currents for the Chalmers VCSEL [6].

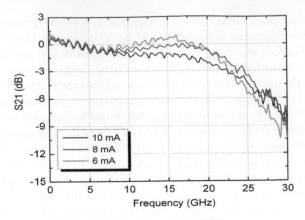

FIGURE 14.15 Measured S21 at 25 °C for Finisar 28 G VCSELs.

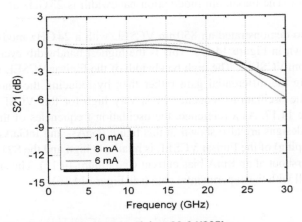

FIGURE 14.16 Simulated S21 at 25 °C for Finisar 28 G VCSEL.

An optical eye diagram at 28 Gb/s with a 47% mask margin is shown in Figure 14.18 for this VCSEL. The higher damping of the S21 has a beneficial effect on the eye opening and mask margin, consequently enabling data transmission at record-high bit rates. Preliminary results demonstrate 55 Gb/s error-free data transmission [8,47].

These devices are fabricated using the standard Finisar process. The capacitance across the native oxide is minimized in two ways: the isolation implant reduces the radius affecting capacitance, and the taper on the oxide increases the thickness of the capacitor at high radii, where the area is large. The parasitic time constants are further reduced using lower mirror resistance and thicker silicon dioxide dielectric under the bond pad. The lower mirror consists of 37 AlGaAs n-DBR pairs and the upper p-DBR consists of 24 AlGaAs pairs. The current confinement is a nominally 6 μm oxide aperture near the second null above the active region. The epitaxial

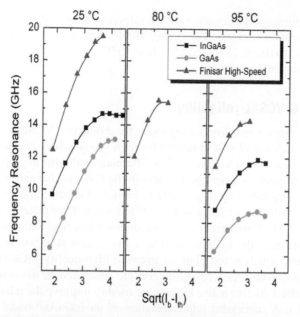

FIGURE 14.17 Measured oscillation frequency (Fr) at 25 °C for Finisar VCSELs.

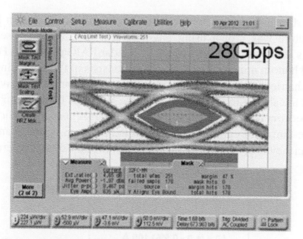

FIGURE 14.18 Finisar VCSEL eye diagram at 28 Gbps.

growth method was metal organic chemical vapor deposition (MOCVD) using an Aixtron system. The process, except for minor modifications to reduce capacitance and accommodate the modified epi, was identical to the standard commercial process. The design methodology used a combination of a 1D optical stack simulator combined with a 2D finite element VCSEL simulator. The use of DC and transient simulations optimized the structure for the relevant DC and AC characteristics. The simulated relaxation oscillation frequency (ROF) from a transient simulation for a 7 mA bias was 20 GHz at 25 °C and 15 GHz at 95 °C, which is in excellent agreement with the experimental results.

14.2.5 28 G VCSEL reliability

In addition to speed, reliability is a key factor for production VCSELs. As previously stated, a high differential gain is needed for high speed, which is achieved by using thin, highly strained InGaAs QWs with a differential gain that is proportional to the In composition. Strong carrier confinement in the QWs is also extremely important and can be accomplished by using AlGaAs barriers. However, the optimum growth temperature is different for AlGaAs and InGaAs when these materials are grown by MBE or MOCVD. High-quality InGaAs requires a lower growth temperature than AlGaAs. In practice, the InGaAs/AlGaAs active region is grown at an intermediate temperature, which is too high for growing high-quality InGaAs and too low for the AlGaAs layers. As a result, defects are generated in the active region. To mitigate this effect Al-free active regions are used to improve the reliability of high-speed VCSELs. A successful implementation of an InGaAsP-based active region is described in [26]. The high material gain and the low transparency carrier concentration make the InGaAsP SC-MQWs a suitable material system for high-speed applications.

The band diagram of the active region is schematically shown in Figure 14.19. The epi structure was grown using low-pressure MOCVD on a semi-insulating (1 0 0) GaAs substrate. The authors use a standard AlGaAs-based DBR structure as the top and bottom mirrors. The active layer consists of three 80A InGaAsP QWs, 100A InGaP barriers, and AlGaAs cladding layers that form a 1L cavity. The layer compositions are shown in Figure 14.19. These VCSELs exhibit kink-free current-light output performance with threshold currents 0.4 mA and slope efficiencies of 0.6 mW/mA. The threshold current change with temperature is less than 0.2 mA, and the slope efficiency drops by less than 30% when the substrate temperature is raised

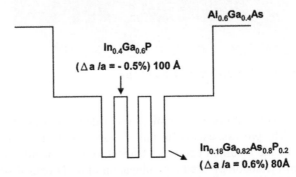

FIGURE 14.19 Band diagram of a VCSEL structure with Al-free active region [26].

from room temperature to 85°C. The small signal modulation response S21 is shown in Figure 14.20 for different bias currents. The 3 dB bandwidth is 14.5 GHz at room temperature, which is a respectable number considering the large thickness of the QWs and barriers.

Current Finisar device results for a VCSEL with an Al-free active region show optical bandwidth at more than 18 GHz and further improvements are expected. Figure 14.21 is a plot of the measured S21 characteristic of a typical VCSEL.

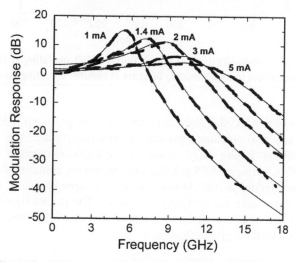

FIGURE 14.20 Normalized S21 versus frequency for a VCSEL from [26].

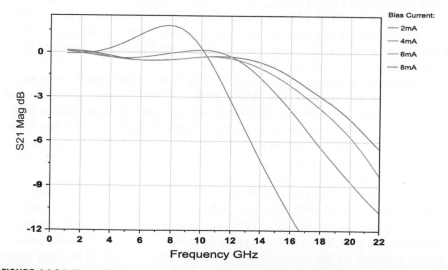

FIGURE 14.21 Normalized S21 at 25 °C measured at various bias currents.

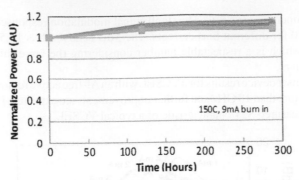

FIGURE 14.22 Typical reliability result of Al-free Finisar VCSEL operating at 150 °C, 9 mA for 300 h, which is equivalent to approximately 50 years at nominal operating conditions.

Simulations of this VCSEL show acceptable eye diagrams up to 95 °C die temperature. We have also conducted preliminary reliability studies on these devices with positive results. Figure 14.22 shows a typical reliability result for a VCSEL operating at 150 °C ambient and 9 mA bias current, which, assuming this is the same device reliability model as our 14 Gb/s devices, is approximately representative of 50 years of life at nominal operating conditions. The results shown here for the Al-free VCSEL are obtained on 8 μm aperture devices with promising reliability, making them practical for productization.

14.3 LONG WAVELENGTH VCSELs (1.3–1.6 μM)

InGaAsP/InP-based Bragg grating and distributed feedback (DFB) lasers have been the sources for long-haul 1.55 μm optical fiber backbone networks for the past three decades, and clearly meet the optical power and dispersion requirements for these applications. However, their price point is still far too high to meet the demands of modern data communication networks.

Low-cost VCSELs operating in the 1.3–1.55 μm wavelength range are an attractive option for high-speed optical MANs, LANs, and FTTH. There is an additional challenge in realizing low-cost, high-bandwidth networks. A larger portion of the low-loss fiber bandwidth must be accessible to enable the utilization of coarse wavelength division multiplexing (CWDM). As a result of these challenges and opportunities, there has been an intense effort to realize low-cost, long-wavelength VCSELs between 1.3 and 1.6 μm over the past decade [27].

Semiconductor lasers operating in the 1.3–1.6 μm region require materials with bandgaps between 0.95 and 0.78 eV. One of the requirements for alloy semiconductors is that they must be reasonably closely lattice matched to readily available binary substrates (GaAs or InP). For many years, InGaAsP on InP was the only materials system available for the communications wavelengths, so most of the long wavelength communications lasers today are fabricated from this system.

However, the InGaAsP/InP system contains several critical limitations that would affect the performance of VCSELs in the long wavelength range. One of the limitations includes the lack of alloy combinations that are lattice matched and produce a large difference in refractive index with reasonable thermal conductivity for the distributed Bragg reflectors (DBRs). Another limitation is the low T_0 (the temperature coefficient of laser threshold current) compared to GaAs-based systems. In addition, the thermal conductivity of the DBR mirror or cladding layers is inferior to GaAs-based structures, resulting in greater junction heating under operation.

The InGaAsP limitations lead to alternative approaches [27], which include InGaAsP/InP quantum well active regions with non-epitaxial DBR mirrors or active gain region materials closely lattice matched to GaAs, combined with GaAs/AlAs DBR mirror technology.

InGaAsP QW-based VCSELs have been fabricated using metal mirrors [28], wafer bonded AlAs/GaAs mirrors, combined InGaAsP/InP and AlAs/GaAs metamorphic mirrors [29], AlGaAsSb/AlAsSb mirrors [30], and dielectric mirrors [31]. GaAs-based VCSEL approaches include InAs quantum dot active regions [32], GaAsSb/InGaAs Type II quantum wells [33], and GaInNAs [27].

The large electronegativity of N and its small covalent radius cause different behaviors in the energy band properties when N is added to GaAs or GaInAs, compared to other III–V alloys. Nitrogen causes formation of bonding and anti-bonding orbital bands, which leads to a decrease of the energy bandgap in GaInAs, thus making it a suitable material for long-wavelength VCSELs [27]. GaInAs provides several advantages over InGaAsP, including better electron confinement, availability of oxide layers (AlAs), and use of the well-developed GaAs/AlGaAs DBR technology.

Several groups were initially successful in demonstrating continuous wave (CW), room temperature, electrically-driven 1.3 μm GaInNAs VCSELs [34–37]. These structures include 850 nm VCSELs with current-carrying n- and p-type DBRs [36], while other groups used either an intracavity-contacted structure [37] or a structure with two n-type DBRs and a tunnel junction [35].

14.3.1 Transceivers with 1.3 μm VCSELs

There has been significant progress in the development of commercial LW VCSELs for use in fiber optic transceiver applications [38]. Our example here uses a 10G SFP + transceiver application [38]. Similar to conventional 850 nm VCSEL technology, these LW VCSELs are GaAs-based devices that use lateral steam oxidation for both current and optical confinement. The active region then uses MBE grown InGaNAs-based quantum wells with an N-type GaAs/AlGaAs DBR. The top DBR is a SiN/SiO$_2$-based structure in order to reduce optical absorption and passivate the die. Like most conventional 850 nm VCSELs, the cathode contact is on the bottom of the die, allowing for smaller die size and better wafer yields.

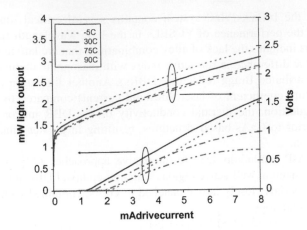

FIGURE 14.23 *L-I-V* characteristics for a 10 Gb/s VCSEL from Ref. [38].

Figure 14.23 shows light and voltage versus drive current at various operating temperatures for a typical 10 Gbps LW VCSEL. Under normal operation, bias current for this device is 5.5 mA at 25 °C and increases to a maximum of 7 mA at 75 °C. Power output is ~1 mW with greater than 30 dB side mode suppression ratio over the entire operating temperature range. Typical differential resistance is 110 Ω. Work to further reduce voltage at 0 °C and 30 °C is ongoing.

Figure 14.24 shows S21 frequency response for a 7 μm aperture LW VCSEL at 25 °C as a function of bias current. 3 dB bandwidth at its normal operating point is 7 GHz. While using smaller oxide aperture devices improves bandwidth, this sacrifices overall reliability. Laser driver and peaking circuit configurations can still be

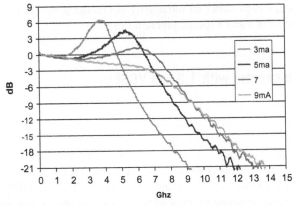

FIGURE 14.24 S21 frequency response for a 7 μm aperture LW VCSEL at 25 °C as a function of bias current [38].

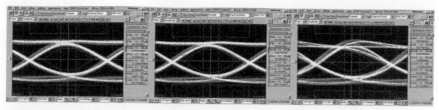

FIGURE 14.25 Sample transmit eye diagrams for 7 μm oxide aperture LW VCSEL SFP+module with commercial driver at 10.3125 Gbps over temperature.

developed to deliver over >20% mask margin over the full 0–75 °C operating temperature range for devices with frequency response comparable to what is shown in Figure 14.20.

Sample eye diagrams over temperature for a commercial driver and module equipped with a 7 μm LW VCSEL are shown in Figure 14.25. Typical rise and fall times are 40 and 50 ps with the 10.3125 Gbps filter engaged, which should give less than 2.4 dB of transmitter and dispersion penalty based on simulations.

The standard Arrhenius model was used to project VCSEL wear-out times under normal use, based on performance under the extreme conditions given in the reliability test matrix below (Figure 14.26). Equation (14.5) gives the Arrhenius model

$$A = (J_{\mathrm{acc}}/J_{\mathrm{op}})^{N}\exp[(E_a/k_B)x(1/T_{\mathrm{jop}} - 1/T_{\mathrm{jacc}})], \tag{14.5}$$

where A is the acceleration factor relative to the normal operating condition, J_{acc} is the current density at the accelerated test condition, J_{op} is the current density at the normal operating condition, and k_B is Boltzmann's constant. The junction temperature (kelvin) at the normal operating condition is T_{jop} and the junction temperature at the accelerated test condition, and is T_{jacc}. The activation energy E_a and the current density exponent N are the two empirically determined parameters used to predict wear-out times.

Condition	Temp	Current (mA)	# Devices
1	70	16	20
2	85	14	20
3	70	14	20
4	100	12	20
5	85	12	20
6	70	12	20
7	125	9	20
8	100	9	20
9	125	7	20

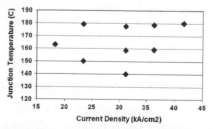

FIGURE 14.26 Drive current and temperature conditions used in reliability test matrix with adjacent figure showing current densities and junction temperatures for a typical wafer under test.

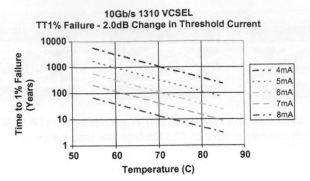

FIGURE 14.27 Time to 1% failure versus temperature and drive current for minimum oxide aperture size [38].

The Arrhenius model was implemented to calculate E_a and N for this sample 10 Gbps LW VCSEL wafer. The 12 mA current condition at oven temperatures of 100 °C, 85 °C, and 70 °C was used to fit E_a at 1.17 eV for the sample wafer. Four drive currents all with a nominal junction temperature of 170 °C are used to fit the current density exponent N at 4.09. These values for E_a and N are typical for oxide confined VCSEL with an InGaNAs-based active region.

For this sample wafer, the predicted time to 1% wear-out failure at 7 mA 70 °C for a 10 Gbps LW VCSEL that meets product specification equaled between 35 and 115 years. The projected time to 1% failure for a 7 μm device as a function of drive current and temperature is shown in Figure 14.27. This plot shows that at a typical operating condition of 55 °C and 6.0 mA, the projected time to 1% VCSEL wear-out failure will be on the order of 700 years. For a transceiver in the field, other failure mechanisms are likely to dominate before this time.

The LW VCSEL technology described here allows development of LR modules that could enable 500 m to 1 km links, which would help fill a distance gap between SW VCSELs (300 m at 10 Gb/s) and LW DFBs (10 km at 10 GB/s). Advantages over long wavelength DFBs are potential reduced cost owing to wafer-scale optical measurements and potential ability to be packaged in a non-hermetic package. In addition, LW VCSELs have been demonstrated to be lower power consumption than current-generation LW DFBs.

14.3.2 1550 nm Long-wavelength VCSELs

Most long-wavelength VCSELs used for the optical communications wavelengths introduced a tunnel junction to eliminate the *p*-doping in the epi-mirror in order to lower the cavity losses [40–44]. Dielectric top DBR mirrors are also used and replace inefficient semiconductor mirrors. A favorable approach for high-speed applications utilizes a buried tunnel junction (BTJ) as a current aperture [44]. Other methods

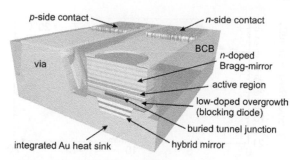

FIGURE 14.28 One thousand five hundred and fifty nanometer VCSEL structure [39].

use GaAs-based DBRs wafer-fused on an InP-based active region [41] or replacing both epi-mirrors with dielectric reflectors [43]. Reliable VCSELs with high data rates were successfully demonstrated using these approaches [39].

A high-speed 1550 nm VCSEL is schematically illustrated in Figure 14.28 [39]. This BTJ VCSEL features a highly strained active region, optimized doping levels, and a low-parasitic chip layout. The BTJ-concept allows effective current-confinement, single-mode operation, low-absorption losses, and consequently, small threshold-current densities. Small mesa-diameters, reduced contact pad size, benzocyclobu-thene-spacers, and modulation-doped n-InP regrowth layers yield high parasitic cut-off frequencies. Embedding the entire chip in an Au pseudo-substrate allows for efficient heat sinking for small back-mirror diameters. The details of the design are further described in [39].

The threshold currents for a device with a 6 μm aperture range from 1.1 mA at 20 °C to 2.9 mA at 80 °C. The output power is above 4.2 mW and 1.3 mW at the respective temperatures. At 20 °C, the threshold voltage is only 0.95 V, indicating a voltage drop of less than 150 mV at the hetero-barriers for a photon energy of 0.8 eV. The differential quantum efficiency (DQE) changes from 45% to 30% over the entire temperature range.

The small signal modulation response of a 5 μm device for different bias-levels at 20 °C is shown in Figure 14.29. Maximum modulation bandwidths exceed 17 GHz. For typical biases applied under large-signal modulation, the relaxation peak is strongly damped, which yields a flat response that is preferred by most applications. The extracted f_{RS} are 18.5 and 7.5 GHz at 20 °C and 90 °C, respectively. This temperature-dependent drop is mainly attributed to the thermal degradation of the differential gain.

Excellent modulation performance is attributed to the novel short-cavity concept with improved electrical, optical, and thermal. Error-free transmission in a back-to-back configuration was achieved at bitrates of 25 Gb/s at 55 °C and 35 Gb/s at 20 °C [97]. In addition, the authors demonstrated error-free transmission over 4.2 km of single-mode fiber. This design makes this device an attractive potential source for optical interconnects and access networks.

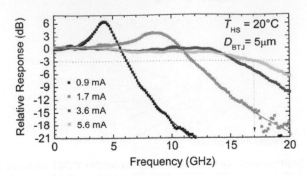

FIGURE 14.29 Small signal modulation response for VCSEL with a 5 μm aperture [97].

14.3.3 Distributed Bragg reflector lasers

Another competing technology at 1310 nm and 1550 nm is offered by distributed reflector (DR) lasers. Demonstrations show that these lasers can be modulated at high data rates and can operate at relatively low, VCSEL-like currents. A small cavity volume and high differential gain of the active region are required to achieve high-speed modulation. This is accomplished by employing highly strained AlInGaAs quantum wells (QWs) and high-reflectivity DBR mirrors [45]. A cross-section of the Fujitsu DR laser is shown in Figure 14.30 [45].

The active region of this laser consists of 12 AlInGaAs compressively strained QWs and a cavity length of 100 μm. Further details of the laser structure are described in [45]. The light-current (*L-I*) characteristics of this DR laser demonstrate a low threshold current of 3.6 mA at 25 °C and 10 mW output power at 42 mA. The threshold current increases to 5.5 mA at 50 °C. This laser is also characterized by a high 3 dB bandwidth of 28.3 GHz at room temperature and over 20 GHz at 85°C, which highly exceeds the best reported speed for long-wavelength VCSELs. Fujitsu used this laser to demonstrate open eye diagrams at ambient temperatures up to 85 °C for a PRBS31 NRZ signal at 40 Gb/s. Further reduction of the cavity volume will lower the threshold and operating currents even more, which will bring the power requirements even closer to those of LW VCSELs.

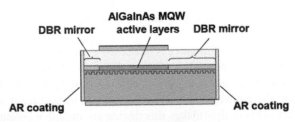

FIGURE 14.30 Cross-section of a Fujitsu DR laser [45].

14.4 DATA RATES >28 GB/S

Optical data links based on directly-modulated VCSELs that operate at 28 Gb/s are becoming a reality. However, the data traffic crossing the network is growing at a pace of 1000 times per decade, demanding transmitters and receivers to operate at even higher rates. To meet a continuously growing demand in data traffic, shorter-distance datacom networks are migrating to even higher speeds. In datacom and access networks, direct modulation is currently predominant due to its cost advantages. However, this is becoming more and more difficult as the required data rate moves toward 56 Gb/s and higher. Robust, high-bit rate sources with ultralow costs are greatly needed for the short distance optical networks such as Local Area Networks (LAN), chip-to-chip, or supercomputer backplane applications. VCSELs are commonly known as the most ideal candidates [12,46,47,99] for these applications for the foreseeable future, but the device performance must meet the system demand.

As mentioned earlier in this chapter, error-free transmission at 56 Gb/s using a directly modulated 850 nm VCSEL has been demonstrated [47,99]. However, directly modulated VCSELs face many challenges as the data rate increases. Since the oscillation frequency is proportional to the square root of the bias current, a fourfold increase in the current density is required in order to double the oscillation frequency. This has very serious reliability implications, as the device lifetime is inversely proportional to the cube of the current density. In addition, as the power density increases the wavelength chirp, dynamic beam degradation and spatial hole burning contribute to a decline of optical transmission quality (e.g. eye mask margin). As a result, various alternative approaches are being investigated and several have already been demonstrated [48–59]. These approaches include VCSELs employing electrooptical modulation, coupled resonant cavities, polarization modulation, and optical injection locking. In addition, we can achieve higher data rates by using advanced modulation techniques (e.g. PAM, CAP, DMT) and parallel optic links, including (C)WDM and spatial multiplexing using multi-core fibers. In this section, we will compare these different methods and describe their advantages, challenges, and tradeoffs.

14.4.1 Eletrooptical modulation

The concept of Electrooptically-Modulated (EOM) filter section was proposed in [48]. According to this concept, an EOM device utilizing resonant interaction between two coupled cavities or waveguides is realized in VCSEL geometry. The device contains a refractive index tunable element (electrooptical modulator) controlled by an applied voltage. Successful realization of the first EOM VCSEL based on two coupled cavities is described in [48]. The structure of this VCSEL is shown in Figure 14.31.

14.4.1.1 Active region

In this structure, the indirect modulation of light is based on the Electrooptic (EO) Effect, which is a change of the refractive index as a result of an applied electric field.

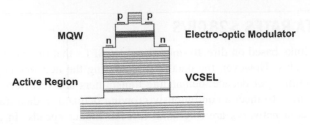

FIGURE 14.31 EOM VCSEL structure [48].

If an electric field is applied perpendicular to the layers in a layered semiconductor structure containing quantum wells (QWs) and barriers, then the conduction and the valence bands of the semiconductor layers tilt due to the potential of the external field. This causes a decrease in the effective band gap energy, and consequently, a shift in the absorption edge to longer wavelengths. This phenomenon is known as the Quantum Confined Stark Effect. Based on the Kramers-Kronig relations, a change in the absorption coefficient "a" results in a change of the refractive index "n." In an optoelectronic device, the QW structure is undoped and placed inside of a p-n junction. The QW composition is designed so that the exciton absorption peak is blue shifted with respect to the lasing wavelength of the VCSEL. When a bias is applied, this creates an electric field across the junction, where both the absorption and the refractive index decrease. The bias range is adjusted to minimize the loss due to the absorption, and also to provide a sufficient change of the refractive index. This means that for efficient laser modulation, the eleoctroptic structure must be characterized by a strong dependence of the refractive index on the electric field. Several approaches based on coupled QWs were proposed [48,60]. One of these concepts is based on the wavelength tuning of the reflectivity dip of the top DBR [48] (see Figure 14.32).

The operation of this VCSEL is based on the stop-band-edge tunable DBR. The principle of operation is illustrated in Figure 14.32, and Figure 14.32a shows the reflectivity spectrum for a zero bias applied to the modulator cavity. The position of the black dot marks the lasing wavelength. Since the reflectivity dip is not aligned with the lasing wavelength, the modulator cavity is not transparent and the output power is zero. Figure 14.32b refers to the resonance state in which a bias is applied to the modulator, the refractive index is changed, and the reflectivity dip of the modulator cavity is shifted toward the resonance with the VCSEL cavity (a shift from the dashed to the solid line in Figure 14.32b). The resonant coupling between both cavities causes a boost in field intensity in the modulator cavity, thus increasing the output power of the laser. The index profiles and the electric field are schematically illustrated in Figure 14.32c and d. It is important to note that the change in the electric field strength between the "on" and "off" states of the device is only significant in the modulator cavity and in the top DBR, but is negligible in the cavity of the VCSEL and in the rest of the device. This ensures that the photon density in the

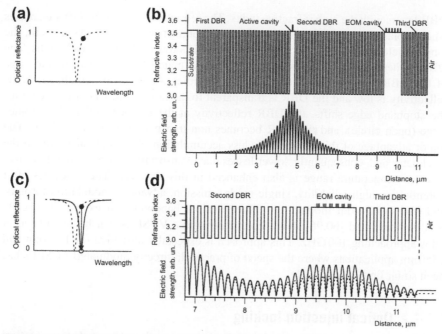

FIGURE 14.32 EOM VCSEL based on coupled resonant cavities [48].

VCSEL cavity is independent of the modulator state (on or off). This approach, however, suffers from limitations related to the accumulation of the optical power in the modulator cavity.

As a result, a different concept of an EOM VCSEL was proposed to address this issue [48,96]. In this approach (see Figure 14.33 [96]), a section of the top DBR is undoped and includes an electrooptic structure, which consists of QWs placed in the

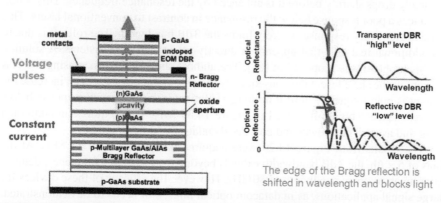

FIGURE 14.33 EOM VCSEL-based stop-band-edge tunable DBR [96].

DBR layers with the high refractive index. The refractive index of the QWs increases when a bias is applied, enhancing the index contrast of the DBR. As a result, the long-wavelength edge of the DBR stopband shifts toward longer wavelengths. The DBR is designed for the lasing wavelength to match the long-wavelength edge of the DBR stopband at a zero bias (filled circle). At this wavelength, the DBR reflectivity is low and the DBR is transparent to the laser light. Under applied bias, the stopband edge shifts, the DBR reflectivity at the lasing wavelength becomes large (open circle), and the DBR becomes non-transparent to the laser light. The major advantage of this approach is less accumulation of the optical power in the modulator and, consequently, lower losses due to absorption. Robust operation over a wider temperature range is also enhanced in this structure. These devices offer potential for high reliability, single-mode emission, advanced modulation formats, and high-bandwidth modulation. Three decibel bandwidth in excess of 35 GHz was demonstrated [60,96], but theoretically, these VCSELs can be modulated at rates approaching 100 GHz. This approach is especially attractive for 1310 nm and 1550 nm applications where the speed of practical, directly modulated VCSELs has been so far limited to 12 GHz.

14.4.2 Optical injection locking

A different approach to high-speed VCSELs is based on the concept of strong Optical Injection Locking (OIL). Using this method, resonance frequency enhancement in excess of 80 GHz was reported [61]. In strong injection locking, the resonance frequency is dominated by the competition between the main locked mode, at the master laser frequency, and the intrinsic cavity mode of the slave laser, which is shifted by the injection [61]. The modulated optical sideband of the locked mode is resonantly enhanced by the cavity mode. Therefore, it is the frequency difference between the two that determines the resonance frequency enhancement. Although the resonance frequency of an injection-locked laser shows a marked enhancement that is significantly higher than that achievable by conventional lasers, the modulation response typically drops sharply before it is enhanced by the resonance frequency. This action produces a poor response below the resonance in contrast to conventional lasers. This type of frequency response severely limits the 3 dB bandwidth. The roll-off is due to an additional real pole that appears in directly-modulated OIL systems. A solution to this problem was proposed in [61]. The authors show that by increasing the bias current of the slave laser, and consequently the photon number, they can increase the low-pass pole frequency until it no longer dominates the frequency response below the resonance. This is illustrated in Figure 14.34 [61]. This requires an additional laser and as such has power and cost disadvantages.

The results from Figure 14.29 were acquired using a 1550 nm VCSEL. At the bias of $5 \times I$th, the 3 dB bandwidth extends beyond the resonance frequency, demonstrating an intrinsic bandwidth of 80 GHz. However, the efficacy of these devices for large signal applications, as in datacom optical links, still needs to be demonstrated. One of the problems is the strong bandwidth dependence on the bias current.

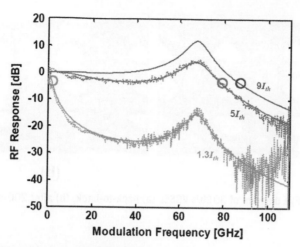

FIGURE 14.34 RF response of OIL VCSEL [61].

14.4.3 Advanced modulation with VCSELs

To satisfy the needs of higher capacity traffic, data communication systems based on VCSELs can use more spectral efficient modulation formats. In recent years, different modulation formats were used with VCSELs. Direct modulation of M-ary pulse amplitude modulation (PAM-M), single-carrier quadrature amplitude phase modulation (QAM), orthogonal frequency division multiplexing (OFDM), and carrierless amplitude phase modulation (CAP) shows the potential to provide data transmission beyond 30 Gb/s with lower bandwidth requirements than NRZ-OOK. However, the higher spectral efficiency of this modulation format comes together with higher requirements in signal-to-noise ratio requirements, and more complex architecture of the transmitter and receiver. In this section, we will review the advanced modulation formats used for direct modulation of VCSELs, along with their advantages and disadvantages.

14.4.3.1 M-ary pulse amplitude modulation

In NRZ-OOK, the modulator has two possible amplitude levels. In a M-ary PAM system, the modulator has M possible amplitude levels. With a signal rate of the system $R_s = 1/T_s$, where T_s is the period of one symbol, the bitrate is $R_b = R_{symb} * \log_2(M)$. Therefore, using M-ary PAM modulation in any given channel bandwidth will transmit $\log_2(M)$ faster. However, the signal-to-noise ratio requirement for a certain bit error ratio (BER) increases 9.54 dB compared to binary modulation for the same baud rate [49]. PAM has relative system simplicity in comparison to other advanced modulation schemes.

PAM-4 modulation of VCSELs was demonstrated at bitrates beyond 30 Gb/s [50–52]. In [50], an 850-nm VCSEL was used to transmit data through 200 m MMF

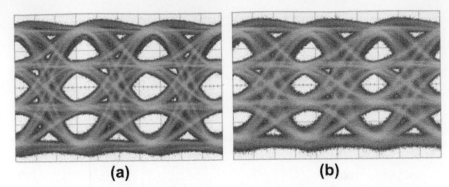

(a) **(b)**

FIGURE 14.35 Eye diagram of 30 Gb/s VCSEL (a) back-to-back, (b) after 200 m MMF.

with real-time error measurement. The eye diagram is shown in Figure 14.35. In [53], 100 Gb/s VCSEL transmissions were presented by using polarization multiplexing a forward-error correction.

14.4.3.2 *Quadrature amplitude modulation*

Quadrature amplitude modulation (QAM) is a modulation scheme that moderates two sinusoidal carriers 90° out-of-phase with each other. The components of each carrier are called inphase and quadrature. Both modulated carriers are summed to result in a signal with amplitude and phase modulation. The spectrum of the QAM signal is centered on the symbol rate frequency. QAM is typically a bandpass transmission; however, for direct modulation of VCSELs, low subcarrier frequency modulation is ideal in order to occupy the maximum spectral bandwidth of the VCSEL.

- **Single-Cycle Subcarrier Modulation**

  Single-Cycle Subcarrier Modulation is a specific case of the Quadrature Amplitude Modulation (QAM) where the carrier is equal to the symbol rate. The first spectrum lobe occupies the spectral band from DC to 2*fsym. 16-QAM direct modulation of a VCSEL at 37 Gb/s was presented in [54]. In [54], the inphase and quadrature signals are generated by a XOR gate, with the baseband data in one input and data clock in the other input. The advantage of this method comes from the simplicity of the high-speed transmitter using a commercial available XOR gate.

- **Half-Cycle Subcarrier Modulation**

  Half-Cycle Subcarrier Modulation uses a carrier that is half the symbol rate. This scheme has a 33% higher spectral efficiency than a single cycle. Direct modulation of a VCSEL at 10 Gb/s was demonstrated [55] with a similar approach as [54].

14.4.3.3 Discrete multitone modulation

Discrete Multitone Modulation (DMT) is a baseband multicarrier modulation that is capable of high-speed transmission with VCSELs. DMT uses multiple orthogonal subcarriers to divide the available bandwidth into many small narrow bands, transmitting the data on each band in parallel. Figure 14.36 shows a representation in time and frequency domain of a DMT signal with four subcarriers. A DMT transmitter scheme is shown in Figure 14.37. QAM is used as a modulation

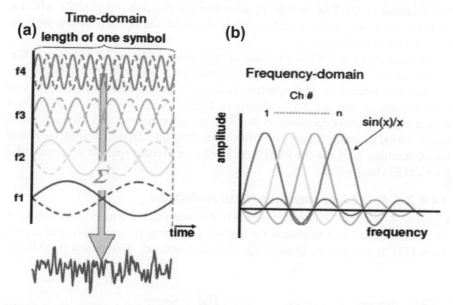

FIGURE 14.36 DMT representation of four subcarriers (a) time domain and (b) frequency domain.

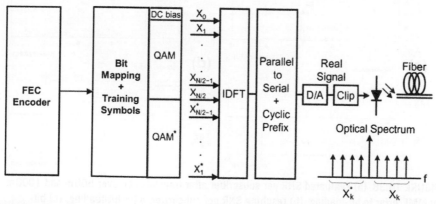

FIGURE 14.37 Block diagram of DMT transmitter.

format for each subcarrier. The DMT transmitter uses an inverse Fourier transform to generate orthogonal subcarriers. At the receiver side, fast Fourier transform is used to obtain the transmitted data from each subcarrier. Orthogonally subcarriers allow for independent subcarrier demodulation. DMT uses cycle prefix to avoid ISI caused by dispersion, and training symbols to estimate and equalize the channel. An important feature of DMT is that it offers flexible modulation order of subcarrier; therefore, it is possible to allocate the number of bits and power per subcarrier, depending on the SNR of each subcarrier (a technique known as bit loading). A disadvantage of DMT is the high Peak to Average Power Ratio (PAPR), which is proportional to the number of subcarriers. Clipping of the signal is used to increase the valuable signal power.

Modulation up to 30 Gb/s was demonstrated using a 850-nm VCSEL over 500 m of multimode fiber [56]. In [56], a 6 GHz bandwidth is used with bit loading implementation to achieve the high spectral efficiency by modulating 150 subcarriers up to 128-QAM for the best SNR subcarriers. Figure 14.38 shows the bit and power allocation that varies depending on the SNR of each subcarrier (bit loading). These results reveal spectral efficiency of seven for the best subcarriers, showing the potential of reaching 100 Gb/s per VCSEL by having an average spectral efficiency of 4 over 25 GHz bandwidth.

14.4.3.4 Carrierless amplitude phase modulation

Carrierless Amplitude Phase Modulation (CAP) was proposed in 1975 by Bell Labs as a viable modulation technique for high-speed communication links over copper wires [57]. It was derived from the Quadrature Amplitude Modulation (QAM) and

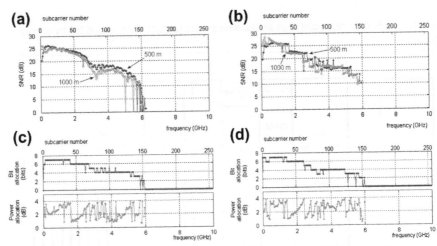

FIGURE 14.38 (a) Evaluated SNR per subscriber after transmission over 500 m and 1000 m of MMF, prior to bit-loading, (b) resulting SNR per subcarrier, after bit-loading, (c) bit-loading for 500 m MMF, and (d) bit-loading parameters for 1000 m MMF.

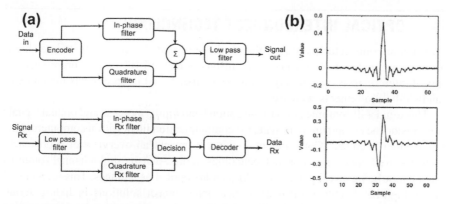

FIGURE 14.39 (a) Block diagram of CAP transmitter and receiver, and (b) waveforms of the in-phase and quadrature.

might be considered its variation, even though there is a fundamental difference in the way the signal is generated.

Figure 14.39 shows the general block diagram of a two-dimensional CAP transmitter and receiver. The incoming data is encoded into multilevel symbols. The multilevel symbols are then shaped by orthogonal filters, resulting in orthogonal waveforms. The output of the filters is summed into a single signal where the result passes through a low-pass filter and is then transmitted. For the two-dimensional form of the modulation, the shaping orthogonal filters are a product of a Square-Root Raised Cosine Filter (SRRC) and a sine or cosine waveform,

$$f_1(t) = h_{SRRC}(t)\cos(2\pi f_c t),$$
$$f_2(t) = h_{SRRC}(t)\sin(2\pi f_c t),$$

where $f_1(t)$ and $f_2(t)$ are the inphase and quadrature filters. At the receiver side, matched filtering is used to recover each component. Higher dimension CAP is achieved by using more than two orthogonal filters. However, the filter design with optimized orthogonality is challenging to obtain for higher order [58]. Therefore, the bits per symbol with CAP modulation is $k = D * \log_2(L)$, where D is the number of dimensions and L is the number of levels in each dimension.

Experimental work was presented up to 40 Gb/s CAP modulation by direct modulation [59]. Using VCSELs, only lower bitrates were demonstrated [60]. However, we believe there is no intrinsic limitation inhibiting the effectiveness of the modulation with VCSELs at higher bitrates and higher order modulations. 2D-CAP16 direct modulation over 25 GHz bandwidth VCSEL could be a potential solution to provide 100 Gb/s per VCSEL.

It should be pointed out, however, that advanced modulation requires high level ICs on the receive side and as such, will have power consumption, cost, and size implications for real applications.

14.5 OPTICAL INTERCONNECT TECHNOLOGY

Optical interconnect technology is already the preferred choice for any "wired" high-speed communication channel covering distances beyond 100 m. The preference for optical interconnects over purely electrical, copper-based interconnects increases as bandwidth requirements increase.

This trend will continue as communication bandwidths increase and optical signals move inside the equipment to provide the communication links within computers. This new frontier is called "Computercom." We will present an overview of short-reach, VCSEL-based optical interconnect technology to high-density multichannel solutions that are now being investigated for multi-terabit/s optical backplanes. This section will survey high-bandwidth, short-reach optical interconnect solutions including serial, parallel optics, and coarse WDM. Since VCSELs are used in Enterprise (<300 m) and Datacenter (<100 m) networks, we will focus on these two networking segments. Historically, storage links were the first to utilize optical interconnects, largely based on the longer distances from computers to storage racks and the relatively modest bandwidth required. In 2005, IBM introduced the ASCI Purple system, which was one of the first systems to utilize optics for rack-to-rack cluster links. Initially, the system deployed all electrical links for rack-to-rack cluster interconnects; however, as the price of optics continued to drop during this period, the longer links were replaced with optics. As the demand for interconnectivity continues to increase to new levels, the communication bottlenecks have moved from inter-rack to board-to-board within a rack, and are now approaching board-level distances [62–67], also see Figure 14.40.

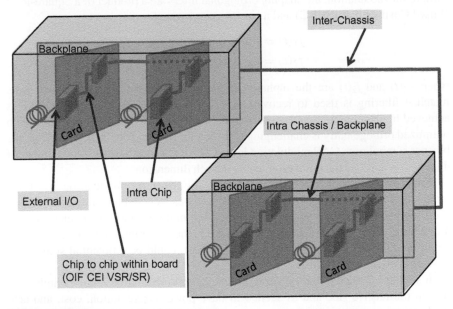

FIGURE 14.40 Next generation interconnect.

For optics to compete with copper at shorter and shorter link distances, there are a number of areas that require additional focus. Increasing channel data rates has improved cost, power and density, and this trend will continue with data rates up to 25 Gbps and beyond. Low-cost optical interconnects for datacom are based on multimode fibers and VCSEL technology [46,68–70]. Rack-to-rack cluster fabric in particular makes effective use of parallel optical modules employing these technologies. In comparison to single-mode technology, which is more commonly used in telecom, multimode technology is more alignment tolerant which translates to lower manufacturing cost. Multimode VCSELs provide an easier and more cost-effective way to test light source than edge-emitting single-mode lasers, as they are easy to make in compact arrays and are much more power efficient. Although the modal dispersion creates distance limitations for multimode fiber, it is the best choice for these primarily short (~100 m) datacom links. Higher data rate VCSELS are being developed with the focus on 25 Gbps technology for 100 Gbps Ethernet (4 × 25 Gbps parallel). Figure 14.41 shows 40, 50, and 55 Gbps transmit eyes of a Finisar VCSEL-based link [47]. An improved version of this link operating at over 56 Gbps will be discussed at OFC 2013 [99].

The use of optical connections over copper cable is determined by the data rate and the distance of the link. As data rates increase, the traditional copper cables struggle to maintain signal integrity over sufficient distance, providing an advantage for deployment of optical links. The typical data rate per lane in datacenters today is 10 Gbps and the average link length is 10 m. Evidently, the product of the Link Bandwidth and Link Distance over the past few decades has remained approximately constant at 100 Gbps × meter. We can use this information to determine future adoption trends for optics, and it implies that when the bandwidth reaches 100 Gbps per link, optics will be used for links as short as 1 m.

Optical transceivers are used at many points within the Datacom network (Figure 14.42), including in switches and on network cards in both the Storage and Local networks. Innovation continues at every level within the network as data rates increase throughout.

This technology will continue to improve. Figure 14.43 shows a prototype highly compact optical module employing flip-chip attachment of the VCSELs and

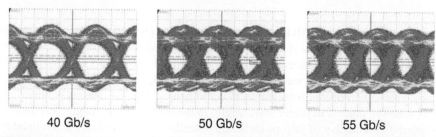

40 Gb/s   50 Gb/s   55 Gb/s

FIGURE 14.41 Optical link eye diagrams for three data rates 40 Gb/s, 50 Gb/s, and 55 Gb/s [47].

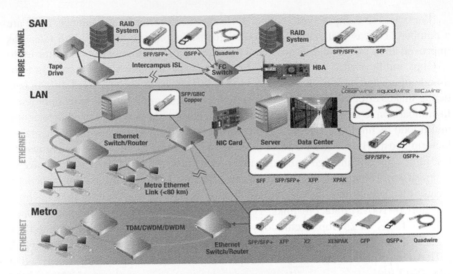

FIGURE 14.42 Optical transceiver applications in datacom.

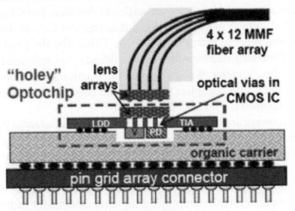

FIGURE 14.43 Compact optical module prototype [71].

photodiode arrays to a CMOS chip with optical vias (holes in the Si substrate) to permit coupling to an optical fiber array [71]. This module provides up to 300 Gbps (24 × 12.5 Gbps in each direction) at 8.2 pJ/bit with a density of 1 Tbps/cm.

Although the 850 nm wavelength for VCSEL links has been the standard for many years, the optimal wavelength is continuously debated. Recently, there has been renewed interest in longer wavelengths in the 900–1100 nm range. This interest is spurred by a number of factors, including potential speed, efficiency and reliability improvements, ease in fabricating backside emitting VCSELs, and the potential for low-cost coarse wavelength division multiplexing (CWDM) transceivers [76–85].

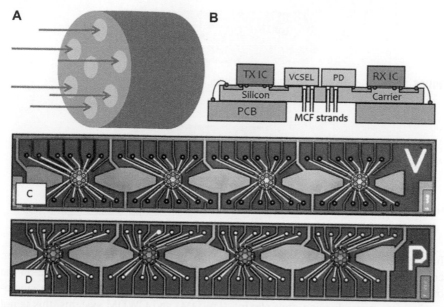

FIGURE 14.44 (a) Transmission through a multicore fiber. (b) Flip-chip board design of [86] and die images of (c) VCSELs, and (d) photodiodes from [86].

CWDM transceivers based on multiple wavelengths can improve the band width (BW) per fiber. This also paves the way for future research that uses different methods, such as multicore fiber and bi-directionality, to achieve the same goals.

Multicore fiber is a promising innovation for parallel applications [86,87]. Up to 120 Gb/s was transmitted through a single fiber consisting of seven graded index multimode cores arranged in a hexagonal lattice pattern, as shown in Figure 14.44a. The cost of these fibers is estimated to be 1.5 times greater than a single fiber alone, compared to equivalent current technology that is seven times greater for seven individual fibers. As shown in Figure 14.44b, arrays of VCSELs and photodiodes can be flip-chip mounted on a silicon substrate with vias designed to couple the multicore fiber strands. Additional optical packaging for an industrial application would have to be designed so the multicore fiber ribbon using a multi-terminal (MT) connector could couple to the VCSEL and photodiode array. This implementation is complex, because it requires precise angular alignment between the VCSEL/PIN arrays and the fiber. Figure 14.44c and d shows the VCSEL and photodiode arrays built in [86]. In this case, they are arranged in a hexagonal array using a common cathode configuration.

Bi-directionality in a single-core multimode fiber may be an additional option to consider, as it offers another way to reduce the number of fibers in the ribbon without adding significant complexity. Figure 14.45 shows how two separate wavelengths can be used to support bi-directional transmission. The use of a DIF (Diffractive

FIGURE 14.45 Bi-directional fiber implementation.

FIGURE 14.46 Bi-directional fiber combined with WDM.

Interference Filter) to reflect the beam emitting from the fiber allows for the beam coupling into the fiber to transmit through. A different DIF relay would be on the other side of the fiber to separate the channels. While commonly used on single-mode dense wave division multiplexing (DWDM) and CWDM systems, this is not commonly implemented in short-wave multimode products.

It is apparent that these improvements could be combined to further increase bandwidth. For example, Figure 14.46 shows how the bi-directional fiber could be combined with wave division multiplexing (WDM) capabilities using a DOE (Diffractive Optical Element) to separate outgoing wavelengths. A total of four wavelengths would be traveling through the fiber (two in each direction). The DIF on each side would be designed to transmit two of them while reflecting the other two. The reflected wavelengths can then be separated using the injection molded DOE. This optical system could be designed in a one piece injection molded material to reduce manufacturing and alignment complexity.

Finally, everything could be combined to use the available bandwidth and fiber most efficiently. Figure 14.47a shows how bi-directional, WDM, and multicore fibers can be combined to provide 14 channels in a single direction through a single fiber at potentially $25\,\text{Gb/s} \times 14\ \text{channels} = 350\,\text{Gb/s}$ through a single fiber in one direction, which is well beyond what was demonstrated in [86]. This is perhaps the most difficult geometry to develop, primarily because the VCSEL/PD layout needs to correspond to the fiber core geometry and the optical system requires precise angular alignment between the fiber and the VCSEL array.

A parallel architecture that is more readily available is shown in Figure 14.47b. A 2D array was demonstrated in [72–75] along with WDM capabilities. By coupling four wavelengths into a single fiber, a parallel product could easily reach 100 Gb/s per fiber, or 400 Gb/s for in a four fiber ribbon. To achieve the density needed for a 2D VCSEL/PD array, a flip-chip bonding scheme would need to be used with either substrate emitting VCSELs, which would preclude the use of 850 nm, or the PCB could be drilled. These are common in research, but have yet to be used extensively

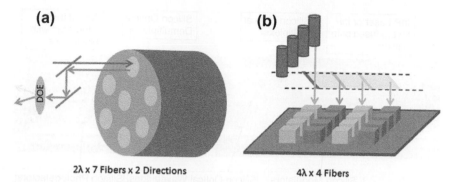

FIGURE 14.47 (a) Bi-directional WDM multicore fiber. (b) 2D array of 4 wavelengths for 16 channels.

in industry. Using a 2D array with only a single fiber ribbon, as opposed to a 2D ribbon, would present significant cost and area savings, given the cost of producing a 2D ribbon and its size.

14.6 COMPARISON OF VCSELS AND SILICON PHOTONICS

Recently, there has been significant research activity in the area of silicon photonics for chip-to-chip communications and other datacenter applications [88–93]. This section will review the cost, performance, and size tradeoffs of silicon photonics, compared to directly modulated VCSELs and InP directly modulated lasers for datacom applications. The term "silicon photonics" has been used to mean many different devices, including silicon waveguides, silicon optical multiplexers and demultiplexers, silicon modulators, and silicon photodetectors, which may be monolithically integrated with one another and/or supporting electronics. It is suggested that the cost of silicon photonics-based optoeletronics devices will be dramatically lower than "traditional" InP and GaAs directly modulated or externally modulated solutions. In addition, since CMOS is associated with low power consumption, there is a perception that a silicon photonics solution will be lower power consumption than "traditional" solutions.

The ultimate vision is to enable chip-to-chip interconnects and some researchers expect silicon photonics to get optics into "Moore's law." Since the early 1990s, researchers have known that it is possible to create high-quality (low loss and single-mode) optical waveguides on silicon-on-insulator (SOI) wafers [92]. It has also been demonstrated that one can attenuate light, and shift its phase, by free carrier injection (or depletion) [94].

Silicon is not able to detect signals at communication wavelengths of 1300 nm and 1550 nm because such photon energies are less than the bandgap; however, these wavelengths represent standard communication bands. This is because optical

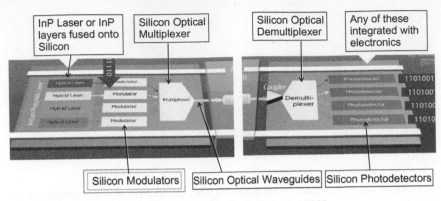

FIGURE 14.48 An example of silicon photonics implementation [98].

fiber, to which devices must eventually interface, has low propagation losses in these bands. Silicon's inability to absorb these wavelengths has been overcome by taking advantage of the small bandgap of germanium grown on silicon [94]. Turning silicon into a laser is far more difficult because of hard limits imposed by nature. Silicon is an inefficient light emitter due partly to its indirect bandgap and partly to free carrier absorption. The vast majority of electron-hole pairs created by electrical or optical injection lose their energy to heat before they can create a photon. In addition, injected free carriers absorb photons through free-carrier absorption. Any light that is created or that enters from the outside will not be amplified because the rate of stimulated emission is far below the rate of absorption by carriers. Amplification is the prerequisite to lasing, which is why it is so difficult to make silicon lase using conventional approaches [95]. As a solution, hybrid approaches that combine InP lasers with Si devices are used. An example of a silicon photonics implementation is shown in Figure 14.48. This hybrid approach combines a continuous wave (CW) InP laser, a Si modulator, s Si optical multiplexer, a Si demultiplexer, and a SiGe photodetector. Substantial research effort is currently focused on modulators, as they are crucial for the generation and transmission of high-speed signals. This is an area where silicon could provide significant advantage over the competing technologies. Modulators with large optical bandwidth, high-speed operation, and good modulation efficiency are required to efficiently send information at high frequencies. Modulators based on a Mach-Zehnder interferometer architecture are of particular interest since they satisfy these criteria. They are a class of interferometric modulators that rely on the relative phase shift of one branch with respect to the other in order to achieve partial-to-full cancelling of the signal at the output, as shown in Figure 14.49.

One fundamental problem with silicon is its lack of a static dipole moment, which is a consequence of its centrosymmetric crystal structure. This means that the linear electrooptic effect, which is the phenomenon that makes LiNbO$_3$ (lithium niobate) and III–V semiconductors good electrooptic materials, is absent

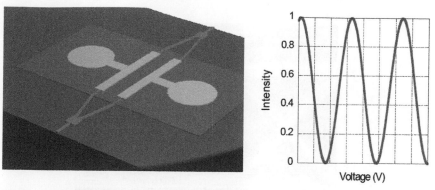

FIGURE 14.49 Mach-Zehnder modulator.

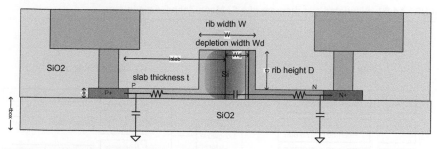

FIGURE 14.50 Carrier depletion Si modulator.

in silicon. Consequently, the only practical way to encode data onto light is by modulating its absorption (and index) via electron and hole injection, or depletion. A waveguide structure of a Si modulator operating on this principle is shown in Figure 14.50.

It has been demonstrated that despite these modulators' limitations, they can operate at 25 Gb/s when they are used in the Mach-Zehnder configuration. Simulations of these structures predict good performance at 56 Gb/s. A simulated eye diagram is shown in Figure 14.51.

Let's consider the advantages and drawbacks of silicon photonics for short-reach (SR) applications when compared to VCSEL-based technology. An example of an implementation of both technologies is shown in Figure 14.52.

In this demonstration, we are comparing multimode VCSELs that are coupled to multimode fibers with a silicon photonics module that contains a single CW InP laser, coupled to four silicon modulators that are aligned to 8-μm single-mode fibers. Based on the current status of both technologies with the data rate limited to 25/28 Gb/s, the shortwave VCSEL transceiver cost is lower than silicon photonics. This is because of the low cost of multi-mode VCSEL packaging, which is much

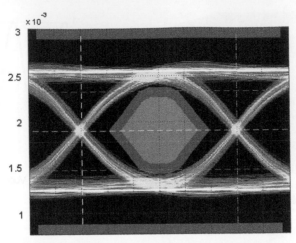

FIGURE 14.51 Simulated eye diagram of a Si Mach-Zehnder modulator.

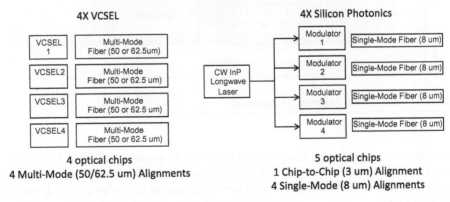

FIGURE 14.52 Short-wave VCSEL versus silicon photonics.

less stringent than single-mode packaging. It is relatively simple to align a VCSEL to a 50 or 62.5 μm multimode fiber; however, much higher precision is required to align a silicon modulator to a single-mode (8 um) fiber. In addition, shortwave VCSELs do not require expensive hermetic packaging currently required by most commercial InP lasers. Integrating silicon photonics with electronics can improve performance by decreasing parasitics, and reducing the size of the optoelectronic chip and assembly costs; however, the existing benefits of integration are still limited for several reasons. The transceiver size is currently determined by the optical and electrical interfaces, so the resulting VCSEL and silicon photonics packages are of the same size. Another limitation is due to the fact that discrete InP lasers must still be used to provide the light source. An additional limitation is that silicon modulators and waveguides do not require the feature sizes required for electronics at 25 Gbps and beyond. Integrating electronics and photonics then unnecessarily requires the photonics process to utilize more expensive fabrication process nodes than required. Also monolithic integration of electronics and photonics requires migrating the photonics process to new fab process nodes every time a new process node is required for higher speed electronics. This is very costly and may be prohibitive for true commercial applications.

If the power budget is analyzed correctly, the shortwave VCSEL will have lower power consumption than silicon photonics for distances less than 100 m at 4×25 Gb/s. Many silicon photonics publications often consider only the switching power of the Mach-Zehnder modulator, with excellent results in the range of 500 fJ/bit to 2 pJ/bit. However, a more practical and appropriate approach is to compare power consumption that is used to send bits error-free from one advanced simulation and computing initiative (ASIC) to another ASIC. Using this approach, the transmitter power consumption is calculated as a function of the link budget, assuming that the power consumed by the pre-drive shaping is the same for both technologies. The results of this analysis are illustrated in Figure 14.53, which clearly show the advantage of the VCSEL-based link.

There are products, however, where silicon photonics might provide advantages, making them an interesting technology to pursue. As discussed earlier in this chapter, direct modulation of VCSELs may not be the most ideal solution for achieving the next speed hurdle of 56 Gb/s. As mentioned earlier, simulation results indicate that Si modulators might show promising performance at these data rates so silicon photonics should be considered along with the other options mentioned in Section 14.4 of this chapter. In addition, high-density single-mode parallel products could be another area where silicon photonics could prove successful. For this application, silicon photonics would compete with long-wavelength VCSELs, reviewed in Section 14.3 of this chapter.

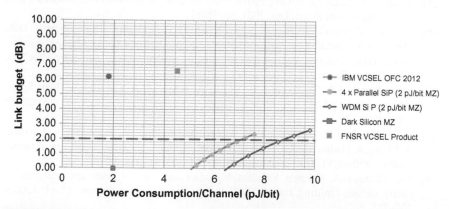

FIGURE 14.53 Power consumption comparison of VCSEL and silicon photonics links.

14.7 CONCLUSIONS

VCSELs have taken the majority of the volume in data link applications due to their cost-competitiveness, low power consumption, and ability to support 10 Gb/s links up to 300 m distance, which is adequate for datacenter applications. Optical data links based on directly-modulated VCSELs that operate at 28 Gb/s per channel have been demonstrated and "hero" data rates are exceeding 56 Gb/s. However, due to the reliability requirements VCSELs utilizing indirect modulation techniques will most likely be used at 56 Gb/s and above. Silicon photonics should also be considered as a viable solution at these data rates especially for high-density single-mode parallel products.

Acknowledgments

The authors would like to acknowledge the contributions of their colleagues from Finisar: Jim Tatum, Luke Graham, Katharine Schmidtke, Frank Flens, Thelinh Nguyen, Gilles Denoyer, Tim Moran, Chris Cole, Roberto Rodes, Steve Joiner, Daniel Mahgerefteh, Gary Landry, Ralph Johnson, Brendan Hamel-Bissell, Greta Light, Theresa Guittard, Ed Shaw, Andy MacInnes, Yasuhiro Matsui, Charlie Roxlo, Ashish Verma, Tsurugi Sudo, and William Wang.

References

[1] LightCounting, Inc. <http://www.lightcounting.com/June2012Forecast.cfm>.

[2] K. Iga, Surface-emitting laser—its birth and generation of new opto-electronics field, IEEE J. Sel. Top. Quant. Electron. 6 (6) (2000) 1201–1215.

[3] J.A. Tatum, VCSEL proliferation, Proc. SPIE 6484 (2007) 648403-1–648403-7.

[4] S.B. Healy, E.P. O'Reilly, J.S. Gustavsson, P. Westbergh, Å. Haglund, A. Larsson, A. Joel, Active region design for high-speed 850 nm VCSELs, IEEE J. Quantum Electron. 46 (4) (2010) 506–512.

[5] J. Tatum, VCSELs for 10 Gb/s optical interconnects, in: Broadband Communications for the Internet Era Symp. Dig. 2001, pp. 58–61.

[6] P. Zhou, J. Cheng, C.F. Schaus, S.Z. Sun, K. Zheng, E. Armour, C. Hains, W. Hsin, D.R. Myers, G.A. Vawter, Low series resistance, high-efficiency GaAs/AlGaAs vertical-cavity surface-emitting lasers with continuously-graded mirrors grown by MOCVD, IEEE Photon. Technol. Lett. 3 (7) (1991) 591–593.

[7] P. Westbergh, J.S. Gustavsson, B. Kögel, Å. Haglund, A. Larsson, A. Joel, Speed enhancement of VCSELs by photon lifetime reduction, Electron. Lett. 46 (13) (2010) 938–940.

[8] P. Westbergh, J.S. Gustavsson, B. Kögel, Å. Haglund, A. Haglund, A. Larsson, A. Mutig, A. Nadtochiy, D. Bimberg, A. Joel, 40 Gbit/s error-free operation of oxide-confined 850 nm VCSEL, Electron. Lett. 46 (14) (2010) 1014–1016.

[9] Y.C. Chang, L.A. Coldren, Efficient, high data rate, tapered oxide aperture vertical-cavity surface-emitting lasers, IEEE J. Sel. Top. Quant. Electron. 15 (3) (2009) 704–715.

[10] S.A. Blokhin, J.A. Lott, A. Mutig, G. Fiol, N.N. Ledenstov, M.V. Maximov, A.M. Nadtochiy, V.A. Shchukin, D. Bimberg, oxide-confined 850 nm VCSELs operating at bit rates up to 40 Gbit/s, Electron. Lett. 45 (10) (2009) 501–503.

[11] N. Suzuki, T. Anan, H. Hatakeyama, K. Fukatsu, K. Yashiki, K. Tokutome, T. Akagawa, M. Tsuji, High-speed 1.1 μm range ingaas-based VCSELs, IEICE Trans. Electron. E92-C (7) (2009) 942–950.

[12] Luke A. Graham, Hao Chen, Deepa Gazula, Timo Gray, James K. Guenter, Bobby Hawkins, Ralph Johnson, Chris Kocot, Andrew N. MacInnes, Gary D. Landry, Jim A. Tatum, The next generation of high-speed VCSELs at Finisar, 2011 Photonics West.

[13] A. Larson, Advances in VCSELs for communication and sensing, IEEE J. Sel. Top. Quant. Electron. 17 (6) (2011)

[14] L. Coldren, S. Corzine, Diode Lasers and Photonic Integrated Circuits, John Wiley & Sons, Hoboken, NJ, 1995.

[15] P. Westbergh, J.S. Gustavsson, A. Haglund, M. Sköld, A. Joel, A. Larsson, High-speed low-current density 850 nm VCSELs, IEEE J. Sel. Top. Quant. Electron. 15 (3) (2009) 694–703.

[16] K.L. Lear, A. Mar, K.D. Choquette, S.P. Kilcoyne, R.P. Schneider, Jr., K.M. Geib, High frequency modulation of oxide-confined vertical-cavity surface-emitting lasers, Electron. Lett. 32 (5) (1996) 457–458.

[17] Y. Wei, J.S. Gustavsson, M. Sadeghi, S.M. Wang, A. Larsson, Dynamics and temperature dependence of 1.3 μm GaInNAs double quantum well lasers, IEEE J. Quant. Electron. 42 (12) (2006) 1274–1280.

[18] P. Zhou, J. Cheng, C.F. Schaus, S.Z. Sun, K. Zheng, E. Armour, C. Hains, W. Hsin, D.R. Myers, G.A. Vawter, Low series resistance, high-efficiency GaAs/AlGaAs vertical-cavity surface-emitting lasers with continuously-graded mirrors grown by MOCVD, IEEE Photon. Technol. Lett. 3 (7) (Dec. 1991) 591–593.

[19] F.H. Peters, M.H. MacDougal, High-speed, high-temperature operation of vertical-cavity surface-emitting lasers, IEEE Photon. Technol. Lett. 13 (7) (2001) 645–647.

[20] W. Hofmann, High speed buried tunnel junction vertical-cavity surface-emitting lasers, IEEE Photon. J. 2 (5) (2010) 802–815.

[21] E.P. O'Reilly, A.R. Adams, band-structure engineering in strained semiconductor lasers, IEEE J. Quant. Electron. 30 (2) (1994) 366–379.

[22] Y.C. Chang, L.A. Coldren, Efficient, high data rate, tapered oxide aperture vertical-cavity surface-emitting lasers, IEEE J. Sel. Top. Quant. Electron. 15 (3) (2009) 704–715.

[23] M. Azuchi, N. Jikutani, M. Arai, T. Kondo, F. Koyama, Multioxide layer vertical cavity surface-emitting lasers with improved modulation bandwidth, in: Proc. 5th Pacific Rim Conf. Lasers Electro-Opt. vol. 1 2003 p.163.

[24] A. Larsson, P. Westbergh, J. Gustavsson, Å. Haglund, B. Kögel, High-speed VCSELs for short-reach communication, Semicond. Sci. Technol. 26 (1) (2011) 014017-1–014017-5.

[25] P. Westbergh, J.S. Gustavsson, B. Kögel, Å. Haglund, A. Larsson, A. Joel, Speed enhancement of VCSELs by photon lifetime reduction, Electron. Lett. 46 (13) (2010) 938–940.

[26] Ya-Hsien Chang, H.C. Kuo, Fang-I. Lai, Yi-An Chang, C.Y. Lu, L.H. Laih, S.C. Wang, Fabrication and characteristics of high-speed oxide-confined VCSELs using InGaAsP–InGaP strain-compensated MQWs, J. Lightwave Technol. 22 (12) (2004) 2828–2833.

[27] J.S. Harris, GaInNAs long-wavelength lasers: progress and challenges, Semicond. Sci. Technol. 17 (8) (2002) 880–891.

[28] H. Soda, K.I. Iga, C. Kitahara, Y. Suematsu, GaInAsP/InP surface-emitting injection lasers, Jpn. J. Appl. Phys. 18 (12) (1979) 2329–2330.

[29] W. Yuen, G.S. Li, R.F. Nabiev, J. Boucart, P. Kner, R.J. Stone, D. Zhang, M. Beaudoin, T. Zheng, C. He, K. Yu, M. Jansen, D.P. Worland, C.J. Chang-Hasnain, High-performance 1.6 µm single-epitaxy, top-emitting VCSEL, Electron. Lett. 36 (13) (2000) 1121–1123.

[30] E. Hall, G. Almuneau, J.K. Kim, O. Sjolund, H. Kroemer, L.A. Coldren, electrically-pumped, single-epitaxial VCSELs at 1.55 µm with sb-based mirrors, Electron. Lett. 35 (16) (1999) 1337–1338.

[31] S. Uchiyama, N. Yokouchi, T. Ninomiya, Continuous-wave operation up to 36 °C of 1.3 µm GaInAsP-InP vertical-cavity surface-emitting lasers, IEEE Photon. Technol. Lett. 9 (2) (1997) 141–142.

[32] J.A. Lott, N.N. Ledentsov, V.M. Ustinov, N.A. Maleev, A.E. Zhukov, A.R. Kovsh, M.V. Maximov, B.V. Volovik, Z.I. Alferov, D. Bimberg, InAs-InGaAs quantum dot VCSELs on GaAs substrates emitting at 1.3 µm, Electron. Lett. 36 (16) (2000) 1384–1385.

[33] O. Blum, J.F. Klem, Characteristics of GaAsSb single quantum well lasers emitting near 1.3 µm, IEEE Photon. Technol. Lett. 12 (7) (2000) 771–773.

[34] T. Sarmiento, H.P. Bae, T.D. O'Sullivan, J.S. Harris, GaAs-based 1.53 m GaInNAsSb vertical-cavity surface-emitting lasers, Electron. Lett. 45 (2009) 978–979.

[35] K.D. Choquette, J.F. Klem, A.J. Fischer, O. Blum, A.A. Allerman, I.J. Fritz, S.R. Kurtz, W.G. Breiland, R. Sieg, K.M. Geib, J.W. Scott, R.L. Naone, Room temperature, continuous wave InGaAsN quantum well vertical-cavity lasers emitting at 1.3 m, Electron. Lett. 36 (2000) 1388–1390.

[36] M.C. Larson, C.W. Coldren, S.G. Spruytte, H.E. Petersen, J.S. Harris, Low-threshold oxide-confined GaInNAs long wavelength vertical cavity lasers, IEEE Photon. Technol. Lett. 12 (2000) 1598–1600.

[37] G. Steinle, H. Riechert, A.Y. Egorov, Monolithic VCSEL with InGaAsN active region emitting at 1.28 m and CW output power exceeding 500 W at room temperature, Electron. Lett. 37 (2001) 93–95.

[38] Luke A. Graham, Jack L. Jewell, Kevin D. Maranowski, Max V. Crom, Stewart A. Feld, Joseph M. Smith, James G. Beltran, Thomas R. Fanning, Melinda Schnoes, Matthew H. Gray, David Droege, Vera Koleva, Mike Dudek, John Fiers, LW VCSELs for SFP+ Applications, SPIE (2007)

[39] W. Hofmann, High-speed buried tunnel junction vertical cavity surface-emitting lasers, IEEE Photon. J. 2 (5) (2010)

[40] N. Nishiyama, C. Caneau, B. Hall, G. Guryanov, M.H. Hu, X.S. Liu, M.-J. Li, R. Bhat, C.E. Zah, Long-wavelength vertical cavity surface-emitting lasers on InP with lattice-matched AlGaInAs–InP DBR grown by MOCVD, IEEE J. Sel. Top. Quant. Electron. 11 (5) (2005) 990–998.

[41] A. Mereuta, G. Suruceanu, A. Caliman, V. Iacovlev, A. Sirbu, E. Kapon, 10-Gb/s and 10-km error-free transmission up to 100 degrC with 1.3–11 m wavelength, wafer-fused VCSELs, Opt. Express 17 (15) (2009) 12981–12986.

[42] M.C. Amann, W. Hofmann, InP-Based Long-wavelength VCSELs and VCSEL arrays, IEEE J. Sel. Top. Quant. Electron. 15 (3) (2009) 861–868.

[43] W. Hofmann, M. Müller, A. Nadtochiy, C. Meltzer, A. Mutig, G. Böhm, J. Rosskopf, D. Bimberg, M.C. Amann, C. Chang-Hasnain, 22-Gb/s Long-wavelength VCSELs, Opt. Express 17 (20) (2009) 17547–17554.

[44] M. Ortsiefer, R. Shau, G. Böhm, F. Köhler, M.C. Amann, Low-threshold index-guided 1.5–11 m long-wavelength vertical cavity surface-emitting laser with high efficiency, Appl. Phys. Lett. 76 (16) (2000) 2179–2181.

[45] T. Yamamoto, A. Uetake, K. Otsubo, M. Matsuda, S. Okumura, S. Tomabechi, M. Ekawa, Uncooled 40-Gbps direct modulation of 1.3-μm-wavelength AlGaInAs distributed reflector lasers with semi-insulating buried-heterostructure, in: 22nd IEEE International Semiconductor Laser Conference (ISLC) 2010, Kyoto, Japan, September 26–30, 2010, pp. 193–194.

[46] R.H. Johnson, D.M. Kuchta, 30 Gb/s directly-modulated 850 nm datacom VCSELs, in: Proc. Conf. Lasers Electro-Optics, San Jose, CA, 2008, Post-deadline Paper CPDB2.

[47] D.M. Kuchta, A.V. Rylyakov, C.L. Schow, J.E. Proesel, C. Baks, C. Kocot, L. Graham, R. Johnson, G. Landry, E. Shaw, A. MacInnes, J. Tatum, 55 Gb/s directly-modulated 850 nm VCSEL-based optical link, in: Proc. IPC, Burlingame, CA, 2012.

[48] V.A. Shchukin, N.N. Ledentsov, J.A. Lott, H. Quast, F. Hopfer, L.Ya. Karachinsky, M. Kuntz, P. Moser, A. Mutig, A. Strittmatter, V.P. Kalosha, D. Bimberg, Ultra highspeed electrooptically-modulated VCSELs: modeling and experimental results, in: Proc. SPIE, Physics and Simulation of Optoelectronic Devices XVI, 2008, pp. 6889–16.

[49] J. Cunningham, D. Beckman, D. Huang, T. Sze, K. Cai, A. Krishnamoorthy, PAM-4 signaling over VCSELs using 0.13 m CMOS, in: Adaptive optics: Analysis and methods/computational optical sensing and imaging/information photonics/signal recovery and synthesis topical meetings on CD-ROM, Technical Digest (Optical Society of America, 2005), Paper IWD3.

[50] K. Szczerba, P. Westbergh, J. Gustavsson, A. Haglund, J. Karout, M. Karlsson, P. Andrekson, E. Agrell, A. Larsson, 30 Gbps 4-PAM transmission over 200 m of MMF using an 850 nm VCSEL, in: Optical Communication (ECOC), 2011 37th European Conference and Exhibition, 18–22 September 2011, pp. 1–3.

[51] J.D. Ingham, R.V. Penty, I.H. White, P. Westbergh, J.S. Gustavsson, A. Larsson, 32 Gb/s multilevel modulation of an 850 nm VCSEL for next-generation datacommunication standards, Lasers and Electro-Optics (CLEO), 2011 Conference, 1–6 May 2011, pp. 1–2.

[52] R. Rodes, J. Estaran, B. Li, M. Muller, J.B. Jensen, T. Gruendl, M. Ortsiefer, C. Neumeyr, J. Rosskopf, K.J. Larsen, M. Amann, I.T. Monroy, 100 Gb/s single VCSEL data transmission link, in: Optical Fiber Communication Conference and Exposition (OFC/NFOEC), 2012 and the National Fiber Optic Engineers Conference, 4–8 March 2012, pp. 1–3.

[53] K. Szczerba, B. Olsson, P. Westbergh, A. Rhodin, J.S. Gustavsson, A. Haglund, M. Karlsson, A. Larsson, P.A. Andrekson, 37 Gbps transmission over 200 m of MMF using single-cycle subcarrier modulation and a VCSEL with 20 GHz modulation bandwidth, in: Optical Communication (ECOC), 2010 36th European Conference and Exhibition, 19–23 September 2010, pp. 1–3.

[54] T. Pham, R. Rodes, J. Jensen, C. Chang-Hasnain, I. Monroy, Half-Cycle QAM Modulation for VCSEL-based optical links, in: Optical Communication (ECOC), 2012 38th European Conference and Exhibition, Paper Mo.1.B.3, 2012.

[55] S.C.J. Lee, F. Breyer, S. Randel, D. Cardenas, H.P.A. van den Boom, A.M.J. Koonen, Discrete multitone modulation for high-speed data transmission over multimode fibers using 850-nm VCSEL, Optical Fiber Communication–includes post deadline papers, 2009, OFC 2009, Conference, 22–26 March 2009, pp. 1–3.

[56] D.D. Falconer, Carrierless AM/PM, Bell Laboratories Technical, Memorandum, 1975.

[57] A.F. Shalash, K.K. Parhi, Multidimensional carrierless AM/PM systems for digital subscriber loops, IEEE Trans. Commun. 47 (11) (1999) 1655–1667.

[58] J.D. Ingham, R.V. Penty, I.H. White, D.G. Cunningham, 40 Gb/s carrierless amplitude and phase modulation for low-cost optical datacommunication links, in: Optical Fiber Communication Conference and Exposition (OFC/NFOEC), 2011 and the National Fiber Optic Engineers Conference, 6–10 March 2011, pp. 1–3.

[59] R. Rodes, M. Wieckowski, T. Pham, J. Jensen, J. Turkiewicz, J. Siuzdak, I. Monroy, Carrierless amplitude phase modulation of VCSEL with 4 bit/s/Hz spectral efficiency for use in WDM-PON, Opt. Express 19 (2011) 26551–26556.

[60] A. Paraskevopoulos, H.J. Hensel, W.D. Molzow, H. Klein, N. Grote, N. Ledentsov, V. Shchukin, C. Möller, A. Kovsh, D. Livshits, I. Krestnikov, S. Mikhrin, P. Matthijsse, G. Kuyt, Ultra-high-bandwidth (>35 GHz) electrooptically-modulated VCSEL, Paper PDP22, OFC/NFOEC 2006, March 5–10, Anaheim, CA, 2006.

[61] E.K. Lau, X. Zhao, H.K. Sung, D. Parekh, C. Chang-Hasnain, M. Wu, OSA 16 (9) (2008) 6609–6618.

[62] A. Benner, M. Ignatowski, J. Kash, D. Kuchta, M. Ritter, Exploitation of optical interconnects in future server architectures, IBM J. Res. Develop. 49 (2005) 755–775.

[63] C. DeCusatis, C.J.S. DeCusatis, Fiber Optic Essentials, Academic Press, New York, 2006.

[64] <http://www-03.ibm.com/systems/power/hard-ware/775/>

[65] B. Arimilli, R. Arimilli, V. Chung, S. Clark, W. Denzel, B. Drerup, T. Hoefler, J. Joyner, J. Lewis, L. Jian, N. Nan, R. Rajamony, The PERCS high-performance interconnect, in: 18th IEEE Annu. Symp. High Perform. Interconnects, 2010, pp. 75–82.

[66] J. Kim, W.J. Dally, S. Scott, D. Abts, Technology-driven, highly-scalable dragonfly topology, in: Proc. 35th, Annu. Int. Symp. Comput. Archit., 2008, pp. 77–88.

[67] A. Gara, M.A. Blumrich, D. Chen, G.L.-T. Chiu, P. Coteus, M.E. Giampapa, R.A. Haring, P. Heidelberger, D. Hoenicke, G.V. Kopcsay, T.A. Liebsch, M. Ohmacht, B.D. Steinmacher-Burow, T. Takken, P. Vranas, Overview of the blue Gene/L system architecture, IBM J. Res. Develop. 49 (2–3) (2005) 195–212.

[68] P. Pepeljugoski, S.E. Golowich, A.J. Ritger, P. Kolesar, A. Risteski, Modeling and simulation of the next-generation multimode fiber, J. Lightwave Technol. 21 (5) (2003) 1242–1255.

[69] P. Pepeljugoski, M.J. Hackert, J.S. Abbott, S.E. Swanson, S.E. Golowich, A.J. Ritger, P. Kolesar, Y.C. Chen, P. Pleunis, Development of system specification for laser-optimized 50-μm multimode fiber for multigigabit short-wavelength LANs, J. Lightw. Technol. 21 (5) (2003) 1256–1275.

[70] D. Kuchta, R. Michalzik, F. Koyama, Progress in VCSEL-Based Parallel Links, VCSELs—Fundamentals, Technology and Applications of Vertical-Cavity Surface-Emitting Lasers, Springer-Verlag, New York, 2011.

[71] F.E. Doany, C.L. Schow, B.G. Lee, R.A. Budd, C.W. Baks, C.K. Tsang, J.U. Knickerbocker, R. Dangel, B. Chan, H. Lin, C. Carver, J. Huang, J. Berry, D. Bajkowski, F. Libsch, J.A. Kash, Terabit/s-class optical PCB links incorporating 360-Gb/s bidirectional 850 nm parallel optical transceivers, J. Lightwave Technol. 30 (4) (2012) 560–571.

[72] B.E. Lemoff et al., Parallel-WDM for multi-Tb/s optical interconnects, 2005 IEEE LEOS Annual Meeting Conference Proceedings, vol. 1, 2005, pp. 359–360.

[73] B.E. Lemoff et al., MAUI: Enabling fiber-to-the-processor with parallel multiwavelength optical interconnects, J. Lightwave Technol. 22 (9) (2004) 2043–2054.

[74] B.E. Lemoff et al., 500-Gbps Parallel-WDM optical interconnect, in: Proceedings Electronic Components and Technology, ECTC 2005, vol. 2, 2005, pp. 1027–1031.

[75] B.E. Lemoff et al., Demonstration of a compact low-power 250-Gb/s Parallel-WDM optical interconnect, IEEE Photon. Technol. Lett. 17 (1) (2005) 220–222.

[76] Y. Okabe, H. Sasaki, Compact Multi/Demultiplexer System Consisting of Stacked Dielectric Interference Filters and Aspheric Lenses Yutaka Okabe and Hironori Sasaki, vol. 4652, pp. 3–9, 2002.

[77] H. Sasaki, Y. Okabe, CWDM multi/demultiplexer consisting of stacked dielectric interference filters and off-axis diffractive lenses, IEEE Photon. Technol. Lett. 15 (4) (2003) 551–553.

[78] L.A. Buckman, B.E. Lemoff, A.J. Schmit, R.P. Tella, W. Gong, Demonstration of a small-form factor WWDM transceiver module for 10-Gb/s local area networks, IEEE Photon. Technol. Lett. 14 (5) (2002) 702–704.

[79] B.E. Lemoff, L.B. Aronson, L.A. Buckman, Zigzag waveguide demultiplexer for multimode WDM LAN, Electron. Lett. 34 (10) (1998) 1014.

[80] J.N. McMullin, R.G. DeCorby, C.J. Haugen, Theory and simulation of a concave diffraction grating demultiplexer for coarse WDM systems, J. Lightwave Technol. 20 (4) (2002) 758–765.

[81] Ondax, Volume holographic gratings (VHG) introduction incidence transmission Bragg condition and selectivity, 2005, pp. 1–7.

[82] D. Yang, H. Wang, X. Guo, J. Zhao, H. Xiang, Wavelength demultiplexing with layered multiple Bragg gratings in LiNbO3:Fe crystal, Appl. Opt. 46 (23) (2007) 5604–5607.

[83] M.W. Beranek, J.C. Bartella, B.A. Capron, D.M. Griffith, D.G. Koshinz, D.L. Livezey, H.P. Soares, Physical design and fabrication of a multiple element slab waveguide spectrograph for multimode fiber-optic WDM systems, IEEE Trans. Comp., Hybrids, Manufact. Technol. 16 (1993) 511–516.

[84] J. Qiao, F. Zhao, R.T. Chen, J.W. Horwitz, W.W. Morey, Athermalized low-loss echelle-grating-based multimode dense wavelength division demultiplexer, Appl. Opt. 41 (31) (2002) 6567.

[85] J. Leuthold, C.W. Joyner, Multimode interference couplers with tunable power-splitting ratios, J. Lightwave Technol. 19 (5) (2001) 700–707.

[86] B.G. Lee et al., End-to-end multicore multimode fiber optic link operating up to 120 Gb/s, J. Lightwave Technol. 30 (6) (2012) 886–892.

[87] B. Zhu, T.F. Taunay, M.F. Yan, M. Fishteyn, G. Oulundsen, D. Vaidya, 70-Gb/s multicore multimode fiber transmissions for optical data links, IEEE Photon. Technol. Lett. 22 (22) (2010) 1647–1649.

[88] R.A. Soref, Silicon-based optoelectronics, Proc. IEEE 81 (12) (1993) 1687–1706.

[89] R. Soref, The past present and future of silicon photonics, IEEE J. Sel. Top. Quant. Electron. 12 (6) (2006) 1678–1687.

[90] D. Guckenberger, S. Abdalla, C. Bradbury, J. Clymore, P. De Dobbelaere, D. Folz, S. Gloeckner, M. Harrison, S. Jackson, D. Kucharski, Y. Liang, C. Lo, M. Mack, G. Masini, A. Mekis, A. Narasimha, M. Peterson, T. Pinguet, J. Redman, S. Sahni, B. Welch, K. Yokoyama, S. Yu, Advantages of CMOS photonics for future transceiver Applications, in: Proc. 36th Eur. Conf. Exhib. Opt. Commun., 2010, pp. 1–6.

[91] Y.A. Vlasov, S. Assefa, W.M.J. Green, M. Yang, C.L. Schow, A. Rylyakov, CMOS Integrated Nanophotonics: enabling technology for exascale computer systems, Presented at the SEMICON, SEMI Technol. Symp., Tokyo, Japan, December 2010.

[92] D. Van Thourhout, Si Photonics, in: Proc. Opt. Fiber Commun. Conf., 2010 [Online]. <http://photonics.intec.ugent.be/download/>

[93] D.A.B. Miller, Optical interconnects, in: Proc. Opt. Fiber Commun. Conf., 2010 [Online]. <http://ee.stanford.edu/~dabm>

[94] B. Jalali, S. Fathpour, Silicon photonics, J. Lightwave Technol. 24 (12) (2006) 4600–4615.

[95] B. Jalali, Making silicon lase, Sci. Am. (2007) 58–65.

[96] J.A. Lott, High-speed VCSELs and subassemblies for short-range optical communications, in: ECOC, September 2010.

[97] M. Muller, M.C. Amann, in: ICTON, 2011.

[98] M. Paniccia, The 50 Gbps SiPhotonics link, intel.com/pressroom.

[99] D.M. Kuchta, A.V. Rylyakov, C.L. Schow, J.E. Proesel, C. Baks, C. Kocot, L. Graham, R. Johnson, G. Landry, E. Shaw, A. MacInnes, J. Tatum, 56 Gb/s directly-modulated 850 nm VCSEL-based optical link, (Invited) in: Proc. OFC, Anaheim, CA, 2013.

[100] D. Cunningham, Example Calculations For 32GFC VCSEL-Based Links, document 11-230v0, <http://www.t11.org>.

[101] J. Tatum, VCSEL Based 32GFC Data links, document 10-268v0, <http://www.t11.org>.

Implementation Aspects of Coherent Transmit and Receive Functions in Application-Specific Integrated Circuits

15

Andreas Leven and Laurent Schmalen

Bell Labs, Alcatel-Lucent, Lorenzstr. 10 70435 Stuttgart, Germany

In this chapter, we will discuss implementation aspects of current and future application-specific integrated circuits (ASICs) for coherent optical transmission systems. Main building blocks for such ASICs are data converters, baseband signal processing, forward error correction and interfacing. We will highlight selected implementation details.

15.1 INTRODUCTION

Coherent line cards are seen as key enablers for high-speed optical communication systems. Today, more than 80% of all shipped 100 G line cards are based on coherent technology [1]. This percentage is expected to increase at least for the long-haul market, with more and more companies adhering to the OIF recommendation laid down in the 100 G Ultra Long-Haul Dense Wavelength Division Multiplexing (DWDM) Framework Document [2]. Today's coherent receivers utilize digital signal processing, mostly implemented in an application-specific integrated circuit (ASIC) or field-programmable gate array (FPGA).

A signal processing device for optical coherent communication can be partitioned as follows:

- Data converters serve as the interface to the optical front-ends. On the receive side, analog-to-digital converters (ADCs) translate the analog waveforms into time and amplitude discrete signals with a given resolution and sampling rate. On the transmit side, digital-to-analog converters (DACs) are utilized if complex waveforms beyond Quadrature Phase Shift Keying (QPSK), possibly software-defined, are desired or if pre-filtering for chromatic dispersion or spectral shaping is required. Data converter designs are called mixed-signal circuits since both analog and digital functions are performed. Typically, a coherent receiver requires four ADCs. If the transmitter comprises data converters, four DACs are required as well.

Optical Fiber Telecommunications VIA. http://dx.doi.org/10.1016/B978-0-12-396958-3.00015-9

- The data is processed in the actual digital signal processor (DSP). The signal processor comprises implementation of all necessary algorithmic processing steps, i.e. chromatic dispersion compensation, timing recovery, polarization demultiplexing, frequency and phase estimation, and so forth on the receive side and pre-filtering for CD compensation and spectral shaping on the transmit side. Typically not only the whole signal processor but also constituent processing blocks are implemented in real time, meaning that for each block, there is a continuous input and output of data. The signal flows continuously as a data stream through the signal processor, with only flip-flops as buffering storage to ensure synchronous timing throughout the whole process.
- The data to be transmitted is encoded in a forward error correction (FEC) encoder (transmit path) and the processed data is verified and corrected in the FEC decoder (receive path). The overall encoder and decoder are also implemented in real-time, but the constituent parts are typically not. As for a block code the FEC decoder needs information about the whole received code word, a full FEC frame (several tens or even hundreds of thousand bits) is stored in a memory. The actual decoder then operates on the stored code word while the subsequently incoming data is stored in another memory block. The amount of time a decoder can operate on a single FEC frame is then given by the total size of the memory divided by the data rate.
- The data has to be transferred to/from another device responsible for framing the data and potentially for secondary FEC encoding/decoding. There are several standardization activities ongoing or finalized that define interfaces with high data throughput of 100 Gbps and beyond. Examples are SFI-S [3], CEI-28G-VSR or the Interlaken Protocol [4]. Interface considerations are outside the scope of this chapter and will not be discussed further.

Figure 15.1 illustrates the interconnection of the signal processor parts as discussed above. In the remainder of this chapter, we will design challenges, technology aspects, and partitioning of these functions. For details on algorithm and code choices for the DSP and FEC part, please refer to Chapters 27 and 28.

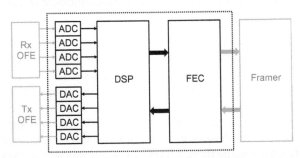

FIGURE 15.1 Block diagram of a typical coherent integrated circuit (hatched box).

15.2 ASIC DESIGN OPTIONS AND LIMITATIONS

Traditionally, there are a number of different devices to choose from for implementing a digital signal processor for communication purposes:

General-purpose digital signal processors (DSP processors) allow for short design cycles as a number of functions, for instance multiply-and-accumulate (MAC) operations, finite impulse response (FIR) filters, or fast Fourier Transform (FFT), are already preconfigured. Unfortunately, today the data throughput achievable with these devices is not sufficient to support multiple tens of gigabits, as required for optical coherent communications.

Field-programmable gate arrays (FPGA) are gaining popularity also for signal processing applications. Benefits of an FPGA design versus an ASIC design are the significant lower development costs, reconfigurability and therefore lower design risks. Several coherent receiver demonstrators have been built using FPGA technology [5,6]. These demonstrators typically operate at a much lower data rate, albeit there have been implementations up to 100 Gbps using FPGA farms [7]. FPGAs are ideally suited for research, early demonstrators, and algorithm verification before implementation. Size, cost, and power consumption considerations prevent their use as a signal processing device in production line cards as of today.

Today, all commercial line cards that are based on coherent reception are using one or multiple ASICs for data conversion, signal processing and error correction. Fully customized for use in coherent optical line cards, ASICs offer the best performance and lowest power consumption at the expense of a complex and costly design cycle.

Referring to Figure 15.1, the different parts of a coherent receiver and transmitter require different design methodologies. The DSP and FEC part are typically designed in a standard cell design technique. Signal processing algorithms and error correction logic are specified and described down to bit equivalency in a higher programming language as C or Matlab. A hardware description language (HDL, for instance VHDL or Verilog) is then utilized for a high-level description of the ASIC, which is called register transfer level (RTL). In a next step, called synthesis, the RTL description is broken down into smaller logic entities, as multiplexers, adders, flip-flops, or basic logic functions. These logic entities are called standard cells. Actual hardware descriptions of the standard cells are provided by the ASIC manufacturer in a cell library. In a final step before tape-out, the standard cells need to be arranged within the chip frame and interconnected. This step is called "place and route" (P&R). Synthesis and P&R are typically automated processes performed by appropriate CAD tools. The description of the ASIC in HDL is still typically a manual process performed by a skilled ASIC designer. Each step, manual or automated, requires extensive verification and analysis to assure a proper translation of the required algorithms into a circuit with reasonable size and power consumption for a given throughput. Since DSP and FEC are purely digital functions, they are best realized in CMOS technology.

Signal converters (ADC and DAC) comprise both analog and digital parts and therefore are called "mixed-signal" circuits. Because of the stringent requirements concerning sampling speed, bandwidth, and signal integrity, they require a different design methodology. Typically, a full custom design procedure is implemented for the design of data converters at speeds considered here. For analog parts and even critical digital parts, the circuit layout is done manually and parasitic extraction is performed to augment the circuit simulation with more realistic models.

High-speed signal converters can be built in CMOS technology or SiGe-based bipolar technology. Due to the high degree of parallelization, CMOS favors high-sampling rate designs, while SiGe bipolar supports higher bandwidths. One possible technology option is to choose a SiGe-based converter design and a CMOS-based design for the digital functionalities. This would probably lead to converters with optimum frequency response and would allow for choosing a more advanced CMOS technology, since for CMOS standard cell libraries typically mature faster than their analog counterparts. Unfortunately, this dual-chip approach would lead to significant challenges with respect to the data transfer in between the converter ASIC and digital ASIC. A typical converter used in coherent optical communications exhibits a physical resolution of 8 bits. This amount of data needs to be transported to/from the digital ASIC. If we assume a power consumption of about 10 mW/GBit for a serializer/deserializer pair and a converter sampling rate of about 60 GSample/s, this would lead to a total power consumption of about 40 W for interconnecting the eight converters shown in Figure 15.1 to the signal processor.

15.3 HIGH-SPEED DATA CONVERTERS

Nowadays, the needs of optical communication systems drive the highest sampling speeds of digital-to-analog (DAC) and analog-to-digital (ADC) converter technology. Converters with sampling rates of 56 GSample/s have been demonstrated [8,9] for optical communication purposes. These rates far exceed what is commonly used in other applications as consumer electronics or mobile communications [10]. In the following, we will briefly review basic converter design approaches, highlight challenges for scaling data converters to sampling rates in the order of 60 GSample/s and beyond and finally we will give an outlook on predicted converter evolution in the light of CMOS technology scaling. Due to space constraints and the more readily available information in the public domain, we will restrict ourselves to discussing ADCs. Nevertheless, a significant portion of the findings discussed here also holds true for DACs.

15.3.1 ADC architectures

One of the most basic ADC architectures is the so-called flash ADC. This architecture is applicable to a range of different technologies. One of the earliest implementations of a flash ADC was implemented in vacuum tube technology [11]. A flash ADC is based on a set of parallel comparators, see Figure 15.2. For a resolution of N bits, $2^N - 1$ comparators are needed. The signal to be digitized is sampled and held in a

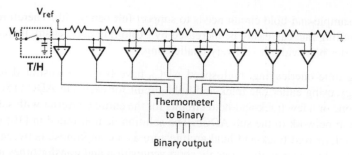

FIGURE 15.2 Simplified schematic of a 3-bit ADC.

storage capacitor (box labeled T/H in Figure 15.2). From there, it is fed in parallel to all comparators, each having a slight offset in threshold. The output signal of this operation is a binary word, with all bits equal to "1" for bits representing a voltage smaller than the input signal, and all others set to "0." This encoding is called thermometer code. A subsequent coding block is required to translate this thermometer code into a binary or gray code.

A significant advantage of this converter structure is its high sampling speed, up to the clocking speed that the respective technology allows for. Therefore, this design is often used for high-speed converters in technologies that permit very high clocking speeds, as for instance SiGe-based converters [12,13]. CMOS technologies do not offer clocking speeds as high as the sampling speeds required in optical communication systems. To overcome this limitation, parallel converter structures have been proposed, with the time interleaved structure being the most commonly implemented architecture [14]. In a time-interleaved ADC, a number of parallel sub-ADCs are employed for sampling a signal at an overall much higher rate than the individual sub-ADCs could support. If N denotes the number of parallel sub-ADCs, each sub-ADC is clocked at $1/N$ of the overall sampling rate. A polyphase clock distribution network needs to be set up in a way that each sub-ADC receives a clock signal that is shifted by $2\pi/N$ with respect to the neighboring sub-ADC. This assures an equal-distance time sampling for the overall ADC with a digitized data stream comprising samples from a series of different sub-ADCs, all running at a much lower sampling rate. Even if this is a very elegant method of designing data converters with very high sampling rate, several challenges need to be considered:

- Even if the sampling clock frequency is lowered by a factor of N, signal integrity of the clock, especially with respect to timing jitter and clock skew, needs to be as high as for a full rate clock. That means that not only the reference clock source needs to be extremely clean, but also the polyphase clock generation and distribution network needs to be designed with great care.
- The first track-and-hold stage will see the full high-speed analog signal and therefore needs to be able to resolve the received signal up to the full bandwidth of the overall converter. Also, the signal distribution network to

each sample-and-hold circuit needs to support full bandwidth. Therefore, even if sampling rate can easily be increased by exploiting parallelism, increasing converter bandwidth is still very challenging.

Using time-interleaving, high-speed ADCs have been demonstrated in CMOS technology using either one track-and-hold circuit for each sub-ADC [15] or alternatively one or a few track-and-hold circuits for the entire converter with subsequent distribution network to the sub-ADCs. The solution demonstrated in [16] utilizes a fourfold interleaved track-and-hold architecture as a compromise between a single track-and-hold device with extremely short acquisition and transfer times and a full array of track-and-hold circuits with its challenges with respect to matching and analog signal distribution. Furthermore, [16] proposes an alternative track-and-hold design: Instead of using a switch and capacitor design as depicted in Figure 15.2, a charge-mode sampling is used. Unfortunately, no further details of the sample architecture are available in the public domain.

The complexity of the circuit depicted in Figure 15.2 increases exponentially with the physical resolution of the converter. Since a high-speed time-interleaved converter running at 56 GSample/s or higher requires a few hundred instances of the base converter, power consumption and area requirements can become significant even if a sub-optimum solution for the base circuitry is chosen. In [16], a successive-approximation-register (SAR) design is proposed for the sub-ADCs to reduce overall circuit complexity and enabling a higher physical resolution than would be possible with a flash design. A simplified schematic of an SAR ADC is depicted in Figure 15.3. As the name suggests, the correct digital output of this converter is successively approximated by adjusting the bits in a test register, which is then converted into an analog signal by a reference DAC, then comparing this signal with the analog signal received from the T/H circuit. The result of this comparison is used to adjust the test register. The operation starts with setting the most significant bit (MSB) to "1," all others to "0." The reference DAC then produces a mid-level analog signal. If this signal is smaller than the analog signal coming from the T/H, the MSB is set to "0," otherwise, it is kept at "1." The process continues with the next highest bit, until all bits have been set correctly. Obviously, this method requires multiple clock cycles. Therefore, the internal digital clock for an SAR runs at a multiple of the sampling rate of the sub-ADC. Also combination of the circuits depicted in Figure 15.2 and Figure 15.3 are possible.

15.4 IMPLEMENTATION OF SIGNAL PROCESSING ALGORITHMS AT HIGH SPEED

Similarly to the converter design, digital signal processing structures have to be implemented for being able to process data at 100 Gbps and beyond. Unfortunately, not all algorithms can be parallelized without modifications and loss of performance.

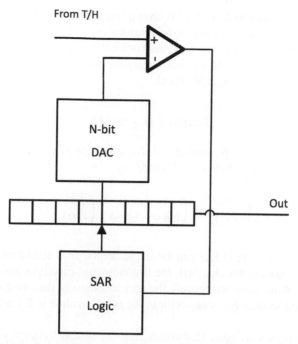

FIGURE 15.3 Simplified block diagram successive-approximation-register ADC.

Algorithms that are time invariant can simply be parallelized by instantiating the circuitry that implements the algorithm multiple times. Often, the complexity can be reduced by sharing resources between multiple instances. An example of a structure that can easily be parallelized is a finite impulse response filter (FIR) with constant coefficients. If the filter coefficients are not constant, e.g. within an adaptive filter structure, there might not be an equivalent parallel structure, e.g. when the update of the filter coefficients is performed once per sampling period. If one is willing to compromise on update speed by accepting an update rate once every clock cycle, with the clock cycle being $1/n$ times the sampling period, with n the parallelization factor, an equivalent structure can be implemented.

The signal processing algorithms necessary for implementing a high-speed coherent receiver are discussed in detail in Chapter 27. In the following, the implementation of some of these algorithms will be discussed in more detail.

15.4.1 Quadrature imbalance compensation

In an initial step, impairments of the optical and electrical front-end are compensated. One example is the correction of quadrature imbalance stemming from imperfect phase control in the optical hybrid. Quadrature imbalance compensation is well known in wireless communications and has been proposed to be used in optical communications as well [17]. If there is no amplitude and offset error of the in-phase and quadrature component of the signal, quadrature imbalance compensation can be performed by first measuring the cross correlation between the in-phase (I) and quadrature (Q) components of the received signal, which is proportional to the sine of the phase error ψ of the optical hybrid:

$$\langle I(t)Q(t)\rangle = \frac{1}{2}\sin(\psi). \tag{15.1}$$

I and Q are assumed to be normalized. The components can then be transformed in corrected orthogonal components I' and Q' by

$$\begin{pmatrix} I'(t) \\ Q'(t) \end{pmatrix} = \begin{pmatrix} 1 & 0 \\ \tan\psi & \sec\psi \end{pmatrix} \begin{pmatrix} I(t) \\ Q(t) \end{pmatrix}. \tag{15.2}$$

Equations (15.1) and (15.2) can be implemented in a feed-forward structure. This has two major drawbacks: first, the trigonometric functions have to be implemented in a look-up table and second, the normalization of the two components has to be performed accurately; as any error in the normalization will lead to quadrature imbalance.

The circuit shown in Figure 15.4 avoids these drawbacks by employing a feedback structure: the cross correlation is measured after the actual compensation, weighted with a convergence factor μ and then integrated. After convergence is achieved, the cross correlation is zero in average and the output of the integrator is constant and, according to Eq. (15.1) proportional to the sine of the phase error. This result is then multiplied with the I tributary and added to the Q-tributary to yield the corrected output. Please note that, according to Eq. (15.2), the result should be divided by the cosine of the phase error as well to yield the correct amplitude. This step can be omitted if a gain control circuit is placed after the quadrature imbalance compensation.

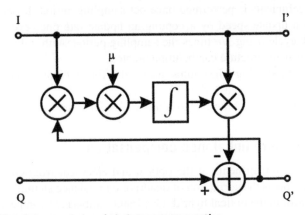

FIGURE 15.4 Circuit for quadrature imbalance compensation.

15.4.2 Chromatic dispersion compensation

It is advisable to split the equalization of the received signal in two steps: first, perform a static or slowly adaptive equalization on each polarization tributary separately and second, perform a fast adaptive joint equalization on both polarization tributaries. The first equalizer is typically chosen to have a much longer impulse response and can be used to compensate for quasi-static effects as chromatic dispersion or frequency response of the optical front-end. The second one having a shorter impulse response but a faster adaptation speed is typically used for polarization tracking, equalization of polarization mode dispersion as well as residual chromatic dispersion not compensated for by the static equalizer.

Typically, equalizers for data rates considered here employ digital block filters [18]. Block filtering involves the calculation of a finite set, or block of output values based on a finite set of input values. This can be performed in time domain or equivalently in frequency domain [19]. Algorithms have been developed for block filtering to achieve identical outputs as sequential filtering, most notably the "overlap-and-save" method and the closely related "overlap-and-add" method.

Let us assume that the input data sequence is partitioned in blocks of length n and that k is the impulse response length of the desired filter function. In case of overlap-and-save, n input samples are concatenated with k symbols from the next block and then convoluted with the impulse response. The first k samples of the output of the convolution are not used while the remaining n samples constitute the correct filter output. Therefore, this method is often also referred to as "overlap-and-dump" or "overlap-and-scrap."

In the case of overlap-and-add, n input samples are padded with k zeros before being convoluted with the impulse response of the desired filter. After convolution, k trailing samples are stored to be added to the k leading samples of the following result of the convolution. Overlap-and-add is typically slightly more efficient in terms of implementation complexity and is often chosen when the filter response is static or changes only slowly in time. For fast adaptive filters, the overlap-and-save method is preferred, because for the overlap-and-save method, each output block is the filtering result of exactly one impulse response function, while in the case of the overlap-and-add method, the portion that is saved from the previous result to be added to the current result might have been calculated with a different impulse response function if the filter function changed in between two clock cycles.

Block filtering can be efficiently implemented in the frequency domain, especially if the impulse response length is comparable to the block length. Complexity estimations that compare frequency domain and time domain implementation can for instance be found in [20].

Frequency domain filtering requires the implementation of discrete Fourier and discrete inverse Fourier transforms. The most commonly used algorithm to implement Fourier transforms in hardware is the Cooley-Tukey FFT algorithm [21]. The basic idea of the Cooley-Tukey algorithm is to break up a transformation of length N in two transformations, each of length $N/2$. This can be done recursively, until one reaches a transform of trivial size (two, four, or eight, for instance). It is possible not only to divide up the FFT in two parts as described before (radix-2). Other very common implementations are also radix-4 implementations, where in each step, the FFT is split in four sub-FFTs, or a mixture of radix-2 and radix-4 (split radix), which are most suitable for hardware implementation.

15.4.3 Polarization tracking

Polarization tracking and Polarization-Mode Dispersion (PMD) equalization are typically performed using a two-in two-out adaptive filter. An adaptive filter can be partitioned in three parts: the actual filter bank, an error estimator, and a device for updating the filter coefficients.

The filter itself has typically a rather short impulse response. Because it needs to follow arbitrary polarization rotation, a rather fast update of the coefficients is required. Therefore, as discussed above, an overlap-and-save implementation is preferable.

In a second step, the error of the signal coming from the filter bank needs to be estimated. There are a number of techniques available for error estimation, namely insertion of training symbols, decision feedback or measuring a known property of the signal. The former have good tracking properties but require the inclusion of carrier synchronization in the feedback loop. The latter one has advantages with respect to loop delay, and therefore potentially offers faster tracking speed. A very popular measure is the constant modulus criterion, see Chapter 27. The constant modulus criterion penalizes deviation of the amplitude of the equalized signal from a desired fixed value. It is obvious that this criterion is optimally suited for Phase Shift Keying (PSK) modulated signals, including Quadrature Phase Shift Keying (QPSK). Several modifications for higher order Quadrature Amplitude Modulated (QAM) modulation have been proposed, see [22] for a comparison.

In a third step, the updated filter coefficients have to be calculated from the estimated error. Several algorithms for this are known in the literature, for instance the method of steepest descent. Most practical from an implementation standpoint is the least-mean-squares algorithm (LMS) [23]. The idea of the LMS algorithm is to estimate the gradient of the error by partial derivatives of the mean squared error with respect to the filter coefficients. The gradient estimates are calculated from instantaneous measures of the error, i.e. the difference between the desired amplitude and the instantaneous signal amplitude after the adaptive filter. In each step, the filter coefficients are updated by adding a small portion proportional to the negative gradient estimate. A weighting factor μ is again utilized for controlling the adaptation speed and residual error of the adapted filter coefficients. The exact formulation of this algorithm and a comparison with a decision feedback structure can for instance be found in [24].

Figure 15.5 shows a simplified circuit diagram of the above described algorithm. From the schematics it seems that the filter coefficient update is more complex than the actual filtering itself. As block processing is utilized, one can contemplate to reduce the complexity by not updating the filter coefficients every sample but only for every block, based on one or multiple measurements of the filter output.

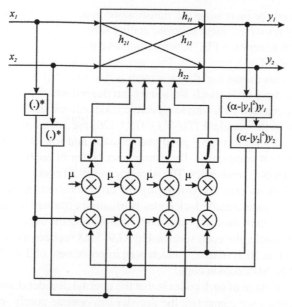

FIGURE 15.5 Circuit for polarization demultiplexing and equalization.

This compromises the adaptation speed and residual noise somewhat, but this might be tolerable depending on the desired adaptation speed.

Often, a coherent receiver uses oversampling to enable better equalization of the received data. Twofold oversampling is simplest in terms of implementation. This means that the adaptive equalizer is fractionally spaced. For the calculation of the error, only baud-spaced output data is utilized. Nevertheless, as fractional-spaced filter coefficients are needed, updating has to be performed on the fractional data.

15.5 SOFT-FEC IMPLEMENTATION AT DATA RATES OF 100 G OR HIGHER

With the advent of coherent detection and the increasing availability of computing resources, soft-decision forward error correction (SD-FEC) has gained widespread attention. Soft-decision forward error correction is considered an important means to improve the reach performance of 100 Gbps fiber optic communication systems and will become essential for novel modulation formats with even higher spectral efficiency. FEC algorithms in optical communications are commonly classified into three generations. First generation FEC schemes, specified in ITU-T G.975 [25], were almost entirely based on the de facto industry standard (255,239) Reed-Solomon code using a hard-decision decoder. This code has a relatively small overhead of 6.7% and realizes a net coding gain of 6.2 dB for an output bit error rate of 10^{-15}.

With the need for increasing the transmission reach and with increasing available computing resources, more efficient coding schemes had to be developed. The so-called second generation FEC schemes improve the net coding gain compared to the (255, 239) Reed-Solomon code. They are almost entirely based on concatenated product codes. This means that the bit stream is first encoded by a simple algebraic code, several of these code words are jointly interleaved and the output is, after partitioning, again encoded by a simple algebraic code. Almost all 7% Overhead (OH) codes specified in the standard ITU-T G.975.1 [26] are such concatenated product codes. These codes achieve very good performance and low error floors with limited implementation complexity. The component codes of such schemes are often BCH and/or Reed-Solomon codes. The advantage of BCH codes is that very efficient closed-form solutions exist for computing the error locations at the decoder [27]. The decoding of concatenated block codes is usually performed iteratively, by alternately decoding each of the component codes, improving the overall result after each iteration. For example, the code specified by G.975.I.3 realizes a net coding gain of 8.99 dB at an output BER of 10^{-15}, which is an improvement of 2.79 dB compared to the (255, 239) Reed-Solomon code.

The main advantage of such codes is that the internal decoder dataflow is small as once the syndromes are computed, the decoder can operate purely in the syndrome domain and only needs to flip some bits and update the affected syndromes. Two very recent developments in second generation HD-FEC codes are the so-called *continuously-interleaved codes* [28] and *staircase codes* [29]. The latter can be considered as some kind of combination of block turbo codes and a convolutional-like structure, with very efficient hard-decision decoding performance. The net coding gain of the staircase code amounts to 9.41 dB at a BER of 10^{-15}, which is only 0.56 dB away from the capacity of the hard-decision equivalent channel [29].

15.5.1 Soft-decision decoding in optical communications

One possibility to realize even higher net coding gains is to use soft-decision instead of hard-decision decoding. Soft decision means that all information originating from the DSP is exploited and no 0/1 decision is made prior to FEC decoding. The 0/1 decision in front of the FEC decoder leads to an unrecoverable loss of information, which should be exploited in order to improve the FEC performance. From an informational theoretical point of view, a performance gain of 1–2 dB (depending on the total FEC overhead) can be achieved by using soft decision.

Figure 15.6 shows the theoretically achievable gain by exploiting soft decision for BPSK and QPSK modulation formats as a function of the total FEC overhead. Thus, especially with the advent of coherent transmission schemes and the utilization of high-resolution ADCs, soft-decision decoding became attractive in coherent receivers to increase the transmission reach. FEC schemes with soft-decision decoding are typically classified as third generation FEC schemes.

Today, there exist mainly two competing classes of codes that allow for soft-decision decoding and that are attractive for an implementation at decoding throughputs of 100 Gbps and above. The first class corresponds to the natural extension of the second

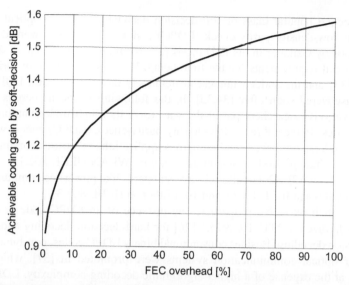

FIGURE 15.6 Achievable (information theoretical) gain by utilizing soft-decision decoding (AWGN noise model).

generation decoding schemes, concatenated product codes. These codes are also often denoted as block turbo codes. Iterative soft-decision decoding of block turbo codes has been pioneered by Pyndiah in [30]. Instead of performing *maximum a posteriori* (MAP) soft-decision decoding of the single component codes, Pyndiah proposed to use a sub-optimal component decoder based on the Chase-II algorithm. In this case, each sub-decoder generates 2^t test vectors by trying all possible bit combinations in the t least reliable positions of the input word (which are determined by the soft information) and the hard decision elsewhere. These test vectors are decoded using a conventional hard-decision decoder, re-encoded, and compared with the received soft information. Using a simple approximation algorithm, these 2^t output vectors are then used to update the soft information for the subsequent decoding steps/iterations.

The advantage of this decoding algorithm is that it is highly parallelizable. All component code words and all test vectors can be computed independently. Furthermore, as only some bits are flipped, the internal data flow of the decoder can be reduced. Another advantage of block turbo codes is that they converge quite rapidly and only require a small number of decoding iterations. As the product code usually has a large minimum distance, the error floor is usually relatively low and has a relatively steep slope. These codes however do not come without disadvantages. The main disadvantage is that large block lengths are required to realize codes with small overhead. This leads to a larger decoding delay. Further disadvantages include a heuristic decoding algorithm, bad predictability of the decoding performance, and inflexibility with respect to varying frame sizes and overheads. The application of block turbo codes with soft-decision decoding in the context of optical communications has been demonstrated for instance in [31].

The second popular class of soft-decision decodable codes in optical communication is low-density parity-check (LDPC) codes. LDPC codes were developed in the 1960s by Gallager in his PhD thesis [32] but then forgotten due to their high computational requirements for the time being. With the discovery of turbo codes in 1993 [33] and the sudden interest in iteratively decodable codes, LDPC codes were rediscovered soon after [34,35]. In the following years, numerous publications by various researchers paved the way for a thorough understanding of this class of codes. In recent years, low-density parity-check (LDPC) codes, which are now well understood, have found numerous applications in various communication standards, including wireless standards like, e.g., WLAN (IEEE 802.11), WiMAX (IEEE 802.16), DVB-S2, but also wireline standards like powerline communications (ITU-T G.9960 and IEEE 1901), and 10 G Ethernet (IEEE 802.3).

Due to their attractive properties, one instance of an LDPC code has already been standardized in ITU-T G.975.1 [26] for hard-decision decoding and later for soft-decision decoding. Besides conventional binary LDPC codes, non-binary LDPC codes for optical communication systems were promoted in [37], which come, however, at the expense of a higher algorithmic decoding complexity. LDPC codes usually suffer from an error floor at BERs of 10^{-8} to 10^{-10}. Therefore, modern high-performance FEC systems are typically constructed using a soft-decision LDPC inner code which reduces the BER to a level of 10^{-3} to 10^{-5} and a hard-decision outer code which pushes the system BER to levels below 10^{-12}, as proposed in [36]. In the remainder of this chapter, we will give an introduction to LDPC codes and present decoding and encoding circuits together with code classes suitable for high-speed operations at throughputs of 100 Gbps and beyond. Finally, we will show how different soft-decision FEC codes can be evaluated using a single database of measurements, which allows for a fast and easy system design.

15.5.2 Low-Density Parity-Check (LDPC) codes

An LDPC code is defined by a sparse binary parity check matrix $\mathbf{H}$ of dimension $M \times N$, where N is the code word length (in bits) of the code and M denotes the number of parity bits. Usually, the number of information bits equals $M - N$. The design rate of the code amounts to $r = (N - M)/N$. The overhead of the code is defined as $\mathrm{OH} = 1/r - 1 = M/(N - M)$. Sparse means that the number of 1s in $\mathbf{H}$ is small compared to the number of zero entries. Practical codes usually have a fraction of 1s that is below 1% by several orders of magnitude. We start by introducing some notation and terminology related to LDPC codes. Each column of the parity check matrix $\mathbf{H}$ corresponds to one bit of the FEC frame. As LDPC codes can also be represented in a graphical structure called Tanner graph (which we will not introduce here), the columns of $\mathbf{H}$ and thus the FEC frame bits are sometimes also called *variable nodes*, referring to the graph-theoretic representation. Similarly, each row of $\mathbf{H}$ corresponds to a parity check equation and ideally defines a parity bit (if $\mathbf{H}$ has full rank). Owing to the Tanner graph representation, the rows of $\mathbf{H}$ are associated to so-called *check nodes*.

LDPC codes are often classified into two categories: regular and irregular LDPC codes. The parity check matrix of regular codes has the property that the number of 1s in each column is constant and amounts to d_v (called variable node degree) and that the number of 1s in each row is constant and amounts to d_c (check node degree). Clearly, $N \cdot d_v = M \cdot d_c$ has to hold, i.e. $r = 1 - d_v/d_c$. On the other hand, irregular LDPC codes [38] have the property that the number of 1s in the different rows and/or columns of **H** is not constant. To ease the implementation of the decoder (as we will see below), it is often assumed that the number of 1s per row (check node degree d_c) is constant and only the number of 1s per column is subject to variation. Such codes are called *check-regular codes*. The irregularity can be characterized by the *degree profile* of the parity check matrix **H**. The degree profile indicates the fraction of columns/rows of a certain degree and $a_{v,i}$ represents the fraction of columns with i ones (e.g. if $a_{v,3} = 1/2$, half the columns of **H** have three 1s). Note that $\sum_i a_{v,i} = 1$ has to hold.

In order to describe the decoder in the remainder of this chapter, we introduce some additional notation. First, we denote by **x** the vector of $K = N - M$ information bits. The single elements of **x** are denoted by x_i, i.e. $\mathbf{x} = (x_1, x_2, \ldots, x_i, \ldots, x_K)^T$. After encoding, the code word $\mathbf{y} = (y_1, y_2, \ldots, y_N)^T$ of length N results. We denote by y_i the single bits of the code word **y**. The code is said to be *systematic* if the information vector **x** is included in the code word, e.g. if $\mathbf{y} = (x_1, \ldots, x_K, p_1, \ldots p_M)^T$, with $\mathbf{p} = (p_1, \ldots, p_M)^T$ denoting the vector of M parity bits. Furthermore, let the set $\mathcal{N}(m)$ denote the positions of the 1s in the mth row of **H**, i.e. $\mathcal{N}(m) = \{i : H_{m,i} = 1\}$. A binary vector **y** is a code word of the code defined by **H** if $\mathbf{Hy} = \mathbf{0}$, with the additions defined over the binary field (addition modulo-2, or XOR, respectively). The set of code words is thus defined to be the null space of **H**. This signifies that the product of each row of **H** and **y** must be zero, or

$$\sum_{j \in \mathcal{N}(m)} y_j = 0, \text{ for all } m \text{ with } 1 \leq m \leq M.$$

EXAMPLE

Let us illustrate the concept of LDPC codes by a toy example, taken from [39]. This example defines the parity check matrix **H** of two concatenated (4, 5) single parity check codes, separated by a 4×5 block-interleaver. The parity check matrix of dimension 9×25 is given by

$$\mathbf{H} = \begin{pmatrix}
1 & 1 & 1 & 1 & & & & & & & & & & & & & & & & & 1 & & & & \\
& & & & & 1 & 1 & 1 & 1 & & & & & & & & & & & & & 1 & & & \\
& & & & & & & & & 1 & 1 & 1 & 1 & & & & & & & & & & 1 & & \\
& & & & & & & & & & & & & 1 & 1 & 1 & 1 & & & & & & & 1 & \\
1 & & & & & 1 & & & & & 1 & & & & & 1 & & & & & 1 & & & & \\
& 1 & & & & & 1 & & & & & 1 & & & & & 1 & & & & & 1 & & & \\
& & 1 & & & & & 1 & & & & & 1 & & & & & 1 & & & & & 1 & & \\
& & & 1 & & & & & 1 & & & & & 1 & & & & & 1 & & & & & 1 & \\
& & & & & & & & & & & & & & & 1 & 1 & 1 & 1 & & & & & & 1
\end{pmatrix}.$$

Thus, $M = 9, N = 20$, and the code has an overhead of 56% (given by $M/(N - M)$) or a rate of $r = 0.64$, equivalently. We have $\mathcal{N}(1) = \{1; 2; 3; 4; 17\}$, $\mathcal{N}(2) = \{5; 6; 7; 8; 18\}, \mathcal{N}(3) = \{9; 10; 11; 12; 19\}, \mathcal{N}(4) = \{13; 14; 15; 16; 20\}$, $\mathcal{N}(5) = \{1; 5; 9; 13; 21\}$, $\mathcal{N}(6) = \{2; 6; 10; 14; 22\}$, $\mathcal{N}(7) = \{3; 7; 11; 15; 23\}$, $\mathcal{N}(8) = \{4; 8; 12; 16; 24\}, \mathcal{N}(9) = \{17; 18; 19; 20; 25\}$. This means for instance that we have (using the first row of **H**) $y_1 + y_2 + y_3 + y_4 + y_{17} = 0$, or, if the code is systematic, $x_1 + x_2 + x_3 + x_4 + p_1 = 0$, i.e. $p_1 = x_1 + x_2 + x_3 + x_4$, defines the first parity bit.

The goal of the LDPC code designer is to find a matrix **H** that realizes a code which yields high coding gains and which possesses some structure facilitating the implementation of the decoder. In this chapter, we will not focus on the first issue and point the interested reader to numerous research articles published on that topic (e.g. [38,40–43]). Instead, we will focus in the remainder of this chapter on code structures that allow for a decoder implementation operating at information throughputs of 100 Gbps and beyond.

15.5.3 Decoding of LDPC codes

The most common decoding algorithm for LDPC codes is the sum-product decoding algorithm and its simplified versions. These algorithms can be described either in a graph structure commonly denoted as Tanner graph, or given directly in a form suitable for implementation. While the former is advantageous for describing and understanding the underlying operations, we only focus on the latter in this chapter. Good descriptions of the Tanner graph and derivations of the decoding algorithm are given in, e.g., [44,27].

In the context of coherent detection, we can usually assume an equivalent Additive White Gaussian Noise (AWGN) channel model. This assumption is justified by the central limit theorem which applies due to the extensive use of filtering in the inner DSP stages. If we denote by y_i the single bits of a code word **y** $(i = 1, \ldots, N)$ as defined above, then the received values at the input of the decoder amount to $z_i = y_i + n_i$, with n_i Gaussian distributed noise samples of zero mean and variance σ^2. Usually, the received noisy samples are converted into the *log-likelihood* (LLR) domain, which leads to a numerically more stable decoder implementation [45]. The LLR of a received sample z_i is defined as

$$L(z_i) = \log\left(\frac{p(z_i|y_i = 0)}{p(z_i|y_i = 1)}\right),$$

with $p(z_i|y_i = k) = \exp\left(-(z_i - (1 - 2k))^2/(2\sigma^2)\right)/\sqrt{2\pi\sigma^2}$ the pdf of the received sample z_i under the AWGN assumption conditioned on the transmitted bit y_i and with bipolar signaling ($y_i = 0 \to +1, y_i = 1 \to -1$). Conveniently, it turns out that $L(z_i) = z_i \cdot 2/\sigma^2 = z_i \cdot L_c$ under the AWGN assumption. Usually, the value $L_c = 2/\sigma^2$ is assumed to be constant and predetermined.

In the following, we describe the row-decoding or layered decoding algorithm, as introduced in [46]. The decoder continuously updates the received LLR values with the goal to compute LLRs that approximate the *maximum a posteriori* (MAP) values. The received vector of LLRs **z** is therefore copied to a memory $\hat{\mathbf{z}}$ of size N, which is continuously updated. The decoding operation in the layered decoder consists of three steps, where the first step prepares the input data, the second step performs the computation of new *extrinsic* data and the final step updates the LLR memory. The decoder carries out the three steps sequentially for each row of the parity check matrix **H**. After all rows have been considered, a single decoding iteration has been carried out. The decoder usually carries out several iterations, where the number depends on the available resources.

In the following, we describe the three steps for a single row m of **H**. The first step consists in computing card $(\mathcal{N}(m)) = d_c$ temporary values $t_{m,i}$, with

$$t_{m,i} = \hat{z}_i - e_{m,i} \quad \text{for all } i \in \mathcal{N}(m),$$

for all non-zero entries (indexed by i) of the mth row of **H**. The value $e_{m,i}$ is the (stored) extrinsic memory for row m and variable $\hat{z}_i$. At the beginning of the decoding of a single frame, all $e_{m,i}$ are initialized by zero and then continuously updated. Note that in total $\sum_m \text{card}(\mathcal{N}(m))$ memory locations are required for storing the $e_{m,i}$. If the code is check-regular, $\sum_m \text{card}(\mathcal{N}(m)) = M d_c$. In the second step, the extrinsic memory is updated using the $t_{m,i}$ according to

$$e_{m,i} = 2 \cdot \tanh^{-1}\left(\prod_{j \in \mathcal{N}(m) \backslash \{i\}} \tanh\left(\frac{t_{m,j}}{2}\right) \right) \quad \text{for all } i \in \mathcal{N}(m)$$

$$= \left[\prod_{j \in \mathcal{N}(m) \backslash \{i\}} \text{sign}\left(t_{m,j}\right) \right] \cdot \phi\left(\sum_{j \in \mathcal{N}(m) \backslash \{i\}} \phi\left(|t_{m,j}|\right) \right),$$

with

$$\phi(\tau) = -\log\left(\tanh\left(\frac{\tau}{2}\right) \right) = \log\left(\frac{e^\tau + 1}{e^\tau - 1}\right),$$

where the second expression may be more suitable for implementation, as the multiplication is replaced by an addition and instead of two functions $\tanh(\cdot)$ and $\tanh^{-1}(\cdot)$, only a single function $\phi(\cdot)$ needs to be implemented (or approximated by a look-up table). The product (or the sum) is carried out over all entries in $\mathcal{N}(m)$ except the one under consideration i. For instance, in the above example, where $\mathcal{N}(5) = \{1; 5; 9; 13; 21\}$, $e_{5,13}$ is computed as

$$e_{5,13} = 2 \cdot \tanh^{-1}\left(\prod_{j \in \mathcal{N}(5) \backslash \{13\}} \tanh\left(\frac{t_{5,j}}{2}\right) \right) = 2 \cdot \tanh^{-1}\left(\prod_{j \in \{1;5;9;21\}} \tanh\left(\frac{t_{5,j}}{2}\right) \right)$$

$$= 2 \cdot \tanh^{-1}\left(\tanh\left(\frac{t_{5,1}}{2}\right) \tanh\left(\frac{t_{5,5}}{2}\right) \tanh\left(\frac{t_{5,9}}{2}\right) \tanh\left(\frac{t_{5,21}}{2}\right) \right).$$

The derivation of the equation for the extrinsic update is beyond the scope of this chapter. We refer the interested reader to, e.g., [44,27]. Usually, if high decoder throughputs shall be achieved, the computation of $e_{m,i}$ should be further simplified. An often employed simplification, which we will consider in the remainder of this chapter, is the so-called *min-sum approximation* which leads to

$$e_{m,i} = \left[\prod_{j \in \mathcal{N}(m)\setminus\{i\}} \text{sign}(t_{m,j}) \right] \min_{j \in \mathcal{N}(m)\setminus\{i\}} |t_{m,j}| \quad \text{for all } i \in \mathcal{N}(m).$$

As this second step computes the output of the parity check node of the check node in the graphical representation of LDPC codes, it is frequently denoted by *check node operation*. Finally, in the last step, the LLR memory is updated according to

$$\hat{z}_i = t_{m,i} + e_{m,i} \quad \text{for all } i \in \mathcal{N}(m)$$

and the decoder continues with decoding row $m + 1$ or, if $m = M$, restarts at $m = 1$ (next iteration). Figure 15.7 shows the simplified block diagram of the check node decoding operation for card $(\mathcal{N}(m)) = 4$ (i.e. **H** contains four 1s at row m). Note that the routing network to access the different $\hat{z}_i$ (which are not necessarily stored in neighboring memory locations) is not shown in the figure.

The update step can be simplified in actual implementation. Let $\mu_m^{[1]} = \min_{j \in \mathcal{N}(m)} |t_{m,j}|$ be the minimum of the absolute value of all involved incoming messages $t_{m,j}$ at row m and let $i_m^{[1]} = \arg\min_{j \in \mathcal{N}(m)} |t_{m,j}|$ be the position (i.e. the index) of this first minimum. Let furthermore $\mu_m^{[2]} = \min_{j \in \mathcal{N}(m)\setminus\{i_m^{[1]}\}} |t_{m,j}|$ be the second minimum of all incoming messages. We further define $s_m = \prod_{j \in \mathcal{N}(m)} \text{sign}(t_{m,j})$. The output message can then be computed by

$$e_{m,i} = s_m \cdot \text{sign}(t_{m,i}) \cdot \begin{cases} \mu_m^{[1]} & \text{if } i \neq i_m^{[1]} \\ \mu_m^{[2]} & \text{if } i = i_m^{[1]} \end{cases}.$$

Thus, the main burden for implementing the check node operation consists in finding the first and second minimum of the incoming messages together with the

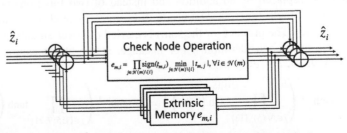

FIGURE 15.7 Simplified description of a check node decoding element in the layered LDPC decoder.

position of the first minimum. High-speed circuits for finding the first two minima have been addressed in, e.g., [47,48]. A simplified implementation utilizing only the first minimum $\mu_m^{[1]}$ together with a simple correction has been presented in [49]. This approach comes, however, at the expense of a slight loss in decoding performance (leading to a smaller coding gain).

The memory requirements of this algorithm are N memory cells for storing the (continuously updated) *a posteriori* values $\hat{z}_i$. Additionally, the storing requirements of the extrinsic memory $e_{m,i}$ amount to the total number of 1s in $\mathbf{H}$: however, due to the simplified implementation using the minimum approximation, the extrinsic memory does not need to store all d_c different values per row, but only the d_c different signs, both minima $\mu_m^{[1]}$ and $\mu_m^{[2]}$ and the position of the first minimum. The values $e_{m,i}$ can then be computed on the fly as required.

As the min-sum algorithm is only an approximation to the full belief-propagation expression, numerous attempts have been made to improve the performance of the min-sum algorithm. Two notable improvements are the normalized min-sum algorithm and the offset min-sum algorithm [50], where the latter is better suited for an implementation. The update rule of the offset min-sum algorithm reads

$$
e_{m,i} = s_m \cdot \operatorname{sign}(t_{m,i}) \cdot \begin{cases} \max(\mu_m^{[1]} - \beta, 0) & \text{if } i \neq i_m^{[1]}, \\ \max(\mu_m^{[2]} - \beta, 0) & \text{if } i = i_m^{[1]}, \end{cases}
$$

where the variable β can either be constant and determined offline or be updated during decoding according to a predefined rule.

Another important aspect with respect to the implementation, which has been left out of the consideration so far, is the fixed-point resolution required within the decoder. The soft-information at the input of the decoder and the data that is processed inside the decoder are usually stored in fixed-point representation. The important parameters are the overall number of bits required for this representation, as well as the number of bits allocated to the fractional part. It is generally difficult to give a precise estimate of these numbers, as they strongly depend on the DSP front-end that is used and of the parameters of the code (variable and check node degrees, number of iterations, decoding algorithm, etc.). Several research papers deal with the analysis of the required number of bits, e.g. [51–53]. A rule of thumb, taken from [52], indicates that 6 bits lead to acceptable decoding performance. However, depending on the circumstances and the code, a decoder with 4 or 5 bit may also produce reasonable results.

15.5.4 Quasi-cyclic LDPC codes

Generally, in the case of LDPC codes, a very easy decoding algorithm exists and in contrast to classical channel coding, the challenge is not to find a good decoder for a given code, but to find a good code given the decoding algorithm. A plethora of design algorithms have been developed for LDPC codes, the two best known being the progressive-edge growth (PEG) [41] and the ACE algorithms [42]. However, these algorithms lead to $\mathbf{H}$-matrices that are often random-like and do not possess any structure suitable for implementation. For this reason, most LDPC codes that are implemented today are so-called Quasi-Cyclic (QC) LDPC codes. They have a parity check matrix with a structure that allows for inherent parallelization of the decoder and leads to an efficient encoder realization. Almost all LDPC codes utilized in practice belong to the class of QC-LDPC codes.

QC-LDPC codes are constructed using a so-called lifting matrix $\mathbf{A}$. The matrix $\mathbf{H}$ is constructed from the lifting matrix by replacing each element of $\mathbf{A}$ with either an all-zero matrix of size $S \times S$ or a cyclically permuted identity matrix of size $S \times S$. We adhere to the following notation: S denotes the lifting factor, i.e. the size of the all-zero or cyclically shifted identity matrix. If the entry of $\mathbf{A}$ at row m and column i, i.e. $A_{m,i} = -1$, then the all-zero matrix of size $S \times S$ is used, and if $A_{m,i} \geq 0$, $A_{m,i}$ denotes how many cyclic right shifts of the identity matrix are performed. If $\dim(\mathbf{H}) = M \times N$, then $\dim(\mathbf{A}) = M' \times N'$, with $M' = M/S$ and $N' = N/S$.

EXAMPLE

Let us illustrate the construction of QC-LDPC codes by a small (artificial) example with a lifting matrix of size $\dim(\mathbf{A}) = 3 \times 5$ and a lifting factor of $S = 5$ (leading to $\dim(\mathbf{H}) = S \cdot \dim(\mathbf{H}) = 15 \times 25$, this corresponds to a code of design rate $r = 0.4$, or an overhead of 150%, respectively).

$$A = \begin{pmatrix} -1 & 0 & 1 & 1 & 2 \\ 2 & -1 & 4 & 2 & 1 \\ 1 & 3 & -1 & 3 & 1 \end{pmatrix} \rightarrow H =$$

Note that for clarity only the 1s are shown in the description of $\mathbf{H}$. The density of this code's parity check matrix is still relatively high as 16% of the entries of $\mathbf{H}$ are 1s. Comparing this for instance with the code of same rate $r = 0.4$ specified in the DVB-S2 standard [54], with $N = 64,800$, $M = 38,880$, $S = 360$, only 0.0093% of $\mathbf{H}$ are 1s. The LDPC code defined by the lifting matrix $\mathbf{A}$ is an irregular (but check-regular) code. Its degree distribution of this code can be easily derived from $\mathbf{H}$ (and $\mathbf{A}$). The first 15 out of 25 variable nodes are connected to two check nodes while the remaining 10 variable nodes are connected to three check nodes. Thus $a_{v,2} = 0.6$ and $a_{v,3} = 0.4$. The code is check-regular as all check nodes connect to four variable nodes; the check node degree thus equals $d_c = 4$.

Before proceeding, we introduce some additional notation which will help us describe the encoder in Section 15.5.5. We can write $\mathbf{H}$ as

$$\mathbf{H} = \begin{pmatrix} \mathbf{P}^{(A_{1,1})} & \cdots & \mathbf{P}^{(A_{1,N'})} \\ \vdots & \ddots & \vdots \\ \mathbf{P}^{(A_{M',1})} & \cdots & \mathbf{P}^{(A_{M',N'})} \end{pmatrix}$$

with $\mathbf{P}^{(A_{m,i})}$ denoting either the $S \times S$ cyclically shifted identity matrix with the shift value defined by $A_{m,i}$ or the all-zero matrix if $A_{m,i} = -1$. $\mathbf{P}^{(A_{m,i})}$ can thus also be expressed as

$$\mathbf{P}^{(A_{m,i})} = \mathbf{I}_1^{A_{m,i}-1} = \underbrace{\mathbf{I}_1 . \mathbf{I}_1 \cdots \mathbf{I}_1}_{A_{m,i}-1 \text{ times}} \text{ with } \mathbf{I}_1 = \begin{pmatrix} & 1 & \\ & & \ddots & \\ & & & 1 \\ 1 & & & \end{pmatrix}$$

if $A_{m,i} > 0$ and $\mathbf{P}^{(0)} = \mathbf{I}$ (identity matrix) by definition. $\mathbf{P}^{(-1)} = \mathbf{0}$ is defined to be the all-zero matrix.

The advantage of such a structured representation is that with the layered decoder, S consecutive rows of $\mathbf{H}$ (i.e. one row of $\mathbf{A}$) can be decoded in parallel, as these operations are fully independent and only access independent memory locations. Figure 15.8 shows the block diagram of a parallelized row decoder implementation suitable for the matrix $\mathbf{A}$ given in the example above. The layered decoder shown in Figure 15.7 is basically repeated S times. If we subdivide the memory $\hat{\mathbf{z}} = \left(\hat{\mathbf{z}}_1^T \ \hat{\mathbf{z}}_2^T \ \hat{\mathbf{z}}_3^T \ \hat{\mathbf{z}}_4^T \ \hat{\mathbf{z}}_5^T \right)^T$ into five chunks of S entries each, with each chunk corresponding to a column of $\mathbf{A}$, then we can describe the layered decoder for QC-LDPC codes applied to the above example as follows. Similarly

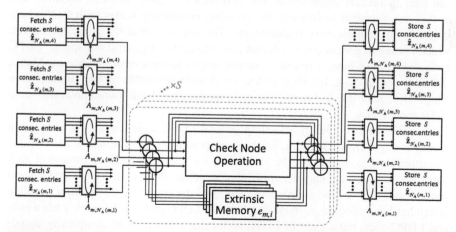

FIGURE 15.8 Simplified description of a check node decoding element for quasi-cyclic (QC) LDPC codes.

to the general case, we define $\mathcal{N}_{\mathbf{A}}(m) = \{i : A_{m,i} \neq -1\}$. In the example, we thus have $\mathcal{N}_{\mathbf{A}}(1) = \{2; 3; 4; 5\}$, $\mathcal{N}_{\mathbf{A}}(2) = \{1; 3; 4; 5\}$, and $\mathcal{N}_{\mathbf{A}}(3) = \{1; 2; 4; 5\}$. Let furthermore $\mathcal{N}_{\mathbf{A}}(m, i)$ be the ith entry of the ordered sequence corresponding to the set $\mathcal{N}_{\mathbf{A}}(m)$.

If the first row of $\mathbf{A}$ shall be decoded, the decoder in Figure 15.8 first needs to fetch the respective $S \cdot \text{card}(\mathcal{N}(1)) = S \cdot d_c = 4S$ memory locations. With the above-mentioned subdivision, the upper block fetches the entry $\hat{\mathbf{z}}_{\mathcal{N}_{\mathbf{A}}(1,4)} = \hat{\mathbf{z}}_5$ and so on. This block of S samples is then cyclically shifted according to the value $A_{1,\mathcal{N}_{\mathbf{A}}(1,4)} = A_{1,5} = 2$. With $S = 5$ as in the example above, this means that $\hat{\mathbf{z}}_5 = (\hat{z}_{5,1} \quad \hat{z}_{5,2} \quad \hat{z}_{5,3} \quad \hat{z}_{5,4} \quad \hat{z}_{5,5})^T = (\hat{z}_{21} \quad \hat{z}_{22} \quad \hat{z}_{23} \quad \hat{z}_{24} \quad \hat{z}_{25})^T$. After the cyclic shift, we get the output vector $(\hat{z}_{5,4} \quad \hat{z}_{5,5} \quad \hat{z}_{5,1} \quad \hat{z}_{5,2} \quad \hat{z}_{5,3})^T = (\hat{z}_{24} \quad \hat{z}_{25} \quad \hat{z}_{21} \quad \hat{z}_{22} \quad \hat{z}_{23})^T$. The cyclic shift can be realized for instance using a barrel shifter circuit. Similarly, the remaining three memory chunks are fetched and shifted accordingly with the respective value defined by $\mathbf{A}$. Then the outputs are fed into one of the S parallel check node decoding blocks and before updating the LLR memory, the blocks have to be grouped again and the shifts need to be reverted using the respective entries of $\mathbf{A}$. This is shown in the right part of Figure 15.8.

A further level of parallelization is possible, especially for large matrices $\mathbf{A}$. If the matrix $\mathbf{A}$ is designed in such a way that no two consecutive rows share common entries (i.e. accesses to the same memory locations), these two rows can be computed in parallel. Note that the order of the rows of $\mathbf{A}$ (and $\mathbf{H}$) does not change the properties of the code. Therefore, the rows of $\mathbf{A}$ can be arranged in such a way that the parallelism is maximized.

15.5.5 Efficient encoding of quasi-cyclic LDPC codes

LDPC codes have appealing properties due to their very easy decoding algorithm and their theoretical foundations. The drawback of LDPC codes is, however, that in their most general realization, the encoding complexity scales with $O(N^2)$ both in processing and memory requirements. The reason is that the generator matrix, which is used to generate the different code words, is in general not sparse. LDPC codes which also have a sparse generator matrix (Low-Density Generator-Matrix (LDGM) codes) exist, but their utilization is often infeasible as they induce quite a high error floor due to low-weight code words. They have found, however, some niche applications.

A method for efficiently encoding general LDPC codes, which is based on a greedy triangulation algorithm, has been proposed in [55]. The application of this algorithm may, however, still be prohibitive for a general implementation at high data rates, especially if encoders operating at a throughput of 100 Gbps and beyond shall be developed. The solution to overcome the encoding complexity problem which has been adopted by most standardized LDPC codes is to employ not a general LDPC code, but a class of LDPC codes that allow for low-complexity encoding while preserving the other advantageous features of general codes. Almost all these codes belong to the class of systematic (irregular) repeat-accumulate (RA) codes [56]. An RA code generates the parity bits by repeating each information bit several times according to a predetermined pattern (which is defined by the degree profile of the code), interleaving the block of repeated bits, processing this block by a simple binary accumulator, and puncturing the output of the accumulator according to a predefined pattern to get the desired number of parity bits. Repeat-accumulate codes as such do not allow for a representation as quasi-cyclic (QC) matrix to realize the aforementioned parallelization of the decoder. However, the class of structured-IRA (sIRA) codes enables such a structure based on cyclic permutation matrices [57], which is a generalization of Tanner's QC-RA codes [58]. In the following, we consider such sIRA codes only. The lifting matrix (and thus the parity check matrix) can in this case be partitioned as $\mathbf{A} = (\mathbf{A}_s\ \mathbf{A}_p)$ with $\mathbf{A}_s$ being the lifting matrix related to the systematic bits and $\mathbf{A}_p$ being the lifting matrix related to the parity check bits. The matrix $\mathbf{A}_p$ is in this case a matrix with a *duo-diagonal* structure. As we will see below, this structure simplifies encoding and leads to other appealing properties (such as degree-2 variable nodes, which are required for good decoding performance, and the avoidance of cycles in the code graph containing only degree-2 nodes). However, note that the introduction of such a structure also changes the properties and the BER performance of the code, which has to be taken into account by the code designer [59]. The matrices $\mathbf{A}$ and $\mathbf{H}$ can be represented as follows (note that the "-1" entries in $\mathbf{A}$ and the zeros in $\mathbf{H}$ are not shown for clarity).

$$
\mathbf{A} = (\mathbf{A}_s\ \mathbf{A}_p) = \begin{pmatrix} \mathbf{A}_s & \begin{matrix} 0 & & & & \\ 0 & 0 & & & \\ & 0 & \ddots & & \\ & & \ddots & 0 & \\ & & & 0 & 0 \end{matrix} \end{pmatrix} \rightarrow \mathbf{H} = \begin{pmatrix} \mathbf{H}_s & \begin{matrix} 1 & & & & & & \\ 1 & 1 & & & & & \\ & 1 & 1 & & & & \\ & & 1 & \ddots & & & \\ & & & \ddots & 1 & & \\ & & & & \ddots & 1 & \\ & & & & & 1 & 1 \end{matrix} \end{pmatrix}
$$

If we recall the definition of LDPC codes, $\mathbf{H}$ corresponds to the parity check matrix of the code, and if the code is a systematic code (i.e. the information bits are part of the code word), then $\mathbf{H}_s$ corresponds to the systematic bits. If we define $\mathcal{N}_s(m) = \{i : H_{s,m,i} \neq 0\}$ with $H_{s,m,i}$ denoting the entry of $\mathbf{H}_s$ at row m and column i, and if we define a code word $\mathbf{y} = (x_1, \ldots, x_{N-M}, p_1, \ldots p_M)^T$ with $p_1, \ldots, p_M$ denoting the parity bits and if we recall that $\mathbf{Hy} = 0$, then the first row of $\mathbf{H}$ completely defines the first parity bit as

$$
p_1 = \sum_{j \in \mathcal{N}_s(1)} x_j,
$$

where we use the signum notation ($\sum$) to denote XOR operations (modulo-2 addition). This readily allows us to compute the first S parity bits, as $p_m = \sum_{j \in \mathcal{N}_s(m)} x_j$, for $m \leq S$. After that, a more general expression has to be employed as the remaining parity bits are also dependent on the previously computed parity bits

$$p_m = p_{m-S} + \sum_{j \in \mathcal{N}_s(m)} x_j \quad \text{for } m > S.$$

Using backward substitution of all previously computed parity bits, the new parity bit p_m can be computed. Note that due to the QC-property of the code, again always blocks of S parity bits can be easily computed in parallel. This approach has, however, the disadvantage that the last S parity bits $p_{M-S+1}, \ldots, p_M$ are connected only to a single parity check equation and thus their *a posteriori* values are only updated once per iteration, which does not lead to any information improvement on these nodes and may not be desirable in all cases. Using the degree distribution notation introduced above, this means that $a_{v,1} \geq S/N$. Often, the code designer wants $a_{v,1} = 0$. One approach to overcome this deficiency is to replace this last column of **A** by a so-called *tailbiting* column in the lifting matrix **A**. We define the tailbiting column to have two identical entries at rows 1 and U and an entry with zero shift at the last row M. Usually, for reasons which will become clear below, the tailbiting column is placed at the first column of $\mathbf{A}_p$. Note that different definitions of the tailbiting column exist in the literature. An example of a matrix **A** with a tailbiting column is

$$
\mathbf{A} = \begin{pmatrix} \mathbf{A}_s & \mathbf{A}_p' \end{pmatrix} = \begin{pmatrix} \mathbf{A}_s & \begin{matrix} 1 & 0 & & & & \\ & 0 & \ddots & & & \\ & & \ddots & \ddots & & \\ 1 & & & \ddots & \ddots & \\ & & & & \ddots & 0 \\ & & & & & 0 & 0 \\ 0 & & & & & & 0 \end{matrix} \end{pmatrix}
\tag{15.3}
$$

This tailbiting approach is used, for example, in the codes defined in the WiMAX and Wireless LAN standards [60,61]. A detailed description of the encoding procedure can be found in Annex G of [61]. An observation of this matrix reveals that (binary, i.e. modulo-2) adding all the rows of the parity check matrix **H** in chunks of S rows leads an $S \times N$ matrix $\widetilde{\mathbf{H}}$ defined by

$$\widetilde{H}_{m,i} = \sum_{j=0}^{M'-1} H_{jS+m,i} \quad \text{for } 1 \leq m \leq S \text{ and } 1 \leq i \leq N.$$

This matrix $\widetilde{\mathbf{H}}$ can be described by the corresponding lifting matrix $\widetilde{\mathbf{A}} = \begin{pmatrix} \widetilde{\mathbf{A}}_s & 0 & -1 & \cdots & -1 \end{pmatrix}$. This lifting matrix immediately defines the first S parity bits, which can be directly computed from the systematic information bits using the sum of all rows multiplied by the information vector. The remaining parity bits can be computed using the first parity bits as described below. Note that the systematic part $\widetilde{\mathbf{A}}_s$ of that corresponding lifting matrix is not constructed using cyclically shifted identity matrices as described above but rather using sums of several cyclically shifted identity matrices.

A simplified encoder circuit for generating the M parity bits for a code based on a lifting matrix as defined in (15.3) is shown in Figure 15.9. From this circuit, it becomes immediately obvious why these codes are called repeat-accumulate codes. First, the input vector $\mathbf{x}$ of size $N-M$ is subdivided into $K' = N' - M' = (N - M)/S$ chunks $\mathbf{x}_i$ of S bits each. The input is then processed chunk-wise along with the different columns of $\mathbf{A}_s$ (remember that each chunk $\mathbf{x}_i$ is assigned to the ith column of $\mathbf{A}_s$). The left part of the circuit shown in Figure 15.9 consists of M' parallel branches, corresponding to the M' rows of $\mathbf{A}_s$. The AND gates in the left part of the circuit describe the routing of the input chunk $\mathbf{x}_i$ to the various branches (if $A_{m,i} = -1$, the chunk $\mathbf{x}_i$ is not considered). As the number of non-zero entries in each column of $\mathbf{H}$ is usually larger than 1, this corresponds to the repetition part of the code. Note that all the data paths in Figure 15.9 are S parallel paths, which are not shown for clarity. After performing the cyclic shift defined by the lifting matrix, the single branches are accumulated, such that after all K' chunks $\mathbf{x}_i$, $1 \leq i \leq K'$ have been processed, the M'

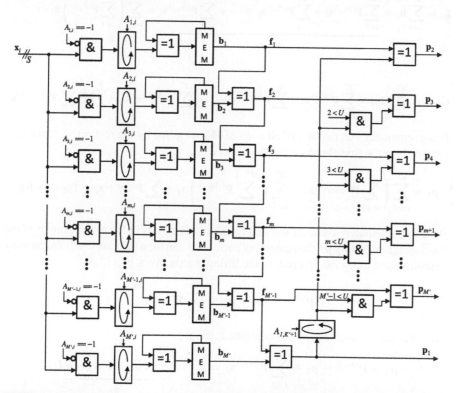

FIGURE 15.9 Simplified block diagram of an encoder circuit for quasi-cyclic (QC) irregular repeat-accumulate (IRA) LDPC codes with tailbiting.

memory blocks contain vectors of size S that correspond to the multiplication of the input vector $\mathbf{x}$ with S consecutive rows of the systematic part of $\mathbf{H}$. The output of the mth branch is denoted as $\mathbf{b}_m$ and defined by

$$\mathbf{b}_m = \sum_{i=1}^{K'} \mathbf{P}^{(A_{m,i})} \mathbf{x}_i.$$

In the following, the different $\mathbf{b}_m$ are accumulated to realize the sum of all rows as required by $\widetilde{\mathbf{A}}_s$ for computing the first parity chunk. The intermediate outputs of this accumulator are denoted as $\mathbf{f}_m$ and are defined by

$$\mathbf{f}_1 = \mathbf{b}_1 = \sum_{i=1}^{K'} \mathbf{P}^{(A_{1,i})} \mathbf{x}_i,$$

$$\mathbf{f}_2 = \mathbf{b}_1 + \mathbf{b}_2 = \sum_{i=1}^{K'} \mathbf{P}^{(A_{1,i})} \mathbf{x}_i + \sum_{i=1}^{K'} \mathbf{P}^{(A_{2,i})} \mathbf{x}_i = \sum_{i=1}^{K'} \left(\mathbf{P}^{(A_{1,i})} + \mathbf{P}^{(A_{2,i})} \right) \mathbf{x}_i,$$

$$\vdots,$$

$$\mathbf{f}_m = \sum_{j=1}^{m} \mathbf{b}_j = \underbrace{\sum_{i=1}^{K'} \mathbf{P}^{(A_{1,i})} \mathbf{x}_i + \sum_{i=1}^{K'} \mathbf{P}^{(A_{2,i})} \mathbf{x}_i + \cdots + \sum_{i=1}^{K'} \mathbf{P}^{(A_{m,i})} \mathbf{x}_i}_{m \text{ times}} = \sum_{i=1}^{K'} \left(\sum_{j=1}^{m} \mathbf{P}^{(A_{j,i})} \right) \mathbf{x}_i,$$

$$\vdots,$$

$$\mathbf{f}_{M'-1} = \sum_{m=1}^{M'-1} \mathbf{b}_m = \sum_{i=1}^{K'} \mathbf{P}^{(A_{1,i})} \mathbf{x}_i + \sum_{i=1}^{K'} \mathbf{P}^{(A_{2,i})} \mathbf{x}_i + \cdots + \sum_{i=1}^{K'} \mathbf{P}^{(A_{M'-1,i})} \mathbf{x}_i = \sum_{i=1}^{K'} \left(\sum_{j=1}^{M'-1} \mathbf{P}^{(A_{j,i})} \right) \mathbf{x}_i.$$

If we remember that the sum of all rows of $\mathbf{H}$ multiplied by the respective input vectors leads to the first chunk of parity bits, i.e.

$$\mathbf{p}_1 = \sum_{i=1}^{K'} \left(\sum_{m=1}^{M'} \mathbf{P}^{(A_{m,i})} \right) \mathbf{x}_i = \sum_{i=1}^{K'} \left(\sum_{m=1}^{M'-1} \mathbf{P}^{(A_{m,i})} \right) \mathbf{x}_i + \sum_{i=1}^{K'} \mathbf{P}^{(A_{M',i})} \mathbf{x}_i = \mathbf{f}_{M'-1} + \mathbf{b}_{M'}$$

We can immediately compute the first chunk of parity bits in the lower branch of the block diagram. Using $\mathbf{p}_1$, the remaining parity bits can then be computed. We have by definition and using the first row of the lifting matrix given by (15.3)

$$\mathbf{P}^{(A_{1,K'+1})} \mathbf{p}_1 + \mathbf{p}_2 = \sum_{i=1}^{K'} \mathbf{P}^{(A_{1,i})} \mathbf{x}_i.$$

With $\mathbf{p}_1$, which has already been computed, we get

$$\mathbf{p}_2 = \mathbf{P}^{(A_{1,K'+1})} \mathbf{p}_1 + \sum_{i=1}^{K'} \mathbf{P}^{(A_{1,i})} \mathbf{x}_i = \mathbf{P}^{(A_{1,K'+1})} \underbrace{\left(\mathbf{f}_{M'-1} + \mathbf{b}_{M'} \right)}_{\mathbf{p}_1} + \mathbf{f}_1.$$

Thus, the second chunk of parity bits $\mathbf{p}_2$ is obtained by summing up $\mathbf{p}_1$ shifted by the value $A_{1,K'+1}$ and $\mathbf{f}_1$. Note that $\mathbf{p}_1$ is present at the bottom branch of the encoder circuit.

The remaining parity bits are generated by the same rule, for instance, $\mathbf{p}_3$ is obtained by using the second row of $\mathbf{A}$ leading to

$$\mathbf{p}_2 + \mathbf{p}_3 = \sum_{i=1}^{K'} \mathbf{P}^{(A_{2,i})} \mathbf{x}_i$$

$$\Rightarrow \mathbf{p}_3 = \mathbf{p}_2 + \sum_{i=1}^{K'} \mathbf{P}^{(A_{2,i})} \mathbf{x}_i = \underbrace{\mathbf{P}^{(A_{1,K'+1})} \mathbf{p}_1 + \sum_{i=1}^{K'} \mathbf{P}^{(A_{1,i})} \mathbf{x}_i}_{\mathbf{p}_2} + \sum_{i=1}^{K'} \mathbf{P}^{(A_{2,i})} \mathbf{x}_i$$

$$= \mathbf{P}^{(A_{1,K'+1})} \mathbf{p}_1 + \sum_{i=1}^{K'} \left(\mathbf{P}^{(A_{1,i})} + \mathbf{P}^{(A_{2,i})} \right) \mathbf{x}_i = \mathbf{P}^{(A_{1,K'+1})} \mathbf{p}_1 + \mathbf{f}_2.$$

This procedure can be repeated until the row U, which contains the second entry (here "1") in the tailbiting column $K' + 1$, is reached. We can then conclude that

$$\mathbf{p}_m = \mathbf{P}^{(A_{1,K'+1})} \mathbf{p}_1 + \mathbf{f}_{m-1} \quad \text{for } 2 \le m \le U.$$

The next parity chunk $\mathbf{p}_{U+1}$ is obtained by

$$\mathbf{P}^{(A_{U,K'+1})} \mathbf{p}_1 + \mathbf{p}_U + \mathbf{p}_{U+1} = \sum_{i=1}^{K'} \mathbf{P}^{(A_{U,i})} \mathbf{x}_i$$

$$\Rightarrow \mathbf{p}_{U+1} = \mathbf{P}^{(A_{U,K'+1})} \mathbf{p}_1 + \mathbf{p}_U + \sum_{i=1}^{K'} \mathbf{P}^{(A_{U,i})} \mathbf{x}_i$$

$$= \left(\underbrace{\mathbf{P}^{(A_{1,K'+1})} + \mathbf{P}^{(A_{U,K'+1})}}_{=0} \right) \mathbf{p}_1 + \mathbf{f}_U$$

$$= \mathbf{f}_U.$$

This means that starting with row U (corresponding to the computation of $\mathbf{p}_{U+1}$), the addition of the shifted version of $\mathbf{p}_1$ has to be disabled (and this continues for the remaining rows). This is achieved by the AND gate in the branch connecting $\mathbf{p}_1$ to the addition with the respective $\mathbf{f}_m$.

Sometimes, repeat-accumulate codes cannot be used as the use of degree-2 nodes may not be desirable (if, e.g., regular LDPC shall be employed). In this case, the techniques presented in [62], which generalize the description given here, may be used to realize circuits for efficient encoding of general quasi-cyclic LDPC codes.

15.6 PERFORMANCE EVALUATION OF DIFFERENT CODING CONCEPTS

Before the implementation of an LDPC code-based FEC scheme in an ASIC, extensive simulations have to be carried out in order to assess its performance. Besides simulations, one often wants to evaluate the performance based on some measurement database that has been carried out over a fiber testbed. For most experimental transmitters it is difficult to generate transmission sequences that are valid code words of a given FEC scheme. It is therefore often assumed that a pre-FEC *bit error rate* BER [36] or *mutual information* (MI) threshold exists [63], for which a subsequent soft-decision FEC in conjunction with an outer hard-decision FEC is able to reduce the overall system BER beyond 10^{-12} or even lower. Often, *pseudo-random bit sequences* (PRBS) or interleaved versions thereof are preferred. These sequences generally do not correspond to code words of common soft-FEC realizations. Furthermore, it is inconvenient to redo involved transmission experiments whenever new coding schemes need to be evaluated. We seek therefore for a possibility to assess the soft-FEC performance of different coding schemes using recorded channel data measurements gained from PRBS data, without the need of redoing the experiments. Commonly, such experiments with coherent reception use offline processing for the required signal processing based on a stored database of measurements originating from a fast sampling scope. In the following, we propose to integrate the evaluation of the soft-FEC component into that signal processing chain to evaluate the performance of different codes [64].

The method we propose is based on the linearity of all practically employed FEC codes. In order to devise a strategy for performance assessment, we assume that the coding scheme to be tested generates one (or several) valid code words. These valid code words are scrambled using a (time-varying) binary scrambler which generates, by modulo-2 addition (i.e. XOR) of a scrambling sequence, the desired transmit bit sequence. This sequence corresponds in general to the PRBS which—after modulation—corresponds to the transmitted symbol sequence used in the experiment. This virtual transmitter (which is actually not implemented) is depicted in the top part of the block diagram given in Figure 15.10. If we exploit the fact that all practically relevant codes are linear codes, we can assume that the code word to be transmitted is just the all-zero code word, such that no actual code words need to be generated.

The corresponding receiver is given in the bottom part of the block diagram in Figure 15.10. The sequence of transmitted symbols is either taken from the measurement database or recovered from the received noisy symbols. The sequence recovery can be performed by either cross-correlating some or all of the received symbols with transmit symbols or, if a PRBS is transmitted, utilizing a simple PRBS synchronization algorithm. With the reconstructed transmit sequence, the scrambling sequence can be computed in a way that—within the virtual transmitter—the scrambler modifies the code words such that, after modulation, the actual transmitted sequence is generated. After demodulation, the scrambler de-scrambles the soft information, which can then be used to perform soft-FEC decoding. After the soft-FEC decoding step, the number of errors is counted by comparison with the assumed code words. If the all-zero code word has been supposedly transmitted, this step corresponds to counting the resulting "1"s after decoding. The main difficulty of this approach lies in the computation of the scrambler, which needs to be recomputed in each frame. The computation of the scrambler requires knowledge of the modulation format, of the mapping of bit patterns to modulation symbols and of a possible interleaver between FEC and modulation. Based on this information and on the reference sequence, which can be obtained using a PRBS synchronizer, the scrambler is

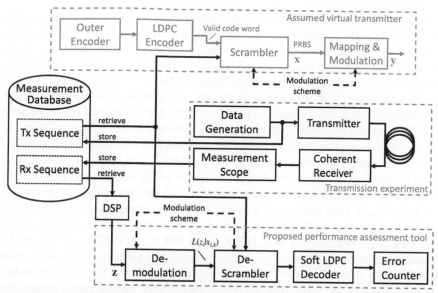

FIGURE 15.10 General setup of an optical transmission experiment with the proposed tool for offline evaluation of the soft-FEC performance.

then computed. As an example, the computation of the scrambler for a differentially encoded 16-QAM mapping has been derived in [64]. In the following, we show some results obtained using this mapping, based on a 16-QAM transmission experiment.

In the experiment, which has been described in detail in [65], 77 optical carriers are transmitted in the C-band with 50 GHz spacing. Odd and even channels are multiplexed separately and modulated independently. The FPGA-based 16-QAM transmitter generates a $2^{15} - 1$ PRBS sequence and subsequently maps the bits to symbols of the 16-QAM symbol stream. The 16-QAM symbols are fed to a digital-analog converter (DAC) with differential outputs. One of the differential output lines is delayed with respect to the other. The two signals are then utilized for modulating the I and Q branch of a nested Mach-Zehnder modulator. The outputs of the IQ-modulators for even and odd channels are separately polarization multiplexed using a delay line emulator. All polarization multiplexed 16-QAM modulated channels are combined in a wavelength selective switch (WSS) to form the 77 channel DWDM signal spaced on a 50 GHz wavelength grid. After amplification and power equalization the signal has been launched into a re-circulating loop composed of a G.654B compliant pure-silica core fiber Corning® Vascade® EX2000. Twelve spans of 50 km length with EDFA only amplification are organized in one roundtrip with 600 km length incorporating a gain equalizer and a polarization scrambler per loop.

As a receiver, a real-time storage oscilloscope with standard coherent frontend is used. The processing algorithm comprises re-sampling up to 64 GS/s, channel skew adjustment, timing recovery, dispersion compensation, polarization demultiplexing, frequency and phase recovery, and two adaptive FIR filters for post-equalization of both recovered polarizations. Polarization demultiplexing is based on a 9 tap fractionally spaced butterfly equalizer with blind adaptation using the multi-modulus algorithm (MMA). The received symbols are demodulated in a two-step demodulator. First, differential detection is performed on the quadrants of the 16-QAM to compute soft information for the first two bits. A soft-input/soft-output module according to [66] can be used for this task. Second, soft information for the other bits is obtained by conventional soft-demodulation, i.e. by computing the differences of the Euclidean distances between the received value and those constellation points where the respective bit is either "0" or "1." Details on how to compute the scrambler are not given here but can be found in [64].

The resulting soft information is de-scrambled using the recovered computed scrambling sequence and then fed to the soft-FEC component. In this work, we evaluate the soft-FEC component based on two different LDPC codes of rate $r = 0.86$ (16.3% OH) and length $N = 25,160$. Both codes are check-regular, irregular codes. The first code (Code 1) with check node degree $d_c = 28$ and variable node degree distribution $\{a_{v,2} = 0.285, a_{v,3} = 0.494, a_{v,7} = 0.022, a_{v,8} = 0.199\}$ is optimized to have a good waterfall behavior, while the second code (Code 2, with $d_c = 29$, $\{a_{v,2} = 0.137, a_{v,3} = 0.651, a_{v,7} = 0.022, a_{v,8} = 0.19\}$) aims at reducing the error floor by optimizing the degree profile such that no cycles comprising only degree-2 variable nodes result. After soft-FEC decoding, either the number of non-zero output bits (errors) can be counted, or the transmitted data bit sequence can be recovered by a subsequent scrambling step. We applied the soft-FEC assessment tool to the recorded data for the entire C-band at 2400 km transmission length. Figure 15.11 shows the BER at the LDPC decoder output vs. the hard-decision

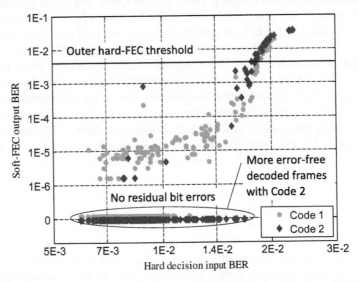

FIGURE 15.11 Simulation results, soft-FEC output BER vs. hard-decision input BER for the investigated LDPC codes.

demodulator output BER for all data and both codes. The waterfall region was found in the range of 1.5×10^{-2} to 2×10^{-2} for the demodulator output BER. We assume that an additional outer FEC decoder (7% OH) is used [36]. This code requires a BER below 4×10^{-3}. At this threshold, the BER at demodulator output is 1.8×10^{-2}.

15.7 CONCLUSION

In this chapter, we have discussed several implementation aspects of current and future application-specific integrated circuits (ASIC) for coherent optical transmission systems. We highlighted the main building blocks for such ASICs, which are data converters, baseband signal processing, and forward error correction.

References

[1] R. Kline, High-Speed Optics: Global 40G/100G Market Outlook, Ovum, Reference Code: OT00063-033, 01/2012.

[2] Optical Internetworking Forum, OIF-FD-100G-DWDM-01.0—100G Ultra Long Haul DWDM Framework Document, 2009.

[3] Optical Internetworking Forum, Scalable Serdes Framer Interface (SFI-S): Implementation Agreement for Interfaces Beyond 40G for Physical Layer Devices, 2008.

[4] Cortina Systems and Cisco Systems, Interlaken Protocol Definition, Revision 1.2, 2008.

[5] A. Leven, N. Kaneda, Y.-K. Chen, A real-time CMA-based 10 Gb/s polarization demultiplexing coherent receiver implemented in an FPGA, in: OFC/NFOEC, 2008, p. OTuO2.

[6] T. Pfau et al., First real-time data recovery for synchronous QPSK transmission with standard DFB lasers, IEEE Photon. Technol. Lett. 18 (18) (2006) 1907–1909.

[7] M. Birk et al., Field trial of a real-time, single wavelength, coherent 100 Gbit/s PM-QPSK channel upgrade of an installed 1800 km link, in: OFC/NFOEC, San Diego, CA, 2010, p. PDPD1.

[8] I. Dedic, 56Gs/s ADC: enabling 100GbE, in: OFC/NFOEC, San Diego, CA, 2010, p. OThT6.

[9] Y.M. Greshishchev et al., A 56GS/S 6b DAC in 65nm CMOS with $256 \times 6b$ memory, in: Solid-State Circuits Conference Digest of Technical Papers, San Francisco, CA, 2011, pp. 194–196.

[10] B.E. Jonsson, A survey of A/D-converter performance evolution, in: Electronics, Circuits and Systems (ICECS) Conference, Athens, Greece, 2010, pp. 766–769.

[11] R.W. Sears, Electron beam deflection tube for pulse code modulation, Bell Syst. Tech. J. 27 (1) (1948) 44–57.

[12] R.A. Kertis et al., A 35 GS/s 5-Bit SiGe BiCMOS flash ADC with offset corrected exclusive-or comparator, in: Bipolar/BiCMOS Circuits and Technology Meeting, 2008, pp. 252–255.

[13] S. Shahramian, S.P. Voinigescu, A.C. Carusone, A 35-GS/s, 4-bit flash ADC with active data and clock distribution trees, IEEE J. Solid-State. Circ. 44 (6) (2009) 1709–1720.

[14] C. Vogel, H. Johansson, Time-interleaved analog-to-digital converters: status and future directions, in: Proceedings ISCAS, 2006, pp. 3386–3389.

[15] Y.M. Greshishchev et al., A 40GS/s 6b ADC in 65nm CMOS, in: ISSCC Solid-State Circuits Conference Digest, 2010, pp. 390–391.

[16] P. Bower, I. Dedic, Highspeed converters and DSP for 100G and beyond, Opt. Fiber Technol. 17 (5) (2011) 464–471.

[17] I. Fatadin, S.J. Savory, D. Ives, Compensation of quadrature imbalance in an optical QPSK coherent receiver, IEEE Photon. Technol. Lett. 20 (2008) 1735–1773.

[18] C. Burrus, Block implementation of digital filters, IEEE Trans. Circuit Theory 18 (1971) 697–701.

[19] G. Clark, S. Parker, S. Mitra, A unified approach to time- and frequency-domain realization of FIR adaptive digital filters, IEEE Trans. Acoust Speech Signal Process 31 (1983) 1073–1083.

[20] B. Spinnler, Complexity of algorithms for digital coherent receivers, in: Proc. ECOC, 2009, Paper 7.3.6.

[21] J. Cooley, J. Tukey, An algorithm for the machine calculation of complex Fourier series, Math. Comput. 19 (1965) 297–301.

[22] I. Fatadin, D. Ives, S.J. Savory, Blind equalization and carrier phase recovery in a 16-QAM optical coherent system, J. Lightwave Technol. 27 (15) (2009) 3042–3049.

[23] B. Widrow, Thinking about thinking: the discovery of the LMS algorithm, IEEE Signal Process Mag. 22 (2005) 100–106.

[24] S.J. Savory, Digital filters for coherent optical receivers, Opt. Express 16 (2008) 804.

[25] ITU-T G.975 Forward error correction for submarine systems, 1996.

[26] ITU-T G.975.1 Forward error correction for high bit-rate DWDM submarine systems, 2004.

[27] T.K. Moon, Error Correction Coding: Mathematical Methods and Algorithms, John Wiley & Sons, 2005.

[28] M. Scholten, T. Coe, J. Dillard, Continuously-Interleaved BCH (CI-BCH) FEC delivers best in class NECG for 40G and 100G metro applications, in: Proc. OFC/NFOEC, Paper NTuB3, San Diego, 2010.

[29] B.P. Smith, A. Farhood, A. Hunt, F.R. Kschischang, J. Lodge, Staircase codes: FEC for 100 Gb/s OTN, IEEE, J. Lightwave Technol. (2012) 110–117.

[30] R.M. Pyndiah, Near-optimum decoding of product codes: block turbo codes, IEEE Trans. Commun. 46 (8) (1998) 1003–1010.

[31] T. Mizuochi et al., Forward error correction based on block turbo code with 3-bit soft decision for 10-Gb/s optical communication systems, IEEE J. Sel. Top. Quant. Electron. 10 (2) (2004) 376–386.

[32] R.G. Gallager, Low-Density Parity-Check Codes, MIT, 1963.

[33] C. Berrou, A. Glavieux, P. Thitimajshima, Near Shannon limit error-correcting coding and decoding: turbo codes, in: International Conference on Communications, Geneva, 1993.

[34] D.J.C. MacKay, R.M. Neal, Near Shannon limit performance of low density parity check codes, Electron. Lett. 32 (1996) 1645–1646.

[35] N. Wiberg, Codes and Decoding on General Graphs. PhD Thesis, Linköping University, Linköping, Sweden, 1996.

[36] Y. Miyata, K. Kubo, H. Yoshida, T. Mizuochi, Proposal for frame structure of optical channel transport unit employing LDPC codes for 100 Gb/s FEC, in: Proc. OFC/NFOEC, Paper NThB2, 2009.

[37] I.B. Djordjevic, B. Vasic, Nonbinary LDPC codes for optical communication systems, IEEE Photon. Technol. Lett. 17 (10) (2005) 2224–2226.

[38] T. Richardson, M.A. Shokrollahi, R. Urbanke, Design of capacity-approaching irregular low-density parity-check codes, IEEE Trans. Inf. Theory 47 (2) (2001) 619–637.

[39] S. ten Brink, A. Leven, L. Schmalen, FEC and soft decision: concepts and directions, in: Proc. OFC/NFOEC, Tutorial OW1H.5, Los Angeles, 2012.

[40] N. Bonello, S. Chen, L. Hanzo, Design of low-density parity-check codes, IEEE Vehicular Technol. Mag 6 (4) (2011) 16–23.

[41] X. Hu, E. Eleftheriou, D.M. Arnold, Regular and irregular progressive edge-growth tanner graphs, IEEE Trans. Inf. Theory 51 (1) (2005) 298–386.

[42] T. Tian, C.R. Jones, J.D. Villasenor, R.D. Wesel, Selective avoidance of cycles in irregular LDPC code construction, IEEE Trans. Commun. 52 (8) (2004) 1242–1247.

[43] G. Liva et al., Design of LDPC codes: a survey and new results, J. Commun. Software Syst. 2 (3) (2006) 191–211.

[44] T. Richardson, R. Urbanke, Modern Coding Theory, Cambridge University Press, 2008.

[45] J. Hagenauer, E. Offer, L. Papke, Iterative decoding of binary block and convolutional codes, IEEE Trans. Inf. Theory 42 (2) (1996) 429–445.

[46] D.E. Hocevar, A reduced complexity decoder architecture via layered decoding of LDPC codes, in: Proc. IEEE Workshop on Signal Processing Systems (SIPS), 2004.

[47] C.-L. Wey, M.-D. Shieh, S.-Y. Lin, Algorithms of finding the first two minimum values and their hardware implementation, IEEE Trans. on Circ. Syst. – I: Regular Papers 55 (11) (2008) 3430–3437.

[48] L.G. Amarù, M. Martina, G. Masera, High speed architectures for finding the first two maximum/minimum values, IEEE Trans. Very Large Scale Integr. (VLSI) Syst. 20 (12) (2012) 2342–2346.

[49] A. Darahiba, A.C. Carusone, F.R. Kschischang, Bit-serial approximate min-sum LDPC decoder and FPGA implementation, in: Proc. IEEE International Symposium on Circuits and Systems (ISCAS), 2006.

[50] J. Chen, A. Dholakia, E. Eleftheriou, M.P.C. Fossorier, X.-Y. Hu, Reduced-complexity decoding of LDPC codes, IEEE Trans. Commun. 53 (8) (2005) 1288–1299.

[51] D. Oh, K. Parhi, Min-sum decoder architecture with reduced word length for LDPC codes, IEEE Trans. Circ. Syst. I 57 (1) (2010) 105–115.

[52] X. Zuo, R.G. Maunder L. Hanzo, Design of fixed-point processing based LDPC codes using EXIT charts, in: Proc. Vehicular Technology Conference, San Francisco, 2011.

[53] Z. Zhang, L. Dolecek, M. Wainwright, V. Anantharam, B. Nikolic, Quantization effects in low-density parity-check decoders, in: Proc. International Conference on Communications (ICC), 2007.

[54] ETSI DVB-S2, ETSI Standard EN 302 307 V. 1.2.1, 2009.

[55] T. Richardson, R. Urbanke, Efficient encoding of low-density parity-check codes, IEEE Trans. Inf. Theory 47 (2) (2011) 638–656.

[56] D. Divsalar, H. Jin, R.J. McEliece, Coding theorems for turbo-like codes, in: Proc. Allerton Conference on Communication, Control, and Computing, 1998.

[57] Y. Zhang, W.E. Ryan, Structured IRA codes: performance analysis and construction, IEEE Trans. Commun. 55 (5) (2007) 837–844.

[58] R.M. Tanner, On quasi-cyclic repeat-accumulate codes, in: Proc. Allerton Conference on Communictions, Control, and Computing, 1999.

[59] L. Schmalen, S. ten Brink, G. Lechner, A. Leven, On threshold prediction of low-density parity-check codes with structure, in: Proc. Conference on Information Sciences and Systems (CISS), Princeton, 2012.

[60] IEEE, 802.11, Wireless LAN Medium Access Control (MAC) and Physical Layer (PHY) Specifications, 2012.

[61] IEEE, 802.16, Air Interface for Broadband Wireless Access Systems, 2009.

[62] Z. Li, L. Chen, L. Zeng, S. Lin, W.H. Fong, Efficient encoding of quasi-cyclic low-density parity-check codes, IEEE Trans. Commun. 54 (1) (2006) 71–81.

[63] A. Leven, F. Vacondio, L. Schmalen, S. ten Brink, W. Idler, Estimation of soft-FEC performance in optical transmission experiments, IEEE Photon. Technol. Lett. 23 (20) (2011) 40–42.

[64] L. Schmalen, F. Buchali, A. Leven, S. ten Brink, A generic tool for assessing the soft-FEC performance in optical transmission, IEEE Photon. Technol. Lett. 24 (1) (2012) 40–42.

[65] K. Schuh et al., 15.4 Tb/s transmission over 2400 km using polarization multiplexed 32-Gbaud 16-QAM modulation and coherent detection comprising digital signal processing, in: Proc. ECOC, Paper We.8.B.4, Lausanne, 2011.

[66] S. Benedetto, D. Divsalar, G. Montorsi, F. Pollara, A soft-input soft-output APP module for iterative decoding of concatenated codes, IEEE Commun. Lett. 1 (1) (1997) 22–24.

All-Optical Regeneration of Phase Encoded Signals

16

Joseph Kakande, Radan Slavík, Francesca Parmigiani,
Periklis Petropoulos, and David Richardson

Optoelectronics Research Centre, University of Southampton, Southampton, Hampshire,
SO17 1BJ, United Kingdom

PHASE SENSITIVE OPTICAL REGENERATION

16.1 INTRODUCTION

The development of the erbium-doped fiber amplifier (EDFA) revolutionized optical communications, eliminating fiber propagation loss as the primary limiting factor to transmission capacity, and enabling mass access to data transfer over inter-continental distances. By adopting wavelength division multiplexing (WDM) of high baud-rate (on-off keyed, OOK) channels, data signaling rates approaching 1 Tbit/s down a single fiber were commonly deployed commercially for long-haul and high-capacity backbone communications by the late 1990s, and the lay perception of the optical fiber was as a channel of effectively limitless bandwidth. However, due to the continued emergence of bandwidth intensive applications and services, e.g. social networking and more recently high definition video streaming, data transmission demands have continued to grow exponentially and there is an ever increasing appreciation that fiber bandwidth is a far from infinite resource: fiber bandwidth will need to be used more efficiently in future communication systems if available capacity is to stay ahead of demand for very much longer. Indeed there are fears in terms of crunches in both capacity and energy requirements ahead without further radical innovation at the physical layer. To this end the use of higher signaling rates, coupled with use of higher spectral efficiency modulation formats, has become a key research challenge.

The ability to signal in phase as well as amplitude has long been exploited in radio telecommunications to boost spectral efficiency. In fiber communications, formats such as differential phase shift keying (DPSK) and differential quadrature phase shifted keying (DQPSK) were rapidly shown to offer significant benefits over amplitude only formats, such as OOK. DPSK, for example, offers considerable advantages in terms of resilience to transmission impairments such as dispersion and non-linearity, as well as an inherent 3 dB improvement in receiver sensitivity over OOK if balanced detection is utilized [1], and combined with sophisticated error correction

Optical Fiber Telecommunications VIA. http://dx.doi.org/10.1016/B978-0-12-396958-3.00016-0

techniques further improvements in system margin and/or reach are enabled. More recently, a resurgence in the use of coherent detection has allowed for an increase in spectral efficiency by the adoption of higher order modulation formats, such as coherent quadrature phase shift keying (QPSK) [2], as well as in reach due to the ability to digitally compensate, in real-time, for both linear and nonlinear transmission impairments [3]. Polarization multiplexed QPSK (PM-QPSK) has recently become the standard for long-haul 100 Gbit/s optical transmission and in the laboratory far more exotic modulation formats, such as 16-QAM are routinely studied with heroic attempts being made to work with much higher complexity variants, e.g. 256-QAM and above, which places great demands on system optical signal-to-noise ratio (OSNR) and on the underpinning component and sub-system technologies.

The reach of fiber links is ultimately limited by noise, generated either linearly due to quantum noise resulting from loss and quantum limited amplification, or nonlinearly due to interactions between the propagating signal and other optical waves in the same fiber [4]. Typically, amplifiers are required every 60–120 km, compensating for around 12–24 dB worth of losses, and it is possible to minimize noise pickup in links by using low noise figure (NF) lumped amplifiers and precompensating for the loss by amplifying before the lossy element. Additional distributed Raman amplification is also often deployed to enhance system performance. While optical amplification can be used to substantially offset the detrimental impact of loss, the presence of optical nonlinearity means that beyond a certain peak power level, amplification degrades, rather than enhances the received signal quality. The primary nonlinear mechanism for signal quality degradation is cross phase modulation (XPM) between channels and amplified spontaneous emission (ASE), which gives rise to phase noise. For further details the interested reader is referred to Ref. [4] for a comprehensive review of nonlinear physical noise generation mechanisms and their impact within optical fiber communication systems.

The solution to such noise pickup is to periodically regenerate the signal, restoring it back to its original quality. Typically, commercial regenerators utilize a process referred to as Optical-Electrical-Optical (OEO) conversion. This involves the noisy optical signal (i) being detected using a photo-receiver thereby converting it into an analog electrical signal, (ii) re-digitizing the analog signal thereby removing the noise, and (iii) re-modulating this digital stream onto a new non-degraded optical carrier. Digital error correction can also be used on the electrical signal, enhancing the performance of the system significantly. While such an approach potentially allows transmission over essentially unlimited distances, the clear drawback is that the OEO conversion and associated electronic processing substantially restricts the speed and latency of the system; moreover there are associated power consumption/heat dissipation issues which are ultimately limiting as the bit rate is increased. The ideal solution to these problems would then be to perform the regeneration function all optically.

Over the years there has been substantial work on the problem of all-optical regeneration with many interesting and impressive results achieved in the laboratory [5]. However, given that most research over the past 30 years has concerned the regeneration of simple intensity modulated signals the majority of this work is unfortunately now largely redundant given the clear and necessary migration to more complex phase-based optical signaling formats. The challenge in the field of optical regeneration has shifted and is now to develop regeneration techniques that can operate on complex, phase-based modulation format signals. Work in this area only really started some 5–6 years ago and the purpose of this chapter is to review the impressive progress made to date.

All-optical phase regeneration can be approached in a number of distinct ways. The most obvious way, perhaps, is to try to exploit all the work done to date on amplitude encoded signals by providing some means to convert an incident phase-encoded signal to an amplitude encoded signal, to regenerate that all-optically, before converting back to the original (regenerated) phase encoded format. To an extent this can be viewed as the phase-to-amplitude converter analog of the OEO regenerator described previously. A more attractive approach, however, is to exploit optical processes which are inherently sensitive to optical phase—of which the Optical Parametric Amplifier (OPA), and more specifically the Fiber OPA (FOPA), is the archetypal example [6]. FOPAs have a number of interesting modes of operation and performance features that make them especially attractive for manipulating advanced modulation format signals.

In conventional Phase Insensitive (PI) mode, in which there is no well-defined phase relationship between the waves involved (Signal, Idler, and Pump(s)), FOPAs can be used to limit the intensity of signals via a gain saturation mechanism [7]. This has been used to regenerate the intensity of DPSK signals [8], reducing the nonlinear pickup of noise in transmission spans following the regenerator, with demonstrations showing reach enhancement [9]. The extinction ratio of OOK signals can also be increased using a higher order four-wave mixing (FWM) effect in FOPAs, allowing for another optical regeneration mechanism [10].

In Phase Sensitive (PS) mode, two pump symmetric (degenerate) FOPAs can directly eliminate phase noise from DPSK signals due to their π-periodic step phase characteristics. When saturated, they can also use the same amplitude limiting function described earlier to eliminate amplitude noise [11]. This combination of effects can be used to perform simultaneous phase and amplitude regeneration of DPSK signals [11,12]. As we shall describe later this phase regeneration capability can be used to extend the reach of DPSK-bearing fiber links. This configuration, however, is limited to two-level PSK signals, and is not applicable to higher order modulation formats such as QPSK that are superseding DPSK. In order to work on higher modulation formats it is necessary to cascade multiple parametric effects to build up the necessary device transfer functions [13].

In this chapter we review the general principles and approaches used to regenerate phase encoded signals of differing levels of coding complexity. We will describe the key underpinning technology and present the current state of the art. This chapter is structured as follows: we first review the different approaches and nonlinear processes that can be used to perform the regeneration of phase encoded signals. Our primary focus is on parametric effects, which as we have explained previously can operate directly on the optical phase. We then proceed to review progress on regenerating the simplest of phase modulation formats, namely DPSK/BPSK and for which the greatest progress has been made to date. We then move on to discuss progress in regenerating more complex modulation format signals: in particular (D) QPSK and other M-PSK signals. Next we review the choice of nonlinear components available to construct phase regenerators. It is not our intention to provide a rigorous and complete overview of all options in this section but rather to describe the various merits and issues associated with each component type and material system. Finally, we review the prospects for regenerating even more complex signals including QAM and mixed phase-amplitude coding variants. We end with a brief summary and conclusions.

16.2 APPROACHES TO REGENERATION OF PHASE ENCODED SIGNALS

For the purpose of this review we classify a PSK regenerator in the broadest of terms as a device that actively reduces the accumulated phase noise between an optical PSK transmitter and receiver. However, before we begin, it is worth noting the whole host of other (passive) ways to achieve the same goal, such as installing lower noise amplifiers [14], using transmission fiber with reduced loss and/or nonlinearity [15], and optimizing the link dispersion map [16]. The regenerators discussed in this chapter post-compensate either for accumulated phase noise, or for amplitude noise that could seed phase noise via nonlinear mechanisms during further propagation [4]. Given the variety of physical mechanisms that can create phase noise within an optical fiber link, multiple approaches have been proposed to actively perform regeneration, each seeking to balance the complexity of implementation with the required level of performance.

It is useful to divide the schemes for optical phase regeneration into two broad categories, *direct* and *indirect*. Direct regenerators operate on the premise that a signal has already been degraded in phase, and therefore signal quality can only be improved by suppressing these phase perturbations. To perform direct phase regeneration, a method of distinguishing between the noise and signal, both of which occur in phase, is required. Indirect regenerators act by suppressing amplitude fluctuations which could seed phase noise further on—eliminating the need to distinguish between the signal, which occupies phase space, and noise, located in the amplitude domain. Note that the process of amplitude noise suppression must not create significant added phase noise, to make it worthwhile.

16.2.1 Indirect PSK regeneration

Indirect phase regeneration is often referred to as phase-preserving amplitude regeneration, which is rather apt as it describes the actual mechanism involved in the regenerator. The devices are also referred to in the literature as *limiters*, as they limit signal power to a narrow range. It is suitable in systems where nonlinear phase noise is the dominant limiting factor in receiver bit error rate (BER) [17]. The historical use of OOK in optical links means that all-optical amplitude regeneration has been extensively investigated in the past, predominantly using nonlinear (power-dependent) optical effects such as nonlinear refraction and parametric gain saturation. Many of the schemes, however, cannot be used for indirect phase regeneration as they intrinsically transform the amplitude noise into the phase (or frequency) domain. It is worth clarifying that every all-optical amplitude regenerator will add some phase noise, although some do so to a lesser extent, and therefore more tolerably, than others. We highlight some key examples here.

16.2.1.1 Nonlinear loop mirror

Nonlinear optical loop mirrors (NOLMs) transfer the optical signal from an input to an output port with a power dependent transfer function, mediated by the interference between two power-imbalanced counter-propagating portions of the signal within a loop comprising a Kerr optical material [18]. The nonlinear transfer function is obtained due to the different amounts of phase modulation induced on the signal via the Kerr effect. A plateau in the output signal power can then be obtained while providing minimum phase distortion [19] by optimizing the device configuration and by defining the input signal dynamic range. Stephan et al. [20] demonstrated one such device, referred to as a nonlinear amplifying loop mirror (NALM) due to the insertion of a gain element (EDFA) inside the loop. By installing the NALM inside a recirculating loop with 100 km transmission per span, they observed significant improvement in launch power margin, Figure 16.1, hence superior link BER [20].

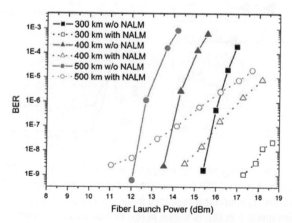

FIGURE 16.1 BER dependence of the fiber launch power for different transmission distances with and without a NALM regenerator every 100 km, by Stephan et al. (reproduced from [20]).

16.2.1.2 Four-wave mixing-based limiting

FWM is another process that can generate a plateau in its amplitude transfer function without significantly distorting the phase [5,21]. In [21], Matsumoto demonstrates a FWM limiter (Figure 16.2a) installed either before, or within a 5 span, 200 km recirculating loop (40 km per span). The setup used for characterizing the regenerative properties of the limiter for 10 GHz return-to-zero DPSK data signals is reproduced in Figure 16.2b. Note that the signal OSNR after the transmitter (before limiting) was deliberately degraded to around 21 dB, to ensure that there was enough amplitude noise on the signal for the limiting to provide a significant improvement. The results obtained (shown in Figure 16.2d) clearly show the improvement obtained with periodic regeneration [21]. Other results presented in [21] confirmed that the improvement was due to the reduction in phase noise buildup along the transmission line.

16.2.1.3 Optical injection locking

The phase and amplitude transfer characteristics of injection-locked semiconductor lasers have also been proposed and demonstrated as a means to perform limiting of PSK signals [22,23]. By carefully choosing the power of the master (input signal), the slave laser can follow all phase changes on the master up to a limit in frequency, while holding the output amplitude constant. Being optoelectronic in nature the

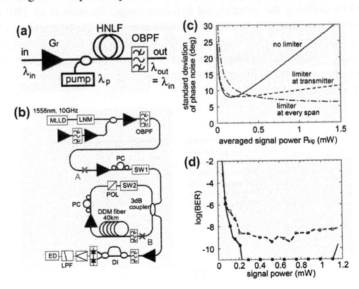

FIGURE 16.2 Phase-preserving amplitude regenerator based on saturated FWM (a) and set-up of 10 Gbit/s short-pulse DPSK transmission using the regenerator (b). Comparison of performance when the regenerator was placed in the recirculating loop (point B in (b)), or at a single position (point A in (b)) is shown in (c). BER performance when regenerator in the recirculating loop (solid) and without the regenerator (dashed) is shown in (d). MLDD: mode-locked diode laser, LNM: modulator, SW1, 2: acousto-optic switches, DI: delay interferometer, LPF: low-pass filter, ED: error detector (taken from [6]).

bandwidth is currently limited to just a few tens of GHz. The technique has the benefit of being very simple in terms of implementation, and lends itself well to integration.

16.2.1.4 Saturable absorption

A novel multiple-quantum-well (QW) semiconductor saturable absorber structure has also been demonstrated to be capable of performing amplitude limiting at over 40 Gbit/s for DPSK input signals [24]. Unlike conventional saturable absorbers, the reflectance of the device decreases when the incident optical power increases, allowing for a power limiting function when operated in reflection. Quang Trung et al. [24] demonstrated the device installed prior to a 100 km transmission span, obtaining an extra 2 dB margin in fiber launch power. The demonstration has also been recently extended to DQPSK [25]. The device is fully passive, eliminating the need for active cooling or electrical biasing, although a practical implementation may require signal preamplification.

16.2.2 Direct phase regeneration

The experiments referred to in Section 16.2.1 all utilized signals degraded with a lot of amplitude noise, but low levels of phase noise, prior to a long transmission. Under those circumstances the limiter does reduce the amplitude-to-phase noise conversion process. However, in cases where phase noise has already been acquired, a direct phase regenerator is required to suppress the noise. We discuss some of the alternatives here.

16.2.2.1 Format conversion-based regeneration

As mentioned earlier, various schemes for amplitude regeneration have been investigated in the past, offering trade-offs between regeneration performance and system complexity. Format conversion-based regenerators leverage these schemes, and rely on optical techniques to convert the optical PSK signal to one or more OOK signals. The OOK signals can then be regenerated in amplitude, without the requirement to be phase preserving, after which the now "clean" amplitude information is transferred back on to the phase of a noise-free optical carrier.

Converting the PSK signal into OOK is done using interference, either differentially or fully coherently. Doing so differentially, in a delay line interferometer (DLI), is the simplest approach, although the logical states of the input signal get changed in the process, albeit in a deterministic fashion, and therefore care must be taken in decoding the signal at the receiver [5]. The conversion can also be done by interfering the signal with a phase-locked local oscillator, eliminating the logic-alteration problem in the differential case, although performing the phase-locking represents a significant technical challenge in itself.

In [21], a delay line interferometer (DLI) was used to convert a DPSK signal into an amplitude-encoded signal before the regenerative step and format reconversion. A self-phase modulation (SPM)-based fiber device was used to regenerate the

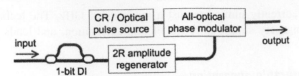

FIGURE 16.3 Block diagram of an all-optical DPSK signal regenerator using a straight-line phase modulator. CR: clock recovery circuit; DI: delay interferometer (reproduced from [5]).

OOK signal which was followed by an all-optical OOK-PSK conversion device using cross-phase modulation (XPM). The general scheme is shown in Figure 16.3. The regenerative capabilities of one such device are presented in [5,21]. It was demonstrated that the regenerator could, at least up to a certain level of nonlinear noise, significantly reduce the receiver power penalty, although for higher levels of nonlinear phase noise some error flooring was observed.

16.2.2.2 Phase sensitive amplification

Phase sensitive amplification (PSA) has been widely recognized as an effective way to regenerate phase encoded signals thanks to its phase squeezing capabilities [26,27]. It has also been shown to regenerate the amplitude of the signal when operated in the saturated regime [27]. While a format conversion-based device requires knowledge of the signal symbol rate, as well as recovery of the signal clock so as to perform amplitude regeneration, PSAs operate directly on the input signal field, and can be incorporated in-line using continuous wave pumps. They also allow the possibility of low noise amplification, although when used as regenerators this feature is not required. These benefits have spurred significant research efforts into PSAs in recent times, and we summarize key results in this field in the following sections.

16.3 PSA-BASED PHASE REGENERATION

Phase sensitive optical amplifiers exhibit gain characteristics that depend on the phase of the input optical signal relative to some local optical reference. The theory of PSAs is more than five decades old, and interest in them first stemmed from a semi-classical realization that knowledge of signal phase allows signal measurements to be made more precisely than the Uncertainty Principle dictates [28], and later from a more in-depth analysis by Caves [29] that showed how this could be used for noiseless amplification. Research efforts into PSA increased significantly due to the new appreciation of their ability to squeeze the noise added to any input light below the quantum limit in one of the two phase quadratures; an excellent review by Slusher and Yurke [30] highlights the diverse applications of squeezed light in communications and sensing. Second order nonlinearity in bulk crystals was used to demonstrate degenerate phase sensitive amplification, both for squeezing [30] and amplification below the 3 dB quantum noise limit [31].

Practical PSAs deployed within optical networks would allow significant benefits—increased receiver OSNR allowing more spectrally efficient communications, increased amplifier spans, and interesting photon correlation characteristics of possible interest for quantum communications. The ability to provide different gains to the two phase quadratures can also be used for all-optical signal processing, including phase regeneration and signal sampling/characterization. Performing PSA using third order nonlinearity in optical fibers rather than in second order crystals was a first step toward practical network applications due to the increased robustness, improved power efficiency, and ease of system integration. Marhic et al. successfully demonstrated the first degenerate PSA by 1991 [32] but progress in the field was quite limited up until the demonstration of amplification with 1.8 dB NF by Imajuku et al. [33]. Since then, PSAs have been demonstrated operating over wide bandwidths [34,35] as well as with a record low noise figure [36].

PSAs principally operate by amplifying one of the input signal's quadratures and de-amplifying the other [32]. This is very convenient for BPSK regeneration because signal information is distinguishable from the noise because they can be made to occupy orthogonal quadratures [11,26]. While PSA can be obtained in various media (as discussed later on in the chapter), we focus on FOPA implementations here as FOPA devices comprise the majority of PSA based phase regenerators in the literature. We discuss the theory behind phase sensitive (PS)-FOPAs in this section and highlight key experimental demonstrations of regeneration.

16.3.1 Binary phase quantization theory

The direct phase regeneration of BPSK signals requires a device capable of amplifying a signal's in-phase component, while de-amplifying the quadrature component. One such device is the PSA. PS-FOPAs can be achieved via interferometric (Figure 16.4, left) or non-interferometric means (Figure 16.4, centre). As experimentally demonstrated

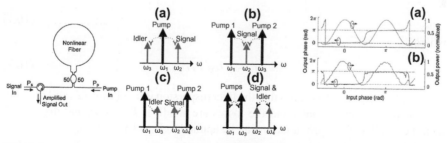

FIGURE 16.4 Left—General configuration of the NOLM-based interferometric PSA. Middle—FWM-based PSAs (a) single pump non-degenerate, (b) dual pump degenerate, (c) dual pump non-degenerate, and (d) wavelength exchanger. Right—(a) analytical and (b) experimental phase-to-amplitude and phase-to-phase transfer functions of a non-degenerate-idler PSA with a maximum of 11 dB of on-off gain, taken from [37].

in [37], conventional PSAs squeeze the input phase to one of two π-separated phase levels, accompanied by a sinusoidal phase-to-amplitude transfer characteristic (Figure 16.4, right). The theory of both interferometric and non-interferometric PSAs is discussed as follows.

16.3.1.1 Interferometric PSA

BPSK regeneration was in fact first proposed and demonstrated using degenerate PSA in a Sagnac loop [26]. Figure 16.4 (left) shows the setup for a typical interferometric PS-FOPA, based around a Sagnac interferometer. A 3 dB splitter is used to combine a strong pump with a signal, both of which are phase locked at the same frequency. The outputs of the splitter are then coupled into the opposite ends of a nonlinear fiber, allowing the combined beam from each splitter port to travel separately from that emerging from the other splitter port. During propagation in the fiber, the two beams experience a nonlinear phase, after which they are recombined in the splitter. An amplified version of the signal can be seen to emerge from the same splitter input port as the original signal and can be retrieved using a circulator as shown, while the pump exits the loop at the other splitter port. A modification to this setup would be to send the two combined outputs through separate fibers and have a second 3 dB splitter to separate the signal and pump, effectively creating a nonlinear Mach-Zehnder interferometer, but this requires two length-matched nonlinear fibers and is susceptible to acoustic and thermally induced phase perturbations and hence is more unstable.

The gain expression for this device can be found in a straightforward fashion. Assuming P_s and P_p to be the input signal and pump powers respectively, γ to be the nonlinear coefficient of the fiber, L to be the effective length of the fiber and ϕ to be the phase difference between signal and pump at the splitter, then the output signal power P_s can be found as [32,38]

$$P_s(\phi) = P_s\cos^2(\Phi_{NL}\cos(\phi)) + P_p\sin^2(\Phi_{NL}\cos(\phi)) - \sqrt{(P_pP_s)}\sin(\phi)\sin(2\Phi_{NL}\cos(\phi)),$$

(16.1)

where

$$\Phi_{NL} = \gamma L\sqrt{P_sP_p}.$$

(16.2)

Note that Eq. (16.1) assumes that the fiber dispersion is negligible (a valid assumption for narrowband signals) and the losses are small. If $P_p \gg P_s$ and $\Phi_{NL} \ll 1$, then the small angle approximation can be applied to Eq. (16.1), giving

$$P_s(\phi) = P_s\left(1 - \frac{\Phi_{NL}^2}{2}\cos^2(\phi)\right) + P_p\Phi_{NL}^2\cos^2(\phi) - 2\Phi_{NL}\sqrt{(P_pP_s)}\sin(\phi)\cos(\phi).$$

(16.3)

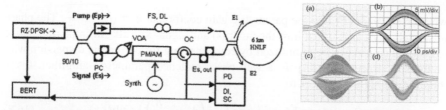

FIGURE 16.5 Experimental setup for RZ-DPSK regeneration (from [12]). FS: fiber stretcher; DL: delay line; VOA: variable optical attenuator; OC: optical circulator; PC: polarization controller; PD: photodiode; DI: delay interferometer; SC: sampling oscilloscope; Synth: synthesizer; BERT: bit-error-ratio tester; PM/AM: noise adding phase and amplitude modulators. Right: eye diagrams after demodulation: (a) RZ-DPSK data, (b) with added AN, (c) with added PN and AN, (d) after regeneration [26].

It can be seen from Eq. (16.3) that the signal gain will vary with phase ϕ with π periodicity. The solutions for the maximum and minimum gain, G_1 and G_2, respectively, are [38]:

$$G_1 = 1 + 2\Phi^2 + 2\Phi\sqrt{1 + \Phi^2}, \tag{16.4}$$

$$G_2 = 1 + 2\Phi^2 - 2\Phi\sqrt{1 + \Phi^2}, \tag{16.5}$$

where

$$\Phi = \frac{\gamma L P_p}{2}. \tag{16.6}$$

Because $G_1 \cdot G_2 = 1$, it can be calculated (not shown here) that the noise figure for such an amplifier would be 0 dB, for a more detailed quantum-mechanical analysis see Shirasaki and Haus [39]. Note that the maximum gain varies with $(\gamma L P_p)^2$; this will be referred to shortly while comparing interferometric and non-interferometric PSAs.

Croussore et al. [26] reported the first experimental demonstrations of DPSK regeneration using an interferometric PSA in the setup shown in Figure 16.5. A DPSK signal was degraded in phase and amplitude using a single frequency tone imparted onto inline electrooptic modulators, and the ability of the device to suppress both forms of noise noted, Figure 16.5 (right). For amplitude regeneration, a plateau in the amplitude transfer function of the loop mirror was used, similar to that described in Section 16.2.1.

16.3.1.2 Non-interferometric PSA

Fiber-based BPSK regeneration in optical fiber can also be obtained using the non-interferometric degenerate PSA, Figure 16.4 (centre, b). For the dual pump (degenerate) PS-FOPA, the gain function can be written as [35,40]

$$G_s(z) = \cosh(gz) + \frac{i}{g}\frac{\kappa}{2}\sinh(gz) + \frac{1}{g}r\exp\left(-i\left(\phi_{\text{rel}} - \frac{\pi}{2}\right)\right)\sinh(gz), \tag{16.7}$$

where $r = 2\gamma\sqrt{P_1 P_2}$ and the parametric gain coefficient $g = \sqrt{r^2 - \frac{\kappa^2}{2}}$. The phase mismatch κ is a measure of how the pump(s) and signal(s) walk-off in phase relative to each other, and can be minimized by compensating for chromatic dispersion with nonlinear refraction [40,41]. In a well-phase matched parametric amplifier, $\kappa \approx 0$, therefore Eq. (16.7) reduces to

$$G_s(z) \approx \cosh(gz) + \exp\left(-i\left(\phi_{\mathrm{rel}} - \frac{\pi}{2}\right)\right)\sinh(gz). \qquad (16.8)$$

By setting the sum of the pumps' phases as the phase reference, then

$$G_s(z) \approx \cosh(gz) + \exp\left(-i\left(2\phi_s - \frac{\pi}{2}\right)\right)\sinh(gz). \qquad (16.9)$$

And by using the identity $\cosh^2 - \sinh^2 = 1$, Eq. (16.9) further reduces to

$$G_s(z) \approx \cosh(gz) + \left[1 + \exp\left(-i\left(2\phi_s - \frac{\pi}{2}\right)\right)\tanh(gz)\right]. \qquad (16.10)$$

Assuming an input analytical signal $\exp(i\phi_s)$, then the output of the PSA can be obtained by multiplying the analytical input with the gain function, hence

$$\exp(i\phi_s)G_s(z) \approx \cosh(gz)\left[\exp(i\phi_s) + \exp\left(-i\left(\phi_s - \frac{\pi}{2}\right)\right)\tanh(gz)\right]. \qquad (16.11)$$

Observe that $\tanh(gz)$ is bounded within the range ± 1, and that $\cosh^2(gz)$ gives the peak power gain.

The first experimental demonstration of such a regenerator was reported by Croussore et al. [12]. As shown in Figure 16.6, two pumps were generated by applying sinusoidal modulation and sideband filtering of the same optical carrier as the signal (tapped off before the modulation). The pumps, 40 GHz on either side of the signal, were subsequently amplified and combined with the noisy data signal. Noise was emulated via amplitude and phase modulation at a single frequency, and the regenerator was found to cope well with both the phase and amplitude noise. For amplitude regeneration, saturation of the FWM process was exploited, similarly to FWM-based

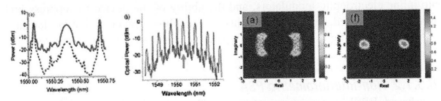

FIGURE 16.6 Left: Spectra for degenerate non-interferometric PSA with 80 GHz spaced pumps. (a) Spectrum before (dashed) and after (solid) PSA. (b) Full output spectrum: red arrow indicates location of the input signal. Right: Constellation plots of the input (a) and output (f) of the regenerator (taken from [12].) (For interpretation of the references to color in this figure legend, the reader is referred to the web version of this book.)

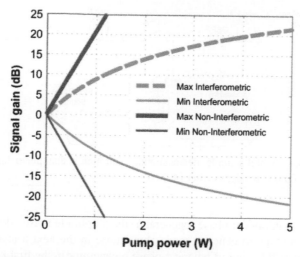

FIGURE 16.7 Analytic comparison (by simulation) of interferometric and non-interferometric PSAs using pump power and signal gain as reference metrics.

amplitude limiters. Constellation diagrams clearly showing regeneration in both the phase and amplitude are also shown in Figure 16.6 (right).

16.3.1.3 Comparison of interferometric and non-interferometric PSAs

From the expressions for the maximum gain of the interferometric PS-FOPA, Eq. (16.4), and that of the phase-matched non-interferometric non-degenerate PS-FOPA, Eq. (16.10), it can be seen that in the case of the former, the gain scales quadratically with pump power as compared to exponentially in the latter. As an example, for a fiber of nonlinear coefficient 12/W/km and effective length 200 m (typical values), the expected gains in interferometric and non-interferometric mode are shown in Figure 16.7, calculated analytically.

For these parameters, achieving 20 dB gain would require 4 W of pump power in the interferometric PSA, as compared to 1 W in the non-interferometric. In addition, the interferometric PSA is inherently single channel, while the non-interferometric device can operate with multiple signals provided that corresponding idlers are presented at the device input. Non-interferometric PSAs are therefore more energetically efficient, and therefore the preferred choice for most recent demonstrations of BPSK regeneration.

16.3.2 Saturation of PSAs for amplitude noise suppression

The phase regeneration function in degenerate PS-FOPAs is typically accompanied by phase-to-amplitude conversion [11]. As a result phase noise at the PSA input is transformed into amplitude noise at the output. Consequently, if such a PSA was

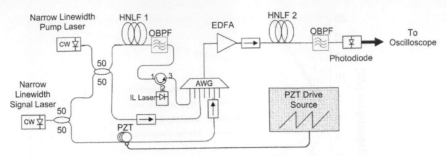

FIGURE 16.8 Experimental setup for demonstrating saturation in dual pump degenerate PS-FOPA. OBPF: Optical Bandpass Filter [42].

used as a mid-span optical phase regenerator, the amplitude noise added within the regenerator would act to generate more phase noise in the next transmission span, reducing the overall benefit of having a phase regenerator in the first place. It therefore becomes crucial when PSA-based phase regenerators are deployed mid-span to saturate the PS-FOPAs (by boosting the input signal power), so as to clamp the maximal signal gain, and minimize the amplitude fluctuations seen on the signal at the PSA output [11].

16.3.2.1 Experimental observation of saturation in degenerate PS-FOPA

In [42], an experiment was performed to study the saturation process in a FOPA and demonstrates the amplitude noise suppression in a PS-FOPA phase regenerator. The setup is shown in Figure 16.8. The PSA was a dual pump degenerate device, with the pumps 200 GHz either side of the signal in frequency. The phase of the signal wave was modulated using a 50 kHz ramp function applied to a piezo-electric fiber stretcher, allowing the phase-to-amplitude conversion to be observed at the PSA output.

The normalized signal gain curves are shown in Figure 16.9. At the lowest signal level assessed, a sinusoidal gain curve was obtained indicating the absence of significant gain saturation. As the signal power increased the gain curve is seen to acquire

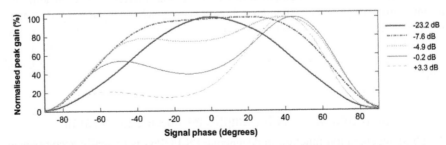

FIGURE 16.9 Phase-to-amplitude transfer characteristics for degenerate 2P PS-FOPA when signal power is boosted high enough to observe gain saturation [42].

a flat-top. Because each curve was captured for a fixed signal power at the PS-FOPA input, this differs from PI saturation (in which the gain varies with input signal power) because a nonlinear dependence on input signal phase exists. In this case, every point on the curve experiences a gain shift due to the alteration of the phase matching condition by the variation in the signal/pump power proportions during propagation. As the signal and pumps propagate down the fiber in which the PS interaction occurs, at every point a nonlinear contribution to the phase matching exists as a result of SPM and XPM respectively. Because the gain is dependent on the signal phase, the absolute powers and therefore the nonlinear phase matching varies with the input phase. For strong signal levels, this phase dependent behavior leads to a modification of the gain characteristic, and ultimately to gain saturation. These results were validated by similar observations in [43].

16.4 BLACK-BOX PSA-BASED BPSK REGENERATION

To advance the work done by Croussore and Li [12], several key issues needed to be addressed. Firstly, a field-installed PSA requires a means of phase locking the pump(s) to the input signal. A temporary work-around of this issue was employed in the demonstrations of Ref. [12] by using the same laser for both pump(s) and signal. Maintaining the phase relationship between local free-running pump(s) and the incoming BPSK signal is challenging in practice because phase encoded signals have their carrier suppressed. Consequently, techniques such as injection locking that work for OOK carrier recovery [44] cannot be used without placing restrictions on the nature of the data pattern applied to the signal [45].

In addition to phase locking, the ability of the device to cope with real-world impairments, such as wideband noise in phase and frequency, residual chromatic and polarization mode dispersion, and slow polarization drifts, needed to be considered. All of these issues are discussed in this section.

16.4.1 Phase locking of pumps and signal

To perform optical carrier recovery from a BPSK signal, a number of options are available. Feedback [46,47] or feed-forward methods [48] both exist. Alternatively the carrier can be transmitted in a separate polarization or frequency channel to the signal [49], or the data coding may be modified to leave some residual component of the carrier in the data spectrum [45]. The feedback methods require short loop delays to achieve reasonable (>1 MHz) bandwidths, something that is ultimately limited by the physical layout of electronic and optical devices. The electrical feed-forward schemes are generally limited to processing signals with modest baud rates (e.g. in [48], 10 Gbaud signals are processed with >10 GHz electronics) although an optically assisted scheme that can use electronics slower than the data baud rate was recently reported [50]. In that method, the inherently ultrafast FWM is used to down-convert the carrier variations to the baseband (<10 GHz) allowing subsequent electronic-based processing before transfer back into the optical domain to obtain the underlying carrier frequency.

In [51,52], a simple method to synthesize the phase-locked pumps is shown, which is described here in detail. If φ_{pump1}, φ_{pump2}, and $\varphi_{Carrier}$ refer to the absolute phases of the two pumps and signal carrier respectively, then "classic" phase locking of all three waves would require the conditions

$$\varphi_{Pump1} = \varphi_{Pump2} \quad \text{and} \quad \varphi_{Carrier} = \varphi_{Pump2}. \tag{16.12}$$

In the case of the regenerator, the requirement is to lock the relative, rather than absolute phases, and therefore the condition is

$$\varphi_{Pump1} + \varphi_{Pump2} = 2\varphi_{Carrier}. \tag{16.13}$$

This condition is much simpler to satisfy than that in Eq. (16.12). First, a free-running pump (Pump 1) is mixed with the input signal of phase φ_{Signal} (which is a combination of the signal carrier as well as the overlying phase modulation) to generate an idler via FWM. The nature of the FWM process means that the idler's phase is [53]

$$\varphi_{Idler} = 2\varphi_{Signal} - \varphi_{Pump1}. \tag{16.14}$$

The BPSK signal comprises a carrier with a phase shift of either 0 or π and thus

$$2\varphi_{Signal} = 2\varphi(0 \, or \, \pi) + 2\varphi_{Carrier} \equiv 2\varphi_{Carrier}. \tag{16.15}$$

Substituting Eq. (16.15) into (16.14) yields

$$\varphi_{Idler} = 2\varphi_{Carrier} - \varphi_{Pump1}, \tag{16.16}$$

which has two important consequences. Firstly, the idler does not bear the data modulation. Secondly, the idler phase is exactly what is required for the phase of Pump 2, as follows from Eq. (16.13). Direct amplification of the idler to serve as Pump 2 would not provide good performance, partly due to its poor OSNR (it carries all the amplitude noise present on the input signal, and it can also be quite weak in power). Also, some of the high frequency phase noise present on the data signal is also present on the idler [50]. Optical injection locking of a semiconductor laser to the noisy idler [50] can be used to address both of these issues for the following reasons. Firstly, injection locking inherently allows regeneration of signal amplitude [22]. By reducing the injected power, the injection locking bandwidth can be reduced, meaning that any high-frequency noise present on the idler is not transferred to the injection locked-laser output. Secondly, an additional benefit of injection locking is the significant amplification of the weak idler (by gains of up to 60 dB or more), as needed for the synthesis of Pump 2.

The practical implementation as demonstrated in [51] is shown in Figure 16.10. Pump 1 was combined with the data signal and sent to a highly nonlinear fiber (HNLF) to generate a modulation-stripped idler that was then used to injection-lock a slave laser and realize Pump 2.

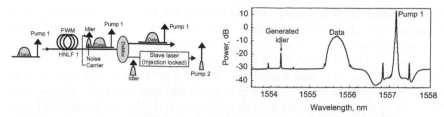

FIGURE 16.10 Phase synchronization of the pumps to the data signal. (a) Set-up;
(b) spectral characteristics at the output of HNLF 1.

16.4.2 Noise suppression capability of the black-box regenerator

As mentioned previously, noise generated inside a real fiber network can be divided
into that generated by linear processes (typically amplified spontaneous emission
(ASE) from EDFAs) and nonlinear processes (e.g. XPM between the signal and ASE
noise). For laboratory experimentation, the former source of noise ("linear noise")
can be emulated simply by directly adding ASE to the signal, and this technique was
used in [51]. The latter ("nonlinear noise") being predominantly phase noise was
emulated in the work described below by driving a large-bandwidth (40 GHz) phase
modulator with broadband electronic noise. This electronic noise was generated by
feeding ASE into a fast photodiode (16 GHz bandwidth) and amplifying the resultant
signal [51]. The set-up of the noise emulator is shown in Figure 16.11.

The full regenerator set-up is shown in Figure 16.12. The input was a noisy DPSK
signal, the polarization of which was automatically aligned to the regenerator's

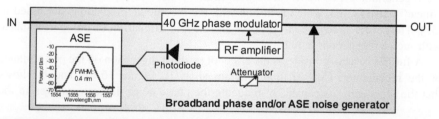

FIGURE 16.11 Noise emulator for adding linear noise (ASE) and nonlinear noise (phase-only
noise) via phase modulator driven by broadband RF noise generated using a fast photodiode
detecting ASE noise.

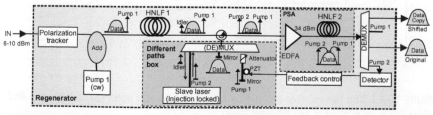

FIGURE 16.12 PSA regenerator set-up.

principal polarization axis using a polarization tracker. For phase synchronization of the pumps, the scheme shown in Figure 16.10 was used. The pumps were 200 GHz away from signal. Pump 1 was a CW laser emitting 17 dBm of power. After HNLF 1, the three waves (Data signal, Pump 1, Idler) were separated in a 4-channel 200-GHz demultiplexer placed behind a circulator. A mirror provided retro-reflection in the data path as well as in the path of Pump 1, which also included a piezoelectric transducer (PZT) fiber stretcher and a variable attenuator. A Discrete Mode (DM) semiconductor laser [54], which was injection locked to the idler, was used to generate Pump 2 in the idler path.

The data signal with the two (phase-locked) pumps was then launched into the PSA which consisted of a high-power EDFA (total power of 31 dBm) and a 180-m sample of a high stimulated Brillouin scattering (SBS)-threshold HNLF (HNLF 2) [55,56]. HNLF 2 had an alumino-silicate core and had a linear strain gradient applied along its length (ranging from 400 to 20 g) to broaden the SBS gain bandwidth and reduce the peak gain. For optimum performance the PSA was operated in deep saturation (see Section 16.3.2).

The BER at the regenerator input and output was recorded at 40 Gbit/s for various levels of noise, generated as shown in Figure 16.11. An optically preamplified receiver and a 1-bit delay interferometer (DLI) for data demodulation was used to detect the data. In Figure 16.13a, it can be seen that no power penalty with respect to the back-to-back is observed when there is no noise at the PSA input. For all levels of phase-only (Figure 16.13a), ASE-only (Figure 16.13b), and combined noise (Figure 16.13c), we see that there is some error flooring for the signal both with and without the regenerator. By comparing Figure 16.13a and Figure 16.13b it can be seen that the regenerator copes better with "nonlinear" (phase-only) phase noise than with "linear" noise (ASE), which is in line with the theoretical predictions discussed in [57]. The reader is referred to [57] for an in-depth discussion of link performance with such a regenerator in line.

A further issue to consider is the effect of residual dispersion on the performance of the regenerator. Constellation diagram analysis, Figure 16.14a and b, shows that the regenerator is capable of reducing the phase noise fluctuations even in the

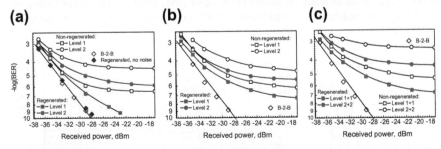

FIGURE 16.13 BER curves when phase-only noise (a), ASE-only noise (b), and phase+ASE noises (c) were applied at two different levels (from [57]).

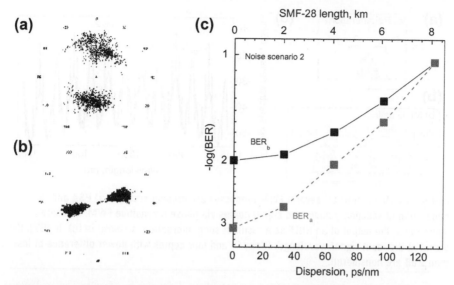

FIGURE 16.14 Calculated constellation diagrams of a 40 Gbit/s DPSK signal (a) before and (b) after the regenerator when the signal has experienced residual dispersion of 80 ps/nm. (c) Calculated BER as a function of the residual dispersion (from [58]).

presence of some residual dispersion (80 ps/nm in the diagrams shown) albeit at the expense of increased amplitude stability. This influences the BER as shown in Figure 16.14c (taken from [58]), where it can be seen that the regenerator can cope well with residual dispersion corresponding to residual dispersion accumulated in about 2 km of SMF-28, but gives no advantage for dispersion accumulated in an SMF-28 length of 8 km. More discussion regarding this topic can be found in [58].

Another potentially useful feature of the regenerator is its inherent capability of wavelength multicasting through cascaded FWM processes, as explained in Figure 16.15. As sketched in this figure, several FWM interactions may occur simultaneously: PSA, which involves the two pumps and the data, is responsible for the regeneration (Figure 16.15a), whereas new (secondary) pumps as well as new copies of the data signal are generated via phase insensitive FWM interactions (Figure 16.15b). In the saturation regime, the high input signal power gives rise to the generation of multiple FWM products which inherit the amplitude and phase properties of the initially interacting waves (Pump 1, Pump 2, and Data). Theoretical and experimental analysis showing that the generated signal copies are also regenerated is presented in [57]. This feature was used for performing simultaneous regeneration and wavelength conversion in a field trial experiment as described later. This feature is shown in the regenerator setup in Figure 16.12 showing two outputs from the regenerator (at the original wavelength and at a wavelength-shifted wavelength). In practice, up to five output copies were reported [57], Figure 16.15c.

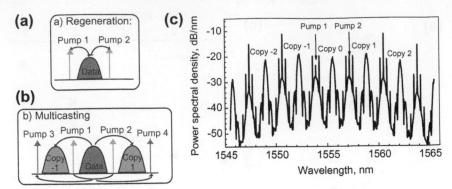

FIGURE 16.15 In an HNLF, several FWM processes are present such as (a) PSA and (b) generation of secondary pump and signal copies via phase insensitive FWM. Measured spectrum at the output of an HNLF as a result of such interaction is shown in (c). In [57], the signal at the input signal wavelength (Copy 0) and four copies with power difference of less than 2 dB are considered.

16.4.3 Field results with DWDM network-generated noise

A field transmission experiment which made use of the wavelength conversion/multicasting feature of the regenerator was carried out [58]. A conceptual outline of the network and its practical implementation are shown in Figure 16.16 a and b respectively. At the transmitter, 37 CW semiconductor lasers were combined on a 100 GHz dense wave length division multiplexing (DWDM) ITU grid and modulated with 40 Gbit/s, 2^{31}–1 pseudo-random bit sequence (PRBS) DPSK data. To de-correlate adjacent channels, the odd/even channels were split in an interleaver, 55 ns of relative delay was introduced, and the channels then recombined. In order to facilitate placing the regenerator at the mid-point of the network and to operate at twice the transmission distance that was physically available, a wavelength shift was incorporated in the regeneration process, Figure 16.16b. Obviously, conversion of the wavelength would not normally be required in most transmission applications, nevertheless, it serves to illustrate another useful functionality of the technology (wavelength conversion/multicasting) as previously reviewed [57,58]. Thirty-seven channels (excluding ITU Channel 23) were sent down the dark fiber link (part of the UK JANET

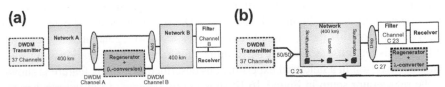

FIGURE 16.16 Model experiment with regenerator as an in-line device (a) and the implemented network that emulates key features of the model network (b).

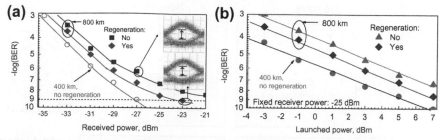

FIGURE 16.17 **BER curves at the output of the 2nd round-trip (800 km) with and without mid-span regeneration: (a) for launched power of 5 dBm; (b) as a function of the launched power for fixed receiver power of −25 dBm. For reference, measurements of the signal at the mid-point (after wavelength conversion) are also shown.**

Aurora Network) that extends from Southampton to London and back again (400 km dispersion-compensated transmission distance, six in-line flat-gain EDFAs with maximum input/output powers of −5/15 dBm, operated in automatic gain control mode with a nominal gain of 20 dB). The maximum total power launched into the link was 7 dBm with a maximum power along the link of 15 dBm. The detailed configuration is shown in [58]. At its output, ITU Channel 27 was converted to Channel 23 [Copy −1] (either with or without regeneration) and sent through the link again together with all the remaining channels. A total of 38 channels occupied the frequency grid at any time.

The regenerator was mounted into a standard telecommunication rack along with other network components and test gear. To enable a comparative study with/without regeneration, conventional wavelength conversion could be carried out by switching off Pump 2, resulting in phase insensitive FWM-based wavelength conversion.

Results obtained after 800 km with the regenerator used "in-line" after 400 km are shown in Figure 16.17. Use of a mid-point regenerator was capable of reducing the BER power penalty by a factor of two (e.g. at BER = 10^{-9}, from 5 to 2 dB). The regenerator also lowered the error floor by one order of magnitude, Figure 16.17a. Varying the input power into the link, Figure 16.17b, showed almost 4 dB power penalty in the second round trip. This value was reduced to 2 dB when the regenerator was used.

16.4.4 Analog error correction in PSA regenerators

PS regenerators can correct errors when differential decoding of PSK is performed. This is counter-intuitive as regenerators typically do not improve BER when placed in front of a receiver. This phenomenon is due to a correctable class of errors specific to a "differential" receiver. Consider an example where two consecutive bits with identical phases of π, Figure 16.18, which during transmission accumulate phase errors of $\pi/3$ and $-\pi/3$, respectively. A DPSK receiver detects the difference in the phases of these consecutive bits as $2\pi/3$. The receiver's decision point is set at $\pi/2$ resulting in an erroneous detection of π as the differential

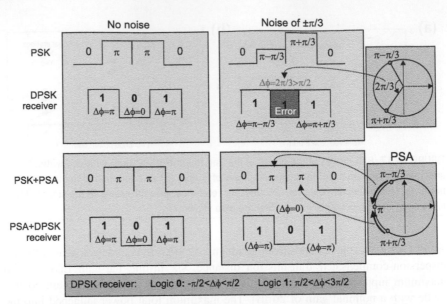

FIGURE 16.18 Schematic explanation of the PSA for BER improvement in differential coherent (DPSK) receivers. A data sequence 0, π, π, 0 is considered.

phase of the two bits rather than the correct value of 0. On the other hand, a PSA corrects the phase of each bit, before the evaluation of their difference in the differential receiver, so such an error would not occur. This ability to correct for differential errors means that after the PSA process the absolute and differential phase noise exhibit almost the same statistics for signals degraded by nonlinear phase noise [57]. This suggests that a DPSK receiver supported by a PSA pre-amplifier could perform as well as an ideal homodyne PSK receiver. Therefore, in principle a homodyne coherent receiver, which is usually implemented with power greedy digital electronics, could be substituted by a simpler DPSK receiver assisted by a PSA regenerator.

16.5 MPSK PHASE REGENERATION

Conventional PSA-based schemes are incapable of directly supporting advanced formats such as quadrature phase shift keying, whose four phase states allow two bits per transmitted symbol. To achieve regeneration of signals with more than two levels, a number of indirect approaches have been suggested. One of these is the use of parallel BPSK regenerators, each regenerating one of the I and Q quadratures in a QPSK signal [59]. Another solution is to perform BPSK to OOK conversion using linear interferometers, to amplitude regenerate the OOK streams in parallel, and then to perform all-optical format conversion to synthesize a clean (regenerated)

QPSK signal [60,61]. The parallel approach of both these schemes, often favored in electronics, is complicated in optics due to the requirement to equalize and stabilize multiple optical paths (often over 100 m in length with silica fiber-based implementations) in phase, polarization and propagation delay, which is both complicated and costly. In addition, linear scaling of component count and power utilization would offset much of the benefit of adapting higher order modulation formats.

In addition to regeneration, for more sophisticated processes such as analog-to-digital conversion (ADC), quantization to more than two levels is an absolute requirement. Research into photonic quantizers for ADCs has been ongoing for four decades. While numerous architectures have been proposed [62], a niche for an ultrafast all-optical quantizer still exists. Photonic quantizers providing bandwidths >100 GHz, effective number of bits (ENOB) of at least 4 and power consumption under 5 W would offer real competitive advantage as compared to the current electronic alternatives [62,63]. With this in mind, effort was made to realize a scheme to quantize phase encoded signals to multiple levels.

16.5.1 PSA-based multilevel phase regeneration

16.5.1.1 Multilevel phase quantization concept

By developing a more general understanding of how phase can be manipulated, it is possible to derive schemes by which more sophisticated phase transfer functions can be synthesized. A simple way to do this is by taking the cosine function as an example. For an input ϕ ranging from 0 to 2π, $\cos(\phi)$ resembles a two-level phase quantization operation as its result is a variable of phase $\phi_s = 0$ for $\pi/2 \leq \phi < 3\pi/2$, and $\phi_s = \pi$ everywhere else. $\cos(\phi)$ can be rewritten as the sum of complex exponentials obtaining

$$|A_s(\varphi)| \cdot \exp(i\varphi_s) = \exp(i\varphi) + \exp(-i\varphi) = 2\cos(\varphi)$$

$$= \begin{cases} |2\cos(\varphi)| \cdot \exp(i \cdot 0), & \text{if } \frac{\pi}{2} \leq \varphi < \frac{3\pi}{2}, \\ |\cos(\varphi)| \cdot \exp(i \cdot \pi), & \text{elsewhere,} \end{cases} \qquad (16.17)$$

where ϕ_s is the two-level quantized phase output, and $A_s(\phi)$ is the accompanying π-periodic amplitude response.

Equation (16.17) suggests that an arbitrarily phase encoded signal $\exp(-i\phi(t))$ can be quantized in phase to two levels by coherently adding to it a phase conjugated version of the input signal, $\exp(-i\phi(t))$. This is in fact what happens in a degenerate PS-FOPA, as can be seen in Eq. (16.11).

The real leap forward enabling multilevel phase quantization is the realization that Eq. (16.17) can be generalized by rewriting it as [13]

$$A_s(\phi) \cdot \exp(i\phi_s) = \exp(i\phi) + \frac{1}{M-1} \exp(-i\phi(M-1)), \qquad (16.18)$$

where M refers to the number of phase levels. Figure 16.19 shows the result of evaluating Eq. (16.18) for three values of M. For $M = 2$, the phase transfer function is very

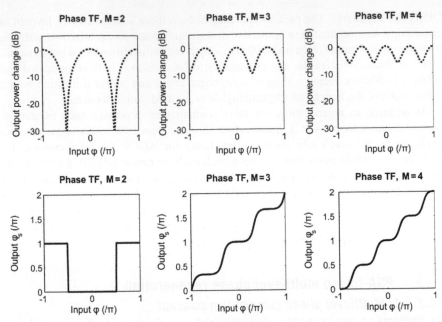

FIGURE 16.19 Evaluation of Eq. (16.13) showing how multilevel phase TFs are achieved as *M* is varied.

sharp, with a perfect π step. This is accompanied by a strong phase-to-amplitude conversion. As the value of M increases, the periodicity of the phase transfer function (TF) matches M as expected, but the step becomes less defined, and the depth of the phase-to-amplitude conversion is also seen to reduce.

The phase transfer functions as derived in Eq. (16.18) are all monotonic with M turning points. While this provides maximum local flatness at 0, $2/M\pi$ $3/M\pi$, etc., it is not necessarily ideal for a quantizer in which the target is to minimize the global phase error variance. One solution is to rewrite Eq. (16.18) by substituting the coefficient of the conjugate term with a variable m,

$$A_s(\phi) \cdot \exp(i\phi_s) = \exp(i\phi) + m \cdot \exp(-i\phi(M-1)). \qquad (16.19)$$

Figure 16.20 shows the result of varying m in Eq. (16.19) for $M=4$. Three values are used, including 0.33, which is the coefficient as calculated from Eq. (16.18). To optimize for m, a misfit factor was designed such that an ideal value can be numerically identified. The misfit factor, as shown in Eq. (16.20), integrates the difference between a given phase transfer function, and an ideal step, hence the smaller the number achieved the better the quantizer

$$\text{Misfit factor} = \log\left(\frac{2}{\pi}\int_{-\pi/M}^{\pi/M}|\phi_s(m,\phi)|\,d\phi\right). \qquad (16.20)$$

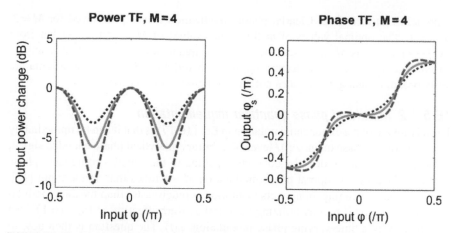

FIGURE 16.20 Evaluation of Eq. (16.19) for *M*=4, showing the transfer functions for various values of *m*. Dotted line is for *m*=0.25, solid line *m*=0.33, and dashed line *m*=0.5.

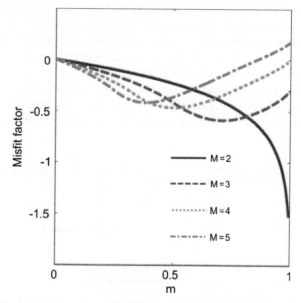

FIGURE 16.21 Evaluation of Eq. (16.20) showing how the misfit factor (as calculated in Eq. (16.15)) as a function of *m* for various values of *M*. Optimum values for *M*=2, 3, 4, and 5 are *m*=1, 0.71, 0.50, and 0.38 respectively.

The impact of varying *m* on the misfit factor is shown in Figure 16.21 for *M* values of 2–5. A misfit factor of 0 is obtained for *m*=0 for all values of *M*, indicating that no phase quantization is achieved, and that the output phase is identical to that at the input. The smaller the misfit factor, the closer the phase transfer function approaches an ideal step. Clearly, phase quantization is easily achieved for $M=2$, at $m=1$. The optimal values of m for higher values of M depart somewhat from $1/(M-1)$. Also, as M increases, the degree of quantization achievable is compromised, as evidenced by the decreasing misfit factor. Note also that the minima are fairly broad, meaning that m does not have to be set very precisely.

16.5.1.2 Multilevel phase quantizer implementation

The advantage of the exponential notation in Eq. (16.17) is that it can be immediately deduced how to phase quantize M levels to arbitrary analytical phase encoded signal, $\exp(i\varphi(t))$. This can be done by coherently adding to the signal a conjugated phase harmonic bearing a temporal phase modulation $M-1$ times that on the input [13]. Provided that the conjugation mechanism used is much faster than the data modulation (this criterion is met by utilizing the ultrafast Kerr effect), φ in Eq. (16.17) can be replaced by a time-varying phase modulation, $\varphi(t)$. The question is then how to synthesize, from an input phase modulated signal, the $(M-1)$th phase harmonic, conjugate it, scale it by m, and coherently add it to the input.

The functions of phase multiplication, conjugation, and coherent addition can be performed using FWM as shown in Figure 16.22. Firstly, the phase harmonic is generated from the signal using a cascaded FWM process (Figure 16.22a). This is done by combining the signal with a strong pump beam at a frequency detuning Δf in a nonlinear medium. By optimizing the phase matching and the strength of the nonlinear interaction, a spectral cascade of FWM products is generated (Figure 16.22a). Because FWM is momentum conserving, the comb of products possesses an overlying phase modulation that is a perfect integer multiple of the modulation present on the signal at the mixer input. Next, a second FWM process is carried out using two

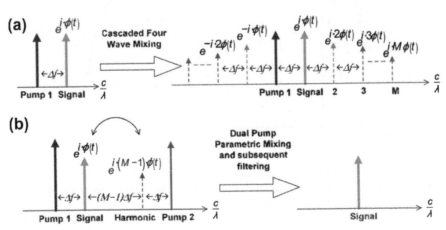

FIGURE 16.22 Illustration of how the M-level staircase transfer function necessary for phase quantization is achieved. (a) By mixing a pump beam with a phase modulated signal in a nonlinear medium higher order phase harmonics of the signal can be generated. (b) The signal is then coherently combined with the $(M-1)$th harmonic using a dual pump parametric process.

pumps located symmetrically around the signal and phase harmonic (Figure 16.22b) to coherently conjugate and add the (M−1)th phase harmonic to the signal.

16.5.1.3 Inline PSA QPSK regenerator

As discussed in Section 16.5.1.2, a single non-degenerate PSA can be directly used for MPSK regeneration provided that the input signal is coherently added to a conjugated phase harmonic (idler) bearing a temporal phase modulation $M - 1$ times that on the input. Figure 16.22 shows how this can conceptually be achieved using FWM in a two-step process. In [64] this was demonstrated using two nonlinear fibers (one for each step). Firstly, the QPSK signal, $\varphi(t)$, was mixed with a pump to generate a FWM comb, including the required idler, $3 \cdot \varphi(t)$. The signal and conjugated harmonic were then coherently added using a dual pump non-degenerate PSA. The first of the two PSA pumps was derived from the free-running laser used in the comb generation stage, and the second by injection locking the fourth FWM harmonic ($4 \cdot \varphi(t)$, which is modulation stripped for a QPSK input) to a semiconductor laser, satisfying the phase locking requirement. The relative powers of the signal and harmonic needed to be optimized to take the PSA power gain G into account; as such a signal-harmonic offset of m_{eff} was used, where $m_{\text{eff}} = m*\sqrt{1 - 1/G}$. Absolute phase deviations of up to 60 degrees per symbol (peak-to-peak) for a QPSK input at 10 Gbaud were reduced by a factor of 2 [64].

In the interests of brevity that experiment is not described in detail here, but rather a modification to the setup that allowed a single HNLF to be used [13,65] is presented. The two stage MPSK regenerator [64] is defined by the distinction between the first HNLF in which the phase multiplication and pump recovery occurs, and the second HNLF in which the conjugated idler is coherently added to the signal, hereby regenerating the phase. An interesting variant of that scheme was discovered [13,65] in which phase sensitive gain is obtained directly from a two-pump (2P) non-degenerate parametric amplifier without an idler at the amplifier input, in apparent contradiction to the expected characteristics of a non-degenerate PSA. This device configuration, in addition to allowing QPSK phase regeneration in a simpler configuration relying on just one nonlinear element, provided significantly enhanced amplitude noise improvement.

The concept is in effect of a fusion of the two separate parametric effects into one distributed action. Provided that the two pumps 1 and 2 as well as the signal are present at the mixer input, the coherent mixing process can then be distributed along the fiber length; initially the cascaded mixing process dominates, and then as the harmonic idler starts to grow in power the coherent conjugated addition takes place. In [13,65], this was experimentally demonstrated.

The output of a CW laser at 1555.7 nm was split into two, with one portion coupled into a 10 GHz comb generator, see Figure 16.23. Comb lines at −190 and +570 GHz detuning were injected into semiconductor lasers, providing two pump beams phase locked to the signal carrier. The rest of the signal light was modulated with a PRBS to generate single polarization QPSK. This was sent through a noise additive module to emulate the effects of linear (related to quantum noise and ASE)

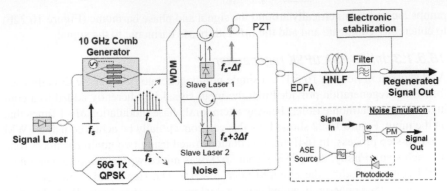

FIGURE 16.23 Regenerator setup, Tx—transmitter, WDM—wavelength division demultiplexer, PZT—piezo-electric fiber stretcher, PM—phase modulator [65].

and nonlinear phase noise (related to nonlinear amplitude-to-phase conversion). This module (shown as an inset in Figure 16.23) comprised an ASE source whose output was split into two, one portion being detected and the resulting electrical white noise being used to drive a LiNbO$_3$ phase modulator through which the signal was passed, and the other portion being optically combined with the signal in a coupler.

The signal was then combined with the pumps and all the waves were amplified in an EDFA, leading to 50 mW of signal power and 250 mW power per pump. They were then sent into an HNLF (OFS, Denmark). The HNLF parameters were length 300 m, nonlinear coefficient 11.6/W/km, zero dispersion wavelength (ZDW) of 1553 nm and 1550 nm dispersion slope (DS) 0.018 ps/nm^2/km. It also had a strain gradient to increase its SBS threshold, allowing the use of continuous wave pumps. Slow thermo-acoustic relative phase drifts were suppressed by monitoring the signal power at the PSA output and controlling the PZT. The signal was then assessed using an EXFO constellation analyzer (PSO-200) based on all-optical sampling capable of operation up to 100 Gbaud.

The input and output spectra to the PSA are shown in Figure 16.24 and Figure 16.25, respectively. At the input to the PSA, there is a very weak component at the idler frequency (+380 GHz detuning) emanating from weak FWM in the high power EDFA, but at −40 dB relative to the signal this does not affect the subsequent parametric interaction as verified by numerical simulations. 7 dB phase sensitive gain variation was obtained at the PSA output as measured with the feedback to the PZT turned off. The spectrum at the output to the PSA Figure 16.25 suggests two separate interactions occur simultaneously. Firstly, the presence of the strong component at +190 GHz detuning indicates coherent phase multiplication via mixing of the signal with the pump at −190 GHz, and secondly, the strong idler at +380 GHz indicates conventional 2P phase insensitive amplification. There are many other parametric interactions that occur due to the strength of the signal and pumps, as shown in the wideband output spectrum, Figure 16.26. These extra mixing products can be used to enhance the functionality of the device, such as being used as a wavelength multicaster by accessing the wavelength-translated copies of the input signal. It could,

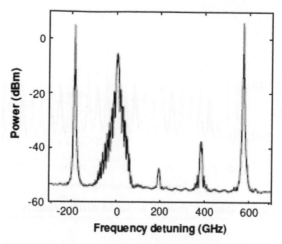

FIGURE 16.24 PSA input spectrum, signal located at 0 GHz detuning.

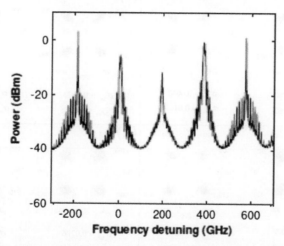

FIGURE 16.25 PSA output spectrum, signal located at 0 GHz detuning, output idler.

however, be viewed as being energetically inefficient, but this transfer of energy to unwanted frequencies can be minimized by careful selection of the fiber parameters.

To characterize the regenerator, signal constellations including data on the phase error variance and normalized variance of amplitude noise are shown in Figure 16.27 [65]. While this statistical information would be more robust if obtained from a homodyne receiver without digital phase compensation for the intradyne local oscillator, it is still useful for quantifying the relative signal improvements derived from the regenerator.

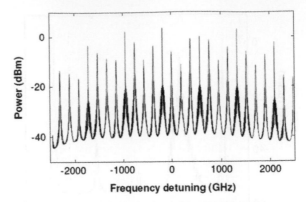

FIGURE 16.26 PSA output spectrum showing wideband mixing products.

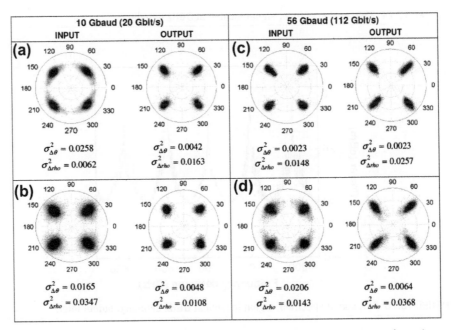

FIGURE 16.27 Regenerator performance at 10 and 56 Gbauds. (a) Input phase noise only; (b) Phase and amplitude noise; (c) No noise; (d) Phase noise only. $\sigma^2_{\Delta\theta}$ is the phase error variance, $\sigma^2_{\Delta\rho}$ is the normalized amplitude noise variance.

The regenerator was first assessed at 10 Gbaud with added phase noise only, emulating the nonlinear regime. Pseudo-Gaussian phase fluctuations would be expected at the input, with an artificial roll-off at the tails of the distribution due to saturation of the photodiode for high ASE levels. The regenerator desirably reduced the phase error variance by a factor of 6, while the amplitude noise variance only increased by 2.6 (Figure 16.27, Cell a), suggesting an overall benefit as the input BER is dominated by the phase noise. This phase noise reduction is comparable to the numerically predicted factor of 5.5 for the parallel BPSK regenerator scheme [59], denoting that the inline compactness of the approach does not come with an associated performance penalty. For QPSK, a comparison with studies on the impact of phase estimation errors on BER suggests that the output phase error variance of $0.0042\,\mathrm{rad}^2$ (Figure 16.27, Cell a) approximately corresponds to an SNR penalty under $0.5\,\mathrm{dB}$ for a BER of 10^{-4}, while the input variance corresponds to a penalty of approximately $4\,\mathrm{dB}$ [66]. This nonlinear phase noise reduction implies that (i) the reach of the transmission span can be increased and (ii) the tolerance to nonlinearity is significantly enhanced; hence higher receiver OSNRs can be envisaged.

In the presence of linear noise emulated by ASE loading (hence degraded OSNR with both phase and amplitude fluctuations), the phase error variance and normalized amplitude variance are simultaneously reduced by approximately 3.2, indicating even better net regenerator performance (Figure 16.27, Cell b). This ability to concurrently reduce both absolute phase and amplitude variance prior to the receiver suggests that the regenerator should provide significant signal quality improvement if deployed before a differential receiver, which normally requires a trade-off between reduced complexity and lower noise tolerance compared to a fully coherent one.

The symbol rate was increased to 56 Gbaud and the measurements repeated. Without added noise, the regenerator preserves the phase quality at the expense of some amplitude noise (Figure 16.27, Cell c). It is believed that this is a result of the wideband parametric interactions occurring in the PSA, as shown in Figure 16.26, that transfer some amplitude noise to the signal. In the presence of phase noise, the phase error variance is reduced by a factor of 3.2, with increased amplitude noise at the output (Figure 16.27, Cell d). For QPSK, however, in which the information is solely contained in the phase, this increased nonlinear phase noise tolerance would translate to increased reach or higher signal launch powers to improve OSNR at the receiver.

16.5.2 Other approaches to QPSK regeneration

16.5.2.1 Interferometric (parallel PSA) QPSK regenerator

Zheng et al. [59] proposed a technique that utilizes two parallel PSAs to separately regenerate the I and Q quadratures of an input BPSK signal (Figure 16.28). These regenerated streams are then combined in parallel in a coupler, thereby re-synthesizing a cleaner QPSK signal. This concept has not been demonstrated experimentally, but numerical simulations showed the potential of the device to suppress phase noise significantly [59]. In principle, the performance of this device should be similar to that of the inline MPSK regenerator presented in Section 16.5.1.2, if correctly implemented. A similar scheme has been proposed by Da Ros and Peucheret [67] exploiting saturation effects in fiber optical parametric amplifiers using a pair of degenerate dual-pump PSAs. These interferometric schemes require one PSA per input quadrature, and therefore have double the component count for QPSK as compared to BPSK regeneration, in contrast to the inline approach (see Figures 16.28–16.30).

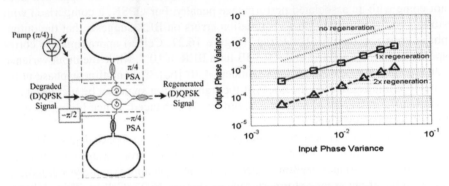

FIGURE 16.28 QPSK regenerator concept using two parallel PSAs. Left—setup. Right—numerical simulation results [59].

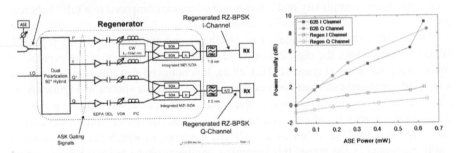

FIGURE 16.29 SOA-based QPSK regenerator scheme by Kimchau et al. Left—experimental setup. Right—power penalty measurements at BER 10^{-9} as ASE is added to the input signal [61].

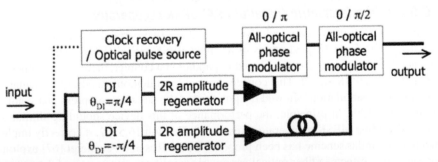

FIGURE 16.30 Block diagram of the all-optical DQPSK signal regenerator [5].

16.5.2.2 *Interferometric format conversion-based QPSK regeneration*

A scheme utilizing the separate regeneration of the I and Q quadratures was recently demonstrated by Kimchau et al. [61] (Figure 16.29). A 90° optical hybrid with homodyne detection was used to convert the phase information of the I and Q components into four OOK signals that were then used to gate two all-optical wavelength converters based on semiconductor Mach-Zehnder interferometers, creating two phase coherent signals from a common local oscillator, each of which bears one of the I and Q streams. Combining these two BPSK signals in quadrature creates the regenerated QPSK signal. A major strength of this scheme is the visible route toward on-chip integration, although the demonstration itself included a separate EDFA for each of the 90° optical hybrid's outputs. Note that implementing it requires the use of RZ input pulses.

A similar scheme to this has been suggested by Matsumoto [5] (Figure 16.30). A DQPSK signal is converted to two parallel OOK signals by a pair of DLIs and the noise on the OOK signals is suppressed by all-optical 2R amplitude regenerators. The regenerated OOK signals are used as control pulses for a pair of all-optical phase modulators to generate phase-remodulated output pulses. This type of regenerator converts the phase difference between consecutive pulses in the input signal to the absolute phase of the outgoing regenerated signal so that the phase data encoded on the regenerated signal is altered by the regenerator. The conversion of the phase data can be undone by suitable encoding or decoding in the electrical domain.

It is worth noting that all these interferometric schemes rely on regenerating the I and Q quadratures separately. This means that as the input signal scales in complexity, for higher values of M, the regenerator component count will scale linearly. Also, these schemes are only capable of operating on modulation formats that can be generated by summing BPSK signals, such as QPSK, while the inline approach can cope with any M value, including odd numbers such as 3 and 5. This makes the latter approach more versatile and significantly easier to implement, as the component count is independent of the complexity of the input signal.

16.5.3 **Flexible MPSK quantization**

The inline PSA-based regeneration approach is inherently multi-format, as discussed in Section 16.5.1. Reconfiguring the regenerator from QPSK to an alternate modulation format such as 8-PSK can be achieved simply by generating a broader FWM comb (by increasing the signal and pump powers), passively selecting the desired harmonic and tuning the injection locked laser to the corresponding frequency. This was experimentally demonstrated in [13].

A narrow linewidth CW fiber laser (Rock laser, NP Photonics) at 1555.7 nm was split into two, with one portion coupled into a 10 GHz comb generator. Two semiconductor lasers were injection locked to comb lines, one representing Pump 1 at approximately −200 GHz detuning from the signal frequency and the other at a frequency $(M-1) * 200$ GHz away, with the value of M depending on the modulation format being assessed. For both lasers the injected power was close to −30 dBm. The rest of the signal light was modulated, either in a LiNbO$_3$ phase modulator with information generated in a 12 GS/s arbitrary waveform generator, or in the case of

QPSK, in a LiNbO$_3$ Mach-Zehnder modulator (MZM) driven by an electrical PRBS pattern, the optical output of which was coherently multiplexed from BPSK to QPSK in a DLI.

The modulated signal was sent through a noise additive module to emulate the effects of linear and nonlinear phase noise. The signal (50 mW) was then combined with the pumps (total power 500 mW) and all the waves were amplified in a polarization maintaining (PM) EDFA. Use of a PM EDFA ensured that the signal and pumps were co-polarized in the subsequent HNLF.

They were then sent into a strained HNLF whose parameters were length 300 m, nonlinear coefficient 11.6/W/km, zero dispersion wavelength 1552 nm ZDW and 1550 nm DS 0.018 ps/nm^2/km. This had a strain gradient to increase its SBS threshold, allowing the use of CW pumps. Optical spectra at the input and output of the HNLF are shown in Figure 16.31b and c respectively. Slow thermo-acoustic relative phase drifts were suppressed by monitoring the signal power at the PSA output and

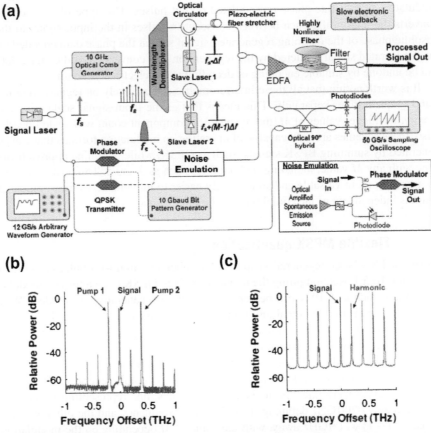

FIGURE 16.31 (a) Schematic of the quantizer. EDFA, erbium-doped fiber amplifier. **(b)** Optical spectrum at the input and **(c)** output of the highly nonlinear fiber. The weak tones (over 35 dB less than pumps) in (b) result from FWM in the EDFA and can be ignored [13].

controlling the PZT. The processed signal was then combined with a local oscillator (tapped off the signal laser prior to data modulation) in a 90° hybrid followed by real-time 50 GS/s sampling of the two hybrid outputs detected in single ended fashion. Off-line post-processing was used to retrieve the phase and amplitude of the signal.

To verify the operating principle, the signal phase was first varied over 2π at a rate of 150 MHz. To switch to M-level quantization, the only requirement is to tune the frequency of laser 2 to $f_s + (M-1)\Delta f$, and optimize the pump and signal powers into the HNLF. This means that from a practical point of view, changing the modulation format or supported signal bandwidth (given by Δf) can be achieved in milliseconds or less without any hardware changes, relying on the rapid tuning capability of semiconductor lasers. The choice of signal and pump powers depends on the fiber parameters, M-level, and application, but as an example, 50 mW and 250 mW respectively proved ideal for QPSK regeneration.

The signal can be represented in constellation diagrams, as shown in Figure 16.32. Quantization to 3, 5, and 6 levels was demonstrated. As expected,

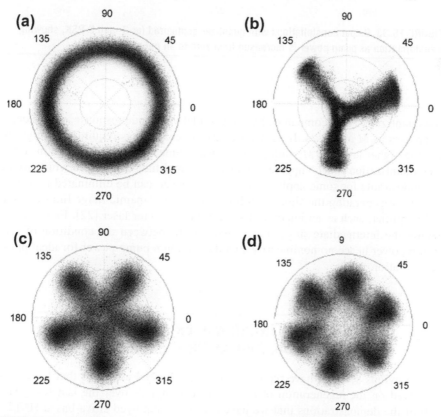

FIGURE 16.32 Signal constellation diagrams. (a) Before the quantizer, signal occupies every phase state. (b) After phase quantizer with Pump 2 at $f_s + 2\Delta f$, giving 3 levels. (c) After phase quantizer with Pump 2 at $f_s + 4\Delta f$, giving 5 levels. (d) After phase quantizer with Pump 2 at $f_s + 5\Delta f$, giving 6 levels.

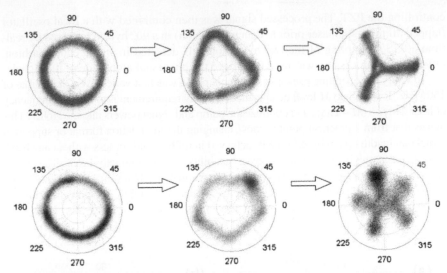

FIGURE 16.33 Output constellations with quantizer configured for 3- and 5-PSK, showing transformation as pump power is increased from zero to an optimum value.

the quantization is accompanied by a sinusoidal amplitude response, whose depth decreases as M increases. If stronger phase squeezing was required, as might be for the higher levels of M, multiple nonlinear elements can be cascaded in series. The quantization is accompanied by phase-to-amplitude conversion which would be undesirable for some applications; this, however, can be eliminated by subsequently regenerating the signal amplitude in a high dynamic range limiting optical amplifier, such as an injection-locked semiconductor laser [22]. Figure 16.33 shows the intermediate stages for 3- and 5-PSK, between the condition of zero pump power hence no nonlinearity, and the optimum pump power for ideal phase regeneration.

16.6 CHOICE OF NONLINEAR MATERIALS AND DESIGNS FOR ALL-OPTICAL SIGNAL PROCESSING

The previous sections have reported on the substantial advances that have been achieved on the regeneration of phase encoded signals over the last few years. All of the demonstrations that we have reviewed employed silica-based HNLFs. While this technology arguably represents the most mature solution in terms of achieving highly controllable nonlinear optical effects, it is often desirable to opt for more compact nonlinear devices that lend themselves to easier sub-system integration. This section briefly reviews the features as well as some of the potential issues associated with each of the various material platforms that have been investigated for all-optical signal processing and the regeneration of phase encoded optical signals in particular. We start with an overview of the considerations that relate to $\chi^{(3)}$-based nonlinear systems, including highly nonlinear optical fibers (both in silica and non-silica glasses) and then present recent progress achieved in periodically poled lithium niobate (PPLN) and semiconductor optical amplifiers (SOAs).

16.6.1 Third-order nonlinearity systems

The observation of broadband nonlinear effects, and especially parametric effects such as those involved in OPAs, requires a combination of both high nonlinearity and suitable dispersion characteristics. The design of optical fibers for nonlinear applications relies on engineering precisely the geometry and refractive index profile of the core-cladding region. It is then possible to attain designs that ensure a tightly confined guided field, thereby resulting in a high nonlinear coefficient, and that at the same time to precisely manipulate the overall chromatic dispersion of the fiber. Germanium-doped silica fibers exhibiting exceptional dispersion properties and low transmission losses (around 1 dB/km) over a broadband wavelength range represent a well-established and reliable nonlinear platform [68]. However, their nonlinear coefficients typically range between 10 and $30 \, \text{W}^{-1} \, \text{km}^{-1}$, therefore in order for nonlinearities of sufficient strength to be observed at realistic power levels, long lengths (hundreds of meters) of fiber need to be used. The amount of launched power into such devices is ultimately restricted by the onset of SBS, which depletes the signal and effectively acts as a power limiter and noise source. The growth of SBS effects can be inhibited by broadening the SBS gain bandwidth and reducing the peak gain, by, for example, applying a gradient either on the core diameter, the temperature or the strain of the fiber [69]. While an order of magnitude increase in SBS threshold can be achieved in this way all of these techniques affect the chromatic dispersion uniformity along the fiber length which often significantly degrades device performance in other ways. Doping the core with aluminium [56,68] which provides for low peak Brillouin gain coefficients has also been shown to be an effective technique, however this comes at the expense of a reduced nonlinear coefficient (typically $\gamma = 7 \, \text{W}^{-1} \, \text{km}^{-1}$) and a higher attenuation (around 15 dB/km). Nevertheless various PSA demonstrations reported in the literature have already benefited from the application of these techniques [57,58,64,65].

More compact fiber-based solutions operating at mW power levels are possible by employing glasses with a higher nonlinear refractive index n_2, such as bismuth oxide, lead silicate, and chalcogenides. These glasses have invariably been used in either step-index [70–73] or photonic crystal fiber [74–76] designs, exhibiting up to three orders of magnitude higher nonlinearity than conventional silica single-mode fibers. Tapering of the waist diameter of such fibers has also been used to boost the nonlinear coefficient still further [77,78]. Moreover, the relative strength of the Kerr to Brillouin nonlinearities in these glasses is much greater than in silica and consequently the maximum Kerr-induced non-linear phase shift before the onset of SBS is often much higher than for silica-based HNLFs [70,79], although in absolute terms the SBS threshold may be much lower. Fibers of this class have been used in the past for the implementation of short (a few meters long) reshaping regenerators for intensity modulated signals [71,80], which can subsequently be used as the building blocks for indirect phase regeneration subsystems based on format conversion. Nevertheless, phase and amplitude regeneration in a FWM-based PSA has already been demonstrated in a 5.6 m-long bismuth oxide HNLF ($\gamma = 1100/(\text{W km})$, $D = -300\,\text{ps}/(\text{nm km})$, $\gamma P_{\text{SBSth}} L_{\text{eff}} = 1.1$) [12], while using 1.56 m of a lead silicate fiber with significantly lower dispersion, phase sensitive amplification over a broad wavelength range has been demonstrated ($\gamma = 820/(\text{W km})$, $D < -5\,\text{ps}/(\text{nm km})$, $\gamma P_{\text{SBSth}} L_{\text{eff}} > 0.83$) [81].

When it is important that the device length should be of the order of cm, fiber solutions have to be abandoned. Indeed, when considering PSA applications for example, any device that could provide sufficient parametric gain could eventually be used for regeneration. Chalcogenide glasses have long been known for their high nonlinear refractive index [82] and recent results have shown their potential for the implementation of cm-long rib waveguides with a nonlinear coefficient of the order of $10^5\,\text{W}^{-1}\text{km}^{-1}$ [83]. Dispersion tailoring as required for different applications, as well as low propagation losses (0.05 dB/cm) are also possible [84–86]. Silicon waveguides have also attracted significant interest recently, both because of their strong nonlinearities and their potential for further integration with electronic integrated circuits. Nonlinear waveguides have been demonstrated employing either pure silicon waveguide structures or silicon-organic hybrids, where the silicon waveguide is covered with organic nonlinear materials [87–89]. The cross-sectional dimensions of the waveguides can be of the order of tens of nanometers and the high optical confinement contributes to achieving an ultrahigh nonlinearity (up to $4 \times 10^5\,\text{W}^{-1}\text{km}^{-1}$) [90–92]. By appropriate engineering of the waveguide dimensions, the dispersion can be tailored to the desired values [90,91], and by adopting new fabrication processes relatively low losses (4.5 dB/cm) are achievable. Impressive recent results have shown the potential of mm-long waveguide devices in ultrahigh-speed optical signal processing at data rates up to 1.28 Tb/s at a relatively low operating power [92]. The amount of power launched into either chalcogenide or silicon waveguides is ultimately limited by the onset of two-photon absorption (TPA). In silicon, in particular, where the effect is stronger, TPA is likely to be an issue for any applications operating below 2 μm (see Table 16.1).

Table 16.1 Summary of typical parameters for the $\chi^{(3)}$-based devices/materials discussed above. Two-photon absorption figure of merit: TPA FOM $=n_2/(\beta\lambda)$, where β is the TPA coefficient. Stimulated Brillouin scattering figure of merit: SBS FOM $=P_{SBSth}\gamma L_{eff}$, where L_{eff} is the effective length.

Device and Material	Nonlinearity (/W/km)	Dispersion (ps/nm/km)	Loss (dB/m)	Device Length (m)	TPA FOM (cm/GW)	SBS FOM (rad)
Ge-Silica HNLF [68]	10–30	Selectable close to zero	10^{-3}	100–1000	High[a]	0.21
Strained Al-Silica HNLF [56]	7.4	Selectable close to zero	0.015	100–300	High[a]	1.56
Soft glass fibers						
Chalc. fiber taper [77,78]	68×10^3	282	18	0.22	2.83	0.85
W-type index profile [81]	820	Selectable close to zero	2.1	2	High[b]	>0.83
Chalc. waveguides	10×10^3	29	60	0.06	60	Predict~1[a]
As$_2$S$_3$ [85] Ge$_{11.5}$ As$_{24}$Se$_{64.5}$ [86]	136×10^3	66	260	0.018	60	Predict~1[a]
Si waveguides [90,91]	3×10^5	Selectable close to zero	400	0.005	0.77	Predict high[b]
SOH waveguides [87–89]	1×10^5	−7000	1500	0.004	2.19	Predict high[b]

[a]No direct measurement found.

16.6.2 Quadratic nonlinearity systems

A different approach to nonlinear processing involves the use of quadratic ($\chi^{(2)}$) nonlinearities. In this case, the material of preference for telecommunications applications is lithium niobate, which can be periodically poled to form highly efficient nonlinear waveguides of a few centimeters length. Since most $\chi^{(2)}$ effects involve the generation of a second harmonic, it is necessary to employ cascaded nonlinearities in order to facilitate the use of pumps and signals in the same wavelength band. Such schemes typically use one pump operating within the phase-matched frequency range

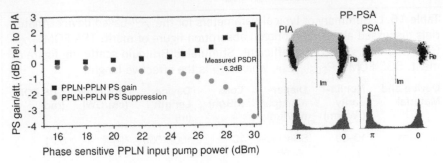

FIGURE 16.34 Characterization of a phase-squeezing PSA based on PPLN devices (taken from [95].)

of the PPLN waveguide for the generation of the second harmonic, which in turn interacts with the signal via, e.g. difference frequency generation. The application of cascaded quadratic nonlinear processes offers a number of potential advantages, such as broadband operation, low-latency, no intrinsic frequency chirp, and SBS resilience [93–100]. However, such devices are ultimately limited by power transfer to the third harmonic wavelength, infrared absorption, and subsequent temperature instability. A non-degenerate PSA was demonstrated in a PPLN waveguide using cascaded quadratic nonlinearities in [93]. Subsequently, this concept was also employed in [95] to demonstrate squeezing of the phase noise of a DPSK signal (Figure 16.34). Interestingly, this demonstration also considered the generation of the phase-locked signal-idler pair in a second PPLN waveguide, further miniaturizing the system. In a different application, phase-regenerative wavelength conversion was demonstrated in [94].

16.6.3 Semiconductor optical amplifiers

Despite the advantages that PPLN-based devices offer both in terms of compactness and overcoming SBS issues, the pump power levels required for the operation of phase regenerators based on such devices, are still comparable to those of fiber-based systems. Semiconductor optical amplifiers (SOAs), on the other hand, are also promising active devices for signal processing as they have high integration ability and low switching energy [101–103]. By employing intersubband transition (ISBT) XPM effects [104] in quantum well or quantum dot material structures [102] SOA-based processing systems that combine a number of nonlinear devices on the same chip have already been demonstrated. In order to achieve operation at repetition rates exceeding 100 Gbit/s the slow response component originating from the carrier energy relaxation and band-to-band carrier recombination can be solved using blue-wavelength filtering [105], or different configurations of symmetric-Mach-Zehnder structures with SOAs in both arms [104,106,107]. However, a trade-off relationship between the optical power density required for absorption saturation and the relaxation time is intrinsic to this technology. The use of a monolithically integrated array of four SOAs inside two parallel Mach-Zehnder structures in a large-scale integrated circuit has been considered for indirect phase regeneration of DPSK signals based on format conversion [108]. Furthermore, the possibility of implementing the PSA operation in SOAs has also been investigated [109] and SOA-based signal regeneration (at least for the DPSK format) should be possible.

16.7 FUTURE TRENDS AND RESEARCH DIRECTIONS

So far this chapter has documented the remarkable progress that has been achieved in the regeneration of phase modulated signals over the last few years. It has to be appreciated, however, that these advances represent only the first bold steps toward the implementation of all-optical processing systems that will address the networks of the future. In the paragraphs that follow, we will review early progress and challenges associated with two distinct directions, namely the simultaneous phase regeneration of more than one channel and the processing of signals modulated using more spectrally efficient modulation formats, such as quadrature amplitude modulation (QAM).

16.7.1 Multiwavelength regeneration

In general, the multiwavelength operation of a nonlinear device is complicated, mainly since each of the co-existing signals can invariably act as a pump to the system. For example, the addition of an extra signal on a device based on $\chi^{(3)}$ nonlinearities can affect its response by the onset of additional effects originating by either cross-phase modulation, four-wave mixing, or even Raman scattering. These impairments are less severe in cases where the signals are only required to propagate in a linear fashion within the nonlinear medium and share a common pump [14], or where the non-linearities are localized within a certain wavelength region [110]. However, even in these cases, the presence of multiple signals drawing simultaneously from the power of the same pump can give rise to crosstalk. Despite this multi-channel PSA demonstrations based on a single pump non-degenerate parametric process have already been reported both in fibers and PPLN waveguides. In the PPLN-based demonstration reported in [110] one common pump was employed to effect phase squeezing of four independent channels through the use of cascaded quadratic nonlinearities. The experiment achieved squeezing of the standard deviation of the phase angle by up to 1.3 times, whereas the power of each of the four channels was ~15dB weaker than the pump to ensure that the parasitic crosstalk components were weaker than 22dB below the generated signal-idler pairs. A fiber-based two-channel phase regenerative PSA based on a two-pump (per channel) degenerate parametric amplifier has also been presented recently (Figure 16.35) [111]. In this experiment, the two data signals shared the nonlinear medium and a common pump, whereas independent second pumps and phase-locking mechanisms for these pumps to each one of the signals

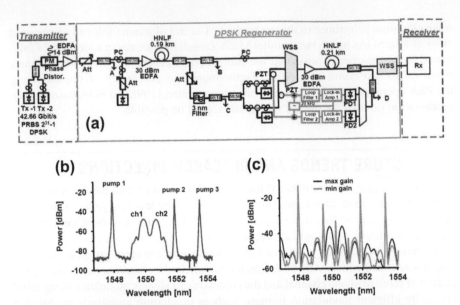

FIGURE 16.35 (a) A PS-FOPA based two-channel phase regenerator; (b) optical spectral trace of the pumps and signals at the input of the PS-FOPA, pumps 1 and 2 amplify channel 1, and pumps 1 and 3 amplify channel 2; (c) optical spectral traces at the output of the PS-FOPA. Adapted from [111].

were used. In order to avoid parasitic four-wave mixing interactions between the various pumps and signals, the signals were at least 10 dB weaker than each of the pumps, and care was taken to ensure that the frequency spacing between the two signals and the common pump was a non-integer multiple of the spacing between the two signals.

The phase-preserving amplitude regeneration scheme based on the application of a nonlinear amplifying loop mirror has also been extended to accommodate multiwavelength operation [112]. In this case, the approach taken was to demultiplex and employ a different nonlinear element per channel inside the loop mirror, thus allowing all other components of the regenerator to be shared between the different channels. Finally, a number of multi-channel intensity regenerators have been developed [113,114] which can be used as building blocks in phase regenerators based on format conversion.

16.7.2 Regeneration of other advanced modulation formats

In general, the operation of signal regenerators relies on developing step-like transfer functions characterizing the input and output parameters of interest (i.e. amplitude and/or phase). The majority of research on signal regeneration has been concerned with binary modulation formats requiring a single-stepped transfer function, whether this is in amplitude or in the 2π phase cycle. Clearly, the processing of more complex modulation formats that encode the information in multiple levels of phase and/or amplitude require the engineering of increasingly more complex transfer functions. This was shown in Section 16.5 for the case of the QPSK format. The same section described the route toward using the principle of phase-sensitive amplification to implement transfer functions that satisfy a larger number of phase-only coding levels. This scaling of the PSA regeneration principle can satisfy certain useful modulation formats, such as 8-PSK. The regeneration of QAM signals in a single device would, however, require the implementation of both intensity and phase transfer characteristics that exhibit more than two levels. It has been shown that it may be possible to achieve intensity transfer characteristics that exhibit more than two levels and operate in a phase-preserving fashion [115] by extending the nonlinear loop mirror-based scheme of [19]. However, the regeneration of the complex field in both phase and amplitude, and indeed in more than just two levels still represents a problem of considerable challenge. It is likely that the solution might lie in combining the respective techniques for phase and amplitude regeneration; however, whether this route is feasible in practice still remains to be seen.

16.8 CONCLUSIONS

In this chapter we have reviewed the need, general principles, and approaches used to regenerate phase encoded signals of differing levels of coding complexity and have described the key underpinning technologies used. We have discussed the evolution of the field and have presented the current state of the art both in terms of the demonstrations of novel concepts that address and extend the range of functionality that can be achieved (e.g. range of advanced modulation format signals that can be addressed and new functions such as phase quantization) and progress in terms of extending practicality (e.g. in terms of providing black-box operation and routes to integration). We have also given a personal perspective in terms of key issues and directions for future work.

From all that we have written it should be clear that this is a fast moving, exciting, and potentially important research field with significant advances made in recent years, enabled in many instances by progress in the underpinning technologies, which include both highly nonlinear fibers and high performance narrow-linewidth lasers. It should, however, be equally clear that many challenges remain and that will need to be overcome if all-optical regeneration is ever to make it from the laboratory to commercial deployment in future optical networks—not least in terms of enabling significant levels of WDM capability and the development of compact, power efficient variants. Much work needs to be done and one can anticipate considerable innovation and interesting results ahead with many other application opportunities (e.g. within test and measurement systems) opening up along the way—we hope that this chapter encourages others to choose to enter and contribute to this interesting and potentially very important field.

References

[1] P.A. Humblet, M. Azizoglu, On the bit error rate of lightwave systems with optical amplifiers, J. Lightwave Technol. 9 (1991) 1576–1582.

[2] M. Salsi, H. Mardoyan, P. Tran, C. Koebele, E. Dutisseuil, G. Charlet, S. Bigo, 155 × 100 Gbps coherent PDM-QPSK transmission over 7,200 km, in: Proceedings of the 35th European Conference on Optical Communication (ECOC), 2009, pp. 1–2.

[3] M. Salsi, C. Koebele, P. Tran, H. Mardoyan, E. Dutisseuil, J. Renaudier, M. Bigot-Astruc, L. Provost, S. Richard, P. Sillard, S. Bigo, and G. Charlet, Transmission of 96 × 100 Gbps with 23 percent super-FEC overhead over 11,680 km, using optical spectral engineering, in: Proceedings of the 2011 Optical Fiber Communication Conference (OFC/NFOEC), 2011, pp. 1–3.

[4] S. Kumar (Ed.), Impact of Nonlinearities in Fiber Optic Communications, Optical and Fiber Communications Reports, Springer, vol. 7, 2011 ISBN 978-1-4419-8139-4. <http://www.springer.com/physics/optics+%26+lasers/book/978-1-4419-8138-7>.

[5] M. Matsumoto, Fiber-based all-optical signal regeneration, IEEE J. Selected Topics Quantum Electron. 18 (2012) 738–752.

[6] C. McKinstrie, S. Radic, Phase-sensitive amplification in a fiber, Opt. Express 12 (20) (2004) 4973–4979.

[7] K. Inoue, Optical level equalisation based on gain saturation in fiber optical parametric amplifier, Electron. Lett. 36 (2000) 1016–1017.

[8] M. Matsumoto, Regeneration of RZ-DPSK signals by fiber-based all-optical regenators, IEEE Photon. Technol. Lett. 17 (2005) 1055–1057.

[9] M. Matsumoto, Performance improvement of phase-shift-keying signal transmission by means of optical limiters using four-wave mixing in fibers, J. Lightwave Technol. 23 (2005) 2696–2701.

[10] C.S. Bres, A.O.J. Wiberg, J.M. Chavez-Boggio, S. Radic, Optical demultiplexing with extinction ratio enhancement based on higher order parametric interaction, in: Proceedings of the 35th European Conference on Optical Communication (ECOC), 2009, pp. 1–2.

[11] A. Bogris, D. Syvridis, RZ-DPSK signal regeneration based on dual-pump phase-sensitive amplification in fibers, IEEE Photon. Technol. Lett. 18 (2006) 2144–2146.

[12] K. Croussore, G. Li, Phase and amplitude regeneration of differential phase-shift keyed signals using phase-sensitive amplification, IEEE J. Selected Topics Quantum Electron. 14 (2008) 648–658.

[13] J. Kakande, R. Slavik, F. Parmigiani, A. Bogris, D. Syvridis, L. Gruner-Nielsen, R. Phelan, P. Petropoulos, D.J. Richardson, Multilevel quantization of optical phase in a novel coherent parametric mixer architecture, Nat. Photon. 5 (2011) 748–752.

[14] Z. Tong, C. Lundstrom, P.A. Andrekson, C.J. McKinstrie, M. Karlsson, D.J. Blessing, E. Tipsuwannakul, B.J. Puttnam, H. Toda, L. Gruner Nielsen, Towards ultrasensitive optical links enabled by low-noise phase-sensitive amplifiers, Nat. Photon. 5 (2011) 430–436.

[15] X. Chongjin, G. Raybon, P.J. Winzer, Transmission of mixed 224-Gb/s and 112-Gb/s PDM-QPSK at 50-GHz channel spacing over 1200-km dispersion-managed LEAF spans and three ROADMs, J. Lightwave Technol. 30 (2012) 547–552.

[16] S. Chandrasekhar, X. Liu, Impact of channel plan and dispersion map on hybrid DWDM transmission of 42.7-Gb/s DQPSK and 10.7-Gb/s OOK on 50-GHz grid, IEEE Photon. Technol. Lett. 19 (2007) 1801–1803.

[17] A. Bononi, N. Rossi, P. Serena, Transmission limitations due to fiber nonlinearity, in: Proceedings of the 2011 Optical Fiber Communication Conference and Exposition (OFC/NFOEC), and the National Fiber Optic Engineers Conference, 2011, pp. 1–3.

[18] N.J. Doran, D. Wood, Nonlinear-optical loop mirror, Opt. Lett. 13 (1988) 56–58.

[19] K. Cvecek, G. Onishchukov, K. Sponsel, A.G. Striegler, B. Schmauss, G. Leuchs, Experimental investigation of a modified NOLM for phase-encoded signal regeneration, IEEE Photon. Technol. Lett. 18 (2006) 1801–1803.

[20] C. Stephan, K. Sponsel, G. Onishchukov, B. Schmauss, G. Leuchs, Nonlinear phase noise reduction in a DPSK transmission system using cascaded nonlinear amplifying loop mirrors, IEEE Photon. Technol. Lett. 21 (2009) 1864–1866.

[21] M. Matsumoto, A fiber-based all-optical 3R regenerator for DPSK signals, IEEE Photon. Technol. Lett. 19 (2007) 273–275.

[22] A. Fragkos, A. Bogris, D. Syvridis, R. Phelan, Amplitude noise limiting amplifier for phase encoded signals using injection locking in semiconductor lasers, J. Lightwave Technol. 30 (5) (2012) 764–771.

[23] A. Fragkos, A. Bogris, D. Syvridis, R. Phelan, Colorless regenerative amplification of constant envelope phase-modulated optical signals based on injection-locked Fabry Perot lasers, IEEE Photon. Technol. Lett. 24 (2012) 28–30.

[24] L. Quang Trung, L. Bramerie, N. Hoang Trung, M. Gay, S. Lobo, M. Joindot, J.L. Oudar, J.C. Simon, Saturable-absorber-based phase-preserving amplitude regeneration of RZ DPSK signals, IEEE Photon. Technol. Lett. 22 (2010) 887–889.

[25] L. Quang Trung, L. Bramerie, M. Gay, M. Joindot, J. Simon, A. O'Hare, N. Hoang Trung, J. Oudar, All-optical phase-preserving amplitude regeneration of 28-Gbaud RZ-DQPSK signals with a microcavity saturable absorber in a recirculating loop experiment, in: Optical Fiber Communication Conference and Exposition (OFC/NFOEC), 2011 and the National Fiber Optic Engineers Conference, 2011, pp. 1–3.

[26] K. Croussore, I. Kim, C. Kim, Y. Han, G.F. Li, Phase-and-amplitude regeneration of differential phase-shift keyed signals using a phase-sensitive amplifier, Opt. Express 14 (2006) 2085–2094.

[27] D. Levandovsky, M. Vasilyev, P. Kumar, Amplitude squeezing of light by means of a phase-sensitive fiber parametric amplifier, Opt. Lett. 24 (1999) 984–986.

[28] H.A. Haus, J.A. Mullen, Quantum noise in linear amplifiers, Phys. Rev. 128 (1962) 2407.

[29] C.M. Caves, Quantum limits on noise in linear amplifiers, Phys. Rev. D 26 (1982) 1817.

[30] R.E. Slusher, B. Yurke, Squeezed light for coherent communications, J. Lightwave Technol. 8 (1990) 466–477.

[31] J.A. Levenson, I. Abram, T. Rivera, P. Grangier, Reduction of quantum noise in optical parametric amplification, J. Opt. Soc. Am. B 10 (1993) 2233–2238.

[32] M.E. Marhic, C.H. Hsia, J.M. Jeong, Optical amplification in a nonlinear fiber interferometer, Electron. Lett. 27 (1991) 210–211.

[33] W. Imajuku, A. Takada, Y. Yamabayashi, Low-noise amplification under the 3 dB noise figure in high-gain phase-sensitive fiber amplifier, Electron. Lett. 35 (1999) 1954–1955.

[34] R. Tang, P. Devgan, V.S. Grigoryan, P. Kumar, Inline frequency-non-degenerate phase-sensitive fiber parametric amplifier for fiber-optic communication, Electron. Lett. 41 (2005) 1072–1074.

[35] R. Tang, J. Lasri, P.S. Devgan, V. Grigoryan, P. Kumar, M. Vasilyev, Gain characteristics of a frequency nondegenerate phase-sensitive fiber-optic parametric amplifier with phase self-stabilized input, Opt. Express 13 (2005) 10483–10493.

[36] Z. Tong, C. Lundstrom, A. Bogris, M. Karlsson, P. Andrekson, D. Syvridis, Measurement of sub-1 dB noise figure in a non-degenerate cascaded phase-sensitive fiber parametric amplifier, in: Proceedings of the 35th European Conference on Optical Communication (ECOC), 2009, pp. 1–2.

[37] C. Lundström, Z. Tong, M. Karlsson, P.A. Andrekson, Phase-to-phase and phase-to-amplitude transfer characteristics of a nondegenerate-idler phase-sensitive amplifier, Opt. Lett. 36 (2011) 4356–4358.

[38] W. Imajuku, A. Takada, Gain characteristics of coherent optical amplifiers using a Mach-Zehnder interferometer with Kerr media, IEEE J. Quant. Electron. 35 (1999) 1657–1665.

[39] M. Shirasaki, H.A. Haus, Squeezing of pulses in a nonlinear interferometer, J. Opt. Soc. Am. B 7 (1990) 30–34.

[40] M.E. Marhic, Fiber Optical Parametric Amplifiers, Oscillators and Related Devices, Cambridge University Press, 2007, ISBN: 9780521861021. <http://www.cambridge.org/gb/knowledge/isbn/item1173531/?site_locale=en_GB>

[41] G. Agrawal, Nonlinear Fiber Optics, Academic Press, 2001.

[42] J. Kakande, F. Parmigiani, R. Slavik, L. Gruner-Nielsen, D. Jakobsen, S. Herstrom, P. Petropoulos, D.J. Richardson, Saturation effects in degenerate phase sensitive fiber optic parametric amplifiers, in: 36th European Conference and Exhibition on Optical Communication (ECOC), 2010, pp. 1–3.

[43] C. Lundstrom, B. Corcoran, T. Zhi, M. Karlsson, P. A. Andrekson, Phase and amplitude transfer functions of a saturated phase-sensitive parametric amplifier, in: 37th European Conference and Exhibition on Optical Communication (ECOC), 2011, pp. 1–3.

[44] M.J. Fice, A. Chiuchiarelli, E. Ciaramella, A.J. Seeds, Homodyne coherent optical receiver using an optical injection phase-lock loop, J. Lightwave Technol. 29 (2011) 1152–1164.

[45] A. Chiuchiarelli, M.J. Fice, E. Ciaramella, A.J. Seeds, Effective homodyne optical phase locking to PSK signal by means of 8b10b line coding, Opt. Express 19 (2011) 1707–1712.

[46] I. Kim, K. Croussore, X. Li, G. Li, T. Hasegawa, N. Sugimoto, All-optical carrier phase and polarization recovery using a phase-sensitive oscillator, in: Proceedings of the 2007 Optical Fiber Communication and the National Fiber Optic Engineers Conference (OFC/NFOEC), 2007, pp. 1–3.

[47] J.M. Kahn, BPSK homodyne detection experiment using balanced optical phase-locked loop with quantized feedback, IEEE Photon. Technol. Lett. 2 (1990) 840–843.

[48] S.K. Ibrahim, S. Sygletos, R. Weerasuriya, A.D. Ellis, Novel carrier extraction scheme for phase modulated signals using feed-forward based modulation stripping, in: Optical Communication (ECOC), 36th European Conference and Exhibition on, 2010, pp. 1–3.

[49] M.J. Fice, A.J. Seeds, B.J. Pugh, J.M. Heaton, S.J. Clements, Homodyne coherent receiver with phase locking to orthogonal-polarisation pilot carrier by optical injection phase lock loop, Conference on Optical Fiber Communication—Incudes Post Deadline Papers, OFC 2009 2009 (2009) 1–3.

[50] R. Slavik, J. Kakande, D. J. Richardson, Feed-forward optical domain carrier recovery from high baud rate PSK signals using relatively slow electronics, in: 37th European Conference and Exhibition on Optical Communication (ECOC), 2011, pp. 1–3.

[51] R. Slavik, F. Parmigiani, J. Kakande, C. Lundstrom, M. Sjodin, P.A. Andrekson, R. Weerasuriya, S. Sygletos, A.D. Ellis, L. Gruner-Nielsen, D. Jakobsen, S. Herstrom, R. Phelan, J. O'Gorman, A. Bogris, D. Syvridis, S. Dasgupta, P. Petropoulos, D.J. Richardson, All-optical phase and amplitude regenerator for next-generation telecommunications systems, Nat. Photon. 4 (2010) 690–695.

[52] R. Weerasuriya, S. Sygletos, S.K. Ibrahim, R. Phelan, J. O'Carroll, B. Kelly, J. O'Gorman, A. D. Ellis, Generation of frequency symmetric signals from a BPSK input for phase sensitive amplification, in: Proceedings of the 2010 Optical Fiber Communication (OFC), 2010, pp. 1–3.

[53] J.A. Armstrong, N. Bloembergen, J. Ducuing, P.S. Pershan, Interactions between light waves in a nonlinear dielectric, Phys. Rev. 127 (1962) 1918.

[54] R. Phelan, B. Kelly, J. O'Carroll, C. Herbert, A. Duke, J. O'Gorman, $-40°C < T < 95°$ C mode-hop-free operation of uncooled AlGaInAs-MQW discrete-mode laser diode with emission at $\lambda = 1.3\,\mu$m, Electron. Lett. 45 (2009) 43–45.

[55] L. Gruner-Nielsen, S. Dasgupta, M.D. Mermelstein, D. Jakobsen, S. Herstrom, M.E.V. Pedersen, E.L. Lim, S. Alam, F. Parmigiani, D. Richardson, B. Palsdottir, A silica based highly nonlinear fiber with improved threshold for stimulated Brillouin scattering, in: 36th European Conference and Exhibition on Optical Communication (ECOC), 2010, pp. 1–3.

[56] L. Gruner-Nielsen, S. Herstrom, S. Dasgupta, D. Richardson, D. Jakobsen, C. Lundstrom, P.A. Andrekson, M.E.V. Pedersen, B. Palsdottir, Silica-based highly nonlinear fibers with a high SBS threshold, in: 2011 IEEE Winter Topicals (WTM), 2011, pp. 171–172.

[57] R. Slavik, A. Bogris, F. Parmigiani, J. Kakande, M. Westlund, M. Skold, L. Gruner-Nielsen, R. Phelan, D. Syvridis, P. Petropoulos, D.J. Richardson, Coherent all-optical phase and amplitude regenerator of binary phase-encoded signals, IEEE J. Selected Topics Quant. Electron. 18 (2012) 859–869.

[58] R. Slavik, A. Bogris, J. Kakande, F. Parmigiani, L. Gruner-Nielsen, R. Phelan, J. Vojtech, P. Petropoulos, D. Syvridis, D.J. Richardson, Field-trial of an all-optical PSK regenerator/multicaster in a 40 Gbit/s, 38 channel DWDM transmission experiment, J. Lightwave Technol. 30 (2012) 512–520.

[59] Z. Zheng, L. An, Z. Li, X. Zhao, X. Liu, All-optical regeneration of DQPSK/QPSK signals based on phase-sensitive amplification, Opt. Commun. 281 (2008) 2755–2759.

[60] M. Matsumoto, All-optical DQPSK signal regeneration using 2R amplitude regenerators, Opt. Express 18 (2010) 10–24.

[61] N.N. Kimchau, K. Tomofumi, M.G. John, P. Henrik, J.B. Daniel, All-Optical 2R Regeneration of BPSK and QPSK Data using a 90 degree optical hybrid and integrated SOA-MZI wavelength converter pairs, in: Proceedings of the 2011 Optical Fiber Communication Conference (OFC/NFOEC), 2011, p. OMT3.

[62] G.C. Valley, Photonic analog-to-digital converters, Opt. Express 15 (2007) 1955–1982.

[63] R.H. Walden, Analog-to-digital converter survey and analysis, IEEE J. Selected Areas Commun. 17 (1999) 539–550.

[64] J. Kakande, A. Bogris, R. Slavik, F. Parmigiani, D. Syvridis, P. Petropoulos, D.J. Richardson, First demonstration of all-optical QPSK signal regeneration in a novel multi-format phase sensitive amplifier, in: Proceedings of the 36th European Conference and Exhibition on Optical Communication (ECOC), 2010, pp. 1–3.

[65] J. Kakande, A. Bogris, R. Slavik, F. Parmigiani, D. Syvridis, P. Petropoulos, D. Richardson, M. Westlund, M. Skold, QPSK phase and amplitude regeneration at

56 Gbaud in a novel idler-free non-degenerate phase sensitive amplifier, in: Proceedings of the 2011 Optical Fiber Communication Conference (OFC/NFOEC), 2011, p. OMT4.

[66] C. Yu, S. Zhang, P.Y. Kam, J. Chen, Bit-error rate performance of coherent optical M-ary PSK/QAM using decision-aided maximum likelihood phase estimation, Opt. Express 18 (2010) 12088–12103.

[67] F. Da Ros, C. Peucheret, QPSK phase regeneration in saturated degenerate dual-pump phase sensitive amplifiers, in: 2011 IEEE Photonics Conference (PHO), 2011, pp. 105–106.

[68] M. Hirano, T. Nakanishi, T. Okuno, M. Onishi, Silica-based highly nonlinear fibers and their application, IEEE J. Selected Topics Quant. Electron. 15 (2009) 103–113.

[69] E. Myslivets, C. Lundstrom, J.M. Aparicio, S. Moro, A.O.J. Wiberg, C.S. Bres, N. Alic, P.A. Andrekson, S. Radic, Spatial equalization of zero-dispersion wavelength profiles in nonlinear fibers, IEEE Photon. Technol. Lett. 21 (2009) 1807–1809.

[70] J.H Lee, T. Nagashima, T. Hasegawa, S. Ohara, N. Sugimoto, K. Kikuchi, Bismuth-oxide-based nonlinear fiber with a high SBS threshold and its application to four-wave-mixing wavelength conversion using a pure continuous-wave pump, J. Lightwave Technol. 24 (2006) 22–28.

[71] F. Parmigiani, S. Asimakis, N. Sugimoto, F. Koizumi, P. Petropoulos, D.J. Richardson, 2R regenerator based on a 2-m-long highly nonlinear bismuth oxide fiber, Opt. Express 14 (2006) 5038–5044.

[72] F. Poletti, X. Feng, G.M. Ponzo, M.N. Petrovich, W.H. Loh, D.J. Richardson, All-solid highly nonlinear singlemode fibers with a tailored dispersion profile, Opt. Express 19 (2011) 66–80.

[73] J.S. Sanghera, L. Brandon Shaw, I.D. Aggarwal, Chalcogenide glass-fiber-based mid-IR sources and applications, IEEE J. Sel. Top. Quant. Electron. 15 (2009) 114–119.

[74] J.Y.Y. Leong, P. Petropoulos, J.H.V. Price, H. Ebendorff-Heidepriem, S. Asimakis, R.C. Moore, K.E. Frampton, V. Finazzi, X. Feng, T.M. Monro, D.J. Richardson, High-nonlinearity dispersion-shifted lead-silicate holey fibers for efficient 1-μm pumped supercontinuum generation, J. Lightwave Technol. 24 (2006) 183–190.

[75] S. Afshar V, W.Q. Zhang, H. Ebendorff-Heidepriem, T.M. Monro, Small core optical waveguides are more nonlinear than expected: experimental confirmation, Opt. Lett., 34 (2009), 3577–3579.

[76] S.D. Le, D.M. Nguyen, M. Thual, L. Bramerie, M. Costa e Silva, K. Lenglé, M. Gay, T. Chartier, L. Brilland, D. Méchin, P. Toupin, J. Troles, Efficient four-wave mixing in an ultra-highly nonlinear suspended-core chalcogenide As38Se62 fiber, Opt. Express 19 (2011) B653–B660.

[77] D.-I. Yeom, E.C. Mägi, M.R.E. Lamont, M.A.F. Roelens, L. Fu, B.J. Eggleton, Low-threshold supercontinuum generation in highly nonlinear chalcogenide nanowires, Opt. Lett. 33 (2008) 660–662.

[78] B.J. Eggleton, B. Luther-Davies, K. Richardson, Chalcogenide photonics, Nat. Photon. 5 (2011) 141–148.

[79] C. Fortier, J. Fatome, S. Pitois, F. Smektala, G. Millot, J. Troles, F. Desevedavy, P. Houizot, L. Brilland, N. Traynor, Experimental investigation of Brillouin and Raman scattering in a 2SG sulfide glass microstructured chalcogenide fiber, Opt. Express 16 (2008) 9398–9404.

[80] L. Fu, M. Rochette, V. Ta'eed, D. Moss, B. Eggleton, Investigation of self-phase modulation based optical regeneration in single mode As2Se3 chalcogenide glass fiber, Opt. Express 13 (2005) 7637–7644.

[81] M.A. Ettabib, L. Jones, J. Kakande, R. Slavik, F. Parmigiani, X. Feng, F. Poletti, G.M. Ponzo, J.A. Shi, M.N. Petrovich, W.H. Loh, P. Petropoulos, D.J. Richardson, Phase sensitive amplification in a highly nonlinear lead-silicate fiber, Opt. Express 20 (2012) 1629–1634.

[82] M. Asobe, T. Kanamori, K. Kubodera, Applications of highly nonlinear chalcogenide glass fibers in ultrafast all-optical switches, IEEE J. Quant. Electron. 29 (1993) 2325–2333.

[83] K. Suzuki, Y. Hamachi, T. Baba, Fabrication and characterization of chalcogenide glass photonic crystal waveguides, Opt. Express 17 (2009) 22393–22400.

[84] M.D. Pelusi, V.G. Ta'eed, F. Libin, E. Magi, M.R.E. Lamont, S. Madden, C. Duk-Yong, D.A.P. Bulla, B. Luther-Davies, B.J. Eggleton, Applications of highly-nonlinear chalcogenide glass devices tailored for high-speed all-optical signal processing, IEEE J. Sel. Top. Quant. Electron. 14 (2008) 529–539.

[85] M.R. Lamont, B. Luther-Davies, D.-Y. Choi, S. Madden, B.J. Eggleton, Supercontinuum generation in dispersion engineered highly nonlinear($\gamma = 10/\text{W/m}$) As2S3 chalcogenide planar waveguide, Opt. Express 16 (2008) 14938–14944.

[86] X. Gai, S. Madden, D.-Y. Choi, D. Bulla, B. Luther-Davies, Dispersion engineered Ge11.5As24Se64.5 nanowires with a nonlinear parameter of $136\,\text{W}^{-1}\,\text{m}^{-1}$ at 1550 nm, Opt. Express 18 (2010) 18866–18874.

[87] T. Vallaitis, S. Bogatscher, L. Alloatti, P. Dumon, R. Baets, M.L. Scimeca, I. Biaggio, F. Diederich, C. Koos, W. Freude, J. Leuthold, Optical properties of highly nonlinear silicon-organic hybrid (SOH) waveguide geometries, Opt. Express 17 (2009) 17357–17368.

[88] J. Leuthold, C. Koos, W. Freude, Nonlinear silicon photonics, Nat. Photon. 4 (2010) 535–544.

[89] C. Koos, P. Vorreau, T. Vallaitis, P. Dumon, W. Bogaerts, R. Baets, B. Esembeson, I. Biaggio, T. Michinobu, F. Diederich, W. Freude, J. Leuthold, All-optical high-speed signal processing with silicon-organic hybrid slot waveguides, Nat. Photon. 3 (2009) 216–219.

[90] M.A. Foster, A.C. Turner, J.E. Sharping, B.S. Schmidt, M. Lipson, A.L. Gaeta, Broad-band optical parametric gain on a silicon photonic chip, Nature 441 (2006) 960–963.

[91] R. Salem, M.A. Foster, A.C. Turner, D.F. Geraghty, M. Lipson, A.L. Gaeta, Signal regeneration using low-power four-wave mixing on silicon chip, Nat. Photon. 2 (2008) 35–38.

[92] L.K. Oxenlowe, J. Hua, M. Galili, P. Minhao, H. Hao, H.C.H. Mulvad, K. Yvind, J.M. Hvam, A.T. Clausen, P. Jeppesen, Silicon photonics for signal processing of Tbit/s serial data signals, IEEE J. Sel. Top. Quant. Electron. 18 (2012) 996–1005.

[93] K.J. Lee, F. Parmigiani, S. Liu, J. Kakande, P. Petropoulos, K. Gallo, D. Richardson, Phase sensitive amplification based on quadratic cascading in a periodically poled lithium niobate waveguide, Opt. Express 17 (2009) 20393–20400.

[94] S. Liu, K.J. Lee, F. Parmigiani, J. Kakande, K. Gallo, P. Petropoulos, D.J. Richardson, Phase-regenerative wavelength conversion in periodically poled lithium niobate waveguides, Opt. Express 19 (2011) 11705–11715.

[95] B.J. Puttnam, D. Mazroa, S. Shinada, N. Wada, Phase-squeezing properties of non-degenerate PSAs using PPLN waveguides, Opt. Express 19 (2011) B131–B139.

[96] F. Gomez-Agis, C.M. Okonkwo, A. Albores-Mejia, E. Tangdiongga, H.J.S. Dorren, 320-to-10 Gbit/s all-optical demultiplexing using sum-frequency generation in PPLN waveguide, Electron. Lett. 46 (2010) 1008–1009.

[97] A. Bogoni, W. Xiaoxia, S.R. Nuccio, A.E. Willner, 640 Gb/s All-optical regenerator based on a periodically poled lithium niobate waveguide, J. Lightwave Technol. 30 (2012) 1829–1834.

[98] H. Hao, R. Nouroozi, R. Ludwig, C. Schmidt-Langhorst, H. Suche, W. Sohler, C. Schubert, Polarization-insensitive 320-Gb/s in-line all-optical wavelength conversion in a 320-km transmission span, IEEE Photon. Technol. Lett. 23 (2011) 627–629.

[99] G.-W. Lu, T. Miyazaki, Optical phase erasure based on FWM in HNLF enabling format conversion from 320-Gb/s RZDQPSK to 160-Gb/s RZ-DPSK, Opt. Express 17 (2009) 13346–13353.

[100] A.E. Willner, O.F. Yilmaz, W. Jian, W. Xiaoxia, A. Bogoni, Z. Lin, S.R. Nuccio, Optically Efficient Nonlinear Signal Processing, IEEE J. Sel. Top. Quant. Electron. 17 (2011) 320–332.

[101] A. Poustie, SOA-based all-optical processing, in: Optical Fiber Communication and the National Fiber Optic Engineers Conference, OFC/NFOEC 2007, 2007, pp. 1–38.

[102] T. Akiyama, M. Sugawara, Y. Arakawa, Quantum-Dot Semiconductor Optical Amplifiers, Proc. IEEE 95 (2007) 1757–1766.

[103] K. Hinton, G. Raskutti, P.M. Farrell, R.S. Tucker, Switching energy and device size limits on digital photonic signal processing technologies, IEEE J. Sel. Top. Quant. Electron. 14 (2008) 938–945.

[104] O. Wada, Recent progress in semiconductor-based photonic signal-processing devices, IEEE J. Sel. Top. Quant. Electron. 17 (2011) 309–319.

[105] E. Tangdiongga, Y. Liu, H. de Waardt, G.D. Khoe, A.M.J. Koonen, H.J.S. Dorren, X. Shu, I. Bennion, All-optical demultiplexing of 640 to 40 Gbits/s using filtered chirp of a semiconductor optical amplifier, Opt. Lett. 32 (2007) 835–837.

[106] S. Nakamura, Y. Ueno, and K. Tajima, Error-free all-optical demultiplexing at 336 Gb/s with a hybrid-integrated symmetric-Mach-Zehnder switch, in Optical Fiber Communication Conference and Exhibit, OFC 2002, 2002, pp. FD3-1–FD3-3.

[107] T. Hirooka, M. Okazaki, T. Hirano, P. Guan, M. Nakazawa, S. Nakamura, All-optical demultiplexing of 640-Gb/s OTDM-DPSK signal using a semiconductor SMZ switch, IEEE Photon. Technol. Lett. 21 (2009) 1574–1576.

[108] M. Bougioukos, C. Kouloumentas, M. Spyropoulou, G. Giannoulis, D. Kalavrouziotis, A. Maziotis, P. Bakopoulos, R. Harmon, D. Rogers, J. Harrison, A. Poustie, G. Maxwell, H. Avramopoulos, Multi-format all-optical processing based on a large-scale, hybridly integrated photonic circuit, Opt. Express 19 (June) (2011) 11479–11489.

[109] R.P. Webb, J.M. Dailey, R.J. Manning, A.D. Ellis, Phase discrimination and simultaneous frequency conversion of the orthogonal components of an optical signal by four-wave mixing in an SOA, Opt. Express 19 (2011) 20015–20022.

[110] B.J. Puttnam, A. Szabo, D. Mazroa, S. Shinada, N. Wada, Multi-channel phase squeezing in a PPLN-PPLN PSA, in: Optical Fiber Communication Conference and Exposition (OFC/NFOEC), 2012 and the National Fiber Optic Engineers Conference, 2012, pp. 1–3.

[111] S. Sygletos, P. Frascella, S.K. Ibrahim, L. Grüner-Nielsen, R. Phelan, J. O'Gorman, A.D. Ellis, A practical phase sensitive amplification scheme for two channel phase regeneration, Opt. Express 19 (2011) B938–B945.

[112] K. Cvecek, K. Sponsel, C. Stephan, G. Onishchukov, R. Ludwig, C. Schubert, B. Schmauss, G. Leuchs, Phase-preserving amplitude regeneration for a WDM RZ-DPSK signal using a nonlinear amplifying loop mirror, Opt. Express 16 (2008) 1923–1928.

[113] C. Kouloumentas, P. Vorreau, L. Provost, P. Petropoulos, W. Freude, J. Leuthold, I. Tomkos, All-fiberized dispersion-managed multichannel regeneration at 43 Gb/s, IEEE Photon. Technol. Lett. 20 (2008) 1854–1856.

[114] F. Parmigiani, L. Provost, P. Petropoulos, D.J. Richardson, W. Freude, J. Leuthold, A.D. Ellis, I. Tomkos, Progress in multichannel all-optical regeneration based on fiber technology, IEEE J. Sel. Top. Quant. Electron. 18 (2012) 689–700.

[115] M. Hierold, T. Roethlingshoefer, K. Sponsel, G. Onishchukov, B. Schmauss, G. Leuchs, Multilevel phase-preserving amplitude regeneration using a single nonlinear amplifying loop mirror, IEEE Photon. Technol. Lett. 23 (2011) 1007–1009.

Ultra-High-Speed Optical Time Division Multiplexing

17

**Leif Katsuo Oxenløwe, Anders Clausen, Michael Galili,
Hans Christian Hansen Mulvad, Hua Ji, Hao Hu, and Evarist Palushani**

DTU Fotonik, the Technical University of Denmark, Kgs. Lyngby, Denmark

17.1 BACKGROUND

The idea of using optical means to temporally multiplex data channels to generate higher baud rates than are otherwise available is a mature idea. In 1968, Bell Labs researchers Kinsel and Denton [1] described a free-space system, where the output from a He-Ne laser was modulated and multiplexed to reach a capacity of 448 Mbit/s. In 1971–1972, detailed studies of demultiplexers were presented [2,3] aiming at total capacities of 6.7 Gbit/s. In 1987, Tucker et al. experimentally demonstrated the highest transmission bit rate at that time, higher than any available electrical bit rate [4] by multiplexing four 4 Gbit/s channels reaching 16 Gbit/s and transmitting it over 8 km of fiber. The attraction of optical time division multiplexing (OTDM) has always been the same—namely the promises of achieving higher bit rates per channel than electronics could provide, thus alleviating the so-called electronic speed bottleneck [5]. The benefits of this would be to capitalize on the inherent high-speed response of some optical devices, the offer of system design flexibility, such as adjustable bandwidth of different channels, and simpler systems with potentially less hardware to manage, e.g. very few lasers. Since 1987, the highest reported OTDM bit rate has always been higher than the highest available ETDM bit rate at any given time, and the same attractive features as given above have fueled a large amount of progress in the field. Especially when utilizing the ultra-fast Kerr effect in nonlinear fibers, as suggested by Morioka in 1987 [6], bit rates began to move into the 100s Gbit/s regime, and eventually 640 Gbit/s demultiplexing and transmission was demonstrated by Nakazawa et al. [7]. In year 2000, polarization multiplexing (pol-MUX) was also employed to generate 1.28 Tbit/s transmission (640 Gbaud per polarization) [8], an idea already suggested in 1968 [1].

For 10 years, 640 Gbaud per polarization would remain the highest reported symbol rate, and endeavors to boost the per-channel bit rate to the Tbit/s regime went toward advanced modulation formats, such as the generation and detection of 2.56 Tbit/s using DQPSK and pol-MUX on a 640 Gbaud RZ pulse train [9], as demonstrated by Weber et al. More recently, the generation and detection of 8PSK at 640 Gbaud with pol-MUX resulted in an error free net bit rate of 3.56 Tbit/s using coherent detection [10]. In 2009, the symbol rate was finally increased to

Optical Fiber Telecommunications VIA. http://dx.doi.org/10.1016/B978-0-12-396958-3.00017-2

1.28 Tbaud in a ingle-polarization single-wavelength channel [11], and this remains the highest symbol rate per polarization to date. 1.28 Tbit/s DPSK on 1.28 Tbaud was transmitted over 50 km of SMF-IDF in 2010 [12]. In [13], 5.1 Tbit/s net data rate in a single-wavelength channel was demonstrated by combining 1.28 Tbaud with DQPSK modulation and pol-MUX and simple direct detection. Recently, 10.2 Tbit/s line rate on 1.28 Tbaud per polarization transmission was demonstrated using 16 QAM, pol-MUX, and coherent detection [14].

Internet traffic is constantly growing [15,16] and this trend is expected to continue [16,17]. The capacity has persistently been upgraded to support this increase, as e.g. seen by the release of the IEEE standardization of 40 and 100 Gbit/s communication in 2010 [18]. Despite these initiatives, predictions indicate that the apparently unlimited fiber capacity will be insufficient long before 2025 [17]. Consequently, an ongoing quest for more capacity, lower power consumption and reduced physical footprints spurs researchers on worldwide to explore the feasibility of a wide variety of potential technologies to achieve these objectives. Several impressive experiments have demonstrated both generation and detection of signals with ultra-high aggregated bit rates. For the purpose of this chapter, the experiments can be divided into multiple carrier and single carrier experiments. In the multiple carrier experiments, a number of Wavelength Division Multiplexing (WDM) channels and/or Orthogonal Frequency Division Multiplexing (OFDM) channels are modulated often utilizing both advanced modulation formats such as higher-order Quadrature Amplitude Modulation (QAM) formats or Quadrature Phase Shift Keying (QPSK) combined with Polarization Division Multiplexing (PDM) reaching bit rates, after subtracting the assumed Forward Error Correction (FEC) overhead, of 20 Tbit/s [19], 101 Tbit/s [20], and 102 Tbit/s [21]. If using several cores in the fiber, Space Division Multiplexing (SDM) can be used to increase the bit rate even further. By using 3 and 19 fiber cores, bit rates of 109 Tbit/s [22] and very impressive 305 Tbit/s [23] have been demonstrated. Despite using different schemes to achieve these results, they all have one thing in common; the symbol rate is defined and limited by the driving electronics, thus the symbol rate used in [19–23] is varying between 5.5 and 43 GBaud.

In OTDM, as sketched in Figure 17.1, the idea is to increase the symbol rate beyond the limits of electronics. When creating such high bit rates, active optical gating is required to extract the individual data signals again.

17.1.1 Ultra-fast nonlinear switching elements and materials

OTDM has now been an active field of research for more than four decades with multi-Tbit/s transmission demonstrated, so why has it not yet transcended into commercial systems. One reason is the state of ultra-fast devices and subsystems that are stable, compact, highly reliable and cost-effective (cost-per-bit and energy-per-bit). In the late 1980s, focus was on electro-optic solutions, which can be compact, cost-and-energy-effective and with potential for monolithic integration into advanced chips. But they rely on charged carrier transport, limiting the speed. It should be noted, though, that filtering-assisted cross-phase modulation

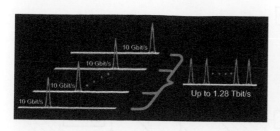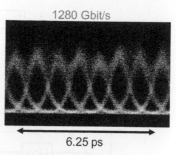

1280 Gbit/s

6.25 ps

FIGURE 17.1 The basic principle of Optical Time Division Multiplexing (OTDM). Left: Schematic showing low rate data signals on short pulses being temporally interleaved to form a serial string of ultra-high bit rate data. Right: An eye diagram of a 1.28 Tbit/s OTDM data signal measured on an optical sampling oscilloscope.

(XPM) in semiconductor optical amplifiers (SOAs) have been shown to operate at 320–640 Gbit/s [24–26]. Four-wave mixing (FWM) in (SOAs) has also been used at 400 Gbit/s [27]. However, these reports tend to include large power penalties and unavoidable pattern effects. The optical Kerr effect in nonlinear waveguides, such as highly nonlinear fiber (HNLF) [6], on the other hand has resulted in a multitude of successful reports on low-penalty ultra-high-speed switching, routinely reaching 640 Gbit/s, e.g. [7–9,28–30] and also 1.28 Tbit/s [11–13]. HNLF does not rely on carrier effects, so it is truly ultra-fast. However, it is bulky and difficult to stabilize temperature. However, stable HNLF-based switches can be made, as this chapter will discuss. A trend up through the 2000s has been to identify compact devices, which rely on the Kerr effect or equally ultra-fast nonlinear optical effects. This has given rise to a large number of demonstrations using e.g. chalcogenide (ChG) non-linear glass waveguides up to 1.28 Tbit/s [31,32]. The last 6 years, a strong focus has also been on the use of nonlinear silicon waveguides, often termed nanowires due to their 300–500 nm cross-sectional width [33]. Si-nanowires have been demonstrated to be able to demultiplex 1.28 Tbit/s [34], do 640 Gbit/s wavelength conversion [35] and perform serial-to-parallel conversion at 640 Gbit/s [36]. Lithium niobate has also proven very fruitful with χ^2-based 640 Gbit/s demonstrations [37–40].

All in all, there are a number of compact materials available now with demonstrated Tbit/s potential offering the possibility to integrate into multi-functional chips.

17.2 THE BASIC OTDM SYSTEM AND ITS CONSTITUENT PARTS

In Figure 17.2 the basic principle of an on-off keying OTDM system is shown.

The pulse source is the main component of the transmitter. It generates a pulse train of Return-to-Zero (RZ) pulses characterized by e.g. shape, temporal Full Width Half Maximum (FWHM), timing jitter, and repetition rate B, labeled the base rate

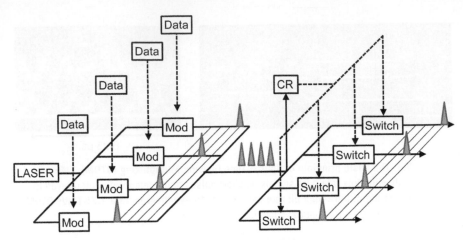

FIGURE 17.2 Principle of an Optical Time Division Multiplexed point-to-point system.

frequency. The pulse train is split into N branches, each including an external modulator, in which intensity modulates the pulse trains with unique data streams. If the pulses are sufficiently narrow, a specifically designed time delay in each branch allows the N data signals to be bit interleaved, i.e. multiplexed, thus generating an aggregated OTDM data signal with a bit rate of NB bit/s. When detecting the signal, the ultra-fast OTDM data signal should be downscaled in bit rate, allowing electronics to process the signal. In the transmitter, it was relatively easy to overcome the potential bit rate limitations induced by the low bandwidth of the electronics. However, in the receiver, the OTDM scheme takes its toll, as detection of the signal is very challenging; both clock recovery and demultiplexing of each tributary channel are required. A part of the OTDM signal is tapped for the Clock Recovery (CR) circuit, which extracts a clock corresponding to the base rate frequency B of the OTDM signal. Because the OTDM signal does not include a distinct frequency component at B, the clock extraction is challenging and modifications of existing commercial clock recovery schemes are required. The base rate clock is applied as a control signal to the switches, used to optically demultiplex each individual time channel from the aggregated data signal. Until recently, no existing switches offered demultiplexing of all channels using only one single switch and consequently an array of switches were envisioned, as sketched in Figure 17.2. Each of the demultiplexed base rate signals can subsequently be detected, by injecting the signals into base rate receivers triggered by the base rate clock. Hence, by using relatively slow electronics, very high-speed optical signals can be generated and detected. By adding multi-level modulation formats and polarization multiplexing, extra data can be imposed to each symbol, thus increasing the total bit rate.

In Figure 17.2, regarding transmission it is tacitly assumed that it is possible to connect the high-speed transmitter and receiver via a fiber transmission span. However, due to the narrow pulses of the transmitter, the spectrum is very broad and is consequently quite sensitive to even small values of uncompensated dispersion. Nevertheless, high-speed transmission has been demonstrated both in the laboratory environment, see e.g. [14], and in deployed fibers [41].

17.2.1 Essential functionalities for point-to-point

In the following, the essential functionalities sketched in Figure 17.2, necessary to build an OTDM system, will be described.

17.2.1.1 Pulse sources and pulse compression

In order to design a functioning OTDM system, a high-quality pulse source fulfilling a number of requirements is indispensable. By inspecting Figure 17.2, it is intuitively clear that the pulse shape, the width of the pulses, the timing jitter, and finally the energy in the pedestals are vital for the OTDM system. Additionally, in [42] other key parameters are listed.

Pulse trains emitted from lasers can typically be described by e.g. Gaussian or Hyperbolic Secant pulse shapes. Typically the emitted pulses will include an additional pedestal, which can be addressed using the parameter Pulse Tail Extinction Ratio (PTER) defined as the relation between the peak power in the pulse and the power in the pedestal. In the receiver the OTDM signal is demultiplexed to the individual base rate channels, before being *O/E* converted and processed electronically. If the electrical fields from the pulses are overlapping, either due to the pulse width or due to a finite PTER, the neighboring channels, upon *O/E* conversion, can deteriorate the demultiplexed channel. The noise terms in the receiver due to this process can be shown to consist of Intersymbol Interference (ISI) and interferometric crosstalk [43]. The interferometric crosstalk terms are dependent on the coherence time of the pulse source in the OTDM system. If the delay between the pulses is larger than the coherence time, the interferometric crosstalk terms will vary fast, and can be regarded as noise, and is denoted as incoherent crosstalk [43]. In [44], the requirements to the FWHM and PTER specifically for OTDM systems based on these considerations are given. A design criterion for the analysis in [44] is the maximum introduction of a 1 dB power penalty for modulating, multiplexing, and demultiplexing the pulses. The listed requirements indicate how strict system design requirements for an OTDM system are.

For Gaussian pulses, the FWHM pulse width should be less than 0.5 times the time slot. For FWHM pulse widths of approximately 0.4 times the time slot, the impact of PTER determines the system performance of the pulses, and consequently the effort for decreasing the pulse width further is not awarded with better performance. From Figure 17.3 [44], the PTER for different aggregated bit rates can be extracted to 27, 33, 37, and 41 dB at 160, 320, 640, and 1280 Gbit/s, respectively, when the base rate is 40 Gbit/s.

From the simulations it can be seen that the increase in PTER requirements for increased number of OTDM channels is not directly scalable, e.g. the increase in PTER from four to eight channels compared to an increase from 8 to 16 channels

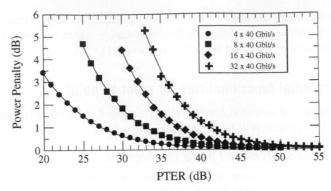

FIGURE 17.3 Power penalty versus PTER for pulse width FWHM equal to 0.4 times the width of the OTDM time slot. The 1 dB penalty is obtained for PTER equal to 27, 33, 37, and 41 dB, for 160, 320, 640, and 1280 Gbit/s respectively [44].

are not identical. This is ascribed to the gradual change of the nonGaussian shape for the probability density function (pdf) of the incoherent crosstalk toward a Gaussian-shaped pdf for an increase in the number of crosstalk terms, i.e. OTDM channels [45].

17.2.1.2 Demultiplexing: single channel

Several switches based on different physical processes and materials have been used successfully to demultiplex an OTDM signal. Generalized requirements to the shape, width, or extinction ratio (ER) of the induced switching window (SW) are presented here (see Figure 17.4).

In [46], a model simulating the impact of the switch parameters on the demultiplexed signal by determining the Eye Opening Penalty (EOP) is presented. This model is used to extract general rule-of-thumb design values for the switching window. Two different SW shapes have been investigated—a bell-shaped and a square-shaped SW represented by a Gaussian and a fifth order Super Gaussian shape, respectively.

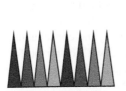
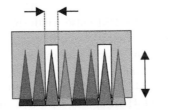
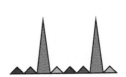

FIGURE 17.4 An OTDM signal shown to the left is injected into a switch (middle). The switch is characterized by the width, extinction ratio and shape of the SW as indicated by the gray box. These parameters will have an impact on the quality of the demultiplexed signal as shown to the right.

Based on the simulations it is concluded that the impact of timing jitter cannot be eliminated by varying the width of the SW. However, a Super Gaussian-shaped SW is better suited to reduce the impact of timing jitter than a Gaussian SW, so a squared shaped SW will be preferred when possible. The suggested design parameters for the demultiplexing switch are summarized in Table 17.1.

An example of a widely used practical implementation of an OTDM demultiplexer is shown in Figure 17.5. This Nonlinear Optical Loop Mirror (NOLM) [47] is capable of demultiplexing a 1.28 Tbit/s data signal. A NOLM is a nonlinear fiber-based Sagnac interferometer. A data pulse is split in two at the NOLM input, a clockwise (CW) and counter-clockwise (CCW) propagating pulse. By adjusting the internal polarization controller in the loop, one may assure destructive interference at the output of the NOLM. A short control pulse co-propagating with the CW data pulse will generate a phase shift on the CW data pulse, due to cross-phase modulation (XPM) in the highly nonlinear fiber (HNLF) in the loop. This phase change will alter the interference conditions at the input/output coupler and ideally give a constructive interference at the output-port. As XPM in HNLF is ultra-fast, the NOLM can be made to open and close on a sub-ps time scale. The NOLM in Figure 17.5 contains a 15 m short HNLF with a zero dispersion at 1544 nm and a slope of 0.015 ps/nm^2 km. With a short HNLF and a zero dispersion wavelength in between the control and data, walk-off is reduced to a negligible level (less than 50 fs). Furthermore, the short HNLF helps to enhance overall stability by reducing the amount of second-order PMD.

Table 17.1 Suggested design parameters for optical demultiplexing switches.

Shape	Min. ER (dB)	Optimum FWHM SW Width (ps)			
		160 Gbit/s	320 Gbit/s	640 Gbit/s	1280 Gbit/s
Gauss	20	4.38	2.19	1.09	0.54
Super Gauss	15	6.25	3.13	1.56	0.78

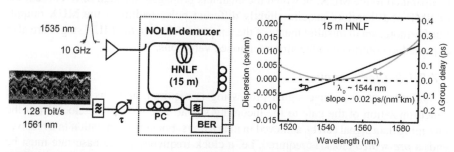

FIGURE 17.5 1.28 Tbit/s demultiplexer. A NOLM with 15 m HNLF with a dispersion minimizing the walk-off between the data and control pulses.

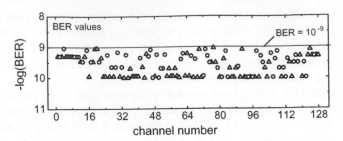

FIGURE 17.6 BER characterization of all 128 tributaries at 10 Gbit/s in the 1.28 Tbit/s data signal. All channels are below BER 10^{-9}. (After [11]. © 2012 IET.)

The sketched NOLM is used to demultiplex the generated 1.28 Tbit/s data signal shown in Figure 17.1. The demultiplexed data channels are characterized in terms of bit error rate (BER) performance. It is very important that there is as low a spectral overlap as possible between the control and data in the NOLM, which is challenging if the system is constrained to remain in the C-band and the spectra are about 15 nm wide each. However, this effect can be minimized by keeping the detrimental SPM broadening at a minimum by careful adjustment of the polarization. This is enabled by a short HNLF used in the NOLM, and by using a moderate control power to get a good SPM/XPM balance (in this case $\sim 0.7\pi$ XPM). The control power used is 15.2 dBm, and the data signal input power is 15.7 dBm.

In order to validate the integrity of the full 1.28 Tbit/s data signal, all 128 channels are characterized. Figure 17.6 shows the BER results for all 128 demultiplexed OTDM tributaries of the 1.28 Tbit/s data signal. The integrity of the 1.28 Tbit/s data signal is tested by verifying that error-free performance can be obtained for all channels after demultiplexing. This test is performed by lowering the attenuation before the receiver sufficiently to obtain a BER $< 10^{-9}$ for each individual channel. Note that the error detector decision level is not adjusted for minimum BER, but only until a BER $< 10^{-9}$ is detected. For all 128 channels, a BER $< 10^{-9}$ can be obtained, confirming the integrity of the 1.28 Tbit/s data signal. The variation in BER performance is attributed to the MUX, in which the channels propagate through different lengths of SMF and therefore acquire slightly different pulse widths at the MUX output. Furthermore, small misalignments of the temporal delays in the MUX stages are also expected to contribute to the aforementioned variations.

17.2.1.3 Clock recovery

In OTDM systems, active signal processing is required, and this in turn requires synchronization of the gate to the incoming signal. Recovering the clock from the incoming data signal can be achieved in many ways. For synchronization in a receiver end, a *pre-scaled* clock is required, i.e. a clock frequency at the base rate must be extracted. This base rate signal is then used as the control for the demultiplexer. The timing jitter of the recovered clock is a critical parameter. The relative timing jitter between the data and the demultiplexing switching window (determined by the recovered clock and the data) should not exceed 8% of the time slot at the specific OTDM line rate [48].

Ultra-high-speed clock recovery has proven to be exceptionally challenging, and only very recently was 640 Gbit/s reached allowing for full transmission system demonstrations [24]. In [24], filtering-assisted cross-phase modulation (f–a XPM) in an SOA was used, as in [25] at 320 Gbit/s. The fast red-shift is filtered out and used as an error signal for a feedback loop to lock on to. The demonstration of 640 Gbit/s clock recovery including full transmission was the first ever successful demonstration, and was obtained at the COBRA Institute at the Technical University of Eindhoven. This was soon followed by a second 640 Gbit/s demonstration at the Technical University of Denmark [40]. Here, a PPLN device was used, relying on the $\chi^{(2)}$-mediated process of sum-frequency generation, which is truly ultra-fast and not depending on any carrier recovery. Several other approaches have appeared since then, such as using a base rate electro-optical phase modulation on the 640 Gbit/s data signal [49] as demonstrated at Tohoku University in Japan, or using a fiber-based single parametric device in a balanced-detection scheme extracting an error-signal from the two sidebands of a FWM idler [50] as demonstrated at UC San Diego. Various other approaches to clock recovery have also been demonstrated over the years at lower bit rates, such as electro-absorption modulator-based mixing up to 320 Gbit/s [51–54], FWM in an SOA up to 400 Gbit/s [27], SOA-based interferometers such as the SLALOM/TOAD (Semiconductor Laser Amplifier in a Loop Mirror/Terahertz Optical Asymmetric Demultiplexer) up to 160 Gbit/s [55,56]. Electro-optical solutions have the advantage of using the local electrical clock directly in the mixing process, not needing to convert it to an optical clock for all-optical mixing. On the other hand, electro-optical switches are inherently slower, i.e. in terms of timing resolution, than for instance parametric all-optical switches, disregarding carrier-mediated nonlinear effects. So using PPLNs or HNLFs or the like allows for synchronization of very high bit rates. Adding a base rate phase mark on one channel and then filter that out for clock recovery has also proven viable at 650 Gbit/s and simultaneously allowing for channel identification [57], see Section 17.2.2.1.

In the following, some key techniques are described in more detail.

17.2.1.3.1 SOA-based clock recovery at 640 Gbit/s (first ever)

Figure 17.7 shows the schematic of the setup used in [24] to accomplish the world's first demonstration of clock recovery at 640 Gbit/s.

The carrier density of an SOA will undergo a rapid decrease as a response to an incoming pump pulse, giving rise to a fast red-shifted frequency chirp [58]. The carrier density will then partially recover due to carrier-carrier scattering and carrier heating [59] and then slowly fully recover by electrical pumping. The recovery part will give rise to a fast and slow blue-shift of a probe wavelength. By the use of clever optical filtering, the fast red-shift response can be emphasized. In the configuration in [24] as shown in Figure 17.7, the data pulses act as pump creating XPM on the clock pulses, and a narrow filter offset to the blue side of the clock is implemented.

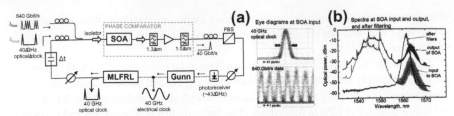

FIGURE 17.7 640 Gbit/s clock recovery using filtering-assisted XPM in a single SOA. Left: Schematic of experimental setup. Middle (a): Input waveforms to the SOA. Right (b): Spectra at the input and output of the SOA. (Adapted from [26]. © 2012 IEEE.)

The spectrum will be red-shifted ultra-fast, giving rise to a dip in transmission through the filter, and the partial-recovery blue-shift coinciding with the partial gain recovery gives rise to a fast closing of the dip. So every time a data pulse (640 Gbit/s) coincides with a clock pulse (40 GHz repetition rate) a fast change in transmission through the filter is observed, yielding a 40 GHz tone mixed with the data signal base rate. If the clock is in sync with the data, a clear 40 GHz tone is observed, if the clock is not in sync, the clock pulses will scan across the data and yield a cross-correlation of the data signal, i.e. a slowly varying sampled version of the data pulses. This can be used as an error signal. In the present configuration, the feedback loop acts like an injection locked loop, where the 40 GHz tone is filtered through a Gunn oscillator and fed to a mode-locked fiber ring laser (MLFRL). This laser now locks its repetition rate to the 40 GHz clock tone, and sends pulses via an appropriate time delay back into the SOA with the data pulses. The system was demonstrated to be able to extract accurately the 40 GHz clock of a 640 Gbit/s data signal and uses it for demultiplexing the data signal after transmission over 50 km of SMF-IDF fiber. In [25], a similar configuration was used, except that a red-shifted filter was employed on the clock pulses, and that a phase-locked loop (PLL) was used for feedback. This was demonstrated up to 320 Gbit/s using only III–V semiconductors for the active components in the loop, making it interesting for future integration to a single chip.

17.2.1.3.2 PPLN-based 640 Gbit/s clock recovery

The approach above using an SOA is interesting for integration reasons. However, in addition to the fast effects in semiconductors, they also have slow recovery times, which will inevitably lead to patterning effects for OOK modulation formats. To avoid this, a truly ultra-fast scheme would be preferable. Nonlinear wave mixing in lithium niobate is one such scheme, as proposed in [60]. In Figure 17.8, a schematic setup of a clock recovery scheme employing a PPLN device as the nonlinear optical mixer is shown. This scheme was used for the second ever demonstration of 640 Gbit/s clock recovery in a full transmission experiment [39,40].

The all-optical mixer here is based on sum-frequency generation in a quasi-phase-matched (QPM) adhered ridge waveguide periodically poled LiNbO$_3$ module (ARW PPLN). A semiconductor external cavity tunable mode-locked laser (TMLL) is used as

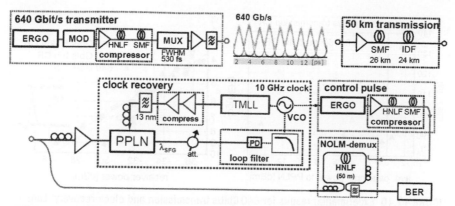

FIGURE 17.8 640 Gbit/s clock recovery using sum-frequency generation in PPLN. (After [39]. © 2009, IEEE.)

the local clock pulse source, and its pulses are compressed from about 1.5 ps to 700 fs FWHM by soliton compression in a two-stage high-power EDFA compressor, before being merged with the data pulses in the PPLN. The voltage controlled oscillator (VCO) drives the TMLL with a 10 GHz tone, and this same clock is used to drive an external cavity solid state mode-locked laser with an erbium glass as gain medium and a saturable absorber mirror as mode-locker (ERGO laser), which produces control pulses for the NOLM demultiplexer. The feedback loop is an active integrating PLL [61]. The data is transmitted over 50 km SMF-IDF with negligible residual dispersion and PMD.

The ARW PPLN used in this experiment, sketched in Figure 17.9, is designed for efficient processes from the 1550 nm wavelength range [62]. Active Mg:LiNbO$_3$ is set on a low-index adhesive, and together with a ridge structure this gives an optical waveguide with a tight optical confinement with a high index difference [62]. The ridge (2.5 μ high and 8 μ wide) ensures good modal overlap between light at the fundamental wavelength (1550 nm) and the second harmonic (780 nm), which will enhance the conversion efficiency. The length of the device is 30 mm and the QPM

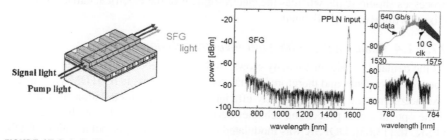

FIGURE 17.9 Left: The principle of SFG in the PPLN. Right: Spectral input and output to the PPLN. (After [39]. © 2009, IEEE.)

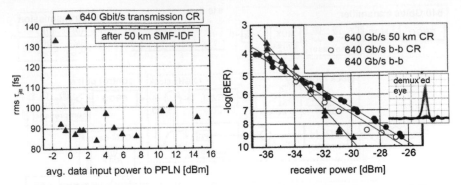

FIGURE 17.10 Experimental results for 640 Gbit/s transmission and clock recovery. Left: Dynamic range for low timing jitter. Right: Bit error rate performance. (After [39]. © 2009, IEEE.)

period is 17 μm, resulting in efficient sum-frequency generation (SFG) between the 1557 nm data and the 1567 nm clock at 782 nm. The normalized conversion efficiency for the 30 mm device is 350%/W for the packaged module, and 900%/W for the naked chip. The experimental spectra are shown in Figure 17.9 (right).

Figure 17.10 (left) shows the clock recovery locking performance in terms of rms timing jitter of the control pulses to the demultiplexer vs. average input data power. The average power of the clock pulses into the PPLN is kept at −4 dBm. The rms jitter is kept below 100 fs over a dynamic range of 16 dB from −1 to +15 dBm. The PPLN thus only requires 1-mW average 640-Gbit/s data power to operate properly. As seen in Figure 17.10 (right), the full 640-Gbit/s system gives error-free performance with no error floor and no pattern dependence and there is less than 1-dB transmission induced power penalty after 50-km fiber. Using the CR without transmission (i.e. b-b CR), a penalty of only 2.7 dB to the reference b-b is incurred, and using the CR after 50 km an additional penalty of only 0.8 dB is added.

In summary, there are various clock recovery schemes demonstrated today, working up to 640 Gbit/s. At present, there is still no 1.28 Tbit/s demonstration, but the PPLN scheme [39] or the mentioned HNLF-scheme [50] relying on truly ultra-fast processes should scale, whereas the electro-optical or the electrical carrier-mediated mixers will be more challenged.

17.2.1.4 Compatibility with multi-level modulation formats

As the demand for higher capacity grows, higher spectral efficiency is required. This goes for all systems, and therefore it is interesting to investigate if OTDM systems can also accommodate highly spectrally efficient higher order modulation formats, and if the optical signal processing schemes derived for OOK are also applicable to e.g. xPSK. In [13,63], DQPSK was modulated on a 1.28 Tbaud per polarization OTDM pulse train, and using pol-MUX this achieved 5.1 Tbit/s capacity on a single laser. This was for direct detection, but using coherent detection allows for even higher order modulation. Various advantages resulting from the combination of coherent detection with OTDM data include: compatibility with spectrally efficient modulation formats, high sensitivities and simultaneous demultiplexing, and demodulation of ultra-high-speed data signals. Detection of certain advanced modulation formats [64] requires the presence of a coherent receiver. Time division demultiplexing performed with a pulsed local oscillator (LO) in a coherent optical receiver was introduced in [65,66]. This principle was later introduced and demonstrated for synchronous digital coherent demultiplexing [65] and demodulation in a digital coherent receiver [67]. The utilization of an OTDM-coherent receiver for demultiplexing and demodulation reduces the bandwidth requirements for the optical-to-electrical conversion and the *A/D*-converters due to the reduced rate of the demultiplexed data signal.

The combination of sub-picosecond pulses, advanced modulation formats (e.g. 16-QAM), polarization multiplexing (pol-MUX), and digital coherent receivers, allows for the generation and transmission of up to 10.2 Tbit/s [14,68] on a symbol rate of 1.28 Tbaud per polarization. In this case, the narrow data pulses were generated in a 200 m DF-HNLF. Figure 17.11a shows the spectra of both data and local oscillator (LO) at the input of the 90°-hybrid (receiver side). Both spectra are Gaussian-like and have the same FWHM of 1.4 THz. The 10.2 Tbit/s signal was sent through 29 km dispersion managed fiber (DMF) composed of 19-km super large area fiber (SLA) and 10-km inversed dispersion fiber (IDF). The residual phase profile of the DMF is also shown in Figure 17.11a. The autocorrelation traces of the resulting 300 fs Gaussian-like pulses (both 10-GHz LO and 1.28-Tbaud 16-QAM signal) in

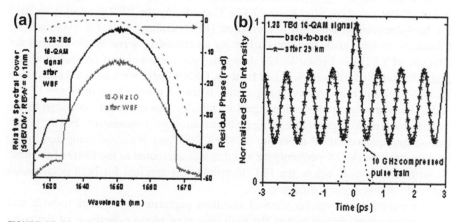

FIGURE 17.11 (a) Optical spectra of the compressed and filtered 10-GHz LO pulses and the 1.28-Tbaud 16-QAM data signal measured at the input of the 90°-hybrid as well as the residual phase profile of the 29 km transmission link. (b) Autocorrelation traces of the compressed 10-GHz LO pulse train (dashed) as well as the modulated and optically multiplexed 1.28-Tbaud RZ-16-QAM data signal (solid black line: back-to-back, red stars: after 29 km transmission). (From [14]. © 2012 IEEE). (Blue lines indicate polarization maintaining fibers). (For interpretation of the references to color in this figure legend, the reader is referred to the web version of this book.)

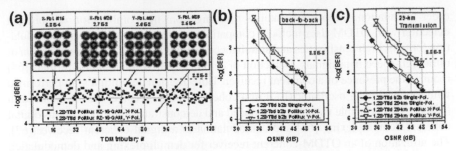

FIGURE 17.12 10.2-Tbit/s BER measurements: (a) All 256 tributaries (X and Y polarizations) of the 10.2-Tbit/s data signal in back-to-back configuration. The insets show constellation diagrams with BER for selected tributaries. (b) BER vs. OSNR for one tributary for 1.28-Tbaud single polarization (5.1 Tbit/s) and after polarization multiplexing (10.2 Tbit/s). (c) BER vs. OSNR for one tributary after 29 km transmission. (From [15]. © 2012 IEEE.)

back-to-back configuration and after transmission, measured at the balanced photo-detectors of the coherent receiver, are shown in Figure 17.11b. The autocorrelation traces in Figure 17.11b show negligible pulse broadening due to transmission.

Figure 17.12 shows the measured BER performance for the back-to-back and transmitted data signals. By scanning the time delay between the OTDM signal and the base rate LO it is possible to measure all 256 tributaries (both polarizations: X-Pol., Y-Pol.) as shown in Figure 17.12a. Considering the BER limit of 3.8×10^{-3} for hard-decision forward error correction (FEC) coding with 7% overhead, all tributaries perform error-free, thus achieving an error-free serial single-channel net data rate of 9.5 Tbit/s. Figure 17.12a shows also some examples of constellation diagrams for selected tributaries. A full BER vs. OSNR curve for a selected channel in back-to-back and transmission configuration is shown respectively in Figure 17.12b and c. For the single-polarization signal no penalty was observed compared to the back-to-back performance, indicating no degradation by transmission. For the polarization multiplexed signal a penalty of about 3 dB was observed compared to the back-to-back pol-MUX performance and this was attributed to the PMD of the link. Nevertheless, a BER below the FEC threshold was reached for both polarization states of the investigated tributary.

The experimental results showed excellent performance in back-to-back and after transmission, which proves the high degree of phase coherence of the compressed pulses.

17.2.2 Advanced network functionalities

In the above sections, point-to-point OTDM systems were assumed, i.e. where the OTDM data is simply generated and transmitted to one destination. It has also been speculated that it may be beneficial to use the individual OTDM tributary channels as routing/switching entities, much like in WDM systems using reconfigurable add/drop multiplexers operating on the individual wavelength tributary channels. The idea was to investigate if OTDM could deliver the same functionalities as WDM networks, i.e. use the OTDM data on a tributary basis. This would first of all require a clear distinction between the individual tributaries, and secondly methods to drop individual channels and add new ones. This would allow for the construction of a serial optical data bus (OTDM data signal), which could then add/drop individual channels when passing network nodes. This section will present some work on this topic, by addressing 640 Gbit/s channel identification, 640 Gbit/s add/drop multiplexing, and finally by addressing a scenario where Ethernet data packets are synchronized to a local node clock and made fully compatible with the OTDM data bus, and subsequently multiplexed into an OTDM data stream. This final scheme is shown to work for up to 1.28 Tbit/s serial data.

17.2.2.1 Channel identification and simultaneous synchronization

Figure 17.13 shows a scheme to obtain channel identification and simultaneous clock recovery. In this scheme, a small spectral feature is added to one optical TDM channel in the transmitter, and this is then simply filtered out in the receiver, allowing for easy identification of that particular channel. This scheme was demonstrated at 650 Gbit/s, i.e. where a marked channel was added to a 640 Gbit/s optical TDM signal [57]. This scheme should be scalable to Tbit/s speeds. The added channel could stem from an Ethernet packet, and several spectral marks could be envisaged to allow for individual labeling of the TDM data packets. As seen in Figure 17.13, a

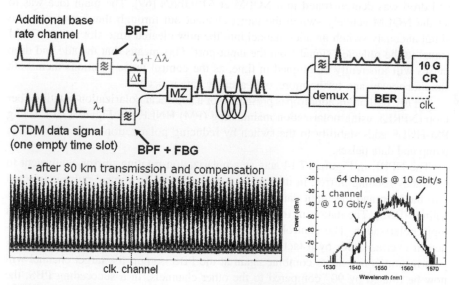

FIGURE 17.13 650 Gbit/s channel identification and simultaneous clock recovery by adding a small spectral feature to one TDM channel. More details in [57].

640 Gbit/s OTDM data signal is filtered with a fiber Bragg grating (FBG) notch filter to carve out a narrow hole in its spectrum. An additional base rate channel is aligned with a purposefully created 65th empty time slot and interleaved, thus creating a 650 Gbit/s OTDM signal. The added channel does not contain a spectral hole, and is also moved a little bit in central wavelength, so there is a good contrast at the hole wavelength between the added channel and the original other 64 OTDM tributary channels. In the receiver, this difference at the hole wavelength is easily detected, and the 10 GHz clock recovery in the receiver simply locks to the added channel. This channel is thus identified, and the remaining 64 channels are then simply known by their distance to the marked channel.

The system was demonstrated in a transmission experiment, where a Mach-Zehnder (MZ) filter was added to compensate for spectral tilt in the used EDFAs. SMF (80 km) and DCF (11 km) were used, and the channel identification and receiver synchronization worked very well, and all 65 channels were error free [57]. In fact, without transmission, there was negligible penalty of using the added channel as clock compared to using the transmitter's own clock signal, the receiver sensitivity of the added channel before and after transmission also had no penalty. So this scheme is quite robust.

17.2.2.2 Add/drop multiplexing

Once the individual tributaries are identified, one may start to consider individual treatment of the channels, such as adding or dropping channels. Add/drop multiplexing has been demonstrated using various switching materials, but at 640 Gbit/s only PPLN [38] and HNLF have been shown to operate. For instance, simultaneous add and drop was demonstrated in a NOLM at 640 Gbit/s [69]. The main idea was to let the NOLM actively switch the target channel out through the output-port and simultaneously switch an add-channel into the now cleared time slot in the OTDM signal to the reflected signal from the input-port. This means that the add and drop pulses will inherently be aligned in time, as the control pulse positions them both simultaneously, and furthermore, only a single gate is needed.

In [70], 640 Gbit/s add/drop is presented for a nonlinear polarization-rotating fiber loop (NPRL), using polarization maintaining (PM) HNLF, see Figure 17.14. Using PM-HNLF adds stability to the switch by reducing polarization walk-off between pump and data pulses.

The NPRL in Figure 17.14 acts as a polarization-rotating mirror, equivalent to a Faraday rotating mirror, in its passive state, i.e. when no control pulse is injected into the loop and the input data signal is aligned to 45° on the polarization beam-splitter (PBS). The data sent into the loop emerges again at 90° with respect to its input polarization. The control pulse will now delay the target data pulse component it overlaps with by a factor of π or 180°. When the two counter-propagating polarization components combine again at the PBS output, the target channel will now be rotated by 90° compared to the other channels. In a succeeding PBS, the target is separated from the other channels, leaving an empty time slot, where an add channel may be inserted by a passive coupler. Figure 17.14 shows cross-correlations

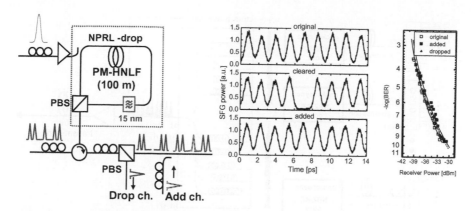

FIGURE 17.14 Example of a 640 Gbit/s add/drop multiplexer using a nonlinear polarization-rotating fiber loop (NPRL). The original, the dropped target and the added channel have similar BER performance.

of the original data, the cleared time slot and the data with the added channel. In the cleared time slot the dropped channel is clearly very well suppressed, which is the key to successful operation. The suppression ratio is on the order of 20 dB, which is adequate for 640 Gbit/s operation [69,70]. Using flat-top control pulses is also useful, as discussed in [70] in order to clear the whole time slot. Flat-top pulses for optical signal processing are discussed more in Section 17.2.3.1.

17.2.2.3 Synchronization of 10 Gbit/s Ethernet packets and OTDM

With channel identification and add/drop multiplexing in place, OTDM networking emerges as a possibility, where the network structure may be designed so as to take advantage of the serial nature of the OTDM data and the direct access to the individual bits.

For instance, it may be beneficial to create a high-speed optical serial data bus to carry Ethernet data packets (frames) in optical TDM time slots, that is, multiplex Ethernet packets into time slots by bit interleaving. Ethernet is basically a TDM-based technology, and when considering the need for future Tbit/s Ethernet [5] and the Tbit/s capacity encountered in massive data centers, it makes sense to explore the potential of optical TDM in conjunction with Ethernet. Figure 17.15 shows an Ethernet compatible optical TDM scenario. As Ethernet packets are asynchronous, and optical TDM systems are very synchronous, it becomes important to synchronize the data packets to a local master clock. This can be carried out using the concept of a time lens [71] by which the packet may be stretched or compressed to fit with the local clock frequency. The heart of this scheme is thus to use the optical time lens effect derived from the space-time duality [71] to slightly stretch or compress the asynchronous Ethernet data packets to synchronize them to a local master clock. When synchronized they may be NRZ-RZ converted, for example, through optical

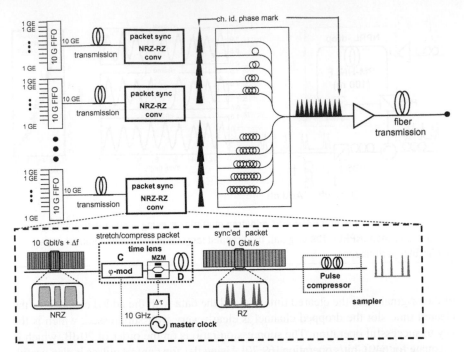

FIGURE 17.15 Envisaged Ethernet compatible optical TDM system with a packet sync-and-mux scheme. Each 10 GE packet is synchronized to the local master clock via a time lens based-sync-unit, essentially stretching or compressing the data packet to fit the individual bits to the local rep-rate. The synchronized bits in the packet are simultaneously individually converted to RZ pulses by pulse carving in the MZM and chirping in the time lens. The synchronized RZ bits are subsequently time domain multiplexed. Thus each packet is ascribed its own time slot. Time slot *N* may also be marked with a channel ID, e.g. by a slight phase modulation, allowing for easy channel identification and sub-rate clock recovery at the receiver.

sampling or pulse compression, and subsequently multiplexed together with other Ethernet packets or optical TDM channels. In this scheme, each Ethernet packet would be assigned an OTDM time slot.

In order to interface between asynchronous Ethernet networks and bit-interleaved synchronous high-speed TDM networks, a number of challenges need to be addressed, such as packet sizes varying from 64 to 1518 bytes (according to Ethernet standard IEEE 802.3), repetition rate variations, timing jitter reduction, non-return-to-zero (NRZ) to return-to-zero (RZ) format conversion, and data pulse compression. According to the protocol of 10 GE WAN PHY, the repetition rate of each Ethernet packet can vary with up to ± 20 ppm of the nominal transmission rate, i.e. ± 200 kHz frequency offset between transmitter and receiver must be tolerated [72,73].

Using the simplified time lens configuration of Figure 17.15, an asynchronous 10 Gbit/s Ethernet packet with the maximum standardized size of 1518 bytes can be synchronized and retimed to a master clock with 200 kHz frequency offset, format converted from RZ to NRZ, and finally optical time division multiplexed (OTDM) with a serial 1.28 Tbit/s RZ-OOK signal having a vacant time slot [74].

The concept of a time lens stems from the time-space duality, which refers to the analogy between the paraxial diffraction of beams through space and the dispersion of narrowband pulses through dielectric media in time [71]. Since a spatial lens can be used to obtain the Fourier transform of a spatial profile at the spatial focus, a time lens can also be used to obtain the Fourier transform of a temporal profile at the temporal focus. In Fourier analysis, any time shift or timing jitter only changes the phase in the frequency domain but does not change the envelope, which can be expressed as $x(t - n\Delta T - \delta t) \leftrightarrow X(\omega)e^{-j\omega(n\Delta T + \delta t)}$. The asynchronicity of the incoming packet can be viewed as a time wandering or time shift of the packet pulses relative to the local master clock. Hence, after the Fourier transformation, these time shifts can be transferred into the frequency domain. If we only detect the envelope of the electrical field in the time domain but discard the phase of the electrical field, the time wandering or time shift can be removed. Therefore, we can use the time lens to cancel the time wandering caused by the asynchronous nature of Ethernet packets and also to reduce the timing jitter. As we will see in later sections, this concept, also referred to as the time domain optical Fourier transformation (OFT) [75], can also be used to perform time-to-frequency conversion (OTDM-to-WDM conversion [36]), frequency-to-time conversion (WDM-to-OTDM conversion [76]), pulse shaping for e.g. flat-top pulse generation [77] or for cleaning up distorted pulses due to dispersion in a transmission link [75].

The time lens in this experiment consists of a cascaded phase modulator and Mach-Zehnder modulator (MZM) followed by a piece of fiber as the dispersive element, as shown in Figure 17.15. The phase modulator is driven by a sinusoidal signal, which locally approximates a quadratic phase modulation. The MZM is used to remove the part of the waveform subjected to the lower part of the sinusoidal phase modulation (corresponding to positive chirp) and to only keep the waveform part overlapped with the upper part of the sinusoidal phase modulation (corresponding to negative chirp), as sketched in Figure 17.16a. The dispersive element (dispersion compensating fiber (DCF) in this experiment) provides the temporal focus in the system. In addition, all the bits in the packet experience negative chirp and can be compressed into short pulses in the DCF, which in turn allows for converting an NRZ signal into an RZ signal.

Figure 17.17 shows results on packet sync-and-MUX up to 1.29 Tbit/s [74]. In this case, the 10 G input asynchronous Ethernet packet with a data rate of 9.9534 Gbit/s is converted into a synchronized Ethernet packet with a data rate of 9.9536 Gbit/s. At the same time, the Ethernet packet is format converted into an RZ signal with a full width at half maximum (FWHM) of 6 ps. The converted RZ signal is further pulse compressed to a FWHM of 400 fs in a 500 m dispersion-flattened highly nonlinear fiber. Figure 17.17 (top left) shows the original NRZ eye diagram and the compressed

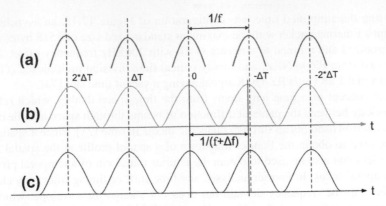

FIGURE 17.16 Operation principle of the time lens-based sync-unit. (a) Phase modulation. (b) Packet data pulses before synchronization. (c) Packet data pulse after synchronization.

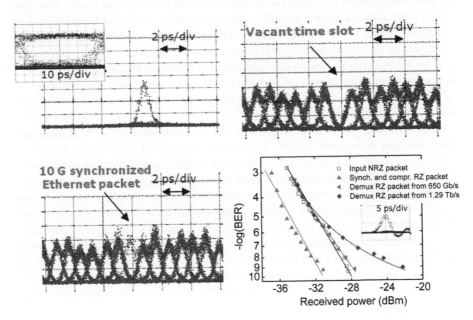

FIGURE 17.17 10 Gbit/s Ethernet packet synchronization and multiplexing to 1.29 Tbit/s. Top left: Original NRZ (inset) and synchronized RZ pulse compressed Ethernet data signal. Top right: Prepared OTDM signal with a vacant time slot. Lower left: Multiplexed signal with synchronized Ethernet channel. Lower right: BER performance. (Adapted from [74]. © 2011 OSA.)

RZ pulse of the synchronized data packet. The latter is measured on an optical sampling oscilloscope with a 900 fs timing resolution. In parallel, a 1.28 Tbit/s OTDM data signal is generated, but with the time slot moved slightly so as to make room for an additional vacant time slot (Figure 17.17 (top right)). The synchronized and compressed Ethernet packet is now interleaved with the OTDM data signal, in the vacant time slot, thus creating a 1.29 Tbit/s OTDM signal (Figure 17.17 (lower left)). The optical sampling oscilloscope traces portray the data pulses wider than they really are, due to its limited resolution, and the data pulses do not really overlap as much as it appears here. To characterize the data quality, the added packet data is characterized in terms of bit error rate. In the receiver, synchronized to the master clock, a NOLM is used to demultiplex the 10 G Ethernet packet from the high-speed serial data stream. The control pulse is at 1533 nm and has a pulse width of 470 fs. Finally, the demultiplexed 10 G Ethernet packet is detected by a 10 Gbit/s receiver and measured by an oscilloscope and an error analyzer, which are both triggered by the master clock.

BER is measured for the input NRZ packet, the synchronized and compressed RZ packet and the demultiplexed RZ packet from the aggregated 1.29 Tbit/s OTDM signal and also from the aggregated 650 Gbit/s OTDM signal. Compared to the input NRZ packet, the synchronized and compressed RZ packet has a 3.1 dB negative power penalty which is the expected benefit from the NRZ-to-RZ format conversion. Compared to the synchronized RZ packet, the demultiplexed RZ packet from the aggregated 650 Gbit/s OTDM signal and 1.29 Tbit/s OTDM signal has an additional power penalty of 3.8 dB and 8.7 dB at the BER of 10^{-9}, after the multiplexing and demultiplexing. This is due to some pulse pedestal from the compressed Ethernet pulses, which overlaps with the neighboring channels and creates some intersymbol interference. These types of pedestals may be removed by various tricks in later generations of the experiment.

In summary, we have described how an Ethernet packet with the maximum standardized packet size of 1518 bytes can be synchronized and retimed to a master clock with 200 kHz frequency offset (corresponding to 20 ppm) and at the same time be format converted from NRZ to RZ and subsequently pulse compressed and multiplexed in time with a 1.28 Tbit/s or a 640 Gbit/s OTDM signal having a vacant time slot. The scheme does not require any packet clock recovery, although an initial alignment between the incoming packet and the master clock is required, and could be achieved in practice by using a packet envelope detector. Error free performance of synchronizing, retiming, multiplexing with a 1.28 Tbit/s or 640 Gbit/s OTDM signal and finally demultiplexing back to 10 Gbit/s of this Ethernet packet is achieved.

17.2.3 Impairment-tolerant switches

OTDM signals at very high bit rates are very sensitive to many timing effects, such as pulse waveform distortions due to dispersion in transmission fiber, phase noise and timing jitter, and synchronization between data and control signals. The higher the bit rate, the stricter the requirements become. Furthermore, as very high bit rate OTDM relies on optical signal processing, where polarization is often an issue, polarization control thus also becomes important. This section will describe some methods that have been demonstrated to increase the tolerance to these impairments, such as flat-top control pulses for increased timing jitter tolerance (up to 30% of the time slot) and polarization-independent demultiplexing (and wavelength conversion).

17.2.3.1 Flat-top pulses for timing jitter tolerant demultiplexing

The introduction of a switching mechanism which is tolerant to timing jitter is highly desirable. In all-optical demultiplexing the switching mechanism is implemented by combining the OTDM data signal with a base rate control signal. In these systems the shape of the control pulse determines the shape of the switching window. To increase the tolerance to timing jitter in ultra-fast optical switches, flat-top pulses have been demonstrated to be very beneficial [78,79]. Different approaches have been proposed for the generation of these types of pulses, such as long-period grating (LPG) filters [80], electro-optic sinusoidal phase modulation [81], super-structured fiber Bragg gratings (SSFBG) [82], or frequency-to-time conversion via the optical Fourier transformation (OFT) [77] technique.

Figure 17.18a shows the cross-correlation trace of a sub-ps FWHM flat-top pulse generated via the LPG filter technique [80]. This pulse was used as control in a 640 Gbit/s NOLM demultiplexing experiment in order to compensate the data timing jitter. By changing the relative time delay between control and data, it is possible to measure the timing tolerance of the demultiplexer. Figure 17.18a also shows a cross-correlation of the 640 Gbit/s data signal. It is clear that the flat-top part of the control pulse overlaps with most of the data pulse, and hence it is expected that moving the control pulse with respect to the data will be more tolerant than using a narrow Gaussian-shaped control pulse. The flat-top pulse determines the shape of the

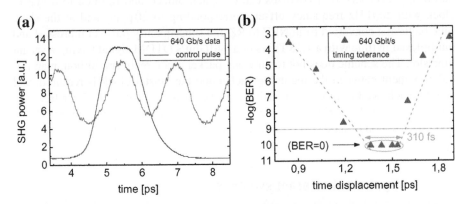

FIGURE 17.18 (a) Cross-correlation traces of the 640 Gbit/s OTDM signal and the flat-top pulse used to demultiplex the data signal. (b) Measured BER timing tolerance for 640 Gbit/s operation. (After [78]. © 2008, IEEE.)

switching window in the NOLM, since the used NOLM has negligible walk-off. As seen in Figure 17.18b, this switching window gives rise to a range of 310 fs, where the BER is $<10^{-9}$ [78]. This is about three times higher tolerance to timing jitter than the 8% of the time slot requirement generally imposed on Gaussian pulses.

17.2.3.2 Polarization-independent switching

The state of polarization (SOP) of transmitted data signals fluctuates in time due to varying fiber birefringence and/or temperature fluctuations. For this reason a polarization-independent (PI) switch, which can process the data signal directly after transmission, with identical performance for any data input SOP, is highly desirable. Depending on the way the PI-switches operate they can be grouped in (a) switches where the PI operation is *inherent* to the switching principle, and (b) *polarization diversity switches*, where the orthogonal components of the incoming data SOP are separately processed. For these two types of switches, both fiber- and SOA-based [83–85] solutions have been demonstrated.

A switch is considered inherently PI when a splitting of the incoming data pulse into its orthogonal polarization components is not required to obtain a PI operation. This type of PI switch can be achieved by exploiting the evolution of the data SOP and/or control signal SOP throughout the nonlinear medium. One example of an inherent PI operation based on XPM is the PI-NOLM [86] operation with flat-top control pulses [84], the latter generated in an interferometric LPG pair, operating in partial destructive interference. The principle of the PI-NOLM is based on the cancelation of the polarization-dependence of the XPM phase shift by careful use of the periodic power transfer function of the NOLM [86]. This requires no structural modifications of the device itself, and is achieved solely through adjustment of the power level and the polarization state of the control pulses. Figure 17.19 shows the BER measurement for the standard and PI-NOLM operation, in which the SOP of the data signal (640 Gbit/s RZ-OOK) is randomized at high speed (113 kHz) in a

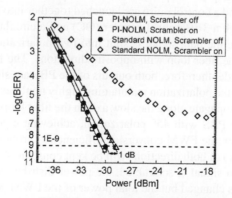

FIGURE 17.19 BER curves for 640 Gbit/s demultiplexing for the standard and PI-NOLM operation. (From [84]. © 2010 IEEE.)

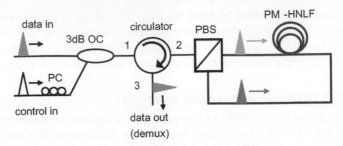

FIGURE 17.20 Polarization diversity optical switch using a polarization-maintaining highly nonlinear fiber (PM-HNLF) loop with bi-directional operation. (Blue lines indicate polarization maintaining fibers.) (For interpretation of the references to color in this figure legend, the reader is referred to the web version of this book.)

polarization scrambler. The standard NOLM operation (polarization sensitive operation) is reached when the control power is 26.5 dBm. When the control power is increased to 27.6 dBm and the control polarization is optimized then the NOLM starts working in PI mode. As it can be seen from Figure 17.19, when the scrambler is off, the power sensitivity (at BER $= 10^{-9}$) is the same for both cases. When the scrambler is turned on, the standard NOLM operation exhibits an error floor above BER $= 10^{-7}$. In contrast, the PI-NOLM shows merely a power penalty of 1 dB.

In polarization diversity demultiplexers, the two random orthogonal polarization components of the incoming data SOP are split, separately switched, and then recombined to obtain the demultiplexed data signal. The two SOP components are switched by separate control pulses but with equal control intensities giving the device the PI property. The switching occurs either by cross-phase modulation (XPM) or four-wave mixing (FWM). Even though these effects are still inherently polarization dependent, the diversity scheme guarantees the overall PI operation.

The operation principle of the polarization diversity switch is shown in Figure 17.20 [83]. A data signal and a control pulse are launched together into a polarization beam splitter (PBS) through a 3 dB optical coupler (OC) and a circulator. In the polarization diversity switch, the data signal is split into two polarization components that propagate through the fiber loop with opposite directions. The fast axis of the PBS is rotated by 90° inside; therefore, both outputs of the PBS are slow axis and aligned with the slow axis of the polarization maintaining highly nonlinear fiber (PM-HNLF). All signals always propagate along the slow axis in the fiber loop. If the control pulse is launched into the PBS with 45° polarization, achieving equal intensity in both directions of the loop, the FWM conversion efficiencies (proportional to the square of the pump intensity) of both directions will be kept constant. When the polarization of the incoming data signal is scrambled, the power distribution of the data signal at the PBS outputs is changed but the total power of the FWM products will be kept constant. The switched signal is recombined in the PBS and output from port three of the circulator.

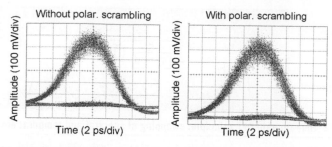

FIGURE 17.21 Eye diagrams of the switched signal with and without polarization scrambling.

The power fluctuation of the switched signal is monitored with a polarization scrambled incoming data signal. The maximum power fluctuation is less than 0.2 dB, effectively demonstrating the polarization insensitivity of the switch. The eye diagrams of the demultiplexed 10 Gbit/s from the 640 Gbit/s data signal with and without polarization scrambling of the incoming data signal are shown in Figure 17.21. The eye diagrams show negligible difference irrespective of whether the scrambler is on or off. The BER performance confirms that there is negligible polarization dependence using this scheme [83], which is also seen for wavelength conversion, see Section 17.2.4.2.

This section has described various approaches to overcome some of the most severe limitations imposed on systems operating with ultrashort pulses. There are thus demonstrated schemes to increase the tolerance to timing jitter and make polarization-independent switches.

17.2.4 Processing of all bits in one device—conversion-type functionalities

In the previous sections, we mostly considered the processing of a single tributary at a time. This implies that for all N tributary channels to be processed, N parallel switches are required. This can result in high power consumption for a whole OTDM system, see Section 17.4. However, for some functionalities, all channels are processed simultaneously in a single device. Such functionalities thus offer energy-efficient systems. In this section, we will address four such functionalities, namely serial-to-parallel conversion (OTDM-to-WDM conversion), wavelength conversion, regeneration, and packet switching.

17.2.4.1 Serial-to-parallel conversion

As mentioned in Section 17.2.2.3 above, the optical time lens effect or the time domain optical Fourier transformation (OFT) technique is a very versatile tool, and it was shown to enable synchronization of optical data packets.

The basic principle of the OFT scheme employed in Section 17.2.2.3 is shown in Figure 17.22. The output time domain waveform in this configuration is the Fourier transform of the input waveform, i.e. the spectral envelope shape at the input [87].

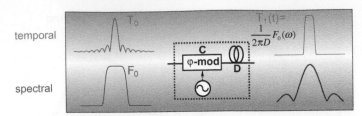

FIGURE 17.22 Time-domain OFT principle: swapping of spectral and temporal envelopes.

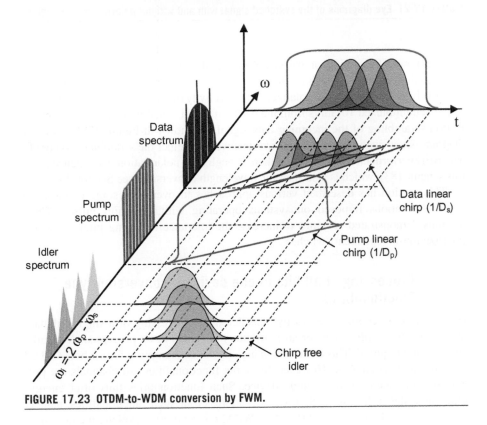

FIGURE 17.23 OTDM-to-WDM conversion by FWM.

Thus, if a rectangular-shaped spectrum is generated, it may readily be transformed to a rectangular (flat-top) pulse, as described in Section 17.2.3.1. Reversing the phase modulation and the dispersion also allows for temporal to spectral transformation. One application of this is to transform the serial OTDM data signal into separate parallel WDM channels. For an efficient solution, the electro-optic phase modulation is exchanged with an optical phase modulation, which may be provided in various ways. Using four-wave mixing (FWM) allows for very strong phase modulations. Figure 17.23 shows a schematic of OTDM-to-WDM conversion using FWM.

The data and pump are first dispersed in standard single-mode fiber (SMF) to create a linear chirp on the pulses. The rectangular pump spectrum is sent through a fiber of dispersion D, starting at its Fourier transform position. After D, it will have acquired a flat-top waveform. To obtain OTDM-to-WDM conversion with spectral compression of the individual WDM channels, allowing for DWDM creation, the data pulses must traverse a fiber with dispersion exactly equal to half of the pump's dispersion, i.e. $D/2$. Now, these two signals are injected into a nonlinear medium, e.g. a HNLF, where they interact via FWM. The pump and data signals create an idler at the optical frequency $\omega_i = 2\omega_p - \omega_s = 2\pi c/\lambda_i$. Because the pump is chirped and covers all the OTDM channels, different channels will overlap in time with different parts of the pump spectrum and thus give rise to idlers at different wavelengths. And because the data is chirped with $D/2$ compared to the pump being chirped by D, each OTDM pulse will be converted to an idler with a compressed spectrum. This makes it possible to place the idlers very densely, i.e. to create a DWDM signal. Each DWDM channel now originates from a specific OTDM channel, and they may simply be filtered out by passive filters. In this way, all OTDM channels can be demultiplexed simultaneously.

Figure 17.24 shows the basic schematic setup for OFT-based OTDM-to-WDM conversion. In principle all OTDM channels can be demultiplexed in one device [88], but in practice it may be difficult to avoid cross-talk among those data pulses which overlap with two neighboring pump pulses. It is therefore more practical to assume that two of these OFT-based demultiplexers or OTDM-DWDM converters in parallel are needed to demultiplex all channels. Still, this is potentially a great reduction in active components in an OTDM receiver. We will see in Section 17.4 that this gives rise to a significant power consumption reduction, rendering an OTDM point-to-point system less power consuming than a WDM point-to-point system for high bit rates.

Figure 17.25 shows results from an experimental demonstration where a 640 Gbit/s OTDM signal is converted into a DWDM signal with only 25 GHz spacing

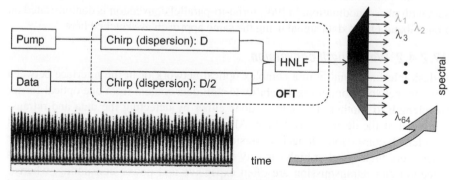

FIGURE 17.24 Schematic setup for OTDM-to-DWDM conversion using the optical Fourier transformation technique (OFT) by four wave mixing in a HNLF. In practice, two OFT units would be preferred.

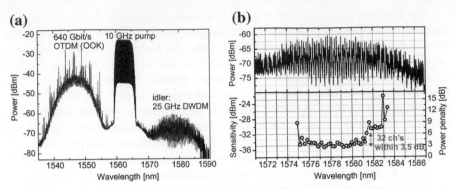

FIGURE 17.25 Experimental demonstration of 640 Gbit/s OTDM-DWDM conversion by OFT using FWM in HNLF. (a) Output spectrum from the HNLF showing the OTDM data, the flat-top pulse and the DWDM idler channels. (b) Zoom in on the DWDM idler spectrum (top) and the receiver sensitivity and power penalty at BER 10^{-9} for the demultiplexed channels (bottom). (After [89]. © 2011 OSA.)

between the DWDM channels [36,89]. Forty of the 64 tributaries are simultaneously demultiplexed with a BER less than 10^{-9}. The channels to the sides of the spectrum correspond to the OTDM channels overlapping with the edges of the pump pulse, and thus have less conversion efficiency, and ultimately lower OSNR, and therefore higher penalty. The most extreme channels cannot get a BER less than 10^{-9}. The receiver sensitivity at BER 10^{-9} is measured for the 40 center-most channels and 32 of these have a sensitivity spread within 3.5 dB. This clearly demonstrates the effectiveness of this scheme. With only two of such OFT units in parallel using temporally detuned pumps, all OTDM channels should be attainable, irrespective of the OTDM bit rate.

As the phase modulation is obtained by FWM, this scheme is also applicable to data with phase modulation. In [90], serial-to-parallel conversion is demonstrated on a 640 Gbaud QPSK data signal using the same approach as sketched here.

17.2.4.2 Wavelength conversion

A desirable functionality in a complex optical network or in the interface between different network domains is the ability to shift data content from one optical carrier wavelength to another. This can be done either by detecting the data and retransmitting it at the desired wavelength. Alternatively it can be performed by optical wavelength conversion. Optical conversion might be desirable especially in the case where very high speed or high capacity data channels are concerned. In these cases detection and retransmission are often expensive and power consuming, whereas optical wavelength conversion can be made transparent to bit rate and/or modulation format. In this section a number of approaches to optical wavelength conversion will be highlighted.

Optical wavelength conversion has been the topic of much interest for numerous years and a variety of physical effects have been demonstrated for optical wavelength conversion. Furthermore, several platforms or materials have been investigated both in the form of fibers and planar waveguides. In this section we will focus on systems and demonstrations relevant to high-speed optical signals.

One of the most popular platforms for performing optical wavelength conversion is HNLF. This is a relatively mature technology which has several important benefits for optical signal processing. HNLF typically has very low propagation loss, less than 1 dB per km, and splicing to standard single-mode fiber can be done with almost no loss. This allows for a very long effective interaction length between optical signals in HNLF. Consequently, conversion efficiencies are typically higher in wavelength converters implemented in HNLF compared to other, more nonlinear but also more lossy platforms for wavelength conversion.

In HNLF, wavelength conversion by e.g. four-wave mixing (FWM) has been applied in a polarization diversity scheme allowing for successful conversion of a polarization multiplexed 1.28 Tbit/s OTDM data signal [91]. This is the highest single wavelength channel wavelength converter demonstrated to date. Figure 17.26 shows the scheme for polarization diversity FWM wavelength conversion. The CW pump is aligned 45° to the main axis of the polarizing beam splitter (PBS) and is thus propagating in both directions through the polarization maintaining HNLF (PM-HNLF). Each polarization of the data signal carrying 640 Gbit/s is thus wavelength converted in opposite propagation directions through the PM-HNLF and the wavelength converted output can be extracted at the circulator output. This is the same scheme as discussed in Section 17.2.3.2.

Figure 17.27 shows the BER results for this polarization diversity wavelength scheme. Less than 0.5 dB polarization sensitivity is obtained. Error free operation for both the wavelength converted 640 Gbit/s RZ-DPSK signal and the wavelength converted 1.28 Tbit/s pol-MUX RZ-DPSK signal are achieved. As discussed in Section 17.2.3.2, polarization independence is very attractive, and now several functionalities are demonstrated.

FWM in a dispersion optimized sample of HNLF has been demonstrated in [92] for multicasting of 320 Gbit/s OTDM data onto a total of eight copies at different wavelengths. In this case, a large FWM bandwidth had been achieved by suppressing dispersion fluctuations through the length of the HNLF by precisely applied strain to

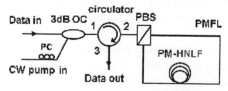

FIGURE 17.26 Scheme for polarization-independent polarization-diversity wavelength conversion.

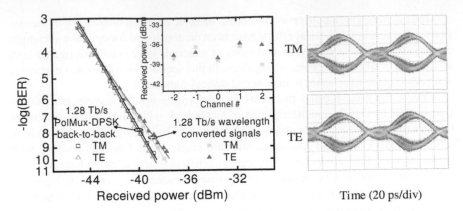

FIGURE 17.27 1.28 Tbit/s polarization multiplexed-polarization independent wavelength conversion. (After [91]. © 2009 OSA.)

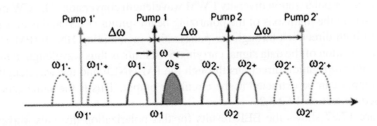

FIGURE 17.28 One-to-eight multicasting in a two pump FWM wavelength converter. (From [92]. © 2009 IEEE.)

segments of the fiber. The multicasting was achieved in a two-pump FWM scheme as illustrated schematically in Figure 17.28.

Wavelength conversion by cross-phase modulation (XPM) between an amplitude modulated data signal and a continuous wave (CW) probe signal in an HNLF has been successfully demonstrated up to 640 Gbit/s single polarization OTDM data [93]. The scheme is shown in Figure 17.29 where an optical filter is used to extract the wavelength converted data signal generated at one of the sidebands to the CW probe.

Figure 17.30 shows experimental results for the Raman-enhanced XPM in a HNLF scheme operating at 640 Gbit/s. The performance is error free with a modest power penalty of less than 3 dB.

A key enabler to obtain ultra-high bit rate conversion is to use sharp notch filters to suppress the CW signal out of the HNLF.

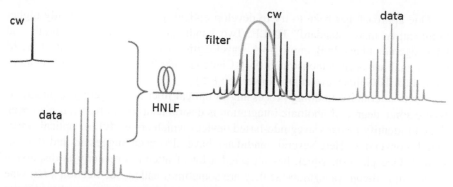

FIGURE 17.29 Wavelength conversion by XPM. Pumping the HNLF with a Raman pump additionally increases the conversion efficiency.

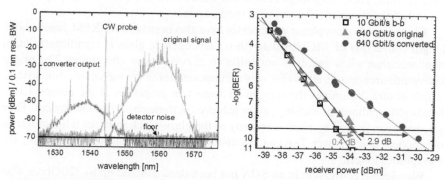

FIGURE 17.30 640 Gbit/s wavelength conversion based on Raman-enhanced XPM in HNLF. (After [93]. © 2008 IEEE.)

HNLF has several significant advantages as a platform for wavelength conversion; however, it also has certain limitations. The silica glass which is the main ingredient in the HNLFs used in the demonstrations mentioned above has a fairly low optical nonlinearity, which can be somewhat enhanced by designing fibers with high confinement of the optical field. Still, in the demonstrations above many meters of HNLF are required to achieve sufficient nonlinear interaction. This in turn creates challenges regarding dispersion and polarization control through such long fibers. The nonlinear effects used for wavelength conversion are typically sensitive to the polarizations of the interacting fields and to dispersion properties of the nonlinear medium—HNLF in the cases above. A large effort has been done by many research groups to try to identify other nonlinear platforms to supplement HNLF for applications where the limitations of HNLF become critical.

One approach has been to try to develop optical fibers with significantly higher nonlinearity than "standard" HNLF. Two candidates here are the Bismuth and Chalcogenide fibers. In Bismuth fiber, wavelength conversion of an 80 Gbit/s OTDM data signal has been demonstrated [94]. Chalcogenide fiber has so far been demonstrated for 10 Gbit/s wavelength conversion [95].

As an alternative to doing processing in optical fiber and targeting applications where some degree of photonic integration is desirable, a large effort has also been done to identify planar waveguide-based devices which are useful for optical wavelength conversion. Here several candidates have also been suggested and demonstrated. One platform which has attracted a lot of attention recently is dispersion engineered silicon waveguides as they are sometimes called, see Section 17.3. One clear attraction of silicon waveguides is the possibility to perform wavelength conversion over record high bandwidths, as demonstrated in [96], where a 40 Gbit/s data signal was converted over 100 nm and in [97], where a 10 Gbit/s signal was converted over 700 nm. These very large conversion bandwidths are enabled by very careful control of the chromatic dispersion in the Si devices.

Chalcogenide glass planar waveguides have also been used for XPM-based wavelength conversion of 40 Gbit/s data [98]. Chalcogenide glass is significantly more nonlinear than silica and does not suffer from two photon absorption. The planar waveguide architecture allows for very accurate control of the waveguide dimensions, which in turn determines the waveguide dispersion properties of the device. Large bandwidth nonlinear processing can be achieved in these chalcogenide devices. They are, however, limited by photosensitivity of the glass which imposes a limitation on the applied optical power. This can to some extent be compensated for by making the waveguides longer.

Wavelength conversion in an SOA has been demonstrated up to 320 Gbit/s with the inherent potential of offering gain to the process. In [99], a combination of cross gain modulation (XGM) and filtering-assisted XPM was used for wavelength conversion. Quantum dot SOAs have also recently become available, and they are very interesting, due to their inherent fast receovery compared to bulk SOAs. In [100], 320 Gbit/s wavelength conversion using FWM in an SOA was demonstrated.

PPLNs have also been demonstrated for wavelength conversion up to 320 Gbit/s based on sum-frequency generation (SFG) and difference-frequency generation (DFG) [101]. In [102], a 320 Gbit/s DQPSK signal was converted using SHG and DFG. Equivalently, yet more challenging and more impressive 640 Gbit/s regeneration using PPLN has also been demonstrated, see Section 17.2.4.3 [37].

17.2.4.3 Regeneration

Using optical regeneration, it is possible to extend the transmission reach of a data signal. It is also possible to expand the information spectral density for a link with a given transmission reach, ultimately allowing for reaching the capacity limit given by Shannon's theorem [103–105] when limited by nonlinear effects in the transmission fiber and ASE noise. This goes for both binary direct detection systems as well as coherent detection systems [105]. A challenge for regenerators is to regenerate multiple wavelength channels without having to split up the individual wavelengths and regenerate them separately, as the latter holds no clear advantage over o/e-3R-e/o. For regenerators, the benefit in terms of energy efficiency comes when many bits are processed simultaneously in the same device, i.e. when fewer power hungry components are required [106]. This could be the case for very high bit rate signals [106] such as multi-100s Gbit/s OTDM signals [107] or for high-order modulation formats such as 16 QAM or 16 PSK [108]. Recently, a number of impressive schemes for regenerating optical phase modulated signals have been conceived and demonstrated, e.g. [108–110], and methods for scaling these to higher order formats are under way. However, over the last 10 years, OTDM regeneration has also been demonstrated in various forms, and merging the two efforts of high symbol rate and high-order modulation regeneration may double the benefit in terms of energy efficiency. As most demonstrated phase-sensitive regenerators rely on parametric effects in highly nonlinear fiber, this should also be applicable to ultra-high-speed OTDM data signals carrying higher order modulation. An example of a parametric effect used for QPSK on a high symbol rate is described in [90], where serial-to-parallel conversion of a 640 Gbaud signal is demonstrated using FWM. As for regeneration, so far only a single demonstration has reached 640 Gbit/s (640 Gbaud), and this is based on the use of a PPLN device [37]. Again, an ultra-fast parametric effect is used, namely sum-frequency generation with pump depletion under the quasi-phase matching condition. In [37], 3R regeneration of a 640 Gbit/s OOK data signal is demonstrated. A 640 GHz clean clock signal is used to generate a SFG product together with the data signal. This process requires one photon from each, and hence both the clock and the original data signal will experience pump depletion. Filtering out the output clock signal, a logically inverted replica of the original data is obtained on a clean clock signal. The power transfer curve of this SFG pump depletion process reveals a step-like shape, such that the zero and one level noise distribution will be redistributed favorably, i.e. yielding a regenerated data signal.

Figure 17.31 shows regeneration results in terms of eye diagrams. The distorted 640 Gbit/s data signal is clearly cleaned up, yielding a significantly opened eye diagram. The biggest noise suppression takes place on the one level, due to the characteristics of the power transfer curve of the regenerator [37]. The sides of the eyes are also cleaned up, due to the data transfer to the clean clock signal with low jitter.

Figure 17.32 shows the BER performance of the regenerator. The leftmost curve shows the original data signal and the regenerated signal, revealing a very small penalty. When the OSNR of the input signal is lowered, the BER curve of the regenerated signal remains pretty much the same, showing an improvement of the BER curve at 12 dB OSNR deterioration. At 20 dB, corresponding to the eye diagrams in Figure 17.31, the deteriorated signal cannot get below BER 10^{-5}, but the regenerated signal is pretty much the same as always, demonstrating a very clear regeneration. The regenerated signal has a sensitivity at BER 10E–9 of about −16 dBm and a variation of about 0.5 dB for all the curves in Figure 17.32. The timing jitter at the output is also kept pretty constant around 40 fs, even when the input timing jitter ranges from 40 to 300 fs. So this scheme is very clearly demonstrated to operate well at up to

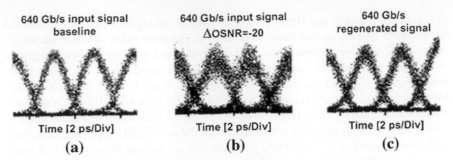

FIGURE 17.31 Eye diagrams at 640 Gbit/s showing the SFG-based pump depletion regeneration of the distorted data signal. (a) Original 640 Gbit/s data signal. (b) Distorted 640 Gbit/s data signal with purposefully added ASE noise to lower the OSNR with 20 dB compared to the original signal. (c) Regenerated signal at 640 Gbit/s with clear noise suppression in the one level and reduced timing jitter. (From [37]. © 2012 IEEE.)

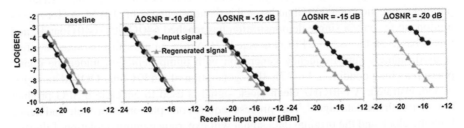

FIGURE 17.32 BER characteristics of the regenerator. Curves show demultiplexed to 40 Gbit/s data signals stemming from the 640 Gbit/s deteriorated signal and the regenerated signal with clear improvement. (From [37]. © 2012 IEEE.)

640 Gbit/s. It would be desirable to keep the signal at the same wavelength, though, so with the above scheme, an additional wavelength converter is required.

Though the above results are the only ones at 640 Gbit/s, several other approaches, which should be scalable, have been demonstrated. These include the following two based on polarization rotation in a Kerr switch based on HNLF [111,112]. In [111], a full 160 Gbit/s 3R regeneration is demonstrated by switching the data onto a clean clock signal in the Kerr switch. The data is first made flat-top-like by transmission through a short piece of highly birefringent fiber. The regenerated signal is now on another wavelength, as above, and a SPM-based wavelength shifter is added to obtain a regenerated signal at the same wavelength as the original data. In [112], retiming is demonstrated keeping the same wavelength in a single device. Again, an HNLF-based Kerr switch is used, and the data is made flat-top by the use of a super-structured fiber Bragg grating (SSFBG) filter. A clean clock signal is used as pump to switch out the central part of the broadened flat-top data pulses, resulting in a retimed output at the same wavelength as the original data.

All in all, today there are several methods and materials available to perform ultra-fast all-optical regeneration. The benefits may be potential power savings with increased transmission lengths and increased channel capacity, especially for high symbol rates and higher order modulation. However, to make these regenerators practical, they must be multi-wavelength operational, i.e. able to handle WDM data, and be cost-effective. This is a challenge for the future, but potentially, using the above-mentioned OFT technique combined with an OTDM regenerator may be a viable path.

17.2.4.4 Packet switching at 640 Gbit/s

Optical packet switching has been researched for a long time, but seems inhibited by the lacking prospects of robust solutions for optical buffering. However, for certain niche applications, in a short range network with a limited number of packets to manage, a central control unit may be able to administer the timing of packet transmission from the various transmitters, so as to ensure a free gate upon time-of-arrival at the packet switch, thus alleviating the need for optical buffering at the switch. In such a scenario, ultra-high bit rate data packets are interesting, since it allows for a high throughput yet keeping the number of packets low, and also allows for sharing the switching energy among more bits, thus reducing the overall power consumption. The switching could be performed like a wavelength conversion, embracing all the various approaches discussed above in Section 17.2.4.2. It could also be performed in an on/off type of switch as discussed here.

The schematic diagram of the optical packet switching (OPS) concept is shown in Figure 17.33 [113]. The system consists of four main blocks: the transmitter, a 50-km dispersion-compensated transmission link, the optical packet switch, and the receiver. In the transmitter, 640 Gbit/s data payload is generated by OTDM. The optical packets have a duration of 153.6 ns consisting of 89.6 ns of data payload

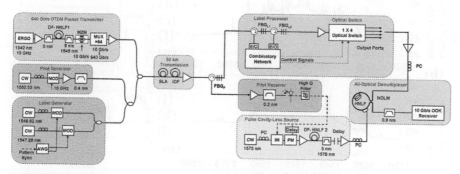

FIGURE 17.33 Experimental setup for 640 Gbit/s optical packet switching. ERGO: erbium glass oscillating pulsed-generating laser, MOD: modulator, AWG: arbitrary waveform generator, DF-HNLF: dispersion-flattened highly nonlinear fiber, SLA: super large area fiber, IDF: inverse dispersion fiber, FGB: fiber-Bragg grating, E/O: electro-optical conversion, PC: polarization controller. (From [113]. © 2011 OSA.)

separated by a 64 ns guard band. In order to distribute the clock in the system, a clock pilot was generated by modulating a CW laser with a master clock at 10 GHz and inserted in-band within the 640 Gbit/s data signal's spectrum [114]. To improve the OSNR of the clock pilot, a programmable Waveshaper, a filter with a 3 dB bandwidth of 0.4 nm and 30 dB suppression of the rejection band centered at 1552.52 nm, is used to carve a portion of the spectrum where the pilot is inserted.

In order to address the destination of the packets, the in-band labeling technique is employed [115]. The packet address is encoded with labels at wavelengths within the bandwidth of the payload and a duration equal to the payload length. Each label has a binary value: "0" means no signal at the label wavelength, "1" means an optical signal at the label wavelength. Thus, 2^N addresses can be encoded by only using N in-band labels. Here two in-band labels at 1546.62 nm and 1547.28 nm are used to encode four addresses. Figure 17.34a and b shows the spectra of the transmitted signal at the receiver side before and after the labels and pilot are extracted. The pilot, located between the first (640 GHz) and second (1.2 THz) spectral line of the signal, is considered in-band due to the minimum number of spectral lines required to preserve an undistorted signal pulse. Nevertheless, the pilot can be placed close to the central wavelength if a narrower (0.25 nm) FBG is employed.

The 640 Gbit/s OTDM packetized data, together with the pilot and labels, is transmitted over a dispersion and dispersion slope compensated fiber span of 50-km optimized at 1545 nm. The transmission link is composed of 25-km SLA (super large area) fiber ($D = 20$ ps/nm/km, $S = 0.06$ ps/nm²/km, and PMD $= 0.04$ ps/km$^{1/2}$) and 25-km IDF (inverse dispersion fiber) ($D = -20$ ps/nm/km, $S = -0.06$ ps/nm²/km, and PMD $= 0.02$ ps/km$^{1/2}$) with a total loss of 12 dB.

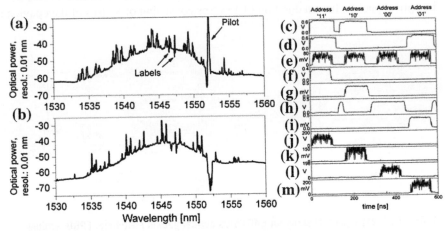

FIGURE 17.34 Spectra of the 640 Gbit/s OTDM signal after transmission and dynamic operation of the optical packet switch before (a) and after (b) labels and pilot extraction; (c and d) labels of the input optical packets; (e) input packet payload; (f and i) control signals; j–m switched packets. (From [113]. © 2011 OSA.)

After the 50 km transmission link, the 640 Gbit/s data packets are fed into the OPS. The OPS extracts the optical labels by cascading two FBGs (FBG$_{L1}$ and FBG$_{L2}$) centered at the label wavelengths and two optical circulators. The FBGs have Gaussian transfer functions with 98% reflectivity and 6 GHz 3 dB bandwidth to avoid significant slicing of the payload spectrum, which might otherwise lead to distortions. The parallel labels are detected and processed by the electrical combinatory network. The combinatory network processes asynchronously and on the fly the parallel labels and can be scaled for larger number of labels without increasing the latency. The combinatory network provides the control signal of the electro-optical switch with a latency <3 ns. Finally, the payload is fed via polarization controllers into a 1×4 electro-optic switch based on LiNbO$_3$ technology. According to the encoded addresses, one switch is enabled at a time and the switched optical packet is evaluated in the optical demultiplexer and receiver.

Figure 17.34c–m shows the dynamic operation of the OPS. The extracted labels and the 640 Gbit/s data packets are shown in Figure 17.34c–e, respectively. According to the packet addresses, the control signals generated by the combinatory network to drive the 1×4 switch are shown in Figure 17.34f and i. Figure 17.34j and m shows the switched packets from port 1 to port 4 with their respective addresses "11," "10," "00," and "01."

The pilot clock is separated via an optical circulator and a FBG (FBG$_P$), which has a central frequency and bandwidth identical to the carved portion of the data spectrum. The extracted pilot is amplified and filtered with a 3 dB bandwidth of 0.2 nm and converted to the electrical domain by a 12 GHz photo-receiver. The electrical signal is subsequently filtered by a high-Q filter (10-MHz) and used to drive a cavity-less pulse source [116], which consists of two RF-driven stages (intensity and phase modulation) followed by pulse-compression to produce pulses with 1 ps duration and 10 GHz repetition rate at 1578 nm. This scheme allows for generating an optical clock pulse on the fly once the RF pilot clock is applied. As the clock and payload are transmitted synchronously and are spectrally in-band, they are affected by the same impairments leading to the same phase drifts. Thus, the relative phase between the clock and data is preserved.

As discussed in more detail in [113], the output from all 4 ports on the switch are error-free with a low switching penalty of about 1 dB. The whole setup even works with transmission, with an additional penalty after 50 km transmission of ~3 dB. So, 640 Gbit/s single-channel packet switching really does work.

17.2.5 Transmission

Transmission of sub-picosecond pulses is challenging due to GVD and PMD. However, there are several demonstrations of transmission over relevant distances. Most notably, a 640 Gbaud data signal was transmitted over 525 km standard fiber [117] using the time-domain optical Fourier transformation (OFT) technique to compensate for linear transmission impairments. In [12], a 1.28 Tbaud data signal was transmitted over a 50 km dispersion managed link consisting of SMF and IDF. Temperature

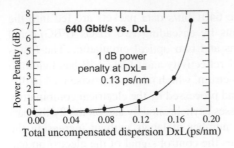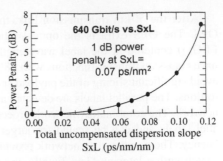

FIGURE 17.35 Dispersion tolerance for a 640 Gbit/s data signal. System power penalty at BER 1OE-9 as a function of accumulated uncompensated dispersion, $D \times L$ (left), and ditto dispersion slope, $S \times L$ (right). The 1 dB power penalty points correspond to about 7.5 m standard SMF.

fluctuations will influence the dispersion, and hence some level of adaptive compensation is needed. This may be realized using e.g. OFT or a spectral pulse shaper (SPS) with adjustable phase [118]. This section will describe transmission challenges and give examples of state-of-the-art solutions to overcome these challenges.

17.2.5.1 Dispersion tolerance

Detailed simulations based on a full 640 Gbit/s OTDM signal transmitted over standard single-mode fiber reveal that an uncompensated dispersion of ± 0.13 ps/nm, corresponding to approximately ± 7.5 m SMF, will introduce a 1 dB power penalty, see Figure 17.35.

The acceptable uncompensated dispersion at 640 Gbit/s can also be estimated using a simple rule of thumb stating that every time the bit rate is increased fourfold, the uncompensated distance is reduced by a factor 16 [119]. Using 40 km SMF at 10 Gbit/s as standardized in [120] as a starting point leads to $\sim \pm 10$ m of uncompensated SMF which would be acceptable at 640 Gbit/s, in good agreement with the detailed simulations. As the spectrum increases for increased bit rate, the impact of dispersion slope S will also affect the transmitted OTDM signal. Therefore, also slope compensation should be included when designing the transmission links for high-speed systems.

17.2.5.2 640 Gbaud transmission experiments

The transmission of ultra-high symbol rates is primarily limited by higher-order chromatic dispersion (CD) and higher-order polarization mode dispersion (PMD), causing distortion of sub-ps optical pulses due to their broad bandwidth requirement. The bandwidth limitation and uneven gain profile of optical amplifiers can also cause distortion of such short optical pulses. Other impairments come from the nonlinear Kerr effects in the transmission fiber and the ASE noise generated in optical amplifiers

[121]. Various methods have been demonstrated for mitigating higher-order CD in the transmission fiber, including various types of dispersion-compensating fibers, phase-modulation, phase-conjugation, time-domain optical Fourier transformation, and spectrally resolved phase-shaping. First-order PMD is typically limited by launching the data signal along a principal state of polarization (PSP) axis of the transmission span. However, higher-order PMD must also be taken into account due to the broad bandwidth of the optical pulses. Pulse shaping by spatial light modulation can potentially compensate all-order PMD, but has not yet been directly employed in an ultra-high-speed OTDM transmission experiment. In the following, we briefly review some of the main high-speed OTDM transmission experiments (640 Gbaud and 1.28 Tbaud), where various techniques for compensation of transmission impairments have been employed.

The first 640 Gbaud transmission experiment was carried out by Nakazawa et al. [7]. Here, 400 fs pulses at 640 Gbit/s were transmitted over a 60 km link consisting of two spans, based on single-mode fiber (SMF) and dispersion-slope compensation fiber (DSCF) and DSF to compensate for second- and third-order CD. The received power penalty at BER 10^{-9} of the demultiplexed 10 Gbit/s data was mainly attributed to the PMD, which was 0.08 ps/km in the first span and 0.11 ps/km in the second span. In 2000, Yamamoto et al. succeeded in extending the transmission distance to 92 km by employing reverse dispersion fiber (RDF) [122]. The RDF has nearly the same dispersion and dispersion slope compared to SMF, but with opposite sign. It therefore results in improved higher-order CD compensation compared to using DSCF, and also has a lower PMD. The 92 km transmission line was divided in two spans, the first consisting of 21 km SMF and 24 km RDF, and the second of 25 km SMF and 20 km RDF. A 2 km DSF was inserted in the second span to compensate for the residual dispersion slope of the entire span. The PMD of the first and second span were 0.04 ps/km and 0.03 ps/km, respectively. As a result, a pulse width of 500 fs could be obtained after the 92 km transmission. In 2000, a polarization-multiplexed 640 Gbaud (per polarization) OOK data signal, yielding 1.28 Tbit/s data capacity, was successfully transmitted over 70 km [8]. The link consisted of a 39.7 km SMF, a 4.6 km DSF, and a 25.1 km RDF. In addition, pre-chirping with a 10 GHz phase modulation was employed to compensate for residual third- and fourth-order dispersion in the span. As a result, a 380 fs pulse was only broadened by 20 fs after transmission.

In 2006, a polarization-multiplexed 640 Gbaud (per polarization) OTDM signal encoded with DQPSK was transmitted over a 160 km link, yielding a data capacity of 2.56 Tbit/s on a single-wavelength channel [9]. The fiber link was composed of two 80 km dispersion-managed fiber spans, each consisting of 53 km super-large area fiber (SLA) and 27 km inverse dispersion fiber (IDF). The SLA and IDF are dispersion-matched in order to compensate both for dispersion and dispersion slope, similarly to the SMF/RDF combination. The pulse width was 0.65 ps at the output of the transmitter, and 0.75 ps after the 160 km transmission. The minimum pulse width was limited by the bandwidth of the cascaded non-gain-flattened EDFAs, fourth order CD, and higher-order PMD. Nevertheless, a BER performance within standard FEC limits was obtained after transmission.

More recently, the transmission reach for 640 Gbaud per polarization data was extended to 525 km, which is the longest reach yet reported for this symbol rate [117]. The link consisted of seven 75 km dispersion-managed spans composed of 50 km SMF and 25 km IDF to provide both dispersion and dispersion slope compensation. The average differential group delay of the fiber link was 1.2 ps. First-order PMD was mitigated by launching in a PSP axis. Importantly, a time-domain optical Fourier transformation (OFT) circuit was employed after demultiplexing of the transmitted 640 Gbaud DPSK data. The OFT circuit is composed of a phase modulator operated in round-trip configuration to achieve a sufficient chirp rate, followed by a dispersive fiber. The OFT can transfer the spectral profile of the input waveform into the temporal intensity profile of the output waveform [75]. Hence, linear transmission impairments such as higher-order CD and PMD, which do not alter the spectrum, can be mitigated by OFT by transferring the undistorted spectrum into the time domain. The pulse width before transmission was 600 fs. After transmission, the pulse width fluctuated between 630 fs and 730 fs, due to time-varying higher-order PMD effects. As a result, the BER for the best-performing channels after demultiplexing was $\sim 10^{-8}$–10^{-9}, and varied with time. However, when the OFT circuit was employed, the corresponding channels could be reduced to a BER below 10^{-9} with a transmission power penalty of 2 dB. All channels performed below the limit for standard forward error-correction (FEC). In a subsequent experiment, the DPSK modulation was replaced with DQPSK in order to achieve a 1.28 Tbit/s single-channel transmission over the 525 km fiber link, with a BER performance within FEC limits [123]. Later, the 640 Gbaud DQPSK was combined with polarization-multiplexing to achieve a single-channel data rate of 2.56 Tbit/s [124]. However, the transmission reach in this case was limited to 300 km (four spans), due to the influence of higher order PMD introducing detrimental crosstalk between the two polarizations. This crosstalk increased from −14 dB after 300 km to −11.3 dB after 525 km transmission. The BER performance for the 2.56 Tbit/s polarization-multiplexed data after 300 km transmission was limited to $\sim 10^{-5}$, which is more than two orders of magnitude worse compared to the 1.28 Tbit/s single-polarization DQPSK case, yet still below the FEC limit.

Another interesting technique for CD compensation is mid-span optical phase-conjugation, which has been used in a 640 Gbaud transmission over 100 km standard SMF [125]. Phase-conjugation of the optical data was achieved by a four-wave mixing (FWM) process using two orthogonal pumps in a HNLF, thus generating an idler copy of the data signal at the same wavelength, but with the orthogonal polarization (allowing for its extraction). The FWM process is located between two identical fiber spans of 50 km SMF. In this way, the even orders of the dispersion which are accumulated in the first span can be compensated by propagation of the FWM-generated idler data copy through the identical, second span. The uneven orders of dispersion must be compensated by other means. In this experiment, the third-order dispersion (dispersion slope) was compensated by high-slope DCFs. The 640 Gbaud OOK data pulses were recovered at the receiver and demultiplexed with error-free BER performance with a 5.3 dB OSNR penalty relative to the back-to-back case.

17.2.5.3 Tbaud transmission and dispersion compensation by spatial light modulation

The first 1.28 Tbaud per polarization transmission experiment was reported in 2010 [12]. The employed data format was DPSK, and the transmission span was 50 km long, consisting of 25 km SLA and 25 km IDF to provide simultaneous dispersion and dispersion-slope compensation. A 10 GHz optical tone for clock recovery was transmitted in the L-band along with the data signal in the C-band. Figure 17.36 (left) shows the optical data spectrum with the dispersion profile of the 50 km span, and pulse autocorrelations (a) before and (b) after transmission. The pulse width is slightly broadened from 420 fs to 470 fs due to the residual higher-order CD and higher-order PMD of the fiber link. Note that PSP-launching was used to mitigate first-order PMD. The BER measurements revealed error-free performance with a transmission power penalty of only 1 dB. In 2011, Richter et al. demonstrated a transmission experiment over a 29 km link of polarization-multiplexed 1.28 Tbaud 16 QAM, yielding a total capacity of 10.2 Tbit/s [14]. The dispersion-managed fiber link was composed of 19 km SLA and 10 km IDF. Additional CD compensation up to fourth order was performed with a liquid crystal on silicon (LCOS) device (as described in the following section). First-order PMD was mitigated by PSP-launching. As a result, 300 fs pulses could be transmitted with negligible broadening. The BER performance after coherent detection was below the hard-decision FEC threshold of 3.8×10^{-3}. No transmission penalty was observed for the single-polarization case, whereas a 3 dB penalty in the polarization-multiplexed case was attributed to the PMD in the setup.

A more advanced technology allowing for compensation of higher-order CD is pulse shaping using spatial light modulation [126], which can be achieved using gratings and e.g. LCOS arrays. Basically, a wavelength grating is used to spatially distribute the different frequency-components of an optical waveform onto a LCOS

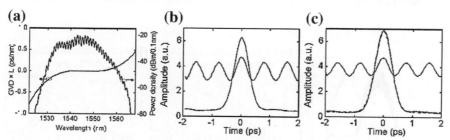

FIGURE 17.36   1.28 Tbaud RZ-DPSK transmission over 50 km fiber link. Left: Group velocity dispersion profile of the 50-km transmission link (blue curve) and spectrum of the 1.28 Tbit/s data signal (red curve). Autocorrelation traces: (b) 320 Gbit/s (red) and 1.28 Tbit/s (blue) before transmission; (c) 320 Gbit/s (red) and 1.28 Tbit/s (blue) after transmission, and 10 GHz control pulse (black dots). (From [12]. © 2010 IEEE.) (For interpretation of the references to color in this figure legend, the reader is referred to the web version of this book.)

array, where the phase (and amplitude) of each component can be individually controlled by each pixel [127]. Hence, an LCOS device enables precise, frequency-resolved control of the phase of the optical pulses. The LCOS technique has been employed for higher-order CD compensation of OTDM data up to 1.28 Tbaud. Paquot et al. have shown compensation of CD up to fourth order for 275 fs pulses using an LCOS array [118]. The LCOS compensation was automatically adjusted using a feedback loop with an optical RF spectrum analyzer to monitor the 1.28 THz tone of the data signal, providing the feedback signal for re-programming the LCOS array. The optical RF spectrum analyzer is based on XPM by the OTDM data on a CW beam in a dispersion-engineered As_2S_3 chip, enabling THz bandwidth RF spectrum measurements. In another experiment, an LCOS array enabled a nearly distortion-less transmission of 610 fs pulses over a 160 km standard SMF link [128]. The main part of the CD was compensated by 22 km DCF, and the residual dispersion including higher-order terms was compensated using the LCOS array. Figure 17.37 (left) shows the spectrum of 610 fs pulses transmitted at either 1542 nm or 1556 nm, on top of the residual dispersion profile of the 160 km SMF-DCF span. As can be seen, the 1542 nm pulse is subjected to a strong dispersion slope (β_3), while the 1556 nm pulse is subjected to a strong dispersion curvature (β_4). The autocorrelations in Figure 17.37a and b show the effect of compensating the higher-order CD terms using the LCOS device. As expected from the dispersion profile, the 1542 nm pulse can be nearly fully recovered by compensating up to third order dispersion, while compensation up to fourth order is required for the 1556 nm pulse. The transmission of even shorter pulses down to 320 fs was limited by higher-order PMD.

The liquid crystal technology also has the potential for compensating higher-order PMD. Several liquid crystal modulator arrays (LCM) can be employed to compensate the frequency-dependent Jones matrix associated with the all-order PMD of a fiber transmission span, which has been demonstrated by Miao et al. [129]. Here, the Jones matrix of a higher-order PMD emulator is firstly evaluated by launching different states of polarization (SOP), and measuring the output SOP spectra using

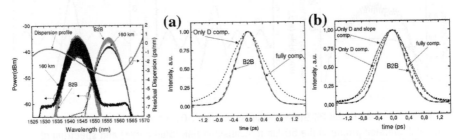

FIGURE 17.37 Distortion-less 610 fs pulse transmission over 16 km SMF-DCF link. Left: Residual dispersion profile of 160 km SSMF-DCF span and spectra of 10 GHz pulses before (B2B) and after transmission (160 km). Autocorrelations for (a) 1542 nm pulse and (b) 1556 nm pulse before (B2B) and after 160 km transmission, showing the effects of compensating higher orders of dispersion using the LCOS device. (From [128]. © 2011 IEEE.)

a wavelength-parallel polarimeter. The inverse matrix can be written as a product of three elementary rotation matrices, which are then programmed into separate layers of a four-layer LCM, positioned after the PMD emulator. As a result, the Jones matrix of the entire system can be corrected to a frequency-independent constant matrix. The system was tested with 800 fs input pulses at 1550 nm. The higher-order PMD emulator had a mean DGD of ~5.5 ps, resulting in a highly distorted output waveform spread over ~10 ps. When the PMD was compensated with the four-layer LCM, the 800 fs could be almost fully recovered. These results indicate that the LCOS technology can potentially mitigate the higher-order PMD limitations in ultra-high-speed OTDM transmission.

17.2.5.4 *640 Gbaud field trial with RZ-NRZ conversion for improved transmission performance*

To demonstrate the robustness of OTDM transmission, it is necessary to perform a field trial, to test if real-world conditions can be overcome. In this section, a presentation of a 640 Gbaud per polarization field trial experiment is given.

Figure 17.38a shows the schematic setup for the field trial of a 1.19 Tbit/s NRZ OTDM signal. It mainly consists of a 640 Gbit/s RZ-OOK OTDM transmitter (L-band), a phase-coherent OTDM generator, a polarization multiplexer, a field transmission link, a polarization demultiplexer, and a 640 Gbit/s receiver. A 640 Gbit/s OTDM RZ signal at 1590 nm is generated by optical time division multiplexing of a 10 GHz short pulse, which has been on-off-keying (OOK) modulated at the 10 Gbit/s base rate with a PRBS $(2^{31}-1)$ signal. The 640 Gbit/s data pulse has a FWHM of 600 fs. To generate a phase-coherent OTDM signal, the original OTDM signal is mapped onto a coherent CW light beam. A coherent signal is needed in order to convert it to NRZ. The coherent OTDM generator is basically a polarization-rotating Kerr switch [130] as shown in Figure 17.38b. The 640 Gbit/s data pulses and coherent CW light beam at 1545 nm are launched into a highly nonlinear fiber (HNLF). At the fiber output a polarizer is placed with its axis (vertical axis in Figure 17.38b), orthogonal to the CW light. The polarization of the data is 45° with respect to the polarizer. The 640 Gbit/s data pump switches the CW light by cross-phase modulation (XPM) induced polarization rotation in the HNLF, which generates a pulse-to-pulse phase-coherent OTDM signal. The HNLF used in the experiment is alumino-silicate strained in order to increase the Stimulated Brillouin Scattering (SBS) threshold [131], and is kindly provided by OFS Fitel Denmark. The average power of the data and the CW light are 25 dBm and 27 dBm, respectively. The zero-dispersion wavelength of the Al-HNLF is at 1560 nm, minimizing the walk-off. The 640 Gbit/s coherent OTDM (Co-OTDM) signal is then polarization multiplexed by the polarization multiplexer, constituting a 1.28 Tbit/s line rate Co-OTDM PDM-RZ signal, which corresponds to a data rate of 1.19 Tbit/s, after subtracting the 7% FEC overhead.

A wavelength selective switch (WSS) is used to implement a filtering function by combining those of a 1.28 THz ring resonator and a 500 GHz Gaussian filter, as shown in Figure 17.39a, in order to convert the RZ format into NRZ format. The spectra of the original 640 Gbit/s OTDM, the generated 640 Gbit/s phase-coherent

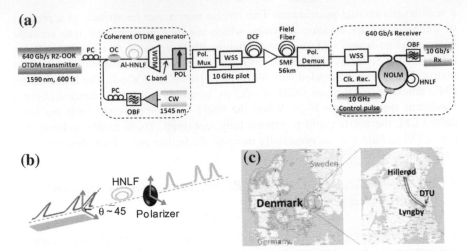

FIGURE 17.38 (a) Schematic setup for the field transmission of a 1.28 Tbit/s pol-MUX-NRZ signal. PC: Polarization controller, OC: Optical coupler, HNLF: Highly nonlinear fiber, WDM: Wavelength splitter (C- and L-band), POL: Polarizer, CW: Continuous wave, OBF: Optical bandpass filter, Pol. Mux: Polarization multiplexer, WSS: Wavelength selective switch, Pol. Demux: Polarization demultiplexer, Clk. Rec.: Clock recovery, NOLM: Nonlinear loop mirror. (b) The principle of the Kerr switch based coherent OTDM generator. (c) Route of the field installed fiber in Denmark. (From [41]. © 2012 OSA.)

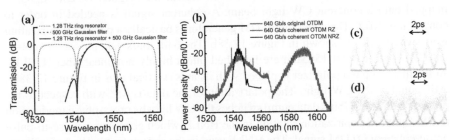

FIGURE 17.39 (a) Filtering function of the WSS; (b) spectra of original 640 Gbit/s L-band OTDM signal, generated 640 Gbit/s coherent OTDM RZ signal and NRZ signal; (c) and (d) optical sampling eye diagrams of the 640 Gbit/s coherent OTDM RZ (c) signal and NRZ (d) signal. (From [41]. © 2012 OSA.)

RZ and the 640 Gbit/s NRZ signals are shown in Figure 17.39b. Optically sampled eye diagrams of the 640 Gbit/s Co-OTDM RZ and NRZ signals are shown in Figure 17.39c–d. Finally, an in-band 10-GHz pilot tone at 1539.6 nm is inserted through the WSS for clock recovery [113].

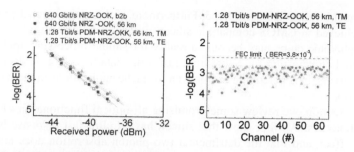

FIGURE 17.40 Field trial BER characterization. Left: BER vs. received power of demultiplexed channels with and without PDM. Right: BER for all 128 tributaries (64 OTDM channels and two polarizations). (From [41]. © 2012 OSA.)

The 1.28 Tbit/s signal is dispersion pre-compensated by a 7 km DCF and then launched into a 56 km field transmission link, which is a loop between the DTU lab in Lyngby and Hillerød in Denmark, Figure 17.38c. The launched power into the field transmission link is 11 dBm, with less than 1 dB nonlinearity penalty when increased to 18 dBm, which is a benefit of the lower peak power of NRZ compared to RZ. The mean differential group delay (DGD) of the field installed fiber is 0.2 ps and has negligible impact on the NRZ transmission.

The RZ-NRZ conversion and field transmission are successful [41]. Figure 17.40 shows BER-based characterization results of the transmitted NRZ data signal. Figure 17.40 (left) shows the BER vs. received power to the receiver for demultiplexed channels for the 640 Gbit/s NRZ back-to-back and after the 56 km field-deployed fiber, and also the transmitted TE and TM signals when both polarizations are transmitted simultaneously. In all cases, nice straight BER curves are observed down to well below the FEC limit. BER values for all the 64 OTDM tributaries in both polarizations are shown in Figure 17.40 (right). All 128 channels show a BER well below the FEC limit of 3.8×10^{-3}, and thus this confirms that this is indeed a full 1.19 Tbit/s post-FEC error-free PDM-NRZ OOK data signal transmitted through this field-deployed fiber.

In summary, it is indeed possible to transmit ultra-high-speed OTDM data signals over relevant distances. There exists various schemes for adaptive dispersion and PMD compensation, and 640 Gbaud OTDM is robust enough for field trial transmission over 56 km of installed fiber.

17.3 SILICON PHOTONICS AND ULTRA-FAST OPTICAL SIGNAL PROCESSING

From a power consumption perspective, all-optical signal processing may be suitable for functionalities where many bits are processed in a few devices [106]. This is the case for e.g. wavelength conversion and all-optical regeneration, especially if very high channel bit rates are considered. Furthermore, if the optical signal processing can be achieved in CMOS-compatible silicon waveguides, then ultra-fast, energy-efficient photonic chips may become a reality for simple processing applications, which may be interesting for future ultra-fast serial data links, e.g. in data centers, between servers, in supercomputers, or even for niche applications in the core transport network.

In this section, we review some promising all-optical functionalities based on silicon photonics. We first show that silicon has an ultra-fast response based on the Kerr effect, and that the detrimental two-photon absorption does not necessarily pose a serious limitation to the use of silicon optical processing, neither for XPM nor for FWM. In particular we use nano-engineered silicon waveguides enabling efficient phase-matched four-wave mixing (FWM) for ultra-high-speed optical signal processing of ultra-high bit rate serial data signals. We show that silicon can indeed be used to control Tbit/s serial data signals [34], perhaps paving the way for future ultra-fast optical chips. From an energy perspective, the most promising functionalities are those that process many bits in few devices, e.g. *conversion-type* functionalities where e.g. a full serial data signal is converted to some other format. In such a scheme, all the bits in the serial signal are processed in the same device. We pay particular attention to such functionalities and we will describe various potentially energy-efficient schemes of conversion, focusing on wavelength conversion [35] and serial-to-parallel conversion [36], all using Si waveguides.

17.3.1 Silicon photonics for ultra-fast optical signal processing

Nonlinear signal processing in silicon by the nonlinear optical Kerr effect has been described and used for many applications over the last 5–10 years, e.g. NRZ-to-RZ conversion [130], regeneration, multi-casting, multiple-wavelength source, monitoring, demultiplexing, and many more, see e.g., [33,96,132–148]. Most of these demonstrations rely on ultrafast FWM or even two-photon absorption (TPA), but SPM and XPM are also useful, as will be seen. Adding other nonlinear materials to Si slot waveguides can induce a high nonlinearity but avoid the detrimental nonlinear absorption of silicon. Such materials could be organic molecules [149,150], recently used to enable a 40 Gbit/s data modulator [151] and demultiplexing 170 Gbit/s [152]. Using silicon nitride instead of pure silicon changes the bandgap, allows for alleviation of TPA and has been used e.g. to generate frequency combs [153]. Recently, the use of amorphous silicon has been demonstrated. Amorphous silicon also has a different bandgap to crystalline silicon, and therefore avoids TPA [154,155], has been demonstrated to generate 26 dB FWM gain and used e.g. to sample a 320 Gbit/s data signal.

Figure 17.41 shows numerical simulations of the phase response of a nonlinear silicon waveguide in a pump-probe configuration, using the model described in reference [156]. A 5 ps pump pulse is injected into the waveguide, and a 5 ps probe pulse is then stepped in time across the pump pulse to monitor the response of the

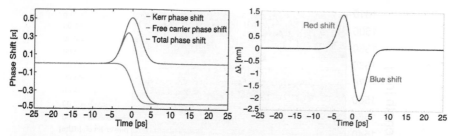

FIGURE 17.41 Phase shift and chirp from a silicon waveguide. Left: Theoretical phase shift consisting of two contributions, the Kerr shift and a free-carrier shift. Right: Associated frequency chirp, revealing an equally fast red- and blue-shift. (Blue- lines indicate polarization maintaining fibers.) (For interpretation of the references to color in this figure legend, the reader is referred to the web version of this book.)

waveguide. The refractive index of the silicon is modified by the high-power pump pulse by two contributions: the optical Kerr effect and a free-carrier effect. The free-carrier effect is slow, whereas the Kerr effect is almost instantaneous. Figure 17.41 (left) shows the individual contributions along with the total phase shift, with a fast initial rise, dominated by the Kerr effect, followed by a fast decline to an almost flat negative phase value, corresponding to a very slow carrier recovery. The recovery time is on the order of 10 ns.

Figure 17.41 (right) shows the corresponding wavelength shift (frequency chirp), and reveals an ultra-fast initial red-shift, owing to the Kerr effect, followed by an equally ultra-fast almost complete recovery by a blue-shift, owing to the almost flat free-carrier recovery. This means that for silicon waveguides, both red- and blue-shifting is ultra-fast and useful for optical signal processing, as confirmed by experiments recently [157].

To evaluate the modeling presented above, a detailed experimental time-resolved spectral analysis was carried out and presented in [158]. A resulting spectrogram is shown in Figure 17.42. The pulses used here are sub-ps (585 fs FWHM), in order to fully verify the ultra-high-speed potential.

The spectrogram in Figure 17.42 contains both dynamic XPM and FWM information. When the pump and probe overlap in time around 0 on the time axis, FWM products are generated on the high and low wavelength sides, and additionally, the probe spectrum around 1539 nm is clearly distorted due to XPM. The top FWM product is seen to have a short tail, and this actually comes from the probe pulse, which suffered from a trailing pulse tail. As FWM is very linear in probe power, a good replica of the probe pulse is obtained, including the tail. The main responses are all on the order of the pulse widths used, i.e. around 830 fs FWHM, corresponding to the cross-correlation of the probe with the pump pulse.

Figure 17.43 shows the XPM-induced chirp on the probe pulse, derived from Figure 17.42 by plotting the central wavelength of the probe over time. It is clearly

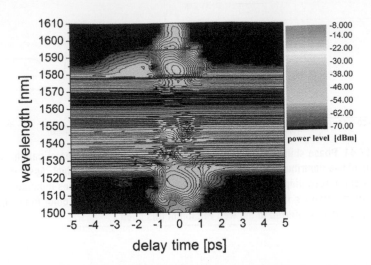

FIGURE 17.42 Experimental spectrographic pump-probe characterization with sub-ps pulses. The pump is at 1562nm, and the probe at 1539nm. When the pump and probe coincide in time, FWM products occur at the top and the bottom of the graph. (From [158]. © 2011 IEEE.)

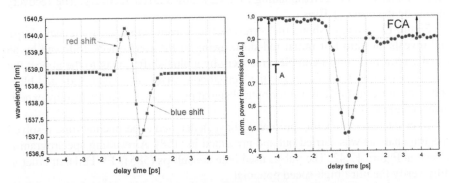

FIGURE 17.43 Left: Experimental wavelength shift due to XPM. A fast red-shift is followed by an equally fast blue-shift. Right: Probe transmission through the silicon waveguide. A fast power reduction and equally fast partial recovery due to TPA between the pump and the probe is followed by a free-carrier absorption (FCA) level with a very slow recovery.

observed that the fast red-shift is followed by an equally fast blue-shift, as expected. The blue-shift is almost entirely recovered to its original 1539 nm value, owing to the very slow carrier recovery.

Figure 17.43 (right) shows the probe transmission. TPA between the pump and the probe creates a fast drop and partial recovery in the probe transmission, and the

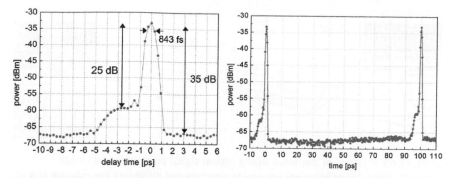

FIGURE 17.44 Measured FWM switching window. Left: 843 fs FWHM wide switching window, corresponding to the cross-correlation between the pump and the probe. There is a very good contrast of 25–35 dB, the lowest owing to the probe pulse used in this experiment. Right: Scanning over two switching windows revealing no memory effect in between.

transmission power eventually ends on a level determined by the number of generated free-carriers giving FCA.

To analyze the temporal response of the FWM, a slice of the FWM idler spectrum at 1585 nm is extracted and shown in Figure 17.44, as a switching window profile on a logarithmic scale. The main peak of this switching window (centered ∼0 ps) in Figure 17.44 (left) resembles a Gaussian profile with a measured FWHM of 826 fs. The response immediately goes down with 35 dB after the pump has left the waveguide, showing no sign of free-carriers. The bulge on the front is due to the tail of the probe pulse itself. Figure 17.44 (right) shows a zoom-out over the 10 GHz repetition rate period of 100 ps, the distance between pump pulses (and probe pulses), and a truly ultra-fast response with no memory effects from the free carriers is observed. As seen in Figure 17.43 (right) there are free carriers generated, but these do not interfere with the FWM temporal response. The effect of FCA is mainly to reduce the optical powers, and hence reduce the conversion efficiency. Thus, FWM in silicon nanowires should lend itself nicely to truly ultra-fast operation.

17.3.2 Demonstrations of silicon-based optical signal processing of OTDM signals

In the following, a series of demonstrations of ultra-high-speed optical signal processing using FWM in nonlinear silicon waveguides is presented. Figure 17.45 summarizes the basic concept. A 640–1280 Gbit/s OTDM data signal is injected into the Si waveguide together with a control signal, being either pulses (as sketched) or continuous wave. Through FWM, an idler is generated containing the information of the data pulse. If a short low-rate control pulse is scanned across the data signal inside the waveguide, the data eye diagrams may be sampled. Figure 17.45

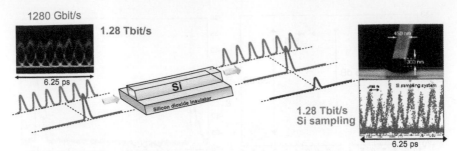

FIGURE 17.45 Principle of four-wave mixing optical signal processing in a silicon waveguide. Right: Image of typical waveguide structure and a sampled eye diagram of a 1.28 Tbit/s data signal using the Si waveguide.

shows an example of a 1.28 Tbit/s sampled eye diagram together with a microscope image of the used waveguide. When comparing the Si-sampled eye with the eye diagram using a commercial HNLF-based sampling oscilloscope (top left corner), it is clear that the Si waveguide system has much higher timing resolution. This is partly due to the short waveguide, resulting in negligible walk-off and pulse broadening. The used Si waveguide [159] has tapered end-sections providing low coupling loss (1.5 dB per facet).

17.3.2.1 1.28 Tbit/s demultiplexing and optical sampling of OOK and DPSK data signals

As mentioned above, when the control is a short pulse at a low repetition rate, one may sample the data signal. If the repetition-rate of the control pulses (16 MHz) is not a sub-rate of the OTDM data signal, the control pulses will scan across the data pulses, and hence give rise to a slowly varying idler signal. This can easily be detected by a sensitive slow photodetector, and when attached to a slow standard sampling oscilloscope, eye diagrams with ultra-high timing resolution can be constructed, cf. Figure 17.46.

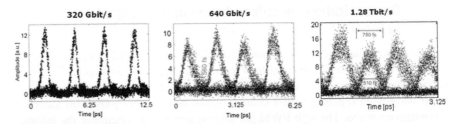

FIGURE 17.46 Ultra-fast optical sampling using FWM in a silicon waveguide. With a timing resolution of about 400 fs, 1.28 Tbit/s data signal can easily be resolved.

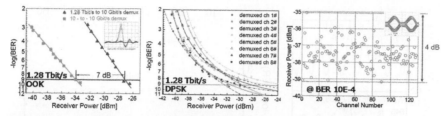

FIGURE 17.47 1.28 Tbit/s demultiplexing in a silicon waveguide. Left: 1.28 Tbit/s OOK BER performance. Middle: 1.28 Tbit/s DPSK BER performance. The error floor is caused by reflections in the particular waveguide for this experiment, being different to the one used in the OOK example. Right: Receiver power at a BER of 10E–4 for all 128 tributaries in the 1.28 Tbit/s DPSK signal.

The pulse width of the control pulse is 750 fs, and in the 1.28 Tbit/s case, the data pulses are 330 fs wide [159]. The measured eyes are a convolution of the data with the square of the pump, and comparing the sampled data pulse width to the autocorrelation of the data pulse width, the timing resolution of the scope can be deduced to about 400 fs [160]. In the 640 Gbit/s case, the data pulses are 500 fs FWHM, but appear roughly the same as the Tbit/s case on the sampled eye, indicating the system is close to its resolution limit. However, the Si sampling systems is clearly able to resolve Tbit/s data.

If the control pulse is operating at a sub-rate of the OTDM signal, i.e. at the base-rate, then demultiplexing can be carried out. Since FWM is phase preserving, the demultiplexing performance is independent of the data format. Figure 17.47 shows BER performance for 1.28 Tbit/s demultiplexing.

Demultiplexing 1.28 Tbit/s to 10 Gbit/s is in both the OOK and DPSK cases error-free, i.e. with a BER reaching 10E–9, demonstrating the strength of the silicon plat-form. In the OOK case, there is no sign of an error floor, but a 7 dB power penalty compared to a 10 Gbit/s back-to-back reference [159]. The penalty is partly due to a relatively low conversion efficiency. In the DPSK case [161], there is an error floor, which is caused by some excessive reflections occurring with this particular used device [162]. In the DPSK case, all 128 channels are demultiplexed and character-ized, however, at BER 10E–4 for ease of measurements. They behave with similar performance within 4 dB.

These results clearly demonstrate that nonlinear silicon waveguides can indeed be used for signal processing of Tbit/s data signals. The control pulse, though, is at 10 GHz in these cases. To really talk about ultra-high bit rate optical signal processing, the processing needs to take place not at the base rate, but at the line rate. In the following we will see examples of this.

17.3.2.2 640 Gbit/s wavelength conversion

Wavelength conversion is an example of signal processing at the aggregate line rate, i.e. the OTDM bit rate. In [35] the first clear demonstration (with BER character-ization) of optical signal processing at >100 Gbit/s was presented in the form of

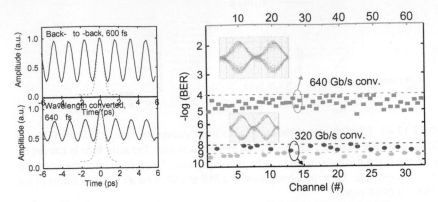

FIGURE 17.48 640 Gbit/s wavelength conversion using FWM in a nonlinear silicon waveguide. Left: Autocorrelation traces of the original 640 Gbit/s DPSK signal and the converted 640 Gbit/s DPSK signal. There is a slight pulse broadening owing to limited filter bandwidth after conversion. Right: BER results for 320 and 640 Gbit/s DPSK wavelength conversion [35] for all-data channels. All results are within FEC requirements, as all-channels are below BER 10E–4.

a 640 Gbit/s wavelength conversion experiment. The data signal was a 640 Gbit/s DPSK signal, and the pump was CW. Figure 17.48 shows 640 Gbit/s DPSK wavelength conversion results together with BER measurements for 320 Gbit/s. This is the highest signal processing speed reported using silicon, and Figure 17.48 shows full BER characterization for all tributary channels. All channels are within the standard FEC requirements with a 7% overhead, corresponding to 595 Gbit/s error-free conversion. This functionality is promising for low energy consumption: we used as low as 110 fJ/bit pump energy for the 640 Gbit/s case. We have previously shown 95 fJ/bit for 640 Gbit/s wavelength conversion using highly nonlinear fiber (HNLF) [91]. Wavelength conversion is more challenging than demultiplexing, as all-time channels need to be switched simultaneously, whereas a demultiplexer only switches one channel at a time. The demonstration of ultra-high-speed wavelength conversion proves that it is possible to isolate the FWM effect from the TPA/FCA effects and allow for ultra-fast processing. Here this was done simply by keeping the CW pump power below the TPA threshold. Note that the BER limitations in the 640 Gbit/s wavelength conversion result are due to OSNR limitations, stemming from the relatively modest conversion efficiency (−30 dB) of the used device and the insertion loss, resulting in low FWM output power.

17.3.2.3 640 Gbit/s serial-to-parallel conversion

As mentioned above, 1.28 Tbit/s demultiplexing has been demonstrated, but a single-channel demultiplexer may not be the best choice in terms of energy efficiency. If the demultiplexing can be based on a conversion scheme like the wavelength conversion, it may be more efficient [106]. With the development of the time lens concept, or

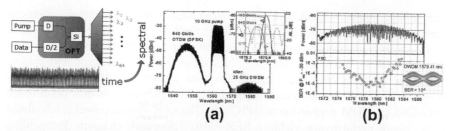

(a)   **(b)**

FIGURE 17.49 **640 Gbit/s serial-to-parallel conversion in a nonlinear silicon waveguide using the optical time lens or optical Fourier transformation (OFT) technique. Left: Schematic setup. Middle (a): Involved spectra, of the 640 Gbit/s DPSK data signal, the 10 GHz flat-top pump and the resulting 25 GHz spaced DWDM channels (inset: zoom-in on DWDM channels and optical tunable filter (OTF) transfer function to filter out individual DWDM channels). Right (b): Converted DWDM channels and corresponding BER performance (inset: received eye diagram at a BER 10E−6). Forty channels are within FEC limits of 2E−3 BER, i.e. more than half of all the 64 channels are simultaneously demultiplexed. (Adapted from [89]. © 2011 OSA.)**

equivalently the optical Fourier transformation (OFT) technique, a versatile new tool for manipulating optical waveforms has been created, as discussed in Sections 17.2.2.3, 17.2.3.1, and 17.2.4.1. As discussed, the technique allows for spectral-to-temporal domain and temporal-to-spectral domain transformations. In [36], it was demonstrated that this could also be done in a Si waveguide. The basic principle is sketched in Figure 17.49 (left). As described in Section 17.2.4.1, the OFT is based on FWM in the Si waveguide by chirped pump and data pulses. If the pump is dispersed twice as much as the data, spectral compression of the converted channels is achieved, and each temporal channel is furthermore converted to a separate WDM channel. In [36], 25 GHz spaced DWDM channels were generated this way from a 640 Gbit/s OTDM data signal. Figure 17.49 (middle) shows the involved spectra. Note that the pump spectrum is made flat-top in order to also create a flat-top temporal waveform to overlap with more than half of the OTDM channels. Figure 17.49 (right) shows the converted WDM channels and their corresponding BER performance. Forty of the 64 OTDM tributaries were simultaneously demultiplexed to individual 25 GHz spaced DWDM channels with BER performance below the FEC limit. This means that only two OFT units would be needed to demultiplex all channels, instead of 64 individual demultiplexers. This type of conversion may thus prove more energy efficient and more practical than single channel demultiplexing.

17.4 ENERGY PERSPECTIVES AND POTENTIAL APPLICATIONS

Historically, increasing the ETDM bit rate by a factor of four has reduced the cost per bit by 40%, partly due to less power consumption and easier management, e.g. by fewer components. For OTDM, the challenge is to maintain this trend, even though optical signal processing is required. Current electronic solutions seem limited to about 120 Gbit/s [163] due to power and heating constraints. In [164], power consumption over capacity is calculated for OTDM and WDM point-to-point systems. No in-line functionalities other than amplification and passive dispersion compensation are considered. Total system power consumption is shown in Figure 17.50, as presented in [164], with the addition of a point based on OTDM-to-WDM conversion at 640 Gbit/s. The total system power is calculated for a system performance yielding BER <10E-9 for each channel, disregarding common equipment (such as data modulators), but including cooling of all active devices, and electrical power to all amplifiers, lasers etc. [164]. Three classical OTDM receivers are investigated assuming N parallel demultiplexers based on SOAs or HNLF: OTDM 1: SOAs + single ctrl pulse source, OTDM 2: SOAs + N ctrl pulse sources, OTDM 3: HNLF + N ctrl pulse sources (worst case scenario). In addition to this, the serial-to-parallel scheme is added, assuming two HNLF-based OFTs, sharing a pump source and using the pump power used in the experimental demonstration described in Section 17.2.4.1.

Generally WDM is less power consuming due to the passive demultiplexing. SOAs are known to require less switching energy than HNLF, and the SOA schemes are indeed less power consuming than the HNLF case. The case with a single pump source is seen to scale favorably with higher bit rates, as more channels will be processed using the same number of pumps. The single pump source case approaches the WDM case for high bit rates. However, this example is based on an SOA as the active switching element, and it is questionable if SOAs will at all be able to operate with low penalty at these high bit rates. The worst case is by far the N parallel HNLFs and

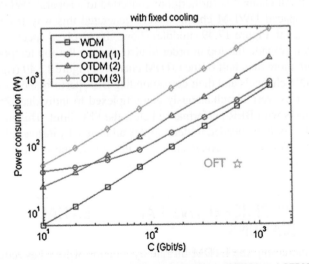

FIGURE 17.50 Total system power consumption comparison of WDM and OTDM systems for point-to-point system. Adapted from [164] with added OFT data point.

pump sources, being an order of magnitude worse than the WDM case. However for serial-to-parallel conversion at high bit rates (here 640 Gbit/s) the OTDM system is an order of magnitude less power consuming than the WDM system, and two orders of magnitude better than the worst-case parallel HNLF receiver. These calculations are realistic in the sense that they are based on the values used in the experimental demonstration.

Assuming two OFTs for all-channels demultiplexed, only 20 W total system power is needed at 640 Gbit/s (independent of the bit rate) compared to 200 W for a 640 Gbit/s WDM system [164]. This agrees well with the general understanding that processing more bits in fewer components is where OTDM has its strongest advantage [106,165]. In fact, for these operations, the energy per bit scales inversely with the bit rate, as most power is used for the control signal. And irrespective of the total OTDM bit rate, only two OFTs will be required, and to a large extent, the same pump power will be needed.

So overall, there may be power benefits in the OTDM transmitter relying on only a single pulse source, there may be power benefits in-line for regeneration, relying on only a single regenerator for all channels, and in the receiver with one to two OFT-based OTDM-to-WDM converters.

So, based on this discussion, we believe that conversion-type functionalities are energetically favorable. For serial-to-parallel conversion, pretty much the same switching power is required for various bit rates, so the same system power consumption is required, apart from the data signal power for increased OSNR of the data signal, and this leads to better energy efficiency for higher bit rates. It is even better than for WDM systems for this simple point-to-point consideration. So the most favorable functionalities supported by OTDM appear to be high-speed: regeneration, wavelength conversion, serial-to-parallel conversion and even packet switching if intelligent network structures are designed to reduce the need for optical buffering. Such functionalities may find applications in short-haul high-data-density networks such as data centres and supercomputers where ultra-high-speed serial data buses may be needed.

17.5 SUMMARY

In this chapter, the state of the art of OTDM systems have been presented, with a focus on experimental demonstrations. In summary, this chapter has especially highlighted demonstrations at 640–1280 Gbaud per polarization based on a variety of materials and functionalities. As the following list clearly shows, there are many technical solutions available today, and this list is not even complete, due to lack of space. The following has been demonstrated:

- Demultiplexing up to 1.28 Tbaud using HNLF, Si, ChG (640 Gbaud also SOA).
- Clock recovery at 640 Gbaud using SOA, PPLN, HNLF, and electro-optic phase modulation.

- Transmission at 640 Gbaud per polarization up to 525 km and at 1.28 Tbaud up to 50 km. Several dispersion compensating schemes are adequate such as dispersion compensating fibers, time domain optical Fourier transformation, spatial light modulation, which is even adaptive.
- Multi-level compatibility: 1.28 Tbaud per polarization with DQPSK reaching 5.1 Tbit/s and 16 QAM reaching 10.2 Tbit/s (including 29 km transmission).
- Channel identification and base rate synchronization at 650 Gbit/s.
- Add/drop multiplexing demonstrated at 640 Gbaud using HNLF and PPLN.
- 10 GbE packet synchronization and OTDM to 1.29 Tbit/s (Tbaud) demonstrated.
- Pulse shaping for flat-top pulse generation used for timing jitter tolerant demultiplexing at 640 Gbaud.
- Polarization-independent demultiplexing and wavelength conversion at 640 Gbaud.
- Wavelength conversion at 640 Gbaud per polarization using HNLF and Si, and 1.28 Tbit/s pol-MUX polarization-independent operation using HNLF.
- Serial-to-parallel conversion of 640 Gbaud using HNLF and silicon. Up to 1.28 Tbit/s using QPSK.
- Regeneration at 640 Gbit/s (Gbaud) using PPLN.
- Packet switching of 640 Gbit/s (Gbaud).
- Format conversion: RZ-to-NRZ conversion at 640 Gbaud.

Many essential network functionalities are available today using a plethora of available materials, so now it is time to look at new network scenarios that take advantage of the serial nature of the data, e.g. try to come up with practical schemes for ultra-high bit rate optical data packets in supercomputers or within data centers.

References

[1] T.S. Kinsel, R.T. Denton, Terminals for a high-speed optical pulse code modulation communication system: II. Optical multiplexing and demultiplexing, Proc. IEEE 56 (2) (1968) 146–154.
[2] F.S. Chen, Demultiplexers for high-speed optical PCM, J. Quant. Electron. 7 (1) (1971) 24–29.
[3] F.S. Chen, T.S. Kinsel, Experimental evaluation of an optical time division demultiplexer for twenty-four channels, Appl. Opt. 11 (1972) 1411–1418.
[4] R.S. Tucker, G. Eisenstein, S.K. Korotky, L.L. Buhl, J.J. Veselka, G. Raybon, B.L. Kasper, R.C. Alferness, 16-Gbit/s fiber transmission experiment using optical time-division multiplexing, Electron. Lett. 23 (1987) 1270–1271.
[5] R.S. Tucker, G. Eisenstein, S.K. Korotky, Optical time-division multiplexing for very high bit-rate transmission, J. Lightwave Technol. 6 (11) (1988) 1737–1749.
[6] T. Morioka, M. Saruwatari, A. Takada, Ultrafast optical multi/demultiplexer utilizing optical Kerr effect in polarization-maintaining single-mode fibres, Electron. Lett. 23 (9) (1987) 453–454.

[7] M. Nakazawa, E. Yoshida, T. Yamamoto, A. Sahara, TDM single channel 640 Gbit/s transmission experiment over 60 km using 400 fs pulse train and walk-off free, dispersion flattened nonlinear optical loop mirror, Electron. Lett. 34 (9) (1998) 907–908.

[8] M. Nakazawa, T. Yamamoto, K.R. Tamura, 1.28 Tbit/s–70 km OTDM transmission using third-and fourth-order simultaneous dispersion compensation with a phase modulator, Electron. Lett. 36 (24) (2000) 2027–2029.

[9] H.G. Weber, S. Ferber, M. Kroh, C. Schmidt-Langhorst, R. Ludwig, V. Marembert, C. Boerner, F. Futami, S. Watanabe, C. Schubert, Single channel 1.28 Tbit/s and 2.56 Tbit/s DQPSK transmission, Electron. Lett. 42 (3) (2006) 178–179.

[10] C. Schmidt-Langhorst, R. Ludwig, D.-D. Groß, L. Molle, M. Seimetz, R. Freund, C. Schubert, Generation and coherent time-division demultiplexing of up to 5.1 Tb/s single-channel 8-PSK and 16-QAM signals, in: Proceedings of the OFC 2009, San Diego, California, USA, Paper PDPC6, 2009.

[11] H.C.H. Mulvad, L.K. Oxenløwe, M. Galili, A. Clausen, L. Grüner-Nielsen, P. Jeppesen, 1.28 Tbit/s single-polarisation serial OOK optical data generation and demultiplexing, Electron. Lett. 45 (5) (2009) 280–281.

[12] H. Hu, M. Galili, L.K. Oxenløwe, J. Xu, H.C.H. Mulvad, C. Peucheret, A.T. Clausen, P. Jeppesen, Error-free transmission of serial 1.28 Tbaud RZ-DPSK signal, in: Proceedings of the ECOC 2010, Turin, Italy, Paper P4.18, 2010.

[13] H.C.H. Mulvad, M. Galili, L.K. Oxenløwe, H. Hu, A. Clausen, J.B. Jensen, C. Peucheret, P. Jeppesen, Error-free 5.1 Tbit/s data generation on a single-wavelength channel using a 1.28 Tbaud symbol rate, in: Procedings of the IEEE PHO Annual, Paper PD1.2, 2009.

[14] T. Richter, E. Palushani, C. Schmidt-Langhorst, M. Nŀolle, R. Ludwig, J. K. Fischer, C. Schubert, Single wavelength channel 10.2 Tb/s TDM-data capacity using 16-QAM and coherent detection, in: Optical Fiber Communication Conference OFC/NFOEC 2011, Los Angeles, California, USA, Paper PDPA9 (postdeadline).

[15] DE-CIX: Traffic Statistics. <http://www.de-cix.net/about/statistics/>, 2012.

[16] M. Mauldin, The state of the global internet, Webinare. <www.telegeography.com>, 2011.

[17] E.B. Desurvire, Capacity demand and technology challenges for lightwave systems in the next two decades, J. Lightwave Technol. 24 (12) (2006) 4697–4710.

[18] IEEE Std 802.3ba-2010: Amendment 4: Media Access control parameters, physical layers and management parameters for 40 Gb/s and 100 Gb/s operation.

[19] D. Hillerkuss, R. Schmogrow, T. Schellinger, M. Jordan, M. Winter, G. Huber, T. Vallaitis, R. Bonk, P. Kleinow, F. Frey, M. Roeger, S. Koenig, A. Ludwig, A. Marculescu, J. Li, M. Hoh, M. Dreschmann, J. Meyer, S. Ben Ezra, N. Narkiss, B. Nebendahl, F. Parmigiani, P. Petropoulos, B. Resan, A. Oehler, K. Weingarten, T. Ellermeyer, J. Lutz, M. Moeller, M. Huebner, J. Becker, C. Koos, W. Freude, J. Leuthold, 26 Tbit/s line-rate super-channel transmission utilizing all-optical fast Fourier transform processing, Nat. Photon. 5 (2011) 364–371.

[20] D.Y. Qian, M.F. Huang, E. Ip, Y.K. Huang, Y. Shao, J.Q. Hu, T. Wang 101.7-Tb/s (370 × 294-Gb/s) PDM-128QAM-OFDM transmission over 3 × 55 km SSMF using Pilot-based phase noise mitigation, in: Proceedings of the OFC 2011, Los Angeles, California, USA, Paper PDPB5, 2011.

[21] A. Sano, T. Kobayashi, S. Yamanaka, A. Matsuura, H. Kawakami, Y. Miyamoto, K. Ishihara, H. Masuda, 102.3-Tb/s (224 × 548-Gb/s) C- and extended L-band all-Raman

transmission over 240 km using PDM-64QAM single carrier FDM with digital pilot tone, in: Proceedings of the OFC 2012, Los Angeles, CA, USA, Paper PDP5C.3, 2012.

[22] J. Sakaguchi, Y. Awaji, N. Wada, A. Kanno, T. Kawanishi, T. Hayashi, T. Taru, T. Kobayashi, M. Watanabe, 109-Tb/s ($7 \times 97 \times 172$-Gb/s SDM/WDM/PDM) QPSK transmission through 16.8-km homogeneous multi-core fiber, in: Proceedings of the OFC 2011, Los Angeles, CA, USA, Paper PDPB6, 2011.

[23] J. Sakaguchi, B.J. Puttnam, W. Klaus, Y. Awaji, N. Wada, A. Kanno, T. Kawanishi, K. Imamura, H. Inaba, K. Mukasa, R. Sugizaki, T. Kobayashi, M. Watanabe, 19-core fiber transmission of $19 \times 100 \times 172$-Gb/s SDM-WDM-PDM-QPSK signals at 305 Tb/s, in: Proceedings of the OFC 2012, Los Angeles, California, USA, Paper PDP5C.1, 2012.

[24] E. Tangdiongga, H.C.H. Mulvad, H.de Waardt, G.D. Khoe, A.M.J. Koonen, H.J.S. Dorren, SOA-based clock recovery and demultiplexing in a lab trial of 640-Gb/s OTDM transmission over 50-km fibre link, in: Proceedings of the ECOC 2007, Berlin, Germany, Paper PD 1.2, 2007.

[25] L.K. Oxenløwe, D. Zibar, M. Galili, A.T. Clausen, L.J. Christiansen, P. Jeppesen, Filtering-assisted cross-phase modulation in a semiconductor optical amplifier enabling 320 Gb/s clock recovery, in: Proceedings of the ECOC 2005, Glasgow, Scotland, Paper We3.5.5, 2005.

[26] H.C.H. Mulvad, E. Tangdiongga, O. Raz, J.H. Llorente, H. de Waardt, H.J.S. Dorren, 640 Gbit/s OTDM lab-transmission and 320 Gbit/s field-transmission with SOA based clock recovery, in: Proceedings of the OFC 2008, San Diego, California, USA, Paper OWS2, 2008.

[27] O. Kamatani, S. Kawanishi, Prescaled timing extraction from 400 Gb/s optical signal using a phase lock loop based on four-wave-mixing in a laser diode amplifier, Photon Technol. Lett. 8 (8) (1996) 1094–1096.

[28] T. Yamamoto, M. Nakazawa, Ultrafast OTDM transmission using novel fiber devices for pulse compression, shaping, and demultiplexing, J. Opt. Fiber Commun. Rep. 2 (3) (2005) 209–225.

[29] L.K. Oxenløwe, K.S. Berg, A. Clausen, J. Seoane, A. Siahlo, P. Jeppesen, M. Schmidt, M. Le Schilling, Q.T. Nghi, Specialty fibers for 160, 320 and 640 Gb/s signal processing, in: Proceedings of the CLEO'04, San Francisco, California, USA, Paper CThQ1, 2004.

[30] A.I. Siahlo, A.T. Clausen, L.K. Oxenløwe, J. Seoane, P. Jeppesen, 640 Gb/s OTDM transmission and demultiplexing using a NOLM with commercially available highly non-linear fiber, in: Proceedings of the CLEO 2005, Baltimore, Maryland, USA, Paper CTuO1, 2005.

[31] M. Galili, J. Xu, H.C.H. Mulvad, L.K. Oxenløwe, A.T. Clausen, P. Jeppesen, B. Luther-Davies, A. Rode, S. Madden, A. Rode, D-Y. Choi, M. Pelusi, F. Luan, B.J. Eggleton, Breakthrough switching speed with an all-optical chalcogenide glass chip: 640 Gbit/s demultiplexing, Opt. Express 17 (4) (2009) 2182–2187.

[32] T.D. Vo, H. Hu, M. Galili, E. Palushani, J. Xu, L.K. Oxenløwe, S.J. Madden, D-Y. Choi, D. Bulla, M.D. Pelusi, J. Schroder, B. Luther-Davies, B.J. Eggleton, Photonic chip based 1.28 Tbaud transmitter optimization and receiver OTDM demultiplexing, in: Proceedings of the OFC 2010, San Diego, California, USA, Paper PDPC5, 2010.

[33] R. Salem, M.A. Foster, A.C. Turner, D.F. Geraghty, M. Lipson, A.L. Gaeta, Signal regeneration using low-power four-wave mixing on silicon chip, Nat. Photon. 2 (2008) 35–38.

[34] H. Ji, H. Hu, M. Galili, L.K. Oxenløwe, M. Pu, K. Yvind, J.M. Hvam, P. Jeppesen, Optical waveform sampling and error-free demultiplexing of 1.28 Tbit/s serial data in a silicon nanowire, in: Proceedings of the OFC 2010, San Diego, California, USA, Paper PDPC7, 2010.

[35] H. Hu, H. Ji, M. Galili, M. Pu, H.C.H. Mulvad, L.K. Oxenløwe, K. Yvind, J.M. Hvam, P. Jeppesen, Silicon chip based wavelength conversion of ultra-high repetition rate data signals, in: Proceedings of the OFC 2011, Los Angeles, California, USA, Paper PDPA8, 2011.

[36] H.C.H. Mulvad, E. Palushani, H. Hu, H. Ji, M. Galili, A.T. Clausen, M. Pu, K. Yvind, J.M. Hvam, P. Jeppesen, L.K. Oxenløwe, Ultra-high-speed optical serial-to-parallel data conversion in a silicon nanowire, in: Proceedings of the ECOC 2011, Geneva, Switzerland, Paper Th.13.A.2, 2011.

[37] A. Bogoni, X. Wu, S.R. Nuccio, A.E. Willner, 640 Gb/s all-optical regenerator based on a periodically poled Lithium Niobate waveguide, J. Lightwave Technol. 30 (12) (2012) 1829–1834.

[38] A. Bogoni, X. Wu, S.R. Nuccio, J. Wang, Z. Bakhtiari, A.E. Willner, Photonic 640-Gb/s reconfigurable OTDM add-drop multiplexer based on pump depletion in a single PPLN waveguide, J. Select. Topics Quantum Electron. 18 (2) (2012) 709–716.

[39] L.K. Oxenløwe, F. Gomez-Agis, C. Ware, S. Kurimura, H.C.H. Mulvad, M. Galili, H. Nakajima, J. Ichikawa, D. Erasme, A.T. Clausen, P. Jeppesen, 640-Gbit/s data transmission and clock recovery using an ultrafast periodically poled lithium niobate device, J. Lightwave Technol. 27 (3) (2009) 205–213.

[40] L.K. Oxenløwe, F. Gomez-Agis, C. Ware, S. Kurimura, H.C.H. Mulvad, M. Galili, H. Nakajima, J. Ichikawa, D. Erasme, A.T. Clausen, P. Jeppesen, 640 Gbit/s data transmission and clock recovery using an ultra-fast periodically poled lithium niobate device, in: Proceedings of the OFC 2008, San Diego, California, USA, Paper PDP22, 2008.

[41] H. Hu, P. Münster, E. Palushani, M. Galili, K. Dalgaard, H.C.H. Mulvad, P. Jeppesen, L.K. Oxenløwe, 640 Gbaud NRZ-OOK data signal generation and 1.19 Tbit/s PDM-NRZ-OOK field trial transmission, in: Proceedings of the OFC 2012, Los Angeles, USA, Paper PDP5C.7, 2012.

[42] L.A. Jiang, E.P. Ippen, H. Yokoyama, Semiconductor mode-locked lasers as pulse sources for high bit rate data transmission, J. Opt. Fiber. Commun. Rep. 2 (1) (2005) 1–31.

[43] P.J. Legg, M. Tur, I. Andonovic, Solution paths to limit interferometric noise induced performance degradations in ASK/direct detection lightwave networks, J. Lightwave Technol. 14 (9) (1996) 1943–1954.

[44] A.T. Clausen, H.N. Poulsen, L.K. Oxenløwe, A.I. Siahlo, J. Seoane, P. Jeppesen, Pulse source requirements for OTDM systems, in: Proceedings of the LEOS 2003, Tucson, USA, Paper TuY2, 2003.

[45] X. Jiang, I. Roudas, K. Jepsen, Asymmetric probability density function of a signal with interferometric crosstalk in optically amplified systems, in: Proceedings of the OFC 2000, vol. 2, San Jose, California, USA, 2000, pp. 151–153.

[46] A.T. Clausen, Experimental and theoretical investigations of systems with potential for terabit capacity, PhD Thesis, DTU Fotonik, Kgs. Lyngby, Denmark, 2006.

[47] N.J. Doran, D. Wood, Nonlinear-optical loop mirror, Opt. Lett. 13 (1) (1988) 56–58.

[48] M. Jinno, Effects of crosstalk and timing jitter on all-optical time-division demultiplexing using a nonlinear fiber Sagnac interferometer switch, J. Quant. Electron. 30 (12) (1994) 2842–2853.

[49] P. Guan, H.C.H. Mulvad, K. Kasai, T. Hirooka, M. Nakazawa, High time-resolution 640-Gb/s clock recovery using time-domain optical Fourier transformation and narrowband optical filter, Photon. Technol. Lett. 22 (23) (2010) 1735–1737.

[50] B.P.P. Kuo, A.O.J. Wiberg, C.S. Brès, N. Alic, S. Radic, Ultrafast clock recovery and sampling by single parametric device, Photon. Technol. Lett. 23 (3) (2011) 191–193.

[51] D.T.K. Tong, K.L. Deng, B. Mikkelsen, G. Raybon, K.F. Dreyer, J.E. Johnson, 160 Gbit/s clock recovery using electroabsorption modulator-based phase-locked loop, Electron. Lett. 36 (23) (2000) 1951–1952.

[52] M. Schmidt, K. Schuh, E. Lach, M. Schilling, G. Veith, 8 × 160 Gbit/s (1.28 Tbit/s) DWDM transmission with 0.53 bit/s/Hz spectral efficiency using single EA-modulator based RZ pulse source and demux, in: Proceedings of the ECOC 2003, Rimini, Italy, Paper Mo3.6.5, 2003.

[53] C. Boerner, C. Schubert, C. Schmidt, E. Hilinger, V. Marembert, J. Berger, S. Ferber, E. Dietrich, R. Ludwig, B. Schmauss, H.G. Weber, 160 Gbit/s clock recovery with electro-optical PLL using bidirectional operated electroabsorption modulator as phase comparator, Electron. Lett. 39 (14) (2003) 1071–1073.

[54] L.K. Oxenløwe, A.I. Siahlo, K.S. Berg, A.T. Clausen, B.M. Sørensen, K. Yvind, P. Jeppesen, K.P. Hansen, K. Hoppe, J. Hanberg, A novel 160 Gb/s receiver configuration including a glass crystal pulsed laser, photonic crystal fibre and a simple dynamic clock recovery scheme, in: Proceedings of the ECOC 2003, Rimini, Italy, Paper Th2.5.3, 2003.

[55] I.D. Phillips, P.N. Kean, N.J. Doran, I. Bennion, D.A. Pattison, A.D. Ellis, Simultaneous demultiplexing, data regeneration, and clock recovery with a single semiconductor optical amplifier-based nonlinear-optical loop mirror, Opt. Lett. 22 (17) (1997) 1326–1328.

[56] L.K. Oxenløwe, C. Schubert, C. Schmidt, E. Hilliger, J. Berger, U. Feiste, R. Ludwig, H.G. Weber, Optical clock recovery employing an optical PLL using cross-phase modulation in a Sagnac interferometer, in: Proceedings of the CLEO 2001, Baltimore, Paper CThU2, 2001.

[57] M. Galili, H.C. Hansen Mulvad, H. Hu, L.K. Oxenløwe, F. Gomez Agis, C. Ware, D. Erasme, A.T. Clausen, P. Jeppesen, 650 Gbit/s OTDM transmission over 80 km SSMF incorporating clock recovery, channel identification and demultiplexing in a polarisation insensitive receiver, in: Proceedings of the OFC'10, San Diego, California, USA, 2010. Paper OWO3, 2010.

[58] J. Mørk, A. Mecozzi, C. Hultgren, Spectral effects in short pulse pump-probe measurements, Appl. Phys. Lett. 68 (4) (1996) 449–451.

[59] A.V. Uskov, J.R. Karin, J.E. Bowers, J.G. McInerney, J.L. Bihan, Effects of carrier cooling and carrier heating in saturation dynamics and pulse propagation through bulk semiconductor absorbers, J. Quant. Electron. 34 (11) (1998) 2162–2171.

[60] F.G. Agis, C. Ware, D. Erasme, R. Ricken, V. Quiring, W. Sohler, 10-GHz clock recovery using an optoelectronic phase-locked loop based on three-wave mixing in periodically poled lithium niobate, Photon. Technol. Lett. 18 (13) (2006) 1460–1462.

[61] F.G. Agis, L.K. Oxenløwe, S. Kurimura, C. Ware, H.C.H. Mulvad, M. Galili, D. Erasme, Ultrafast phase comparator for phase-locked loop-based optoelectronic clock recovery systems, J. Lightwave Technol. 27 (13) (2009) 2439–2448.

[62] S. Kurimura, Y. Kato, M. Maruyama, Y. Usui, H. Nakajima, Quasi-phase-matched adhered ridge waveguide in LiNbO$_3$, Appl. Phys. Lett. 89 (2006) 191123-1–191123-3.

[63] H.C.H. Mulvad, M. Galili, L.K. Oxenløwe, H. Hu, A.T. Clausen, J.B. Jensen, C. Peucheret, P. Jeppesen, Demonstration of 5.1 Tbit/s data capacity on a single-wavelength channel, Opt. Express 18 (2) (2010) 1438–1443.

[64] P.J. Winzer, R.J. Essiambre, Advanced modulation formats for high-capacity optical transport networks, J. Lightwave Technol. 24 (12) (2006) 4711–4728.

[65] F. Ito, Interferometric demultiplexing experiment using linear coherent correlation with modulated local oscillator, Electron. Lett. 32 (1) (1996) 14–15.

[66] F. Ito, Demultiplexed detection of ultrafast optical signal using interferometric cross-correlation technique, J. Lightwave Technol. 15 (6) (1997) 930–937.

[67] C. Zhang, Y. Mori, K. Igarashi, K. Katoh, K. Kikuchi, Ultrafast operation of digital coherent receivers using their time-division demultiplexing function, J. Lightwave Technol. 27 (3) (2009) 224–232.

[68] T. Richter, E. Palushani, C. Schmidt-Langhorst, R. Ludwig, L. Molle, M. Nölle, J.K. Fischer, C. Schubert, Transmission of single-channel 16-QAM data signals at Tbaud symbol rates, J. Lightwave Technol. 30 (4) (2012) 504–511.

[69] H.C. Hansen Mulvad, M. Galili, L.K. Oxenløwe, A.T. Clausen, L. Grüner-Nielsen, P. Jeppesen, 640 Gbit/s optical time-division add-drop multiplexing in a non-linear optical loop mirror, IEEE Lasers and Electro-Optics Society Winter Topical Meeting, Innsbruck, Austria, 2009, Paper MC4.4, 2009.

[70] H.C. Hansen Mulvad, M. Galili, L. Grüner-Nielsen, L.K. Oxenløwe, A.T. Clausen, P. Jeppesen, 640 Gbit/s time-division add-drop multiplexing using a non-linear polarisation-rotating fibre loop, in: Procedings of the ECOC 2008, Brussels, Belgium, Paper Tu.3.D.6, 2008.

[71] B.H. Kolner, Space-time duality and the theory of temporal imaging, J. Quant. Electron. 30 (8) (1994) 1951–1963.

[72] C. Meirosu, P. Golonka, A. Hirstius, S. Stancu, B. Dobinson, E. Radius, A. Antony, F. Dijkstra, J. Blom, C. de Laat, Native 10 Gigabit Ethernet experiments over long distances, Future Gener. Comput. Syst. 21 (4) (2005) 457–468.

[73] J. Areal, H. Hu, C. Peucheret, E. Palushani, R. Puttini, A. Clausen, M. Berger, A. Osadchiy, L.K. Oxenløwe, Analysis of a time-lens based optical frame synchronizer and retimer for 10G Ethernet aiming at a Tb/s optical router/switch design, in: Proceedings of the ONDM 2010, Kyoto, Japan, Paper P-3, 2010.

[74] H. Hu, J. Laguardia Areal, H.C. Hansen Mulvad, M. Galili, K. Dalgaard, E. Palushani, A.T. Clausen, M.S. Berger, P. Jeppesen, L.K. Oxenløwe, Synchronization, retiming and time-division multiplexing of an asynchronous 10 Gigabit NRZ Ethernet packet to terabit Ethernet, Opt. Express 19 (26) (2011) B931–B937.

[75] M. Nakazawa, T. Hirooka, F. Futami, S. Watanabe, Ideal distortion-free transmission using optical Fourier transformation and Fourier transform-limited optical pulses, Photon. Technol. Lett. 16 (4) (2004) 1059–1061.

[76] H.C.H. Mulvad, H. Hu, M. Galili, H. Ji, E. Palushani, A.T. Clausen, L.K. Oxenløwe, P. Jeppesen, DWDM-to-OTDM conversion by time-domain optical Fourier transformation, in: Proceedings of the ECOC 2011, Geneva, Switzerland, Paper Mo.1.A.5, 2011.

[77] E. Palushani, L.K. Oxenløwe, M. Galili, H.C. Hansen Mulvad, A.T. Clausen, P. Jeppesen, Flat-top pulse generation by the optical Fourier transform technique for ultrahigh speed signal processing, J. Quant. Electron. 45 (11) (2009) 1317–1324.

[78] L.K. Oxenløwe, R. Slavík, M. Galili, H.C.M. Mulvad, A.T. Clausen, Y. Park, J. Azaña, P. Jeppesen, 640 Gbit/s timing jitter tolerant data processing using a long-period fiber grating-based flat-top pulse shaper, J. Sel. Top. Quant. Electron. 14 (3) (2008) 566–572.

[79] F. Parmigiani, L.K. Oxenløwe, M. Galili, M. Ibsen, D. Zibar, P. Petropoulos, D. Richardson, A.T. Clausen, P. Jeppesen, All-optical 160 Gbit/s RZ data retiming system incorporating a pulse shaping fibre Bragg grating, in: Proceedings of the ECOC 2007, Berlin, Germany, 2007, pp. 219–220.

[80] R. Slavík, Y. Park, J. Azaña, Long-period fiber-grating-based filter for generation of picosecond and sub-picosecond transform-limited flat-top pulses, Photon. Technol. Lett. 20 (10) (2008) 806–808.

[81] K.N. Berger, B. Levit, B. Fischer, J. Azaña, Picosecond flat-top pulse generation by low-bandwidth electro-optic sinusoidal phase modulation, Opt. Lett. 33 (2) (2008) 125–127.

[82] P. Petropoulos, M. Ibsen, A.D. Ellis, D.J. Richardson, Rectangular pulse generation based on pulse reshaping using a superstructured fiber Bragg grating, J. Lightwave Technol. 19 (5) (2001) 746–752.

[83] H. Hu, H.C. Hansen Mulvad, M. Galili, E. Palushani, J. Xu, A.T. Clausen, L.K. Oxenløwe, P. Jeppesen, Polarization-insensitive 640 Gb/s demultiplexing based on four wave mixing in a polarization-maintaining fibre loop, J. Lightwave Technol. 28 (12) (2010) 1789–1795.

[84] E. Palushani, H. Hu, M. Galili, H.C.H. Mulvad, R. Slavík, L.K. Oxenløwe, A.T. Clausen, P. Jeppesen, 640 Gbit/s polarisation-independent demultiplexing in a standard nonlinear-optical-loop-mirror using a cascaded long-period grating pulse shaper, in: IEEE Photonics Society 23rd Annual Meeting, Denver, Colorado, USA, pp. 203–204, Paper TuM2, 2010.

[85] C. Schubert, J. Berger, U. Feiste, R. Ludwig, C. Schmidt, H. Weber, 160-Gb/s polarization insensitive all-optical demultiplexing using a gain-transparent ultrafast nonlinear interferometer (GT-UNI), Photon. Technol. Lett. 13 (11) (2001) 1200–1202.

[86] H.C.M. Mulvad, M. Galili, L.K. Oxenløwe, A.T. Clausen, L.G.-Nielsen, P. Jeppesen, Polarization-independent high-speed switching in a standard non-linear optical loop mirror, in: Proceedings of the OFC 2008, San Diego, California, USA, Paper OMN3, 2008.

[87] E. Palushani, L.K. Oxenløwe, M. Galili, H.C.H. Mulvad, A.T. Clausen, P. Jeppesen, Flat-top pulse generation by the optical Fourier transform technique for ultra-high-speed signal processing, J. Quant. Electron. 45 (11) (2009) 1317–1324.

[88] K.G. Petrillo, M.A. Foster, Scalable ultrahigh-speed optical transmultiplexer using a time lens, Opt. Express 19 (15) (2011) 14051–14059.

[89] H.C.H. Mulvad, E. Palushani, H. Hu, H. Ji, M. Lillieholm, M. Galili, A.T. Clausen, M. Pu, K. Yvind, J.M. Hvam, P. Jeppesen, L.K. Oxenløwe, Ultra-high-speed optical serial-to-parallel data conversion by time-domain optical Fourier transformation in a silicon nanowire, Opt. Express 19 (26) (2011) B825–B835.

[90] E. Palushani, T. Richter, R. Ludwig, C. Schubert, H.C. Hansen Mulvad, A.T. Clausen, L.K. Oxenløwe, OTDM-to-WDM conversion of complex modulation formats by time-domain optical Fourier transformation, in: Proceedings of the OFC 2012, Los Angeles, California, USA, Paper OTh3H.2, 2012.

[91] H. Hu, E. Palushani, M. Galili, H.C. Hansen Mulvad, A.T. Clausen, L.K. Oxenløwe, P. Jeppesen, 640 Gbit/s and 1.28 Tbit/s polarisation insensitive all optical wavelength conversion, Opt. Express 18 (10) (2010) 9961–9966.

[92] C.S. Bres, A. Wiberg, B.P.P. Kuo, N. Alic, S. Radic, Wavelength multicasting of 320-Gb/s channel in self-seeded parametric amplifier, Photon. Technol. Lett. 21 (14) (2009) 1002–1004.

[93] M. Galili, L.K. Oxenlowe, H.C.H. Mulvad, A.T. Clausen, P. Jeppesen, Optical wavelength conversion by cross-phase modulation of data signals up to 640 Gb/s, J. Sel. Top. Quant. Electron. 14 (3) (2008) 573–579.

[94] J.H. Lee, T. Nagashima, T. Hasegawa, S. Ohara, N. Sugimoto, K. Kikuchi, Bismuth oxide nonlinear fibre-based 80 Gbit/s wavelength conversion and demultiplexing using cross-phase modulation and filtering, Electron. Lett. 41 (22) (2005) 1237–1238.

[95] V.G. Ta'eed, L. Fu, M. Pelusi, M. Rochette, I.C.M. Littler, D.J. Moss, B.J. Eggleton, Error free all optical wavelength conversion in highly nonlinear As-Se chalcogenide glass fiber, Opt. Express 14 (22) (2006) 10371–10376.

[96] B.G. Lee, A. Biberman, A.C. Turner-Foster, M.A. Foster, M. Lipson, A.L. Gaeta, K. Bergman, Demonstration of broadband wavelength conversion at 40 Gb/s in silicon waveguides, Photon. Technol. Lett. 21 (3) (2009) 182–184.

[97] N. Ophir, R.K.W. Lau, M. Menard, X. Zhu, K. Padmaraju, Y. Okawachi, R. Salem, M. Lipson, A.L. Gaeta, K. Bergman, Wavelength conversion and unicast of 10-Gb/s data spanning up to 700 nm using a silicon nanowaveguide, Opt. Express 20 (6) (2012) 6488–6495.

[98] V. Ta'eed, M.D. Pelusi, B.J. Eggleton, D.-Y. Choi, S. Madden, D. Bulla, B. Luther-Davies, Broadband wavelength conversion at 40 Gb/s using long serpentine As2S3 planar waveguides, Opt. Express 15 (23) (2007) 15047–15052.

[99] Y. Liu, E. Tangdiongga, Z. Li, H.D. Waardt, A.M.J. Koonen, G.D. Khoe, X. Shu, I. Bennion, H.J.S. Dorren, Error-free 320-Gb/s all-optical wavelength conversion using a single semiconductor optical amplifier, J. Lightwave Technol. 25 (1) (2007) 103–108.

[100] M. Matsuura, O. Raz, F. Gomez-Agis, N. Calabretta, H.J.S. Dorren, 320 Gbit/s wavelength conversion using four-wave mixing in quantum-dot semiconductor optical amplifiers, Opt. Lett. 36 (15) (2011) 2910–2912.

[101] A. Bogoni, X. Wu, I. Fazal, A.E. Willner, 320 Gb/s nonlinear operations based on a PPLN waveguide for optical multiplexing and wavelength conversion, in: Proceedings of the OFC 2009, San Diego, California, USA, Paper OThS5, 2009.

[102] H. Hu, R. Nouroozi, R. Ludwig, B. Huettl, C. Schmidt-Langhorst, H. Suche, W. Sohler, C. Schubert, Polarization-insensitive all-optical wavelength conversion of 320 Gb/s RZ-DQPSK signals using a Ti:PPLN waveguide, Appl. Phys. B 101 (4) (2010) 875–882.

[103] C.E. Shannon, Bell Syst. Technol. J. 27 (1948) 379–423 623–656.

[104] C.E. Shannon, W. Weaver, The Mathematical Theory of Communication, University of Illinois Press, 1963.

[105] A.D. Ellis, J. Zhao, Channel capacity of non-linear transmission systems, in: S. Kumar (Ed.), Impact of Nonlinearities on Fiber Optic Communications, Optical and Fiber Communications Reports, vol. 7, Springer, 2011.

[106] R.S. Tucker, K. Hinton, Energy consumption and energy density in optical and electronic signal processing, IEEE Photon. J. 3 (5) (2011) 821–833.

[107] L.K. Oxenløwe, H.C. Hansen Mulvad, H. Hu, H. Ji, M. Galili, M. Pu, E. Palushani, K. Yvind, J.M. Hvam, A.T. Clausen, P. Jeppesen, Ultrafast nonlinear signal processing in silicon waveguides, in: Proceedings of the OFC 2012, Los Angeles, California, USA, Paper OTh3H.5, 2012.

[108] J. Kakande, R. Slavik, F. Parmigiani, P. Petropoulos, D. Richardson, All-optical processing of multi-level phase shift keyed signals, in: Proceedings of the OFC 2012, Los Angeles, California, USA, Paper OW1I, 2012.

[109] R. Slavík, F. Parmigiani, J. Kakande, C. Lundström, M. Sjödin, P.A. Andrekson, R. Weerasuriya, S. Sygletos, A.D. Ellis, L. Grüner-Nielsen, D. Jakobsen, S. Herstrøm,

R. Phelan, J. O'Gorman, A. Bogris, D. Syvridis, S. Dasgupta, P. Petropoulos, D.J. Richardson, All-optical phase and amplitude regenerator for next-generation telecommunications systems, Nat. Photon. 4 (2010) 690–695.

[110] P. Andrekson, Phase sensitive fiber optic parametric amplifiers, in: Proceedings of the ECOC 2011, Geneva, Switzerland, Paper Th.11, 2011.

[111] S. Watanabe, F. Futami, R. Okabe, Y. Takita, S. Ferber, R. Ludwig, C. Schubert, C. Schmidt, H.G. Weber, 160 Gbit/s optical 3R-regenerator in a fiber transmission experiment, in: Proceedings of the OFC 2003, Atlanta, Georgia, USA, vol. 3, Paper PD16-P1-3 2003.

[112] F. Parmigiani, L.K. Oxenløwe, M. Galili, M. Ibsen, D. Zibar, P. Petropoulos, D.J. Richardson, A.T. Clausen, P. Jeppesen, All-optical 160 Gbit/s retiming system using fiber grating based pulse shaping technology, J. Lightwave Technol. 27 (9) (2009) 1135–1141.

[113] F. Gomez-Agis, H. Hu, J. Luo, H.C.H. Mulvad, M. Galili, N. Calabretta, L.K. Oxenløwe, H.J.S. Dorren, P. Jeppesen, Optical switching and detection of 640 Gbits/s optical time-division multiplexed data packets transmitted over 50 km of fiber, Opt. Lett. 36 (17) (2011) 3473–3475.

[114] F. Gomez-Agis, N. Calabretta, A. Albores-Mejia, H.J.S. Dorren, Clock-distribution with instantaneous synchronization for 160 Gbit/s optical time domain multiplexed systems packet transmission, Opt. Lett. 35 (19) (2010) 3255–3257.

[115] N. Calabretta, W. Wang, T. Ditewig, O. Raz, F. Gomez-Agis, S. Zhang, H. de Waardt, H.J.S. Dorren, Scalable optical packet switches for multiple data formats and data rates packets, Photon. Technol. Lett. 22 (7) (2010) 483–485.

[116] H. Hu, H.C. Hansen Mulvad, C. Peucheret, M. Galili, A. Clausen, P. Jeppesen, L.K. Oxenløwe, 10 GHz pulse source for 640 Gbit/s OTDM based on LiNbO3 modulators and self-phase modulation, Opt. Express 19 (26) (2011) B343–B349.

[117] T. Hirano, P. Guan, T. Hirooka, M. Nakazawa, 640-Gb/s/channel single-polarization DPSK transmission over 525 km with ultrafast time-domain optical fourier transformation, Photon. Technol. Lett. 22 (14) (2010) 1042–1044.

[118] Y. Paquot, J. Schröder, J.V. Erps, T.D. Vo, M.D. Pelusi, S. Madden, B. Luther-Davies, B.J. Eggleton, Single parameter optimization for simultaneous automatic compensation of multiple orders of dispersion for a 1.28 Tbaud signal, Opt. Express 19 (25) (2011) 25512–25520.

[119] S. Bigo, Technologies for global telecommunications using undersea cables, in: I.P. Kaminow, T. Li, A.E. Willner (Eds.), Optical Fiber Telecommunications V B Systems and Networks, Academic Press, California, USA, 2008, pp. 561–611.

[120] IEEE Std 802.3-2008—IEEE Standard for Information Technology—Telecommunications and Information Exchange Between Systems—Local and Metropolitan Area Networks—Specific Requirements Part 3: Carrier Sense Multiple Access With Collision Detection (CSMA/CD) Access Method and Physical Layer Specifications—Section Four.

[121] H.-G. Weber, R. Ludwig, Ultra-high-speed OTDM transmission technology, Optical Fiber Telecommunications V B: Systems and Networks, Elsevier Inc., 2008, pp. 201–232.

[122] T. Yamamoto, E. Yoshida, K.R. Tamura et al., 640 Gbit/s optical TDM transmission over 92 km through a dispersion managed fiber consisting of single-mode fiber and reverse dispersion fibre, Photon. Technol. Lett. 12 (3) (2000) 353–355.

[123] P. Guan, H.C. Hansen Mulvad, Y. Tomiyama, T. Hirano, T. Hirooka, M. Nakazawa, Single-channel 1.28 Tbit/s-525 km DQPSK transmission using ultrafast time-domain

optical Fourier transformation and nonlinear optical loop mirror, IEICE Trans. Commun. E94-B (2011) 430–436.

[124] P. Guan, T. Hirano, K. Harako, Y. Tomiyama, T. Hirooka, M. Nakazawa, 2.56 Tbit/s/ch polarization-multiplexed DQPSK transmission over 300 km using time-domain optical Fourier transformation, Opt. Express 19 (26) (2011) B567–B573.

[125] B.P.-P. Kuo, E. Myslivets, A.O.J. Wiberg, S. Zlatanovic, C.-S. Brès, S. Moro, F. Gholami, A. Peric, N. Alic, S. Radic, Transmission of 640-Gb/s RZ-OOK channel over 100-km SSMF by wavelength-transparent conjugation, J. Lightwave Technol. 29 (4) (2011) 516–523.

[126] A.M. Weiner, Femtosecond pulse shaping using spatial light modulators, Rev. Sci. Instrum. 71 (5) (2000) 1929–1960.

[127] M.A.F. Roelens, S. Frisken, J.A. Bolger, D. Abakoumov, G. Baxter, S. Poole, B.J. Eggleton, Dispersion trimming in a reconfigurable wavelength selective switch, J. Lightwave Technol. 26 (1) (2008) 73–78.

[128] G. Tologlou, H.C.H. Mulvad, K. Dalgaard. M. Galili, A.T. Clausen, P. Jeppesen, L.K. Oxenløwe, Distortion-less 610 fs pulse transmission over 160 km SSMF-DCF using wavelength selective switch for compensation of chromatic dispersion, in: Proceedings of the Photonics Conference (PHO), 2011, Arlington, USA, 2011, pp. 829–830.

[129] H. Miao, A.M. Weiner, L. Mirkin, P.J. Miller, Broadband all-order polarization mode dispersion compensation via wavelength-by-wavelength Jones matrix correction, Opt. Lett. 32 (16) (2007) 2360–2362.

[130] Y. Ding, H. Hu, M. Galili, J. Xu, L. Liu, M. Pu, H.C.H. Mulvad, L.K. Oxenløwe, C. Peucheret, P. Jeppesen, X. Zhang, D. Huang, H. Ou, Generation of a 640 Gbit/s NRZ OTDM signal using a silicon microring resonator, Opt. Express 19 (7) (2011) 6471–6477.

[131] L. Grüner-Nielsen, S. Dasgupta, M.D. Mermelstein, D. Jakobsen, S Herstrøm, M.E.V. Pedersen, E.L. Lim, S. Alam, F. Parmigiani, D. Richardson, B. Pálsdóttir, A silica based highly nonlinear fibre with improved threshold for stimulated brillouin scattering, in: Proceedings of the ECOC 2010, Torino, Italy, Paper Tu.4.D.3, 2010.

[132] S.F. Preble, Q. Xu, M. Lipson, Changing the colour of light in a silicon resonator, Nat. Photon. 1 (2007) 293–296.

[133] F. Li, M. Pelusi, D.-X. Xu, A. Densmore, R. Ma, S. Janz, D.J. Moss, Error-free all-optical demultiplexing at 160 Gb/s via FWM in a silicon nanowire, Opt. Express 18 (4) (2010) 3905–3910.

[134] H. Hu, H. Ji, M. Galili, M. Pu, H.C.H. Mulvad, L.K. Oxenløwe, K. Yvind, J.M. Hvam, P. Jeppesen, 320 Gb/s phase-transparent wavelength conversion in a silicon nanowire, in: Proceedings of the OFC 2011, Los Angeles, California, USA, Paper OWG6, 2011.

[135] B. Corcoran, C. Monat, M. Pelusi, C. Grillet, T.P. White, L. O'Faolain, T.F. Krauss, B.J. Eggleton, D.J. Moss, Optical signal processing on a silicon chip at 640 Gb/s using slow-light, Opt. Express 18 (8) (2010) 7770–7781.

[136] A. Biberman, N. Ophir, A.C. Turner-Foster, M.A. Foster, M. Lipson, A.L. Gaeta, K. Bergman, On-chip wavelength multicasting of 3 × 320-Gb/s pulsed-RZ optical data, in: Proceedings of the ECOC 2010, Torino, Italy, Paper We.7.E.3, 2010.

[137] M.A. Foster, A.C. Turner, J.E. Sharping, B.S. Schmidt, M. Lipson, A.L. Gaeta, Broadband optical parametric gain on a silicon photonic chip, Nature 441 (2006) 960–963.

[138] M.A. Foster, R. Salem, D.F. Geraghty, A.C. Turner-Foster, M. Lipson, A.L. Gaeta, Silicon-chip-based ultrafast optical oscilloscope, Nature 456 (2008) 81–84.

[139] J.S. Levy, A. Gondarenko, M.A. Foster, A.C. Turner-Foster, A.L. Gaeta, M. Lipson, CMOS-compatible multiple-wavelength oscillator for on-chip optical interconnects, Nat. Photon. 4 (2010) 37–40.

[140] D.J. Moss, L. Fu, I. Littler, B.J. Eggleton, Ultrafast all-optical modulation via two-photon absorption in siliconon-insulator waveguides, Electron. Lett. 41 (6) (2005) 320–321.

[141] T.K. Liang, L.R. Nunes, T. Sakamoto, K. Sasagawa, T. Kawanishi, M. Tsuchiya, Ultrafast all-optical switching by crossabsorption modulation in silicon wire waveguides, Opt. Express 13 (19) (2005) 7298–7303.

[142] Ö. Boyraz, P. Koonath, V. Raghunathan, B. Jalali, All optical switching and continuum generation in silicon waveguides, Opt. Express 12 (17) (2004) 4094–4102.

[143] R.L. Espinola, J.I. Dadap, Jr R.M. Osgood, C-band wavelength conversion in silicon photonic wire waveguides, Opt. Express 13 (11) (2005) 4341–4349.

[144] H. Rong, Y.-H. Kuo, A. Liu, M. Paniccia, High efficiency wavelength conversion of 10 Gb/s data in silicon waveguides, Opt. Express 14 (3) (2006) 1182–1188.

[145] K. Yamada, H. Fukuda, T. Tsuchizawa, T. Watanabe, T. Shoji, S. Itabashi, All-optical efficient wavelength conversion using silicon photonic wire waveguide, Photon. Technol. Lett. 18 (9) (2006) 1046–1048.

[146] Q. Lin, J. Zhang, P.M. Fauchet, G.P. Agrawal, Ultrabroadband parametric generation and wavelength conversion in silicon waveguides, Opt. Express 14 (11) (2006) 4786–4799.

[147] R. Dekker, A. Driessen, T. Wahlbrink, C. Moormann, J. Niehusmann, M. Först, Ultrafast Kerr-induced all-optical wavelength conversion in silicon waveguides using 1.55 μm femtosecond pulses, Opt. Express 14 (18) (2006) 8336–8346.

[148] Y.-H. Kuo, H. Rong, V. Sih, S. Xu, M. Paniccia, O. Cohen, Demonstration of wavelength conversion at 40 Gb/s data rate in silicon waveguides, Opt. Express 14 (24) (2006) 11721–11726.

[149] J. Leuthold, C. Koos, W. Freude, Nonlinear silicon photonics, Nat. Photon. 4 (2010) 535–544.

[150] C. Koos, P. Vorreau, P. Dumon, R. Baets, B. Esembeson, I. Biaggio, T. Michinobu, F. Diederich, W. Freude, J. Leuthold, Highly-nonlinear silicon photonic slot waveguide, in: Proceedings of the OFC, San Diego, California, USA, Paper PDP25, 2008.

[151] L. Alloatti, D. Korn, D. Hillerkuss, T. Vallaitis, J. Li, R. Bonk, R. Palmer, T. Schellinger, A. Barklund, R. Dinu, J. Wieland, M. Fournier, J. Fedeli, P. Dumon, R. Baets, C. Koos, W. Freude, J. Leuthold, 40 Gbit/s silicon-organic hybrid (SOH) phase modulator, in: Proceedings of the ECOC 2010, Torino, Italy, Paper Tu.5.C.4, 2010.

[152] C. Koos, P. Vorreau, T. Vallaitis, P. Dumon, W. Bogaerts, R. Baets, B. Esembeson, I. Biaggio, T. Michinobu, F. Diederich, W. Freude, J. Leuthold, All-optical high-speed signal processing with silicon-organic hybrid slot waveguides, Nat. Photon. 3 (2009) 216–219.

[153] Y. Okawachi, K. Saha, J.S. Levy, Y.H. Wen, M. Lipson, A.L. Gaeta, Octave-spanning frequency COMB generation in a silicon nitride chip, Opt. Lett. 36 (17) (2011) 3398–3400.

[154] B. Kuyken, S. Clemmen, S.K. Selvaraja, W. Bogaerts, D. Van Thourhout, P. Emplit, S. Massar, G. Roelkens, R. Baets, On-chip parametric amplification with 26.5 dB gain at telecommunication wavelengths using CMOS compatible hydrogenated amorphous silicon waveguides, Opt. Lett. 36 (4) (2011) 552–554.

[155] B. Kuyken, H. Ji, S. Clemmen, S.K. Selvaraja, H. Hu, M. Pu, M. Galili, P. Jeppesen, G. Morthier, S. Massar, L.K. Oxenløwe, G. Roelkens, R. Baets, Nonlinear properties of and nonlinear processing in hydrogenated amorphous silicon waveguides, Opt. Express 19 (26) (2011) B146–B153.

[156] Q. Lin, O.J. Painter, G.P. Agrawal, Nonlinear optical phenomena in silicon waveguides: modeling and applications, Opt. Express 15 (25) (2007) 16604–16644.

[157] H. Ji, C.S. Cleary, J.M. Dailey, J. Wang, H. Hu, R.P. Webb, R.J. Manning, M. Galili, P. Jeppesen, M. Pu, K. Yvind, L.K. Oxenløwe, Two-copy wavelength conversion of an 80 Gbit/s serial data signal using cross-phase modulation in a silicon nanowire and detailed pump-probe characterisation, in: Proceedings of the ECOC 2012, Amsterdam, Netherlands, Paper We.2.E.3, 2012.

[158] M. Ma, M. Galili, L.K. Oxenløwe, H. Ji, H. Hu, M. Pu, H.C. Hansen Mulvad, P. Jeppesen, Detailed time-resolved spectral analysis of ultra-fast four-wave mixing in silicon nanowires, in: Proceedings of the IPC 2011, Arlington, Virginia, USA, Paper TuV2, 2011.

[159] H. Ji, M. Pu, H. Hu, M. Galili, L.K. Oxenløwe, K. Yvind, J.M. Hvam, P. Jeppesen, Optical waveform sampling and error-free demultiplexing of 1.28 Tb/s serial data in a nanoengineered silicon waveguide, J. Lightwave Technol. 29 (4) (2011) 426–431.

[160] P.A. Andrekson, M. Westlund, Nonlinear optical fiber based high resolution all-optical waveform sampling, Laser Photon. Rev. 1 (3) (2007) 231–248.

[161] H. Ji, M. Galili, H. Hu, M. Pu, L.K. Oxenløwe, K. Yvind, J.M. Hvam, P. Jeppesen, 1.28-Tb/s demultiplexing of an OTDM DPSK data signal using a silicon waveguide, Photon. Technol. Lett. 22 (23) (2010) 1762–1764.

[162] L.K. Oxenløwe, H. Ji, M. Galili, M. Pu, H. Hu, H.C.H. Mulvad, K. Yvind, J.M. Hvam, A.T. Clausen, P. Jeppesen, Silicon photonics for signal processing of Tbit/s serial data signals, J. Sel. Top. Quant. Electron. 18 (2) (2012) 996–1005 (special issue).

[163] R. Driad, R.E. Makon, V. Hurm, K. Schneider, F. Benkhelifa, R. Lösch, J. Rosenzweiget, InP DHBT-based ICs for 100 Gbit/s data transmission, in: Proceedings of the IPRM 2008, Versailles, France, Paper TuB2.2, 2008.

[164] J. Xu, C. Peucheret, P. Jeppesen, Power consumption comparison between point-to-point WDM and OTDM systems, in: Proceedings of the ICTON 2010, Munich, Germany, Paper Th.A1.1, 2010.

[165] K. Hinton, G. Raskutti, P.M. Farrell, R.S. Tuckeret, Switching energy and device size limits on digital photonic signal processing technologies, J Sel. Top. Quant. Electron. 14 (3) (2008) 938–945.

Technology and Applications of Liquid Crystal on Silicon (LCoS) in Telecommunications

18

Stephen Frisken, Ian Clarke, and Simon Poole

All with Finisar Australia Ltd., Australia

18.1 INTRODUCTION

The use of liquid crystal technology in telecommunications is now well established. Initially, applications were limited to attenuation and simple binary switches, but more recently we have seen the adoption of liquid crystal into gain equalizers, wavelength blockers, and then into wavelength switches. Within these switches, the control of light on a pixel-by-pixel basis has been enabled by developments in the display technologies and in particular by Liquid Crystal on Silicon. (LCoS). This chapter seeks to clarify the basic operating principles of LCoS in existing wavelength-switching applications, and to give an outline of the broad scope of new opportunities that are arising from the intrinsic performance and flexibility of LCoS as a switching medium.

To do this, we will first provide some context for this chapter by taking a high level look at how reconfigurable optical networks have developed, before diving deep into the principles of operation of liquid crystals, focusing in particular on how they operate within an LCoS chip. We then explain in detail the design and operation of an LCoS-based Wavelength Selective Switch (WSS), with particular emphasis on the key optical parameters which determine performance in an optical communications network. In the final two sections we look at how LCoS-based WSSs will fit into future network architectures and also briefly review the myriad of additional applications that the flexibility of the LCoS-based WSS architecture has enabled in recent years.

18.2 ROADMS AND RECONFIGURABLE OPTICAL NETWORKS

Wavelength Division Multiplexed (WDM) optical communication networks have evolved over the past 20 years from simple point-to-point transmission paths [1], to the highly interconnected mesh networks currently being deployed [2]. This has required a significant evolution in the technologies for multiplexing, routing, and demultiplexing the individual transmission wavelengths (Figure 18.1).

Optical Fiber Telecommunications VIA. http://dx.doi.org/10.1016/B978-0-12-396958-3.00018-4

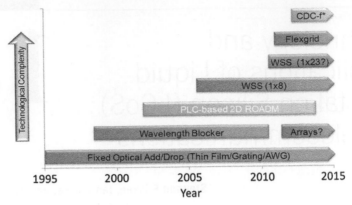

FIGURE 18.1 Evolution of multiplexing, routing, and demultiplexing technologies for WDM networks.

Early WDM networks used a combination of multilayer dielectric filters [3], fiber Bragg grating filters [4] and Arrayed Waveguide Routing (AWG) [5] technologies to multiplex and demultiplex the wavelengths while any reconfiguration was achieved through manual patching of individual demultiplexed wavelengths.

The use of multiple channels in an optical amplifier presented new challenges to the network as the increasing number of channels affected the amplifier's spectral gain. Optimization of the amplified transmission links was enabled through the introduction of spectrally dynamic telecom components [6] which could respond to the varying spectral gain of the amplifier to compensate for changes to the traffic load, and hence ensure a consistent optical signal-to-noise ratio. Such optical Dynamic Gain Flattening Filters (DGFF) were initially broadband devices which acted on the spectrum in a continuous manner. However, it was recognized that increased spectral resolution would allow channel-by-channel gain modification and, perhaps more importantly, provide the ability to block a channel (by providing greater than 35 dB attenuation within a single channel) without affecting the neighboring channel. These channel equalizers are based on dispersing the light with a bulk grating and then individually attenuating the light in each channel with an actuator array—typically either liquid crystal [7] or Micro-Electro-Mechanical Systems (MEMS) [8]. Blocking the optical power on a channel-by-channel basis created what became known as a wavelength or channel blocker [9].

This in turn led to the development of the first Reconfigurable Optical Add/Drop Multiplexers (ROADMs) based on a broadcast-and-select architecture utilizing wavelength blockers and passive splitters and combiners (typically AWGs) as shown in Figure 18.2. However, the requirement for multiple wavelength blockers, together with the large losses associated with the passive splitter and combiner couplers, means that these architectures do not scale well beyond a degree 2 node (e.g. East/West Connectivity) so interconnecting a large number of directions of traffic at a node was complex and prohibitively expensive.

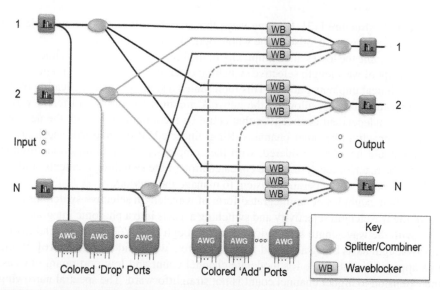

FIGURE 18.2 A 1 × N broadcast-and-select reconfigurable wavelength cross-connect utilizing wavelength blockers between passive splitters and combiners. The local drop (left) and add (right) ports at the bottom are "colored" as only one wavelength is coupled to each port of the AWG and additional amplification stages (not shown) are required as the number of directions increases.

While wavelength routing on early blocker-based ROADMs could be reconfigured affecting the existing traffic (hitless), the limitation was that each add or drop port was colored (i.e. had a pre-defined wavelength routed through it), with each new transceiver requiring provisioning at the time of deployment. Nevertheless, these colored ROADM architectures provided an initial starting point for network reconfigurability and developments in laser tuning technology [10] provided the next stimulus for the development of a colorless architecture, which would be important for a truly hands-off remote reconfiguration.

One approach to creating truly dynamically reconfigurable networks relied on the capability of AWGs to break out each separate wavelength channel on to individual fibers, which could then be routed through a photonic cross-connect switch with a port count equivalent to the product of the number of channels and the number of fibers in a wavelength cross-connect switch [11]. While enormous amounts of effort were expended on addressing this architecture it was spectacularly unsuccessful in delivering even a single commercial network, as it was plagued by the complexity of the fiber structure and the poor scalability of manufacturing yields within the MEMS switching chips as the number of ports increased into the hundreds. One alternative large-scale cross-connect switch was known as the "bubble switch," as it relied on total internal reflection of light at an interface created by injection of a bubble into a waveguide structure [12]. In some ways this switch epitomized the euphoria and subsequent temporary demise of the whole optical networking space.

To overcome the losses and complexity of the broadcast-and-select architecture, a new concept of wavelength selective switching emerged [13–15] with a range of different technical implementations. This represented a significant simplification of the switching functionality, allowing the network architect to consider different wavelengths as independent connections that could be reconfigured through the network without impacting the other channels. Ring, linear add/drop, and true mesh architectures could now be considered, with the complexity of the connections limited only by the reach of the system, the concatenation of the switching elements, and the ability of the network operating system to manage the network.

The first examples of an integrated form of wavelength selective switching combined the wavelength selectivity and switching aspects into a photonic integrated circuit, with the wavelength selectivity created through an AWG and the switching on a wavelength basis being achieved through discrete waveguide switches [16]. While this approach is attractive for moderate channel count and low port count devices, the extension to higher channel count is not straightforward. The spectral narrowing effects of the limited bandwidth of the AWG (two per node) limit the transmission spectral bandwidth when signals must pass through multiple nodes. In addition to the narrowed spectrum it is difficult to maintain the other important filter characteristics such as group delay and polarization effects across the whole band.

Supporting this vision to reconfigure the demultiplexed channels, advances in MEMS technology (especially two-axis mirror arrays) and liquid-crystal switching were combined with the wavelength dispersive techniques developed for wavelength blockers. This offered advantages in being able to reconfigure not only the physical wavelength connection (by varying the MEMS' angle orthogonal to the dispersive axis) but also being able to block unwanted wavelengths and independently control the power (by varying the MEMS' angle in the dispersive axis) of each of the channels.

By using a disperse-and-switch approach the channel clear pass band is only limited by the optical resolution of the system, which in turn (in common with other optical monochromators) is determined by the number of lines of grating which the optical beam addresses. Even with reasonably compact designs it was possible to achieve broad flat pass bands at first for 100 GHz channel spacing and later for the technically more demanding 50 GHz spacing.

A now-common architecture employing WSSs is illustrated in Figure 18.3, which employs a broadcast section at the ingress to a wavelength cross-connect node, and a WSS associated with each direction at the egress to select from the various broadcast wavelength channels, and importantly, to block channels in the directions not selected.

The loss of a WSS is now typically around 4 dB—much reduced from the loss of greater than 10 dB, that would be incurred using blockers and combiners, so large-port-count wavelength cross-connects can be achieved without amplification. However, as multiple channels can be present at the same wavelength, achieving high extinction of the blocked channels is of particular importance because the coherent

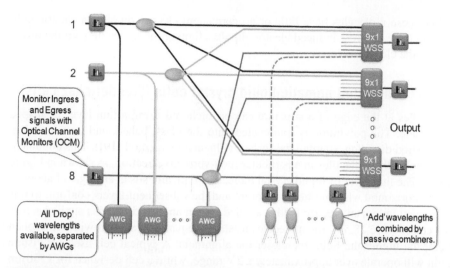

1

2

Monitor Ingress
and Egress
signals with
Optical Channel
Monitors (OCM)

8

9x1
WSS

9x1
WSS

o
o   Output
o

9x1
WSS

All 'Drop'
wavelengths
available, separated
by AWGs

AWG   AWG   ∘ ∘ ∘   AWG

'Add' wavelengths
combined by
passive combiners.

FIGURE 18.3 A broadcast-and-route ROADM, variants of which are commonly employed in the industry.

beating of the interference terms lies within the electrical bandwidth of the receiver and scales with the electric field vector.

More recently LCoS-based WSSs have been developed [17,18] and commercialized and the operating principles, design, and performance of these devices are outlined in the following sections.

18.3 BACKGROUND AND TECHNOLOGY OF LCOS

Liquid crystals, first observed in 1888, are liquids in which the molecules retain some order. In the smectic phase they sit in regular layers. In the nematic phase, the regular layering is lost, but the molecules still have a preferred direction (the "director"). Heating a liquid crystal in the nematic phase beyond the "clearing temperature" (the temperature at which the liquid crystal loses its long distance order and the bulk liquid becomes transparent) will result in a further transition to a totally random (isotropic) liquid. In the remainder of this chapter we will only consider the nematic phase.

Commercial liquid crystal materials, designed to operate in the nematic phase usually consist of a mixture of several different organic molecules with rod-like structures. An example of such a liquid crystal molecule is shown in Figure 18.4.

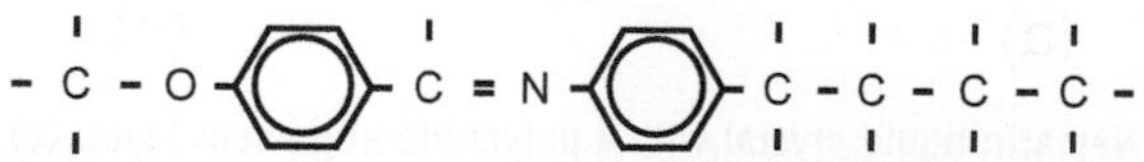

$$-C-O-\bigcirc-C=N-\bigcirc-C-C-C-C-$$

FIGURE 18.4 MBBA molecule (4-methoxybenzylidene-4-butylaniline), perhaps the most studied liquid crystal molecule, first discovered by Hans Kelker [20].

While these molecules have little or no permanent electric dipole, an electric field will produce a strong-induced electric dipole, aligning the molecule with the direction of the electric field.

18.3.1 Untwisted nematic liquid crystal cells (Fréedericksz cells)

Molecules at the edge of a structure can be anchored using a thin layer of shaped polyimide. The polyimide is spin coated onto the glass, baked, and then buffed in the desired direction, shaping the surface (Chatelain's method [19]). This alignment layer holds the molecules in a particular (non-vertical) direction, as shown in Figure 18.5a. The first layer of liquid crystal molecules typically sits at an angle of about 3° to the horizontal when no field is present and the subsequent layers conform to this alignment unless an electric field is applied. When an electric field is applied across the cell (Figure 18.5b), the molecules rotate to be partially aligned with the applied field; the higher the field, the closer the alignment. A typical cell used for visible light will operate over approximately a 2V range, while a cell designed for 1550 nm light will require approximately twice the voltage range and twice the cell thickness. Once the applied electric field exceeds 10^6 V/m, the molecules are nearly fully aligned with applied field and there is little further change in the birefringence [21]. In a Fréedericksz [22] cell the director is essentially uniform for the molecules, independent of their distance from the base of the cell. This is in contrast to the twisted nematic cells used in most liquid crystal displays.

In practice, an applied voltage will gradually be canceled by charge build-up within the cell due to the migration of free ions in the solution. To counteract this, the amplitude of the applied voltage is held constant but the sign of the field is switched at a rate much faster than the time taken for the molecules to realign (typically at 1 kHz or more). The orientation of the liquid crystal molecule stays essentially constant with only the induced electrical dipole swapping direction. However, if the electric field amplitude in the two directions is not equal, either because the applied

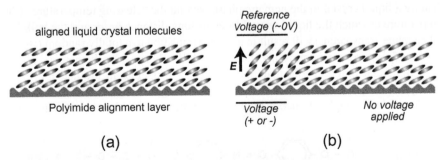

(a)   (b)

FIGURE 18.5 (a) Nematic liquid crystal with a polyimide alignment layer. (b) The addition of a voltage across the liquid crystal causes the molecules to rotate, increasingly aligning themselves with the direction of the electric field as the field increases.

voltages are different or because of a non-symmetric ion build-up in the alignment layer, the LCoS image will flicker.

18.3.2 Cell construction

A sketch of a Liquid Crystal on Silicon (LCoS) cell is shown in Figure 18.6. The CMOS chip controls the voltage on plates buried just below the chip surface, each controlling one pixel. For example, chips with XVGA resolution will have 1024×768 plates, each with an independently addressable voltage.

A common voltage for all the pixels is supplied by an indium tin oxide layer on the cover glass. This layer should be optimized for the wavelength range being controlled. Layers optimized for visible light are often absorbing in the near infrared. Not shown in Figure 18.6 are the spacer balls or glass rod used to ensure that a uniform thickness of liquid crystal is maintained across the cell.

18.3.3 Driving a cell

The variable voltage needed to control each pixel may be applied via analog ramp voltage which is disconnected from a pixel at a time corresponding to the desired voltage level. This voltage source must be resilient to changes in current draw ("stiff") because the load on the voltage source will change as pixels reach their desired charge level and are disconnected from the supply. A lack of stiffness in the source will result in the voltage jumping as the load is removed, increasing the voltage applied to cells still connected to the source.

Alternatively, the variable voltage may be simulated using pulse-width modulation. This has the advantage of being purely digital, but the major disadvantage of introducing an extra source of flicker in the image which is unacceptable for most telecommunications applications. The transfer function of phase versus applied voltage is highly non-uniform (Figure 18.7)—a factor which must be taken into account when determining the appropriate drive voltage for any given application.

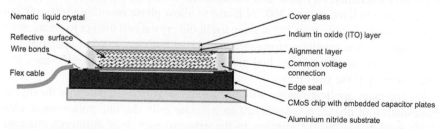

FIGURE 18.6 A conceptual drawing of a Liquid Crystal on Silicon cell with the vertical dimension exaggerated to reveal the different layers. Typical cells are about $3\,cm^2$ and about $2\,mm$ thick.

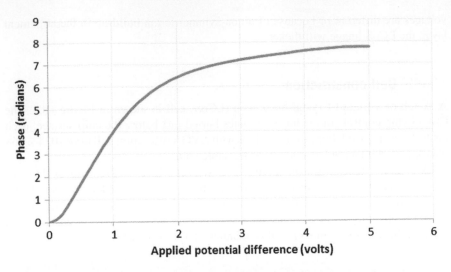

FIGURE 18.7 A typical applied voltage versus phase shift in a Fréedericksz liquid crystal cell.

18.3.4 Spatial light modulators

In a Fréedericksz cell, the polarization orthogonal to the director is minimally affected by any voltage applied to the cell (the "ordinary" axis of the liquid crystal). However, the alignment of the liquid crystal strongly affects the phase velocity of the extraordinary polarization of any light wave passing through. This can be understood intuitively, since the molecule is polarizable along its length and so interacts strongly with any electric field with a component in that direction. If the molecule is parallel to the electric field associated with a light wave (hence perpendicular to both the light-wave direction of travel and the associated magnetic field) it interacts strongly, producing a high refractive index and corresponding phase shift, as shown in Figure 18.7. We can consider the LCoS as a phase (or amplitude if used with cross polarizers) matrix where pixel-by-pixel addressing of the local phase is possible. In general it is desirable to have at least 360° of phase to allow phase resetting (360°=0°) in creating a desired phase map, but making the liquid crystal cell thicker to create more phase can have detrimental performance in terms of both the speed of operation and also the sharpness of the individual pixels which can in turn affect efficiency and light scattering at the reset points.

For the many applications we will discuss below, which are based on the digital manipulation of phase, it is sufficient to consider only the one polarization state which is addressed by the changes to the extraordinary axis. Although attempts have been made [23] to create polarization independent LCoS, in general, polarization diversity techniques (Section 4.4) have been employed in commercial products which use LCoS technology to enable low-loss, polarization-independent operation.

18.3.5 **Holographic Fourier processing**

The use of phase-based spatial light modulators for switching of light was demonstrated as early as 1988 [24]. Much of the early work in holographic steering [25] was focused on demonstrating the principles of holographic interconnects between optical fibers. The spatial light modulator can be thought of digitally recreating the desired wavefront to couple the light to a given fiber or location based on the wavefront of the incident beam of light. It is important, of course, that the hologram intersects efficiently the point where the beams intersected. There are many ways that this can be achieved with an optical scheme. One of the simplest ways is to use a lens to create the Fourier transform where parallel beams of light will intersect at the focal length of the lens. This Fourier processing of the optical beam can be used for both routing and attenuation. Other interesting capabilities of LCoS switching were explored in the 1990s including multicasting through writing multiple images into a single hologram and sharing the power between different locations [26].

18.3.6 **Holographic beam steering**

An LCoS can mimic most optical devices through pixel-by-pixel control of the wavefront. Cylindrical lens, spherical lens, beam splitters are all possible. An intuitive understanding of LCoS routing can be obtained by considering that to steer a beam we need to emulate a simple angled mirror. An angled mirror can be emulated by angling the wavefront. We could do this by applying a linear ramp to the wavefront, however, we would quickly run out of range since the maximum phase shift we can create is a little over 2π. Instead we can borrow a concept from a Fresnel lens and reset the phase back to zero after it reaches 2π, as sketched in Figure 18.8.

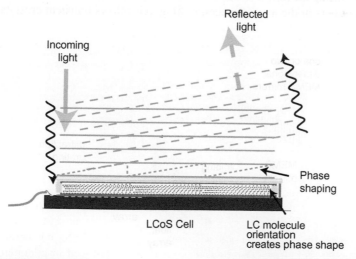

FIGURE 18.8 Beam steering using a Liquid Crystal on Silicon cell configured with an image to create a tilted reflected wavefront.

While the phase shaping resets, the reflected wavefront is continuous. In practice, a perfect hard transition is impossible and some scattering results from these reset points. Because of the periodic nature of these resets, this scattering is not random and so can be eliminated from the optical path through careful design. In this case the desired phase transformation is a linear ramp—but as the angles become larger the required phase ramp becomes more wavelength dependent. Alternatively for a given hologram the efficiency and also the steering angle are functions of wavelength. This creates an issue for wavelength independent switching and so precludes the applications of holographic steering in broadband cross-connects. However, for a WSS, each pixel of an LCoS always sees the same wavelength and so this intrinsic switching limitation does not impact the performance in any way.

18.3.7 **Comparison with other switching technologies**

While LCoS-based WSSs are one commercially successful method of selecting and switching wavelengths, they were certainly not the first, and several other methods are in common use, each bringing its own advantages and limitations. Before looking in detail at the design considerations of an LCoS WSS it is instructive to review some of the other switching technologies that have been employed.

18.3.7.1 *MEMS optical switching*

The simplest and earliest commercial WSSs [27] were based on movable mirrors using Micro-Electro-Mechanical Systems (MEMS). The concept is sketched in Figure 18.9. The incoming light is nearly collimated by the lens array (in fact, it will focus on, or near, the MEMS mirror). The light is then broken into a spectrum by a diffraction grating. Each frequency channel then focuses on a separate MEMS mirror. By tilting the mirror in one dimension, the channel can be directed back into any of the fibers in the array. A second tilting axis allows transient cross-talk to be

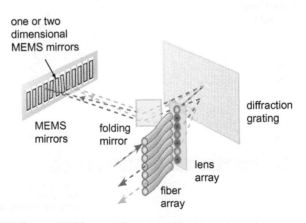

FIGURE 18.9 MEMS-based WSS concept.

minimized, otherwise switching from port 1 to port 3 will always involve passing the beam across port 2. The second axis provides a means to attenuate the signal without increasing the coupling into neighboring fibers.

This technology has the advantage of a single steering surface, not necessarily requiring polarization diversity optics. It works well in the presence of a continuous signal, allowing the mirror tracking circuits to dither the mirror and maximize coupling. MEMS-based WSS typically produces good extinction ratios, but poor open-loop performance for setting a given attenuation level. The main limitations of the technology arise from the channelization that the mirrors naturally enforce. During manufacturing, the channels must be carefully aligned with the mirrors, complicating the manufacturing process. Post-manufacturing alignment adjustments have been mainly limited to adjusting the gas pressure within the hermetic enclosure. This enforced channelization has also proved, so far, an insurmountable obstacle to implementing flexible channel plans where different channel sizes are required within a network. Additionally the phase of light at the mirror edge is not well controlled in a physical mirror so artifacts can arise in the switching of light near the channel edge due to interference of the light from each channel.

18.3.7.2 Binary liquid crystal switching

Binary liquid crystal switching avoids both the high cost of small volume MEMS fabrication and potentially some of its fixed channel limitations. The concept is illustrated in Figure 18.10.

Light, at the input, generally consists of up to 96 separate modulated frequency channels spaced at 50 GHz. Diffraction grating 1 breaks these signals up horizontally into a spectrum. Polarization diversity optics (in practice, usually located before the grating) spatially separate the horizontally and vertically polarized components

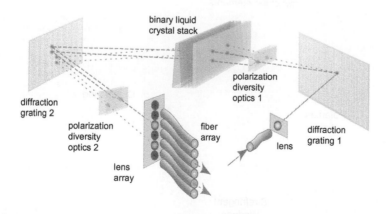

FIGURE 18.10 Binary liquid crystal switching schematic.

and rotate one component by 90°. A software-controlled binary liquid crystal stack, described later, individually tilts each optical channel. A second grating (or a second pass of the first grating) is used to spectrally recombine the beams. The offsets created by the binary liquid crystal stack cause the resulting spectrally recombined beams to be spatially offset, and hence to focus, through the lens array, into separate fibers.

The binary liquid crystal stack is shown schematically in Figure 18.11. Polarized light enters the stack and passes through the first liquid crystal switching element where its polarization is switched to either the fast or slow axis of the subsequent birefringent wedge. On leaving the wedge, the angle of refraction of the light will depend on whether it traveled along the fast or slow axis. Subsequent liquid crystal/birefringent wedge pairs create additional switching elements [28]. A 1×8 switch, as sketched here, requires three switching elements. Usually another element is added to provide attenuation control.

This technology has the advantages of relatively low cost parts, simple electronic control, and stable beam positions without active feedback. It is capable of configuring to a flexible grid spectrum by the use of a fine pixel grid. The inter-pixel gaps must be small compared to the beam size, to avoid perturbing the transmitted light significantly. Furthermore each grid must be replicated for each of the switching stages creating the requirement of individually controlling thousands of pixels on different substrates so the advantages of this technology in terms of simplicity are negated as the wavelength resolution becomes finer.

The main disadvantage of this technology arises from the thickness of the stacked switching elements. Keeping the optical beam tightly focused over this depth is difficult and has, so far, limited the ability of high port count WSSs to achieve very fine (12.5 GHz or less) granularity.

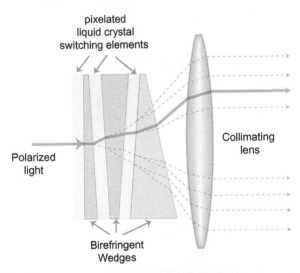

FIGURE 18.11 Binary liquid crystal switching stack schematic.

18.4 LCOS-BASED WAVELENGTH-SELECTIVE SWITCHING

As noted above, WSS-based ROADMs enable efficient routing of multiple wavelength traffic in Ring, Long-Haul, and Mesh networks. A WSS comprises a switching array that operates on light that has been dispersed in wavelength without the requirement that the dispersed light be physically demultiplexed into separate ports. This is termed a "disperse-and-switch" configuration. For example, an 88 channel WDM system can be routed from a "common" fiber to any one of N fibers by employing 88 $1 \times N$ switches. This represents a significant simplification of a demux and switch and multiplex architecture that would require (in addition to $N+1$ mux/demux elements) a non-blocking switch for 88 $N \times N$ channels [29] which would test severely the manufacturability limits of large-scale optical cross-connects for even moderate fiber counts.

LCoS is particularly attractive as a switching mechanism in a WSS because of the near continuous addressing capability, enabling much new functionality. In particular the bands of wavelengths which are switched together (channels) need not be preconfigured in the optical hardware but can be programmed into the switch through the software control. This is advantageous from a manufacturability perspective, with different channel plans being able to be created from a single platform and even different operating bands (such as C and L) being able to use an identical switch matrix. Additionally, it is possible to take advantage of this ability to reconfigure channels while the device is operating. Products have been introduced allowing switching between 50 GHz channels and 100 GHz channels, or a mix of channels, without introducing any errors or "hits" to the existing traffic.

LCoS technology has now enabled the introduction of more flexible wavelength grids which, we shall see later, help to unlock the full spectral capacity of optical fibers. Even more surprising features rely on the phase matrix nature of the LCoS switching element. Features in common use include such things as shaping the power levels within a channel or broadcasting the optical signal to more than one port.

In the previous section, we reviewed various switching mechanisms and in the following we will focus on LCoS and in particular how the switch matrix and optical train cooperate to achieve the specification required for a WSS to be valuable in a switching network. We will consider initially the application to a Broadcast-and-Select ROADM architecture or what is often termed an "Add" WSS, which is in general the most challenging specification. We will then contrast the different requirements for a Route-and-Select ROADM in which two passes of a WSS are used for each switching node of the network.

Figure 18.12 shows schematically the operation of a disperse-and-switch WSS. The various incoming channels of a common port are dispersed continuously onto a switching element which then directs and attenuates each of these channels independently to the N switch ports. The dispersive mechanism is generally based on holographic or ruled diffraction grating similar to those used commonly in spectrometers. It can be advantageous, for achieving resolution and coupling efficiency, to employ a combination of a grating and a reflective or transmissive grating—known

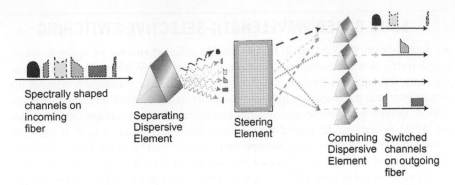

Spectrally shaped channels on incoming fiber

Separating Dispersive Element

Steering Element

Combining Dispersive Element

Switched channels on outgoing fiber

FIGURE 18.12 Internal architecture of a disperse-and-switch WSS. Wavelength channels with optical bandwidth determined by the width of switching elements are independently routed to one of *N* output ports. Power equalization of the channels is achieved through attenuation.

as a GRISM. The operation of the WSS can be bidirectional so the wavelengths can be multiplexed together from different ports onto a single common port. To date, the majority of deployments have used a fixed channel bandwidth of 50 or 100 GHz and 9 output ports are typically used. A practical implementation of this concept is shown in Figure 18.13.

Key target specifications and cost points of a new technology need to be comparable with incumbent and competing switching technologies, and achieving this has led to a broad adoption of LCoS in network applications that do not (yet) use the advanced functions. Key performance criteria that must be considered in the design of a WSS include:

- Number of switch ports.
- Attenuation range and setting accuracy.
- Open-loop or closed-loop control.
- Wavelength range and channel shape (resolution).
- Wavelength setting resolution, calibration, and accuracy.
- Thermal dependence (wavelength tracking).
- Polarization independence.
- Port-to-port isolation and channel extinction
- Phase response of filter (CD and DGD)
- Temporal response (switching speed and transient responses)

18.4.1 Switching between ports

The first studies of LCoS spatial light modulators assumed that to achieve efficiency the switching states were quantized (integral number of pixels per reset)—but it was later shown [30] that with arbitrary pixels per reset that high efficiency can be

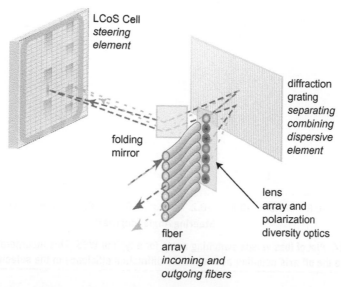

LCoS Cell
*steering
element*

diffraction
grating
*separating
combining
dispersive
element*

folding
mirror

lens
array and
polarization
diversity optics

fiber
array
*incoming and
outgoing fibers*

FIGURE 18.13 Schematic of an LCoS-based WSS showing use of a folding mirror and single dispersive element for separating and combining the different frequencies.

achieved—and use of non-integral pixels per reset results in a continuous steering range without loss discontinuities. The range of steering angles is much more limited for LCoS than MEMS as the loss will increase as a function of the number of resets—particularly for the relatively thick liquid crystal cells required to achieve 2π phase angle in the near infrared. The steering angle is given (in air) by $\theta = \sin^{-1}\left[\frac{\lambda}{lx}\right]$ where λ is the wavelength of the incident light, x is the average number of pixels per reset, and l is the pixel dimension. We note that for a fixed number of pixels the switching angle is a function of wavelength which can be a significant issue for broadband switches but, as noted above, in a WSS the wavelength associated with a switch position is a very well-defined function.

The number of resolvable switch states is a function of the beam dimensions (neglecting other practicalities); a Gaussian beam of radius $= r$ will require an angular separation of approximately $1.6/r$ radians (where r is the mode field radius measured in microns) to achieve 45 dB isolation between neighboring ports. By consideration of the available beam dimensions we can choose the number of ports that can be employed without excess loss. For a typical high port count WSS, Figure 18.14 shows the distribution of loss versus switching angle. One of the other factors that influence the number of ports is the distribution of the diffraction orders. The ports may be arranged spatially to minimize the impact of diffraction orders by avoidance of positions associated with a particular symmetry-related diffraction. For example, it is possible to align the fibers such

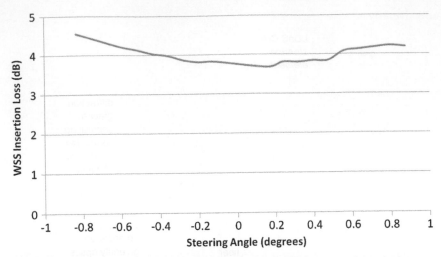

FIGURE 18.14 Plot of loss versus switching angle for a typical WSS. This incorporates losses due to the off-axis coupling as well as the diffraction efficiency of the selected port.

that the symmetric order falls between fibers and is not coupled into any port—though we shall see later that other techniques can also be employed to mitigate these issues.

18.4.2 Optical power control

WSSs have been employed in architectures where they are relied upon not only to direct the light but to balance the optical power between the different channels. In an "Add" architecture (Figure 18.3) this power balancing can not only account for the variations in gain of optical amplifiers based on different loading of the channels, but can also account for link and power level mismatches between the different directions. Although the power levels will often be controlled with a feedback comprising an Optical Channel Monitor (OCM) at the point where the signals are combined, it is clearly advantageous to be able to control the attenuation set point in an open-loop fashion and to ensure a well-defined monotonic function with resolution through a large operating range of attenuation set states. Typical systems today employ between 15 and 20 dB of attenuation with the higher portion of the range usually required to introduce new channels gracefully.

In many optical components an attenuation term can be introduced as an analog response to a single control variable. For example, in liquid crystal attenuators the polarization of the light is controlled by the applied AC voltage, while in a MEMS attenuator the light is steered away from the receiving fiber via a control voltage. Any variation of the response over operating conditions or lifetime (for example, a small shift in mirror angle setting) can lead to an uncertainty in the desired function. One effective way of introducing a highly deterministic attenuation function into an LCoS switch is to direct part of the optical power into null ports—i.e. where no output fiber is present. This capability of LCoS to multicast directed light [31] is dependent upon the pattern or image associated with the channel and so the resultant attenuation level is largely insensitive to small variations in optical coupling or the electrooptic response of the liquid crystal material due to thermal or mechanical perturbations. This allows the use of open-loop control of the attenuation state—i.e. the attenuation level can be set via a look-up table rather than relying on a feedback loop from the system, simplifying the overall control.

There are of course many different ways to attenuate light—for example, it would be possible to just reduce the "strength" of the image by scaling the reset points to be less than 2π. This approach is often problematic as it results in a very strong non-diffracted signal or unwanted coupling into any ports which are related by symmetry to the directed port.

18.4.3 Spectral resolution

An important characteristic of any wavelength switch is the ability of the switch to resolve the wavelength components over a band of interest. Most WSS have been designed to operate over the C-band (1525–1568 nm) or the L-band (1570–1610 nm) but combined C + L band devices have been demonstrated [32]. One of the important advantages of LCoS is that different wavelength bands can be addressed without a redesign of either the CMOS chip or the liquid crystal cell. Of course the resolution of the system is limited eventually by the number of control elements, but in a WSS the optical resolution is likely to be the dominant factor. An interesting problem arises in that for a typical spectrometer an increase in resolution is achieved by reducing the focal-spot size at the switching plane. This condition is the opposite of the requirement for being able to switch through a large number of ports using LCoS and even though anamorphic imaging may result in some expansion in the switching axis, this is still far from being a useful device. One solution to this has been to move away from the traditional 4-f spectrometer designs and to only employ 4-f imaging in the dispersive axis using cylindrical optics. This permits a 2-f switching system in the switching axis, which not only allows a very simple way to have a large beam required for high port count switches but simply transforms a linearly ramped phase to a fiber displacement.

More generally, the properties of LCoS are such that the only requirement is to overlap physically the mode projection of the different ports at the LCoS and a phase image can be constructed to ensure effective coupling between the ports. This is a form of digital holography where the phase front required to couple to a given port is constructed from the input wavefront.

The spectral resolution is now defined by the cylindrical imaging system of the spectrometer and many traditional optical design techniques [33] can be employed to create a flat field and tightly focused point. Once the spot size has been defined and the dispersion of the wavelength across the LCoS determined, the spectral response as a function of frequency, $S(f)$, can be approximately characterized by two parameters [34]:

$$S(f) = \frac{1}{2}\sigma\left[\text{erf}\left(\frac{B/2 - f}{\sqrt{2}\sigma}\right) - \text{erf}\left(\frac{-B/2 - f}{\sqrt{2}\sigma}\right)\right]$$

where B is the width of the aperture opened on the LCoS array mapped to optical frequencies in f and σ is the standard deviation of the Gaussian spot at the array, also mapped to optical frequency.

Another aspect of the spectral response that must be considered is the exact location of the channel edge. Although an LCoS has a large number of columns, the spectral quantization associated with using these columns as channel edges is too limiting for many high resolution applications. Various techniques which rely on the two-dimensional nature of the switching matrix have been developed to ensure a smooth transition of band-edge as illustrated in Figure 18.15. Channel edges can then be calibrated and stored for each required channel or spectral slice, and these can be adjusted for any environmental factors such as ambient temperature.

18.4.4 Polarization-independent operation

Polarization dependence of response is critical for telecommunications operations. It is important to remember that the liquid crystal phase response act only on the extraordinary polarization state and so techniques must be employed to ensure the device behaves independently of polarization state. These techniques are broadly known as polarization diversity, wherein two polarization states are treated separately, and are commonly used in other common components such as circulators and isolators. In the case of wavelength switches the polarization splitting can be angle or displacement based and can be in the plane of either the switching or the dispersion axis. If the polarization states are caused to recombine at the switching element then the paths can be made to be nearly symmetrical (Figure 18.16) and impairments such as differential group delay (DGD), which is a measure of time delay between orthogonal polarizations, and polarization dependent loss (PDL) can be kept to a minimum.

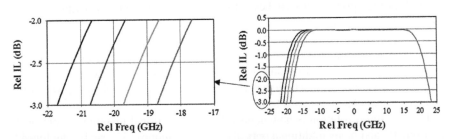

FIGURE 18.15 Spectral response of a switched WSS illustrating tuning of a channel edge with sub-GHz resolution and reconfiguration of channels to variable bandwidth.

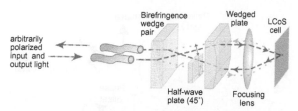

FIGURE 18.16 Principle of polarization diversity where the orthogonal polarization states are equalized and cross at a common point of the LCoS for all wavelengths.

The polarization response of any optical system can be fully quantified by the phase and amplitude response of each polarization state over wavelength and so there is no hidden penalty for employing a polarization diversity technique and in fact the polarization dependence can be superior to other switching or filtering approaches which can be impaired due to angle of reflection or a polarization dependent frequency shift. Experiments [35] have shown that even for demanding optical transmission formats such as DP-QPSK, transmission through a large number of wavelength switches is feasible.

18.4.5 Port isolation and extinction

There are a number of considerations that impact the design of an optical switch but port isolation is perhaps the most significant—particularly for larger port count devices. Port isolation can be considered as misdirected light and in an optical transmission system can result in beat noise if a dropped signal is not sufficiently blocked. In an LCoS switch the light could arise from scattering at the reset points of the reconstructed wavefront. Fortunately there are good techniques to calculate and avoid or even to cancel out [36] these unwanted scattering terms. Most of these techniques remain unpublished but some examples of approaches that attempt to frustrate the constructive scattering have been recently published [37]. Use of a selection of calibration techniques has resulted in commercially available 1 × 9 wavelength switches with port isolation in excess of 35 dB over a whole range of operating conditions, including varying attenuation states.

There is an industry trend to specify cumulative port isolation to reflect the "real-world" worst-case condition where an unwanted channel is present on all ports of a WSS being used as an "Add" device. In this case, the "port isolation" for a given channel is simply the ratio of the channel power in the selected port to the linear summation of the cross-talk power in that channel from each of the other fibers, assuming that a channel of equal power is present on all ports. This "port isolation" is generally expressed in dB. The design considerations for a WSS to be employed in a "route-and-select" architecture (see Figure 18.21) can be relaxed in this respect as the signal traverses two WSSs and hence the channel blocking is effectively doubled.

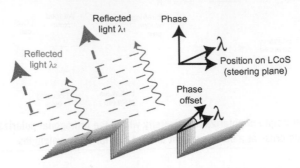

FIGURE 18.17 Phase adjustment through translation of the switching image. Offsetting the ramps creates a phase delay that varies with wavelength producing a group delay. If the rate of offset change is not constant with wavelength then the group delay itself varies with wavelength creating chromatic dispersion.

18.4.6 Phase response

One of the remarkable features of an LCoS-based WSS is the ability to control the relative phase of the directed light at each wavelength. It is this fact that is relied upon in destructive cancellation of the unwanted orders to achieve high port isolation. Functional applications of this will be dealt with in more detail later, but it is worth recognizing that this arises from the digital holographic nature of the switched wavelength. A simple way to consider this is as a shift of the reset position of a linear beam steering function. A small shift in displacement (less than a wavelength) of the created wavefront will result in a shift of the corresponding reflected wavefront and hence the position of the phase resets (Figure 18.17).

As well as the ability to control the phase of light it is encouraging to note that the phase response of filters generated by an LCoS switch—for example, in an interleaved port pattern—is very flat compared to other devices such as interleavers or AWGs. This results in very low system penalties due to dispersion or group delay ripple.

18.4.7 Switching time

Liquid crystal is indeed a dynamically controlled "liquid" cell so it is no surprise that the response of the liquid crystal is a function of temperature. To ensure that switching speeds are maintained, it is therefore advantageous to operate with some form of thermal control. For example, when heated to around 45 °C the liquid crystal will have an optical response of typically a few tens of milliseconds depending on the details of the particular mixture. However, when calculating the overall temporal response of the device (i.e. how fast it switches or attenuates), it is necessary to include the time taken by the driving electronics to address the pixels, together with any algorithmic requirements to transition through multiple intermediate states. This is often necessary to avoid scattering light during a switching or attenuation event.

Understanding the transient response (the optical properties of the device during a switching event) is important. Controlling this transient response is a significant contributor to the drive complexity of an LCoS device. The other transient response that needs to be considered is the response of the liquid crystal to the AC voltage which is applied as described in Section 3.3 above. It is important to set the DC bias accurately over lifetime to reduce any ripple on the transmission over all attenuation levels. The temporal and amplitude response characteristics of the liquid crystal, the design of the backplane, and managing these components' behavior over the lifetime of the device, are all important factors in achieving a reliable telecom grade component.

18.4.8 Device specification calibration and characterization

Although the manufacturing tolerances for a WSS may be reduced by the use of an LCoS-based switching matrix, the task of ensuring specification compliance now lies with the algorithm which drives the LCoS and the calibration of the device subsequent to manufacture. For example, the setting of the channel edges—which in alternative switching technologies may be determined by a final adjustment of the optical elements or gas composition—can now be set after all the packaging is complete. It is also possible to adjust this as a function of ambient temperature, if required. However, the measurement and calibration of such a large number of parameters requires a high level of sophistication and automation in the calibration processes. For example, in a typical calibration process, in excess of 7000 full spectral images will be sent to the LCoS and the device spectral response analyzed, to obtain the calibrated performance required for use in an optical communications network.

The specification and final test is also intensive as a typical specification will incorporate the performance over temperature of say 96 channels over 10 switching states each of which can be attenuated over a range of 150 attenuation states and over 20 parameters will be measured incorporating the loss, group delay, optical isolation, polarization dependence, and spectral performance. The multiplicative effect of all these parameters is overwhelming even without considering different perturbations of the states so the measurement and storage of data needs to be capable of revealing any trends and correlations.

18.5 FUTURE NETWORKS
18.5.1 Overcoming capacity constraints

The next generation of optical networks needs to address, for the first time, the limitations of the available spectrum in the amplification window of optical fiber. This limitation, also known as the Non-Linear Shannon Limit [38,39], is caused by the nonlinearity of the optical fiber and imposes an upper power limit of transmission before nonlinear interactions degrade the Optical Signal-to-Noise Ratio (OSNR) of the transmitted signal. This is significant as the available OSNR is determined by the optical components within a link and hence there is a corresponding limit to spectral density that can be obtained at an error rate which can be compensated for by the Forward Error Correction (FEC) systems available.

Both multimode (refer Chapter 11) or multicore (refer Chapter 12) fibers have been proposed to introduce extra capacity, but these require a complete overhaul of the existing infrastructure and so it is likely that solutions which can squeeze more capacity out of the existing fiber infrastructure will become increasingly important in the coming years. Improvements to fiber linearity, loss, and amplification (e.g. Raman) can increase the potential capacity while being compatible with existing networks and may be introduced in parallel with the approaches to be discussed in this section to deliver a significant increment of available capacity.

Any discussion of next generation transmission is not complete if only point-to-point transmission is analyzed. Consideration must also be given to how a wavelength switched network can be implemented to take advantage of the transmission improvements.

18.5.2 Beyond the ITU grid

A significant portion of spectrum is currently unavailable due to the limitations of the wavelength-switching and routing infrastructure that is in place—in particular the spectral guardbands required between channels. These guardbands are transitional spectral regions which occur between adjacent channels in a WSS, one of which is in a transmit state and the other in a block state. The width of the guardband is determined by the optical performance of the WSS and must allow for worst case scenarios of filter concatenation and accumulated leakage of blocked channels requiring a margin (penalty) to the required OSNR.

One of the ways that this bandwidth could be released is to move to higher data rates, so that an individual channel occupies a proportionately larger spectrum, thus reducing the proportional impact of the guardband. Proposals for increasing the data rate have focused on two alternatives; firstly by increasing the signal constellation size using multi-level phase and amplitude modulation techniques [38] and secondly by employing multiple carriers to carry the data—a concept known as superchannels [40,41].

An example combining both these approaches (Figure 18.18) shows how a 1 Tbit/s transmission system on five carriers each carrying 200 Gbit/s of data could be located in a 175 GHz wide channel, compared to the best currently achievable commercial capacity (10×100 Gb/s using a 50 GHz channel spacing) which would require 500 GHz of spectrum. The superchannel provides an improvement in spectral density of $500/175 = 2.8$, despite having the same guardband spectral allocation at the edge of the channel.

With an LCoS WSS it is possible to create a new functionality—for example, a different grid structure or superchannels—by changing only the software (and possibly calibration requirements) for the WSS without any changes to the optical hardware (and subsequent component requalification). Traditionally, qualification of the

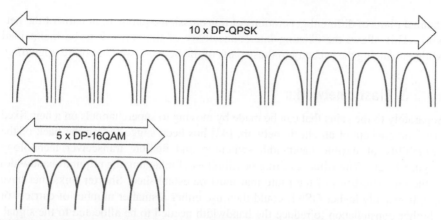

FIGURE 18.18 Packing 1 Tbit/s data into a 175 GHz superchannel showing the 2.8 × improvement in spectral density which can be achieved.

switching element has been a major hurdle to introduction of new products as, particularly for MEMS switching element, a design and qualification cycle can run into millions of development dollars.

An example of this new functionality is the introduction of intrachannel attenuation profiling and variable bandwidth channels (such as shown in Figure 18.19) which has recently been developed for products initially introduced as standard 50 GHz 1 × 9 WSS. Intrachannel attenuation is valuable for accommodating variable OSNR requirements of different carriers within a superchannel, or for compensating nonlinear spectral gain in long-haul transmission.

Finally, an LCoS-based WSS also allows a system to maintain part of its traffic as legacy traffic and introduce new transmission formats gracefully by supporting

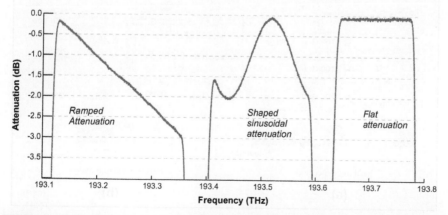

FIGURE 18.19 Measured spectrum of an LCoS-based WSS showing intra-channel attenuation profiling and variable channel bandwidths.

channel plans of different generations of transmission equipment in parts of the network. This allows the existing interconnectivity of the network to be maintained while enhancing the capacity in those parts of the network which require upgrading.

18.5.3 Elastic networks

Separately to the gains that can be made by moving to superchannels on a new fixed grid, the concept of an elastic network [42] has been suggested on the back of the availability of flexible bandwidth switching and flexible transceiver technology (Figure 18.20). This offers exciting possibilities of matching the required bandwidth to the characteristics of the route that must be established. Shorter links that have an intrinsically higher OSNR could then use either a smaller number of carriers or a higher constellation to reduce the bandwidth needed to be allocated to the signal, While links with higher optical impairments can be allocated a higher bandwidth as necessary to obtain suitable OSNR.

Assessing the relative advantage in terms of spectral efficiency of elastic networks can be problematic as assumptions about traffic distribution need to be made in addition to benchmarking, clearly the case that it is being compared against. Comparison of an Elastic Network against an existing 50 GHz network may be reasonable, but it should be noted that an Elastic Network could be implemented without utilizing flexible grid (e.g. by utilizing multiple adjacent 50 GHz channels) but may fail to benefit from the spectral efficiency gains which arise from removal of the intra-channel guardbands. Conversely a flexible grid WSS can be configured to upgrade the grid of a network without requiring the overhead of managing an Elastic Network structure.

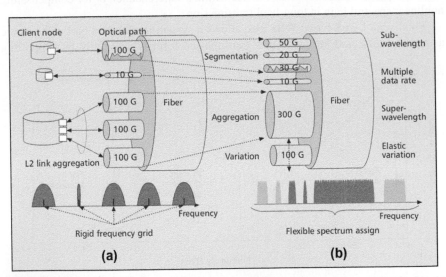

FIGURE 18.20 Spectral assignment in an optical network: (a) Conventional optical path network; (b) Elastic Network (from [43]).

Despite these constraints, many studies have been made quantifying these advantages in different circumstances [43]. In addition to the benefits of wider channels spectral efficiency can be enhanced in elastic networks by choosing the appropriate transmission format depending on the traffic mix and the path length dependent OSNR, allowing a proportion of higher OSNR links to be upgraded to higher constellations [44]. The gain in network capacity is then a function of the link lengths, available OSNR and any overhead required to manage the spectrum. It is in this context that there has been considerable discussion on the role of network defragmentation [45] and the complexity of managing the spectrum—though this latter issue has, of course, been successfully addressed in the RF domain.

18.5.4 **Colorless, directionless, contentionless ROADMs**

A limitation of reconfigurable networks installed to date is that while the core of the network has been able to be reconfigured without affecting other traffic flowing through it, the options have been much more limited for the local add/drop wavelengths. It is conceptually simple to conceive of a solution involving existing network architectures and a large photonic switch to configure any local add/drop wavelength to any available port (channel and direction) [46]. However, the scaling of this to multidirectional nodes involves very large-port-count switches, which are currently not commercially feasible.

Furthermore such architectures, which generally use an Arrayed Waveguide Grating [5] for wavelength selectivity in the drop section, do not address the need for flexible bandwidth required for elastic networks as outlined above. The development of flexible grid multidirectional multiplexing and demultiplexing is therefore an area of active research and the network elements derived from this have come to be known as colorless, directionless, contentionless (CDC) [2] ROADMs. There is also considerable interest in simpler architectures, where the contention management is handled through approaches such as client side fiber cross-connects [47].

Figure 18.21 shows an example of a CDC ROADM architecture known as "Route and Select" in which the back-to-back configuration of high port count WSS provides (a) the core routing capability with, in this case, seven of the ports configured to route the core traffic between directions and (b) the add/drop capability using the remaining ports. The add/drop ports are then available to be connected to mux/demux modules associated with "shared" transceiver banks.

In the Route and Select architecture, the filtering arising from an add/drop operation is in general the sum (in dB) of the filter response of each stage of the wavelength switch. This in turn means that the requirement on concatenation is in fact increased as a typical link will now have twice the number of transits of a WSS. To achieve this special care should be paid to achieving the optimal optical resolution and it is possible to now consider sculpting of the amplitude response of the filter to achieve an enhanced concatenated 3 dB bandwidth as described in Section 18.5.2. This can be done without sacrificing the extinction bandwidth (clear bandwidth centered about the ITU grid that achieves a required blocking ratio) as the blocking ratio is also doubled because of the use of two WSSs.

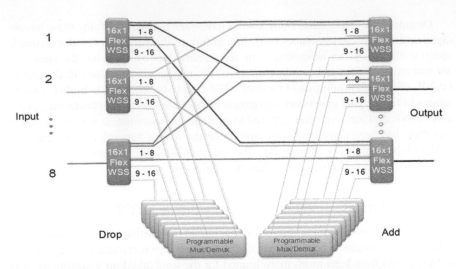

FIGURE 18.21 One proposal for an eight-degree CDC-f network node where the programmable multiplexing can be either multicast and switch or multidirectional wavelength selective switching. In this case Flexgrid WSSs are used at both ingress and egress.

It is likely that in future the two WSS would use the same optical module utilizing different wavelength processing regions of a single matrix switch such as LCoS [48,49], provided that the issues associated with device isolation are able to be appropriately addressed. Channel selectivity ensures only wavelengths required to be dropped locally (up to the maximum number of transceivers in the bank) are presented to the mux/demux module through each fiber, which in turn reduces the filtering and extinction requirements on the mux/demux module.

An architecture that takes advantage of this reduced filtering requirement is a multicast switch module [27]. A recent commercial incarnation of this [50] employs no wavelength selectivity but integrates eight 1×12 broadcast splitters with 12 8×1 switches and therefore allows each of 12 transceivers to be connected to eight WSS devices in an eight-degree node, and hence the signal from each transceiver can leave the node in one of the eight selectable directions, without an issue of contention. The use of broadcast splitters/combiners in a multicast module requires the use of additional amplification stages to overcome the intrinsic losses associated with this architecture. Coherent transmission systems provide a natural filtering between the remaining wavelengths from the local oscillator of a heterodyne receiver, thus reducing system cost and complexity. However, for direct detection systems, an array of tunable filters is still required. One solution to this that has been proposed employs an array of wavelength blockers on a single LCoS backplane [51] to provide a high-density tunable filter array.

Looking further out, a potential solution is to employ wavelength selectivity directly in the multiplexer stages [52]. This requires multiple independent wavelength processing regions on the LCoS chip and is a generalization of the use of two wavelength processing zones discussed above. Most importantly, wavelength selectivity in the form of a multidirectional wavelength selective mux/demux removes the significant multicast loss penalty and brings additional filtering suitable for direct detection also. In these cases the filtering must be low loss and address the requirement for variable optical bandwidths, but as it is traversed only once some of the concatenation characteristics are less.

More generally, there is still much investigation required to determine the optimum architectures and trade-offs for adding and dropping superchannels in the core network. Many of the factors such as loss budgets and amplification requirements are specific to the requirements of particular proprietary systems and the level of flexibility required. In particular, it is unclear whether, if individual carriers of a superchannel are to be separately processed, that the capacity gains anticipated can be achieved in practice. However, where a superchannel is to be added or routed as a whole then there is a significant opportunity to reduce the loss budget through wavelength selectivity in multiplexing or demultiplexing to a transceiver bank.

18.6 EMERGING APPLICATIONS OF LCOS

18.6.1 Fourier domain pulse shaping and generation

The ability of an LCoS-based WSS to independently control both the amplitude and phase of the transmitted signal was discussed previously and leads to the more general ability to manipulate the amplitude and/or phase of an optical pulse through a process known as Fourier domain pulse shaping [53,54]. This process, which requires full characterization of the input pulse in both the time and spectral domains, is shown schematically in Figure 18.22.

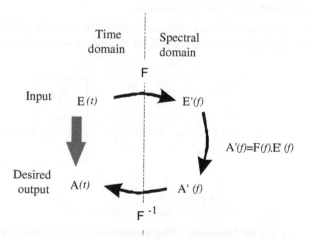

FIGURE 18.22 Schematic representation of Fourier domain pulse shaping.

As an example, an LCoS-based Programmable Optical Processor [55] (POP) was used to broaden a mode-locked laser output into a 20 nm supercontinuum source, while a second such device was used to compress the output to 400 fs, transform-limited pulses [56,57].

Similarly, a multiport LCoS-based POP was demonstrated to demultiplex a 40 GHz pulse train into several different pulse trains at separate wavelengths, with customizable burst numbers [58]. This approach was combined with a nonlinear optical loop mirror to demultiplex 40 Gbit/s signals out of a 160 Gbit/s signal stream [59].

Passive mode-locking of fiber lasers has been demonstrated at high repetition rates, but inclusion of an LCoS-based POP allows tailoring of the dispersion profile across the spectrum. For example, adjusting the phase content of the spectrum has been shown to change the pulse train of a passively mode-locked laser from bright to dark pulses [60].

A similar approach uses spectral shaping of optical frequency combs to create multiple pulse trains. For example, a 10 GHz optical frequency comb was shaped by the POP to generate dark parabolic pulses and Gaussian pulses, at 1540 nm and 1560 nm, respectively [61].

18.6.2 Microwave signal processing

Optoelectronic architectures are well suited for many functions in radar and electronic warfare systems which require signal processing of microwave signals. Early approaches required arrays of gratings [62] to provide the appropriate spectral and temporal response and were therefore limited in terms of their reconfigurability. More recently, an LCoS-based POP provides a one-device solution that allows amplitude shaping of each tap with precise control (Figure 18.23). Implementations using this configuration have produced FIR filters with over 100 taps [63].

18.6.3 Phase-sensitive amplification

Phase sensitive Amplifiers (PSAs) offer the potential to provide amplification with a noise figure below the 3 dB theoretical limit of a phase-insensitive amplifier such

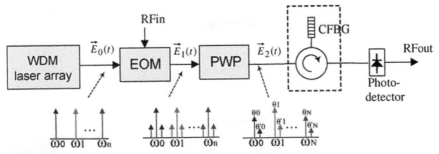

FIGURE 18.23 Multi-tap FIR filter using a Programmable Optical Processor (designated here as a PWP) from [67].

as an EDFA. One approach to implementing PSAs requires phase locking of wavelengths for four-wave mixing, where an LCoS POP has been used to provide the functions of filtering, attenuation, and phase shifting for the pump, signal and idler waves. Using this technique, saturated gain of over 33 dB has been achieved [64] with noise figure as low as 1.1 dB [65].

18.6.4 Programmable interferometer

Optical signal-to-noise ratio (OSNR) is a critical characteristic of a network link, but this quantity is often difficult to measure. By examining the coherence properties of a signal compared to the underlying spontaneous emission noise, it is possible to estimate the true OSNR value. An LCoS POP has been used to dynamical sweep the delay in an interferometer configuration [66,67]. A specific example of this capability is the simultaneous demodulation and demultiplexing of multi-channel DPSK signals [68].

18.6.5 Advanced SLMs

The technical maturity of spatial light modulators based on consumer-driven applications has been highly advantageous to their adoption in the telecommunications arena. There are developments in MEMS phased arrays [69] and other electro-optic spatial light modulators that could be envisaged in the future to be applicable to telecom switching and wavelength processing, perhaps bringing faster switching or having an advantage in simplicity of optical design through polarization-independent operation. For example, the design principles developed for LCoS could be applied to other phase-controllable arrays in a straight-forward fashion if a suitable phase stroke (greater than 2π at 1550 nm) can be achieved. However, the requirements for low electrical crosstalk and high fill factor over very small pixels required to allow switching in a compact form factor remain serious practical impediments to achieving these goals [70].

18.6.6 Modal switching

One of the interesting applications of LCoS switching that has arisen is the ability to transform between modes of optical fibers [71] where these modes can be made to overlap significantly in the far field. For example, a single-mode fiber can have its wavefront transformed to enable selective coupling into just one of the modes of a multimode fiber (see Figure 18.24). To do this with high modal selectivity requires carefully calculating the amplitude and phase response. As in the case of attenuation in WSS the unwanted light can be directed away from the fiber core in the near field by steering or leaving in the zero order (undiffracted) spot which is physically separated from the fiber. This would enable the multiplexing and reconfiguration of the transmission modes of a multimode transmission system. Such systems have been proposed [72] as the basis of higher capacity transmission systems in the future.

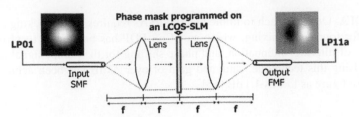

FIGURE 18.24 Modal transformation of an incoming mode of a single-mode fiber to a higher order mode of a multimode fiber (from [73]).

Similarly the use of multicore fiber transmission systems known as Space Division Multiplexing (Chapters 11 and 12) will offer new challenges and opportunities for which the flexibility of LCoS is well suited.

18.6.7 Tunable lasers

The use of LCoS as a filtering technique and hence a tuning mechanism for active fiber and diode lasers has been demonstrated. In early research [73] it was shown that the dispersion of a grating and the digital selection of a wavelength through holographic techniques could be combined in an active cavity and, more recently, linear cavity tunable fiber lasers have been demonstrated [74]. Compact lasers which use a normal-incidence reflective sub-wavelength resonant grating waveguide on silicon substrate [75] can be tuned by varying the liquid crystal refractive index above the grating in an analog tuning mechanism relying on a single control variable.

18.6.8 Programmable high resolution

Applications are emerging where it would be advantageous to have finer spectral resolution of the phase and amplitude response of an LCoS wavelength processor. Examples of this are in pulse shaping, dispersion compensation, and in manipulating the spectrum of advanced transmission formats. The use of the second axis of the LCoS to enhance the spectral resolution through incorporation of a highly dispersive Virtually Imaged Phase Array (VIPA) orthogonal to the principal wavelength dispersive axis has been demonstrated [76], achieving GHz level resolution and wavelength addressing in a filtering application.

18.6.9 Other applications

The application of LCoS in medical, metrology, astronomy, and scientific applications is beyond the scope of this chapter but it is interesting to note that applications of LCoS outside of its current bases in consumer display and telecommunications applications will ensure that it continues to be an area of active research for many years to come.

References

[1] C.A. Brackett, Dense wavelength division multiplexing networks: principles and applications, IEEE J. Sel. Area. Commun. 8 (6) (1990) 948–964.

[2] S. Gringeri et al., Flexible architectures for optical transport nodes and networks, IEEE Commun. Mag. 48 (7) (2010) 40–50.

[3] H.A. Macleod, Thin Film Optical Filters, Institute of Physics, London, 2001 641

[4] G. Meltz, W.W. Morey, W.H. Glenn, Formation of Bragg gratings in optical fibres by a transverse holographic method, Opt. Lett. 14 (1989) 823–825.

[5] C. Dragone, An $N*N$ optical multiplexer using a planar arrangement of two star couplers, IEEE Photon. Technol. Lett. 3 (9) (1991) 812–815.

[6] S. Frisken, D. Abakoumov, A. Bartos, Low-loss polarisation-independent dynamic gain-equalisation filter, in: Optical Fiber Communication Conference, 2000, vol. 2, pp. 251–253.

[7] J. Kelly, Application of liquid crystal technology to telecommunication devices, in: Conference on Optical Fiber Communication and the National Fiber Optic Engineers Conference, OFC/NFOEC 2007, 2007, pp. 1–7.

[8] R. Giles et al., Low-loss channelized WDM spectral equalizer using lightwave micromachines and autonomous power regulation, in: Optical Fiber Communication Conference, 1999, and the International Conference on Integrated Optics and Optical Fiber Communication, OFC/IOOC '99, Technical Digest, vol. Supplement IEEE, 1999, Suppl. PD31/1–PD31/3.

[9] A.R. Ranelli, B.A. Scott, J.P. Kondis, Liquid crystal based wavelength selectable cross-connect, in: Proc. Eur. Conf. on Opt. Commun., 1999, pp. 68–69.

[10] Jens Buus, Markus-Christian Amann, Daniel J. Blumenthal, Tunable laser diodes and related optical sources, John Wiley & Sons, Hoboken, NJ, 2005.

[11] S. Suzuki, A. Himeno, M. Ishii, Integrated multichannel optical wavelength selective switches incorporating an arrayed-waveguide grating multiplexer and thermooptic switches, J. Lightwave Technol. 16 (4) (1998) 650–655.

[12] J.E. Fouquet, Compact optical cross-connect switch based on total internal reflection in a fluid-containing planar lightwave circuit, in: Optical Fiber Communication Conference, 2000, vol. 1, pp. 204–206.

[13] K.W. Cheung, United States Patent No 4, 906, 064: Switch for Selectively Switching Optical Wavelengths, n.d.

[14] G. Hill, A wavelength routing approach to optical communication networks, Brit. Telecom Technol. J. 6 (3) (1988) 24–31.

[15] J.E. Ford et al., Wavelength add-drop switching using tilting micromirrors, J. Lightwave Technol. 17 (5) (1999) 904–911.

[16] B.P. Keyworth, ROADM subsystems and technologies, in: Optical Fiber Communication Conference, 2005, Technical Digest, OFC/NFOEC, vol. 3, IEEE, 2005.

[17] G. Baxter et al., Highly programmable wavelength selective switch based on liquid crystal on silicon switching elements, in: Optical Fiber Communication Conference, 2006 and the 2006 National Fiber Optic Engineers Conference, OFC 2006, IEEE, 2006.

[18] S. Frisken et al., High performance "drop and continue" functionality in a wavelength selective switch, in: Optical Fiber Communication Conference, 2006 and the 2006 National Fiber Optic Engineers Conference, OFC 2006, pp. 1–3.

[19] Par Pieére Chatelain, Sur L'Orientation Des Cristaux Liquides Par Les Surfaces Frottees, Bulletin De La Société 66 (1944) 105–130.

[20] H. Kelker, B. Scheurle, B. Scheurle, H. Kelker, A liquid-crystalline (nematic) phase with a particularly low solidification point, Angew. Chem. Int. Ed. 8 (11) (1969) 884.

[21] Birefringence is the relative velocity of the two orthogonal polarization states. Here the two states are the linear state orthogonal to the molecule (the "ordinary axis"), and the state partly aligned with the molecule (the "extraordinary axis").

[22] V. Zolina, V. Fréedericksz, Forces causing the orientation of an anisotropic liquid, Trans. Faraday Soc. 29 (1933) 919–930.

[23] G.D. Love, Liquid-crystal phase modulator for unpolarized light, App. Opt. 32 (1993) 2222–2223.

[24] M.J. Ranshaw, D.G. Vass, Richard M. Sillitto, Phase or amplitude modulation by birefringence in a liquid-crystal spatial light modulator, in: Optical Computing 88 (Book of Summaries), Toulon, France, n.d.

[25] R.J. Mears et al., Telecommunications applications of ferroelectric liquid-crystal smart pixels, IEEE J. Sel. Top. Quant. Electron. 2 (1) (1996) 35–46.

[26] W.S. Hu, Q.J. Zeng, Multicasting optical cross connects employing splitter-and-delivery switch, IEEE Photon. Technol. Lett. 10 (7) (1998) 970–972.

[27] Robert Anderson, US Patent 6.542, 657: Binary Switch for an Optical Wavelength Router, April 1, 2003.

[28] H. Washburn, M. Xue, US Patent 7,909,958: Apparatus and method for optical switching with liquid crystals and birefringent wedges.

[29] D.J. Bishop, C.R. Giles, G.P. Austin, The Lucent LambdaRouter: MEMS technology of the future here today, IEEE Commun. Mag. 40 (3) (2002) 75–79.

[30] Xinghua Wang et al., Liquid Crystal on Silicon (LCOS) wavefront corrector and beam steerer, Proc. SPIE 5162 (1) (2003) 139–146.

[31] S. Frisken et al., High performance "drop and continue" functionality in a wavelength selective switch, in: Proceedings of the Conference on Optical Fibre Communications, 2006, pp. 1–3.

[32] Finisar WaveShaper 4000/X.

[33] J.F. James, Spectrograph Design Fundamentals, Cambridge University Press, Cambridge, UK, 2007.

[34] C. Pulikkaseril et al., Spectral modeling of channel band shapes in wavelength selective switches, Opt. Express 19 (9) (2011) 8458–8470.

[35] S. Gringeri et al., Real-time 127-Gb/s Coherent PM-QPSK transmission over 1000 km NDSF with >10 cascaded 50 GHz ROADMs, in: 36th European Conference and Exhibition on Optical Communication (ECOC), 2010, pp. 1–3.

[36] S. Frisken, G. Baxter, H. Zhou, D. Abakoumov, Optical Calibration System and Method, US Patent 7457547.

[37] Brian Robertson et al., The use of wavefront encoding to reduce crosstalk in a multicasting fiber telecom switch, in: Optical Fiber Communication Conference, OSA Technical Digest, Paper OM2J.6, Los Angeles, 2012.

[38] R.-J. Essiambre et al., Capacity limits of optical fiber networks, J. Lightwave Technol. 28 (4) (2010) 662–701.

[39] A.D. Ellis, Jian. Zhao, D. Cotter, Approaching the non-linear shannon limit, J. Lightwave Technol. 28 (4) (2010) 423–433.

[40] G. Bosco et al., On the performance of Nyquist-WDM terabit superchannels based on PM-BPSK, PM-QPSK, PM-8QAM or PM-16QAM subcarriers, J. Lightwave Technol. 29 (1) (2011) 53–61.

[41] A. Sano et al., No-guard-interval coherent optical OFDM for 100-Gb/s long-haul WDM transmission, J. Lightwave Technol. 27 (16) (2009) 3705–3713.

[42] M. Jinno et al., Spectrum-efficient and scalable elastic optical path network: architecture, benefits, and enabling technologies, IEEE Commun. Mag. 47 (11) (2009) 66–73.

[43] O. Rival, G. Villares, A. Morea, Impact of inter-channel nonlinearities on the planning of 25–100 Gb/s elastic optical networks, J. Lightwave Technol. 29 (9) (2011) 1326–1334.

[44] Y.-K. Huang et al., 21.7Tb/s field trial with 22 DP-8QAM/QPSK optical superchannels over 1,503 km of installed SSMF, in: Optical Fiber Communication (OFC), Collocated National Fiber Optic Engineers Conference, 2012 Conference on (OFC/NFOEC), vol. Post Deadline Papers, IEEE, 2012, PDP5D.6.

[45] G. Wellbrock, Preparing for the future, in: 36th European Conference and Exhibition on Optical Communication (ECOC), 2010, pp. 1–10.

[46] V. Kaman et al., Comparison of wavelength-selective cross-connect architectures for reconfigurable all-optical networks, in: International Conference on Photonics in Switching, PS '06, 2006, pp. 1–3.

[47] M.D. Feuer et al., Intra-node contention in dynamic photonic networks, J. Lightwave Technol. 29 (4) (2011) 529–535.

[48] Steven James Frisken, United States Patent: 7397980—Dual-Source Optical Wavelength Processor, July 8, 2008.

[49] P. Evans et al., LCOS-based WSS with true integrated channel monitor for signal quality monitoring applications in ROADMs, in: Conference on Optical Fiber Communication/National Fiber Optic Engineers Conference, OFC/NFOEC 2008, 2008, pp. 1–3.

[50] NEL Ltd., NXN non-blocking optical matrix switch, n.d. <http://www.ntt-electronics.com/en/products/photonics/nxn_n_o_m_s.html>.

[51] Y. Sakurai et al., LCOS-based wavelength blocker array with channel-by-channel variable center wavelength and bandwidth, IEEE Photon. Technol. Lett. 23 (14) (2011) 989–991.

[52] N. Fontaine et al., $N \times M$ wavelength selective crossconnect with flexible passbands, in: Optical Conference on Optical Fiber Communication (OFC), Collocated National Fiber Optic Engineers Conference, OFC/NFOEC 2012, Post Deadline Papers, IEEE, PDP5B.2, 2012.

[53] A.M. Weiner, Femtosecond pulse shaping using spatial light modulators, Rev. Sci. Instrum. 71 (2000) 1929–1960.

[54] A.M. Weiner, Ultrafast optical pulse shaping: a tutorial review, Opt. Commun. 284 (15) (2011) 3669–3692.

[55] Finisar WaveShaper Family of Programmable Optical Processors, n.d., <http://www.finisar.com/WaveShaper>.

[56] A.M. Clarke, D.G. Williams, M.A.F. Roelens, M.R.E. Lamont, B.J. Eggleton, Parabolic pulse shaping for enhanced continuum generation using an LCoS-based wavelength selective switch, in: 14th OptoElectronics and Communications Conference (OECC), 2009.

[57] M. Clarke, D.G. Williams, M.A.F. Roelens, B.J. Eggleton, Reconfigurable optical pulse generator employing a Fourier-domain programmable optical processor, J. Lightwave Technol. 28 (1) (2010) 97–103.

[58] M.A.F. Roelens et al., Multi-wavelength synchronous pulse burst generation with a wavelength selective switch, Opt. Express 16 (14) (2008) 10152–10157.

[59] M.A.F. Roelens et al., Flexible and reconfigurable time-domain demultiplexing of optical signals at 160 Gb/s, IEEE Photon. Technol. Lett. 21 (10) (2009) 618–620.

[60] Jochen Schroeder et al., Dark and Bright Pulse Passive Mode-locked Laser with In-cavity Pulse-shaper, Opt. Express 18 (22) (2010) 22715–22721.

[61] T.T. Ng et al., Complete temporal optical fourier transformations using dark parabolic pulses, in: 35th European Conference on Optical Communication, ECOC '09, 2009, pp. 1–2.

[62] D.B. Hunter, R.A. Minasian, Reflectively tapped fibre optic transversal filter using in-fibre Bragg gratings, Electron. Lett. 31 (12) (1995) 1010–1012.

[63] Xiaoke Yi, T.X.H. Huang, R.A. Minasian, Tunable and reconfigurable photonic signal processor with programmable all-optical complex coefficients, IEEE Trans. Microwave Theory Tech. 58 (11) (2010) 3088–3093.

[64] C. Lundstrom et al., Experimental comparison of gain and saturation characteristics of a parametric amplifier in phase-sensitive and phase-insensitive mode, in: 35th European Conference on Optical Communication, ECOC '09, 2009, pp. 1–2.

[65] Z. Tong et al., Towards ultrasensitive optical links enabled by low-noise phase-sensitive amplifiers, Nat. Photon. 5 (7) (2011) 430–436.

[66] Jochen Schroeder et al., Simultaneous multi-channel OSNR monitoring with a wavelength selective switch, Opt. Express 18 (21) (2010) 22299–22304.

[67] O. Brasier et al., OSNR monitoring of Pol-mux Signals using a wavelength selective switch, in: 16th OptoElectronics and Communications Conference (OECC), 2011, IEEE, 2011, p. 28.

[68] Pegah Seddighian, Victor Torres-Company, Lawrence R. Chen, Simultaneous demodulation and demultiplexing of multi-rate WDM DPSK signals using a programmable wavelength-selective switch, Opt. Express 18 (11) (2010) 11657–11663.

[69] A. Gehner et al., Recent progress in CMOS integrated MEMS AO mirror developments, Adaptive Optics for Industry and Medicine: Proceedings of the Sixth International Workshop, National University of Ireland, Ireland, 12–15 June 2007, Imperial College Press, 2008. pp. 53–58.

[70] Jonathan Dunayevsky, David Sinefeld, Dan Marom, Adaptive spectral phase and amplitude modulation employing an optimized MEMS spatial light modulator, in: Optical Fiber Communication Conference, OSA Technical Digest, OM2J.5. <http://www.opticsinfobase.org/abstract.cfm?URI=OFC-2012-OM2J.5>.

[71] M. Salsi et al., Mode-division multiplexing of 2 × 100 Gb/s channels using an LCOS-based spatial modulator, J. Lightwave Technol. 30 (4) (2012) 618–623.

[72] An Li et al., Reception of mode and polarization multiplexed 107-Gb/s CO-OFDM signal over a two-mode fiber, in: Optical Fiber Communication Conference and Exposition (OFC/NFOEC), 2011 and the National Fiber Optic Engineers Conference, 2011, pp. 1–3.

[73] M.C. Parker, R.J. Mears, Digitally tunable wavelength filter and laser, IEEE Photon. Technol. Lett. 8 (8) (1996) 1007–1008.

[74] Feng Xiao, Kamal Alameh, Tong Tak Lee, Opto-VLSI-based tunable single-mode fiber laser, Opt. Express 17 (21) (2009) 18676–18680.

[75] A.S.P. Chang et al., A novel low-cost tunable laser using a tunable liquid-crystal subwavelength resonant grating filter, in: Conference on Lasers and Electro-Optics, CLEO '03 2003, 2003, p. 2.

[76] V.R. Supradeepa, D.E. Leaird, A.M. Weiner, A 2-D VIPA-grating pulse shaper with a Liquid Crystal On Silicon (LCOS) spatial light modulator for broadband, High Resolution, Programmable Amplitude and Phase Control, in: 23rd Annual Meeting of the IEEE Photonics Society, 2010, pp. 494–495.

Index

Printed and bound by CPI Group (UK) Ltd, Croydon, CR0 4YY

03/10/2024

01040320-0002